VOLUME FOUR HUND

METHODS IN
ENZYMOLOGY

Guide to Yeast Genetics:
Functional Genomics, Proteomics,
and Other Systems Analysis,
2nd Edition

Academic Press is an imprint of Elsevier
525 B Street, Suite 1900, San Diego, CA 92101-4495, USA
30 Corporate Drive, Suite 400, Burlington, MA 01803, USA
32 Jamestown Road, London NW1 7BY, UK

First edition 1991
Second edition 2010

For information on all Academic Press publications
visit our website at elsevierdirect.com

ISBN: 978-0-12-375172-0 (Hardback)
ISBN: 978-0-12-375171-3 (Paperback)
ISSN: 0076-6879

Contents

CONTENTS

Section VI. Other Fungi 735

Pascal Braun
Department of Cancer Biology, and Center for Cancer Systems Biology (CCSB), Dana-Farber Cancer Institute; Department of Genetics, Harvard Medical School, Boston, Massachusetts, USA

Donna Garvey Brickner
Department of Biochemistry, Molecular Biology and Cell Biology, Northwestern University, Evanston, Illinois, USA

Jason H. Brickner
Department of Biochemistry, Molecular Biology and Cell Biology, Northwestern University, Evanston, Illinois, USA

Jeffrey L. Brodsky
Department of Biological Sciences, University of Pittsburgh, Pittsburgh, Pennsylvania, USA

J. Ross Buchan
Department of Molecular and Cellular Biology and Howard Hughes Medical Institute, University of Arizona, Tucson, Arizona, USA

Christopher Buser
Department of Molecular and Cell Biology, University of California, Berkeley, California, USA

Andrew P. Capaldi
Department of Molecular and Cellular Biology, University of Arizona, Tucson, Arizona, USA

Cheryl D. Chun
University of California, San Francisco, California, USA

Sean R. Collins
Department of Chemical and Systems Biology, Stanford University School of Medicine, Stanford, California, USA

Michael Costanzo
Banting and Best Department of Medical Research and Department of Molecular Genetics, Terrence Donnelly Center for Cellular and Biomolecular Research, University of Toronto, Toronto, Ontario, Canada

Christopher A. Crutchfield
Department of Chemistry, and Lewis-Sigler Institute for Integrative Genomics, Princeton University, Princeton, New Jersey, USA

Christina A. Cuomo
Broad Institute of MIT and Harvard, Cambridge, Massachusetts, USA

Michael E. Cusick
Center for Cancer Systems Biology (CCSB), and Department of Cancer Biology, Dana-Farber Cancer Institute; Department of Genetics, Harvard Medical School, Boston, Massachusetts, USA

Scott Dixon
Department of Biological Sciences, Columbia University, New York, New York, USA

Matija Dreze
Department of Cancer Biology, and Center for Cancer Systems Biology (CCSB), Dana-Farber Cancer Institute; Department of Genetics, Harvard Medical School, Boston, Massachusetts, USA; Facultés Universitaires Notre-Dame de la Paix, Namur, Belgium

Maitreya J. Dunham
Department of Genome Sciences, University of Washington, Seattle, Washington, USA

Po Hien Ear
Département de Biochimie, Université de Montréal, C.P. 6128, Succursale Centre-ville, Montréal, Québec, Canada

Elke Ericson
Department of Pharmaceutical Sciences, and Donnelly Centre for Cellular and Biomolecular Research, University of Toronto, Toronto, Ontario, Canada; Department of Bioscience, AstraZeneca R&D Mölndal, Mölndal, Sweden

Gerald R. Fink
Whitehead Institute for Biomedical Research, Cambridge Center, Cambridge, Massachusetts, USA

Susan L. Forsburg
Department of Molecular and Computational Biology, University of Southern California, Los Angeles, California, USA

Susan M. Gasser
Friedrich Miescher Institute for Biomedical Research, Basel, Switzerland

Guri Giaever
Department of Pharmaceutical Sciences, and Donnelly Centre for Cellular and Biomolecular Research; Department of Molecular Genetics, University of Toronto, Toronto, Ontario, Canada

Lutz R. Gehlen
Friedrich Miescher Institute for Biomedical Research, Basel, Switzerland

Xue Li Guan
Department of Biochemistry, Yong Loo Lin School of Medicine, National University of Singapore, Singapore

Randal Halfmann
Whitehead Institute for Biomedical Research, and Department of Biology, Massachusetts Institute of Technology, Cambridge, Massachusetts, USA

Franz S. Hartner
Department of Chemical Engineering, Massachusetts Institute of Technology, Cambridge, Massachusetts, USA

Franklin A. H

Department of Biochemistry and Biophysics, University of California at San Francisco, San Francisco, California, USA; Current address: Department of Biochemistry and Molecular Biology, University of Oklahoma Health Sciences Center, Oklahoma City, Oklahoma, USA

Aaron D. Hernday

Department of Microbiology and Immunology, University of California at San Francisco, San Francisco, California, USA

Philip Hieter

Michael Smith Laboratories, Department of Medical Genetics, University of British Columbia, Vancouver, British Columbia, Canada

David E. Hill

Department of Cancer Biology, and Center for Cancer Systems Biology (CCSB), Dana-Farber Cancer Institute; Department of Genetics, Harvard Medical School, Boston, Massachusetts, USA

Shawn Hoon

Stanford Genome Technology Center, Stanford University, Palo Alto, California, USA, and Molecular Engineering Lab, Science and Engineering Institutes, Agency for Science Technology and Research, Singapore

Maki Inada

Department of Molecular Biology and Genetics, Cornell University, Ithaca, New York, USA, and Present address: Department of Biology, Ithaca College, Ithaca, New York, USA

Nicholas T. Ingolia

Department of Cellular and Molecular Pharmacology, Howard Hughes Medical Institute, University of California, San Francisco, and California Institute for Quantitative Biomedical Research, San Francisco, California, USA

Alexander D. Johnson

Department of Biochemistry and Biophysics, and Department of Microbiology and Immunology, University of California at San Francisco, San Francisco, California, USA

Véronique Kalck

Friedrich Miescher Institute for Biomedical Research, Basel, Switzerland

Nevan J. Krogan

Department of Cellular and Molecular Pharmacology, California Institute for Quantitative Biomedical Research, University of California, San Francisco, California, USA

Christian Landry

Département de Biologie, Institut de Biologie Intégrative et des Systèmes, PROTEO-Québec Research Network on Protein Function, Structure and Engineering, Pavillon Charles-Eugène-Marchand, Université Laval, Québec, Canada

Felix H

Department of Chemical Engineering, Massachusetts Institute of Technology, and Whitehead Institute for Biomedical Research, Cambridge Center, Cambridge, Massachusetts, USA

Philippe Lefrançois
Department of Molecular, Cellular and Developmental Biology, Yale University, New Haven, Connecticut, USA

William Light
Department of Biochemistry, Molecular Biology and Cell Biology, Northwestern University, Evanston, Illinois, USA

Susan Lindquist
Whitehead Institute for Biomedical Research, and Department of Biology, Massachusetts Institute of Technology; Howard Hughes Medical Institute, Cambridge, Massachusetts, USA

Wenyun Lu
Department of Chemistry, and Lewis-Sigler Institute for Integrative Genomics, Princeton University, Princeton, New Jersey, USA

Claire Lurin
Unité de Recherche en Génomique Végétale (URGV), Evry Cedex, France

Hiten D. Madhani
University of California, San Francisco, California, USA

Mohan K. Malleshaiah
Département de Biochimie, Université de Montréal, C.P. 6128, Succursale Centre-ville, Montréal, Québec, Canada

Matthias Mann
Proteomics and Signal Transduction, Max Planck Institute of Biochemistry, Martinsried, Germany

Kent McDonald
Electron Microscope Laboratory, University of California, Berkeley, California, USA

Peter Meister
Friedrich Miescher Institute for Biomedical Research, Basel, Switzerland

Vincent Messier
Département de Biochimie, Université de Montréal, C.P. 6128, Succursale Centre-ville, Montréal, Québec, Canada

Eugene Melamud
Department of Chemistry, and Lewis-Sigler Institute for Integrative Genomics, Princeton University, Princeton, New Jersey, USA

Stephen W. Michnick
Département de Biochimie, Université de Montréal, C.P. 6128, Succursale Centre-ville, Montréal, Québec, Canada

Quinn M. Mitrovich
Department of Biochemistry and Biophysics, University of California at San Francisco, San Francisco, California, USA

Dario Monachello
Department of Cancer Biology, and Center for Cancer Systems Biology (CCSB), Dana-Farber Cancer Institute, Boston, Massachusetts, USA; Unité de Recherche en Génomique Végétale (URGV), Evry Cedex, France

Chad L. Myers
Department of Computer Science and Engineering, University of Minnesota-Twin Cities, Minneapolis, Minnesota, USA

Corey Nislow
Department of Molecular Genetics, and Donnelly Centre for Cellular and Biomolecular Research; Banting and Best Department of Medical Research, University of Toronto, Toronto, Ontario, Canada

Tracy Nissan
Department of Molecular Biology, Umeå University, Umeå, Sweden

Suzanne M. Noble
Department of Microbiology and Immunology, University of California at San Francisco, San Francisco, California, USA

Jesper V. Olsen
Novo Nordisk Foundation Center for Protein Research, Faculty of Health Sciences, University of Copenhagen, Denmark

Xuewen Pan
Verna and Marrs McLean Department of Biochemistry and Molecular Biology, Department of Molecular and Human Genetics, Baylor College of Medicine, Houston, Texas, USA

Roy Parker
Department of Molecular and Cellular Biology and Howard Hughes Medical Institute, University of Arizona, Tucson, Arizona, USA

Jeffrey A. Pleiss
Department of Molecular Biology and Genetics, Cornell University, Ithaca, New York, USA

Joshua D. Rabinowitz
Department of Chemistry, and Lewis-Sigler Institute for Integrative Genomics, Princeton University, Princeton, New Jersey, USA

Arjun Raj
Department of Bioengineering, University of Pennsylvania, Philadelphia, Pennsylvania, USA

Oliver J. Rando
Department of Biochemistry and Molecular Pharmacology, University of Massachusetts Medical School, Worcester, Massachusetts, USA

Aviv Regev
Howard Hughes Medical Institute, Broad Institute of MIT and Harvard, and MIT Department of Biology, Cambridge, Massachusetts, USA

Howard Riezman
Department of Biochemistry, University of Geneva, Switzerland

Isabelle Riezman
Department of Biochemistry, University of Geneva, Switzerland

Zygy Roe-Zurz
Membrane Protein Expression Center, University of California at San Francisco, San Francisco, California, USA

Assen Roguev
Department of Cellular and Molecular Pharmacology, California Institute for Quantitative Biomedical Research, University of California, San Francisco, California, USA

Sarah A. Sabatinos
Department of Molecular and Computational Biology, University of Southern California, Los Angeles, California, USA

Robert H. Singer
Department of Anatomy and Structural Biology and The Gruss-Lipper Biophotonics Center, Albert Einstein College of Medicine, Bronx, New York, USA

Michael Snyder
Department of Molecular, Cellular and Developmental Biology, Yale University, New Haven, Connecticut, USA

Robert P. St. Onge
Stanford Genome Technology Center, Stanford University, Palo Alto, California, USA

Gregory Stephanopoulos
Department of Chemical Engineering, Massachusetts Institute of Technology, Cambridge, Massachusetts, USA

Robert M. Stroud
Department of Biochemistry and Biophysics, and Membrane Protein Expression Center; Center for the Structure of Membrane Proteins, University of California at San Francisco, San Francisco, California, USA

Motomasa Tanaka
Tanaka Research Unit, RIKEN Brain Science Institute, Hirosawa, Wako, and PRESTO, Japan Science and Technology Agency, Kawaguchi, Saitama, Japan

Kurt Thorn
Department of Biochemistry and Biophysics, University of California at San Francisco, San Francisco, California, USA

Benjamin P. Tu
Department of Biochemistry, University of Texas Southwestern Medical Center, Dallas, Texas, USA

Alexander van Oudenaarden
Department of Physics, and Department of Biology, Massachusetts Institute of Technology, Cambridge, Massachusetts, USA

Marc Vidal
Department of Cancer Biology, and Center for Cancer Systems Biology (CCSB), Dana-Farber Cancer Institute; Department of Genetics, Harvard Medical School, Boston, Massachusetts, USA

Franco J. Vizeacoumar
Banting and Best Department of Medical Research and Department of Molecular Genetics, Terrence Donnelly Center for Cellular and Biomolecular Research, University of Toronto, Toronto, Ontario, Canada

Elisa Varela
Friedrich Miescher Institute for Biomedical Research, Basel, Switzerland, Present address: Fundacion Centro Nacional de Investigaciones Oncologicas Carlos IIII, Melchor Fernandez Almagro 3, Madrid, Spain

Tobias C. Walther
Organelle Architecture and Dynamics, Max Planck Institute of Biochemistry, Martinsried, Germany

Ilan Wapinski
Howard Hughes Medical Institute, Broad Institute of MIT and Harvard, Cambridge, and Department of Systems Biology, Harvard Medical School, Boston, Massachusetts, USA

Markus R. Wenk
Department of Biochemistry, and Department of Biological Sciences, Yong Loo Lin School of Medicine, National University of Singapore, Singapore

Gregg B. Whitworth
Department of Biology, Grinnell College, Grinnell, Iowa, USA

Hyun Youk
Department of Physics, Massachusetts Institute of Technology, Cambridge, Massachusetts, USA

Daniel Zenklusen
Department of Anatomy and Structural Biology and The Gruss-Lipper Biophotonics Center, Albert Einstein College of Medicine, Bronx, New York, USA

Wei Zheng
Department of Molecular, Cellular and Developmental Biology, Yale University, New Haven, Connecticut, USA

PREFACE

It has been 8 years since the last volumes on yeast genetics appeared in *Methods in Enzymology* (Guide to Yeast Genetics and Molecular and Cell Biology; Volumes 350, 351). At that time, *Saccharomyces cerevisiae* was already acknowledged to be the most advanced system for the exploration of basic questions in cell biology. The existence of a small, well-annotated genome, together with simple and facile genetics and a wealth of functional tools, was propelling new discoveries at an unprecedented rate. Predictably, the number of completely uncharacterized genes has since dwindled from roughly one-third to just a few handfuls. Harder to foresee then, however, was how these new tools would qualitatively change the way one could attack biological questions, rapidly ushering in the "postgenomic" era. Suddenly the experimental world became virtually finite. Interested in how the ER makes a particular lipid? One no longer had to carry out an open-ended search hoping to find a few key players and bootstrap one's way through the pathway. Instead, one could immediately focus on the systematic exploration of several hundred genes whose products localize to that organelle. This targeted exploration could in turn be enormously facilitated by the ready availability of comprehensive collections of null or compromised alleles and tagged genes, exploiting rapid methods for genetic crosses and quantitative phenotypic characterization of the candidates. Of comparable impact is the ability to then transition from a function-centered point of view (i.e., the identification of players important for a cellular process of interest) to one in which the full sets of processes that a given gene impacts can be systematically explored, derived from the functional and physical interactions with other gene products. This new paradigm has already pointed to many novel and unanticipated cellular functions.

In the present volume, we have documented many of the major experimental and analytical advances that have catalyzed this change. Additionally, we have highlighted a variety of powerful improvements in biochemical and cytological approaches. Finally, the volume concludes with basic primers on other yeasts, including *Schizosaccharomyces pombe* as well as the clinically relevant *Cryptococcus neoformans* and *Candida albicans*. Yeast is the best characterized eukaryotic cell and will likely remain so for the foreseeable future. The next exciting frontier to consider is how the availability of a complete wiring diagram will inform our understanding of how the cell functions as an integrated machine.

METHODS IN ENZYMOLOGY

VOLUME 71. Lipids (Part C)
Edited by JOHN M. LOWENSTEIN

VOLUME 72. Lipids (Part D)
Edited by JOHN M. LOWENSTEIN

VOLUME 73. Immunochemical Techniques (Part B)
Edited by JOHN J. LANGONE AND HELEN VAN VUNAKIS

VOLUME 74. Immunochemical Techniques (Part C)
Edited by JOHN J. LANGONE AND HELEN VAN VUNAKIS

VOLUME 75. Cumulative Subject Index Volumes XXXI, XXXII, XXXIV–LX
Edited by EDWARD A. DENNIS AND MARTHA G. DENNIS

VOLUME 76. Hemoglobins
Edited by ERALDO ANTONINI, LUIGI ROSSI-BERNARDI, AND EMILIA CHIANCONE

VOLUME 77. Detoxication and Drug Metabolism
Edited by WILLIAM B. JAKOBY

VOLUME 78. Interferons (Part A)
Edited by SIDNEY PESTKA

VOLUME 79. Interferons (Part B)
Edited by SIDNEY PESTKA

VOLUME 80. Proteolytic Enzymes (Part C)
Edited by LASZLO LORAND

VOLUME 81. Biomembranes (Part H: Visual Pigments and Purple Membranes, I)
Edited by LESTER PACKER

VOLUME 82. Structural and Contractile Proteins (Part A: Extracellular Matrix)
Edited by LEON W. CUNNINGHAM AND DIXIE W. FREDERIKSEN

VOLUME 83. Complex Carbohydrates (Part D)
Edited by VICTOR GINSBURG

VOLUME 84. Immunochemical Techniques (Part D: Selected Immunoassays)
Edited by JOHN J. LANGONE AND HELEN VAN VUNAKIS

VOLUME 85. Structural and Contractile Proteins (Part B: The Contractile Apparatus and the Cytoskeleton)
Edited by DIXIE W. FREDERIKSEN AND LEON W. CUNNINGHAM

VOLUME 86. Prostaglandins and Arachidonate Metabolites
Edited by WILLIAM E. M. LANDS AND WILLIAM L. SMITH

VOLUME 87. Enzyme Kinetics and Mechanism (Part C: Intermediates, Stereo-chemistry, and Rate Studies)
Edited by DANIEL L. PURICH

VOLUME 88. Biomembranes (Part I: Visual Pigments and Purple Membranes, II)
Edited by LESTER PACKER

FUNCTIONAL GENOMICS

CHAPTER ONE

Analysis of Gene Function Using DNA Microarrays

Andrew P. Capaldi

Contents

Abstract

This chapter provides a guide to analyzing gene function using DNA microarrays. First, I discuss the design and interpretation of experiments where gene expression levels in mutant and wild-type strains are compared. I then provide a detailed description of the protocols for isolating mRNA from yeast cells, converting the RNA into dye-labeled cDNA, and hybridizing these samples to a microarray. Finally, I discuss methods for washing, scanning, and analyzing the arrays. Emphasis is placed on describing approaches and techniques that help to minimize the artifacts and noise that so often plague microarray data.

Department of Molecular and Cellular Biology, University of Arizona, Tucson, Arizona, USA

Methods in Enzymology, Volume 470
ISSN 0076-6879, DOI: 10.1016/S0076-6879(10)70001-X

1. INTRODUCTION AND EXPERIMENTAL DESIGN

DNA microarrays are a powerful tool for studying the function of signaling proteins, transcription factors, and the networks that they comprise. The basic premise of the approach is simple; by analyzing the global gene expression profile of mutant and wild-type (WT) strains it should be possible to deduce the function of any protein or genomic element perturbed. In practice, however, the design, execution, and interpretation of these experiments require strict attention to detail. This is particularly true if the data is to be interpreted at a quantitative level.

1.1. Single-mutant analysis

Perhaps the most straightforward approach to interrogating gene function using microarrays is to compare the mRNA levels in a strain with a gene deleted to those in the WT parental strain. This approach has now been used to analyze the function of hundreds of *Saccharomyces cerevisiae* genes leading to genome-wide maps of signaling pathways, cell cycle control, and the DNA damage response (Hughes *et al.*, 2000; Ideker *et al.*, 2001; Roberts *et al.*, 2000; Workman *et al.*, 2006). This approach has also been applied to other yeast species such as *Candida albicans* and *Saccharomyces pombe*, leading to interesting insights into the evolution of signaling systems and the molecular determinants of pathogenicity (Enjalbert *et al.*, 2006; Smith *et al.*, 2002; Tsong *et al.*, 2006; Tuch *et al.*, 2008). The difficulty with such experiments, however, is that gene deletion often leads to secondary changes that make the microarray results difficult to interpret. For example, if removing the gene for one transcription factor also leads to changes in the expression of several other transcription factors, it will not be possible to draw simple conclusions about the genes regulated by the factor that is knocked out. Secondary effects can also be created through downstream changes in metabolite concentrations or the activity of signaling molecules, and are therefore a potential problem in almost all mutant strains. Computational approaches can be used to dissect out the direct and indirect effects of such deletions (Workman *et al.*, 2006) but, in general, this deconvolution is difficult to achieve.

To avoid confounding secondary effects, microarray experiments should therefore be designed around the conditional removal of gene function. In many cases, this can be achieved simply by growing the cells in conditions where the gene product is inactive and then probing mRNA levels shortly after activation by appropriate stimuli. For example, in a recent study of the Hog1 network, array analysis showed that deletion of network components had little or no effect on gene expression in standard growth conditions (Capaldi *et al.*, 2008). However, when the same strains were examined shortly after osmotic stress (10–20 min later), dramatic changes were observed in the expression of hundreds of genes. Correlation with genome-wide transcription factor binding data from chromatin immuno-precipitation (ChIP; described in detail in Chapter 4 of this volume) showed that these effects are direct. By contrast, when transcription factor binding data was compared to expression data collected an hour after exposure to osmotic stress, the correlations were weak, demonstrating a buildup of secondary effects.

In those instances where proteins are constitutively active, alternative approaches can be taken to avoid or minimize secondary effects. For example, several studies have taken advantage of analog-sensitive kinase alleles to conditionally block kinase activity (Carroll *et al.*, 2001; Papa *et al.*, 2003; Zaman *et al.*, 2009). Other approaches to conditional perturbation include utilization of inducible promoters, temperature-sensitive alleles, or drug-regulated protein degradation. In cases where conditional perturbations are not possible, secondary effects need to be kept in mind and analysis limited to global effects and correlations.

A highly related approach to examining gene function using DNA microarrays is to measure gene expression in strains with mutations that eliminate sites of posttranslational modification or modify DNA binding sites. Such studies have been used to build detailed models of regulatory mechanisms (Leber *et al.*, 2004; Springer *et al.*, 2003; Wang *et al.*, 2004), but the same caveats apply. If the mutation leads to constitutive changes or the expression changes are examined long after activating conditions are applied, secondary effects can complicate the interpretation.

In either gene deletion or gene mutation experiments, the resulting data can be examined at several levels. First, it is usually possible to gain insight into the function of the element perturbed by examining the biological role of the genes up- or downregulated in the mutant; for example, by looking for gene ontology (GO) terms enriched in the regulated gene-set (Chu *et al.*, 1998; DeRisi *et al.*, 1997). Second, it is often possible to identify the pathway(s) and/or proteins that are affected by your mutation by looking for correlations with previous datasets. As proteins in the same or interacting pathways regulate highly overlapping gene-sets, comparison with previously acquired KO data (e.g., the Hughes compendium; Hughes *et al.*, 2000) can be highly informative (Marion *et al.*, 2004; Segal *et al.*, 2003). It is also possible to indentify transcription factors regulated by, or that interact with, the gene under study by looking for significant overlap with target genes identified using genome-wide ChIP analysis (Bar-Joseph *et al.*, 2003). Finally, the transcription factors regulated directly or indirectly by the genetic element under investigation can be identified using motif analysis (Beer and Tavazoie, 2004; Roth *et al.*, 1998; Wang *et al.*, 2005). Further details of the computational methods used to perform such analyses are provided in Chapter 2 of this volume.

1.2. Double-mutant analysis

While much can be gained by analyzing a strain with a single mutation/deletion under a single condition, the full power of microarray analysis is only realized when multiple related experiments are compared. By examining expression in a range of conditions it is possible to determine when a protein or pathway of interest is activated and how its function depends on signal level and/or type. Furthermore, by examining expression in a variety of strains, a quantitative genome-wide interaction map can be constructed. This approach is particularly powerful when double mutants are examined, as described below.

A direct or indirect interaction between two factors, A and B, can be inferred when microarray data reveal a significant overlap in the gene-sets that A and B regulate. However, such data cannot be used to determine if the factors act independently, cooperatively, or partially cooperatively to regulate these genes (Fig. 1.1A). To complicate matters further the interaction type may vary from gene to gene. Therefore, to distinguish between these mechanisms, gene expression data in the double-mutant strain must also be examined. If the factors A and B act cooperatively to regulate a gene, the expression defect in the single- (AΔ and BΔ) and double (AΔBΔ)-mutant strains will be identical (Fig. 1.1B, middle panel). By contrast, if the factors act independently, the defect in the double mutant will be the sum of the defects in the single mutants (Fig. 1.1B, top panel). In cases where the interaction is partially cooperative, the expression defect in the double mutant will be somewhere in-between the values expected for a fully independent or fully cooperative interaction (Fig. 1.1B, bottom panel).

The strength of double-mutant analysis is the ability to determine how the interaction of any two proteins (or mutations) affects each and every gene in the genome. This analysis not only makes it possible to build detailed network or circuit diagrams but also provides clues as to the precise mechanism of the identified interaction. For example, where all or most of the genes regulated by factors A and B are influenced by a cooperative interaction, it is highly likely that the interaction occurs at the signaling level. By contrast, where only a subset of genes depends on a cooperative interaction, it is more likely that the interaction occurs at the level of transcription factor activation or through another downstream affector. Importantly, however, the ability to build such detailed models is limited by the noise in the microarray analysis; noise that is compounded when single and double mutants are compared. Therefore, multiple measurements need to be made and statistics applied to distinguish between the possible regulatory mechanisms (Fig. 1.1) at each gene. To allow such error analysis, it is critical to first break down the data describing the interaction at each gene into its fundamental components. Following the example in Fig. 1.1,

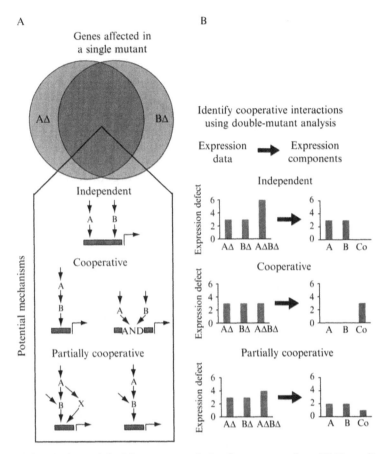

Figure 1.1 Single- and double-mutant analysis of gene expression. (A) Venn diagram summarizing the overlap in genes with a significant defect in gene expression due to deletion of gene A (AΔ) or gene B (BΔ). The wiring diagrams indicate the possible ways factors A and B can interact to regulate expression of overlapping sets of genes. (B) Schematic illustrating the application of the double-mutant approach to analyzing transcriptional network structure and function. The bar graphs on the left show the defects expected in AΔ, BΔ, and AΔBΔ strains for each of the three sample mechanisms. The bar graphs on the right show the values of the three expression components (A, B, and Co) determined by fitting the expression data for the AΔ, BΔ, and AΔBΔ strains (see the text for details).

these components are the induction/repression from A alone, the induction/repression from B alone, and the influence that the interaction between A and B has on expression (or the cooperative component; Fig. 1.1B). The values of these components can be determined simply by comparing the expression defects in microarrays examining the single and double mutants. The array comparing expression in AΔ to the WT reports the value of A + Co for each gene, the array comparing expression in BΔ to the WT reports the value of B + Co for each gene, while the data for the double mutant AΔBΔ compared to the WT reports the value of A + B + Co. In this case, the error for each expression component for each gene can be estimated by propagating the errors through the calculation and using a simple *t*-test (using log expression values). Once this is done it is possible to cluster data based on the type of interaction (pattern of significant A, B, and Co components) and identify groups of genes regulated by a common mechanism. The double-mutant approach, its application, and the analysis of the associated errors are described in more detail in Capaldi *et al.* (2008). Statistical analysis using these or related methods can be carried out using the free software package R (http://www.r-project.org) or MATLAB (http://www.mathworks.com/).

2. METHODS

Once an experiment or series of experiments examining gene function is outlined, it must be translated into a detailed procedure designed to measure mRNA levels while limiting noise (biological or otherwise) in the data. In all cases this means using two color DNA microarrays. The outline of the procedure (developed by DeRisi *et al.*, 1997) is as follows: Cells are grown under the appropriate conditions and mRNA is extracted and purified. The mRNA is then converted into cDNA using reverse transcription and labeled with one of two fluorophores (Cy3 or Cy5). Two cDNA samples, an experimental sample labeled with Cy5 and a control labeled with Cy3, are then hybridized to an array consisting of thousands of different DNA fragments spotted onto a glass slide, where each fragment is complimentary to a single gene. These arrays are then washed to remove any cDNA that binds nonspecifically and analyzed using a laser scanner to measure the Cy5 and Cy3 fluorescence. Finally, after data normalization, the ratio of Cy5 to Cy3 fluorescence at each spot is calculated and used to determine the difference in the mRNA expression levels in the two samples.

2.1. Experimental design

As each microarray compares the expression levels in the two samples, the first step in designing a microarray procedure is deciding which samples will be compared on a given array. The idea is to accurately measure the parameters of interest using the smallest total number of arrays. For large experiments, identifying the best experimental design can be complicated (Kerr and Churchill, 2001), but for more limited experiments some simple principles apply. In most cases where a mutant is being analyzed, it is best to directly compare the WT and mutant strains, grown under the conditions of interest, on the same array. This way the influence that the mutant has on gene expression is determined with the errors from a single array. An alternative approach is to measure gene induction after stimulus in the WT strain on one array (e.g., WT + stress versus WT in no stress) and the gene induction after stimulus in the mutant strain (e.g., AΔ + stress versus AΔ in no stress) on a separate array. In this commonly used scheme the influence that a mutant has on gene expression can only be calculated by dividing the values from the two arrays and thus the errors from each individual array are multiplied. However, in the case where mutant effects are measured on a single array, it is still important to examine expression in the WT strain (e.g., WT + stress versus WT in no stress) to ensure that the genes influenced by mutation are also regulated in the WT background. Where double mutants are examined it is best to compare the expression levels directly to the single mutant(s) on the same array. This reduces the magnitude of the change in gene expression seen on the array and thus the overall noise as its major component is proportional to signal.

2.2. Cell growth

Once the experiments are designed, cells need to be grown and harvested carefully to limit unwanted sample-to-sample variation. First, strains that are going to be compared on a single array should be grown at the same time and in identical medium as variation in nutrient levels, temperature, and other parameters can introduce substantial noise. Second, the strains should grow for at least two doublings to wash out any differences in the overnight cultures. Third, cells should be harvested at the same optical density (OD) and this OD should be selected so that the cells are approximately one doubling away from any transition induced by nutrient depletion. For example, if cells are studied in log growth phase they should be harvested at a density substantially below that of the diauxic shift to ensure that small variations in cell number do not affect the gene expression pattern (for *S. cerevisiae* $OD_{600} = 0.6$ works well). Finally, it is best to harvest cells by filtration using a 0.2 μm 90 mm filters (Millipore). This filter is then rolled, placed in a 50 ml conical tube and submerged in liquid nitrogen. This ensures rapid harvesting (<1 min) and little sample-to-sample variation. For the protocol described below it is best to harvest between 75 and 150 OD_{600}/ml units. Once harvested the cells can be stored at $-80\ °C$ for several weeks before the RNA is extracted.

2.3. Total RNA isolation and purification

To extract the mRNA from the frozen cells, first add 12 ml of AE buffer (50 mM sodium acetate (pH 5.2) and 10 mM EDTA) to the 50 ml conical tube and then rotate/shake the tube to remove the cells from the filter.

When this is done for each of the tubes in the set (no more than eight at a time, but always prepare samples to be compared on the same array at the same time), transfer the cells and buffer to a 50 ml centrifuge tube and add 800 μl of 25% (w/v) SDS to each of the samples. Finally, add 12 ml of 65 °C acid phenol (Fluka #77608, note that this is not standard buffer-saturated phenol) to each tube and incubate for 10 min in a 65 °C water bath, vortexing every 30 s or so. When this procedure is finished, cool the samples on ice for 5 min and then centrifuge for 20 min at 12,000×g and 4 °C. Carefully decant the phenol–buffer mix from these tubes into prespun (5 min at 1500 rpm) phase lock tubes (5-Prime) and then add 13 ml of chloroform to each tube, mix vigorously, and spin at 3000 rpm for 10 min. After centrifugation the RNA will be in the 10–12 ml aqueous layer at the top of the tube, separated from the organic phase by the gel in the phase lock tubes. Pour this solution into a clean centrifuge tube and add 1 ml of 3 M sodium acetate (pH 5.2) and 10 ml of isopropanol; mix by inverting the tube several times and spin for 45 min at 17,000×g and 10 °C. At this point the total RNA will be present in an approximately 1 cm white pellet. Carefully decant the isopropanol and add 10 ml of 70% ethanol to each tube, without disturbing the pellet, and spin again at 17,000×g and 10 °C but this time for 20 min. Decant as much of the ethanol as possible and then spin the tube at 17,000×g again for 1 min to collect any remaining liquid at the bottom of the tube and remove it carefully with a pipette. Finally, let these pellets dry on the bench until they are translucent (30–60 min).

Once the total RNA pellets are dry, resuspend them in 800 μl of RNase-free water and measure the absorbance at 260 and 280 nm. The sample should have >2 mg of RNA and an A_{260}/A_{280} >2.0. It is also useful, especially in the first few RNA preparations or if problems are encountered, to check the integrity of the RNA sample using an Agilent Bioanalyzer or a similar device (agarose gels are only useful for detecting severe degradation). Here a good quality sample should have distinct rRNA and tRNA bands with well-defined edges. If the sample fails in any of the quality controls it should be discarded.

2.4. Purification of poly-A RNA

To isolate mRNA from the total RNA sample, cellulose resin with a poly-deoxythimidine oligomer attached (oligo-dT cellulose) is used to purify transcripts with a poly-A tail. This purification should be done on the same day as the total RNA purification to limit degradation. First, wash 60 mg of cellulose resin three times with 750 μl of NETS buffer (0.6 M NaCl, 10 mM EDTA, 10 mM Tris–HCl (pH 8.0), 2% (w/v) SDS) in a 2 ml screw cap tube. Here the resin should be spun at 3000 rpm for 1 min on a benchtop centrifuge between washes and the buffer removed by aspiration. At this stage, incubate 750 μl of 2–4 mg total RNA at 65 °C for 10 min, and then add to a 2 ml tube along with 750 μl of 2× NETS buffer also at 65 °C. Each tube should then be left to mix on a rotator for 1 h at room temperature. After incubation, apply this sample to a disposable column (BioRad #732-6008) that has been washed once with NETS buffer. Once the resin has settled in the column, wash it three times with 750 μl NETS buffer and then elute the mRNA with 650 μl of 65 °C ETS buffer (NETS buffer without the NaCl) by injecting it directly into the column bed. Finally, add 65 μl of 3 M sodium acetate and 650 μl of isopropanol to these tubes, mix well by inversion, and incubate at -20 °C overnight. The next morning spin the sample at full speed in a benchtop centrifuge for 1 h at 4 °C. When the spin is complete, remove the isopropanol buffer mix from the tube and add 250 μl of 70% ethanol to the sample, taking care not to disturb the pellet, and spin at full speed for 20 min at room temperature. Carefully remove the ethanol from these samples and allow them to air dry completely (residual ethanol will inhibit the reverse transcription in the next step). When dry, the pellets will become white and powdery. Resuspend these pellets in 20 μl of RNase-free water and then spin for 1 min at full speed to remove the cellulose fragments that were not trapped by the column. Remove the supernatant and measure the absorbance and A_{260}/A_{280} ratio. This is best done on a nanodrop spectrophotometer (Thermo Scientific) due to the small volume. The yield should be between around 20 μg and the A_{260}/A_{280} ratio again greater than 2.0.

2.5. Reverse transcription and dye labeling

On the same day as the mRNA is purified it should be converted into cDNA by reverse transcription. Combine four micrograms of poly-A RNA with 5 μg of an oligo-dT primer (T_{20}) and 5 μg of a random primer (N_9) in a total volume of 15.5 μl and incubate at 70 °C for 8 min before cooling on ice. Once cool, perform cDNA synthesis using AffinityScript reverse transcriptase (Stratagene) by adding the RNA and primer mix to 2 μl of enzyme, 3 μl of AffinityScript buffer, 3 μl of 100 mM DTT, 5.9 μl of water, and 0.6 μl of 50× aa-dNTP mix and incubate at 42 °C for 2 h. Here the aa-dNTP mix is made up of 1 mg of aminoallyl-dUTP, 20 μl of water, 30 μl of 100 mM dTTP, and 50 μl of 100 mM A, C, and GTP (store this nucleotide mix at -20 °C in single use aliquots). This reaction will result in a cDNA library where approximately 1/10 bases have a free amine group that can be labeled by Cy5 or Cy3. To degrade the RNA template, add 4 μl of 1 M NaOH and 8 μl of 50 mM EDTA and incubate at 65 °C for 10 min. Finally, neutralize the solution using 40 μl of 1 M HEPES, pH 7.0. The cDNA can then be purified using a Clean and Concentrator-5 Kit (Zymo Research) following the manufacturer's instructions except that the cDNA–HEPES buffer mix should be mixed with 1 ml of binding buffer before applying it to the column. Elute the DNA from the column in 12 μl of water and then determine the yield and A_{260}/A_{280} ratio using the nanodrop spectrophotometer; they should be ≥ 2 μg and 1.8, respectively. The cDNA samples can be stored at -20 °C for many weeks before labeling and hybridization.

Once the cDNA is synthesized it needs to be labeled with either Cy5 (typically the sample) or Cy3 (typically the control). To do this, add 1 μl of 1 M sodium bicarbonate buffer (pH 9.0) to 10 μl of the cDNA solution and add 1 μl of N-hydroxysuccinimidyl ester Cy5 or Cy3 (GE Biosciences) that has been resuspended in DMSO and incubate at room temperature for 4 h. Each Cy dye pack has enough dye to label 4–8 samples. After labeling, purify the samples using the Clean and Concentrator-5 kit, following the manufacturer's instructions, and again measure the concentration and A_{260}/A_{280} ratio using the nanodrop spectrophotometer. The purified and labeled cDNA should have visible color after labeling. If necessary, these samples can be snap frozen in liquid nitrogen and stored at -80 °C.

2.6. Hybridization

The labeled cDNA samples are now ready to hybridize to a DNA micro-array. Many types of microarrays are available for gene expression analysis and each has its advantages and disadvantages. The most common of these are the printed arrays where PCR products or DNA oligomers are spotted onto a polylysine-coated slide and commercial arrays from companies such as Agilent and Roche Nimblegen where oligomers are synthesized directly on the slide. The advantage of printed arrays are that, when thousands of arrays are used, the cost can be as low as $20 per array. However, printed arrays have several distinct disadvantages. First, there tends to be significant variation in the quality and size of the DNA spots. This makes it difficult to accurately determining the expression ratio for some genes and contributes substantially to replicate variation. Second, these arrays have to be post-processed to neutralize the otherwise highly charged lysine surface. This postprocessing leads to imperfections on the surface of the slide and sub-stantial background variation. Finally, the polylysine coated surface is deli-cate and is often damaged in the printing and hybridization procedure leading to further background noise. By contrast, the commercial arrays are printed with high accuracy and on more stable substrates, resulting in very low background noise. Moreover, the stability of the slides surface means that the sample applied to these arrays can be vigorously mixed during hybridization. This dramatically increases the signal-to-noise ratio and means that far less cDNA needs to be applied to the array during hybridization. The cost of such arrays is presently greater than $100 a piece, but this continues to fall. Here, I will describe hybridization to Agilent arrays (see the manufactures website for further details), but the same samples can be hybridized to arrays from other companies following the manufacturer's instructions, or to printed arrays using protocols from Derisi Lab (DeRisi et al., 1997).

The eight array format from Agilent (eight, 15,000 spot arrays per slide) works well when 100 ng of Cy5- and 100 ng of Cy3-labeled cDNA is hybridized to each array. To do this, bring 125 ng of Cy5- and 125 ng of Cy3-labeled sample to 25 μl total volume. Heat this sample to 98 °C for 2 min. Cool the sample by centrifugation for 1 min and then add 25 μl of 2× Hi-RPM hybridization buffer (Agilent) to the sample. Carefully apply 40 μl of the sample to the gasket slide positioned in the base of SureHyb chamber, repeat seven more times with different samples to fill each position, and place the microarray face down onto this gasket slide. Add the top to the hybridization chamber, tighten the screw, and then rotate the chamber slowly to ensure that the large bubble in the chamber moves freely (this bubble is critical for mixing during the hybridization) and that no smaller bubbles are stuck to the array surface. If any bubbles remain fixed to the array, gently tap the chamber on a hard surface until they are dislodged and then place in the rotating oven (Agilent) at 65 °C for 17 h.

2.7. Microarray washing

Once the sample is hybridized to the array, excess sample and cDNA that is bound nonspecifically must be washed off and the array dried. Listed here is a protocol that works well for Agilent arrays; those working with other types of arrays should select the appropriate alternative protocol.

Fill two slide staining dishes with wash buffer I (6× SSPE, 0.005% N-Lauryl sarcosine where SSPE is 150 mM NaCl, 10 mM sodium phosphate, and 1 mM ETDA, pH 7.4), a third chamber with wash buffer II (0.06× SSPE, 0.005% N-Lauryl sarcosine), and a fourth chamber with ozone protection and drying solution (Agilent). Place a slide rack in the second chamber and set chambers 2–4 up on stir plates at a medium setting, ensuring that the stir bars used are small enough to remain below the bottom of the slide rack. Submerge the first array in chamber 1 and pry it open using plastic tweezers so that the gasket slide falls away. Gently move the array from side to side while submerged in the chamber to remove any bubbles from the surface and then place it in the rack in chamber 2 (only handle the label on the array). Repeat this process until all the slides are in chamber 2 and then leave the arrays in this low-stringency wash for 1 min. At this stage transfer the entire slide rack to chamber 3, ensuring that the slides spend minimal time exposed to the air and that little of the wash buffer I is transferred into this new chamber. The slides should then be allowed to sit in this high-stringency wash for exactly 1 min before transferring the rack to the drying solution for 30 s. At this time slowly lift the rack out of the chamber, ensuring that droplets do not form on the surface of the array. Now the array is dry and ready to scan.

2.8. Array scanning

To quantify the amount of Cy5- and Cy3-labeled DNA hybridized to each probe (spot on the array), the washed array is analyzed with a laser scanner that excites both Cy5 (at 635 nm) and Cy3 (at 532 nm) and then measures the emitted light at the appropriate wavelength (570 and 670 nm, respectively) using a photomultiplier. It is best to perform the scan immediately after washing the arrays, but arrays can be stored in a dry ozone-free area for a day or more before scanning if necessary. The most common scanners are the Axon 4000B from Molecular Devices and the DNA microarray Scanner from Agilent.

To get the best quality data from the arrays, the scanner needs to be set up appropriately. First, the lasers should be focused onto the surface of the slide that is spotted with the probes. This focal plane is easily identified as the position where a scan gives the highest overall signal. Next, keeping the laser power at 100%, the voltage applied to the photomultiplier tube should be set to ensure that the highest Cy5 and Cy3 signals measured are just below the maximum of the digitization range (65,536 for a 16-bit A to D converter). This ensures the highest possible signal-to-noise ratio without losing data at some probes due to saturation. Often it is necessary to scan arrays several times to find such settings. This does not present a problem as the Cy dyes are photostable.

The resulting image should show clear spots, each with little pixel-to-pixel variation in the Cy5 and Cy3 ratio, surrounded by a uniform background with low signal (40–50 units on 65,000 unit scale, Fig. 1.2A). Poor quality array images generally indicate a problem with sample labeling or with the wash and hybridization steps. If the signal/noise is uniformly low in a single color, the problem is likely to be poor labeling. This is often due to degraded dye. If the signal/noise is poor in both colors (Fig. 1.2B and C), the problem likely lies in the wash (low signal; overly stringent washing, high background; poor washing). However, poor labeling in both channels can lead to similar problems. Large regions without signal can be caused by bubbles on the surface of the array or by leakage during hybridization (Fig. 1.2D). Finally, speckles on the array are often caused by dust or precipitate in the hybridization or wash buffers. Array images should be saved as a high-resolution tif image file for further analysis

2.9. Gridding and normalization

Once a microarray image has been collected the precise position and identity of each probe (or "spot") must be identified. This is done through a process known as gridding. First a grid file must be assembled; this establishes the identity and expected location of each spot on the array. The grid is then overlaid onto the array and any variation in spot location or

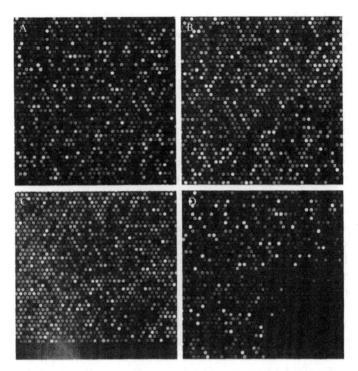

Figure 1.2 High-resolution DNA microarray images showing common hybridization artifacts. (A) A high-quality array as defined by a high signal-to-noise ratio, a low (background level) signal at the negative control spots, and an absence of washing or hybridization artifacts. (B) A poorly washed array showing the characteristic high background signal. (C) An array with variable background signal caused by cDNA precipitation and poor washing. Such precipitation is often caused by loading too much cDNA onto the chip. (D) An array with nonuniform hybridization due to leakage from the hybridization chamber.

size adjusted for through a probe-by-probe alignment. The gridding software that comes with most scanners does this automatically but in many cases some manual adjustments need to be made. At the end of the process, spots that overlap with artifacts on the array, or are highly irregular, should be flagged to ensure that they do not affect the downstream analysis. At this stage, the Cy5 and Cy3 signal intensity at each probe is determined within each spot using the same software. While this data represents the gene expression changes measured on the array, normalization is required before detailed analysis can be performed. First, any systematic difference between the Cy5 and Cy3 signals, due to differences in the quantum yield and the amount of cDNA loaded on the array, must be corrected. For printed arrays this can be accomplished by multiplying the Cy5 and Cy3 signals by a constant so that their average, across all spots, is the same. Care must be taken, however, to ensure that this is an appropriate normalization.

For example, when a strain with the gene for a global repressor deleted is compared to the WT strain, the average Cy5 and Cy3 signals are expected to differ systematically. In this case a set of spike-in control RNAs, or an appropriate subset of genes, can be used to normalize the Cy5 and Cy3 signals. For commercial arrays, where the DNA in each spot is at high-density and precisely aligned, the Cy3-to-Cy5 ratio tends to be nonlinear as a function of signal intensity, due to dye quenching and other effects, and thus the more sophisticated locally weighted scatterplot smoothing (LOWESS) normalization procedure should be used. Finally, the data for the spots with weak intensity need to be thrown out or at least weighted appropriately. One simple way to do this is to eliminate all data where both the Cy3 and Cy5 signals are less than 1.5-fold above the background. For printed arrays the background is determined by the signal around each spot, while in commercial arrays it is determined by the signal at negative control spots printed at various positions on the array. The former method is critical for subtracting away a variable background signal, while the later method is better where the surface chemistry inside and outside the spot are different and background variation is negligible. As an alternative to throwing out data with low signal, pixel-to-pixel variation in the background and spot intensity can be used to calculate an error range for each spot and these error values propagated through the analysis. Such data filtering and normalization can be carried out in wide number of databases (e.g., the Stanford Microarray Database or Rosetta Resolver). Such databases are also extremely useful for storing microarray results and images and building tables of data from multiple arrays. These tables can then be fed into one or more of the wide range of microarray analysis packages available, or programs such as R and MATLAB (The MathWorks), for detailed analysis as described further in Chapter 2 of this volume.

REFERENCES

Bar-Joseph, Z., Gerber, G. K., Lee, T. I., Rinaldi, N. J., Yoo, J. Y., Robert, F., Gordon, D. B., Fraenkel, E., Jaakkola, T. S., Young, R. A., and Gifford, D. K. (2003). Computational discovery of gene modules and regulatory networks. *Nat. Biotechnol.* **21,** 1337–1342.

Beer, M. A., and Tavazoie, S. (2004). Predicting gene expression from sequence. *Cell* **117,** 185–198.

Capaldi, A. P., Kaplan, T., Liu, Y., Habib, N., Regev, A., Friedman, N., and O'Shea, E. K. (2008). Structure and function of a transcriptional network activated by the MAPK Hog1. *Nat. Genet.* **40,** 1300–1306.

Carroll, A. S., Bishop, A. C., DeRisi, J. L., Shokat, K. M., and O'Shea, E. K. (2001). Chemical inhibition of the Pho85 cyclin-dependent kinase reveals a role in the environmental stress response. *Proc. Natl. Acad. Sci. USA* **98,** 12578–12583.

Chu, S., DeRisi, J., Eisen, M., Mulholland, J., Botstein, D., Brown, P. O., and Herskowitz, I. (1998). The transcriptional program of sporulation in budding yeast. *Science* **282,** 699–705.

DeRisi, J. L., Iyer, V. R., and Brown, P. O. (1997). Exploring the metabolic and genetic control of gene expression on a genomic scale. *Science* **278**, 680–686.

Enjalbert, B., Smith, D. A., Cornell, M. J., Alam, I., Nicholls, S., Brown, A. J., and Quinn, J. (2006). Role of the Hog1 stress-activated protein kinase in the global transcriptional response to stress in the fungal pathogen *Candida albicans. Mol. Biol. Cell* **17**, 1018–1032.

Hughes, T. R., Marton, M. J., Jones, A. R., Roberts, C. J., Stoughton, R., Armour, C. D., Bennett, H. A., Coffey, E., Dai, H., He, Y. D., Kidd, M. J., King, A. M., *et al.* (2000). Functional discovery via a compendium of expression profiles. *Cell* **102**, 109–126.

Ideker, T., Thorsson, V., Ranish, J. A., Christmas, R., Buhler, J., Eng, J. K., Bumgarner, R., Goodlett, D. R., Aebersold, R., and Hood, L. (2001). Integrated genomic and proteomic analyses of a systematically perturbed metabolic network. *Science* **292**, 929–934.

Kerr, M. K., and Churchill, G. A. (2001). Experimental design for gene expression microarrays. *Biostatistics* **2**, 183–201.

Leber, J. H., Bernales, S., and Walter, P. (2004). IRE1-independent gain control of the unfolded protein response. *PLoS Biol.* **2**, E235.

Marion, R. M., Regev, A., Segal, E., Barash, Y., Koller, D., Friedman, N., and O'Shea, E. K. (2004). Sfp1 is a stress- and nutrient-sensitive regulator of ribosomal protein gene expression. *Proc. Natl. Acad. Sci. USA* **101**, 14315–14322.

Papa, F. R., Zhang, C., Shokat, K., and Walter, P. (2003). Bypassing a kinase activity with an ATP-competitive drug. *Science* **302**, 1533–1537.

Roberts, C. J., Nelson, B., Marton, M. J., Stoughton, R., Meyer, M. R., Bennett, H. A., He, Y. D., Dai, H., Walker, W. L., Hughes, T. R., Tyers, M., Boone, C., *et al.* (2000). Signaling and circuitry of multiple MAPK pathways revealed by a matrix of global gene expression profiles. *Science* **287**, 873–880.

Roth, F. P., Hughes, J. D., Estep, P. W., and Church, G. M. (1998). Finding DNA regulatory motifs within unaligned noncoding sequences clustered by whole-genome mRNA quantitation. *Nat. Biotechnol.* **16**, 939–945.

Segal, E., Shapira, M., Regev, A., Pe'er, D., Botstein, D., Koller, D., and Friedman, N. (2003). Module networks: Identifying regulatory modules and their condition-specific regulators from gene expression data. *Nat. Genet.* **34**, 166–176.

Smith, D. A., Toone, W. M., Chen, D., Bahler, J., Jones, N., Morgan, B. A., and Quinn, J. (2002). The Srk1 protein kinase is a target for the Sty1 stress-activated MAPK in fission yeast. *J. Biol. Chem.* **277**, 33411–33421.

Springer, M., Wykoff, D. D., Miller, N., and O'Shea, E. K. (2003). Partially phosphorylated Pho4 activates transcription of a subset of phosphate-responsive genes. *PLoS Biol.* **1**, E28.

Tsong, A. E., Tuch, B. B., Li, H., and Johnson, A. D. (2006). Evolution of alternative transcriptional circuits with identical logic. *Nature* **443**, 415–420.

Tuch, B. B., Li, H., and Johnson, A. D. (2008). Evolution of eukaryotic transcription circuits. *Science* **319**, 1797–1799.

Wang, Y., Pierce, M., Schneper, L., Guldal, C. G., Zhang, X., Tavazoie, S., and Broach, J. R. (2004). Ras and Gpa2 mediate one branch of a redundant glucose signaling pathway in yeast. *PLoS Biol.* **2**, E128.

Wang, W., Cherry, J. M., Nochomovitz, Y., Jolly, E., Botstein, D., and Li, H. (2005). Inference of combinatorial regulation in yeast transcriptional networks: A case study of sporulation. *Proc. Natl. Acad. Sci. USA* **102**, 1998–2003.

Workman, C. T., Mak, H. C., McCuine, S., Tagne, J. B., Agarwal, M., Ozier, O., Begley, T. J., Samson, L. D., and Ideker, T. (2006). A systems approach to mapping DNA damage response pathways. *Science* **312**, 1054–1059.

Zaman, S., Lippman, S. I., Schneper, L., Slonim, N., and Broach, J. R. (2009). Glucose regulates transcription in yeast through a network of signaling pathways. *Mol. Syst. Biol.* **5**, 245.

CHAPTER TWO

AN INTRODUCTION TO MICROARRAY DATA ANALYSIS AND VISUALIZATION

Gregg B. Whitworth

Contents

Department of Biology, Grinnell College, Grinnell, Iowa, USA

Methods in Enzymology, Volume 470
ISSN 0076-6879, DOI: 10.1016/S0076-6879(10)70002-1

Abstract

Microarray experiments offer a potential wealth of information but also present a significant data analysis challenge. A typical microarray data analysis project involves many interconnected manipulations of the raw experimental values, and each stage of the analysis challenges the experimenter to make decisions regarding the proper selection and usage of a variety of statistical techniques. In this chapter, we will provide an overview of each of the major stages of a typical yeast microarray project. We will focus on providing a solid conceptual foundation to help the reader better understand each of these steps, will highlight useful software tools, and will suggest best practices where applicable.

1. Introduction

1.1. Overview

Figure 2.1 illustrates a typical data analysis scheme for a microarray experiment. The first challenge in a microarray project is to develop a clear definition of the biological question of interest, as all subsequent tasks will be guided by the goals of the study. Given the material and time costs associated with microarray experiments, it is well worth designing a data analysis scheme in tandem with your experimental approach, to ensure that the data you produce will be sufficient to rigorously address your question of interest. Once an overall goal for the project is set, the key experimental design choices include the array platform, probe construction and sequence, and experimental protocol. As these aspects of a microarray experiment are covered elsewhere in this series,[1] below we will focus our discussion on the experimental design decision which has the greatest impact on data analysis: single-sample versus competitive hybridization. Once an experiment has been performed, data analysis begins with the acquisition of digital images describing the signal intensity associated with each "spot" or probe on an array. Image analysis results in a table linking each probe on the array with a quantitative response value. In the preprocessing stage, these raw response values are subjected to quality control and are normalized to allow for a fair comparison of measurements between probes on a single array and among different arrays. Once a high-quality set of comparable values has been

---

[1] See Chapter 1 by Andrew Capaldi and Chapter 3 by Maki Inada and Jeff Pleiss in this volume.

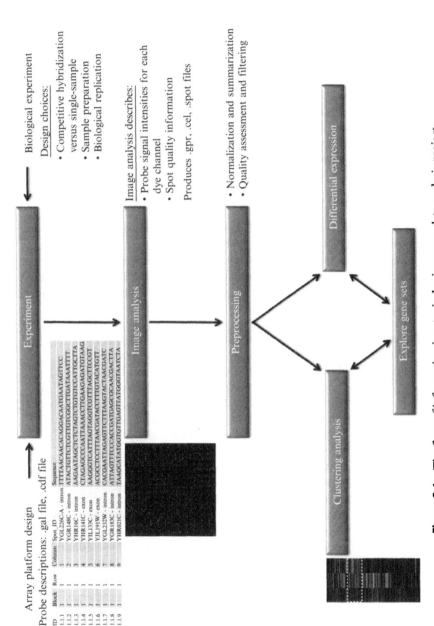

Figure 2.1 The flow of information in a typical microarray data analysis project.

obtained which describe the behavior of each unique target in each experimental sample, results can be visualized with a variety of clustering tools or subjected to statistically rigorous methods for identifying targets that are affected by each experimental treatment. Finally, after we have identified an interesting set of mRNAs, genes, or other targets, there are a number of ways to explore biologically meaningful associations among members of the set.

In this chapter, we will walk through each of these major steps in the chronology of a microarray data analysis scheme. The goals of this chapter are to introduce common terminology, provide a conceptual basis for understanding the major design concerns, and suggest useful software and statistical tools where applicable.

1.2. Commonly used terms in microarray data analysis

Microarray platforms and experimental designs have grown increasingly varied as new uses for microarray technologies have emerged. The terminology used to discuss microarray data analysis procedures is often confusing to newcomers because the vocabulary must be sufficiently abstract to describe a wide range of technologies and assays. Here are some useful working definitions:

Microarray	A microarray can be broadly defined as a high-density grid of probes attached to a two-dimensional surface.
Platform	Platform specifies a particular array technology. This includes the choice of surface material (glass, silicon), probe material (PCR product, oligonucleotide), probe targets (ORFs, SNPs, tiled genome-sequence), and manufacturing process (robotically printed, photolithography, ink-jet).
Probe	A nucleic acid bait sequence, usually a DNA oligonucleotide or PCR product, designed to hybridize to a specific target in samples loaded onto the surface of an array. Arrays may include more than one probe sequence for each target.
Spot, feature	A location on an array where a specific probe has been printed or synthesized. Many array designs contain replicate spots, meaning that a given probe is represented by more than one spot on the array.
Biological sample	The starting material from a biological experiment, such as total RNA or DNA. In some designs this is directly labeled and loaded onto the microarray, in others it is first converted (e.g., from RNA to cDNA).

Target	A molecular species in the sample for which there is a specific probe on the array.
Label	Labels are attached to targets to allow for the detection of material hybridized to each spot on the array. Usually labels are fluorescent dyes, such as Cy3 and Cy5, which are excited by a laser and detected using dye-specific filter sets.

1.3. A simple case study

Consider a simple yeast experiment which we will use to illustrate different aspects of a microarray data analysis project throughout this chapter. Suppose we are interested in performing a common class of microarray experiment: identification of genes which are differentially expressed when yeast cells are exposed an environmental stress. Our biological samples in this example might be total RNA isolated from cells before the treatment and at various time points following the treatment. Our material preparation protocol could involve a reverse transcription reaction in which we synthesize a cDNA copy of isolated RNA molecules, followed by a labeling reaction in which we attach a fluorescent dye to our cDNA targets. Our platform would be a yeast ORF array, containing DNA oligonucleotide probes which are designed to hybridize to sequences specific to each gene in the yeast genome.

Our goal in designing this microarray experiment and analyzing our results will be to identify genes which are differentially expressed in the pre- and posttreatment samples.

2. EXPERIMENTAL DESIGN: SINGLE-SAMPLE VERSUS COMPETITIVE HYBRIDIZATION

From a data analysis perspective, one of the most important decisions you will need to make when planning your microarray experiment is how to setup your sample hybridizations. In single-sample, or "one-channel," experiments each biological sample being assayed is individually hybridized on an array. In a competitive hybridization, or "two-channel" design, two or more biological samples with unique dye-labels are mixed and hybridized together on an array. Typically, your choice of array platform will be tied to your preferred hybridization strategy.

2.1. Single-sample hybridization

A single-sample design can be appealing because it is both conceptually simple and analogous to other common DNA or RNA detection procedures (e.g., northern blot analysis). In these experiments, target molecules from each biological sample are labeled with fluorescent dye and hybridized to the surface of a single array. The intensity of the label associated with each probe is quantified using digital image analysis and these intensity values are used in all subsequent manipulations to represent the quantity of each target in the starting sample. In order for label intensities to be comparable between probes, the labeling procedure must operate at a similar efficiency on each target probed by the array. This means that labeling must be independent of the sequence length or base composition of the target. To meet this requirement, end-labeling schemes are commonly used, whereby a single fluorescent molecule is covalently linked to targets. The major advantage of single-sample designs is that there is a direct, and under optimal conditions potentially linear, relationship between the concentration of each target in the original biological sample and the label intensity measured on the associated probe.

If we were to analyze the biological samples in our simple case study above using a single-channel scheme we would hybridize end-labeled cDNA copies of each of our original biological RNA samples onto individual yeast ORF arrays. The absolute dye intensity imaged on each spot would be used to represent the quantitative expression level of the associated gene target. We could then compare expression levels between samples mathematically. For example, we might ask what the fold-change in expression levels of each transcript is between a pre- and posttreatment sample by comparing probe intensities observed on two different arrays.

2.2. Competitive hybridization

While single-sample designs attempt to measure the *absolute* levels of probed DNA or RNA species, competitive hybridization designs measure the *relative* concentration of each probed species in two or more samples. In standard competitive hybridization designs, two biological samples, usually a "reference" or control sample and an "experiment" or treatment sample, are labeled with two different dyes. The samples are then mixed and hybridized on the surface of an array. Two digital images are acquired for each array using a wavelength filter specific to the emission spectrum of each of the two dyes. During image analysis, the ratio of the two dye intensities is calculated for each spot. This ratio value is then used in all subsequent analysis steps to represent the relative level of targets in the two starting samples.

Competitive hybridization designs can be more difficult for those new to microarrays to understand, because each measurement is inherently relative, rather than absolute. However, competitive hybridization also offers a number of significant advantages over single-sample designs. Competitive hybridization schemes tend to be insensitive to differences in the relative efficiency with which different target sequences pass through the sample preparation protocol and are detected on the array. In a typical gene expression experiment, for example, we can imagine that the efficiency of RNA isolation, cDNA synthesis, dye-labeling, and probe hybridization could each be affected by the length and base composition of each target. Because competitive hybridization arrays measure the relative levels of the same target sequence from two different starting biological samples on each probe, differences in the performance characteristics of individual target sequences are not as relevant to the measurements being made.[2]

One disadvantage to competitive hybridization designs is that, with two samples used per array, there is twice the material cost for each. A second consideration to be aware of is that no two dyes will perform identically in an array experiment. In two-color experiments using Cy3 and Cy5, for example, there is often a "green" or Cy3-shift among low-intensity spots. For this reason, it is important to flip the association of dyes and samples in replicate array experiments ("dye-flipped" replicates), to ensure that observations are not the result of a dye-intensity bias. In Section 4, we will also discuss computational methods which allow us to assess and mitigate dye-intensity bias.

Competitive hybridization designs are not limited to two samples. Conceptually the only limit to the number of samples that can be put on a single array is the number of unique dye wavelengths that can be reliably differentiated. Several commercial scanners now support up to four dyes and several data analysis packages are already prepared to handle arbitrarily large numbers of dye-channels.

2.3. Choosing the best approach

Both of these hybridization strategies have been used in the microarray field to produce accurate and reproducible data. A good litmus test for which approach is best suited to your study may be to consider whether or not your question of interest is *comparative*. In the case study introduced above, our biological question is indeed comparative: we are interested in measuring changes in the relative abundance of transcripts in an experimental and control sample. We have a clear reference, the pretreatment sample, to which we want to compare transcript abundance in posttreatment samples.

---

[2] This is not to suggest that probe intensity is not a consideration in competitive hybridization experiments. As we discuss below, for example, it is important to be aware of dye-specific intensity biases. However, with proper normalization, competitive hybridization designs can be extremely robust across a very wide range of probe intensities.

In this example, the competitive hybridization approach allows us to perform this comparison directly, rather than as mathematical manipulation, thereby simplifying our data analysis approach and avoiding the potential propagation of error.

The advantage of competitive hybridization designs for comparative studies becomes even stronger when we consider investigations involving more than one experimental factor. Suppose we are interested in testing both a wild-type (WT) and mutant strain in our example stress experiment. In a first pass experiment, we might hybridize a pre-stress sample against a post-stress sample for each of these two strains, on two different arrays. How would we interpret our results if we observed subtle differences between the two strains, which could plausibly be either biologically meaningful or simply due to random variation? In a competitive hybridization design we can add a third array into the mix which makes this comparison directly, hybridizing poststress samples from the WT and mutant strains on the same array. Because at least one of two original biological samples has been hybridized onto each of these three arrays we have a powerful tool for ensuring that the observations made on each array are directly comparable.[3] Linear analysis packages such as Limma, discussed in Section 6, allow us to use competitive hybridization designs such as this to rigorously assess the statistical significance of apparent changes in expression level of a given target between two samples.

3. IMAGE ANALYSIS

The computational phase of a microarray experiment begins with the analysis of digital images. Image analysis usually encompasses the following steps:

(1) Identify regions in the image which represent spots where probes have been printed.
(2) Calculate the average intensity of pixels within each spot (foreground pixels).
(3) Calculate the average intensity of pixels which lie outside of spots (background pixels).
(4) Associate spots with platform annotations (probe identifiers).

In this section, we will consider the information held in image files and how it is used to describe the quantity of target material hybridized to probes on an array.

---

[3] In performing this experiment, we would actually use at least six arrays to obtain data from dye-flipped replicates of each experimental contrast. We would probably also perform a fourth type of array, comparing prestressed WT and mutant samples to isolate the effect of the mutation alone.

3.1. What is a digital image?

To fully understand the processing of microarray images, it is important to understand how information is stored in a digital image. The simplest digital image formats store a table of intensity values corresponding to each pixel position in the image. The resolution of an image is equal to the dimensions of this table, usually expressed as the "number of columns × the number of rows," for example, "1024 × 1024." Microarray images are gray-scale, meaning that each pixel has a single intensity value associated with it. The range of intensity values for each pixel depends on the bit-depth of the image. Typically, microarray images are 16-bit, meaning that pixel intensities can range in value from 0 (black) to 65,535 (white).[4] Do not be alarmed if you open a microarray image in your favorite image viewer and it appears to be blank. Microarray images often fail to render properly in photo applications because these software packages are usually designed to handle only 8-bit images.

When working with microarrays, it is important to be aware of the file format being used to store array images. Microarray images should always be kept in loss-less formats. Compressed image formats, such as JPEG, discard pixel information during compression in the interest of decreasing overall file size. Although it might be tempting to choose a compressed format for long-term storage, because microarray image files can be quite large, storing images in a lossy format is the digital equivalent to throwing away data.

The tagged image file format (TIFF) is a commonly used to store microarray images. By default these files store data uncompressed. Each TIFF can hold an arbitrary number of images, called layers. This feature is used in competitive hybridization experiments to save the images from both dye-channels in a single file. As the name suggests, TIFFs can also hold a set of user-defined tags. Microarray scanner software will often save information about the image acquisition session to these files such as the scanner temperature and laser settings. Other common file formats which offer loss-less compression include PNG (portable network graphics) and GIF (graphics interchange format).

3.2. Data files

Input. Image analysis consumes two types of data: the digital images themselves and platform-specific probe annotations. Probe annotations are usually saved in text files which associate spot addresses on the array with information about the probes printed on each spot. Examples include

---

[4] Each bit in the file has one of two values (binary), so with 16-bits per pixel this is 2^{16} possible values.

GenePix GAL files and Affymetrix CDF files. These files allow us to link each observation in the array data to:

(1) a spot location, or set of locations, on the surface of the array (information that can be used in quality assessment)
(2) a probe sequence (necessary for MIAME[5] compliance).

Output. Image analysis produces a table of information describing various properties of each spot on the array. Usually these files have as many rows as there are spots on the array. Example formats include GenePix GPR files, Affymetrix CEL files, and SPOT files. Average intensities of the pixels associated with each spot are extracted from these files in the preprocessing step. Information stored in these files often describes the shape of each spot and variation in pixel intensities, both of which can be used to flag low-quality spots during spot filtering.

3.3. Software tools

3.3.1. Commercial packages

Most commercial microarray scanners are sold with accompanying licenses for image analysis software. For example, GenePix is licensed with Axon scanners and offers an integrated image analysis solution, including control of the scanner to acquire images, import of probe information, spot finding, and pixel intensity analysis. One advantage of commercial image analysis packages is that they tend to offer integrated, user-friendly interfaces. Unfortunately, it can be quite costly to obtain a license for these packages in the absence of a hardware purchase. Also, as with any closed-source software, it is usually not possible to obtain detailed information about the implementation of image analysis algorithms.

3.3.2. Open-source and freely available packages

Although none of the free solutions have reached the maturity of most commercial packages, there are several projects worthy of note:

ScanAlyze (http://rana.lbl. gov/EisenSoftware.htm)	An open-source microarray image analysis package written and maintained by Michael Eisen. (Windows only)
MicroArray_Profile (http:// www.optinav.com/imagej. html)	An ImageJ (http://rsbweb.nih.gov/ij/) plug in that can be used to analyze microarray images. MicroArray_ Profile allows you to define a spot

---

[5] See Section 8.2.

	grid, automatically adjust spot diameters and export labeled mean pixel intensity values. (Cross-platform)
Spot (http://www.cmis.csiro. au/iap/Spot/spotmanual. htm)	An R package that can be used to perform microarray image analysis, originally written by Yang *et al.* (2001). Installation instructions and a step-by-step user guide are both available on the web site. (Cross-platform)

 4. PREPROCESSING

Data preprocessing can encompasses a number of different manipulations of the raw probe intensity values obtained from image analysis. In general, the goals of preprocessing are to ensure that:

(1) Observations are comparable between different probes on a single array.

(2) Observations are comparable between arrays.

(3) Low-quality observations are removed from the dataset.

Preprocessing procedures are dependent on the array platform and hybridization scheme. In the sections that follow we will focus our discussion on preprocessing data from competitive hybridization experiments. In general, a preprocessing workflow will consume data from image analysis files (GPR, CEL, SPOT) and ultimately produce a table which associates a single expression value with each target across each unique experimental condition. This table then serves as the starting point for higher order analysis such as hierarchical clustering or differential expression analysis.

4.1. Software tools

The preprocessing procedure is often the most computationally intensive and data rich stage of a microarray analysis scheme, usually encompassing several transformations of the primary data. Because of this, it is important to establish a set of best practices for your lab which ensure that the preprocessing of data are both consistent between arrays and well documented.

Here we review several useful software tools which can be used to perform preprocessing steps. Section 8 at the end of this chapter addresses data persistence tools which can be used store and replicate the input to, and output from, your preprocessing procedure.

4.4.1. Spreadsheets

Common spreadsheet applications include Microsoft Excel, OpenOffice Calc, and Google Docs Spreadsheets. Spreadsheet applications are appealing working environments because they are familiar to most researchers and microarray data structures fit neatly into two-dimensional tables. Spreadsheet applications usually feature some basic declarative programing facilities, for example, allowing the user to calculate values using predefined formula based on the data held within a range of cells in a table. There are several common pitfalls of spreadsheet software; however, which should be considered before committing to using spreadsheets for the bulk of a data analysis scheme. First, many spreadsheet applications (with the exception of Google Docs) allow the user to perform sorting and filtering operations on arbitrary subsets of rows or columns based on the current user selection. In a single "click" this can lead to disastrous consequences, such as jumbling ratio values and probe labels. Second, spreadsheet packages usually do not build a long-term record of changes made to the data by the user (again, with the exception of Google Docs), placing the onus of manually recording each and every data manipulation on the researcher. Finally, many spreadsheet applications impose hard limitations on the number of columns and rows present in each table. Before committing to a particular software package it is important to verify that the software will support the dimensions of your array platform and the number of arrays you intend to perform.

4.4.2. Commercial statistics packages

Commercial statistical packages such as SPSS or Minitab offer a step-up from basic spreadsheet applications. These software packages are usually designed to efficiently handle large datasets and offer more robust facilities for describing the structure of a tabular dataset than most spreadsheet software. These packages also implement a broader range of statistical algorithms and usually support more sophisticated programing capabilities. Disadvantages of these software packages include the expense of licenses and, because they are closed-source, a lack of access to details about the implementation of the statistical methods they feature.

4.4.3. [R] and Bioconductor

In the domain of microarray data analysis one open-source statistical environment is worthy of special note: R.[6] R is a freely available, open-source derivative of the S-Plus system and has attracted the attention of wide range

---

[6] http://www.r-project.org/

of users in both the academic and private sectors. There are several features which make R appealing as a data analysis environment. First, it is entirely open-source. This means that the software is both free to obtain and that the implementation details of all statistical methods found in R are freely available and open to scrutiny. Second, R has emerged as one of the most accommodating environments for statistical research, leading to the establishment of an active and innovative community. Finally, R implements a full-featured programming language which can be used to abstract and automate sophisticated analysis procedures.

Because of these features, R was chosen as the host environment for the Bioconductor[7] suite of microarray data analysis software (Gentleman *et al.*, 2004). Bioconductor boasts an impressive array of microarray analysis tools which support a wide variety of platforms and address both preprocessing and higher order analysis. Although R and Bioconductor offer an ideal set of microarray data analysis tools, the initial learning curve can be quite steep, especially for those unfamiliar with command-line environments or scripting languages. However, if your microarray project extends beyond a small number of arrays, spending the time to learn how to use R and Bioconductor at the start of a project can save you many hours of work down the road. In contrast to working in a spreadsheet application, once you have designed a data analysis procedure in R you will never have to manually perform it again: any set of commands can be saved in a script file and run on new input data.

To get started with R, one-click installers are available for Windows, Mac OSX, and Linux platforms and can be found in the "download" section of the R-project.org web site. When you launch R on your system you will be presented with an "interactive interpreter" window in which you can enter commands. R comes packaged with a number of documents in PDF format aimed at the new user. The best place to start is with the introductory "R-intro.pdf." On Windows, users can find this document from within the "Rgui" application (found in the Start menu after installation) by opening the "Help" menu, selecting "Manuals (in PDF)," and then "An Introduction to R." The exercises in this document should take new users a few hours to work through and will introduce you to all of the key features and concepts needed to implement data analysis schemes in R.

Installing Bioconductor from within the R environment is easy. To perform a standard installation, enter the following commands at the R prompt (denoted below as a ">"):

```
> source("http://bioconductor.org/biocLite.R")
> biocLite()
```

---

[7] http://www.bioconductor.org/

Additional installation options and instructions for installing nonstandard packages are available on the Bioconductor web site.[8] Once installed, you can load a specific Bioconductor package with the library( ) function. To load the "marray" suite of preprocessing functions, for example, enter:

> library(marray)

The R help() function can be used to obtain information about the use of methods implemented in Bioconductor packages. For example, the following command will display information about the usage of the main array normalization function ("maNorm") in the "marray" package:

> help(maNorm)

Many Bioconductor packages also come packaged with tutorial style documents in PDF format called "Vignettes." To access the vignettes from within R:

> library(Biobase)
> openVignette( )

You can then enter the number corresponding to the vignette of interest and the associated PDF should open on your system. For microarray normalization procedures the "marray" and "Limma" vignettes are great places to get started.

4.2. Calculating ratio values

In a competitive hybridization experiment, ratios are calculated for each spot from the average pixel intensities in each channel. There are several methods which can be used to calculate the average pixel intensity for a spot. The first option to consider is which descriptive statistic will be used, usually the mean or median. On a high-quality spot, the values of the mean and median pixel intensities should be very similar. One advantage to choosing the median pixel intensity over mean is that it will be more robust to small numbers of outliers, which can help to mitigate the effects of small scratches (aberrantly low-intensity pixels) or dust (aberrantly high-intensity pixels). The second consideration is whether the channel ratios are calculated before or after averaging. For example, one can first average the intensity of pixels in the red and green channels independently and then calculate the ratio of these two averages ("ratio of the means" or "ratio of the medians"). Alternatively, one can calculate the ratio of the intensity of the red and green channel for each pixel and then average the set of ratios

---

[8] http://www.bioconductor.org/docs/install/

("mean of ratios" or "median of ratios"). Again, for a high quality spot, we would expect these four values to be similar.

By convention, we usually assign the "red" (Cy5; 635 nm filter) channel to the experimental treatment and "green" (Cy3; 536 nm filter) channel to the reference sample, so that this ratio is greater than 1 when a gene target increases in abundance in response to a treatment, and the ratio is less than 1 when a target decreases in abundance. It is important to remember to "flip" values, or calculate the inverse of the ratios obtained from image analysis, for arrays in which the experimental and reference channels are arranged in the reverse orientation.

Finally, it is also common practice to transform ratio values to a linear scale. A $\log_2$ transformation makes ratios particularly convenient to work with because it is simple to conceptualize the fold-change in a ratio given a $\log_2$ value. For example: $\log_2(2/1) > 1$, $\log_2(1/1) > 0$, and $\log_2(1/2) > -1$.

4.3. Normalizing ratio values

In a competitive hybridization experiment, the total signal intensities of the red and green channels will never be perfectly balanced. As described in Chapter 3 in this volume, consideration is given to signal balance in both the sample preparation and image acquisition stages. However, a final mathematical manipulation of the data is required to account for any remaining array-specific biases in signal intensity in order for ratio values to be fairly compared between different arrays. Many normalization strategies also include adjustments for technical bias to improve comparison between probes on the same array.

It is important to carefully consider both the structure of your data and the underlying biological question of interest when choosing a normalization scheme. For best practices, it is recommended that plots be made which will allow you to assess the overall distribution of ratio values on each array before and after normalization. Histograms or density plots such as those illustrated in Fig. 2.2 are easy to make in spreadsheet software and statistical packages. Boxplots are useful when you want to compare the distribution of ratio values across a large number of arrays on a single axis (Fig. 2.3). Finally, scatter plots comparing ratio values to spot pixel intensity (in Bioconductor, "MA" plots) are useful for assessing the degree to which spot ratios have been influenced by a dye-intensity bias.

4.3.1. Global mean or median normalization

The simplest form of ratio normalization is a global mean or median adjustment. As illustrated in Fig. 2.2, in this normalization procedure each ratio is adjusted by a constant value to center the mean or median of the distribution of ratios observed on the array. In this example, the ratios on

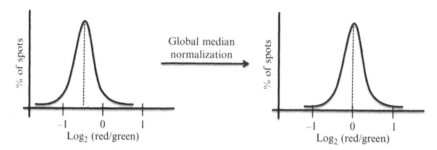

Figure 2.2 A hypothetical distribution of $\log_2$(ratio) values before and after global median normalization. Before normalization this is a "green" array, where the average ratio of red/green is <1. After normalization we have shifted the distribution to the right, moving the median $\log_2$(ratio) value to 0 and preserving the shape of the distribution.

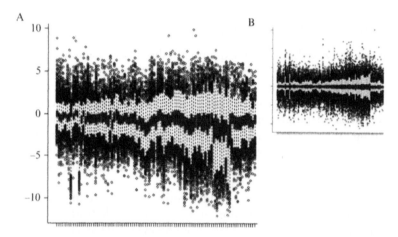

Figure 2.3 Boxplots showing example distributions of $\log_2$(ratio) across a large number of arrays. $\log_2$(ratio) values are shown on the y-axis and data from each array is plotted in a single boxplot along the x-axis. Distributions are shown before (A) and after (B) normalization. Notice that we can easily identify two low-quality arrays as outliers on the left-hand side of these graphs.

our array showed a clear "green" bias ($\log_2$(ratio) < 0), which we can account for by adding a constant value to the $\log_2$ transformed ratios (or by multiplying by a constant value if we are working with raw ratios).

A global adjustment is appropriate if we can assume that the total amount of input material from the experimental and reference samples *should be* equivalent. In most cases a median adjustment is preferable to a mean adjustment, because it will be insensitive to outliers. This adjustment is easy to calculate in a spreadsheet or statistical package.

4.3.2. Normalizing to spike-in controls or housekeeping genes

In cases where we expect a large proportion of targets to show a biologically relevant change in expression levels between the experimental and reference samples, a global median normalization may not be appropriate. The alternative is to normalize features on the array based on the behavior of a small subset of spots we expect to not show expression level changes between the two samples. One option is to choose probes which target "housekeeping" genes which we can reasonably assume will not be affected by the treatment. A second strategy is to design probes which do not target any of the DNA/RNA species in the original biological sample, but rather hybridize to a "spike-in" control which can be added to the samples at a standard concentration. One disadvantage to both of these schemes is they rely heavily on the behavior of a relatively small number of spots, and the consistency of the underlying concentration of material probed by those spots.

4.3.3. Adjusting for intensity bias

A global normalization, based either on the median ratio value or control spots, adjusts all ratios using a constant value, irrespective of the signal intensity observed on each spot. However, most competitive hybridization arrays exhibit a bias in ratio value across different signal intensities. As mentioned previously, one common cause for this bias can be variability in the behavior of different dyes at different levels of intensity. LOESS (local polynomial regression fitting) is a regression-based approach which can be used to adjust ratio values based on the observed relationship between spot ratio and intensity (Cleveland *et al.*, 1992; Yang *et al.*, 2002). An excellent application of this procedure to microarray data is provided by Bioconductor "marray" package.[9]

4.3.4. Adjusting for spatial or print-tip bias

Another source of technical error in microarray data which normalization can be used to minimize is bias in ratios arising from the physical location of spots on the array. Depending on the array platform, location bias can be caused by the manufacturing process (e.g., printing efficiencies of different print-tips in robotically pinned arrays), the hybridization process or inconsistencies in scanner alignment. Although it is possible to manually implement spatially aware normalization procedures using common spreadsheet software, Bioconductor offers a number of well developed and convenient tools. The marray package, for example, can be used to apply LOESS smoothing to ratios based on either printing block (for pinned arrays) or the two-dimensional spatial bias in the local neighborhood of each spot.

---

[9] See documentation for the "maNorm" wrapper function.

The Bioconductor "arrayQuality"[10] package also offers powerful visualization tools for assessing spatial bias across the surface of an array.

4.3.5. More aggressive approaches

If your biological question of interest and the structure of your data conform to a stricter set of assumptions than standard expression arrays, there are a variety of more aggressive normalization techniques which can be used. For example, if you can assume that the variation in ratio distributions should be equivalent between arrays you may consider performing scale or quantile normalizations (Dudoit and Yang, 2002).

4.4. Quality assessment, filtering, and handling replicates

A data analysis procedure is only as strong as the quality of the data fed into it. It is, therefore, important to assess the quality of your array data and implement an effective filtering scheme. Quality assessments can be both qualitative and quantitative, and can be used to filter individual data points or entire arrays.

4.4.1. Spot quality

The first stage of spot filtering occurs during image analysis. Spots which cannot be found by the image analysis software should be flagged for downstream filtering. Spots should also be flagged if they are smeared, continuous with another spot, significantly marred by dust, or scratched. Depending on your array platform, it may also be useful to automatically flag spots which are bigger or smaller than certain reasonable size cutoffs or for which there are extremely high variances in the pixel intensities in either dye-channel. Flags can be added to image analysis data tables using a calculated column in a spreadsheet or with logical index vectors in R. Generally, flagged spots are excluded from ratio normalization calculations.

4.4.2. Array quality

As a general rule of thumb, if an array "looks bad" then it probably is. The Bioconductor arrayQuality[10] package implements a number of plotting techniques which can be used to identify spatial bias across the surface of an array. For example, a heatmap of ratio-value ranks on an array will readily reveal any spatial bias in spot ratios. If your platform design includes replicate spots (identical probes printed in multiple locations) which are distributed across different locations on the array it is possible to assess whether or not variation in ratio values can be explained by surface position. Replicate spots

---

[10] http://bioconductor.org/packages/2.4/bioc/html/arrayQuality.html

also allow you to calculate an average variance among replicate groups, a parameter which can provide a useful measure of array quality.

4.4.3. Utilizing spot and array replicates

There are a number of ways in which technical replication can be used to enhance the power of microarray-based investigations. The central question of how to best make use of technical replication comes down to when in the analysis procedure replicates should be averaged and how to utilize information about the variability underlying each unique measurement. Spot replicate averaging can be accomplished by taking the mean or median. Whenever you average values, it is also important to calculate a measurement of variation among the observations contributing to each average. Variance and standard deviation are used most commonly. As noted above, spot replicate and array replicate variation can be used as a quantitative measure of array quality. Information about the variation among replicates can also be used to "weight" observations in most clustering algorithms (see Section 5), allowing higher quality observations to have a stronger influence on the structure of the resulting graph. Finally, replicate variation is also used in differential expression analysis (see Section 6). When submitting datasets to public repositories (see Section 8.2), it is important to include preaveraged data, so that interested parties can independently recreate these data analysis steps.

4.5. Preprocessing Affymetrix arrays

Although the goals for preprocessing single-sample arrays are similar to those for competitive hybridization arrays, the preprocessing approaches differ significantly. Single-sample arrays, such as the Affymetrix GeneChip platform, often include a series of "Perfect match" (PM) and "Mismatch" (MM) probes for each gene target. Preprocessing of these arrays involves summarizing PM and MM probe set behaviors. There are a number of software packages available which implement different preprocessing algorithms for Affymetrix arrays. Two mature and popular options are:

Expression console	Affymetrix offers their expression array analysis software free of charge for registered users on their web site (http://www.affymetrix.com/support/technical/software_downloads.affx). The latest generation of this analysis suite implements several preprocessing algorithms including MAS, PLIER, and RMA.
Bioconductor	The Bioconductor project includes a number of packages which can be used to normalize data from Affymetrix arrays, including: "affy," "gcrma," and "affyPLM" (Bolstad *et al.*, 2003; Irizarry *et al.*, 2003).

5. Visualizing Data Using Cluster Analysis

One of the most popular ways to explore microarray datasets is with clustering analysis. Microarray datasets may contain information about the behaviors of $\sim 10k-1M$ different probes across dozens or hundreds of different experiments, resulting in a grid of data which is far too large to simply "browse." Clustering techniques are used to order the data according to the behavior of probed targets (genes), experimental factors (arrays), or both. Visualization tools are then used to explore the resulting data structure.

5.1. Hierarchical clustering

The most common clustering algorithm used in the microarray field is hierarchical clustering. Hierarchical clustering is an uncensored machine learning method, meaning that it is used to order a dataset based on the data alone rather than a predefined model.

5.1.1. An example use-case

Let's consider our microarray case study from the introduction. After we have analyzed our array images, normalized and $\log_2$ transformed our ratio values, and filtered out low-quality data we will be left with a table of information which describes the behavior of target genes in each of our biological sample hybridizations (Table 2.1). We can use hierarchical clustering of this table of values to help us identify genes which exhibit similar behaviors across this stress time course. In this case, we would only want to cluster the gene or target axis, because the experiments already have an obvious biologically meaningful order.

5.1.2. How hierarchical clustering works

Hierarchical clustering begins by examining a list of elements, in this example a list of genes, to identify the two elements which are most similar. The results of a hierarchical clustering run are highly dependent on the way in which "similar" is defined. Pearson correlation is a straightforward similarity metric to imagine in this context: the similarity between any

Table 2.1 Example data structure produced by preprocessing which can be used for cluster analysis

	Array 1 (stress time point 1/control)	Array 2 (stress time point 2/control)	...
Gene1	-0.1503	-0.3861	...
Gene2	0.3857	0.2168	...
...	...	...	...

two genes can be calculated as the Pearson correlation of $\log_2$(ratio) values observed across the arrays in the time course. The absolute value of the Pearson correlation could be used if we wanted genes which showed large changes at the same time point to score as similar, irrespective of whether expression levels went up or down. Nonparametric measures such as rank order or Euclidean distances can also be used to evaluate similarity. Choosing a different similarity metric can dramatically affect the structure of a cluster, so it can be beneficial to spend some time exploring your options.

Once the two most similar genes have been identified in the starting list, they are associated on a single branch of a tree. The original data associated with these two genes is then removed from the list of elements and replaced with a new entry that represents the "average" behavior of both. The method by which the "average" behavior of a branch is represented in subsequent similarity calculations (the linkage method) is the second major parameter which controls the behavior of hierarchical clustering.

5.1.3. Common pitfalls

There are several features of hierarchical clustering algorithms which can lead new users astray in the interpretation of their results. First, although hierarchical clusters are often used to "group" genes or experiments into discrete subsets, grouping is not the goal of hierarchical clustering *per se*. Commonly used hierarchical clustering algorithms only operate on the pairwise distances between elements in a list; they do not consider the larger structure of the data. As such, these algorithms are "bottom–up" approaches and do not necessarily create a tree in which the average distance between all elements is fully minimized, nor do they suggest a best cutoff level in the resulting tree to produce meaningful subsets of elements. There are, however, a number of techniques which have been developed to help overcome these limitations of hierarchical clustering, notably the tree "pruning" algorithms implemented in the Bioconductor "hopach" library (van der Laan and Pollard, 2003).

When exploring a hierarchical cluster, it is important to remember that each node has two equivalent orientations and that the orientation chosen when a tree is initially rendered is effectively arbitrary. Flipping the orientation of a node can dramatically change the visual appearance of a cluster, because of changes in the linear order of the gene or array axis, but has no effect on the overall pair–wise distances between genes. When considering the relatedness of elements on a clustered graph it is important to pay close attention to the actual distance between elements in the tree rather than the relative order on the plot.

5.1.4. Software

The open-source Cluster 3.0 package features an easy-to-use interface and a host of clustering procedures (de Hoon *et al.*, 2004). Cluster 3.0 can read data from tab-delimited text files produced in the preprocessing step.

Depending on the setup, clustering runs produce files with some or all of the following extensions:

.cdt	These files hold the original table of data, with targets (genes) listed in each row and sample comparisons (experiments/arrays) listed in each column. The CDT tables may also have a gene weight (GWEIGHT) column and experiment weight (EWEIGHT) row which describe the clustering weight of each gene and experiment, respectively. As mentioned above, it can be useful to draw weights for a clustering run from measures of variance among technical replicates.
.gtr	Gene-tree files contain tables which describe each branch in the tree as an association of two child genes/nodes (1st and 2nd column) and a parental gene/node (3rd column), as well as the distance from children to parent (4th column).
.atr	Array-tree files are identical to gene-tree files, but describe array clustering.

The information saved in these output files can be visualized using the open-source, cross-platform, Java TreeView package (Saldanha, 2004). Java TreeView draws a heatmap of ratio values from data saved in the CDT file and associated gene- or array-trees if similarly named GTR or ATR files are found in the same directory.

5.2. Partitioning and network-based approaches

In addition to hierarchical clustering, there are a wide variety of other types of clustering methods which can be used to explore microarray datasets. Several commonly used partitioning methods include: self-organizing maps (SOMs) (available in Cluster 3.0 and the "som" R package), k-means clustering (available in Cluster 3.0), and Prediction Analysis for Microarrays[11] (PAM) (available as the "PAM" R package and as an Excel plug in). Like hierarchical clustering, these partitioning algorithms make use of distance metrics and linkage methods to cluster closely associated genes or arrays. Unlike hierarchical clustering, partitioning algorithms are designed to construct defined subsets of element, although for some approaches, like SOMs, the use of stochastic parameters means multiple runs on the same dataset may not always produce the same result.

Among network-based approaches, the BioLayout Express[12] implementation of the Markov Cluster Algorithm (MCL) is particularly worthy of note (Freeman et al., 2007). MCL belongs to a class of algorithms which

---

[11] http://www-stat.stanford.edu/~tibs/PAM/
[12] http://www.biolayout.org/

optimize the overall distances between nodes across the entire set of elements being clustered. Network-based approaches are much more flexible than hierarchical clustering in the kinds of paths between nodes which can be drawn, and can be more conducive to a visual exploration of gene groups. BioLayout Express is an open-source, cross-platform software package with excellent documentation and a well-developed user-interface.

6. ASSESSING THE STATISTICAL EVIDENCE FOR DIFFERENTIAL EXPRESSION

Microarray data are often used to identify changes in gene expression in response to an experimental treatment. Conceptually, we would like to be able to extract a unique list of genes from a microarray dataset which are differentially expressed in one biological sample compared to another. The challenge comes when deciding how to draw a cutoff in ratio values. Although a blanket cutoff of a "twofold change" has been used in the microarray field in the past, this is not a statistically rigorous approach nor is it a reasonable approximation for many datasets.

Our goal in assessing the evidence for differential expression of targets in a microarray dataset can be conceptualized in the same terms as any canonical hypothesis testing problem. In this case, the null hypothesis, which we want to know whether or not to reject, is that the observed $\log_2$(ratio) value for a particular target is actually no different than 0. Given the variability in the measurements observed among replicates, and the average expression change across the dataset, we want to calculate statistics which will allow us to assess the probability that accepting or rejecting this null hypothesis will result in a false-positive (type I error; identifying genes as differentially expressed which are not) or false-negative (type II error; identifying genes as not differentially expressed which are) determination. However, calculating meaningful significance levels for each gene in a large dataset is a complex problem. For example, if we test this null hypothesis for 15,000 probes on an array using a classical t-test, the high level of multiple testing shifts the scale of meaningful p-values away from what we are normally used to considering.

Here we briefly discuss two statistical approaches to this problem which are implemented in mature software packages and have been used to great effect in yeast studies:

6.1. Significance analysis of microarrays

Significance analysis of microarrays (SAM) uses gene-specific t-tests, calculated using a nonparametric statistic, to provide an estimate of the false discovery rate at a given ratio value cutoff (Tusher *et al.*, 2001). SAM is flexible enough to handle most common experimental designs, but is less versatile in this regard than Limma. SAM is available as both an [R] library and a Microsoft Excel plug in. It is free to download for academic users.

6.2. Limma

Limma is a Bioconductor software package which uses linear models to analyze microarray data and assess evidence for differential gene expression (Smyth, 2004). Limma can be used to model virtually any experimental design, accounting for sample measurements taken across complex sets of competitive hybridization pairs. An extensive user manual and step-by-step walkthrough are provided in the Limma documentation.[13]

7. EXPLORING GENE SETS

Having identified an interesting group of targets (e.g., genes) using cluster analysis or differential expression assessment, the next question we usually want to address is: what is similar about the members of each group? Here we review some common approaches to this question.

7.1. Gene Ontology term mapping

The Gene Ontology (GO) project is cross-species gene annotation effort which defines a controlled vocabulary of terms describing a gene product's function, biological process, or cellular component (Ashburner et al., 2000). When presented with a subset of genes which show a similar pattern of expression in a microarray dataset, it can be useful to determine whether or not the GO annotations associated with those genes suggest a common biological function. In GO, terms are related to one another through a branched hierarchy.[14] Genes can be annotated with any number of terms from any level of this hierarchy. Conceptually the complex structure of GO terms is appealing, because it allows for a high degree of flexibility in gene labeling. Unfortunately, this structure also makes it difficult to appropriately estimate the statistical relevance of a potentially overrepresented GO term in a list of genes. The GO Slim Mapper[15] hosted at the Saccharomyces Genome Database offers a Web-based solution which assesses the statistical likelihood that a GO term is meaningfully overrepresented in a given set of genes, plotting the results on a GO term graph. The Bioconductor

---

[13] See "userguide.pdf" in the "/doc" subdirectory of the limma library or enter "> library(limma); limma UsersGuide( )" at the R prompt.
[14] More properly, the GO topology is an acyclic graph as child terms can be associated with multiple parents.
[15] http://www.yeastgenome.org/cgi-bin/GO/goSlimMapper.pl

"GOSemSim" package[16] also provides methods for estimating GO semantic similarities in gene sets. Finally, the GeneMAPP software package offers an excellent set of tools for drawing pathways based on GO terms which can then be highlighted based on gene behavior in a microarray dataset (Dahlquist *et al.*, 2002).

7.2. Motif searching

Motif searching can be used to identify potential *cis*-regulatory elements associated with coregulated gene products. MEME and AlignACE are two popular software packages which each offer Web-based submission (Bailey *et al.*, 2009; Hughes *et al.*, 2000). An in-depth discussion of motif searching algorithms by Hao Li appears in "The Guide to Yeast Genetics and Molecular Biology, Part B" (Li, 2002).

7.3. Network visualization

There are a number of tools available which allow one to visualize interaction networks, integrating microarray data, proteomic data and other sources of gene annotations. The open-source Cytoscape[17] project is particularly worthy of note for its excellent documentation, accessible learning curve, and active community.

7.4. Graphing array data on genome tracks

Analysis of tiling microarray data usually involves visualizing probe behaviors on genomic tracks. The "genome browser" tool in the Gaggle project[18] offers an easy-to-learn software solution for genomic visualization of microarray data. Once probe identifiers are associated with a chromosome number and chromosomal coordinates, microarray data can be visualized with heatmaps, scatter plots, or line graphs on top of genome tracks drawn from the UCSC genome browser. A note of caution for those who usually use SGD as the source of genomic coordinates: the UCSC genome browser draws from the October, 2003 assembly, while SGD has continued to update the reference sequence. This means that there are slight inconsistencies between the genome coordinate systems in these two databases. The SGD Genome Browser (GBrowse) offers a Web-based alternative[19] which uses the current SGD coordinates.

---

[16] http://bioconductor.org/packages/2.4/bioc/html/GOSemSim.html
[17] http://www.cytoscape.org/
[18] http://gaggle.systemsbiology.net/docs/geese/genomebrowser/
[19] http://www.yeastgenome.org/cgi-bin/gbrowse/scgenome

8. Managing Data

8.1. Data persistence and integrity

One important, and easily overlooked, consideration when setting up a new microarray data analysis project in your lab is the best way to handle storage of microarray data and analysis results. Individual image and data files can be quite large, and data analysis pipelines typically produce a number of files across several stages. Ultimately, published array data will be archived in a public, off-site, MIAME-compliant database (see below), but how should the data associated with moderate- or large-scale microarray projects be handled within the lab?

One approach is to use a relational database to store array data and provide a Web-based front-end for queries and data-submission (e.g., the UCSF NOMAD[20] project). These solutions allow a working group to store array data in a centralized location, facilitating backups and maintenance. Modern, open-source, relational databases such as MySQL[21] are capable of handling extremely large chunks of data, storing array images alongside data tables (such as image analysis files, probe information, clustering results, etc.), and provide efficient data querying facilities.

Often, however, it is more convenient to work with microarray data as files on the local file system, rather than as tables stored in a relational database. Moving data to and from a database for use in other software packages can be cumbersome and confusing for those unfamiliar with these technologies. In my own work I have settled a convenient solution that works well for most types of projects: version control systems.

Version control systems such as CVS and Subversion[22] were originally developed to facilitate software development projects involving many developers. Using Subversion to manage your microarray data files is fairly simple: you work with your microarray data files in a set of folders saved to your local machine and then synchronize the state of these folders with a Subversions server after each major change or manipulation. Version control systems such as Subversion offer several facilities which are extremely useful in a microarray data analysis project. Below we review some of these features, all of which should be taken into consideration when choosing a data storage strategy for you lab.

[20] http://sourceforge.net/projects/ucsf-nomad/
[21] http://www.mysql.com/
[22] http://subversion.tigris.org/

8.1.1. Data replication

In Subversion parlance the files and folders you keep on your local machine are a "working copy" of a "repository" kept on the server. You can make as many working copies of your files on different computers as you need, each of which can be kept synchronized with the central repository. The concept here is similar to that of a fileserver, but Subversion is more useful in several ways. First, the working copy is actually a local copy of all of your files, so you do not need to be connected to a network to access your data. Second, it is up to you explicitly decide when you want commit local changes to the repository. This allows you to organize sets of changes into logical transactions.

8.1.2. Multiuser support

Many people can work with the microarray data files simultaneously using their own working copies of the archive. When users commit the changes they have made locally back to the repository, Subversion detects conflicts (cases where mutually exclusive changes have been made to a file) and offers a rich set of tools that help you to merge changes into a new version of the file. Subversion servers also allow you to setup user authentication (unique user names and passwords for different member of you lab) and set read/write permissions on different portions of the archive.

8.1.3. Transactions and logs

Subversions servers save *changes* to files in the repository instead of copies of the files themselves. Each time you commit local changes in your working copy to the server, a log entry is created tracking all of the changes that were made. This means that it is possible to revert a working copy, or the central repository itself, to any previous version of the archive. Inevitably, when working on a data analysis project there will come a time when you are unsure of whether or not you have performed an analysis with the intended parameters or when you accidentally overwrite an important set of files. The version control system makes solving these problems straightforward: you can simply revert the repository to the last point in time when you know it was in a sane state. Subversion allows you to save a text log entry alongside every change to the contents of the repository. I have found this to be an incredibly powerful way to keep a record of data analysis projects. Log viewers are available which allow you to browse each of your log entries alongside the associated changes to files in the repository.

8.1.4. Web-based access

The Subversion community has developed a mature Apache web server module which makes it easy to "publish" your Subversion repository on the Web. I have found this to be an extremely convenient way to share data analysis files with off-site collaborators.

There are a number of version control systems available, each with slightly different performance characteristics and feature sets. Subversion is a well rounded, open-source, cross-platform solution which I have found works quite well for data analysis projects. Subversion is comprised of two software components: a *client* that needs to be installed on each computer where you want to make a "working copy" and a *server* that should be setup on a machine with reliable internet connectivity and a regular backup schedule. If you are using Linux, chances are quite good that both the Subversion client and server are already installed. Installers and binary files for other platforms, including Windows and Mac OSX, can be found on the Subversion web site. For windows users I would highly recommend both TortoiseSVN,[23] which integrates Subversion client facilities into Windows explorer, and VisualSVN[24] which allows you to setup a Subversion server and web server using a simple installer.

8.2. Public data repositories and MIAME compliance

Community standards, and many journal submission agreements, require that microarray data presented in publications be submitted to public databases. Before the development of centralized microarray data hosting centers, many researchers posted microarray files on private lab web sites. This solution is problematic for several reasons. First, Web URLs are not a reliable resource: institutions often reorganize the URL structure of their web sites and labs can change institutional affiliation. Second, the structure and completeness of the available data can be extremely varied, making it difficult for interested third parties to recreate the published analysis or use the data in new studies.

In response to inconsistencies in data-sharing practices in the microarray community, the Microarray Gene Expression Data Society[25] (or MGED) was formed to develop a standard describing the Minimal Information About a Microarray Experiment, or MIAME (Brazma *et al.*, 2001). The MIAME standard takes on the daunting task of articulating a formal set of data structures which describe a wide variety of different micro-array experiments and platforms. Because of this, the standard itself is extensive and relies on a relatively abstract nomenclature to remain platform agnostic.

Fortunately there are a number of public, curated, MIAME-compliant databases designed to guide experimenters through the data annotation process. Three popular choices are:

[23] http://tortoisesvn.tigris.org/
[24] http://www.visualsvn.com/server/
[25] http://www.mged.org/

GEO	The Gene Expression Omnibus (GEO) (http://www. ncbi.nlm.nih.gov/projects/geo/) project developed and hosted at the NCBI is a free, full featured, database maintained by a responsive curatorial staff. GEO supports submission of a wide variety of gene expression datasets, offers a user-friendly Web-based submission system, and scales well to handle large submissions. GEO supports timed public release of datasets and the creation of private URLs which can be provided in the peer-review process. Finally, datasets stored in GEO are automatically searched when users enter terms in the "all databases" search box on the NCBI home page.
ArrayExpress	ArrayExpress (http://www.ebi.ac.uk/microarray-as/ae/), developed and hosted at the EBI, is similar to GEO in scope and sophistication. There are some minor differences between the two sites in the searching facilities offered and the batch data upload file formats that are supported.
SMD	The Stanford Microarray Database (http://smd.stanford. edu/index.shtml) offers several Web-based data analysis packages not available from other repositories. However, SMD charges a significant usage fee to labs outside of Stanford University.

For projects with a small number of arrays, submitting data to GEO is as simple as creating an account and following the Web-based guide to upload data files and provide MIAME-compliant annotations. GEO submissions are composed of three types of records:

Platform	The platform record describes your microarray. This includes annotations describing the array substrate, manufacturing process, and probe sequences. If you are working with a common commercial array platform there is a good chance that a record has already been submitted for your array. If this is the case, you do not need to create a duplicate entry; you can simply skip this step and link your samples to the preexisting platform record. It is useful to attach your platform-specific probe/spot annotation table (GAL file, CDF, etc.) as a supplementary file on these records.
Sample	Sample records describe each *hybridization* event in your microarray experiment. For single-sample designs you will create one sample record for each individual array in your

(continued)

experiment. For competitive hybridization designs, you will also create one sample record for each array in your experiment, but in this case each sample record will actually describe *two* biological samples. The sample record combines annotations describing the origin and preparation of the biological sample hybridized on the array, along information about the hybridization conditions and data acquisition procedure. If your data preprocessing scheme involves a normalization step, it is common to provide the results of this normalization in the data table associated with each sample record. Attaching your platform-specific image analysis results (GPR file, CEL, SPOT, etc.) as a supplementary file will allow interested users to explore alternative normalization and preprocessing approaches.

Each sample record points to a single platform record.

Series   The series record describes your study and the relationships among the associated sample records. This record is the main entry point for users interested in your dataset. The series record will allow you to describe the unique experimental factors examined in your study and the structure of your experimental replicates. Finally, the series page provides users with a set of links which allow them to download your data in a variety of formats.

Each series record points to one or more sample records.

All GEO submissions are reviewed by a curator before they are made public. This ensures that the submitted data and annotations meet a set of minimum standards.

If your microarray project involves a large number of arrays, it may be too cumbersome and time consuming to manually submit data to GEO through the Web-based forms. For these cases, GEO supports a number of batch deposit formats. The simplest of these to use and understand is the SOFT format. SOFT files are plain text files which associate annotations, in header rows, with data, in tab-delimited tables. Each of the GEO record types (Platform, Sample, and Series) can be described in a SOFT file and each field that appears on the Web-based forms is given a corresponding label for use in SOFT file header rows. These files are relatively easy to produce with spreadsheet software or using simple scripts. Extensive documentation is available on the GEO web site.[26]

---

[26] http://www.ncbi.nlm.nih.gov/projects/geo/info/soft2.html

ACKNOWLEDGMENTS

I thank M. Bergkessel, C. Guthrie, M. Fitzpatrick, and R. Chande for their critical feedback on this manuscript.

REFERENCES

Ashburner, M., Ball, C. A., Blake, J. A., Botstein, D., Butler, H., Cherry, J. M., Davis, A. P., Dolinski, K., Dwight, S. S., Eppig, J. T., Harris, M. A., Hill, D. P., *et al.* (2000). Gene ontology: Tool for the unification of biology. The Gene Ontology Consortium. *Nat. Genet.* **25,** 25–29.

Bailey, T. L., Boden, M., Buske, F. A., Frith, M., Grant, C. E., Clementi, L., Ren, J., Li, W. W., and Noble, W. S. (2009). MEME SUITE: Tools for motif discovery and searching. *Nucleic Acids Res.* **37,** W202–W208.

Bolstad, B. M., Irizarry, R. A., Astrand, M., and Speed, T. P. (2003). A comparison of normalization methods for high density oligonucleotide array data based on variance and bias. *Bioinformatics* **19,** 185–193.

Brazma, A., Hingamp, P., Quackenbush, J., Sherlock, G., Spellman, P., Stoeckert, C., Aach, J., Ansorge, W., Ball, C. A., Causton, H. C., Gaasterland, T., Glenisson, P., *et al.* (2001). Minimum information about a microarray experiment (MIAME)-toward standards for microarray data. *Nat. Genet.* **29,** 365–371.

Cleveland, W. S., Grosse, E., and Shyu, W. M. (1992). Local regression models. *In* "Statistical Models in S," (J. M. Chambers and T. J. Hastie, eds.), Chapman & Hall.

Dahlquist, K. D., Salomonis, N., Vranizan, K., Lawlor, S. C., and Conklin, B. R. (2002). GenMAPP, a new tool for viewing and analyzing microarray data on biological pathways. *Nat. Genet.* **31,** 19–20.

de Hoon, M. J. L., Imoto, S., Nolan, J., and Miyano, S. (2004). Open source clustering software. *Bioinformatics* **20,** 1453–1454.

Dudoit, S., and Yang, Y. H. (2002). Bioconductor R packages for exploratory analysis and normalization of cDNA microarray data. *In* "The Analysis of Gene Expression Data: Methods and Software," (G. Parmigiani, E. S. Garrett, R. A. Irizarry, and S. L. Zeger, eds.), Springer.

Freeman, T. C., Goldovsky, L., Brosch, M., van Dongen, S., Mazière, P., Grocock, R. J., Freilich, S., Thornton, J., and Enright, A. J. (2007). Construction, visualisation, and clustering of transcription networks from microarray expression data. *PLoS Comput. Biol.* **3,** 2032–2042.

Gentleman, R. C., Carey, V. J., Bates, D. M., Bolstad, B., Dettling, M., Dudoit, S., Ellis, B., Gautier, L., Ge, Y., Gentry, J., Hornik, K., Hothorn, T., *et al.* (2004). Bioconductor: Open software development for computational biology and bioinformatics. *Genome Biol.* **5,** R80.

Hughes, J. D., Estep, P. W., Tavazoie, S., and Church, G. M. (2000). Computational identification of cis-regulatory elements associated with groups of functionally related genes in *Saccharomyces cerevisiae. J. Mol. Biol.* **296,** 1205–1214.

Irizarry, R. A., Bolstad, B. M., Collin, F., Cope, L. M., Hobbs, B., and Speed, T. P. (2003). Summaries of Affymetrix GeneChip probe level data. *Nucleic Acids Res.* **31,** e15.

Li, H. (2002). Computational approaches to identifying transcription factor binding sites in yeast genome. *In* "Methods in Enzymology: Guide to Yeast Genetics and Molecular Biology," Part B, (C. Guthrie and J. Fink, eds.), pp. 484–975, Academic Press.

Saldanha, A. J. (2004). Java Treeview—Extensible visualization of microarray data. *Bioinformatics* **20,** 3246–3248.

Smyth, G. K. (2004). Linear models and empirical Bayes methods for assessing differential expression in microarray experiments. *Stat. Appl. Genet. Mol. Biol.* **3,** Article 3.

Tusher, V. G., Tibshirani, R., and Chu, G. (2001). Significance analysis of microarrays applied to the ionizing radiation response. *Proc. Natl. Acad. Sci. USA* **98,** 5116–5121.

van der Laan, M. J., and Pollard, K. S. (2003). A new algorithm for hybrid hierarchical clustering with visualization and the bootstrap. *J. Stat. Plan. Inference* **117,** 275–303.

Yang, Y. H., Buckley, M. J., and Speed, T. P. (2001). Analysis of cDNA microarray images. *Brief. Bioinform.* **2,** 341–349.

Yang, Y. H., Dudoit, S., Luu, P., Lin, D. M., Peng, V., Ngai, J., and Speed, T. P. (2002). Normalization for cDNA microarray data: A robust composite method addressing single and multiple slide systematic variation. *Nucleic Acids Res.* **30,** e15.

CHAPTER THREE

GENOME-WIDE APPROACHES TO MONITOR PRE-mRNA SPLICING

Maki Inada*,[1] *and* Jeffrey A. Pleiss*

Contents

Abstract

Pre-mRNA processing is an essential control-point in the gene expression pathway of eukaryotic organisms. The budding yeast *Saccharomyces cerevisiae* offers a powerful opportunity to examine the regulation of this pathway. In this chapter, we will describe methods that have been developed in our lab and others to examine pre-mRNA splicing from a genome-wide perspective in yeast. Our goal is to provide all of the necessary information—from microarray design to experimental setup to data analysis—to facilitate the widespread use of this technology.

* Department of Molecular Biology and Genetics, Cornell University, Ithaca, New York, USA
[1] Present address: Department of Biology, Ithaca College, Ithaca, New York, USA

Methods in Enzymology, Volume 470
ISSN 0076-6879, DOI: 10.1016/S0076-6879(10)70003-3

1. INTRODUCTION

In the 30 years since Sharp and Roberts independently demonstrated the presence of split genes (Berget *et al.*, 1977; Chow *et al.*, 1977), it has become abundantly clear that pre-mRNA splicing and its regulation play an essential role in regulating gene expression in eukaryotic organisms (House and Lynch, 2008). By regulating the efficiency of splicing of specific transcripts during development, in specific tissues or in response to external stimuli, the expression levels of particular genes can be controlled. Similarly, splicing can control proteomic diversity by regulating splice site choice via the process known as alternative splicing. Remarkably, whereas the number of genes predicted to be encoded by the human genome has been steadily decreasing over the last 15 years, the fraction of genes known to be alternatively spliced has gradually increased over this same time and is now thought to be over 90% of genes in humans (Wang *et al.*, 2008). While great progress has been made in understanding the mechanistic details of splicing, many questions remain about the pathways used to regulate this process.

It was recognized early on that the components of the spliceosome and the basic mechanisms of splicing are highly conserved from yeast to humans. As such, the budding yeast *Saccharomyces cerevisiae* has played a pivotal role as a model organism for elucidating mechanisms of pre-mRNA splicing. While the underlying machinery is highly conserved between humans and yeast, the genome-wide distribution of introns is in fact quite different. Whereas over 90% of human genes are interrupted by at least one intron, only about 5% of *S. cerevisiae* genes contain a functional intron. In spite of this simplified architecture, one of the earliest recognized examples of regulated splicing was demonstrated in *S. cerevisiae*. In an elegant set of experiments, Roeder's group demonstrated that the Mer2 protein specifically modulates the splicing of the MER1 transcript during meiosis (Engebrecht *et al.*, 1991). The observation that a specific protein can play a pivotal role in the efficient splicing of a distinct set of transcripts highlights both the utility of splicing as a regulator of gene expression and also the need for genome-wide tools to assess global changes in splicing.

This chapter focuses on methods for studying genome-wide changes in pre-mRNA splicing in yeast. The last several years have seen the development of several distinct but related microarray platforms that allow for global analysis of splicing (Clark *et al.*, 2002; Juneau *et al.*, 2007; Pleiss *et al.*, 2007a,b; Sapra *et al.*, 2004; Sayani *et al.*, 2008). In this chapter, we will describe one such methodology using short oligonucleotide sequences that specifically detect each of the different splicing isoforms. The first oligonucleotide-based microarrays that were used to specifically probe changes in splicing status were developed by the Ares lab (Clark *et al.*, 2002).

We have used this approach to identify splicing responses to environmental stress and to evaluate effects of mutations in core spliceosomal components (Pleiss *et al.*, 2007a,b). The goal of this chapter is to guide you through the platform that we are currently using in our lab. While the details presented here are specific to our particular platform, we expect that the protocols can be readily adapted to meet the requirements of other platforms.

2. MICROARRAY DESIGN

The fundamental philosophy underlying splicing-sensitive microarrays is no different from standard gene expression microarrays in that short, complementary oligonucleotides are used to probe the abundance of a given RNA species. Whereas gene expression microarrays have oligonucleotides that target only coding regions of genes, the splicing-sensitive microarrays that we use include additional probes to both intron regions and the junction of the two ligated exons to distinguish changes in pre- and mature mRNA levels, respectively (Fig. 3.1). Using tools that are described below, we have designed sequences that target approximately 6000 genes, ~300 introns, ~300 junctions, tRNAs, snRNAs, snoRNAs, and other functional noncoding RNAs in the yeast genome. The sequences are readily available at the NCBI Gene Expression Omnibus (http://www. ncbi.nlm.nih.gov/geo/, Accession number GPL8154) and can be downloaded and used by anyone to order microarrays from any of several different vendors. In this chapter, we will describe our work using custom Agilent microarrays which contain eight identical hybridization zones each printed with approximately 15,000 of these oligonucleotides. However, we see no reason why these sequences would not be transferrable to other commercial or homemade platforms.

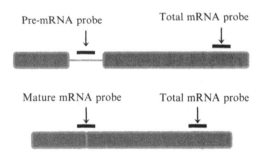

Figure 3.1 Microarray probe design. For each intron-containing gene, a minimum of three probes are designed. One measures changes in total RNA level by hybridizing to a region of the exon. A second measures changes in pre-mRNA level by hybridizing to a region of the intron. The third probe measures changes in mature mRNA levels by hybridizing to the junction of the two ligated exons.

If you are interested in learning how these microarrays are designed, continue reading the rest of this section. If not, order your microarrays and skip ahead to Section 3. As with all microarrays, a key component in the design of splicing-sensitive microarrays is the ability to probe the RNA of interest without cross-reacting with off-target or nonspecific RNAs. For probes that target the coding regions of genes (both intron-containing and intronless genes) this task is made relatively easy by numerous computational programs that have been specifically designed for this purpose. We have used the OligoWiz program (http://www.cbs.dtu.dk/services/OligoWiz/) to design 60 nt (nucleotide) probes to all ~6000 protein-coding genes in the yeast genome (Wernersson et al., 2007). For intron-containing genes, where possible, probes were designed that target regions in both exons 1 and 2. However, because most yeast introns are located at the 5' end of the gene, and exon 1 tends to be very short, probes have been designed to target a region only in exon 2 for most intron-containing genes (Fig. 3.1).

For intron-containing genes, OligoWiz can also be used to design probes targeting intronic regions. Because many yeast introns are short, we hoped to use shorter probe lengths to target the pre-mRNA species. However, it was unclear whether these probes would provide sufficient hybridization capacity. In our initial experiments with the Agilent platform, we tested both long (60 nt) and short (35 nt) probes targeted to the introns of all ~300 intron-containing genes and observed no significant loss of signal intensity when using the shorter probes. Therefore, all of the pre-mRNA specific probes on our microarray target a 35 nt region of the intron of interest. Likewise, because many functional RNAs like tRNAs and snoRNAs are also small, we designed probes to functional RNAs using 35 nt sequences.

From the perspective of specificity, the most difficult oligonucleotides to design are those that target the mature mRNA species by hybridizing to the junction between ligated exons. Whereas the exon-targeting probes can be optimized by moving the targeted region anywhere within the coding sequence, by definition the oligonucleotides which probe changes in mature mRNA levels by targeting the junctions of ligated exons are restricted to the discrete sequences at the end and beginning of those neighboring exons. In designing these probes, we sought to identify the shortest length oligonucleotide which was sufficient to efficiently capture spliced mRNAs. Because of the varying sequence content in the exons of different intron-containing genes, we chose to vary the length such that the sequences upstream and downstream of the junction are energetically balanced (Fig. 3.2). Our initial experiments tested a variety of thermodynamic stabilities for every exon–exon junction in the genome. In these experiments, the best compromise between signal intensity and signal specificity was found for those junction probes that had $\Delta G°$ values

YML056c

$\Delta G°$	Exon 1	Exon 2	$\Delta G°$
10.2	CCAGTTACTG	AAGACGGTAA	10.8
13.8	TCCCAGTTACTG	AAGACGGTAAGT	13.8
16.5	CTTCCCAGTTACTG	AAGACGGTAAGTGT	17.0
18.8	GCTTCCCAGTTACTG	AAGACGGTAAGTGTC	18.5
22.6	TGGCTTCCCAGTTACTG	AAGACGGTAAGTGTCCA	22.3

YML085c

$\Delta G°$	Exon 1	Exon 2	$\Delta G°$
11.2	TATTAGTATTAATG	TCGGTCAAG	10.4
13.9	GTTATTAGTATTAATG	TCGGTCAAGCT	14.2
16.6	AAGTTATTAGTATTAATG	TCGGTCAAGCTG	15.9
19.6	AGAAGTTATTAGTATTAATG	TCGGTCAAGCTGGT	19.5
22.6	AGAGAAGTTATTAGTATTAATG	TCGGTCAAGCTGGTTG	22.4

YML017w

$\Delta G°$	Exon 1	Exon 2	$\Delta G°$
9.4	GAAGAAATGG	GAACAAATAATAC	11.2
13.6	GGGAAGAAATGG	GAACAAATAATACAT	13.8
16.4	CGGGAAGAAATGG	GAACAAATAATACATCT	16.8
19.1	AACGGGAAGAAATGG	GAACAAATAATACATCTAAT	19.8
22.7	GGAACGGGAAGAAATG	GAACAAATAATACATCTAATAAT	22.8

Figure 3.2 Design scheme for junction probes. For each intron-containing gene, a series of probes was created such that the hybridization energy derived from interactions with the upstream and downstream exons were thermodynamically balanced. For some genes, like YML056c, this yielded a nearly equal number of base pairs on either side of the junction. However, because of variable sequence content surrounding the junctions, other genes required longer base pairing regions either upstream (YML085c) or downstream (YML017w) of the exon–exon boundary. The boxed sequences correspond to the best performing probes in test hybridizations, and are included in the final microarray design.

closest to 17 kcal/mol on each side of the junction (Sugimoto *et al.*, 1996). While these parameters were used to design junction probes specific to the *S. cerevisiae* genes, the thermodynamic properties are such that these parameters are likely to be the optimal parameters for the design of junction probes for any organism.

Finally, the architecture of the Agilent microarray platform is such that 60 nt probes are printed with their 3′ ends covalently linked to the glass slide surface. Because the lengths of intron and functional RNA probes are fixed at 35 nt and our junction-specific probes vary from 24 to 36 nt, we included a stalk region at the 3′ end of these oligonucleotides to move the "targeting region" of the probes away from the glass in hopes of making them more readily accessible for hybridization. Several different stalk designs were tested. We settled on a sequence designed by Agilent to have low cross-reactivity with any genomic sequence. As expected, our initial experiments comparing probes containing stalks with those lacking them indicated that the stalks provided improved signal intensity with little or no loss in probe specificity.

3. Sample Preparation

The goal of this section is to describe all the steps needed to go from experimental design to hybridizing a sample on a microarray. Obviously, the details of experimental design will vary. For the purposes of this chapter we will describe a specific experiment comparing a yeast strain containing a temperature-sensitive mutation in a canonical splicing factor to a matched wild-type strain, which will be referred to as the experimental and reference strains, respectively, from here on. In our experience, the data from this type of experiment are most easily understood when a time course is followed after shifting the strains to the nonpermissive temperature. Defects in pre-mRNA processing can often be detected within minutes in an experiment like this. Below are the protocols for each of the major steps in the pathway: cell collection, RNA isolation, cDNA synthesis, fluorescent labeling, and microarray hybridization and washing.

3.1. Cell collection

The first step in an experiment is to collect appropriate cells from the experimental and reference strains. In our experience, microarrays are exceptionally sensitive assays that are able to detect subtle differences in growth and handling of samples. As such, we always collect actively grow-ing cells in early to mid-log phase. Likewise, we work to standardize all experimental conditions to have equivalent volumes, flask sizes, growth media, shift conditions, etc. for both the experimental and reference strains. Our preferred method for harvesting cells is by vacuum filtration using mixed cellulose ester filters (Millipore Cat.#: HAWP02500, or equivalent) and a vacuum manifold apparatus (Millipore Cat.#: XX1002500, or equiv-alent). After collection, the filters can be placed in a 15 ml conical tube and immediately frozen in liquid nitrogen. This method provides a fast and simple mechanism for collecting rapid time points during a time course. We have also collected cells by centrifugation at $5000 \times g$ for 5 min, but disfavor this method because of the time involved in getting cells from growth condition to frozen cells.

The quantity of cells required for an experiment will depend upon your experimental conditions. In general, our protocol requires 40 μg of total RNA for both the experimental and reference samples. We routinely recover 20 μg of total cellular RNA from a single milliliter of cells grown in YPD with an optical density (OD) equal to 0.5 ($\sim 5 \times 10^6$ cells). Our yields of total RNA are typically twofold less for cells grown in synthetic media to the same OD.

3.1.1. Protocol for cell collection

(1) On day 1, start a 5 ml culture of both experimental and reference strains in YPD and allow them to grow overnight at the permissive temperature (25 °C).

(2) On day 2, for both strains, use the 5 ml overnight cultures to inoculate a fresh 50 ml culture to a starting OD of 0.1. Allow these to continue growing at 25 °C.

(3) When the two cultures have reached the appropriate cell density (between an OD of 0.5 and 0.75, ~3–4 h), collect 10 ml of cells by filtration. Immediately transfer the filter into a 15 ml conical tube, cap the tube tightly, and plunge it into liquid nitrogen.

(4) Transfer both the experimental and reference culture flasks to a shaking water bath at the nonpermissive temperature (37 °C). Collect additional 10 ml aliquots as described above after 5, 15, and 30 min. Cells can be stored at − 80 °C until ready for RNA isolation.

3.2. RNA isolation

In our experience, there are two factors that are crucial for efficiently isolating high-quality RNA for use in microarray experiments. The first critical factor is achieving efficient cell lysis. In our protocol, we affect cell lysis by using a combination of heat, exposure to phenol, exposure to SDS, and physical agitation. In our experience the most common reason for obtaining poor RNA yields results from insufficient vortexing during heating (step 2 below). The second crucial factor is maintaining the integrity of the RNA. In this regard, we focus on two aspects: temperature and time. During all of the steps listed below, the samples should be handled on ice. Likewise, where possible, all centrifugation steps should be performed in a refrigerated centrifuge. Additionally, we strive to minimize the amount of time that elapses between taking the cells out of the − 80 °C freezer (step 1) and adding isopropanol to precipitate the RNA (step 8). Every effort should be made to move as expeditiously as possible until this point to ensure high-quality RNA. After the addition of isopropanol, the samples can be stored at − 20 °C indefinitely.

An important improvement to both the reproducibility and integrity of our RNA preparations has come from the use of tubes containing Phase Lock Gel (5 Prime, http://www.5prime.com) to facilitate the separation of the organic and aqueous phases. Use of these tubes allows for nearly quantitative recovery of the aqueous phase and removes the inconsistency associated with manual aspirations at the interphase. Our protocols indicate $3000 \times g$ spins to separate the aqueous and organic phases when using Phase Lock Gel, however, faster spins will give better separation. We routinely spin our samples in a Beckman X-15R centrifuge with a SX4750A rotor at top speed ($5250 \times g$) at 4 °C for 5 min.

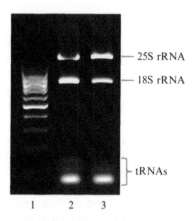

Figure 3.3 Typical banding pattern for purified total RNA. Two independent preparations of total cellular RNA have been separated by gel electrophoresis on a 1% agarose gel run in 1× TAE buffer. Loaded in lane 1 is 0.5 μg of GeneRuler 100 bp DNA Ladder (Fermentas). Lanes 2 and 3 are the total RNA samples. The locations of the 25S rRNA, 18S rRNA, and tRNA species are indicated.

The end of the RNA isolation procedure is the first opportunity to assess the quality of your experiment. We assess RNA quality in several ways. First, we determine the quantity of RNA isolated using a spectrophotometer and compare the results to the expectations described above. If the RNA isolation yield is significantly less than expected, we tend to discard this material, collect new cells and repeat the RNA isolation. Second, we examine the integrity of the isolated RNA by visualization on an agarose gel. While the quality of the mRNAs in the total RNA preparation cannot be directly assessed by the agarose gel, we use the bands corresponding to the ribosomal RNAs as a proxy for the integrity of the mRNAs. Figure 3.3 shows a typical banding pattern seen when 1 μg of RNA is separated on a 1% agarose/1× TAE gel. As an alternative, a higher resolution analysis of mRNA quality can be obtained using instruments such as the Agilent Bioanalyzer. It is perhaps worth a brief note that typical precautions should be used when handling RNA so as to avoid contamination with ribonucleases.

3.2.1. Materials for RNA isolation

15 ml Phase Lock Gel Heavy tubes (5 Prime Cat.#: 2302850)
AES buffer (50 mM sodium acetate (pH 5.3), 10 mM EDTA, 1% SDS)
Acid–phenol:chloroform (5:1) (pH < 5.5) (Ambion Cat.#: AM9720, or equivalent)
Phenol:chloroform:IAA (25:24:1) (Ambion Cat.#: AM9730, or equivalent)

Chloroform
3 M sodium acetate (pH 5.3)
Isopropanol
70% ethanol

3.2.2. Protocol for RNA isolation

(1) Remove conical tubes containing filters from the $-80\ ^\circ$C freezer and place on ice. Immediately add 2 ml of acid–phenol:chloroform, then add 2 ml of AES buffer and vortex well. Cells will be easily removed from the filters by vortexing.

(2) Transfer the tubes to a 65 °C water bath and incubate for 7 min. Vortex thoroughly (5 to 10 seconds) once every minute.

(3) Transfer the tubes to ice and incubate for 5 min. During incubation, prepare one 15 ml Phase Lock Gel Heavy tube for each sample by spinning briefly at $\geq 3000 \times g$.

(4) Transfer the entire organic and aqueous contents to a prespun 15 ml Phase Lock Gel Heavy tube. Leave the filters behind if possible, but do not worry if they do transfer. Once the material is in the Phase Lock Gel Heavy tube, do not vortex. This fragments the Phase Lock Gel. Spin at $\geq 3000 \times g$ at 4 °C for 5 min.

(5) In the same 15 ml Phase Lock Gel Heavy tube, add 2 ml of phenol:chloroform:IAA to the supernatant. Mix by shaking, but do not vortex. Spin at $\geq 3000 \times g$ at 4 °C for 5 min.

(6) In the same 15 ml Phase Lock Gel Heavy tube, add 2 ml of chloroform to the supernatant. Mix by shaking, but do not vortex. Spin once again at $\geq 3000 \times g$ at 4 °C for 5 min.

(7) Prepare a new 15 ml conical tube with 2.2 ml of isopropanol and 200 μl of 3 M sodium acetate.

(8) Pour the supernatant from step 6 into the 15 ml conical tube with isopropanol. Mix by inverting several times.

(9) Transfer 2 ml of the isopropanol slurry into a 2 ml microcentrifuge tube and spin the RNA at top speed in a microcentrifuge ($\geq 14{,}000 \times g$) at 4 °C for 20 min. (The remainder of the RNA slurry can be stored at $-20\ ^\circ$C for future use as this is the most stable storage method for RNA.)

(10) Carefully pour off the supernatant from the 2 ml tube so as not to disrupt the pellet. Add 2 ml of 70% ethanol to the pellet and mix by inverting several times. Spin again at top speed in a microcentrifuge ($\geq 14{,}000 \times g$) at 4 °C for 5 min.

(11) Repeat step 10 once.

(12) Carefully pour off the supernatant from the 2 ml tube so as not to disrupt the pellet, then briefly dry the RNA in a SpeedVac. Do not heat or overdry the samples.

(13) Dissolve the RNA in 50 μl of water. Determine the actual concentration using a spectrophotometer. Use a conversion factor of 1 $A_{260} = 40\ \mu$g/ml RNA. This should give a concentration of approximately 2 mg/ml.

3.3. cDNA synthesis

After successful isolation of total RNA, the next step is to fluorescently label the experimental and reference samples for microarray hybridization. Several different methods exist, including direct RNA labeling (Wiegant et al., 1999), generation of cRNA (Gelder et al., 1990), or generation of cDNA (DeRisi et al., 1997). It is worth noting that Agilent's protocols for gene expression microarrays take advantage of a linear amplification method using T7 RNA polymerase to generate direct-labeled cRNAs. While there are advantages and disadvantages to each of these methods, for reasons described below our protocols are instead designed and optimized for cDNA synthesis.

While traditional expression microarrays (and Agilent's gene expression protocols) use oligo-dT sequences to prime their cDNA (and cRNA) reactions, we instead use random 9-mer oligonucleotides. There are two main reasons for choosing this method. First, we are interested in looking at RNAs independent of their poly-(A) tail status. We presume that many pre-mRNA species may lack fully developed poly-(A) tails, and such species may be undetectable when priming with oligo-dT. The second reason is related to the fact that most *S. cerevisiae* introns are located at the 5' end of their transcripts. Therefore, to distinguish between pre- and mature mRNA species, cDNAs must be produced that correspond to the 5' end of the transcript, the efficiency of which is greatly reduced if all priming events take place at the poly-(A) tail. While random priming helps to alleviate this issue, the potential downside of random priming is the production of cDNAs corresponding to the highly abundant ribosomal RNA species. For a given intron-containing gene there are on the order of one million copies of rRNA for every single copy of pre-mRNA, thereby highlighting the need for highly specific probe design. Nevertheless, we find that the combination of our probe design and priming strategy does produce highly specific data. For example, as a measure of the potential cross-reactivity of the ribosomal cDNA species, we have examined microarrays comparing wild-type strains to strains containing complete deletions of intron-containing genes. Whereas robust signal is detected on the intron-specific probes for the wild-type strain, we find the signal intensity for the deletion strains are significantly reduced to levels near background. This suggests that there is minimal cross-hybridization of the ribosomal cDNAs to our specific probes.

In general, there are two different methods by which fluorescent dyes can be incorporated into cDNA: either by inclusion of a fluorescently labeled nucleotide analog which can be directly incorporated by reverse transcriptase, or by inclusion of a derivatized nucleotide analog containing a reactive chemical group which can be used to covalently attach fluorescent dyes in subsequent reactions. Our protocols utilize the latter method, including aminoallyl-dUTP in the reverse transcription reaction, which is highly reactive to N-hydroxysuccinimidyl (NHS) ester derivatized fluorophores. The advantage to this method is that the aminoallyl-dUTP is efficiently incorporated by reverse transcriptase, whereas fluorescently labeled nucleotide analogs are often poor substrates for polymerases. Because cellular RNAs contain modified nucleotides with potentially reactive primary amines, our protocol includes an RNA hydrolysis step after cDNA synthesis to facilitate their removal. As will be described later, the cDNA purified from this protocol can be efficiently reacted with fluorescent dyes.

The total amount of fluorophore incorporated into a sample is an important factor for optimal microarray signal and can be controlled by adjusting the ratio of aminoallyl-dUTP to dTTP in the cDNA synthesis reaction. Our initial experiments using Agilent microarrays indicated that the specific fluorescent activity of the hybridized sample needed to be significantly lower than what has been typically used for spotted microarrays (DeRisi et al., 1997; Pleiss et al., 2007b). At high concentrations of aminoallyl-dUTP, we observed optical interaction between the two fluorophores, presumably because of the overlap between their emission and absorption spectra. This problem was alleviated by reducing the ratio of aminoallyl-dUTP to dTTP. The different requirements for Agilent and spotted microarrays presumably reflect the differences in the density and orientation of the oligonucleotide probes in these two formats.

For both the experimental and reference samples, a single splicing microarray requires 20 μg of total RNA as starting material. We always perform our microarray experiments as technical repeats where the orientation of the dyes is reversed, resulting in so-called "dye-flipped" replicates. As indicated earlier, this means that 40 μg of total RNA are needed for both experimental and reference samples for a replicate set of hybridizations. We set up a single cDNA synthesis reaction for the entire 40 μg of total RNA, which will be divided later for fluorescent labeling and hybridization. Important considerations in setting up the cDNA reactions are the concentration of total RNA and random primers. Our best results have been achieved using a final concentration of total RNA at or below 0.5 mg/ml and random primers at 0.25 mg/ml. At RNA concentrations higher than this, the efficiency of cDNA synthesis drops off significantly.

For the protocol listed below we purify recombinant MMLV reverse transcriptase and make our own buffers. However, commercial enzymes can also be used. In considering different commercial enzymes it is important to use an enzyme like Superscript Reverse Transcriptase (Invitrogen) that has the RNaseH activity disrupted. However, most commercial enzymes are packaged with a reaction mix that contains buffer, salt and MgCl$_2$ together. When we hybridize our primers to our RNA, we find that it is important to leave out the MgCl$_2$ to avoid stabilizing RNA structures and therefore recommend using homemade buffers for this step. Under the conditions described below we typically achieve cDNA synthesis yields that are between 30% and 50% conversion by mass (12–20 μg of cDNA from 40 μg of starting RNA). As this is a second opportunity for quality control, yields significantly below this expectation warrant a repeat of the cDNA synthesis step or there will be insufficient signal for the microarray experiment.

After a successful cDNA synthesis reaction it is important to purify the products away from any unincorporated aminoallyl-dUTP prior to fluorescent labeling. In the protocol below we use a commercial kit from Zymo Research that is designed to purify oligonucleotides from unincorporated dNTPs. An alternative that we have found to be both cost-effective and high quality is to use 96-well Glass Fiber DNA binding plates. Such plates are made by several manufacturers and are widely available. We have commonly used plates from Nunc (Cat.#: 278010) along with homemade cDNA binding buffer (5 M guanidine–HCl, 30% isopropanol, 90 mM KOH, 150 mM acetic acid) and wash buffer (10 mM Tris–HCl (pH 8.0), 80% ethanol).

3.3.1. Materials for cDNA synthesis

Total RNA for experimental and reference samples
10× RT buffer = 0.5 M Tris–HCl (pH 8.5), 0.75 M KCl
10× dN$_9$ = 5 mg/ml dN$_9$ oligonucleotides
10× MgCl$_2$ = 30 mM MgCl$_2$
10× DTT = 0.1 M DTT
10× dNTP's (+aa-dUTP) = 10 mM ATP, 10 mM CTP, 10 mM GTP, 9.8 mM TTP, 0.2 mM aminoallyl-dUTP
Reverse transcriptase
RNA hydrolysis buffer = 0.3 M NaOH, 0.03 M EDTA
Neutralization buffer = 0.3 M HCl
DNA Clean and Concentrator—25 kit (Zymo Research, Cat.#: D4006)

3.3.2. Protocol for cDNA synthesis
Using the volumes in the protocol listed in Table 3.1 as a guide, but adjusted for actual RNA concentrations, do the following:

(1) Anneal the primers to the total RNA by heating in a 60 °C water bath for 5 min. (Note that at this stage the RNA and primer concentrations (1 and 0.5 mg/ml, respectively) are twice the values that they will be

Table 3.1 Experimental setup for cDNA synthesis reaction

Reagent	Stock concentration	Volume used (μl)	Volume used per 8 rxns (μl)
RNA/primer mix			
RT buffer	10×	5	
dN$_9$	5 mg/ml	5	
Total RNA	2 mg/ml	20	
Water		20	
Total		50	
Enzyme/dNTP mix			
RT buffer	10×	5	44
MgCl$_2$	10×	10	88
DTT	10×	10	88
dNTP's (+aa–dUTP)	20×	5	44
Reverse transcriptase	20×	5	44
Water		15	132
Total		50	440

in the final cDNA synthesis reaction. Note also that MgCl$_2$ is omitted from this step, but buffer and salt are included, which we find increases the yield by about 20% relative to annealing in water alone.)

(2) Immediately after heating the samples, transfer them onto ice for an additional 5 min to allow the primers to anneal.

(3) While the RNA/primer mix is cooling, make the enzyme/dNTP mix as described in Table 3.1.

(4) Add 50 μl of enzyme/dNTP mix to the annealed RNA/primer mix (the RNA should now be at its final concentration of 0.5 mg/ml). Briefly vortex and spin the tubes, then allow them to incubate at 42 °C. In our experience the reaction is >90% complete after 2 h; however, we routinely incubate the samples overnight.

(5) To hydrolyze RNA prior to purification of the cDNA, add 50 μl of RNA hydrolysis buffer, vortex, and spin down. Place this mix in a 60 °C water bath for 15 min, then transfer to ice.

(6) Neutralize the solution by adding 50 μl of neutralization buffer. Vortex and spin down.

(7) Each column in a DNA Clean and Concentrator—25 kit can bind a maximum of 25 μg of cDNA. Therefore, a single cDNA reaction, which starts with 40 μg of total RNA and yields about 20 μg of cDNA, can be purified with a single column. Follow the manufacturer's instructions for purification of single stranded cDNA until the elution step. Proceed with elution as described in step 8.

(8) Transfer the column to a clean 1.7-ml tube. Add 35 μl of water directly onto the filter. Wait 30 s, then spin the samples in the

microfuge at top speed ($>14{,}000{\times}g$) for 1 min (spinning at $10{,}000{\times}g$ is effective for elution and can help to prevent the lids of the collection tubes from snapping off).

(9) Add a second 35 μl aliquot of water directly onto the filter, wait another 30 s, and spin the samples in the microfuge at top speed ($>14{,}000{\times}g$) for 1 min.

(10) After discarding the column, vortex the eluate well and spin down. Quantitate cDNA yield using a spectrophotometer. Although this is a cDNA sample, we continue to use a conversion factor of 1 A_{260} = 40 μg/ml cDNA. Using this conversion factor, expect to recover between 12 and 20 μg of cDNA from 40 μg of starting RNA.

(11) Split the eluate into two equal aliquots ($\sim$33 μl) in 1.7 ml tubes and dry in the SpeedVac. These samples will subsequently be labeled with the two different dyes to be used as matched replicates for the microarray.

3.4. Fluorescent labeling of cDNA

Many different vendors sell fluorescent dyes which have both the appropriate spectral properties and the appropriate derivatizations to react with the primary amine of the aminoallyl modified nucleotide. We have used both Cy dyes from GE Healthcare (Cy3 and Cy5) and Alexa dyes from Invitrogen (Alexa 555 and 647), and have found largely overlapping results. We presume that similar dyes from other vendors could also be used. A major concern for both choosing and handling these fluorescent dyes is that there is ample evidence in the literature that significant oxidation of both Cy5 and Alexa 647 can result from the levels of ozone that are commonly present in the air. Methods for mitigating ozone levels will be discussed in Section 3.6. A possible alternative is to use a new dye from GE Healthcare called Hyper5 which is a modified version of Cy5 that is reportedly stable to ozone (Dar et al., 2008).

A single tube of NHS ester derivatized Cy3 or Cy5 contains a sufficient amount of fluorophore to label 16 cDNA samples. Because the NHS ester is highly unstable we do not store opened dye packages. Therefore, the actual volume of DMSO that we use to dissolve a single dye aliquot is determined by the number of samples. For example, in this protocol where eight different hybridizations are being performed (a four-point time course with dye-flipped replication), a single dye pack should be dissolved in 42 μl of DMSO (enough for eight 5 μl aliquots).

3.4.1. Materials for fluorescent labeling of cDNA

0.1 M sodium bicarbonate (pH 9.0)
DMSO (Fluka, Cat.#: 41647)

Cy3 NHS ester (GE, Cat.#: PA23003)
Cy5 NHS ester (GE, Cat.#: PA23005)
DNA Clean and Concentrator—25 kit (Zymo Research, Cat.#: D4006)

3.4.2. Protocol for fluorescent labeling of cDNA

(1) Dissolve dried down cDNA in 5 μl of 0.1 M sodium bicarbonate. To ensure that all of the cDNA is resuspended in this small volume, vortex well and spin down several times.

(2) Dissolve the Cy3 and Cy5 dyes in 42 μl of DMSO. Because it is difficult to see whether all of the dye is dissolved, vortex well and spin down several times.

(3) To one of the two tubes of experimental cDNA, add 5 μl of Cy3 dissolved in DMSO. To the other tube of experimental cDNA add 5 μl of Cy5 dissolved in DMSO. Do the same for your reference cDNA samples.

(4) Incubate the reactions in the dark in a 60 °C water bath for 1 h. While many dye labeling protocols incubate at room temperature, we observe a significant increase in labeling efficiency at elevated temperatures.

(5) To the 10 μl labeling reaction add 100 μl of DNA binding buffer from the DNA Clean and Concentrator kit, and then proceed according to manufacturer's instructions until the elution step. Proceed with elution as described in step 6.

(6) Transfer the column to a clean 1.7 ml tube. Add 35 μl of water directly onto the filter. Wait 30 s, then spin the samples in the microfuge at top speed ($>14,000 \times g$) for 1 min (spinning at $10,000 \times g$ is effective for elution and can help to prevent the lids of the collection tubes from snapping off).

(7) After discarding the column, quantitate cDNA yield using a spectrophotometer. Continue to use a conversion factor of 1 $A_{260} = 40$ μg/ml cDNA. Using this conversion factor, expect to recover $\geq 50\%$ of the cDNA that was included in the labeling reaction. Note that it is in theory possible to monitor Cy3 and Cy5 incorporation at this step; however, with this protocol the expected absorption levels for these dyes are close to background.

(8) Pool the appropriate Cy3- and Cy5-labeled samples for each hybridization, and dry in the SpeedVac. *Important note*—make sure that the appropriate samples are combined at this step. This is the easiest moment to ruin a great experiment. For example, the Cy3-labeled experimental sample should be combined with the Cy5-labeled reference sample, and *vice versa*.

(9) After the samples have dried in the SpeedVac, it is important to resuspend the samples in water quickly, because Cy5 is highly sensitive to ozone. Resuspend each pellet in 25 μl of water. Vortex well to

ensure proper mixing. We typically proceed immediately to hybridization. But if necessary, samples can be flash frozen in liquid nitrogen and
stored in the dark at $-80\ ^{\circ}\text{C}$ indefinitely.

3.5. Microarray hybridization

The Agilent Custom $8\times15\text{K}$ Microarrays are a single piece of 1×3 in.
microscope glass with eight different hybridization zones. The individual
hybridizations are kept separated by use of a matched gasket slide which,
when sandwiched with the microarray slide, creates eight distinct hybridization compartments. All eight compartments need to be simultaneously
hybridized. A detailed description of the microarray architecture and
instructions for their use accompanies each microarray order and should
be used as a supplement to the protocols described here.

Because dust is highly fluorescent, it is important to keep dust to a
minimum during the hybridization procedure. We minimize the time that
the microarray surfaces are exposed to air and always work on clean surfaces.
Likewise, gloves should always be worn to avoid contamination of the glass.
When possible the glass slides should be handled with tweezers. When this
is not possible, the glass slides should only be handled by their edges.

3.5.1. Materials for hybridization

Agilent Custom $8\times15\text{K}$ Microarray
$2\times$ hybridization buffer (Agilent, Cat.#: 5190-0403)
Eight chamber gasket slides (Agilent, Cat.#: G2534-60014)
Hybridization chamber (Agilent, Cat.#: G2534A)
Hybridization oven (Agilent, Cat.#: G2545A)

3.5.2. Protocol for hybridization

(1) Heat samples to 95 °C for 2 min. Place in a drawer or dark box for
    5 min to cool, then spin down briefly.
(2) To each of the samples add 25 μl of $2\times$ Agilent hybridization buffer.
    Mix by gently pipetting up and down. DO NOT VORTEX as this
    introduces bubbles that are problematic for the hybridization.
(3) Place a gasket slide in a hybridization chamber with the gaskets facing up.
(4) Load 40 μl of each sample into the appropriate gasket section. Avoid
    pipetting bubbles. To reduce the evaporation of the samples that are
    loaded first, the time to load eight samples should be minimized.
(5) After all eight sections have been loaded, carefully place the microarray
    slide onto the gasket slide. Note that it is important to ensure that the
    printed side of the microarray slide is exposed to the sample (see Agilent
    protocol).

(6) Close the hybridization chamber. Once assembled, rotate the entire chamber two or three times to wet the gasket linings. Ensure that any bubbles within the hybridization compartments can rotate freely. Gently tapping the chamber may help to release any stuck bubbles.

(7) Hybridize in a rotating oven at 60 °C for 16 h. In our experience hybridization times ranging from 14 to 18 h produce similar results.

3.6. Microarray washing

Prior to scanning the microarray, unhybridized cDNAs must be washed off of the glass surface. The design of the probes on the microarray is optimized for specificity at 60 °C. Because nonspecific species will begin to cross-hybridize at lower temperatures, an important consideration during the washing steps is minimizing the time between removing the microarray from the hybridization oven and washing away any unbound cDNAs.

A second important consideration during microarray washing is the capacity of atmospheric ozone to oxidize Cy5 dyes. This oxidation potential is most acute on a dried microscope slide after hybridization and washing. Several different mechanisms have been developed to mitigate the effects of ozone. The solution we chose was to create a chamber where ozone can be specifically removed from the air. Prebuilt chambers are commercially available which are of sufficient size to house a microarray scanner (Scigene, NoZone Workspace). Alternatively, ozone scavenging filters are available (Ozone Solutions, NT-40) which can be used to remove ozone from any chamber. We have built a simple chamber in our lab using Plexiglass which has sufficient working space for wash dishes and our microarray scanner (15 cubic feet). If changing the infrastructure in your lab is not a possibility, other options exist. For example, Agilent has developed a stabilizing wash solution (Agilent, Cat.#: 5185-5979). Likewise, Genisphere has developed a coating solution that stabilizes the optical properties of the Cy5 dye (Genisphere, Cat.#: Q500500).

3.6.1. Materials for microarray washing

Glass washing dishes with slide racks (Thermo, Shandon Complete Staining Assembly 121)
Wash I = 6× SSPE and 0.005% sarcosyl (or Agilent, Cat.#: 5188-5325)
Wash II = 0.06× SSPE and 0.005% sarcosyl (or Agilent, Cat.#: 5188-5326)

3.6.2. Protocol for microarray washing

(1) Prepare three glass wash dishes, two with Wash I (one without a slide rack, one with a slide rack), and one with Wash II (with a slide rack).

(2) Remove the hybridization chamber from the hybridization oven. Open the chamber such that the sandwich of the microarray and gasket slides remains intact. Quickly transfer sandwiched slides to the Wash I dish with no rack. With the sandwich completely submerged and supported with one hand, use tweezers to gently pry the two slides apart, allowing the gasket slide to drop to the bottom of the glass dish. Hold the microarray slide with your fingers being careful to touch only the stickers or sides of the glass.

(3) Keep the microarray slide submerged in Wash I and swish the microarray slide back and forth two or three times to remove most of the unhybridized sample.

(4) Quickly transfer the microarray slide into the rack in the second Wash I dish, taking care to minimize exposure of the microarray to air.

(5) Vigorously agitate the rack up and down for 1 min.

(6) Transfer the Wash I dish with microarray into the ozone-free chamber. Place Wash II glass dish with rack in ozone-free chamber. Quickly transfer the microarray slide from Wash I to Wash II.

(7) Gently agitate slide rack up and down, ensuring that all bubbles are washed off the microarray slide.

(8) With a pair of tweezers, grab the microarray slide by a corner with the stickers. Slowly remove the slide from Wash II. If this is done slowly enough (over 10 s) the microarray will come out dry because of the sheeting qualities of the wash solution.

(9) The microarray can either be scanned immediately or can be put into a dark box and protected from ozone until ready to be scanned.

4. Microarray Data Collection

The Agilent microarrays are printed on standard microscope glass (1×3 in.) and can be analyzed using any scanner than can accommodate this format. The features on an Agilent Custom 8×15K Microarray are approximately 60 μm in diameter. We get the highest quality data when we scan at a pixel size of 5 μm; this yields about 100 pixels of data for each feature. There are several companies that manufacture microarray scanners, including Agilent, Molecular Devices, and Tecan, which are capable of scanning at this resolution. We use an Axon 4000B for all of our data collection (Molecular Devices). The instruments listed above all use lasers to excite the Cy3 and Cy5 dyes and photomultiplier tubes (PMTs) to quantitate fluorescence intensity at each spot. Below are some general guidelines to facilitate microarray scanning, but because each of these instruments has different parameters, user guides should be consulted for specific details.

An important assumption underlying microarray experiments is that the total global signal is unchanged between the experimental and reference samples with only a small subset of features showing change in expression behavior. As such, an important consideration when scanning the microarrays is to equalize the total fluorescent signal intensity from the Cy3 and Cy5 samples. In the early days, this required a painstaking process of manually adjusting the PMT gain settings. Fortunately, newer versions of scanner control software allow the user to largely automate this step.

Whereas new software has improved the ability to equalize Cy3 and Cy5 signal, another important consideration is maximizing total signal intensity. For most microarray scanners the dynamic range is limited to about three orders of magnitude. By comparison, the difference in abundance of a rare pre-mRNA species and an abundant mature mRNA species can easily be greater than three orders of magnitude, meaning that no single scanning condition can generate reliable data for every RNA species. Agilent's microarray scanners automate the process of scanning each microarray twice; once using the lasers at full power and once at reduced power settings. The Agilent software then integrates the data from these two scans to increase the dynamic range of the experiment. In our experience, we have empirically identified conditions that maximize data collected from just a single scan. On our Axon 4000B, using the built in software, we set a saturation tolerance level equal to 0.1%. Using these settings and with the PMT gain values between 500 and 600 for both channels, only the most highly abundant RNA species in the cell are oversaturated, yet robust signal can be detected for the rare pre-mRNA species.

5. MICROARRAY DATA ANALYSIS

In this section, we will divide the tasks for data analysis into two general parts: one is the technical details for processing the scans from the previous section, and the second is deriving biological meaning from an experiment. The first step is to extract quantitative measurements for both experimental and reference samples for each of the $\sim 15,000$ features for all eight hybridizations. For this step in the data analysis pathway, processing a splicing-sensitive microarray is no different than processing a standard gene expression microarray. A description of the tools necessary for extracting data from a microarray experiment can be found in Chapter 2 of this volume, and should be consulted for this and subsequent sections. Included in that chapter are both the descriptions of the software that can be used and instructions for their implementation.

5.1. Data normalization

Having successfully extracted quantitative information for all of the features on the microarray, the next step is to mathematically normalize the data derived from the experimental and reference samples. As described in Chapter 2 of this volume, the Bioconductor package of data analysis software (Gentleman *et al.*, 2004) provides several different options for normalizing microarray data. We have compared the output when analyzing a single microarray using several different normalization algorithms and have found that the LOWESS normalization algorithm does the best job at addressing nonlinear behavior often seen in microarrays (Yang *et al.*, 2002). Therefore, we use the maNorm package within Bioconductor to implement LOWESS normalization across all of our data. The output from this analysis is a single value corresponding to the $\log_2$ transformed ratio of the Cy5 intensity to the Cy3 intensity for all 15,000 features.

5.2. Replication

Of the $\sim$15,000 features present on our splicing-sensitive Agilent microarrays there are $\sim$7000 unique features, each of which is replicated at either two or three distinct locations on each microarray. The next step in analyzing the data is to collapse these replicate data into a single averaged value. A spreadsheet program such as Microsoft Excel can be used to calculate averages as well as coefficients of variation for these measurements. Having compressed the data to a single value for each unique feature, the next task is to compare and average the values determined between the dye-flipped replicate experiments. It is important here to repeat that the standard output from the maNorm package is always presented as a ratio of Cy5 to Cy3 intensity. According to the design of our experiment, one microarray compares Cy5-labeled experimental sample with Cy3-labeled reference sample, whereas its corresponding replicate compares Cy5-labeled reference sample with Cy3-labeled experimental sample, which for the purposes of this section we will refer to as "forward" and "flip" experiments, respectively. Because of their orientations, the data output from the maNorm package for the "forward" and "flip" experiments are expected to be negatively correlated. Therefore, to determine the average behavior described by the dye-flipped experiments the "flip" value must be multiplied by -1 prior to averaging. At this point, the value associated with each of the $\sim$7000 unique features represents a composite value incorporating replication within a single microarray and between dye-flipped microarrays and constitutes as many as six independent measurements for each time point.

5.3. Splicing specific data

While no further steps are necessary to analyze the behavior of intronless genes, analyzing intron-containing genes requires additional steps. Whereas just one value is needed to describe the behavior of an intronless gene, even the simplest intron-containing gene has a minimum of three values associated with it, corresponding to the total, pre- and mature mRNA probes. None of these metrics by themselves is sufficient to understand the splicing behavior of a given transcript, but rather they must be considered in relation to one another. In their original experiments, the Ares group addressed this issue by creating splice junction and intron accumulation indexes; they divide the values for both the mature probe and the pre-mRNA probe by the value of the total mRNA probe, respectively (Clark *et al.*, 2002). By comparison, we have chosen to analyze the data by concurrently examining the behaviors of each of the three individual metrics (Pleiss *et al.*, 2007a,b). Database software such as Microsoft Access can be used to associate the total, pre- and mature mRNA specific probes with one another for a given intron-containing gene.

5.4. Extracting biological meaning

Congratulations! Having successfully collected experimental samples, isolated RNAs, converted RNA to cDNA, labeled and hybridized the cDNA to microarrays, extracted and normalized the data, you have now made it to the hard part—understanding the biology underlying your experiment. For traditional gene expression experiments, this is rather straightforward in the sense that one can look for transcripts that show either increased or decreased abundance. Those transcripts showing increased abundance either reflect genes whose transcription has increased or whose degradation has decreased. From the perspective of splicing, it is more difficult to describe what a defect should look like. A simple expectation might be that a defect in splicing would lead to accumulation of pre-mRNA levels with a concomitant decrease in mature mRNA levels. This expectation, however, presumes nearly equal steady state levels of pre- and mature mRNA. For genes that are efficiently spliced, where the mature mRNA level is much greater than the pre-mRNA level, a defect in splicing could be expected to show an accumulation of pre-mRNA with little or no change in mature mRNA. Our experience in analyzing these types of data demonstrates that both of these profiles can be seen among the different intron-containing genes in *S. cerevisiae*. As such, several different descriptors exist which may be used to identify a transcript whose splicing is altered in response to an experimental condition.

With these ideas in mind, the next challenge for any experiment is finding the important biological changes among the sea of data collected in a microarray experiment. Without a doubt, this is the most challenging part of the microarray experiment because no single approach to interrogating the data will identify all of the genes whose behavior is modified. Rather, we find the best way to identify these genes is to look at the data from many different perspectives. For example, in our experience comparing data across an experimental time course is a powerful way to identify genes that are responding to an experimental condition. Two important software packages we use to organize and visualize our data, Cluster (Eisen et al., 1998) and Java Treeview (Saldanha, 2004), are described in the Chapter 2 of this volume. We find that Cluster works particularly well for organizing splicing-specific information. For example, Fig. 3.4 shows the results of a time course examining a mutation in the canonical splicing factor Prp16 versus a wild-type reference. By concurrently examining the behavior of the total, pre-, and mature mRNA, transcripts can be identified whose splicing is affected by the experimental condition. This figure demonstrates the variety of behaviors possible for different transcripts, as described above, demonstrating the global patterns that result from a defect in splicing.

6. FUTURE METHODOLOGIES

Splicing-sensitive microarrays are a powerful tool for examining genome-wide changes in pre-mRNA splicing. However, as with all microarray technologies, the advent of high-throughput, short-read sequencing technologies promises to change the way splicing is studied from a genome-wide perspective (Wold and Myers, 2008). In theory, these short-read sequencing methodologies have an advantage over microarray technologies in that they take an unbiased approach to the experiment. Because microarrays require probes be designed to target-specific RNAs, they are by nature poor at discovering previously uncharacterized species. By directly sequencing total cellular RNA, short-read sequencing methodologies should be able to identify both previously uncharacterized RNAs and novel splicing events. Nevertheless, many of the same challenges that the splicing-sensitive microarray community faced must now be resolved in the context of short-read sequencing methodologies. For example, the most widely used current methods for sequencing cellular RNAs utilize poly-(A) selection schemes to remove ribosomal RNAs from the pool of sequenced samples. For the same reasons described at the beginning of this chapter we think it is likely that many of the interesting RNA processing events happen independent of the poly-(A) status of the RNA. Until such time as these methodologies have been developed for the sequencing technologies,

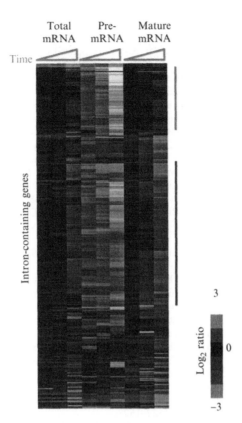

Figure 3.4 Genome-wide changes in pre-mRNA splicing. Results are presented from an experiment comparing a strain containing a cold-sensitive *prp16-302* mutation with a matched wild-type strain as both were shifted to the nonpermissive temperature. Data are shown from unshifted samples (grown at 30 °C), as well as after 10 and 60 min of incubation at 16 °C. Each horizontal line represents the behavior of a single intron-containing gene during this time course. Notice that some genes (indicated with a red bar) show a dramatic increase in pre-mRNA level with very little change in mature mRNA level, whereas other genes (indicated with a green bar) show a strong increase in pre-mRNA level concomitant with a strong decrease in mature mRNA level. (See Color Insert.)

splicing-sensitive microarrays will continue to be a fast, cost-efficient, and effective way to examine genome-wide changes in pre-mRNA splicing.

ACKNOWLEDGMENTS

We thank A. Manfredo, N. Sabet, L. Bud, A. Awan, M. Gudipati, and M. Bergkessel for critical feedback on the manuscript. Thanks also to the students in our Research Practicum Course for testing this protocol prior to publication.

REFERENCES

Berget, S. M., Moore, C., and Sharp, P. A. (1977). Spliced segments at the 5′ terminus of adenovirus 2 late mRNA. *Proc. Natl. Acad. Sci. USA* **74,** 3171–3175.

Chow, L. T., Gelinas, R. E., Broker, T. R., and Roberts, R. J. (1977). An amazing sequence arrangement at the 5′ ends of adenovirus 2 messenger RNA. *Cell* **12,** 1–8.

Clark, T. A., Sugnet, C. W., and Ares, M. (2002). Genomewide analysis of mRNA processing in yeast using splicing-specific microarrays. *Science* **296,** 907–910.

Dar, M., Giesler, T., Richardson, R., Cai, C., Cooper, M., Lavasani, S., Kille, P., Voet, T., and Vermeesch, J. (2008). Development of a novel ozone- and photo-stable HyPer5 red fluorescent dye for array CGH and microarray gene expression analysis with consistent performance irrespective of environmental conditions. *BMC Biotechnol.* **8,** 86.

DeRisi, J. L., Iyer, V. R., and Brown, P. O. (1997). Exploring the metabolic and genetic control of gene expression on a genomic scale. *Science* **278,** 680–686.

Eisen, M. B., Spellman, P. T., Brown, P. O., and Botstein, D. (1998). Cluster analysis and display of genome-wide expression patterns. *Proc. Natl. Acad. Sci. USA* **95,** 14863–14868.

Engebrecht, J. A., Voelkel-Meiman, K., and Roeder, G. S. (1991). Meiosis-specific RNA splicing in yeast. *Cell* **66,** 1257–1268.

Gelder, R. N. V., Zastrow, M. E. V., Yool, A., Dement, W. C., Barchas, J. D., and Eberwine, J. H. (1990). Amplified RNA synthesized from limited quantities of heterogeneous cDNA. *Proc. Natl. Acad. Sci. USA* **87,** 1663–1667.

Gentleman, R. C., Carey, V. J., Bates, D. M., Bolstad, B., Dettling, M., Dudoit, S., Ellis, B., Gautier, L., Ge, Y., Gentry, J., *et al.* (2004). Bioconductor: Open software development for computational biology and bioinformatics. *Genome Biol.* **5,** R80.

House, A. E., and Lynch, K. W. (2008). Regulation of alternative splicing: More than just the ABCs. *J. Biol. Chem.* **283,** 1217–1221.

Juneau, K., Palm, C., Miranda, M., and Davis, R. W. (2007). High-density yeast-tiling array reveals previously undiscovered introns and extensive regulation of meiotic splicing. *Proc. Natl. Acad. Sci. USA* **104,** 1522–1527.

Pleiss, J. A., Whitworth, G. B., Bergkessel, M., and Guthrie, C. (2007a). Rapid, transcript-specific changes in splicing in response to environmental stress. *Mol. Cell* **27,** 928–937.

Pleiss, J. A., Whitworth, G. B., Bergkessel, M., and Guthrie, C. (2007b). Transcript specificity in yeast pre-mRNA splicing revealed by mutations in core spliceosomal components. *PLoS Biol.* **5,** e90.

Saldanha, A. J. (2004). Java Treeview—extensible visualization of microarray data. *Bioinformatics* **20,** 3246–3248.

Sapra, A. K., Arava, Y., Khandelia, P., and Vijayraghavan, U. (2004). Genome-wide analysis of pre-mRNA splicing: Intron features govern the requirement for the second-step factor, Prp17 in *Saccharomyces cerevisiae* and Schizosaccharomyces pombe. *J. Biol. Chem.* **279,** 52437–52446.

Sayani, S., Janis, M., Lee, C. Y., Toesca, I., and Chanfreau, G. F. (2008). Widespread impact of nonsense-mediated mRNA decay on the yeast intronome. *Mol. Cell* **31,** 360–370.

Sugimoto, N., Nakano, S., Yoneyama, M., and Honda, K. (1996). Improved thermodynamic parameters and helix initiation factor to predict stability of DNA duplexes. *Nucleic Acids Res.* **24,** 4501–4505.

Wang, E. T., Sandberg, R., Luo, S., Khrebtukova, I., Zhang, L., Mayr, C., Kingsmore, S. F., Schroth, G. P., and Burge, C. B. (2008). Alternative isoform regulation in human tissue transcriptomes. *Nature* **456,** 470–476.

Wernersson, R., Juncker, A. S., and Nielsen, H. B. (2007). Probe selection for DNA microarrays using OligoWiz. *Nat. Protoc.* **2,** 2677–2691.

Wiegant, J. C., van Gijlswijk, R. P., Heetebrij, R. J., Bezrookove, V., Raap, A. K., and Tanke, H. J. (1999). ULS: A versatile method of labeling nucleic acids for FISH based on a monofunctional reaction of cisplatin derivatives with guanine moieties. *Cytogenet. Cell Genet.* **87,** 47–52.

Wold, B., and Myers, R. M. (2008). Sequence census methods for functional genomics. *Nat. Methods* **5,** 19–21.

Yang, Y. H., Dudoit, S., Luu, P., Lin, D. M., Peng, V., Ngai, J., and Speed, T. P. (2002). Normalization for cDNA microarray data: A robust composite method addressing single and multiple slide systematic variation. *Nucleic Acids Res.* **30,** e15.

CHAPTER FOUR

ChIP-Seq: Using High-Throughput DNA Sequencing for Genome-Wide Identification of Transcription Factor Binding Sites

Philippe Lefrançois, Wei Zheng, *and* Michael Snyder

Contents

Abstract

Much of eukaryotic gene regulation is mediated by binding of transcription factors near or within their target genes. Transcription factor binding sites (TFBS) are often identified globally using chromatin immunoprecipitation (ChIP) in which specific protein–DNA interactions are isolated using an antibody against the factor of interest. Coupling ChIP with high-throughput DNA sequencing allows identification of TFBS in a direct, unbiased fashion; this

Department of Molecular, Cellular and Developmental Biology, Yale University, New Haven, Connecticut, USA

Methods in Enzymology, Volume 470
ISSN 0076-6879, DOI: 10.1016/S0076-6879(10)70004-5

technique is termed ChIP-Sequencing (ChIP-Seq). In this chapter, we describe the yeast ChIP-Seq procedure, including the protocols for ChIP, input DNA preparation, and Illumina DNA sequencing library preparation. Descriptions of Illumina sequencing and data processing and analysis are also included. The use of multiplex short-read sequencing (i.e., barcoding) enables the analysis of many ChIP samples simultaneously, which is especially valuable for organisms with small genomes such as yeast.

1. INTRODUCTION

The number of completely sequenced genomes has increased dramatically with improvements in DNA sequencing technologies, and the determination of coding sequences both *in vivo* and *in silico* has identified novel genes within these newly sequenced genomes (Aparicio *et al.*, 2002). First, understanding gene regulation requires more than just knowing their genomic sequence. Second, one must identify the repertoire of transcription factors present. Coding sequences of transcription factors are often conserved in the course of evolution (Borneman *et al.*, 2006), allowing their discovery in many cases by comparison to homologous transcription factors from closely related organisms (Frazer *et al.*, 2004). Third, it is crucial to establish a list of regulated genes (target genes) for each transcription factor. Computational searches for particular transcription factor DNA binding motifs upstream of putative target genes have been useful tools to obtain such a list, although these predictions require experimental validation *in vivo* (Tompa *et al.*, 2005). Moreover, the presence of a consensus binding motif is not always directly linked to transcription factor binding, as many perfect motifs are not bound by a transcription factor whereas some imperfect motifs are bound under the same environmental conditions (Borneman *et al.*, 2007; Martone *et al.*, 2003). Fourth, transcription factors can regulate genes subjugated to multiple cellular environments and stresses (Harbison *et al.*, 2004). Global characterization of binding sites of a single transcription factor, therefore, demands multiple experiments in different conditions as well as profiling across the whole genome, making such studies labor-intensive. Large international consortiums, such as ENCODE in humans (Birney *et al.*, 2007) and modENCODE in *Drosophila melanogaster* and *Caenorhabditis elegans* (Celniker *et al.*, 2009), aim to characterize every functional DNA element across the whole genome, and the study of transcription factor binding represents a major part of these efforts.

In *Saccharomyces cerevisiae*, there are approximately 200–300 described transcription factors (TFs) among the ~6000 predicted ORFs (Costanzo *et al.*, 2000). Direct analysis of transcription factor binding upstream of target genes was performed initially using DNase footprinting (Axelrod and Majors, 1989) and/or PCR quantification of DNA associated with an immunoprecipitated transcription factor, a procedure called chromatin immunoprecipitation (ChIP) (Kuo and Allis, 1999; Orlando et al., 1997). These methods could only analyze a few promoters at a time, making the comprehensive discovery of unexpected, novel TF-bound DNA elements unrealistic. The development of DNA microarrays technology has provided the field of gene regulation with a powerful tool for genome-wide characterization of transcription factor binding. This technique, termed ChIP-chip, relies on the immunoprecipitation of a transcription factor of interest with its associated DNA, followed by hybridization to a DNA microarray (Horak and Snyder, 2002). In addition, C-terminal protein tagging with an exogenous well-defined epitope (e.g., Myc, HA) circumvents the need for raising native antibodies against every transcription factor (Janke et al., 2004; Longtine et al., 1998). The advantages of epitope tagging include the use of commercially available antibodies, the ability to tag multiple DNA-binding proteins in a high-throughput fashion and a lower occurrence of nonspecific immunoprecipitation and cross-reaction of chromatin, ultimately resulting in decreased noise.

Novel high-throughput sequencing technologies, such as 454/Roche, Solexa/Illumina and ABI/SOLiD, have revolutionized genomic studies by allowing for large-scale sequence analysis through the generation of millions of short sequencing reads. For example, new transcripts and splice variants have been discovered in multiple organisms using RNA-Seq (Lister et al., 2008; Mortazavi et al., 2008; Nagalakshmi et al., 2008; Wilhelm et al., 2008). Transcription factor binding studies have also benefited from ultra-throughput sequencing via the development of ChIP-Sequencing (ChIP-Seq) (Johnson et al., 2007; Robertson et al., 2008). Instead of hybridizing the ChIP DNA sample to a microarray, each sample is processed directly into a DNA library for sequencing and analyzed separately after sequencing. The improved sensitivity and reduced background of ChIP-Seq is replacing the array-based ChIP-chip in mammalian studies aiming to characterize transcription factor binding. Typically, two to four times more transcription factor binding sites (TFBS) are determined using ChIP-Seq in comparison with ChIP-chip; the accuracy and resolution of the data are higher as well (Robertson et al., 2007). ChIP-Seq studies have been used to characterize transcription factor binding during cell growth and a stress response (Johnson et al., 2007; Robertson et al., 2007), enabled the establishment of a regulatory network (Chen et al., 2008) as well as helped to determine epigenetic changes (Marks et al., 2009). ChIP-Seq is widely used by the ENCODE and modENCODE consortia for mapping TFBS in humans, C. elegans and D. melanogaster. In humans, it has been used to examine nucleosome positioning (Schones et al., 2008) and, in yeast, our group has characterized the distribution of three DNA-binding proteins, Cse4p, RNA polymerase II, and Ste12p, using this procedure (Lefrancois et al., 2009).

Recently, efforts have been made to develop a multiplexing scheme for Illumina sequencing, allowing many DNA samples to be sequenced simultaneously (Craig *et al.*, 2008; Cronn *et al.*, 2008; Lefrancois *et al.*, 2009). As a typical flowcell lane currently yields approximately 8 or more million uniquely mapped sequence reads, the number of mapped reads far exceeds the minimal number required for mapping binding sites in yeast, flies, and worms (Lefrancois *et al.*, 2009; Zhong *et al.*, 2010). We have therefore developed a barcoded ChIP-Seq strategy that enables the accurate sequencing and analysis of multiple yeast ChIP samples in the same flowcell lane (Lefrancois *et al.*, 2009). Data generated in this fashion have identified binding sites for Ste12p and RNA PolII, and novel noncentromeric binding sites for Cse4p. We have also characterized the distribution of a reference sample, for example, input DNA. Input DNA, consisting of nonimmuno-precipitated, sonicated, cross-linked DNA, has great importance in ChIP-Seq studies as ChIP DNA samples are normally scored against it for TFBS identification (Auerbach *et al.*, 2009; Rozowsky *et al.*, 2009). The following protocols describe yeast ChIP-Seq, from ChIP to sequence data analysis. We also include our modifications to Illumina sequencing library preparation for generation of barcoded DNA libraries or standard, nonbarcoded DNA libraries.

Computationally, high-throughput sequencing involves handling and analysis of terabytes of sequencing data. Illumina sequencing is a four-color sequencing-by-synthesis approach where incorporation of a reversible terminator nucleotide generates a fluorescence signal detected by a high-sensitivity camera for A, C, G, and T during each cycle. The fluorescent dye is cleaved and the next base is incorporated. Typically, preliminary sequence data analyses are performed using built-in software supplied with the instrument. Fluorescent images of DNA clusters are first analyzed with a module called Firecrest to map cluster location while base-calling is performed with Bustard, which determines the probability of a given nucleotide using fluorescence intensities from the images. Finally, Gerald rapidly aligns 32 bases from the sequence reads to the reference genome using an algorithm called Eland, typically allowing for a maximum of two mismatches. These selected parameters effectively map sequence reads back to the deeply sequenced yeast reference genome. For ChIP-Seq, determination of binding sites from the sequence data is a challenge that has been tackled by different groups with various algorithms (Fejes *et al.*, 2008; Ji *et al.*, 2008; Johnson *et al.*, 2007; Jothi *et al.*, 2008; Nix *et al.*, 2008; Rozowsky *et al.*, 2009; Valouev *et al.*, 2008; Xu *et al.*, 2008; Zhang *et al.*, 2008). ChIP-Seq analysis and the algorithms applied will be described in detail after the protocol section. Conceptually, sequencing reads (or tags) are compiled and genomic regions with an increased number of sequence tags compared to the tags from a control sample are considered as putative TFBS. Next, statistical filtering criteria are used to determine if these putative sites represent true binding sites. After obtaining a preliminary set of TFBS, further bioinformatic analyses are necessary to further analyze the data. These may include analysis of the location of binding sites relative to nearby potential target genes, comparison with gene expression information and gene ontology (GO) analyses of potential targets.

2. PROTOCOLS

2.1. Chromatin immunoprecipitation

DNA–protein complexes formed *in vivo* can be reversibly cross-linked through the application of formaldehyde, and specific DNA–protein interactions are isolated from covalently bound populations using an antibody specific to the transcription factor of interest. Figure 4.1 from Horak and Snyder (2002) summarizes the principal steps of ChIP. We suggest, when possible, tagging the transcription factor of interest with a Myc or HA epitope and performing the immunoprecipitation using commercial antibodies against this epitope; these antibodies generally give little background. As an experimental control, it is possible to IP an untagged version of the same strain and to follow the same protocol. The DNAs from the tagged and untagged strain can be used for qPCR enrichment analysis of selected binding sites prior to proceeding toward sequencing library generation. We adapted this protocol from Aparicio *et al.* (2004, 2005).

(1) Grow 500 ml of yeast cells to exponential mid-log phase ($OD_{600} = 0.6–1.0$). We suggest performing ChIP experiments in biological triplicates.

(2) Treat cells with 14 ml 37% formaldehyde for 15 min, with occasional swirling every 5 min. This allows cross-linking of protein–DNA complexes.

(3) Quench cross-linking reaction by adding 27 ml of 2.5 M glycine for 10 min, with occasional swirling every 5 min.

(4) Collect cells by filtration and wash cells twice with 100 ml of sterile Milli-Q (Millipore, Billerica, MA) water. Rinse the filter with 2 × 20 ml of sterile Milli-Q water to collect cells in a 50 ml Falcon tube. Spin down cells at 4000 rpm for 10 min and discard supernatant. Resuspend the cells in 1 ml water and divide them equally in two 2 ml screw-cap tubes. Repeat this step. Spin down cells at top speed for 3 min, remove the supernatant and put on ice. Measure cell weight. Add 1 ml zirconium beads. One can continue forward to cell lysis or freeze cells at −70 °C for long-term storage.

(5) Resuspend cells in lysis/IP buffer (50 mM Hepes/KOH [pH 7.5], 140 mM NaCl, 1 mM EDTA, 1% Triton X-100, and 0.1% sodium

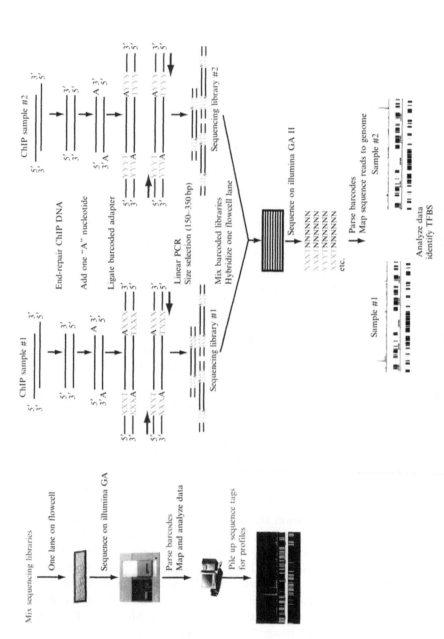

Figure 4.1 Example of a multiplex ChIP-Seq workflow highlighting the principal steps of Illumina sequencing library generation. XXXT and YYYT represent different index sequences. Nonbarcoded ChIP-Seq can be performed by substituting the barcoded adapters with standard Illumina genomic DNA adapters.

deoxycholate) with 1 mM PMSF (Fluka, Buchs, Switzerland) and protease inhibitors (one tablet of Roche Complete protease inhibitor cocktail/50 ml lysis/IP buffer) and lyse them with zirconium beads using a FastPrep Machine (MP Biomedical, Irvine, CA; five times 60 s at a speed of 6.0 m/s).

(6) Recover lysates in a 5-ml snap-cap tube from one 2-ml screw-cap tube by centrifugation at 1500 rpm for 3 min, add 0.5 ml of lysis/IP buffer to the microfuge tube and centrifuge again. Pool the lysate from the other 2 ml screw-cap tube the same way. Add 1 ml of lysis/IP buffer prior to sonication.

(7) Sonicate cell lysates using a Branson Digital 450 Sonifier (Branson, Danbury, CT) to shear DNA. Each sample was sonicated five times 30 s with amplitude of 50%. Between each round of sonications, samples were put on ice for 2 min. The sonicated lysates should be clarified twice, first by a first centrifugation in a Sorvall centrifuge for 5 min at 3000 rpm and then by a second centrifugation in Eppendorf microfuge at 14,000 rpm for 10 min.

(8) Save 250 μl of clarified, sonicated lysate prior to immunoprecipitation to generate input DNA for Illumina sequencing (see next protocol).

(9) Add 2 ml of lysis/IP buffer to each sample, the total volume should be around 6 ml.

(10) Prewash the entire bottle of antibody-coupled beads using lysis/IP buffer. Remove the beads with a broadened 1 ml pipette and transfer to a 15 ml Falcon tube. Wash three times the bottle with 1 ml of fresh lysis/IP buffer to collect all the beads in the 15 ml tube. Vortex briefly and spin 2 min at 2000 rpm in a 4 °C centrifuge. Remove supernatant. Repeat three times with 4–5 ml fresh lysis/IP buffer. Resuspend the beads in an equal volume of lysis/IP buffer (1 ml). For Myc- or HA-tagged strains, we use Sigma EZview anti-Myc affinity gel (Sigma, St. Louis, MO) and Sigma EZview anti-HA affinity gel (Sigma). One antibody bottle can be used for 12 samples.

(11) Add 150–300 μl of prewashed beads to each sample. Immunoprecipitate overnight (12–16 h) on a rocker in the cold room.

(12) After incubation, fill Falcon tube with fresh lysis/IP buffer, pellet antibody beads by spinning 5 min at 3000 rpm in a cold centrifuge and discard supernatant.

(13) Wash the immunoprecipitated samples with 10 ml of appropriate buffer for 5–10 min on a rocker in the cold room. Between washes, spin down the beads in cold centrifuge at 2000 rpm for 2 min. Wash twice with lysis/IP buffer, once with IP/500 mM NaCl buffer (18 ml 5 M NaCl added to 232 ml of lysis/IP buffer), twice with IP wash buffer (10 mM Tris–HCl, 0.25 M LiCl, 0.5% NP-40, 0.5% sodium deoxycholate, and 1 mM EDTA), and once with 1× TE (50 mM Tris–HCl, 10 mM EDTA, pH 8.0). Following the last wash in TE, keep some buffer to transfer beads to a 1.5-ml tube.

(14) Transfer beads from the 15 ml Falcon tube to a 1.5-ml microcentrifuge tube using a broadened 1 ml pipette. Transfer the remaining beads with an additional 0.5 ml of $1\times$ TE. Spin down the beads for 3 min at 14,000 rpm and remove all TE.

(15) Elute the immunoprecipitate from the beads by adding 100–150 μl $1\times$ TE/1% SDS and incubating for 15 min at 65 °C. After 10 min, mix samples briefly. Pellet the beads at 14,000 rpm for 1 min and transfer eluate to a new 1.5 ml tube. Add 150–200 μl of $1\times$ TE/0.67% SDS to the beads and incubate for 10 min at 65 °C. Pellet the beads and pool with previous eluate. Spin down at top speed for 2 min the pooled eluates to remove all beads, as their presence will reduce cross-linking reversal efficiency and therefore ChIP DNA recovery. Transfer eluates to a 2-ml screw-cap tube, avoiding the last 10 μl and the beads at the bottom of the tube.

(16) Reverse protein–DNA cross-links by incubating at 65 °C overnight or for 6–8 h.

(17) Treat samples with proteinase K to remove all proteins from samples. Dilute 20 mg/ml proteinase K (Ambion, Austin, TX) 50-fold in $1\times$ TE. Add 250 μl of diluted proteinase K solution per sample. Incubate between 37 and 50 °C for 2–4 h.

(18) Precipitate DNA with ethanol. Add 3 μl of 20 mg/ml glycogen, 2–5 μl of pellet paint (Novagen, San Diego, CA), 45 μl of 5 M LiCl, and 1 ml of 100% ethanol. Mix thoroughly and incubate at -20 °C overnight or at least several hours. Put samples 1 h at -70 °C. Spin in a cold centrifuge for 20 min at top speed and remove supernatant. The DNA pellet should be slightly pink due to pellet paint. Wash with 1 ml 70% ethanol for 5 min, spin in a cold centrifuge for 10 min at top speed and remove supernatant. Air dry for 10 min. Resuspend in 100 μl $1\times$ TE.

(19) Purify DNA using MinElute PCR purification kit (Qiagen, Valencia, CA). We recommend processing the ChIP DNA sample in two MinElute spin columns. Elution is done in 21 μl EB per column and the two eluates from the same sample are pooled. Samples are stored in a -20 °C freezer.

This procedure typically yields 100–300 ng of ChIP DNA. DNA concentrations can be measured using a Nanodrop spectrophotometer (Thermo Scientific, Waltham, MA) or PicoGreen dsDNA quantification assay (Invitrogen, Carlsbad, CA). qPCR analysis should be performed prior to generation of sequencing library. We typically compare ChIP samples from a tagged strain (experimental sample) versus an untagged strain (control sample) for enrichment in the experimental sample at three known binding sites and at a genomic locus where the transcription factor is not expected to bind (negative control). We have found that ChIP efficiency is the most critical step for success of the entire procedure. Here we suggest steps in the protocol for quality control as well as parameters that can be modified:

(a) *Formaldehyde cross-linking*: Changing the concentration of formaldehyde and the duration of cross-linking can modify the extent of cross-linking. Too much cross-linking can mask the HA or Myc epitopes on tagged transcription factor while too little cross-linking will decrease the immunoprecipitation of the associated DNA.

(b) *Cell lysis*: Five 1-min burst of FastPrep machine typically lyse over 95% of cells. Breaking cells using a paint shaker for 30 min yields about 40% of lysed cells.

(c) *Sonication*: Chromatin should be sheared to a median size of 450–500 base pairs (bp), as measured by gel electrophoresis in a 2% agarose gel. After step 7, take 250 μl of clarified lysate and add an equal volume of 1× TE/1% SDS. Follow the aforementioned protocol from steps 16 to 18, without purification through a MinElute spin column. Then load on a 2% agarose gel for electrophoresis. Ideally, a smear between 100 and 1000 bp should be present, with a median size of 450–500 bp (stronger smear intensity).

(d) *Antibody*: Prior to performing a ChIP experiment with a new tagged strain, a Western blot should be performed to confirm the correct insertion of the epitope. Antibody quantities can also be optimized by preliminary IP experiments with various amounts of antibody.

2.2. Input DNA preparation

Input DNA serves as an important reference sample for ChIP-Seq experiments (Lefrancois *et al.*, 2009; Robertson *et al.*, 2007; Rozowsky *et al.*, 2009). It is used during the scoring process where TFBS are determined based on the sequence reads obtained from a ChIP sample in comparison to input DNA. Input DNA consists of sonicated cross-linked chromatin that is processed in parallel to a ChIP sample, but lacking the immunoprecipitation step. Recent reports have suggested that input DNA represents breaks in chromatin regions of increased accessibility (Auerbach *et al.*, 2009; Teytelman *et al.*, 2009). However, there is currently a debate whether input DNA, normal IgG or, in the case of yeast, untagged strain should be the control sample for scoring ChIP-Seq data. Here we present our laboratory's protocol for isolation of input DNA, which starts at step 8 of the previous ChIP protocol.

(20) Combine 250 μl of 1× TE/1% SDS to the reserved 250 μl of clarified sonicated lysate (from step 8, ChIP protocol) in a 2-ml screw-cap tube.

(21) Reverse cross-links overnight by incubating at 65 °C.

(22) Treat samples with proteinase K as described (step 17, ChIP protocol).
(23) Extract input DNA three times with phenol:chloroform:isoamyl alcohol (25:24:1) (Fluka) followed by a single extraction with chloroform alone. In each case, keep the upper aqueous phase.
(24) Precipitate DNA with ethanol by adding 50 μl of 5 M LiCl and 1 ml of 100% ethanol to the upper aqueous phase from the last chloroform extraction. Enhance precipitation by transferring at − 20 °C for 1 h. Centrifuge samples at top speed for 20 min and discard supernatant. Wash with 1 ml of 70% ethanol in the cold room for 5 min, spin down DNA at top speed for 10 min, discard ethanol and air dry for 10 min. Resuspend DNA in 1× TE (pH 8.0).
(25) RNase-treat the input DNA sample. Add 2 μl of 10 mg/ml DNase-free RNase A (Roche, Indianapolis, IN) and incubate for 30 min at 37 °C.
(26) Purify DNA using a MinElute PCR purification column (Qiagen). Elution is done in 21 μl of EB.

The amount of input DNA recovered using this procedure is much greater than that of a ChIP sample. We recommend the use of one-fifth of each input DNA sample for Illumina sequencing library preparation. If the upper phase seems unclear and the interphase is still very cloudy after the three phenol:chloroform:isoamyl alcohol extractions, an additional extraction should be performed. The phase-lock gel system (5 Prime, Gaithersburg, MD) can be used to perform safer extractions with higher recovery of DNA due to the organic phase and the interphase being sequestered physically at the bottom. This facilitates the removal of the upper, aqueous phase containing DNA.

2.3. Illumina sequencing DNA library generation

ChIP samples must be converted into DNA libraries for sequencing. Protocols differ depending of the sequencing platform used; 454/Roche, Solexa/Illumina, SOLiD/ABI, and Helicos each use different strategies to create a library representing the population of short DNA fragments selected by ChIP. Analysis of TFBS by sequencing technologies does not require very long sequencing reads; large numbers of short reads (e.g., 35 bp) are sufficient for mapping binding sites in most organisms. Therefore, Illumina/Solexa and ABI/SOLiD have been favored over Roche/454 because they both generate millions of very short reads (about 35 bases/read) whereas Roche/454 generate less reads but of longer length (200–300 bases/read). Currently, most ChIP-Seq studies have been performed on the Illumina platform and a few have used SOLiD. Here we describe our procedure to generate standard, nonbarcoded Illumina libraries. We have optimized the ChIP-Seq protocol used in mammalian cell lines experiments

(Robertson *et al.*, 2007) to the yeast context, which follows the manufacturer's guidelines. During Illumina library generation, oligonucleotide adapters are introduced at the ends of the small ChIP DNA fragments that were bound previously by the transcription factor of interest. These adapters allow hybridization of the sample to a flowcell containing a lawn of primers which is used for subsequent cluster generation and sequencing-by-synthesis.

During Illumina library preparation, the sheared ChIP DNA is end-repaired. A single adenosine base ("A") is added to the 3′ end of both strand followed by annealing and ligation to the double-stranded adapter containing a "T" overhang. A short PCR amplification (15–17 cycles) with primers annealing to the adapter sequence is performed to generate a population of adapter–ChIP DNA fragments termed the library. Size selection on a 2% agarose gel allows isolation of the amplified DNA library between 150 and 350 bp. This is the optimal range of fragment size for hybridization to the flowcell and cluster generation according to Illumina's recommendations.

According to bioinformatic simulations based on the yeast genome, only 260,000 uniquely mapped reads would be sufficient to determine at least 95% of the TFBS from a typical punctual TF if these binding sites are enriched at least fivefold in the ChIP sample (Lefrancois *et al.*, 2009). This is very low when compared to the human genome, where 12 M mapped reads is usually used (Rozowsky *et al.*, 2009). A single Illumina flowcell lane generates about 8 M mapped reads so multiple yeast ChIP-Seq samples can be sequenced simultaneously using multiplex Illumina sequencing (Lefrancois *et al.*, 2009). As shown in Fig. 4.1, to generate barcoded Illumina libraries, one can substitute Illumina's genomic DNA adapters for custom-made adapters that contain the adapter sequence from Illumina genomic DNA adapter followed by a nucleotide tag of at least two bases (called the barcode or index; we usually use three bases) and terminated by a single "T" for annealing and ligation to the end-repaired DNA containing an "A" overhang (Craig *et al.*, 2008; Cronn *et al.*, 2008; Lefrancois *et al.*, 2009). Standard Illumina genomic DNA PCR primers are used and the rest of the procedure is intact. ABI/SOLiD has established an indexing strategy since the commercial launch of their platform.

(27) Perform gel electrophoresis on a 2% agarose gel with at least 100 ng of ChIP DNA from step 19 (or input DNA) and size select the DNA smear between 100 and 700 bp. For ChIP, we usually use between 15 and 35 μl of MinElute-purified DNA from step 19. For input, due to its higher DNA concentration, one can apply a lower volume of MinElute-purified DNA from step 26 on the gel (5–10 μl) or gel-purify the same volume as for ChIP but use only 20–25% of the gel-purified input DNA for the next steps. A 100-bp DNA ladder

should be included during gel electrophoresis. Samples should not migrate too much on the agarose gel to allow for isolation of the 100–700 bp smear in a relatively small gel volume. Samples are run typically ~20 min at 100–110 V. The Qiagen QIAquick gel extraction kit is used (Qiagen) and elution is done in 34 μl EB. Although this size selection step is optional, we recommend it for exclusion of very short fragments and longer fragments which are not suitable for Illumina sequencing. For input DNA, the intensity of the smear should be high while for ChIP DNA, the smear should be visible although much fainter.

(28) End-repair DNA for 45 min at room temperature using End-It DNA end-repair kit (Epicentre, Madison, WI). DNA fragments are blunted by end-repair and all 5′ ends are phosphorylated. DNA is purified after end-repair using a QIAquick PCR purification column (Qiagen) and eluted in 34 μl EB.

(29) Add a single adenosine nucleotide ("A") to the 3′ blunted ends of end-repaired DNA fragments (in 34 μl EB). Perform a reaction on eluted sample from step 28 with 10 μl 1 mM dATP, 5 μl 10× NEB buffer 2 and 1 μl Klenow fragment (3′ → 5′ exo minus) (NEB, Ipswich, MA). Mix all components in a PCR plate and cover with a sealing microfilm. Reaction is performed at 37 °C for 30 min in a PCR machine, without the use of a heated lid. Aliquots of 1 mM dATP should be prepared from a 100-mM dATP stock solution (Invitrogen) and frozen at −20 °C. Freeze–thaw should be avoided. The low concentration of dATP permits the single addition of an "A." A MinElute PCR purification column (Qiagen) is used to purify the reaction and DNA is eluted in 10 μl EB.

(30) Ligate Illumina genomic DNA adapters (Illumina, San Diego, CA) or barcoded adapters to the sample for 15 min at room temperature. Mix 10 μl of sample from step 29, 1 μl of diluted oligonucleotide adapters, 1.5 μl of LigaFast T4 DNA Ligase (3 units/μl; Promega, Madison, WI), and 12.5 μl of Rapid Ligation Buffer (Promega). The dilution of Illumina genomic DNA adapters depends of the nature of the sample. For input DNA, Illumina nonbarcoded genomic DNA adapters are diluted 1:20 with Gibco RNase-free, DNase-free water (Invitrogen); for ChIP DNA, Illumina adapters are diluted 1:40. After the 15 min reaction, ligation products are purified with a MinElute PCR purification column and eluted in 10 μl EB. These adapters contain an unpaired "T" overhang which anneals to the 3′ "A" on the sample DNA. Barcoded adapters must have been annealed before being added to the end-repaired DNA. The concentration of diluted barcoded adapters for ligation to input DNA or ChIP DNA samples should mimic that of standard genomic DNA adapters. Barcoded adapter design and annealing will be described in the next section.

(31) Perform a gel electrophoresis on a 2% agarose gel and size select the DNA smear between 150 and 500 bp. We have found that this gel purification prior to PCR amplification has increased the quality of the sequencing libraries. More importantly, it decreases the intensity and occurrence of adapter–adapter dimerization after the PCR amplification. Adapter–adapter dimers amplify preferentially during the following PCR step and appear as a compact bright band around 100–120 bp. This compact band can totally or partially replace the normal smear indicative of a successful library. For this reason, at this step, DNA fragments below 150 bp should be excluded. We recommend using a 2% agarose E-Gel to separate adequately samples during loading and migration. Load 20 μl of a 1:10 diluted Track-It 50 bp DNA ladder (Invitrogen). Add 3 μl of a 1:10 diluted Track-It Cyan/Orange loading buffer (Invitrogen) to each sample. Samples should be separated by at least two empty wells. Load 20 μl of Gibco RNase-free, DNase-free water to all empty wells. Perform gel electrophoresis for 20 min. Recover DNA using the QIAquick gel extraction kit (Qiagen) and elute ligated samples in 28 μl EB. At this step, input DNA libraries should be visible but rather faint while ChIP DNA libraries are fainter than input DNA ones and even sometimes cannot be seen. The lack of a visible ChIP DNA smear at this step does not prevent generation of successful and high-quality libraries.

(32) Amplify the sequencing library by PCR using Illumina genomic DNA primers 1.1 and 2.1. In a PCR plate, mix 28 μl of eluted DNA sample from step 32, 1 μl of 1:1 diluted Illumina genomic DNA primer 1.1, 1 μl of 1:1 diluted Illumina genomic DNA primer 2.1, and 30 μl of Phusion Master Mix with HF Buffer (NEB). Use the following PCR settings with a heated lid: denaturation at 98 °C for 30 s, 17 cycles of amplification (10 s at 98 °C, 30 s at 65 °C, and 30 s at 72 °C), an extra amplification at 72 °C for 5 min and a cool down to 4 °C. Remove enzymes and buffer using a MinElute PCR purification column (Qiagen) and elute in 10 μl EB.

(33) Size select the Illumina sequencing library between 150 and 350 bp by gel electrophoresis on a 2% agarose gel. These size specifications meet the manufacturer's guidelines for cluster generation, optimal at a median fragment size of about 230 bp. The use of a 2% agarose E-Gel is preferable. Loading of samples and ladder are identical to step 31. Run gel electrophoresis for 20 min. A picture of the final library on the gel should be taken. At this step, a medium-to-high intensity smear over 150 bp and under 500 bp should be easy to visualize, suggesting the sequencing library preparation was successful. If there is a faint well-defined band at around 100–120 bp, extreme care should be taken during gel excision to avoid completely this adapter–adapter dimer band. The presence of adapter–adapter dimers

during sequencing will greatly decrease the overall mappability of sequencing reads. The number of uniquely mapping reads could then be very low. Gel extraction is done using a MinElute gel extraction kit (Qiagen) and elution is done in 20–25 μl EB.

(34) Measure DNA concentration and the $Abs_{260 \, nm/280 \, nm}$ ratio using a Nanodrop spectrophotometer (Thermo Scientific). Good quality libraries have an A260/280 ratio between 1.7 and 2.0. Lower values indicate poor quality. The minimal DNA concentration to proceed toward Illumina sequencing is 5.0 ng/μl. Libraries with lower DNA concentrations should be discarded. We typically obtain DNA concentrations over 8.0 ng/μl for ChIP DNA libraries and over 15.0 ng/μl for input DNA libraries.

(35) Store Illumina sequencing libraries at $-70\,^{\circ}$C until they are processed for sequencing.

Samples are now ready for the sequencing step of the ChIP-Seq procedure. They are compatible with Illumina Genome Analyzer and Genome Analyzer II. Generation of barcoded libraries follows an identical procedure except barcoded adapters are added at step 30 instead of Illumina genomic DNA adapters. Prior to the sequencing of barcoded DNA libraries, they must be mixed together in an equimolar ratio using DNA concentrations obtained from Nanodrop. A more precise method to measure DNA concentrations such as the PicoGreen dsDNA quantification assay (Invitrogen) could also be used, although we have obtained good barcode representation with Nanodrop concentrations (less than twofold difference in the number of mapped reads between the least abundant and the most abundant barcoded sample). Here are a few considerations for Illumina sequencing DNA library generation:

(e) *ChIP efficiency*: An insufficient amount of starting DNA material (in this case, ChIP DNA) is the most important cause of failure in library preparation as noted by the complete absence of a smear on the final agarose gel in step 33, the presence of a single intense adapter–adapter dimer band at 100–120 bp or the cooccurrence of a strong adapter–adapter dimer band and of a very faint library smear. Scaling up the ChIP protocol is a solution to generate more DNA as well as the use of tagged strains and/or of ChIP-grade antibodies.

(f) *Adapter dilution*: It is crucial to dilute barcoded adapters to the working concentration of the diluted standard Illumina genomic DNA adapters. If the concentration of barcoded adapters is too high, it may favor the ligation of adapter to other adapters, resulting in the formation of a strong adapter–adapter dimer band during the final gel extraction (step 33) or in the absence of a DNA smear indicative of a successful library. Optimization of concentrations should be first performed on input DNA. Similarly, if problems occur using Illumina genomic DNA adapters, optimization should also be performed on input DNA and then on ChIP DNA.

(g) *PCR amplification*: PCR amplification using Illumina genomic DNA primers 1.1 and 2.1 should stay in the linear range to avoid overrepresentation of some genomic areas among the sequencing library. Sequencing reads would then be very high for these overrepresented regions, creating a bias during data analysis. The manufacturer recommends no more than 18 PCR cycles. One can perform less cycles. The common range for PCR amplification of Illumina DNA libraries lies between 13 and 18 cycles.

2.4. Barcode design and adapter annealing

This section applies specifically to barcoded ChIP-Seq on an Illumina platform. Multiplex sequencing-by-synthesis has been accomplished through the introduction of indexed (or barcoded) adapters (Craig *et al.*, 2008; Cronn *et al.*, 2008; Lefrancois *et al.*, 2009). These strategies have allowed multiplex sequencing and analysis of HapMap loci from different individuals (Craig *et al.*, 2008), chloroplast genomes from different species (Cronn *et al.*, 2008), and yeast ChIP samples (Lefrancois *et al.*, 2009), without the introduction of barcode-induced errors or artifacts. In all cases, a barcode was introduced after the Illumina adapter sequence required for PCR amplification and hybridization to Illumina's flowcell. The resulting sequencing reads first contain the index followed by the sequenced DNA sample. The barcode must contain a final "T" for pairing and ligation to the end-repaired DNA with an "A" overhang. We have used four indexes for barcoded ChIP-Seq: ACGT, CATT, GTAT, and TGCT. We have created a three base index where no barcode contained the same base at each position mainly for two reasons. First, these barcodes have a balanced nucleotide composition in compliance with manufacturer's guidelines. Second, one- or two-base sequencing errors would not result in a barcode being assigned to an erroneous sample as the remaining index base would not match another sample. In our work, the barcode must be intact in all sequencing reads assigned to a sample. With the new Illumina Genome Analyzer II, the increase in read length and the decrease in error rates at later sequenced bases permit generation of longer barcodes. This, coupled to an expected increased number of sequencing reads, could significantly increase the level of multiplexing in yeast ChIP-Seq studies as well as ChIP-Seq studies in small genome organisms. Here we present the procedure for oligonucleotide design and annealing to generate barcoded adapters.

(36) Synthesize oligonucleotides at a 0.05 μmol scale with HPLC purification from MWG/Operon (Eurofins MWG Operon, Huntsville, AL). Oligonucleotide sequences are given in Table 4.1. Note that the

Table 4.1   Oligonucleotide sequences for barcoded ChIP-Seq

Barcode	Forward/reverse	Sequence $(5' \rightarrow 3')^a$
ACGT	Forward[b]	ACACTCTTTCCCTACACGACGCTC TTCCGATCTACGT
	Reverse[c]	CGTAGATCGGAAGAGCTCGTATG CCGTCTTCTGCTTG
CATT	Forward[b]	ACACTCTTTCCCTACACGACGCTC TTCCGATCTCATT
	Reverse[c]	ATGAGATCGGAAGAGCTCGTATGC CGTCTTCTGCTTG
GTAT	Forward[b]	ACACTCTTTCCCTACACGACGCTC TTCCGATCTGTAT
	Reverse[c]	TACAGATCGGAAGAGCTCGTATGC CGTCTTCTGCTTG
TGCT	Forward[b]	ACACTCTTTCCCTACACGACGCTC TTCCGATCTTGCT
	Reverse[c]	GCAAGATCGGAAGAGCTCGTATGC CGTCTTCTGCTTG

[a] From Lefrançois *et al.* (2009).
[b] No modification.
[c] 5' ends are phosphorylated.

    forward primer contains the index and the final "T" at the 3' end while the reverse primer is phosphorylated at the 5' end and the reverse-complement index sequence is found at the 5' end.

(37) Resuspend each primer in annealing buffer (10 mM Tris [pH 7.5], 50 mM NaCl, 1 mM EDTA) to 200 μM.

(38) Mix the forward and reverse primers for each index pair in equal volumes to a final concentration of 100 μM.

(39) Heat to denature in a wet heat block at 95 °C for 5 min.

(40) Remove heat block to room temperature and let primers cool down during 45 min to promote annealing.

(41) Keep on ice for a few minutes and store barcoded adapters at -20 °C.

(42) Dilute barcoded adapters with Gibco RNase-free, DNase-free water (Invitrogen) to the working concentrations of Illumina genomic DNA adapters for generation of input DNA and ChIP DNA libraries (previous protocol). Annealed indexed adapters have different concentrations from each other and differ in the dilutions to obtain the adequate working concentrations. As an example, with our barcoded adapters given in Table 4.1, we have diluted all four adapters 1:30 to generate barcoded input DNA libraries while we have diluted differently adapters for barcoded ChIP DNA libraries: 1:750 for ACGT, 1:450 for CATT, 1:500 for GTAT, and 1:330 for TGCT.

2.5. Illumina sequencing

We follow manufacturer's protocols and guidelines. Detailed protocols for operating the cluster station and sequencing using Genome Analyzer II are available from Illumina's web site. Here we will only briefly describe the various steps of Illumina sequencing. First step is cluster generation on Illumina's cluster station. The cluster station uses microfluidics to physically attach DNA from the sequencing library (step 34) onto one lane of an Illumina flowcell. An Illumina flowcell contains eight lanes and each lane has a lawn of primers with a sequence corresponding to the complement of the Illumina adapter sequence. Samples are denatured and each single-stranded ChIP DNA fragment is connected to the lawn primer via one adapter end. A solid-phase bridge amplification replicates the template DNA fragment from the paired adapter–primer. After denaturation of this double-stranded bridge of DNA, the initial template DNA is washed away and the flowcell-attached replica of the template can undergo successive rounds of bridge amplification to generate a cluster. A cluster contains about 1000 copies from an identical initial template. There are typically 100–120,000 clusters on a single tile, with 100 tiles per flowcell lane. This can give rise to 10–12 M reads per lane. DNA loaded on the flowcell should be at a concentration between 3 and 5 pM and optimal library size should be between 150 and 350 bp. If the DNA library is smaller, too many clusters of smaller size will be present due to the lesser reach of bridge amplification, giving rise to a fewer number of clusters passing quality metrics and a lower number of mapped reads. On the other hand, if the smear size of the library is bigger, fewer clusters of bigger size will be generated due to the greater reach of bridge amplification, resulting to a decreased number of clusters and mapped reads. Just prior to sequencing, a sequencing primer is annealed. The flowcell is then transferred from the cluster station to the Genome Analyzer II for sequencing of DNA clusters. Illumina employs a four-color sequencing-by-synthesis method. Fluorescently labeled reversible terminator ddNTPs are added simultaneously and one base is incorporated per cluster. Laser excitation and fluorescence allows the detection of the first base. The fluorescent dye is cleaved and the first base is unblocked to add the second base using the same reagents. This process is continued for 34–36 cycles by sequencing one base at a time. Each read starts with the template DNA sequence or, in the case of barcoded ChIP-Seq, with the 4-bp barcode. The following section focuses on sequencing reads analysis.

3. SEQUENCING DATA MANAGEMENT

Illumina uses a massively parallel sequencing-by-synthesis approach. A typical run on the Illumina instrument lasts 2–3 days, and generates at least 1 terabyte of data, which poses a big challenge on data storage system and data transfer method. Analyzing these raw image data to get biolog-ically meaningful sequences is also a computationally intensive task. We use Genome Analysis Pipeline (GAP) software (Illumina) to analyze the sequencing data. The minimum system requirement for running this soft-ware is a dual-processor, dual-core computer. If multiprocessor facilities are available, the data analysis time can be greatly reduced by parallelization. The outputs produced by GAP are stored in a hierarchical directory struc-ture called the "run folder." Currently our run-folder resides on the Yale biomedical high-performance computing cluster, which consists of 170 Dell PowerEdge 1955 nodes, and each node contains two dual core 3.0 GHz EM64T Intel CPUs and 16 GB RAM.

It is very important to have an IT infrastructure with sufficient compu-tation capacity, data storage and transfer abilities to support Illumina Genome Analyzer. Depending on the scale of sequencing runs, a laboratory can also consider commercially available Laboratory Information Manage-ment System (LIMS), such as WikiLIMS (BioTeam).

4. GENOME ANALYSIS PIPELINE

Since detailed instructions of installing and running the GAP are available from Illumina, we will only briefly introduce the functionality of pipeline modules related to ChIP-Seq data analysis. Users can refer to the GAP documentation for more details. The documentation files can be obtained with Genome Analyzer machine setup, or browsed through publicly accessible domains. The link to the documentation files at Yale University is http://sysg1.cs.yale.edu:3443/pDir/GAP-1.1.0-docs/.

There are three main modules in the GAP. The first module Firecrest is an image-analysis module. The images are generated from sequencing-by-syn-thesis at hundreds of thousands of clusters. At each cluster, the sequencing machine records four images of added nucleotides (A, G, C, or T) at each synthesis cycle. Firecrest analyzes images captured by the sequencing machine, and remaps cluster positions. In an updated version, Illumina introduced the Integrated Primary Analysis and Reporting (IPAR) Software, which processes images and performs quality control in real time. IPAR removes the need of storing raw images. The second module, Bustard, per-forms base-calling. From the four images captured for A, G, C, and T at each round of synthesis, Bustard calculates the occurrence probability of a certain nucleotide at each cluster, and after 30–34 cycles of synthesis, it concatenates a chain of nucleotides with highest occurrence probabilities into a short sequence tag with length equal to the number of synthesis cycles. It has also some built-in quality control mechanism to determine the confidence of its base-calling. The third module, Gerald, aligns short sequence reads to a reference genome. The alignment software (Eland) in the Gerald module runs very fast and accurately aligns sequence tags of less than 32 bp to a reference genome. Several other open-source programs can also align a large number of short sequences, such as SOAP (Zhang *et al.*, 2008) and MAQ (Li *et al.*, 2008). SOAP has a unique feature of aligning sequences across small gaps in the genome, which is helpful in sequencing transcriptomes. MAQ has a dedicated module to call SNPs and *de novo* genome assembly.

5. EXAMINING DATA QUALITY AND PARSING BARCODED DATA

Before we can use the ChIP-Seq data to answer biological questions, we need to make sure the data quality is of sufficient quality. There are several summary statistics to examine after a sequencing run: % Error (multiplexing runs usually have higher % error than nonbarcoded runs, around 5%), % Phasing ($< 1\%$), total reads (GAII can reach 12–14 million) and cluster density ($\sim 100{,}000$). Some other statistics to verify alignment percentage of short sequences to the genome are Total No Match, % No Match, Total QC Fail, % QC Fail, R0 Multiple Match, R1 Multiple Match, R2 Multiple Match, Total Multiple Match, % Multiple Match, U0 Unique Match, U1 Unique Match, U2 Unique Match, Total Unique Match, and % Unique Match.

For multiplexing runs, users need to parse the Eland query file by matching the first several nucleotides with the barcode sequences, remove the barcode sequences from the sequencing reads, and rerun Eland separately for each parsed data set. In our case, the barcodes are GTAT, CATT, ACGT, TGCT (Lefrancois *et al.*, 2009). Users do not need to check alignment statistics for the entire lane because the alignment uses sequence tags with barcodes at the 5′ end. Instead, users should check these statistics for parsed data with removed barcode sequence. A typical run of yeast multiplexing sequencing with four barcodes has $\sim 15\%$ Total Multiple Match, and $\sim 60\%$ Total Unique Match.

Some sample Perl scripts to perform the barcode parsing and Eland rerunning tasks can be found at http://pantheon.yale.edu/~wz4/Homepage.html. Since the scripts call the Eland program, they need to be run on the same server as the GAP resides, and modified according to user-specific directory structures. If automatic barcode parsing and Eland alignment are desired, please consult with IT support to integrate these functions into the GAP.

6. VISUALIZATION IN GENOME BROWSER

After aligning short sequencing tags onto a reference genome, we can load the data into a Genome Browser to directly visualize how the short tags are distributed across the genome, and whether there is any enrichment near the regions of interest. There are many versions of Web-based genome browsers available for the yeast community, including Gbrowse, UCSC Genome Browser. One can upload data to the server in a format that can be recognized by the genome browser, and visualize the signal track along with other annotations on the host webpage of the specific genome browser. Here we will provide a step-by-step guide to visualize ChIP-Seq data on a local machine using Integrated Genome Browser (IGB) developed by Affymetrix. First, download and launch IGB following the instructions at: http://www.affymetrix.com/partners_programs/programs/developer/tools/download_igb.affx.

To load *S. cerevisiae* annotations, click File → Access DAS/1 Servers; in the pop-up window named "DAS/1 Feature Loader," choose "UCSC" in "DAS Server" pull-down menu, "sacCer1" in "Data Source" pull-down menu, and "1 (or any other chromosome you want to see)" in "Sequence" pull-down menu, and check any interested annotations in the "Available Annotations" window.

To load ChIP-Seq data into IGB, one needs to transform the Eland results into a format that can be recognized by IGB. A sample Perl script for this purpose can be found at http://pantheon.yale.edu/~wz4/Homepage.html. To run this script, simply type the following command in a command line shell:

perl create_sgr_file.pl "eland_result_folder"

The "eland_result_folder" is the folder containing Eland results files parsed into chromosomes, with names "eland_results_chr★.txt."

This script transforms Eland results file into .sgr format, which is compatible with Affymetrix's IGB. The format for each line of an sgr file is:

Chromosome Start_Position Score

where "Score" is the number of overlapping ChIP fragments from the current Start_Position to 1 bp upstream of the next Start_Position. Before counting overlapping ChIP fragments for each genomic position, create_sgr_file.pl also extends sequencing tags in the $3'$ direction to 200 bp since Illumina sequencing tags only represent one end of ChIP fragments, and the average ChIP fragment size in the sequencing library is 200 bp.

6.1. Low-level analysis

The Genome Browser can help scientists visualize and roughly determine sequence-tag enriched regions. To precisely identify TFBS across the genome, we need more rigorous peak-scoring algorithm. Since the emergence of ChIP-Seq technology, a bunch of peak-scoring algorithms have been developed for mammalian genomes, and most of them can also be used to analyze yeast ChIP-Seq data. Here we briefly describe the basic flowchart of peak-scoring algorithms, and summarize the major features of several popular peak-scoring algorithms in Table 4.2. A peak-scoring algorithm usually compares sequencing data from a ChIP experiment with simulated background, or sequencing data from a control experiment (nontagged strain, IgG ChIP, or input DNA). To determine regions with enriched sequence-tag distribution, the scoring algorithm normalizes the ChIP-Seq data and control sequencing data to the same scale, and then uses proper statistical tests (e.g., Binomial, Poisson, Normal) to compare distributions of sequence tags in the two data sets. Significance level is adjusted to control the false discovery rate.

Because the Illumina sequencing platform only reads 30–32 bp from one end of the ChIP DNA fragment, the most enriched regions (peak centers) of sequencing tags may not overlap exactly with the most enriched regions (peak centers) of ChIP DNA fragments, the latter ones corresponding to potential binding sites meaningful to biologists. Several different methods were proposed to convert sequencing tag position to peak center position in published peak-scoring algorithms. The most straightforward way is to extend the sequencing tags to the length of original ChIP DNA fragment, which is about 200 bp due to size selection on agarose gel in sequencing library construction (Rozowsky *et al.*, 2009; Xu *et al.*, 2008). More sophisticated methods include estimating the length of original ChIP DNA fragments with triangle or bell shaped distribution centered at ~200 bp (Fejes *et al.*, 2008), or separating sequencing reads aligned onto Watson and Crick strands, and using the distances between peak center on Watson strand and peak center on Crick strand to estimate the length of original ChIP DNA fragments (Ji *et al.*, 2008; Jothi *et al.*, 2008; Zhang *et al.*, 2008).

The biggest distinction among existing peak-scoring algorithms is the method of extracting background from ChIP-Seq data. Due to the high cost of ChIP-Seq procedure, earlier peak-scoring algorithms often considered one-sample analysis, which compares ChIP data with a null background generated from random permutation or estimated from a Poisson model. One-parameter Poisson model (Feng *et al.*, 2008; Marson *et al.*, 2008; Robertson *et al.*, 2007) has been widely used in these peak-scoring algorithms. Another popular method to estimate background is Monte Carlo sampling (Bhinge *et al.*, 2007; Chen *et al.*, 2008; Fejes *et al.*, 2008; Johnson *et al.*, 2007; Mikkelsen *et al.*, 2007; Robertson *et al.*, 2007; Zhang *et al.*, 2008). Later studies found out that Poisson model with a fixed λ is not good enough to describe nonrandom fluctuations as observed in the input control. To alleviate this problem, CisGenome (Ji *et al.*, 2008) used Negative Binomial instead of Poisson to model the background. MACS (Zhang *et al.*, 2008) used dynamic Poisson parameters. Both studies recognized that the random sampling process had different sampling rates at different positions in the genome, and tried to capture the nature of changing parameters in the underlying Poisson model. Nonetheless, it becomes clear that two-sample

Table 4.2 Comparison of popular ChIP-Seq peak-scoring algorithms

Algorithm	Tag2 peak	Model	1 or 2 sample/ scaling	Unique feature and other notes	References
ChIP-Seq PeakFinder	Use 25-nt tags directly	min_reads > 13 (default 20 in program), max-gap <75, minratio > 5	1 or 2, NA	qPCR to find threshold, MEME motif finding	Johnson *et al.* (2007)
FindPeaks3.1	Three methods: extend fixed length, triangle with user-specified average length, adaptive	FDR using effective genome size and tag number (Monte Carlo background)	1	Directional: fragment after a peak are removed as "noise," subpeak: user-specified valley, peak trimming	Fejes *et al.* (2008)
CisGenome	Boundary refinement, single-strand filtering (detect peaks for +/− data separately)	Sliding window passing user-specified cutoff, one-sample FDR negative binomial, two-sample FDR conditional binomial	1 or 2, use window read counts <n, conditional binomial (n, $c/1 + c$)	Negative binomial model underestimate FDR when window read count is high	Ji *et al.* (2008)

SISSRs	Sep+/−, estimate average fragment length F from data	Net tag counts in overlapping (10 bp) sliding window (20 bp). Use sense–antisense read counts transition point as binding site, user-selected cutoff E, R	1 or 2, NA	One-sample FDR Poisson, two-sample no FDR, use control sample to estimate sensitivity: specificity:empirical p-values	Jothi et al. (2008)
QuEST	Sep+/−, peak shift estimated from position shift between +/− data and most significant peaks	21 bp sliding window, cutoff determined by eFDR, difference between neighboring windows <0.9H of higher peak window, control H < threshold or ChIP/control ratio > threshold	2, extract same number of control data as experimental data, and use the other half of control data to estimate empirical FDR	Gaussian KDE, bandwidth 30, ±3b, FDR for two-sample, at least three user-specified cutoff	Valouev et al. (2008)
U-Seq	Sep+/−, estimate mean fragment length, and shift read position	Overlapping sliding window defined by read position and max size 350 bp	2, trim to equal number of reads	Spike in simulation in input control data compare sum, diff, normdiff, and binomial. Normdiff outperforms	Nix et al. (2008)

(continued)

Table 4.2 (*continued*)

Algorithm	Tag2 peak	Model	1 or 2 sample/ scaling	Unique feature and other notes	References
ChipDiff	Shifting tag by 100 bp	1. Putative enrich region by normalized window read counts 2. fold change or HMM (better)	2, normalize to total reads	Used in histone methylation detection	Xu *et al.* (2008)
MACS	Given bandwidth and fold enrichment, sample 1000 high-quality peaks, sep+/−, calculate *d* between modes, shift tag position by *d/2*	Use 2D windows, Poisson with dynamic parameter to model local background, and calculate *p-value*, eFDR calculated by swapping control/ChIP	1 or 2, linearly scale with total read counts, remove duplicate reads from the same position	eFDR definition different from others, motif occurrence within 50 bp of peak center, average distance from peak center to motif are better than competitors	Zhang *et al.* (2008)
PeakSeq	Extend 200 bp	Binomial (n, 1/2) after scaling	2, Simple Linear Regression using window counts	Two-pass comparison, first to random shuffled background, second to input control data	Rozowsky *et al.* (2009)

analysis is superior because in certain genomic regions, the sequencing tag distribution in an input control experiment shows nonrandom enrichment. Sometimes the same enrichment pattern is also observed in ChIP experiment (Nix et al., 2008; Rozowsky et al., 2009; Zhang et al., 2008). Such enrichment is not likely to be caused by TFBS; instead it probably represents fluctuations due to systematic biases. There are many sources of systematic biases. Known sources include technical reasons such as the method of DNA fragmentation, biased amplification in PCR, error in the sequencing and/or the alignment processes; biological reasons such as the degree of genome repetitiveness, open chromatin structure; statistical reasons such as the dependency among observations from neighboring positions on a chromosome. In both one- and two-sample analyses, it is assumed that the number of sequencing tags observed in a small window of the genome comes from random sampling process. Binomial or Poisson model are often used in two-sample analyses to compare the number of reads in windows of two samples (Feng et al., 2008; Ji et al., 2008; Jothi et al., 2008; Nix et al., 2008; Rozowsky et al., 2009; Valouev et al., 2008; Zhang et al., 2008).

6.2. High-level analysis

Once TFBS are identified from the ChIP-Seq data, one can carry out more high-level analysis to answer biological questions, such as motif analysis, association of TFBS with neighboring genes, and comparison with ChIP-chip data. One can also study the positions of TFBS relative to genome annotation features, such as intragenic versus intergenic binding and binding in 5' or 3' untranslated regions. These analyses are all implemented in an integrative open-source software, CisGenome (Ji et al., 2008). It has many functions varying from low-level analysis to high-level analysis, and its graphic interface under Windows OS is user-friendly for bench scientists. It can be downloaded from the following web site: http://www.biostat. jhsph.edu/~hji/cisgenome/.

6.3. Troubleshooting

If the sequencing run yields many reads, but the percentage of matched reads after alignment is low, a possible explanation for this phenomenon is sample overloading. If too much DNA is loaded onto the flowcell, there will not be enough separation between neighboring clusters and base-calling error rates will be high. By checking the summary statistics "cluster density" and "% Error," one can find deviations from optimal values, and adjust sample concentration accordingly. If summary statistics for sequencing runs are adequate, one still needs to search for technical problems in ChIP and library construction procedures.

A common problem in IGB visualization is that ChIP-Seq data is shown in a separate window from other annotations. The reason is that Eland result files sometimes have different chromosome naming system (e.g., chr01, chr02, ..., chrmt) from that in IGB (e.g., chr1, chr2, ..., chrM). In this case, one needs to rename all the chromosomes in the IGB system to visualize ChIP-Seq data along with other annotations in the same window. If there is not enough memory to load data into IGB, one can load data for one chromosome at a time.

7. CONCLUSION AND FUTURE DIRECTIONS

ChIP-Seq has emerged as a highly sensitive and cost-effective method for genome-wide mapping of TFBS at a high resolution. Barcoded ChIP-Seq enables multiplex short-read sequencing and offers a higher throughput and lower cost per sample. An ongoing debate in the ChIP-Seq field concerns the nature of the control DNA used for scoring ChIP-Seq experiments. Although most groups use input DNA, it is still unsettled whether input DNA, normal IgG DNA or, in the case of yeast, ChIP DNA from an untagged strain is the preferable reference sample for ChIP-Seq. With the read length, read quality, and read quantity improvements of high-throughput DNA sequencing technologies, it will be possible to obtain an increased total of reads with longer sequence lengths. For yeast ChIP-Seq, this will allow an increased multiplex capability. Computational challenges of data handling and long-term data storage require high-performance computing clusters and are much more complex than ChIP-chip analyses that could be performed by most users. The protocols developed for yeast, such as barcoded ChIP-Seq, can be readily extended to lower eukaryotes and eventually to higher eukaryotes with the advent of higher capacity DNA sequencers.

ACKNOWLEDGMENTS

Past and present members of the M. Snyder laboratory have helped to develop these protocols. We are particularly grateful to Jennifer Li-Pook-Than for insightful comments on this manuscript, Ghia M. Euskirchen for pioneer ChIP-Seq protocol development and Christopher M. Yellman for yeast ChIP protocol optimization in our laboratory. This work has been supported by NIH grants. P. Lefrançois has been supported by a master's fellowship from FQRNT and a doctoral fellowship from NSERC during this work.

REFERENCES

Aparicio, S., *et al.* (2002). Whole-genome shotgun assembly and analysis of the genome of Fugu rubripes. *Science* **297,** 1301–1310.

Aparicio, O., *et al.* (2004). Chromatin immunoprecipitation for determining the association of proteins with specific genomic sequences *in vivo*. *Curr. Protoc. Cell Biol.* Chapter 17, Unit 17 7.

Aparicio, O., *et al.* (2005). Chromatin immunoprecipitation for determining the association of proteins with specific genomic sequences *in vivo*. *Curr. Protoc. Mol. Biol.* Chapter 21, Unit 21 3.

Auerbach, R. K., *et al.* (2009). Mapping accessible chromatin regions using Sono-Seq. *Proc. Natl. Acad. Sci. USA.* Epub ahead of print.

Axelrod, J. D., and Majors, J. (1989). An improved method for photofootprinting yeast genes *in vivo* using Taq polymerase. *Nucleic Acids Res.* **17**, 171–183.

Bhinge, A., *et al.* (2007). Mapping the chromosomal targets of STAT1 by Sequence Tag Analysis of Genomic Enrichment (STAGE). *Genome Res.* **17**, 910–916.

Birney, E., *et al.* (2007). Identification and analysis of functional elements in 1% of the human genome by the ENCODE pilot project. *Nature* **447**, 799–816.

Borneman, A. R., *et al.* (2006). Target hub proteins serve as master regulators of development in yeast. *Genes Dev.* **20**, 435–448.

Borneman, A. R., *et al.* (2007). Divergence of transcription factor binding sites across related yeast species. *Science* **317**, 815–819.

Celniker, S. E., *et al.* (2009). Unlocking the secrets of the genome. *Nature* **459**, 927–930.

Chen, X., *et al.* (2008). Integration of external signaling pathways with the core transcriptional network in embryonic stem cells. *Cell* **133**, 1106–1117.

Costanzo, M. C., *et al.* (2000). The yeast proteome database (YPD) and *Caenorhabditis elegans* proteome database (WormPD): Comprehensive resources for the organization and comparison of model organism protein information. *Nucleic Acids Res.* **28**, 73–76.

Craig, D. W., *et al.* (2008). Identification of genetic variants using bar-coded multiplexed sequencing. *Nat. Methods* **5**, 887–893.

Cronn, R., *et al.* (2008). Multiplex sequencing of plant chloroplast genomes using Solexa sequencing-by-synthesis technology. *Nucleic Acids Res.* **36**, e122.

Fejes, A. P., *et al.* (2008). FindPeaks 3.1: A tool for identifying areas of enrichment from massively parallel short-read sequencing technology. *Bioinformatics* **24**, 1729–1730.

Feng, W., *et al.* (2008). A Poisson mixture model to identify changes in RNA polymerase II binding quantity using high-throughput sequencing technology. *BMC Genomics* **9**, S23.

Frazer, K. A., *et al.* (2004). VISTA: Computational tools for comparative genomics. *Nucleic Acids Res.* **32**, W273–W279.

Harbison, C. T., *et al.* (2004). Transcriptional regulatory code of a eukaryotic genome. *Nature* **431**, 99–104.

Horak, C. E., and Snyder, M. (2002). ChIP-chip: A genomic approach for identifying transcription factor binding sites. *Methods Enzymol.* **350**, 469–483.

Janke, C., *et al.* (2004). A versatile toolbox for PCR-based tagging of yeast genes: New fluorescent proteins, more markers and promoter substitution cassettes. *Yeast* **21**, 947–962.

Ji, H., *et al.* (2008). An integrated software system for analyzing ChIP-chip and ChIP-seq data. *Nat. Biotechnol.* **26**, 1293–1300.

Johnson, D. S., *et al.* (2007). Genome-wide mapping of *in vivo* protein-DNA interactions. *Science* **316**, 1497–1502.

Jothi, R., *et al.* (2008). Genome-wide identification of *in vivo* protein-DNA binding sites from ChIP-Seq data. *Nucleic Acids Res.* **36**, 5221–5231.

Kuo, M. H., and Allis, C. D. (1999). *In vivo* cross-linking and immunoprecipitation for studying dynamic Protein:DNA associations in a chromatin environment. *Methods* **19**, 425–433.

Lefrancois, P., *et al.* (2009). Efficient yeast ChIP-Seq using multiplex short-read DNA sequencing. *BMC Genomics* **10**, 37.

Li, H., *et al.* (2008). Mapping short DNA sequencing reads and calling variants using mapping quality scores. *Genome Res.* **18**, 1851–1858.

Lister, R., *et al.* (2008). Highly integrated single-base resolution maps of the epigenome in Arabidopsis. *Cell* **133**, 523–536.

Longtine, M. S., *et al.* (1998). Additional modules for versatile and economical PCR-based gene deletion and modification in *Saccharomyces cerevisiae*. *Yeast* **14**, 953–961.

Marks, H., *et al.* (2009). High-resolution analysis of epigenetic changes associated with X inactivation. *Genome Res.* **19**, 1361–1373.

Marson, A., *et al.* (2008). Connecting microRNA genes to the core transcriptional regulatory circuitry of embryonic stem cells. *Cell* **134**, 521–533.

Martone, R., *et al.* (2003). Distribution of NF-kappaB-binding sites across human chromosome 22. *Proc. Natl. Acad. Sci. USA* **100**, 12247–12252.

Mikkelsen, T. S., *et al.* (2007). Genome-wide maps of chromatin state in pluripotent and lineage-committed cells. *Nature* **448**, 553–560.

Mortazavi, A., *et al.* (2008). Mapping and quantifying mammalian transcriptomes by RNA-Seq. *Nat. Methods* **5**, 621–628.

Nagalakshmi, U., *et al.* (2008). The transcriptional landscape of the yeast genome defined by RNA sequencing. *Science* **320**, 1344–1349.

Nix, D. A., *et al.* (2008). Empirical methods for controlling false positives and estimating confidence in ChIP-Seq peaks. *BMC Bioinformatics* **9**, 523.

Orlando, V., *et al.* (1997). Analysis of chromatin structure by *in vivo* formaldehyde cross-linking. *Methods* **11**, 205–214.

Robertson, G., *et al.* (2007). Genome-wide profiles of STAT1 DNA association using chromatin immunoprecipitation and massively parallel sequencing. *Nat. Methods* **4**, 651–657.

Robertson, A. G., *et al.* (2008). Genome-wide relationship between histone H3 lysine 4 mono- and tri-methylation and transcription factor binding. *Genome Res* **18**, 1906–1917.

Rozowsky, J., *et al.* (2009). PeakSeq enables systematic scoring of ChIP-seq experiments relative to controls. *Nat. Biotechnol.* **27**, 66–75.

Schones, D. E., *et al.* (2008). Dynamic regulation of nucleosome positioning in the human genome. *Cell* **132**, 887–898.

Teytelman, L., *et al.* (2009). Impact of chromatin structures on DNA processing for genomic analyses. *PLoS One* **4**, e6700.

Tompa, M., *et al.* (2005). Assessing computational tools for the discovery of transcription factor binding sites. *Nat. Biotechnol.* **23**, 137–144.

Valouev, A., *et al.* (2008). Genome-wide analysis of transcription factor binding sites based on ChIP-Seq data. *Nat. Methods* **5**, 829–834.

Wilhelm, B. T., *et al.* (2008). Dynamic repertoire of a eukaryotic transcriptome surveyed at single-nucleotide resolution. *Nature* **453**, 1239–1243.

Xu, H., *et al.* (2008). An HMM approach to genome-wide identification of differential histone modification sites from ChIP-seq data. *Bioinformatics* **24**, 2344–2349.

Zhang, Y., *et al.* (2008). Model-based analysis of ChIP-Seq (MACS). *Genome Biol.* **9**, R137.

Zhong, M., *et al.* (2010). Genome-wide identification of binding sites defines distinct functions for *C. elegans* PHA-4/FOXA in development and environmental response. *PLoS Genet.* (In press).

CHAPTER FIVE

Genome-Wide Mapping of Nucleosomes in Yeast

Oliver J. Rando

Contents

Abstract

The packaging of eukaryotic genomes into chromatin has wide-ranging influences on all DNA-templated processes, from DNA repair to transcriptional regulation. The repeating subunit of chromatin is the nucleosome, which comprises 147 bp of DNA wrapped around an octamer of proteins. Positioning of nucleosomes relative to underlying DNA is a key factor in the regulation of gene transcription by chromatin, as DNA sequences between nucleosomes are more accessible to regulatory factors than are DNA sequences within nucleosomes. Here, I describe protocols for mapping nucleosome positions across the yeast genome.

1. Introduction

The yeast genome, like that of all eukaryotes, is packaged into a nucleoprotein complex known as chromatin. The repeating subunit of chromatin is the nucleosome, which consists of 147 base pairs (bp) of DNA wrapped around an octamer of basic histone proteins. The positioning of nucleosomes on underlying DNA has consequences for gene regulation—nucleosomal occlusion of protein-binding sites is generally thought to

Department of Biochemistry and Molecular Pharmacology, University of Massachusetts Medical School, Worcester, Massachusetts, USA

Methods in Enzymology, Volume 470
ISSN 0076-6879, DOI: 10.1016/S0076-6879(10)70005-7

inhibit protein binding. Nucleosomal positioning has also recently been shown to affect sequence evolution, with point mutations fixing at lower rates in linker DNA relative to nucleosomal DNA (Sasaki et al., 2009; Washietl et al., 2008). Though there are clearly DNA sequences that favor and disfavor nucleosomal incorporation (Ioshikhes et al., 1996, 2006; Iyer and Struhl, 1995; Kaplan et al., 2008; Segal et al., 2006; Sekinger et al., 2005; Yuan and Liu, 2008), essentially any DNA sequence can wrap around the histone octamer to form a nucleosome. Furthermore, protein complexes such as Isw2 regulate gene expression by moving nucleosomes away from their thermodynamically favored positions (Whitehouse and Tsukiyama, 2006; Whitehouse et al., 2007). Nucleosome positioning affects signal processing during gene regulation—nucleosomal occlusion of key promoter sequences has been shown to change the regulatory logic of the β-interferon promoter from an OR to an AND gate for three regulatory inputs (Lomvardas and Thanos, 2002), and to separate signaling threshold from dynamic range at the PHO5 promoter in yeast (Lam et al., 2008). Thus, understanding where nucleosomes are located in the genome has implications for a wide range of interesting and important aspects of genome biology.

A number of features distinguish nucleosomal DNA from the linker DNA that intervenes between adjacent nucleosomes. Most notably, nucleosomal DNA is relatively protected from cleavage by a variety of nucleases. The most common nuclease used for assaying nucleosome positions in micrococcal nuclease, which has little (but measureable) sequence preference on naked DNA, but has a dramatic preference for linker DNA over nucleosomal DNA. Thus, identification of DNA protected from micrococcal nuclease digestion has been a mainstay for mapping of nucleosome positions in vivo.

In the past few years, protocols have been developed enabling the mapping of nucleosome positions across the entire genome (Albert et al., 2007; Field et al., 2008; Mavrich et al., 2008a,b; Schones et al., 2008; Weiner et al., 2009; Yuan et al., 2005). I will not describe mapping methods, such as chromatin immunoprecipitation with an antihistone antibody (Bernstein et al., 2004), or formaldehyde-assisted isolation of regulatory elements (Lee et al., 2004), that do not have single-nucleosome resolution. The current state-of-the-art protocols for nucleosome mapping all rely on protection from micrococcal nuclease. Isolated DNA is characterized by tiling microarray or by "deep" sequencing, providing whole-genome, high-resolution data on the packaging state of the yeast genome. The protocols described were developed and validated in Saccharomyces cerevisiae, but we have used them successfully with over 10 different Ascomycete species (not shown), so they can be applied broadly.

2. ISOLATION OF MONONUCLEOSOMAL DNA

• *Solutions (make beforehand)*:

Buffer Z:

1 M sorbitol
50 mM Tris–HCl, pH 7.4

NP buffer:

1 M sorbitol
50 mM NaCl
10 mM Tris, pH 7.4
5 mM MgCl$_2$
1 mM CaCl$_2$

Glycine:

2.5 M glycine

Spermidine:

250 mM in water; ~500 μl aliquots can be stored at $-20\,^{\circ}$C, and can be repeatedly freeze/thawed.

MNase:

We always obtain MNase from Worthington. MNase is resuspended at 20 units/μl in 10 mM Tris, pH 7.4, and stored in ~50 μl aliquots at $-80\,^{\circ}$C—can be freeze/thawed several (at least 2–3) times, store in $-20\,^{\circ}$C after first use of an aliquot.

Proteinase K:

20 μg/μl; ~500 μl aliquots can be stored at $-20\,^{\circ}$C, and repeatedly freeze/thawed.

• *6× Orange G loading buffer*:

60% glycerol
1.5 mg/ml Orange G (Sigma O-1625)

• *Day 0*:

Inoculate a 2–5 ml culture of YPD (or any other media) with the cells of interest—typically S288C derivatives BY4741 (MATa) or BY4742 (MATalpha).

• *Day 1*:

Late PM: Count # cells/ml, calculate inoculation (BY4741 has a generation time of ~90 min in YPD) to have proper OD the following AM.

Inoculate 444 ml YPD in a 2-l flask, grow shaking 200 rpm at 28 °C. We would typically inoculate $\sim$7 μl of saturated BY4741 into this 444 ml culture at 7 p.m. for a mid-log OD the next day around 11 a.m.

We prefer to use shaking water bath incubators when possible, since water bath incubation is less susceptible to temperature variability than incubation in air incubators. We have found an approximately twofold difference in doubling time between different clamp positions in a typical shaking incubator using blown air for heating (cultures near the fan generally grow faster, and are likely warmer), while no such difference is seen between different positions in a shaking water bath.

- *Day 2:*

Check to make sure all water baths are at the correct temperature and have enough water in them.

Grow yeast culture to desired OD_{600} (usually $\sim$0.8).

Add 24 ml formaldehyde (37% as purchased)—to a final of 2%. We have found little difference between 1% and 2% formaldehyde in nucleosome positioning results, but we find better yields from experiments with 2%. This may be related to the spheroplasting step, but whatever the reason, we find optimal nucleosomal yields with 2%. Also, we always use formaldehyde within a month of purchase.

Incubate 30 min, 28 °C/shaking. In general, whatever the growth condition was before fixation, fix cells in those conditions.

Pour culture into a 1-l centrifuge bottle containing 24 ml 2.5 M glycine (to a final conc. of $\sim$125 mM) to react with remaining formaldehyde. If going straight to spheroplasting (i.e., no waiting for other cultures) the glycine can be omitted.

Spin (4000 rpm, SLA-3000 rotor, 5 min). Pour off supernatant and resuspend cell pellet in $\sim$50 ml ddH$_2$O in a 50 ml conical. Pellet cells again (table top Sorvall, 3700 rpm, 2 min).

Cells can now be left on ice as long as necessary (we have tested up to 3 h, with no change in positioning data) for other cells to "catch up," etc.

If necessary make Buffer Z.

Make zymolyase solution (10 mg/ml in Buffer Z). Zymolyase is not particularly soluble at this concentration, so shake well right before adding to yeast to evenly resuspend. We obtain zymolyase from Seigaku, distributed by Cape Cod Associates in the United States. Zymolyase solution lasts roughly a week, so should be made up fresh more or less every experiment.

Resuspend each cell pellet (from 450 ml of cells) in 39 ml Buffer Z. Add 28 μl β-mercaptoethanol (14.3 M, final conc. 10 mM) to each conical. Cap conical, and vortex cells to resuspend—it is important to get cell pellet fully resuspended. Add 1 ml Zymolyase (10 mg/ml). Incubate at 28 °C, shaking, for 30–35 min.

Note that the spheroplasting step is somewhat dependent on growth conditions, and on the strain/species. For example, *S. cerevisiae* in stationary phase are harder to spheroplast than mid-log yeast. Success in this step can be assessed directly by microscopy (spheroplasts will lyse in water, intact yeast will not), but will also be readily apparent when nucleosomal DNA is isolated—unspheroplasted cells will result in a band of undigested large fragments of genomic DNA. If a small fraction of cells are not spheroplasted, this does not seem to be a problem beyond lost yield, as mononucleosomal DNA is gel-purified away from this genomic contamination in any case. But we prefer to have $>90\%$ spheroplasted cells for experiments we care about (i.e., nontroubleshooting).

Meanwhile, add sensitive components to NP buffer: to 5 ml NP buffer add 10 μl 250 mM spermidine (final conc. 500 μM), 3.5 μl of β-mercaptoethanol diluted 1:10 in water (final conc. 1 mM), and 37.5 μl 10% NP-40 (final conc. 0.075%).

During spheroplast spin (below), aliquot micrococcal nuclease (MNase) to three or four Eppendorf tubes. Put a dot of MNase on the side of the tube (halfway down tube).

Suggested concentration range (1 flask of mid-log cells, 444 total ml, $OD_{600} > 0.8$): 3, 6, and 10 μl of MNase. If doing four titration steps, 1, 1.8, 3.5, and 6 μl would be our starting MNase levels (since spheroplasts will be more dilute if dividing into four aliquots). Information from the first titration will guide further titrations.

After zymolyase digestion, pellet cells (4900 rpm in a Sorvall RC3 centrifuge, or $\sim 7000 \times g$, 10 min, 4 °C), or if absolutely necessary on tabletop Sorvall, 3700 rpm, 10 min. Aspirate supernatant—be very careful, as spheroplast pellets are fluffy and will suck into the aspirator. We generally try to get almost all the supernatant while losing a small amount of the spheroplasts (a small amount of supernatant will not affect the MNase reaction). Resuspend cells in ~ 2 ml (600 μl for three titration steps + 200 μl) NP buffer by pipetting up and down several times. Make sure pellet is resuspended.

Add 600 μl of cells to each Eppendorf already carrying micrococcal nuclease. Add the cells to the tubes directly over the spot of nuclease, and try to be even and quick between tubes. I generally start with the most dilute sample and use the same P1000 tip to add to each successive tube to minimize tip changing times.

As soon as all the cells have been added, start the timer, close the tubes, invert them once to mix, and incubate at 37 °C (water bath) for 20 min.

About 5 min before the reaction ends, make 5× STOP buffer—to make 1 ml add 400 μl water, 500 μl 10% SDS, and 100 μl 0.5 M EDTA. STOP buffer needs to be made fresh every time since it precipitates during storage.

Stop the reaction after 20 min by adding 150 μl of (5×) STOP to each digestion.

Add 6 μl proteinase K (20 μg/μl). Invert tubes to mix. Incubate at 65 °C overnight to digest proteins and to reverse formaldehyde cross-links.

• *Day 3*:

Remove tubes from 65 °C.

Phenol:chloroform (P:C) extract once. We do this as below to maximize DNA yield, but any P:C extraction should work here.

Add ∼1 volume (800 μl) phenol:chloroform:IAA, mix, centrifuge 5′ in Eppendorf "heavy" phase lock tubes.

Phase lock tubes—spin at 16,000×*g* for 30 s to bring down gel. Add phenol/aqueous mix, invert repeatedly to form homogeneous solution (do not vortex). Spin at 16,000×*g* for 5 min. Use P200 pipette to recover DNA-containing aqueous layer above phase lock gel.

Add 1/10 volume (75 μl) 3 *M* sodium acetate, pH 5.5.

Add 1 volume isopropanol (fill to top) to precipitate DNA, freeze at −20 °C for 30 min, spin at max speed in microcentrifuge for 10 min. Aspirate supernatant—pellet may be loose, wash with 500 μl ice-cold 70% ethanol, spin 5 more min. Aspirate—be especially careful about pellet in this step. Air dry.

Resuspend pellet in 60 l NEB buffer 2 (dilute commercial stock 1:10!), incubate pellet at 37 °C for 1 h (or leave pellet at 4 °C overnight).

Add 1 l DNase-free RNase (10 μg/μl stock, Roche), incubate at 37 °C for 1 h.

Run products on 1.8% agarose gel to find which part of titration worked the best. Loading ∼1 μl is usually sufficient. Use 6× Orange G loading buffer for all nucleosomal DNA, as the dye front in standard DNA loading buffer runs right on top of the mononucleosomal band in this percentage gel.

Take gel images—see Fig. 5.1. Beware of the bright fuzzy band that runs with the Orange dye front—it is not mononucleosomes!

Take best titration (∼80% mononucleosomal DNA with ∼20% dinucleosome and a ghost of trinucleosome—see below) and load entire fraction onto approximately three lanes of a fresh 1.8% gel. Using a brand new razor blade, gel purify mononucleosomal band away from dinucleosomes, and away from "puff" of degraded DNA/RNA (see Fig. 5.1). As always with gel purification, trim band of excess agarose, but also attempt to minimize UV exposure of DNA.

If there is a lot of degraded DNA (running < 100 bp), first purify the DNA with a Qiagen cleanup column to remove the small pieces. Otherwise, the gel may not run correctly (bands will not separate). Resuspend entire samples (combined titrations) into 45 μl of elution buffer—will fit into two lanes.

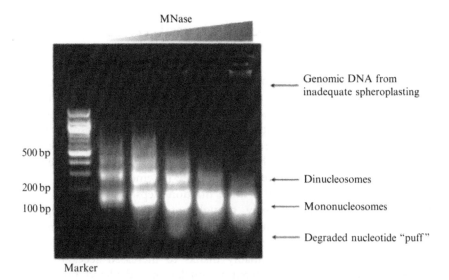

MNase

Genomic DNA from
inadequate spheroplasting

500 bp

200 bp
100 bp

Dinucleosomes

Mononucleosomes

Degraded nucleotide "puff"

Marker

Figure 5.1 Typical MNase titration series. Agarose gel analysis of digestion products isolated from yeast treated with varying amounts (triangle, top) of micrococcal nuclease (MNase). Left lane is 100 bp marker. Features of interest are indicated with arrows.

Use BioRad Freeze-N-Squeeze tubes to purify DNA from gel. We use this method as typical gel purification kits get abysmal yields from DNA in this size range. Mince band, freeze for 5 min at $-20\,^{\circ}$C, spin at $13,000 \times g$ for 3 min. We typically add another 50–100 μl of TE on top of the squished gel in the tube and spin again to increase yield, but this is not necessary.

To clean up DNA, add 1 volume phenol:chloroform:IAA, vortex, remove the aqueous phase to a new tube. Repeat (P:C extract twice).

Add 1 μl of glycogen (20 mg/ml, from *Mytilus edulis*, Sigma G1767) to assist in precipitation. Add 1/10 volume of 3 M Na acetate (pH 5.5), 2.5–3 volumes 100% EtOH. Incubate at $-20\,^{\circ}$C for 20–30 min, spin at max speed for 10 min, gently aspirate off supernatant, being careful not to disturb pellet.

Wash with 500 μl of ice-cold 70% EtOH to remove salt.

Spin max speed for 5–10 min. Aspirate supernatant, and air dry pellet.

Resuspend pellet in 25 μl of H_2O. Pellet may not always completely dissolve—possibly due to glycogen? In that case, brief spin and take supernatant above slurry.

Run 1 μl of mononucleosome prepared on a 1.8% gel after cleanup to confirm the DNA is there, clean, etc.

This DNA can now be labeled for microarray analysis, or ligated to linkers to generate a deep sequencing library.

3. VARIATION IN TITRATION LEVEL USED FOR NUCLEOSOME PURIFICATION

For years we have used nucleosomal DNA from a titration step with ~10–20% dinucleosome. The rationale here is that overdigestion of mono-nucleosomal DNA leads to a slow, progressive trimming of nucleosomes, so the digestion level of a titration step with only mononucleosomal DNA is hard to judge, and hard to reproduce. Using deep sequencing, we have more recently characterized the nucleosomal populations from over and underdigested chromatin as well—see Fig. 5.2 (Weiner *et al.*, 2009).

Nucleosome maps from the three titration steps broadly agree, with differences between them largely, though not entirely, being changes in occupancy rather than positions of nucleosomes. For example, the first and last nucleosomes in coding regions are often relatively highly occupied in the underdigested map, presumably thanks to increased accessibility to MNase of exposed DNA adjacent to nucleosome-depleted promoters. Moreover, so-called "nucleosome-free regions," especially the longer ones characteristic of very highly expressed genes, tend to be partially occupied in underdigested chromatin, suggesting that these regions are in fact occupied by loosely bound histones that are readily digested away.

What does this mean for mapping studies? Most importantly, when com-paring two samples it is crucial to record the gels from which the nucleosomal DNA was isolated. Artifacts are a potential pitfall in data analysis—an accidental difference in digestion levels between wild-type and some mutant would give the false impression that promoter nucleosome occupancy was affected by the mutation of interest. Another way to determine whether chromatin has been

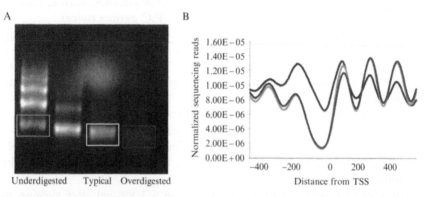

Figure 5.2 Effect of digestion level on nucleosome maps. A. Gel, as in Figure 5.1, showing an MNase titration from which 3 mononucleosome bands have been excised (indicated with boxes), corresponding to under-, well, and over-digested chromatin, from left to right. B. Chromatin maps differ depending on digestion level. Deep seqeucn-ing data for the three nucleosome preps in A were normalized, and data for all genes aligned by transcription start site (TSS) are averaged for each dataset. (See Color Insert.)

digested to similar extents after sequencing is to average nucleosomal data for all genes by the transcriptional start site (Fig. 5.2B), since in this view under- and overdigested chromatin have characteristic patterns that differ somewhat from our preferred titration level.

Finally, it is worth noting that this variation in occupancy at different digestion levels could potentially be utilized as a structural probe for chromatin *in vivo*, and some experimenters may find mapping over a range of digestion levels a valuable tool in their studies.

4. LABELING OF MONONUCLEOSOMAL DNA FOR TILING MICROARRAY ANALYSIS

Note that this protocol describes labeling and hybridization on "homemade" tiling microarrays. For commercial microarrays, labeling, and hybridization should be carried out according to manufacturer's protocols. This is particularly true for Affymetrix microarrays. On the other hand, we have used this protocol quite successfully with Nimblegen and Agilent microarrays, with differences only occurring after cleanup of the labeled material. After cleanup, we add the blocking and hybridization solutions recommended by the manufacturer, then carry out hybridization as described in their protocols.

- *Solutions (make beforehand)*:

2.5× random primer mix (can be made, or can use the buffer from Invitrogen's "BioPrime Klenow" labeling kit):

125 mM Tris, pH 6.8
12.5 mM MgCl$_2$
25 mM β-mercaptoethanol
750 μg/ml random octamers

10× dNTP mix:

1.2 mM dATP, dGTP, dTTP
0.6 mM dCTP
10 mM Tris (pH 8.0), 1 mM EDTA

tRNA:

5 mg/ml in water
Store aliquots at -80 °C

Poly-A RNA:

5 mg/ml in water
Store aliquots at -80 °C

4.1. Protocol

Add 2–3 μg of nucleosomal DNA to a 0.5-ml Eppendorf, and bring the
volume up to 21 μl. In another tube, do the same with 2–3 μg of sheared
genomic DNA.

Add 20 μl 2.5× random primer mix.

Boil 5 min (we use a heat block at 95 °C with water in the tube holders),
place on ice.

After 5 min on ice, add:

- 5 μl dNTP mix
- 3 μl Cy3-dCTP (to nucleosomes) or Cy5-dCTP (to gDNA)—the
  colors can be swapped, but we typically run experiments in this
  direction
- 1 μl high-concentration Klenow.

Place tubes at 37 °C for 1 h.

Add another 1 μl of Klenow.

Incubate another hour at 37 °C.

Stop reactions with 5 μl 0.5 M EDTA.

Mix the two colors, add 400 μl TE (pH 8.0 or 7.4) to stopped reactions, and
add to Microcon 30 filter.

Spin 10–11 min at 10,000 rpm in microcentrifuge, until ∼30–50 μl remain,
solution should be darkly colored.

Add another 400 μl TE.

Add 100 μg yeast tRNA (Sigma).

Add 20 μg poly-A RNA.

Spin 12–13 min at 10,000 rpm, until volume is less than 40 μl.

Recover labeled DNA by inverting Microcon filter in a fresh collection
tube, and spinning 10,000 rpm for 1 min.

Measure volume of recovered DNA and transfer to a 0.5-ml Eppendorf
tube.

Bring volume up to 40 μl with water.

Add 8.5 μl 20× SSC.

Add 1.5 μl 10% SDS (make sure not to add any more SDS than this—we
typically touch the edge of the pipette tip against a clean plastic surface to
wipe away any SDS stuck to the outside of the pipette tip).

Boil hybridization mix for 2 min (we use a heat block at 95 °C with water in
the wells to ensure good heat transfer).

Remove tubes and let them sit at room temperature (in the dark, either in
foil or in a drawer) for 10–15 min.

Give the tube a quick spin, apply to microarray, and hybridize for 12–16 h
at 65 °C.

Wash microarray and scan.

5. Generation of Nucleosomal DNA Libraries for Deep Sequencing

Over the past few years, ultrahigh-throughput "deep" sequencing methods have been effectively used for the analysis of mononucleosomal samples. As of this writing, two major commercial machines have been used for deep sequencing—454 (Roche) and Solexa (Illumina). We have experience with Solexa sequencing and so present a protocol for Solexa library construction, but use of other sequencing methodologies will simply require creating libraries per the instructions of the manufacturer. We use Solexa because provides several million reads 36 bases in length—36 bases is long enough to uniquely identify the vast majority of sequences in the yeast genome, and given that there are ~60,000 nucleosomes in yeast, 3 million reads provides 50× coverage of each nucleosome, allowing occupancy changes to be assessed between conditions. Solexa machine upgrades allow paired-end reads, and longer reads, and neither of these requires a change in the basic protocol.

Start with DNA from MNase titration, after RNase treatment, but prior to gel purification.

Clean up MNase titration with Qiagen MinElute (elute in 50 μl EB).

Treat DNA with alkaline phosphatase (CIP—NEB M0290L) for 1 h at 37 °C.

Gel purify mononucleosomal DNA on 1.8% agarose gel, as described above. Use BioRad Freeze-N-Squeeze tubes to purify DNA from gel. Mince band, freeze for 5 min at −20 °C, spin at 13,000×g for 3 min. We typically add another 50–100 μl of TE on top of the squished gel in the tube and spin again to increase yield, but this is not necessary.

Repair DNA ends using End-It DNA End-Repair Kit (Epicentre Biotechnologies ER0720):

DNA	150 ng
End-It buffer	5 μl
dNTP mix (2.5 mM)	5 μl
ATP (10 mM)	5 μl
End-It enzyme mix	1 μl
	Q/S to 50 μl

Incubate at room temperature for 1 h.

Clean up reaction with Qiagen MinElute column, elute in 30 μl EB.

Klenow exo- (Epicenter Biotechnologies KL06041K):

DNA	30 μl
Klenow buffer	5 μl
dATP (10 mM)	1 μl
Water	14 μl
Klenow exo-	1 μl

Incubate at room temperature for 45 min.

Clean up reaction with Qiagen MinElute column, elute in 20 μl EB.

Dry DNA down to about 10 μl in a Speed Vac, then bring to exactly 10 μl with water.

Ligate Illumina adapters to polished mononucleosomal DNA using Fast-Link DNA Ligation Kit (Epicenter Biotechnologies LK0750H):

DNA	10 μl
Fast-Link Ligation Buffer	1.5 μl
ATP (10 mM)	0.75 μl
Ligase	1 μl
Genomic adapters	2 μl

Incubate at room temperature for 1 h, then add:

Water	7.5 μl
Fast-Link Ligation Buffer	1 μl
ATP (10 mM)	0.5 μl
Ligase	1 μl

Incubate at 16 °C overnight.

Clean up reaction with Qiagen MinElute column, elute in 30 μl EB.

Amplify library with Pfx polymerase (Invitrogen 11708-039):

DNA	30 μl
Pfx buffer	10 μl
Illumina genomic primers (1.1 and 2.1)	1 μl each
dNTPs (10 mM)	3 μl
MgSO$_4$ (50 mM)	2 μl
Water	53 μl
Pfx	1 μl

Step number	Temperature (°C)	Time
1	94	2 min
2	94	15 s
3	65	1 min
4	68	30 s
5	Go to step 2	18 times
6	68	5 min
7	4	Forever

Gel-purify library on 1.5% agarose gel. Band should run ~250 bp. We sometimes see two bands here, and in our experience the smaller band corresponds to primer dimer, and if it is an issue it can be eliminated by gel-purifying DNA after the primer ligation step above, prior to PCR.

Freeze-N-Squeeze purify mononucleosome band from gel as above.

Clone a small portion (~100 ng) of the library using TOPO cloning, and transform into BL21DE3 competent *Escherichia Coli*.

The next day, isolate 20 colonies, and isolate plasmids by miniprep method of your choice. Send these 20 library inserts out for sequencing (or sequence by hand). Inserts should average ~120 bp for our typical MNase digestion level, should map to the yeast genome, and should contain no mitochondrial sequences. If we get more than three inserts which are simply primer dimers, we remake the library from mononucleosomal DNA.

If library looks OK, submit 30 μl of 10 nM sample in EB to your Solexa operator.

REFERENCES

Albert, I., Mavrich, T. N., Tomsho, L. P., Qi, J., Zanton, S. J., Schuster, S. C., and Pugh, B. F. (2007). Translational and rotational settings of H2A.Z nucleosomes across the Saccharomyces cerevisiae genome. *Nature* **446,** 572–576.

Bernstein, B. E., Liu, C. L., Humphrey, E. L., Perlstein, E. O., and Schreiber, S. L. (2004). Global nucleosome occupancy in yeast. *Genome Biol.* **5,** R62.

Field, Y., Kaplan, N., Fondufe-Mittendorf, Y., Moore, I. K., Sharon, E., Lubling, Y., Widom, J., and Segal, E. (2008). Distinct modes of regulation by chromatin encoded through nucleosome positioning signals. *PLoS Comput. Biol.* **4,** e1000216.

Ioshikhes, I., Bolshoy, A., Derenshteyn, K., Borodovsky, M., and Trifonov, E. N. (1996). Nucleosome DNA sequence pattern revealed by multiple alignment of experimentally mapped sequences. *J. Mol. Biol.* **262,** 129–139.

Ioshikhes, I. P., Albert, I., Zanton, S. J., and Pugh, B. F. (2006). Nucleosome positions predicted through comparative genomics. *Nat. Genet.* **38,** 1210–1215.

Iyer, V., and Struhl, K. (1995). Poly(dA:dT), a ubiquitous promoter element that stimulates transcription via its intrinsic DNA structure. *EMBO J.* **14,** 2570–2579.

Kaplan, N., Moore, I. K., Fondufe-Mittendorf, Y., Gossett, A. J., Tillo, D., Field, Y., Leproust, E. M., Hughes, T. R., Lieb, J. D., Widom, J., and Segal, E. (2008). The DNA-encoded nucleosome organization of a eukaryotic genome. *Nature* **458,** 362–366.

Lam, F. H., Steger, D. J., and O'Shea, E. K. (2008). Chromatin decouples promoter threshold from dynamic range. *Nature* **453,** 246–250.

Lee, C. K., Shibata, Y., Rao, B., Strahl, B. D., and Lieb, J. D. (2004). Evidence for nucleosome depletion at active regulatory regions genome-wide. *Nat. Genet.* **36,** 900–905.

Lomvardas, S., and Thanos, D. (2002). Modifying gene expression programs by altering core promoter chromatin architecture. *Cell* **110,** 261–271.

Mavrich, T. N., Ioshikhes, I. P., Venters, B. J., Jiang, C., Tomsho, L. P., Qi, J., Schuster, S. C., Albert, I., and Pugh, B. F. (2008a). A barrier nucleosome model for statistical positioning of nucleosomes throughout the yeast genome. *Genome Res.* **18,** 1073–1083.

Mavrich, T. N., Jiang, C., Ioshikhes, I. P., Li, X., Venters, B. J., Zanton, S. J., Tomsho, L. P., Qi, J., Glaser, R. L., Schuster, S. C., Gilmour, D. S., Albert, I., *et al.* (2008b). Nucleosome organization in the Drosophila genome. *Nature* **453,** 358–362.

Sasaki, S., Mello, C. C., Shimada, A., Nakatani, Y., Hashimoto, S., Ogawa, M., Matsushima, K., Gu, S. G., Kasahara, M., Ahsan, B., Sasaki, A., Saito, T., *et al.* (2009). Chromatin-associated periodicity in genetic variation downstream of transcriptional start sites. *Science* **323,** 401–404.

Schones, D. E., Cui, K., Cuddapah, S., Roh, T. Y., Barski, A., Wang, Z., Wei, G., and Zhao, K. (2008). Dynamic regulation of nucleosome positioning in the human genome. *Cell* **132,** 887–898.

Segal, E., Fondufe-Mittendorf, Y., Chen, L., Thastrom, A., Field, Y., Moore, I. K., Wang, J. P., and Widom, J. (2006). A genomic code for nucleosome positioning. *Nature* **442,** 772–778.

Sekinger, E. A., Moqtaderi, Z., and Struhl, K. (2005). Intrinsic histone-DNA interactions and low nucleosome density are important for preferential accessibility of promoter regions in yeast. *Mol. Cell* **18,** 735–748.

Washietl, S., Machne, R., and Goldman, N. (2008). Evolutionary footprints of nucleosome positions in yeast. *Trends Genet.* **24,** 583–587.

Weiner, A., Hughes, A., Yassour, M., Rando, O. J., Friedman, N. (2009). High-resolution nucleosome mapping reveals transcription-dependent promoter packaging. *Genome Res.* Nov 25. [Epub ahead of print].

Whitehouse, I., and Tsukiyama, T. (2006). Antagonistic forces that position nucleosomes in vivo. *Nat. Struct. Mol. Biol.* **13,** 633–640.

Whitehouse, I., Rando, O. J., Delrow, J., and Tsukiyama, T. (2007). Chromatin remodelling at promoters suppresses antisense transcription. *Nature* **450,** 1031–1035.

Yuan, G. C., and Liu, J. S. (2008). Genomic sequence is highly predictive of local nucleosome depletion. *PLoS Comput. Biol.* **4,** e13.

Yuan, G. C., Liu, Y. J., Dion, M. F., Slack, M. D., Wu, L. F., Altschuler, S. J., and Rando, O. J. (2005). Genome-scale identification of nucleosome positions in *S. cerevisiae*. *Science* **309,** 626–630.

CHAPTER SIX

Genome-Wide Translational Profiling by Ribosome Footprinting

Nicholas T. Ingolia

Contents

Abstract

We present a detailed protocol for ribosome profiling, an approach that we developed to make comprehensive and quantitative measurements of translation in yeast. In this technique, ribosome positions are determined from their nuclease

Department of Cellular and Molecular Pharmacology, Howard Hughes Medical Institute, University of California, San Francisco, and California Institute for Quantitative Biomedical Research, San Francisco, California, USA

Methods in Enzymology, Volume 470
ISSN 0076-6879, DOI: 10.1016/S0076-6879(10)70006-9

footprint on their mRNA template and the footprints are quantified by deep sequencing. Ribosome profiling has already enabled highly reproducible measurements of translational control. Because this technique reports on the exact position of ribosomes, it also revealed the presence of ribosomes on upstream open reading frames and demonstrated that ribosome density was higher near the beginning of protein-coding genes. Here, we describe nuclease digestion conditions that produce uniform $\sim$28 nucleotide (nt) protected fragments of mRNA templates that indicate the exact position of translating ribosomes. We also give a protocol for converting these RNA fragments into a DNA library that can be sequenced using the Illumina Genome Analyzer. Unbiased conversion of anonymous, small RNAs into a sequencing library is challenging, and we discuss standards that played a key role in optimizing library generation. Finally, we discuss how deep sequencing data can be used to quantify gene expression at the level of translation.

1. INTRODUCTION

Gene expression is now measured routinely to characterize the physiological state of cells and to determine the molecular basis of cellular function and dysfunction. Gene expression profiling typically uses mRNA abundance, which can be measured easily, as a proxy for protein production, which is the ultimate effect of gene expression. However, these measurements of mRNA abundance are blind to regulation of protein translation, and there is clear interest in approaches for making comprehensive measurements of protein synthesis. Translational control plays a major role in cellular stress responses in yeast (Hinnebusch, 2005), and homologous pathways are important in mammals as well (Holcik and Sonenberg, 2005). Translation is also regulated in development and differentiation (Sonenberg and Hinnebusch, 2009), including the establishment of mother/daughter asymmetry in yeast (Chartrand et al., 2002; Gu et al., 2004) as well as the filamentous growth response (Gilbert et al., 2007).

Microarrays (Brown and Botstein, 1999), and more recently deep sequencing (Mortazavi et al., 2008; Nagalakshmi et al., 2008), have allowed rapid and comprehensive measurements of mRNA levels. Polysome profiling emerged as a technique for measuring translation with microarrays (Arava et al., 2003; Johannes et al., 1999; Zong et al., 1999) by fractionating transcripts according to the number of bound ribosomes and analyzing the distribution of mRNAs in the resulting fractions. Increases or decreases in the amount of protein synthesized per mRNA will be reflected in the number of ribosomes bound, so translational regulation will shift the distribution of a message between different fractions. This polysome profiling approach has provided measurements of genome-wide translation in yeast, most notably in response to starvation (Preiss et al., 2003; Smirnova et al., 2005). However, the imprecision in polysome fractionation, especially for large numbers of ribosomes, limit the quantitative resolution of polysome profiling. More fundamentally, this approach cannot distinguish ribosomes that are translating protein-coding genes from those on upstream open reading frames (uORFs) (Arava *et al.*, 2005).

In this chapter, we present a protocol for a ribosome profiling, a technique for making quantitative and high-resolution data on all cellular translation. We have used ribosome profiling to measure basal translational efficiency as well as translational regulation and to characterize known and novel sites of noncanonical translation (Ingolia *et al.*, 2009). Ribosome profiling combines the classic observation that the nuclease digestion footprint of a ribosome on an mRNA message indicates its exact position (Steitz, 1969; Wolin and Walter, 1988) with recent advances in ultra high-throughput sequencing (Bentley *et al.*, 2008) that allow the analysis of millions of footprints in parallel. We describe the generation of ribosome footprints as well as the techniques for converting them into a deep sequencing library. We also discuss the analysis of ribosome footprint sequencing data, focusing on measurements of gene expression.

 ## 2. RIBOSOME FOOTPRINT GENERATION AND PURIFICATION

Ribosome profiling requires the preparation of cell extracts containing mRNA-bound ribosomes. One major concern in preparing these extracts is ensuring that the polysomes recovered in the extract reflect the physiological status of translation in the living yeast. However, yeast alter translation very quickly in response to environmental changes such as removal of nutrients (Ashe *et al.*, 2000; Barbet *et al.*, 1996), and cells must be removed from growth media to prepare extracts. To minimize perturbations of *in vivo* translation profiles, polysomes are stabilized by adding cycloheximide to cells immediately before they are harvested. Cycloheximide is a translation elongation inhibitor that immobilizes ribosomes (Godchaux *et al.*, 1967; McKeehan and Hardesty, 1969), thereby preserving a snapshot of their location at the time of cycloheximide addition. Following cycloheximide addition, care is also taken to freeze cells in liquid nitrogen as quickly as possible. If cells are harvested and frozen quickly, ribosome footprinting can be performed on cycloheximide-free extracts. Footprinting in these drug-free extracts shows a marked depletion of ribosome footprints from the 5' region of protein-coding genes (Ingolia *et al.*, 2009), consistent with the idea that initiation is rapidly blocked during cell harvesting and that cycloheximide prevents ribosome run-off during extract preparation.

Ribosome footprints are generated by nuclease digestion of polysomes in cell extracts. Nuclease treatment degrades the mRNA that links together polysomes, freeing individual ribosomes (Fig. 6.1). These ribosomes, bound to $\sim$28 nucleotide (nt) protected mRNA fragments, are isolated by sucrose density gradient centrifugation. The footprint mRNA fragments themselves are then purified through two size selection steps.

Figure 6.1 Schematic of footprint fragment generation. Polysomes, consisting of actively translating ribosomes bound to mRNA templates, are nuclease digested to degrade transcripts. Ribosomes protect ~28 nt mRNA footprint from nuclease digestion. After digestion, footprint-bound ribosomes are isolated and footprint fragments are recovered.

2.1. Extract preparation

Grow a 25-ml overnight starter culture in YEPD and prewarm growth media to 30 °C. Prepare 750 ml prewarmed YEPD in a 2800-ml baffled flask and inoculate it with yeast from the starter culture to an initial OD_{600} of roughly 0.03. Grow cells at 30 °C with 250 rpm shaking to a final OD_{600} of 0.6–0.7, which requires 7–8 h for the standard laboratory strain S288C. The growth conditions can be modified to measure the translational effects of different perturbations provided that the same final quantity of cells is available.

To freeze cells as quickly as possible, prepare for cell harvesting before adding cycloheximide to the culture. Make the polysome lysis buffer (see Section 5.1 for all solutions) and chill it on ice. Fill a 50-ml conical tube with liquid nitrogen and leave it partly immersed in liquid nitrogen. Pierce the cap of the tube several times with a 20-gauge needle to allow nitrogen vapor to escape.

Add 1.5 ml cycloheximide from a 50-mg/ml stock in ethanol to reach a final concentration of 100 μg/ml. Continue cell growth for 2 min, mixing well, to allow the cycloheximide to act. Harvest cells by filtration, which provides rapid and complete removal of media while minimizing perturbations before cells are actually frozen. Use a 90-mm cellulose nitrate filter with a 0.45 μm pore size (Whatman 7184-009) in a vacuum filtration apparatus with a fritted glass support (Kontes 953755-0090). Prewarm the funnel with a small amount of prewarmed growth media and then harvest cells by filtration. As soon as the last liquid has passed through the filter, remove the funnel and scrape the cells from the filter with a metal spatula, taking care not to tear the membrane. Resuspend the cells into a slurry in 2.5 ml of ice-cold polysome lysis buffer as quickly as possible, using a 1000-μl pipette to mix and then pipette cells. Drip the cell slurry into the liquid nitrogen-filled 50 ml conical and cap it, taking care to ensure that the cap is pierced as described above. Place the conical tube upright in a −80 °C freezer to allow the liquid nitrogen to evaporate.

Lyse yeast by cryogenic grinding in a mixer mill (Retsch MM301). The principal concern throughout the lysis is to avoid thawing the extract, and this is addressed by chilling all equipment in liquid nitrogen. Begin by opening a stainless steel grinding chamber and then immersing it and the grinding ball in liquid nitrogen until the nitrogen stops boiling vigorously. Remove the chamber using tongs and pour out all residual liquid nitrogen. Place the frozen yeast pellets and the grinding ball into the chamber and screw it shut tightly. Return the sealed chamber to liquid nitrogen until it is fully rechilled and boiling again stops. Remove the chamber, loosen it one-quarter turn, and mount it on the mixer mill. Grind the sample for 3 min at 15 Hz, then quickly remove the chamber, retighten the seal, and chill it in liquid nitrogen until the nitrogen stops boiling around the chamber. Repeat this process for six total cycles of grinding. After the last grinding cycle, chill the chamber as well as a metal scoop. Fill a new 50 ml conical tube with liquid nitrogen and use the chilled metal scoop to transfer cell powder from the chamber into the conical tube. Keep the scoop cold and, if the recovery of the powder takes too long, reseal and rechill the sample chamber as well. Again place the conical tube upright at -80 °C to allow liquid nitrogen to evaporate. Use caution, as the powder is much more easily dispersed by boiling liquid nitrogen than the frozen cell droplets.

Thaw the cell powder gently and, as soon as it is fully thawed, spin the tube 5 min at $3000 \times g$ in a 4 °C centrifuge to collect unbroken cells and large debris. Remove the supernatant to chilled 1.5 ml microfuge tubes on ice and clarify the lysate by spinning 10 min at $20,000 \times g$ in a 4 °C centrifuge. Recover the supernatant, avoiding both the pellet and the lipid layer at the top of the tube. Find the concentration of the extract by measuring the A_{260} of a 200-fold dilution. Dilute the extract with lysis buffer to achieve an undiluted A_{260} of 200. At this point, aliquots of the extract can be frozen in liquid nitrogen and stored for at least 6 months at -80 °C.

2.2. Nuclease digestion and monosome purification

Nuclease digestion must fully remove unprotected mRNA to produce footprints that precisely define the position of the ribosome. Excessive digestion can degrade the ribosome, which contains an essential RNA component, potentially resulting in the loss of footprint fragments. Tests of commercially available nucleases showed that *Escherichia coli* RNase I was best able to produce ribosome footprint fragments of uniform size. The concentration of RNase I in the digestion reactions was optimized to produce ~ 28 nt footprint fragments (Ingolia *et al.*, 2009). The RNase inhibitor SUPERase·In is effective against RNase I, and for this reason it is used to stop the footprinting digestion.

Gently thaw a 250-μl aliquot of cell extract and add 7.5 μl RNase I 100 U/μl (Ambion AM2294). In parallel, add 5.0 μl SUPERase·In

(20 U/μl; Ambion AM2694) to another aliquot. Incubate the extract samples for 1 h at room temperature with gentle rotation. Stop digestion by adding 5.0 μl SUPERase·In to the RNase-treated sample and keep both samples on ice. Load samples onto sucrose gradients for monosome purification as soon as possible after stopping the digestion.

Prepare 10% and 50% (w/v) sucrose solutions in polysome gradient buffer. Chill the ultracentrifuge rotor at 4 °C. Prepare gradients in Sw41 14 × 89 mm ultracentrifuge tubes. Mark the tubes according to the marker block and fill with 10% sucrose solution to the mark. Use the cannula to underlay 50% sucrose solution to fill the tube entirely. Gently insert the rubber cap into the top of the tube at an angle, with the hole entering the liquid last to allow air to escape. Form the gradient using a Gradient Master (BioComp Instruments) with rotation at 81.5 °C, speed 16, for 1:58. Load the gradients into the Sw41 buckets and balance them by adding additional 10% sucrose solution to the lighter gradient, then return the buckets with gradients to 4 °C. Load extract samples onto sucrose density gradients and seal the buckets well. Centrifuge gradients for 3 h at 35,000 rpm at 4 °C. Retrieve the rotor buckets and store them at 4 °C before fractionation.

Fractionate sucrose gradients on the Gradient Station (BioComp Instruments) at 0.2 mm/s. Measure the A_{260} of the collected material using a continuous UV monitor (such as the BioRad EM-1 Econo UV Monitor) to identify the 80S ribosome peak. Analyze the undigested sample first to ensure that it contains intact polysomes and to determine the approximate location of the 80S ribosome peak. Then, fractionate the digested sample, which has a much larger 80S monosome peak than the undigested sample. Collect this peak, which typically has a volume of 0.8 ml, freeze it, and store at − 80 °C.

2.3. Footprint fragment purification

Purify RNA from the monosome fraction using the SDS/phenol method. Heat acid phenol–chloroform (5:1, pH 4.5; Ambion AM9720) to 65 °C in a fume hood. Place the monosome fraction sample at 65 °C as well and add SDS to a final concentration of 1% (w/v). Add 1 volume of hot acid phenol–chloroform and incubate 5 min at 65 °C, vortexing frequently. Place samples on ice for 5 min, then spin 2 min at full speed in a tabletop microfuge. Recover the aqueous phase, which should be the upper phase if phenol–chloroform was used, but which may be the lower phase if only phenol was used due to the high concentration of sucrose in the aqueous phase. Phase lock gel should be avoided as the sucrose may make the aqueous phase more dense than the gel.

Reextract the aqueous phase with 1 volume of acid phenol–chloroform, vortex for 5 min at room temperature, and spin 2 min at full speed in a tabletop microfuge to separate the phases. Carefully recover the aqueous phase, add 1 volume of chloroform–IAA (24:1), and vortex 1 min at room temperature. Spin 1 min at full speed in a tabletop microfuge to separate and then recover the aqueous phase.

Purify footprint fragments by filtration through a Microcon YM-100 microconcentrator (Millipore 42413). Dilute the aqueous phase to 0.50 ml with 10 mM Tris–Cl (pH 7) and add 2.0 μl SUPERase·In. Prepare a microconcentrator collection tube by placing 2.0 μl SUPERase·In in the tube and then insert the YM-100. Load the sample on the microconcentrator and spin at 500×g, room temperature, until 400–425 μl flow-through has been recovered. This typically takes 20–30 min. Further centrifugation, or centrifugation at higher force, will dramatically increase the amount of high-molecular-weight RNA that passes through the filter. Precipitate RNA from the flow-through (see Section 5.3), using a coprecipitant because the total RNA concentration at this point is quite low.

Resuspend the RNA in 10 μl 10 mM Tris–Cl, pH 7. At this point, dephosphorylate the RNA if it is to be converted into a sequencing library. If randomly fragmented mRNA is prepared (see Section 2.4), dephosphorylate that sample in parallel. Fragments produced by RNase digestion or alkaline hydrolysis (see Section 2.4) have either 3′ phosphoester or 2′, 3′ cyclic phosphodisester termini (delCardayré and Raines, 1995; Markham and Smith, 1952). The phosphatase activity of T4 polynucleotide kinase can dephosphorylate either terminus to leave a free 3′ hydroxyl (Amitsur et al., 1987; Cameron and Uhlenbeck, 1977). Empirically, polynucleotide kinase converts a larger fraction of RNase and alkaline hydrolysis fragments into suitable substrates for polyadenylation than does alkaline phosphatase. To dephosphorylate RNA fragments, denature RNA 2 min at 80 °C, then place on ice. Add 2.0 μl 10× T4 polynucleotide kinase reaction buffer without ATP, 0.5 μl SUPERase·In, 6.5 μl water, and 1.0 μl T4 polynucleotide kinase (10 U/μl; New England Biolabs M0201S). Incubate 1 h at 37 °C.

Perform polyacrylamide gel purification for size selection of mRNA footprint fragments from the mix of digested rRNA. Add 20 μl 2× denaturing gel loading buffer (Invitrogen LC6876) to the dephosphorylated footprint sample. Prepare samples of 1.0 μl 10 bp DNA ladder (1 μg/μl, Invitrogen 10821-015), and of 1.0 μl 20 μM marker oligo, each diluted to 10.0 μl, with 10.0 μl 2× denaturing gel loading buffer. Denature all samples 2 min at 80 °C and place on ice. Load a 15% polyacrylamide denaturing gel and perform electrophoresis to separate fragments of roughly 30 nt. For precast polyacrylamide mini-gels (Invitrogen EC6875BOX), 65 min at 200 V resolves the footprint fragments well. Stain the gel for 3 min in SYBR Gold (diluted from 10,000× in 1× TBE; Invitrogen S11494), then visualize by UV transillumination. Excise the region near the 28 nt marker oligonucleotide (Fig. 6.2) and recover RNA from the gel slice (see Section 5.4). Excise the marker oligonucleotide as well and carry it through the library generation protocol in parallel with the footprint sample. It will serve as a

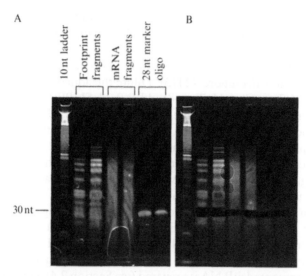

Figure 6.2 Size selection of footprint fragments. (A) A denaturing 15% polyacryl-amide gel was loaded with denatured 10 bp DNA ladder, two nuclease footprinting samples, two samples of randomly fragmented mRNA, and two samples of the 28 nt marker oligonucleotide. After electrophoresis, the gel was stained with SYBR Gold. The footprinting samples contain an array of characteristic rRNA fragments as well as the ribosome-protected mRNA fragments. (B) The gel depicted in (A), after the footprint fragment region from 25 to 31 nt was excised from all samples, guided by the marker oligo and the ladder.

positive control for subsequent steps and provide a size marker for gel purifications.

Resuspend the recovered RNA, typically 50–200 ng, in 10.0 μl 10 mM Tris–Cl, pH 7. A significant fraction of this RNA is digested rRNA. Quantify the size-selected RNA using the Agilent BioAnalyzer Small RNA assay. Dilute 1.0 μl of the size-selected footprint sample with 4.0 μl RNase-free water, and make a second serial dilution of 1.0 μl into 4.0 μl water. One of these dilutions should lie within the range of RNA concentrations for accurate quantitation. Load the 1:25 dilution before the 1:5 dilution, as overloading in earlier samples can distort measurements in later samples.

2.4. Fragmented mRNA preparation

Deep sequencing measurements of fragmented mRNA provide a valuable control and comparison for ribosome footprint density. Both mRNA and ribosome measurements are needed to distinguish translational regulation from control that operates at the level of mRNA transcription and stability. Many standard methods exist to purify mRNA from yeast. These typically involve lysis and total RNA extraction followed by mRNA purification via recovery of polyadenylated messages or removal of rRNA. The following mRNA isolation protocol (Köhrer and Domdey, 1991) is presented as one good approach, but alternatives can doubtlessly be substituted.

Resuspend a small fraction of harvested cells in 600 μl total RNA lysis buffer rather than polysome lysis buffer. Cells may be frozen in lysis buffer and stored at $-80\,^\circ$C indefinitely. Lyse cells and extract RNA using the hot acid–phenol method. Heat aliquots of acid phenol–chloroform and total RNA samples to 65 $^\circ$C. Add SDS to RNA samples to a final concentration of 1% (w/v) and then add 1 volume of hot acid phenol. Proceed with phenol extraction as described above for recovering RNA from the purified monosome fraction (see Section 2.3). The total RNA sample will contain much more debris than the ribosome fraction and may require additional rounds of phenol extraction. After a final chloroform extraction, precipitate the RNA (see Section 5.3). The RNA yield will be quite high and a coprecipitant is not needed. Determine the yield and purity of RNA by spectrophotometry.

Purify mRNA from total RNA using oligo-dT-coated magnetic beads (Dynabeads mRNA purification kit; Invitrogen 610-06) essentially as described by the manufacturer. Prepare 220 μl 1$\times$ binding buffer by diluting it from the 2$\times$ binding buffer stock. Resuspend the magnetic beads by vortexing and take 150 μl beads to a nonstick tube. Collect beads by placing the tube on a magnetic rack for 30 s and carefully pipetting away the storage buffer. Immediately resuspend beads in 100 μl 1$\times$ binding buffer. Repeat this procedure to wash beads again in 1$\times$ binding buffer and leave them in binding buffer. Take 150 μg total RNA and dilute to a final volume of 50 μl with RNase-free water. Add 50 μl 2$\times$ binding buffer and denature 2 min at 80 $^\circ$C, then return immediately to ice and add 1.0 μl SUPERase·In. Remove the binding buffer from the magnetic beads and resuspend them in the RNA sample. Incubate 5 min at room temperature with gentle rotation to allow mRNA to bind to the beads. Return the tube to the magentic rack to collect the beads for 30 s and remove the unbound sample. Wash the beads twice in 100 μl wash buffer B. Resuspend the beads in 20 μl 10 mM Tris–Cl (pH 7) and elute the RNA by heating the beads to 80 $^\circ$C for 2 min. Immediately place the tube on the magnetic rack for 30 s and remove the eluate to a new tube.

Prepare the mRNA fragmentation reaction by mixing 20 μl of RNA with 20 μl of 2$\times$ alkaline fragmentation buffer. Incubate the fragmentation reaction for 40 min at 95 $^\circ$C and return the reaction to ice immediately. Prepare 560 μl stop/precipitation solution in a nonstick microfuge tube on ice and add the fragmentation reaction to the stop solution. Add at least 600 μl isopropanol, then precipitate (see Section 5.3). The stop solution already contains 300 mM salt.

Resuspend the fragmented mRNA in 10 μl 10 mM Tris–Cl (pH 7) and then dephosphorylate the fragments and perform gel size selection in parallel with the ribosome footprint sample (see Section 2.3). All mRNA fragments will be quite small, so there is no need to filter the fragmentation reaction through a microconcentrator.

3. SEQUENCING LIBRARY PREPARATION

Short ribosome footprint RNA fragments must be converted into a DNA library suitable for deep sequencing. Sequencing libraries for the Illumina Genome Analyzer are pools of DNA molecules with constant linker sequences on both sides of the query fragment (Bentley et al., 2008). Creating this library requires attaching linkers to both ends of the RNA fragment as well as reverse transcription. Any sequence preferences in library generation will distort the measured abundance of these fragments. There are many approaches to library generation, and a standard RNA fragment sample is needed to compare between them. A complex but well-characterized pool of small RNAs can be obtained by partial alkaline hydrolysis of yeast mRNA (see Section 2.4). This high-temperature chemical treatment causes relatively uniform fragmentation of RNA, and different fragments of the same transcript should be present at roughly equal abundance in the sample. Deviations from uniform sequencing read coverage across each transcript indicate distortions introduced by library generation, so different protocols can be compared directly by quantifying the uniformity of sequencing coverage.

Extensive optimization has resulted in the library generation protocol depicted in Fig. 6.3. The constraints on generating a sequencing library from ribosome footprint fragments are similar to those that arise when working with endogenous small RNAs such as microRNAs (Berezikov et al., 2006). It is necessary to introduce a known sequence on the $3'$ terminus of the RNA fragment to serve as a primer site for reverse transcription. For RNA fragments produced by nuclease digestion or alkaline hydrolysis, this first requires dephosphorylation of the $3'$ terminus to produce a $3'$ hydroxyl that can serve as a substrate for further manipulation. Various RNA ligases have been used to attach known linker sequences to microRNAs, but these enzymes have significant sequence preferences. Polyadenylation of the $3'$ terminus using E. coli poly-(A) polymerase (Fu et al., 2005) produced more uniform libraries from fragmented RNA than any ligase-based approach tested. Among ligase-based protocols, the use of truncated Rnl2 to attach an adenylylated linker (Ho et al., 2004; Lau et al., 2001; Pfeffer et al., 2005) produces substantially better results than Rnl1 with a phosphorylated linker.

Attachment of the $5'$ linker requires ligation, but this ligation can be performed as an efficient intramolecular reaction of first-strand cDNA using an ssDNA ligase. Intermolecular ligation of a second linker to either RNA

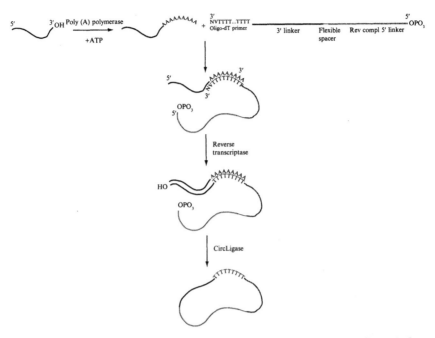

Figure 6.3 Schematic of sequencing library generation protocol. Small RNA fragments are polyadenylated to provide a primer-binding site for reverse transcription. A custom oligonucleotide containing an anchored oligo-d(T) primer as well as linker sequences and a flexible spacer is used to prime reverse transcription. After reverse transcription, the first-strand cDNA is circularized to generate a PCR template with linker sequences on both sides of the target RNA fragment.

or first-strand DNA dramatically distorts mRNA coverage and requires additional gel purification to recover the ligation product. Circularized ssDNA can serve directly as a sequencing template on the Genome Analyzer, though limited PCR amplification to produce a conventional linear dsDNA library does not introduce significant distortions in the relative abundance of sequences in the library. Because the yield of circular ssDNA is relatively low and quantification is more difficult, PCR amplification should be performed unless there is specific evidence that it distorts the contents of the sequencing library.

3.1. Polyadenylation

It is desirable to add a 25–30 nt poly-A tail on each RNA fragment, to ensure that the tail is long enough to serve as a primer-binding site but not dramatically longer. The extent of polyadenylation is controlled by using relatively low amounts of enzyme and limiting the duration of the reaction. Under the reaction conditions below, poly-(A) addition to a synthetic RNA oligonucleotide substrate is directly proportional to the reaction time and proceeds uniformly across the substrate pool (Feng and Cohen, 2000). The tail length is relatively insensitive to the concentration of RNA at or below 500 nM, the concentration used here, but somewhat longer reactions may be required for higher concentrations (Fig. 6.4).

Prepare 10 pmol RNA and dilute to 9.0 μl with RNase-free water. Prepare 2× polyadenylation buffer with ATP as well as polyadenylation enzyme mix on ice. Denature samples 2 min at 80 °C and return immediately to ice. Add 9.0 μl 2× polyadenylation buffer with ATP and 2.0 μl enzyme mix and incubate 10 min at 37 °C. Stop the reaction by adding 80 μl 5 mM EDTA and precipitating the RNA (see Section 5.3).

3.2. Reverse transcription

Resuspend the polyadenylated RNA samples in 10 μl 10 mM Tris–Cl, pH 7. Prepare a template mix with 9.0 μl RNA sample, 1.0 μl dNTPs 10 mM each, 1.0 μl primer 50 μM, and 2.5 μl water. Denature 5 min at 65 °C, then return to ice. Add 4.0 μl 5× FSB, 1.0 μl SUPERase·In, 1.0 μl 0.1 M DTT, mix well, and add 1.0 μl SuperScript III. Incubate 30 min at 48 °C. Add 2.3 μl 1 N NaOH and incubate 15 min at 98 °C.

Gel purify to separate extended reverse transcriptase products from the unextended primer. Prepare a 10% denaturing polyacrylamide gel. Add 22.5 μl 2× denaturing loading dye to each reverse transcription reaction. In parallel, prepare samples with 1.0 μl 10 bp ladder and with 0.5 μl primer 50 μM, dilute to 11.3 μl with water, and add 11.3 μl 2× denaturing loading dye. Denature samples 1 min at 95 °C and return immediately to ice. Load samples on the gel, using two lanes for each reverse transcription reaction. Run the gel under conditions that optimize separation of the unextended primer from the footprint samples, which are all between 90 and 130 nt. When using a precast denaturing polyacrylamide mini-gel (Invitrogen EC6865BOX), 65 min at 200 V will resolve reverse transcription products. Stain the gel 3 min in SYBR Gold and visualize by UV transillumination. Excise the extended reverse transcriptase product band (Fig. 6.5), taking care to avoid the unextended primer. Recover DNA from the gel slice (see Section 5.4).

3.3. Circularization

Resuspend the gel extraction products in 15.0 μl 10 mM Tris–Cl, pH 8. Add 2.0 μl 10× CircLigase buffer, 1.0 μl 1 mM ATP, 1.0 μl 50 mM MnCl$_2$, mix well, and add 1.0 μl CircLigase (Epicentre Biotechnologies CL4111K). Incubate 1 h at 60 °C, then heat inactivate at 80 °C for 10 min. It is not necessary to further purify circularized first-strand cDNA before using it as a PCR template.

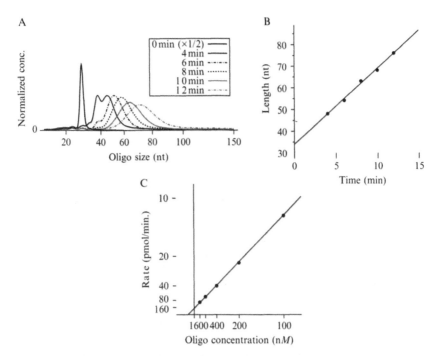

Figure 6.4 Kinetics of polyadenylation reaction. (A) Distribution of oligonucleotide sizes from a time-course of polyadenylation. Polyadenylation reactions were prepared as described, using 8 pmol control RNA oligo per reaction as a substrate. Reactions were carried out at 37 °C for the indicated time, quenched, precipitated, and analyzed using the BioAnalyzer Small RNA assay. (B) Quantification of average product length from (A). The population of polyadenylated products was selected based on a length threshold that excluded 95% of RNA in the unadenylated sample. The average length of the polyadenylated products is plotted against the reaction time along with a linear fit (33.8 nt + 3.5 nt/min). (C) Quantification of average product length as a function of substrate concentration. As in (B), for reactions with different substrate concentrations in an 8 min reaction. The average length of polyadenylation is calculated by subtracting the actual length of the substrate molecule, and the total amount of polyadenylation is calculated by multiplying the average length per molecule by the quantity of substrate in the reaction. The rate, computed from the total amount of polyadenylation, is plotted on a double reciprocal plot against the concentration of substrate, along with a linear fit (K_M = 1500 nM, v_{max} = 180 pmol/min). Note that while total adenylation displays hyperbolic kinetics, tail length is given by adenylation per substrate molecule and is thus independent of substrate concentration in the low-concentration regime and inversely proportional to substrate concentration in the high-concentration regime.

3.4. PCR amplification

Prepare PCR mixes for five reactions, each of 16.7 μl, for each circularized template. Use 16.7 μl 5× Phusion HF buffer, 1.7 μl dNTPs 10 mM each, 0.8 μl library primers 50 μM each, 58.4 μl water, and 5.0 μl circularized

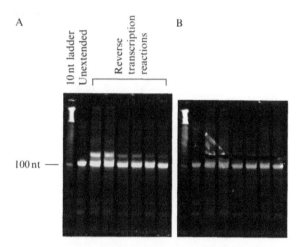

Figure 6.5 Gel purification of reverse transcriptase products. (A) A denaturing 10% polyacrylamide gel was loaded with denatured 10 bp DNA ladder and with the reverse transcriptase primer as well as six lanes containing reverse transcription reactions. The unextended primer band is visible in all reactions at around 100 nt, and the reverse transcription products produce a discrete band at 125–130 nt. (B) The gel depicted in (A), after the reverse transcription products were excised.

DNA, followed by 0.8 μl Phusion polymerase. Set up four PCR tube strips and make one 16.7 μl aliquot of PCR mix into each tube strip. Perform PCR amplification as follows: 30 s at 98 °C; 12 cycles of 10 s at 98 °C, 10 s at 60 °C, and 5 s at 72 °C. Remove one strip tube at the end of the 6th, 8th, and 10th amplification cycle, leaving the last strip tube in the thermal cycler for all 12 cycles. Add 3.4 μl 6× gel loading dye to each reaction.

Prepare an 8% nondenaturing polyacrylamide gel in 1× TBE. Prepare a sample of 1.0 μl 10 bp ladder with 15.7 μl water and 3.4 μl 6× loading dye. Load PCR samples on the gel and run the gel under conditions that optimize separation between 90 and 120 bp fragments. When using a precast polyacrylamide mini-gel (Invitrogen EC6215BOX), 40 min at 180 V separates the sequencing library band from other products. Stain the gel for 3 min in SYBR Gold and then visualize by UV transillumination. A product band should be visible at roughly 120 bp and it should be more intense in samples subjected to more rounds of amplification (Fig. 6.6). There may also be a product band at 90 bp, corresponding to unextended reverse transcription primer. Finally, samples subjected to more rounds of PCR amplification may show additional products much larger than 120 bp. Excise the 120 bp product band from one or two reactions for each sample. Select reactions where amplification has not reached saturation, as judged by increasing intensity of the product band, and where higher molecular weight products are not prominent. Recover DNA from the excised gel slice (see Section 5.4).

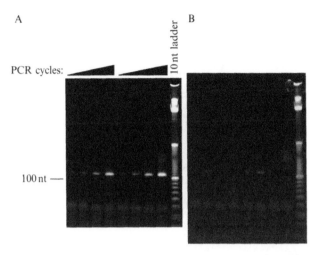

Figure 6.6 Gel purification of PCR-amplified sequencing libraries. (A) A nondenaturing 8% polyacrylamide gel was loaded with sets of PCR reactions from two input templates along with a 10-bp DNA ladder. Each template was amplified in four parallel reactions with increasing numbers of amplification cycles. The prominent 120 nt product is the sequencing library, while the faint 90 nt band represents circularized but unextended reverse transcriptase primer. (B) The gel depicted in (A), after the sequencing library product bands were excised.

Resuspend the gel-purified sequencing library in 20.0 μl 10 mM Tris–Cl, pH 8. Quantify DNA in the library using the Agilent BioAnalyzer DNA 1000 assay. Typically, the recovered DNA will have a concentration of 5–25 nM and there will be no significant peak at 90 bp. This gel-purified library is suitable for sequencing on the Illumina Genome Analyzer.

 ## 4. DATA ANALYSIS

Deep sequencing of ribosome footprints provides a rich data set for studying translation, particularly when it is accompanied by mRNA abundance measurements. The first step in analyzing footprint sequence data is to map the sequencing reads against a reference sequence. Mapping sequencing reads may reveal unanticipated regions of translation, such as uORFs, as well as translation of annotated protein-coding genes. Once the reads have been mapped, expression can be quantified by calculating the density of sequencing reads in a specific gene. The significance of variation between read density in different experimental conditions can be estimated empirically with biological replicates.

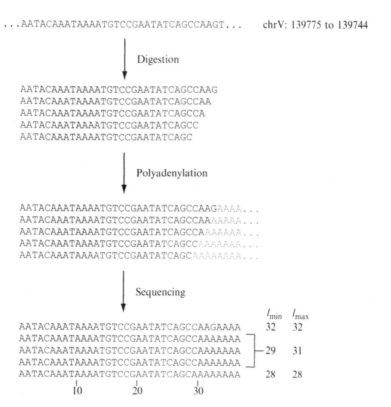

Figure 6.7 Ambiguities in alignments of sequencing reads due to polyadenylation. A sample genomic sequence from the start of the *GCN4* gene is shown along with five possible footprinting products that differ only in the position of their 3′ terminus. Polyadenylation adds poly-(A) tails, shown in gray, but sequencing cannot distinguish which nucleotides occurred in the original RNA fragment and which were added. The middle three sequencing reads are all identical, whereas the 3′ termini of the top and bottom read can be uniquely distinguished. A table of inferred reference alignment lengths l_{min} and l_{max} is given for each sequencing read.

4.1. Mapping polyadenylated sequences

Polyadenylation of footprint fragments introduces difficulties in mapping a sequence to its genomic origin. Ribosome footprint fragments are shorter than the 36 nt sequencing read length provided by the Genome Analyzer, so some poly-A sequence will be present at the end of most sequencing reads. It is not always possible to know how many terminal As in a sequencing read were added during polyadenylation and how many were derived from the RNA fragment (Fig. 6.7). Furthermore, sequencing errors are more likely near the end of the read, making it more difficult to identify the exact extent of the terminal poly-A region. One approach that avoids these difficulties is to align a "seed" from the beginning of each sequencing read to the reference sequence and then to determine the full extent of the alignment. The Bowtie short sequence alignment program is fast, robust, and well-suited to performing these seed alignments (Langmead *et al.*, 2009). Furthermore, it reports alignments to targets in the order in which they appear in the query file, which facilitates postprocessing of seed alignments. Finally, it is capable of reporting multiple alignments as well as reporting on degenerate and unaligned query sequences.

Begin by processing the FastQ format file of sequencing reads to generate a FastQ format file of seed sequences comprising the first 23 nt of the full read. Use Bowtie to align these seed sequences against a reference database consisting of the full yeast genomic sequence. Use the "–v 2" command-line option to allow up to two mismatches in the seed alignment. Use "–m 16" to suppress reporting of seeds with 16 or more matches, which are typically degenerate poly-A sequencing reads that can result from unextended reverse transcription primer, and are uninformative in any case. Use "–unfa" and "–maxfa" to generate files containing the unaligned and degenerate seed sequences.

Process the seed alignments by iterating through the input and alignment files in parallel. When an input sequence is absent from the alignment, skip it and proceed to the next input sequence. When one or more seed alignments are present for an input sequence, extend them find the full reference alignments. Use the coordinates of the seed alignment to extract the reference region corresponding to the full sequencing read. For each possible footprint length l, find the alignment score $s(l)$ by adding the number of mismatches between the sequencing read and the genomic alignment over nucleotides 1 through l and the number of mismatches between the sequencing read and the poly-A linker sequence over nucleotides $l + 1$ through L, the full length of the sequencing read. Determine the best alignment score, $s^\star$, corresponding to the fewest mismatches, and find the set of fragment lengths $\ell^\star$ that give this optimal alignment score. Finally, find the minimum and maximum alignment lengths l_{min} and l_{max}. When the Bowtie results indicate multiple seed alignments, select the best extended reference alignments. In many cases, there will still be multiple, equally high-scoring extended alignments. Next, process the file of unaligned seeds to extract the full-length unaligned sequencing reads.

4.2. Reference databases

A significant fraction of sequencing reads are derived from digested rRNA present in the monosome sample. First, align sequences to a database consisting of just the processed rRNA transcripts (*RDN25-1*, *RDN18-1*, *RDN58-1*, and *RDN5-1*) and proceed with further alignments using only the reads that do not align in the initial rRNA alignment.

Alignments can be performed against the full yeast genomic sequence or against a collection of yeast protein-coding genes. The choice of reference database depends on the purpose of the ribosome profiling experiment. Protein-coding sequences may be the most appropriate reference when the principal goal of an experiment is gene expression measurements. They will include the spliced form of each transcript, so it is not necessary to account further for sequencing reads that overlap splice junctions. In preparing a protein coding reference database, include 15 nt on each end of the coding sequence itself, as the ribosome footprint extends roughly 15 nt to either side of the active codon. Alignments against a full genomic reference sequence are needed to detect novel sites of translation, including uORFs. Footprint sequences that overlap splice junctions substantially will not correspond to any chromosomal sequence in a genomic database. In budding yeast, where splicing is limited and well-characterized, it is possible to correct for the effect of splice junctions on measurements of ribosome occupancy. Another option is to augment the genomic reference with a collection of splice junction sequences.

4.3. Selecting high-quality alignments

Many sequencing reads contain errors, and in some cases these errors compromise the mapping of the RNA fragment to the reference sequence. In other cases, the fragment is atypically short, presumably because of degradation during sample preparation. Noncoding RNA fragments also have a length distribution different than that seen for ribosomal footprints. Yeast ribosome footprints are typically between 27 and 30 nt, and sequencing reads that could not possibly have been derived from an appropriately sized fragment should be excluded. However, the ambiguity in fragment length introduced by polyadenylation makes it impossible to determine the exact length of some fragments. Select sequencing reads that have: (1) two or fewer mismatches total in the alignment; (2) a maximum reference alignment length of at least 27 nt ($l_{max} \geq 27$), excluding alignments that are demonstrably shorter than the minimum footprint length; and (3) similarly, a minimum reference alignment length of no more than 30 nt ($l_{min} \leq 30$).

4.4. Quantifying gene expression

The number of ribosome footprint fragments derived from a gene provides a measurement of that gene's expression level, but two normalization factors are required to make quantitative comparisons. Ribosomes are believed to translate proteins at a similar rate, but this means that a ribosome will spend more time translating a longer gene than a shorter one. Because of this effect, a longer gene will produce more footprint fragments than a shorter gene expressed at the same level. Thus, the read density on a protein-coding gene, expressed in reads per kilobase, provides a directly comparable measurement of expression. Divide the number of sequencing reads for a gene by the size of the gene; if chromosomal alignments are used, correct for any splice junctions that would result in a ~ 28 nt window where reads would not align to the genomic reference sequence. Although 30 additional nucleotides are added to coding sequences when constructing a reference database, do not correct the length for these added nucleotides, as the number of possible 30 nt fragments in the extended sequence is the same as the length of the actual gene.

Samples will differ in the total number of sequencing reads available and the fraction of true footprint fragments, as opposed to rRNA contamination. To account for this difference, divide the fragment density by the total number of true footprint fragments—sequencing reads that align to the reference genome with the appropriate size—to determine reads per kilobase per million (rpkM) (Mortazavi et al., 2008).

It is particularly important to properly characterize expression measurement errors and determine the statistical significance of observed changes when performing genome-wide measurements. Deep sequencing expression measurements are derived from a count of discrete sequencing reads, which introduces errors that follow a well-understood statistical model. When the absolute number of sequencing reads contributing to a measurement is small, statistical sampling error will contribute a large variance to observed expression ratios between two samples. Use a statistical test such as the chi-square test to determine whether the observed distribution of sequencing reads between two samples differs significantly from an expectation derived from the median ratio of reads for well-expressed genes. The statistical variation in expression measurements becomes negligible when the total number of sequencing reads is large; 128 reads is a good threshold for many analyses. Biological variability in samples or in library generation predominates under these circumstances, and the coefficient of variation remains constant with increasing numbers of sequencing reads. Biological replicates prepared independently from matched samples provide the best direct assessment of variability. Determine the distribution of log ratios of sequencing read counts between replicates for a set of genes with similar expression levels. This distribution can be used to assess the likelihood of a measured difference between biologically distinct samples. Typically, the distribution is roughly normal, though as with many biological measurements the tails may contain more highly variable genes than expected. Regardless, it can be used to estimate the fraction of genes whose fragment count ratio would exceed a given threshold by chance. The ratio of the number of genes whose ratio exceeds the threshold in a biological experiment to the number between biological replicates estimates the false discovery rate at that threshold.

Even in the relatively compact yeast genome, there are duplicated and degenerate regions where ribosome footprint sequences cannot be uniquely mapped to a genomic origin. The simplest approach to deal with this problem is to exclude genes with significant nonunique regions, which would affect a relatively small fraction of the yeast genome. A more directed approach would be to exclude only the degenerate regions, ignoring sequencing reads that lie in these areas and correcting the length of the coding sequence for the suppressed region. Finally, it would be possible to quantify read density in degenerate regions and then partition the read density between different chromosomal loci based on the read density in adjoining unique regions (Mortazavi *et al.*, 2008). All of these approaches would fail in the case of perfect or near-perfect paralogues such as the duplicate histone genes in the yeast genome. These duplicate genes should be handled separately, by making a single, combined measurement of their expression and, if necessary, specifically quantifying the presence of signature variations that distinguish between them.

 ## 5. SOLUTIONS AND COMMON PROCEDURES

5.1. Solutions

Alkaline fragmentation solution (2×): 2 mM EDTA, 100 mM Na·CO$_3$, pH 9.2. This solution is prepared by mixing 15 parts 0.1 M Na$_2$CO$_3$ to 110 parts 0.1 M NaHCO$_3$. It will equilibrate with gaseous CO$_2$ to raise the pH over time and thus should be stored in tightly capped, single-use aliquots at room temperature.

Alkaline fragmentation stop/precipitation (540 μl/600 μl): 60 μl 3 M NaOAc (pH 5.5), 2.0 μl GlycoBlue 15 mg/ml (Ambion AM9515), 500 μl RNase-free water.

DNA gel extraction buffer: 300 mM NaCl, 10 mM Tris–Cl (pH 8), 1 mM EDTA.

Polyadenylation enzyme mix: 5.0 μl 2× polyadenylation buffer with ATP, 4.0 μl RNase-free water, 1.0 μl *E. coli* poly-(A) polymerase 5 U/μl (New England Biolabs M0276S). The final concentration of enzyme is 1 U/2 μl.

Polyadenylation buffer with ATP (2×): 2× poly-(A) polymerase buffer, 2 mM ATP, 1 U/μl SUPERase·In. Prepare, for example, 5.0 μl 10× poly-(A) polymerase buffer, 5.0 μl 10 mM ATP, 1.25 μl SUPERase·In, and 13.8 μl RNase-free water.

Polysome gradient buffer: 20 mM Tris–Cl (pH 8), 140 mM KCl, 5.0 mM MgCl$_2$, 100 μg/ml cycloheximide, 0.5 mM DTT, 20 U/ml SUPERase·In.

Polysome lysis buffer: 20 mM Tris–Cl (pH 8), 140 mM KCl, 1.5 mM MgCl₂, 100 μg/ml cycloheximide, 1% (v/v) Triton X-100.

RNA gel extraction buffer: 300 mM NaOAc (pH 5.5), 1 mM EDTA, 0.1 U/μl SUPERase·In.

Total RNA extraction buffer: 50 mM NaOAc (pH 5.5), 10 mM EDTA.

5.2. Oligonucleotides

Reverse transcription primer:

5′-/5Phos/GATCGTCGGACTGTAGAACTCT-
GAACCTGTCGGTGGTCGCCGTATCATT/iSp18/CACTCA/
iSp18/CAAGCAGAAGACGGCATACGATTTTTTTTTTTTTTTT
TTTTVN

This primer is 5′ phosphorylated, to allow circularization, and contains a binding site for the Illumina small RNA sequencing primer as well as one Illumina library primer. The flexible linker consists of a 6-nt spacer sequence flanked by two Spacer 18 linkers, which should provide a complete block to polymerases. Following the linker is the second library primer sequence and a $(dT)_{20}$ primer anchored to the end of the poly-(A) site by degenerate V (A, C, or G) and N (A, C, G, or T) bases.

Library primers:

5′-AATGATACGGCGACCACCGA
5′-CAAGCAGAAGACGGCATACGA

These correspond to the A and B primers present on the Illumina single-read flowcell.

Sequencing primer:

5′-CGACAGGTTCAGAGTTCTACAGTCCGACGATC

This corresponds to the Illumina small RNA sequencing primer.

Control oligonucleotide (RNA):

5′-AUGUACACGGAGUCGACCCGCAACGCGA

This 28 nt RNA oligonucleotide is used as a standard for size selection during library generation. It is an arbitrary sequence which does not resemble any sequence in the yeast genome.

5.3. Nucleic acid precipitation

Perform precipitation of nucleic acids in nonstick tubes, which promote the formation of a tight pellet at the bottom of the tube. Add NaOAc (pH 5.5), for RNA, or NaCl, for DNA, to raise the concentration of Na^+ to 300 mM. Where noted, the protocol includes steps that achieve this salt concentration and no further salt is needed. Add a coprecipitant such as 2.0 μl GlycoBlue whenever precipitating small quantities of nucleic acids. GlycoBlue in particular creates visible pellets but interferes with spectrophotometic quantitation. Add at least 1 volume isopropanol and incubate at least 30 min at − 20 °C. Pellet the sample by spinning 30 min at full speed in a microfuge at 4 °C. Remove as much of the supernatant as possible. Pulse spin the tube to collect the residual liquid at the bottom of the tube, then use a very thin pipette tip such as a 10-μl tip or a gel loading tip to remove as much liquid as possible. Air dry the pellet by leaving the tube open on its side for 5–10 min.

5.4. Nucleic acid gel extraction

Pierce a 0.5-ml microfuge tube with a 20-gauge needle. Nest it inside a 1.5-ml nonstick microfuge tube and label the side of the tube, as the lid may snap off during centrifugation. Place the gel slice in the inner tube and spin the tubes 2 min at full speed in a tabletop microfuge to extrude the gel through the needle-hole into the outer tube. Remove the inner tube, invert it over the outer tube, and tap it to collect any remaining gel debris. Add 400 μl RNA or DNA gel extraction buffer to the gel and incubate overnight at 4 °C with gentle agitation. Cut the tip off of a 1000-μl pipette tip and collect the buffer and gel debris. Load it onto a Spin-X cellulose acetate filter column (Corning 8162) and centrifuge 1 min at full speed in a microfuge. Collect the flow-through and precipitate the recovered nucleic acid (see Section 5.3). Note that both gel extraction buffers already contain 300 mM salt, but add a coprecipitant, as low amounts of nucleic acid are recovered in the gel extractions.

ACKNOWLEDGMENTS

This work was supported by a Ruth L. Kirschstein National Research Service Award from the National Institutes of Health (GM080853) (N. T. I.) and by the Howard Hughes Medical Institute (J. S. W.). We greatly appreciate support and advice from J. Weissman at all stages of this project as well as feedback from G. Brar, S. Churchman, J. Dunn, E. Oh, and S. Rouskin during the optimization of the protocol as well as in preparing the manuscript. S. Ghaemmaghami and J. Newman provided helpful advice based on preliminary ribosome footprinting experiments. We thank D. Bartel, H. Guo, D. Herschlag, J. Hollien, S. Luo, and G. Schroth for valuable discussions of RNA and ribosome methods. We thank P. Walter and T. Aragon for the use of a sucrose density gradient fractionator and C. Chu, J. deRisi, and K. Fischer for help with sequencing.

REFERENCES

Amitsur, M., Levitz, R., and Kaufmann, G. (1987). Bacteriophage T4 anticodon nuclease, polynucleotide kinase and RNA ligase reprocess the host lysine tRNA. *EMBO J.* **6,** 2499–2503.

Arava, Y., Wang, Y., Storey, J. D., Liu, C. L., Brown, P. O., and Herschlag, D. (2003). Genome-wide analysis of mRNA translation profiles in Saccharomyces cerevisiae. *Proc. Natl. Acad. Sci. USA* **100,** 3889–3894.

Arava, Y., Boas, F. E., Brown, P. O., and Herschlag, D. (2005). Dissecting eukaryotic translation and its control by ribosome density mapping. *Nucleic Acids Res.* **33**, 2421–2432.

Ashe, M. P., De Long, S. K., and Sachs, A. B. (2000). Glucose depletion rapidly inhibits translation initiation in yeast. *Mol. Biol. Cell* **11**, 833–848.

Barbet, N. C., Schneider, U., Helliwell, S. B., Stansfield, I., Tuite, M. F., and Hall, M. N. (1996). TOR controls translation initiation and early G1 progression in yeast. *Mol. Biol. Cell* **7**, 25–42.

Bentley, D. R., *et al.* (2008). Accurate whole human genome sequencing using reversible terminator chemistry. *Nature* **456**, 53–59.

Berezikov, E., Cuppen, E., and Plasterk, R. H. A. (2006). Approaches to microRNA discovery. *Nat. Genet.* **38**, S2–S7Suppl.

Brown, P. O., and Botstein, D. (1999). Exploring the new world of the genome with DNA microarrays. *Nat. Genet.* **21**, 33–37.

Cameron, V., and Uhlenbeck, O. C. (1977). 3′-Phosphatase activity in T4 polynucleotide kinase. *Biochemistry* **16**, 5120–5126.

Chartrand, P., Meng, X. H., Huttelmaier, S., Donato, D., and Singer, R. H. (2002). Asymmetric sorting of ash1p in yeast results from inhibition of translation by localization elements in the mRNA. *Mol. Cell* **10**, 1319–1330.

delCardayré, S. B., and Raines, R. T. (1995). The extent to which ribonucleases cleave ribonucleic acid. *Anal. Biochem.* **225**, 176–178.

Feng, Y., and Cohen, S. N. (2000). Unpaired terminal nucleotides and 5′ monophosphorylation govern 3′ polyadenylation by *Escherichia coli* poly(A) polymerase I. *Proc. Natl. Acad. Sci. USA* **97**, 6415–6420.

Fu, H., Tie, Y., Xu, C., Zhang, Z., Zhu, J., Shi, Y., Jiang, H., Sun, Z., and Zheng, X. (2005). Identification of human fetal liver miRNAs by a novel method. *FEBS Lett.* **579**, 3849–3854.

Gilbert, W. V., Zhou, K., Butler, T. K., and Doudna, J. A. (2007). Cap-independent translation is required for starvation-induced differentiation in yeast. *Science* **317**, 1224–1227.

Godchaux, W., Adamson, S. D., and Herbert, E. (1967). Effects of cycloheximide on polyribosome function in reticulocytes. *J. Mol. Biol.* **27**, 57–72.

Gu, W., Deng, Y., Zenklusen, D., and Singer, R. H. (2004). A new yeast PUF family protein, Puf6p, represses ASH1 mRNA translation and is required for its localization. *Genes Dev.* **18**, 1452–1465.

Hinnebusch, A. G. (2005). Translational regulation of GCN4 and the general amino acid control of yeast. *Annu. Rev. Microbiol.* **59**, 407–450.

Ho, C. K., Wang, L. K., Lima, C. D., and Shuman, S. (2004). Structure and mechanism of RNA ligase. *Structure* **12**, 327–339.

Holcik, M., and Sonenberg, N. (2005). Translational control in stress and apoptosis. *Nat. Rev. Mol. Cell Biol.* **6**, 318–327.

Ingolia, N. T., Ghaemmaghami, S., Newman, J. R. S., and Weissman, J. S. (2009). Genome-wide analysis *in vivo* of translation with nucleotide resolution using ribosome profiling. *Science* **324**, 218–223.

Johannes, G., Carter, M. S., Eisen, M. B., Brown, P. O., and Sarnow, P. (1999). Identification of eukaryotic mRNAs that are translated at reduced cap binding complex eIF4F concentrations using a cDNA microarray. *Proc. Natl. Acad. Sci. USA* **96**, 13118–13123.

Köhrer, K., and Domdey, H. (1991). Preparation of high molecular weight RNA. *Methods Enzymol.* **194**, 398–405.

Langmead, B., Trapnell, C., Pop, M., and Salzberg, S. (2009). Ultrafast and memory-efficient alignment of short DNA sequences to the human genome. *Genome Biol.* **10**, R25.

Lau, N. C., Lim, L. P., Weinstein, E. G., and Bartel, D. P. (2001). An abundant class of tiny RNAs with probable regulatory roles in *Caenorhabditis elegans*. *Science* **294,** 858–862.

Markham, R., and Smith, J. D. (1952). The structure of ribonucleic acid. I. Cyclic nucleotides produced by ribonuclease and by alkaline hydrolysis. *Biochem. J.* **52,** 552–557.

McKeehan, W., and Hardesty, B. (1969). The mechanism of cycloheximide inhibition of protein synthesis in rabbit reticulocytes. *Biochem. Biophys. Res. Commun.* **36,** 625–630.

Mortazavi, A., Williams, B. A., McCue, K., Schaeffer, L., and Wold, B. (2008). Mapping and quantifying mammalian transcriptomes by RNA-Seq. *Nat. Methods* **5,** 621–628.

Nagalakshmi, U., Wang, Z., Waern, K., Shou, C., Raha, D., Gerstein, M., and Snyder, M. (2008). The transcriptional landscape of the yeast genome defined by RNA sequencing. *Science* **320,** 1344–1349.

Pfeffer, S., Lagos-Quintana, M., and Tuschl, T. (2005). Cloning of small RNA molecules. *Curr. Protoc. Mol. Biol.* Chapter 26, Unit 26.4, p. 26.4.1–26.4.18.

Preiss, T., Baron-Benhamou, J., Ansorge, W., and Hentze, M. W. (2003). Homodirectional changes in transcriptome composition and mRNA translation induced by rapamycin and heat shock. *Nat. Struct. Biol.* **10,** 1039–1047.

Smirnova, J. B., *et al.* (2005). Global gene expression profiling reveals widespread yet distinctive translational responses to different eukaryotic translation initiation factor 2B-targeting stress pathways. *Mol. Cell. Biol.* **25,** 9340–9349.

Sonenberg, N., and Hinnebusch, A. G. (2009). Regulation of translation initiation in eukaryotes: Mechanisms and biological targets. *Cell* **136,** 731–745.

Steitz, J. A. (1969). Polypeptide chain initiation: Nucleotide sequences of the three ribosomal binding sites in bacteriophage R17 RNA. *Nature* **224,** 957–964.

Wolin, S. L., and Walter, P. (1988). Ribosome pausing and stacking during translation of a eukaryotic mRNA. *EMBO J.* **7,** 3559–3569.

Zong, Q., Schummer, M., Hood, L., and Morris, D. R. (1999). Messenger RNA translation state: The second dimension of high-throughput expression screening. *Proc. Natl. Acad. Sci. USA* **96,** 10632–10636.

SYSTEMATIC GENETIC ANALYSIS

CHAPTER SEVEN

Synthetic Genetic Array (SGA) Analysis in *Saccharomyces cerevisiae* and *Schizosaccharomyces pombe*

Anastasia Baryshnikova,* Michael Costanzo,* Scott Dixon,[†] Franco J. Vizeacoumar,* Chad L. Myers,[‡] Brenda Andrews,* *and* Charles Boone*

Contents

* Banting and Best Department of Medical Research and Department of Molecular Genetics, Terrence Donnelly Center for Cellular and Biomolecular Research, University of Toronto, Toronto, Ontario, Canada
[†] Department of Biological Sciences, Columbia University, New York, New York, USA
[‡] Department of Computer Science and Engineering, University of Minnesota-Twin Cities, Minneapolis, Minnesota, USA

Methods in Enzymology, Volume 470   © 2010 Elsevier Inc.
ISSN 0076-6879, DOI: 10.1016/S0076-6879(10)70007-0   All rights reserved.

Abstract

A genetic interaction occurs when the combination of two mutations leads to an unexpected phenotype. Screens for synthetic genetic interactions have been used extensively to identify genes whose products are functionally related. In particular, synthetic lethal genetic interactions often identify genes that buffer one another or impinge on the same essential pathway. For the yeast *Saccharomyces cerevisiae*, we developed a method termed synthetic genetic array (SGA) analysis, which offers an efficient approach for the systematic construction of double mutants and enables a global analysis of synthetic genetic interactions. In a typical SGA screen, a query mutation is crossed to an ordered array of $\sim$5000 viable gene deletion mutants (representing $\sim$80% of all yeast genes) such that meiotic progeny harboring both mutations can be scored for fitness defects. This approach can be extended to all $\sim$6000 genes through the use of yeast arrays containing mutants carrying conditional or hypomorphic alleles of essential genes. Estimating the fitness for the two single mutants and their corresponding double mutant enables a quantitative measurement of genetic interactions, distinguishing negative (synthetic lethal) and positive (within pathway and suppression) interactions. The profile of genetic interactions represents a rich phenotypic signature for each gene and clustering genetic interaction profiles group genes into functionally relevant pathways and complexes. This array-based approach automates yeast genetic analysis in general and can be easily adapted for a number of different genetic screens or combined with high-content screening systems to quantify the activity of specific reporters in genome-wide sets of single or more complex multiple mutant backgrounds. Comparison of genetic and chemical-genetic interaction profiles offers the potential to link bioactive compounds to their targets. Finally, we also developed an SGA system for the fission yeast *Schizosaccharomyces pombe*, providing another model system for comparative analysis of genetic networks and testing the conservation of genetic networks over millions of years of evolution.

1. INTRODUCTION

A genetic interaction refers to an unexpected phenotype not easily explained by combining the effects of individual genetic variants (Bateson *et al.*, 1905). Importantly, genetic interactions organize into complex networks that may underlie the relationship between an organism's genotype and its phenotype (Waddington, 1957). Thus, an unbiased, systematic analysis of these networks and the interactions that comprise them is required in order to understand the genetic basis underlying disease. Given the complexity of the human genome (Levy *et al.*, 2007), determining how different alleles and polymorphisms combine to manifest a phenotype is a daunting task. Researchers have, therefore, embraced inbred model systems as well as isogenic populations of cultured cells derived from fruit flies and mammals, as platforms to map genetic interactions in a systematic, unbiased, and comprehensive fashion (Dixon *et al.*, 2009).

Large-scale genetic interaction mapping studies were pioneered in *Saccharomyces cerevisae* and focused on the identification of a specific type of interaction termed synthetic lethality (Tong *et al.*, 2001, 2004). Synthetic lethal or sick interactions, in which a combination of mutations in two genes results in cell death or reduced fitness, respectively, has been used extensively in different model organisms to identify genes whose products buffer one another and impinge on the same essential biological process (Fay *et al.*, 2002; Hartman *et al.*, 2001; Lucchesi, 1968). Genome-wide synthetic lethal analysis in yeast revealed the importance of systematic genetic interaction maps for assessing the biological roles of genes *in vivo* and uncovering new components of specific pathways (Pan *et al.*, 2006; Tong *et al.*, 2004). Recently, genetic mapping technologies combined with quantitative phenotypic analyses have enabled further dissection of genetic interactions into different types and classes offering the potential to define protein complex membership and infer order of gene function within specific biochemical pathways (Collins *et al.*, 2006; St Onge *et al.*, 2007).

The first enabling reagent set for large-scale genetic interaction screens in budding yeast was derived from the deletion mutant project, in which each known or suspected open reading frame is deleted and replaced with the dominant drug-resistance marker, *kanMX* (Giaever *et al.*, 2002; Winzeler *et al.*, 1999). The international consortium responsible for this landmark analysis identified ~ 1000 essential genes and constructed ~ 5000 viable haploid deletion mutants. The introduction of molecular tags or barcodes, a unique 20-bp DNA sequence at either end of the deletion cassette, acts as a unique mutant strain identifier enabling the fitness of a particular mutant to be assessed within a population using a barcode microarray (Giaever *et al.*, 1999). Additional libraries have subsequently been developed in which each of the ~ 1000 essential genes are altered in such a way as to produce either conditional alleles (Ben-Aroya *et al.*, 2008; Mnaimneh *et al.*, 2004) or hypomorphic (partially functional) alleles compatible with viability (Schuldiner *et al.*, 2005). Combining the viable deletion mutant and essential gene mutant collections provides the first opportunity for systematic genetic analysis in yeast and the potential for examining the complete genome-wide set of ~ 18 million different double mutants for synthetic genetic interactions.

Synthetic genetic array (SGA) analysis enables the systematic construction of double mutants (Tong *et al.*, 2001), allowing large-scale mapping of synthetic genetic interactions (Costanzo *et al.*, submitted for publication; Tong *et al.*, 2004). A typical SGA screen involves crossing a "query" strain to the array of ~ 5000 viable deletion mutants, but the array may also include essential gene mutants, and through a series of replica-pinning procedures, the double mutants are selected and scored for growth.

Applying SGA analysis on a genome-wide scale to ~1700 query mutations has enabled us to generate a genetic interaction network containing ~170,000 interactions, with functional information associated with the position and connectivity of a gene on the network (Costanzo *et al.*, submitted for publication).

The SGA methodology is versatile because any genetic element (or any number of genetic elements) linked to a selectable marker(s) can be manipulated similarly. In this regard, SGA methodology automates yeast genetics generally, such that specific alleles of genes, including point mutants and temperature-sensitive alleles, or plasmids can be crossed into any ordered array of strains providing systematic approaches to genetic suppression analysis, dosage lethality, dosage suppression or plasmid or reporter shuffling. In this chapter, we describe the steps of SGA analysis in detail and we hope to encourage laboratories from a broad spectrum of fields to adopt this methodology to suit their specific interests.

2. METHODOLOGY

SGA analysis first requires a relatively simple set up, generating the query strains and the array strains, and then the procedure itself basically involves several replica-plating steps, which are amenable to either manual or robotic manipulation. Each step of the procedure is described in detail below.

- SGA query strain construction
- Pin tool sterilization procedures
- Constructing a 1536-density deletion mutant array (DMA)
- SGA procedure
- Double mutant array image acquisition and processing
- Quantitative scoring of genetic interactions using colony size-based fitness measurements
- Interpretation and analysis of genetic interactions

2.1. SGA query strain construction

As mentioned above, SGA enables high-throughput construction and isolation of haploid double mutants by mating a "query" strain of interest harboring SGA reporters to an ordered array of mutant strains. This section describes different approaches for constructing SGA query strains.

2.1.1. Nonessential query strain: PCR-mediated gene deletion

1. Synthesize two gene-deletion primers, each containing 55 bp of sequence at the 5′ end that is specific to the region upstream or downstream of the gene of interest (*Gene X*), excluding the start and stop codons and 22 bp of sequence at the 3′ end that is specific for the amplification of the *natMX4* cassette (Goldstein and McCusker, 1999). The *MX4* cassette amplification sequences include the forward amplification primer (5′-ACATG-GAGGCCCAGAATACCCT-3′) and the reverse amplification primer (5′-CAGTATAGCGACCAGCATTCAC-3′).

2. Set up a 100 μl PCR reaction (68.7 μl H$_2$0, 10 μl 10× PCR buffer, 2 μl 10 mM dNTPs, 2 μl 50 μM forward primer, 2 μl 50 μM reverse primer, ∼0.1 μg p4339 DNA template in 10 μl, 5 μl DMSO, 0.3 μl 5 U/μl Taq polymerase) to amplify the *natMX4* cassette flanked with 55-bp target sequences from p4339 (pCRII-TOPO::*natMX4*) with the gene-deletion primers designed in step 1. Plasmid p4339 serves as a DNA template to amplify the *natMX4* cassette.

3. Cycle as follows in a thermocycler with a heated lid: 95 °C 5 min, 95 °C 30 s, 55 °C 30 s, 68 °C 2 min, repeat 30 times, 68 °C 10 min, hold at 4 °C. PCR products can be stored at −20 °C.

4. Transform the PCR product into the SGA starting strain, Y7092 (*MATα can1Δ::STE2pr-Sp_his5 lyp1Δ ura3Δ0 leu2Δ0 his3Δ1 met15Δ0*) using standard procedures (Winzeler *et al.*, 1999). Y7092 harbors reporters and markers necessary for SGA haploid strain selection following meiotic recombination. In particular, the *MAT*a-specific reporter [*STE2pr-Sp_his5*, composed of the *Saccharomyces cerevisiae STE2* promoter driving the *Schizosaccharomyces pombe his5* gene, which complements *S. cerevisiae his3Δ1* (Tong and Boone, 2006)] was integrated at the *CAN1* locus. Loss of *CAN1* confers canavanine resistance. Y7092 also carries a *lyp1* marker, which confers resistance to thialysine. The *can1* and *lyp1* mutations serve as recessive counterselectable markers designed to remove unwanted heterozygous diploids from the population. Select transformants on YEPD + clonNAT medium (see Section 3).

2.1.2. Nonessential query strain: Gene deletion marker switch method

1. Obtain the deletion strain of interest (*xxxΔ::kanMX4*) from the *MAT*a deletion collection (OpenBiosystems), mate with Y8205 (*MATα can1Δ::STE2pr-Sp_his5 lyp1Δ::STE3pr-LEU2 ura3Δ0 leu2Δ0 his3Δ1*) and isolate diploid zygotes by micromanipulation. Transform the resulting diploid with *Eco*RI-digested p4339 (see above) using standard yeast transformation protocols (Winzeler *et al.*, 1999). This procedure switches the gene deletion marker from *kanMX4* to *natMX4*. Select transformants on YEPD + clonNAT medium (see Section 3).

2. Transfer the resultant diploids to enriched sporulation medium (see Section 3) and incubate at 22 °C for 5 days.
3. A *MATα*-specific reporter (*STE3pr-LEU2*) was integrated at the *LYP1* locus in Y8205 to provide a convenient method for selecting *MATα* meiotic progeny. Resuspend a small amount of spores in sterile water and grow on SD − Leu/Lys/Arg + canavanine/thialysine (see Section 3) to select *MATα* meiotic progeny. Incubate at 30 °C for ∼2 days. To facilitate the selection of *MATα* meiotic progeny we aim to plate ∼200–300 colonies.
4. Replica plate to YEPD + clonNAT (see Section 3) to identify the *MATα* meiotic progeny that carry the query deletion marked with *natMX4* (*xxxΔ::natMX4*).

2.2. Pin tool sterilization procedures

The SGA procedure involves sequential transfer of yeast colonies onto different selective media in order to isolate haploid double mutant strains. Yeast colony transfer is accomplished manually using hand-held pin tools or automatically using robotically controlled pin tools. The following section describes pin tool sterilization procedures for both manual and robotic devices.

2.2.1. Manual pin tools (for low throughput: ∼1–2 genome-wide screens/month)

The following manual pin tools can be purchased from V&P Scientific, Inc. (San Diego, CA): 96 floating pin E-clip style manual replicator (VP408FH), 384 floating pin E-clip style manual replicator (VP384F), Registration accessories: Library Copier™ (VP381), Colony Copier™ (VP380), Pin cleaning accessories: plastic bleach or water reservoirs (VP421), pyrex alcohol reservoir with lid (VP420), pin cleaning brush (VP425).

1. Set up the wash reservoirs as follows: three trays of sterile water of increasing volume—30, 50, and 70 ml, one tray of 40 ml of 10% bleach, one tray of 90 ml of 95% ethanol. To ensure that pins are cleaned properly and to avoid contamination in the wash procedure, the volume of wash liquids in the cleaning reservoirs is calculated to cover the pins sequentially in small increments. For example, only the tips of the pins should be submerged in water in the first step. As the pins are transferred to subsequent cleaning reservoirs and the final ethanol step, the lower halves of the pins should be covered.
2. Let the replicator sit in the 30 ml-water reservoir for ∼1 min to remove the cells on the pins.
3. Place the replicator in 10% bleach for ∼20 s.

4. Transfer the replicator to the 50 ml–water reservoir and then to the 70 ml–water reservoir to rinse the bleach off the pins.
5. Transfer the replicator to 95% ethanol.
6. Let excess ethanol drip off the pins, then flame.
7. Allow replicator to cool.
8. The manual replicator tools are compatible with OmniTray (Nunc) petri dishes. We found that ~35 ml of media in OmniTrays yield optimal results.

2.2.2. Singer RoToR bench top robot (for high-throughput SGA analysis)

The Singer RoToR can be purchased from Singer Instruments (Somerset, UK; www.singerinst.co.uk). This system uses disposable plastic replicators (RePads, Singer Instruments, UK) and thus does not require any sterilization procedures, making it simple and rapid to use. The Singer RoToR HDA bench top robot also uses PlusPlate (Singer Instruments) petri dishes that have a larger surface area but the same external footprint dimensions as OmniTray (Nunc). These larger plates facilitate replica-plating with the disposable RePads. We found that ~50 ml of media in PlusPlates yield optimal results.

2.2.3. BioMatrix robot (for ultrahigh-throughput SGA analysis)

The BioMatrix Colony Arrayer Robot can be purchased from S&P Robotics, Inc. (Toronto, ON; www.sprobotics.com). Use the following procedure to clean and sterilize the replicator pins prior to use of the robot:

1. Fill the sonicator bath with 390 ml sterile distilled water.
2. Clean the replicator pins in the sonicator for 5 min.
3. Remove the water and fill the sonicator bath with 390 ml 70% ethanol.
4. Sterilize the replicator in the sonicator for 20 s per cycle, repeat the cycle twice.
5. Let the replicator sit in a tray of 100 ml of 95% ethanol for 5 s.
6. Allow the replicator to dry over the fan for 20 s.

Use the following procedure to sterilize the pins at the end of each replica pinning step:

1. Set up the wash reservoirs as follows: Program water bath to automatically fill with sterile distilled water from bottle supply source, manually fill brush station with 320 ml sterile distilled water, fill sonicator with 390 ml 70% ethanol and basin with 100 ml 95% ethanol.
2. Let the replicator sit in the water bath for 10 s per cycle, repeat the cycle for four additional times to remove residual cells from replicator pins.
3. Clean replicator pins further at the brush station for three cycles.

4. Sterilize the replicator in the 70% ethanol-sonicator bath for 20 s per cycle, repeat twice.
5. Let the replicator sit in the 95% ethanol reservoir for 5 s.
6. Allow the replicator to dry over the fan for 20 s.
7. The BioMatrix robot can be used in conjunction with OmniTray (Nunc) petri dishes for the replica pinning steps involved in SGA analysis.

2.3. Constructing a 1536-density DMA

The collection of *MAT*a deletion strains is available in stamped 96-well agar or frozen stocks in 96-well plates from various sources (Invitrogen, American Type Culture Collection, EUROSCARF, and Open Biosystems). However, deletion strains are arrayed at a higher density for SGA analysis. Specifically, each SGA array plate consists of 384 mutant strains arrayed in quadruplicate resulting in an array density of 1536 yeast colonies/plate. The following section describes how to assemble high-density SGA arrays from low-density (96 colonies/plate) source arrays available from different suppliers. The procedure described here employs a BioMatrix Colony Arrayer Robot but high-density DMAs can also be assembled using manual pin tools or the Singer RoToR instrument.

1. Peel off the foil coverings slowly on the frozen 96-well microtiter plates.
2. Let the plates thaw completely on a flat surface.
3. Mix the glycerol stocks gently by stirring with a 96-pin replicator.
4. Replicate the glycerol stocks from the 96-well plates onto YEPD + G418 agar plates using the Library Copier^{TM} with the pair of one-alignment holes on the front frame. Take extreme caution that the pins do not drip liquid into neighboring wells.
5. Reseal the 96-well plates with fresh aluminum sealing tape and return to $-80\,^{\circ}C$.
6. Let cells grow at room temperature for $\sim$2 days.
7. Condense four plates of 96-format into one plate of 384-format using the BioMatrix Colony Arrayer Robot 96-pin replicator and the accompanying BioMatrix replicator software.
8. Incubate 384-density arrays at room temperature for $\sim$2 days.
9. Replicate each 384-density DMA in quadruplicate onto a single plate using the BioMatrix robot 384-pin replicator and accompanying BioMatrix replicator software. This will generate a DMA consisting of 384 mutant strains and 1536 yeast colonies.
10. Incubate at room temperature for $\sim$2 days, to generate a 1536-density DMA working copy.

2.4. SGA procedure

The following section provides a detailed description of the six steps involved in SGA-mediated double mutant isolation. The media required for various selections are described in detail in Section 3. A schematic illustrating the SGA procedure is shown in Fig. 7.1.

2.4.1. Query strain and DMA

1. Grow the query strain in a 5 ml YEPD overnight culture.
2. Pour the query strain culture onto a YEPD plate, use the replicator to transfer the liquid culture onto two fresh YEPD plates, generating a source of newly grown query cells for mating to the DMA in the density of 1536 colonies. Pinning the query strain in a 1536-format on an agar plate is advantageous as cells are evenly transferred to subsequent mating steps. One query plate should contain a sufficient amount of cells for mating with eight plates of the DMA. Allow cells to grow at 30 °C for 2 days.
3. Replicate the DMA to fresh YEPD + G418 media. Allow cells to grow at 30 °C for 1 day. The DMA can be reused for three to four rounds of mating reactions.

2.4.2. Mating the query strain with the DMA

1. Pin the 1536-format query strain onto a fresh YEPD plate.
2. Pin the DMA on top of the query cells.
3. Incubate the mating plates at room temperature for 1 day.

2.4.3. *MAT*a/α diploid selection and sporulation

1. Pin the resulting *MAT*a/α zygotes onto YEPD + G418/clonNAT plates.
2. Incubate the diploid selection plates at 30 °C for 2 days.
3. Pin diploid cells onto enriched sporulation medium.
4. Incubate the sporulation plates at 22 °C for 5 days.

2.4.4. *MAT*a meiotic progeny selection

1. Pin spores onto SD − His/Arg/Lys + canavanine/thialysine plates.
2. Incubate the haploid selection plates at 30 °C for 2 days.

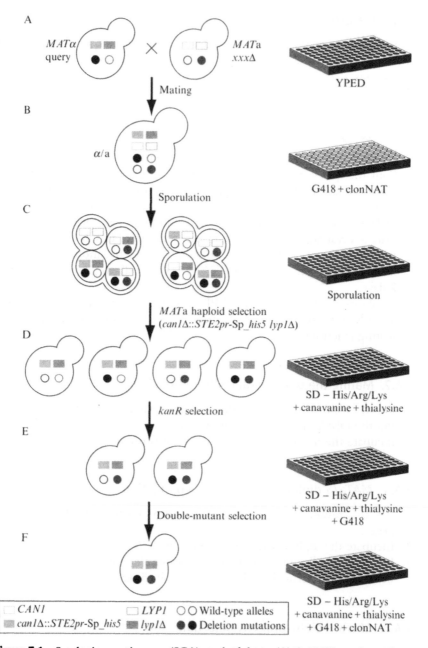

Figure 7.1 Synthetic genetic array (SGA) methodology. (A) A *MAT*α strain carries a query mutation linked to a dominant selectable marker (filled black circle), such as the nourseothricin-resistance marker, *natMX4*, and the SGA reporter, *can1Δ::STE2pr-*Sp_*his5* (in which *STE2pr-*Sp_*his5* is integrated into the genome such that it deletes the open reading frame (ORF) of the *CAN1* gene, which normally confers sensitivity

2.4.5. *MATa-kanMX4* meiotic progeny selection

1. Pin the *MATa* meiotic progeny onto (SD/MSG) — His/Arg/Lys + canavanine/thialysine/G418 plates.
2. Incubate the *kanMX4*-selection plates at 30 °C for 2 days.

2.4.6. *MATa-kanMX4-natMX4* meiotic progeny selection

1. Pin the *MATa* meiotic progeny onto (SD/MSG) — His/Arg/Lys + canavanine/thialysine/G418/clonNAT plates.
2. Incubate the *kanMX4/natMX4* selection plates at 30 °C for 1–2 days.
3. Image double mutant array plates and score for fitness defect. The barcode microarrays can be used as an alternative method to score double mutants for fitness defects. Since each of the deletion mutants is tagged with two unique oligonucleotide barcodes, their growth rates can also be monitored within a population of cells (Decourty *et al.*, 2008; Pan *et al.*, 2004).

2.5. Double mutant array image acquisition and processing

Initial SGA screens focused on detecting severe synthetic sick or synthetic lethal interactions via visual inspection of double mutant colonies and comparison to wild-type controls (Tong *et al.*, 2001, 2004). However, quantitative measurement of genetic interactions offers the potential for constructing high-resolution genetic networks (Collins *et al.*, 2006; St Onge *et al.*, 2007). We developed an automated computational pipeline for acquiring and processing yeast colony data to extract precise fitness and genetic interaction measurements from yeast double mutant colony size (Fig. 7.2) (Baryshnikova *et al.*, manuscript in preparation; Tong *et al.*, 2004).

---

to canavanine). The query strain also lacks the *LYP1* gene. Deletion of *LYP1* confers resistance to thialysine. This query strain is crossed to an ordered array of *MATa* deletion mutants (*xxxΔ*). In each of these deletion strains, a single gene is disrupted by the insertion of a dominant selectable marker, such as the kanamycin-resistance (*kanMX4*) module (the disrupted gene is represented as a filled red circle). (B) The resulting heterozygous diploids are transferred to a medium with reduced carbon and nitrogen to induce sporulation and form haploid meiotic spore progeny. (C) Spores are transferred to a synthetic medium that lacks histidine, which allows selective germination of *MATa* meiotic progeny owing to the expression of the SGA reporter, *can1Δ:: STE2pr-Sp_his5*. To improve this selection, canavanine and thialysine, which select *can1Δ* and *lyp1Δ* while killing *CAN1* and *LYP1* cells, respectively, are included in the selection medium. (D) The *MATa* meiotic progeny are transferred to a medium that contains kanamycin which selects single mutants equivalent to the original array mutants and double mutants. (E, F) An array of double mutants is selected on a medium that contains both nourseothricin and kanamycin. (See Color Insert.)

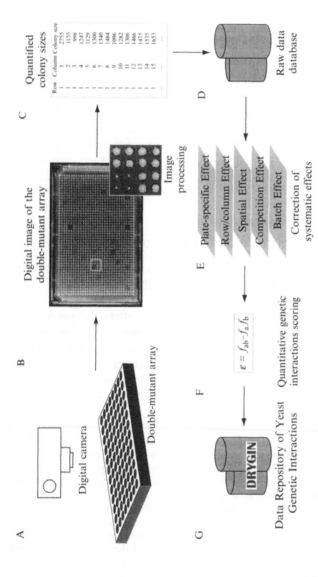

Figure 7.2 Computational pipeline for processing SGA data. (A) Double mutant array plates are photographed by a high-resolution digital camera. (B, C) Digital images of the double mutant array plates are processed by a custom-developed image processing software that identifies the colonies and measures their areas in terms of pixels. (D) Quantified double mutant colony sizes are stored in the database for further manipulation and analysis. (E) To identify quantitative genetic interactions, the yeast colony data is retrieved from the database and a series of normalizations are applied to correct for numerous systematic experimental effects. (F) Genetic interactions are measured by combining the corrected double mutant fitness and the fitnesses of the two single mutants. (G) Genetic interaction data is made available via the DRYGIN web database system.

Double mutant array plates are photographed in a light controlled environment using a high-resolution digital imaging system developed by S&P Robotics, Inc. (Toronto, ON). Digital images are processed using custom-developed image-processing software that measures colony area in terms of pixels (Tong *et al.*, 2004). Colony sizes are then stored in a PostgreSQL database for further manipulation and analysis. Thorough quality control procedures are applied to the data to ensure correct plate identity, screen quality, and proper image processing.

2.6. Quantitative scoring of genetic interactions using colony size-based fitness measurements

Quantitative genetic interactions define double mutant combinations that deviate from an expected phenotype (Bateson *et al.*, 1905). Mutations in independent genes often combine in a multiplicative manner and the resulting double mutant phenotype should be equivalent to the product of the two individual mutations (Mani *et al.*, 2008). The extent of the genetic interaction is consequently measured as $\varepsilon = f_{ab} - f_a \cdot f_b$, where f_a, f_b, and f_{ab} are quantitative fitness measures of the two single and the double mutant, respectively (Elena and Lenski, 1997). Negative genetic interactions ($\varepsilon < 0$) refer to double mutants showing a more severe fitness defect than expected, with the extreme case being synthetic lethality. Positive genetic interactions ($\varepsilon > 0$) refer to double mutants with a less severe fitness defect than expected and include interactions such as epistasis and suppression (Collins *et al.*, 2006; Mani *et al.*, 2008; Segre *et al.*, 2005).

To measure positive and negative genetic interactions quantitatively, precise estimates of both single and double mutant fitness are required. One method for measuring genetic interactions from yeast colony data derives single mutant fitness estimates from the average of all double mutants carrying the same query or array mutation (Collins *et al.*, 2006). Double mutant colony sizes are subsequently normalized by single mutant fitness and experimental variance estimates resulting in a single quantitative measure, termed S-score, which reflects both the strength and confidence of the genetic interaction (Collins *et al.*, 2006).

Although quantitative, the S-score does not represent a true fitness-based measure of genetic interaction. Furthermore, this approach does not account for several experimental effects associated with SGA technology that introduce strong systematic biases in large-scale genetic interaction datasets and are particularly pronounced for high-density arrays. These experimental artifacts adversely affect genetic interaction measurements resulting in increased false-positive rates and reduced sensitivity. Thus, we developed a novel method to process raw yeast colony sizes and derive true measures of fitness and genetic interaction (Baryshnikova *et al.*, manuscript in preparation).

To do so, we first identified several systematic experimental factors that contribute the vast majority of the observed variance in colony size and seriously interfered with our ability to detect and measure genetic interactions. Appropriate normalization of these experimental effects is critical to ensure accurate and reliable genetic interaction measurements. The systematic effects and normalization methods are summarized below and described in detail elsewhere (Baryshnikova *et al.*, manuscript in preparation).

Systematic effects and normalization procedures:

1. *Plate-specific effect correction*: normalizes differences in growth between plates due to varying incubation times and query single mutant fitness.
2. *Row/column effect correction*: normalizes differences in colony growth caused by differential exposure to nutrients due to plate position.
3. *Spatial effect correction*: normalizes differences in colony growth due to gradients in media thickness and/or relative proximity to heat sources.
4. *Competition effect correction*: normalizes differences in colony growth due to reduced fitness of neighboring mutant strains that results in reduced competition for nutrients.
5. *Batch effect normalization*: corrects for the striking similarity between genome-wide SGA screens conducted in parallel. We find that the set of screens completed by the same person, on the same robot and at the same time tend to share a common nonbiological signature. In other words, these screens share a set of unusually small or large colonies that could be confused with true genetic interactions, unless considered in the context of other screens performed in the same period of time.

To estimate double mutant fitness from colony size measurements we developed an approach that models colony size as a multiplicative combination of double mutant fitness, systematic experimental factors, and random noise. Specifically, for a double mutant deleted for genes a and b, we modeled colony size as $C_{ab} = f_{ab} \cdot s_{ab} \cdot e$, where C_{ab} is the colony area, f_{ab} is the double mutant fitness, s_{ab} is the combination of all systematic factors, and e is log-normally distributed error. In addition to double mutant fitness, we also obtained accurate single mutant fitness estimates (f_a and f_b) by averaging colony sizes for a given mutant across numerous control experiments using different arrays consisting of *kanMX4*- and *natMX4*-marked deletion mutants whose array positions have been randomized. Genetic interactions are subsequently measured by assuming the multiplicative model for independent genes, that is, $f_{ab} = f_a \cdot f_b + \varepsilon_{ab}$, where ε_{ab} represents the genetic interaction term between genes a and b (Baryshnikova *et al.*, manuscript in preparation).

Along with correcting for systematic effects and estimating the genetic interaction factor, we derived an accurate estimate of variance for the interaction. Our variance estimate is reported as a *p*-value that reflects both the local variability of replicate colonies (based on 4 colonies/plate for each double mutant) and the variability of double mutants sharing the same query or array mutation.

Accounting for systematic effects and estimating genetic interactions and confidence levels separately provide a boost in data quality and capacity for predicting gene function from the resulting genetic interactions (Fig. 7.3). For example, functionally related genes are known to be enriched for genetic interactions and to share similar genetic interaction patterns. Using Gene Ontology coannotation as a measure of functional relatedness, genetic interactions measured using our newly developed score (SGA score) can predict 100 functionally related gene pairs with 62% precision(Fig. 7.3B, i). Failing to account for systematic experimental effects results in a significant reduction in precision (Fig. 7.3B, i). Analogously, similarity of genetic interaction profiles, as measured by Pearson correlation coefficients, computed using SGA scores predicts 100 functionally related gene pairs with 90% precision compared to only 15% precision when normalization procedures are not applied (Fig. 7.3B, ii).

Access to genetic interaction scores and confidence values is provided via a web database system (DRYGIN) which facilitates retrieval of interactions and their analysis (Koh *et al.*, 2010).

2.7. Interpretation and analysis of genetic interactions

Existing genetic interaction datasets in yeast have provided significant insight into the general principles of genetic network connectivity. For example, genes with related biological functions are connected by genetic interactions more often than expected by chance (Tong *et al.*, 2004) (Fig. 7.4). The position and the connectivity of the gene on the genetic interaction network is predictive of function. Moreover, because the genetic network is a small world network (Tong *et al.*, 2004), genes within the same neighborhood of the network tend to interact with one another and thus a sparsely mapped network is predictive of genetic interactions (Fig. 7.4B). Furthermore, synthetic lethal (negative) genetic interactions among nonessential genes generally do not correspond to physical interactions between the corresponding gene products. It has been observed that negative genetic interactions are more frequent between genes lying in different pathways, whereas physical interactions are more frequent among gene products functioning within the same pathway (Bader *et al.*, 2004; Collins *et al.*, 2007; Kelley and Ideker, 2005; Tong *et al.*, 2004; Ye *et al.*, 2005) (Fig. 7.4C). However, when a pathway or complex contain at least one essential gene, they are often enriched for so-called within-pathway synthetic lethal interactions, indicating that a subset of the negative genetic interactions for essential genes overlap with protein–protein interactions (Bandyopadhyay *et al.*, 2008; Boone *et al.*, 2007). Positive genetic interactions can connect members of the same protein complex or pathway. These positive within-pathway interactions may reflect that both the single and

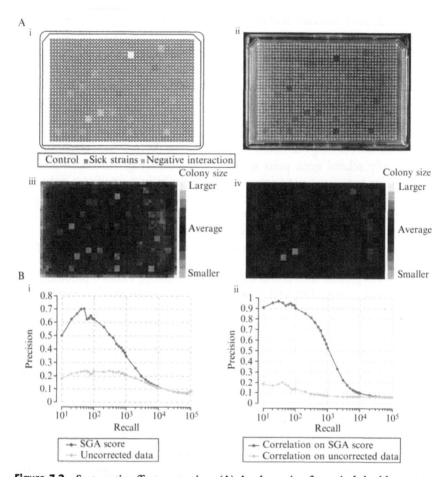

Figure 7.3 Systematic effect correction. (A) A schematic of a typical double mutant array plate shows the systematic biases affecting colony sizes. (i, ii) A typical double mutant array plate contains control spots (gray circles), strains with low fitness (blue circles), negative interactions (red circles), and positive interactions (not shown). On visual inspection, all three cases appear as small colonies or empty spots. (iii) Quantification of colony areas shows a distinctive spatial pattern affecting opposite sides of the plate (bigger colonies on the right, smaller colonies on the left) that was not obvious on visual inspection. Failure to correct for this spatial pattern will result in false-positive interactions. (iv) Corrects spatial patterns, eliminates false positives, and highlights true genetic interactions. (B) Precision–recall curves on genetic interaction scores (i) and genetic profile similarity (ii) show the increased functional prediction capacity of genetic data after correcting for systematic biases. A set of 1712 genome-wide SGA screens (Costanzo *et al.*, in press) were processed using the SGA score (Baryshnikova *et al.*, manuscript in preparation) and a version of the SGA score without systematic effect correction. Both direct genetic interactions and genetic profile similarities, as measured by Pearson correlations, were assessed for function by calculating precision and recall of functionally related gene pairs as described in the study of Myers *et al.* (2006). As a measure of functional relatedness, we used coannotation to the same Gene Ontology term. (See Color Insert.)

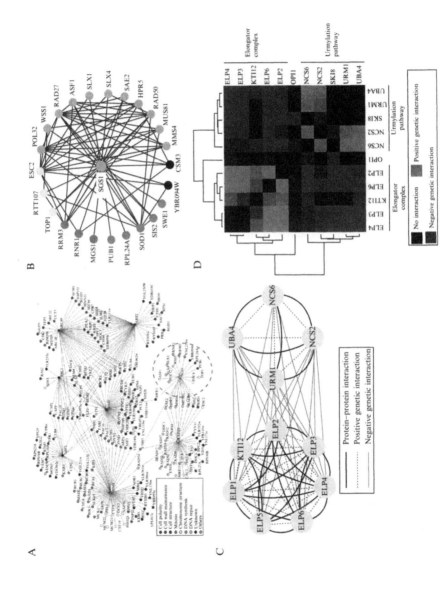

Figure 7.4 (*Continued*)

double mutants of nonessential linear pathways have the same fitness defect and therefore do not show the expected fitness defect associated with the multiplicative model. However, a more recent analysis of the global yeast physical interaction network as defined by affinity purification-mass spectrometry, yeast two-hybrid protocol, or protein-fragment complementation assay (PCA), showed that roughly an equivalent number of physical interactions overlap with negative and positive genetic interaction pairs: ~7% of protein-protein interacting pairs shared a negative genetic interaction, whereas ~5% shared a positive interaction (Costanzo *et al.*, in press). Conversely, only a small fraction of gene pairs that show a genetic interaction (0.4% negative and 0.5% positive) are also physically linked (Costanzo *et al.*, in press). These findings therefore suggest that the vast majority of both positive and negative interactions occurs between, rather than within, complexes and pathways, connecting those that presumably work together or buffer one another, respectively.

The synthetic lethal or negative genetic interaction profile for a particular query gene provides a rich phenotypic signature reflecting the function of the query as it contains genes involved in pathways that buffer the query. On average, negative interactions are approximately twofold more prevalent than positive interactions (Costanzo *et al.*, in press). However, we found subsets of genes that showed a strong bias in interaction type (up to 8–16-fold more negative than positive interactions or vice versa) (Costanzo *et al.*, in press). While the genetic interaction profile composed of negative genetic interactions carries more functional information than that composed only of positive interactions, the profile composed of both positive and negative interactions is most informative about gene function (Fig. 7.4). Clustering of genes according to their genetic interaction profiles is a simple yet very powerful tool for gene function prediction via a "guilt-by-association" approach (Figs. 7.4– 7.5). Open source clustering software, such as Cluster 3.0 (http://bonsai.ims. u-tokyo.ac.jp/~mdehoon/software/cluster/software.htm), reassort rows

---

Figure 7.4   Properties of genetic interactions. (A) Example of a yeast synthetic lethal network. The synthetic lethal network is a sparse network, indicating that genetic interactions are rare. The frequency of true synthetic lethal interactions (blue lines) is less than 1%. A detailed description of how this initial network was generated can be found elsewhere (Tong *et al.*, 2001). (B) Functional neighborhood corresponding to indicated region (dashed gray circle) in (A). Despite being rare, synthetic lethal interactions (blue lines) occur frequently among genes that are functionally related, such as those involved in DNA replication and repair shown here. The frequency of synthetic lethal interaction between functionally related genes ranges from 18% to 25%. (C) Orthogonal relationships. Negative interactions tend to occur between nonessential complexes and pathways. Positive interactions overlap significantly with physical interactions and tend to connect members of the same pathway or complex. (D) Grouping genes according to patterns of genetic interactions revealed a functional relationship between the elongator complex and the urmylation pathway, which act in concert to modify specific tRNAs. (See Color Insert.)

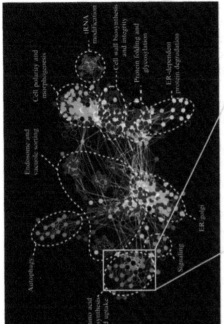

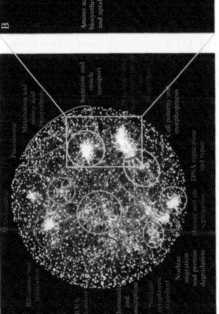

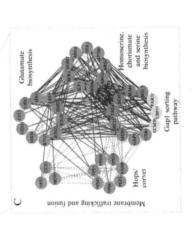

Figure 7.5 (*Continued*)

(query genes) and columns (array genes) of a genetic interaction matrix to place genes sharing similar genetic profiles next to each other. On the resulting clustergram, that is easily visualized with an open source application like Java Treeview (http://jtreeview.sourceforge.net), functionally related genes, including members of the same protein complex or pathway, are normally located in close proximity to each other and coclustering of genes of known function with poorly characterized genes provides strong evidence for cofunction (Tong *et al.*, 2004).

An alternative approach for clustering genetic interaction profiles consists in computing pair-wise correlation coefficients among all genes in a dataset (Costanzo *et al.*, in press; Kim *et al.*, 2001). The resulting data can be visualized as a network (using network visualization software such as Cytoscape (Shannon *et al.*, 2003)) where two genes are connected if their correlation exceeds a chosen threshold. The network can then be reorganized by a force-directed network layout in which highly correlated genes attract each other, while less correlated genes are repelled. Applying such an approach to the correlation-based genetic interaction network generates readily discernable clusters corresponding to distinct biological processes (Fig. 7.5A). This highly structured organization of the genetic map is maintained at increasing levels of resolution. For example, in one region of the global network, the related processes of endoplasmic reticulum

---

Figure 7.5 The genetic landscape of the cell. (A) A correlation-based network connecting genes with similar interaction profiles (Costanzo *et al.*, in press). Genetic profile similarities were measured for all gene pairs by computing Pearson correlation coefficients (PCC) from the complete genetic interaction matrix. Gene pairs whose profile similarity exceeded a PCC > 0.2 threshold were connected in the network. An edge-weighted, spring-embedded network layout, implemented in Cytoscape (Shannon *et al.*, 2003), was applied to determine node position based on genetic profile similarity. This resulted in the unbiased assembly of a network whereby genes sharing similar patterns of genetic interactions are proximal to each other in two-dimensional space, while less-similar genes are positioned further apart. Circled regions correspond to gene clusters enriched for the indicated biological processes. (B) Magnification of the functional map resolves cellular processes with increased specificity and enables precise functional predictions. A subnetwork corresponding to the indicated region of the global map is shown. Node color corresponds to a specific biological process; amino acid biosynthesis and uptake (dark green); signaling (light green); ER/Golgi (light purple); endosome and vacuole sorting (dark purple); ER-dependent protein degradation (yellow); protein glycosylation, cell wall biosynthesis and integrity (red); tRNA modification (fuchsia); cell polarity and morphogenesis (pink); autophagy (orange); uncharacterized (black). (C) Individual genetic interactions contributing to genetic profiles revealed by (B). A subset of genes belonging to the amino acid biosynthesis and uptake region of the network in (B). Nodes are grouped according to profile similarity and edges represent negative (red) and positive (green) genetic interactions. Nonessential (circles) and essential (diamonds) genes are colored according to the biological process indicated in (B) and uncharacterized genes are depicted in yellow. (See Color Insert.)

(ER)/Golgi traffic, endosome/vacuole protein sorting, cell polarity, morphogenesis, cell wall integrity, protein folding, glycosylation, and ER-dependent protein degradation all cluster into well delineated groups (Fig. 7.5B). Furthermore, closer interrogation of the genetic map allows to distinguish uncharacterized genes located next to known functional clusters, suggesting previously unanticipated roles for the uncharacterized genes in the process (Fig. 7.5C).

2.8. *S. pombe* SGA

S. pombe or fission yeast is separated from *S. cerevisiae* by hundreds of millions of years of evolution and therefore provides an excellent system for exploring the conservation of genetic interactions. Moreover, *S. pombe* never underwent an ancient genome duplication and thus unlike *S. cerevisiae*, where functional complementation by paralogous genes can sometimes obscure genetic interactions, the corresponding *S. pombe* ortholog often shows a rich genetic interaction profile (Dixon and Boone, unpublished data). In addition, several gene classes and mechanisms are also absent in *S. cerevisiae* but present in *S. pombe* and other eukaryotes (Aravind *et al.*, 2000). For example, *S. pombe* has an RNA interference (RNAi) pathway similar to that seen in other eukaryotes, while baker's yeast does not (White and Allshire, 2008). Thus, analysis of genetic interactions in *S. pombe* should provide not only complementary information to that obtained in *S. cerevisiae*, but also insight into processes that are inaccessible in *S. cerevisiae*.

The sequencing of the *S. pombe* genome (Wood *et al.*, 2002) and subsequent availability of a genome-wide deletion collection from the commercial company Bioneer (http://pombe.bioneer.co.kr/) has driven the development of SGA-like techniques for this organism (Dixon *et al.*, 2008; Roguev *et al.*, 2007). For *S. pombe* SGA (*Sp*SGA) (Dixon *et al.*, 2008), we developed a protocol that enabled efficient isolation of double mutant haploids through use of the high-density arraying capabilities of the Singer RoToR system.

A schematic illustrating the *Sp*SGA procedure is shown in Fig. 7.6.

2.8.1. *Sp*SGA procedure
Preparing the RoToR for use

1. Turn on the RoToR software and hardware and sterilize the interior using the integral UV lamp. From the main menu, select: LAMP and then set the timer for a minimum of 30 min.

Preparation of a 384-formatted query plate

1. Grow query strain in 8 ml YES liquid media, with shaking, overnight at 30 °C (or appropriate temperature for sensitive strains).

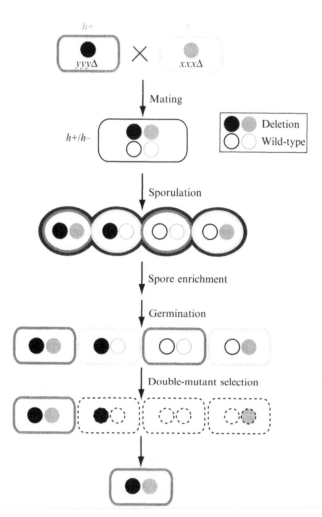

Figure 7.6 Outline of the *Sp*SGA method. Cells of opposite mating type (h+, h−) are mated on minimal SPA media and allowed to sporulate for 3 days at 26 °C. Then, to enrich for spores, mating plates are transferred to 42 °C for 3 days—a treatment that kills unmated haploid cells. Following spore enrichment, cells are transferred to rich medium to allow for germination, then transferred again to double-drug medium to select for recombinant double-mutant progeny. *S. pombe* haploids do not mate on rich medium; therefore, selection for a specific haploid mating type is not required.

2. Pour liquid culture into an empty PlusPlate dish, ensuring the entire surface area is covered.
3. Place the source dish containing the liquid culture in the black position. Place the YES (agar) target plate in the red position. With the RoToR, select: LIQUID HANDLING → BATH → BATH-96. As prompted, load the hopper with 96 long pin pads. Under OPTIONS check

ECONOMY and REVISIT SOURCE. Start the program. This will transfer a droplet of the culture to each position on the array in four steps, generating a 384 array (4 × 96). *Caution.* When in use, the RoToR robot arm and turntable move very quickly. Ensure that the safety cover is in place during operation and that hands and loose clothing are kept well free of the moving parts.

4. Grow the query plate in an incubator at 30 °C for 2 days or until sufficient growth is obtained. With larger colonies, a single query plate can be used for multiple matings.

Preparation of a 384-formatted array plate

1. Plates obtained from Bioneer are arrayed in a 96-position format. These can be rearrayed 1:4, giving four copies of each query or rearrayed at higher density (possibly could use the RoToR 4→1 arraying protocol, although we have not tried that ourselves).
2. Grow the array plate in an incubator at 30 °C for 2 days or until sufficient growth is obtained.

Mating of query to array

1. Place the YES query source plate in the blue position, the YES array source plate in the red position, and the target SPA mating plate in the black position. With the RoToR, select: MATE → 384. Load the hopper with the appropriate number of 384 short pin pads. Start the program.
2. Remove the SPA mating dish and set aside for further processing. The query and/or array source dishes can continue to be used as needed. *Critical step.* It is essential that the query and array source plates are freshly grown (typically 2 days growth is optimal). Plates used after storage at 4 °C do not mate as efficiently. Depending on the amount of cell growth, a single 384-format query plate can be mated to 3–4 separate array plates. However, further attempts may result in a reduced number of transferred cells, interfering with mating efficiency.
3. Following the transfer of cells onto the SPA plate, use the RoToR to transfer a drop of sterile H_2O onto the mated cells and mix them together. Load a PlusPlate dish containing sterile H_2O into the black position and a freshly mated SPA plate in the red position. On the RoToR select: LIQUID HANDLING → BATH → BATH-96. Load 96 long pin pads. Under advanced options, select the Agar Mix tab and set a mix diameter of 0.2 mm. Start the program. *Critical step.* A droplet of water together with the punching of the agar is thought to facilitate access of each yeast to the other. This greatly enhances the mating efficiency, compared to no mixing conditions, and thereby

enhances sporulation substantially. This in turn reduces the number of unmated haploids that need to be eliminated in subsequent steps.

4. Put SPA plates at 26 °C and allow cells to proceed through conjugation and sporulation for 3 days.

5. Put SPA dishes at 42 °C for 3 days. *Critical step.* Ensure that the incubator is maintained at 42 °C. Higher or lower temperatures are not as efficient at eliminating unmated haploids. *Note:* significant temperature gradients may be observed within certain incubators, so it is advisable to measure the temperature at several places within the enclosure.

6. Transfer spores to YES plates as follows. Load the SPA source plate into the red position and the YES target plate into the blue position. Using the RoToR, select: AGAR-AGAR → REPLICATE → REPLICATE ONE → 384-1536. Load the hopper with 384 short pin pads and start the program using default advanced options with "revisit source" enabled.

7. *Pause point.* Once spores have been transferred to YES plates, it is advisable to store the SPA mating plates at 4 °C. Spores remaining on this plate can be subsequently stored up to several months and used later as a source of spores for further confirmatory experiments, such as random spore analysis.

8. Put YES plates at 30 °C and allow cells to germinate for at least 2 days.

9. Transfer germinated cells from source YES plates loaded in the red position to YES + G418 + Nat target plates loaded in the blue position. With the RoToR, select: AGAR-AGAR → REPLICATE → REPLICATE ONE → 1536-1536. Load the hopper with 1536 short pin pads and start the program using default advanced options.

10. Transfer YES + G418 + Nat plates to 30 °C to allow cells to grow for 2 days.

11. *Pause point.* Once yeast cells have grown up on YES + G418 + Nat plates they can be further processed right away, or stored at 4 °C for up to several months for subsequent processing.

12. Image plates to acquire colony size measurements or carry on with additional analysis of recombinant haploids.

3. Media and Stock Solutions

3.1. SGA media and stock solutions

1. *G418 (Geneticin, Invitrogen)*: Dissolve in water at 200 mg/ml, filter sterilize, and store in aliquots at 4 °C.

2. *clonNAT (nourseothricin, Werner BioAgents, Jena, Germany)*: Dissolve in water at 100 mg/ml, filter sterilize, and store in aliquots at 4 °C.

3. *Canavanine (L-canavanine sulfate salt, Sigma, C-9758)*: Dissolve in water at 100 mg/ml, filter sterilize, and store in aliquots at 4 °C.

4. *Thialysine (S-(2-aminoethyl)-L-cysteine hydrochloride, Sigma, A-2636)*: Dissolve in water at 100 mg/ml, filter sterilize, and store in aliquots at 4 °C.

5. *Amino acids supplement powder mixture for synthetic media (complete)*: Contains 3 g adenine (Sigma), 2 g uracil (ICN), 2 g inositol, 0.2 g para-aminobenzoic acid (Acros Organics), 2 g alanine, 2 g arginine, 2 g asparagine, 2 g aspartic acid, 2 g cysteine, 2 g glutamic acid, 2 g glutamine, 2 g glycine, 2 g histidine, 2 g isoleucine, 10 g leucine, 2 g lysine, 2 g methionine, 2 g phenylalanine, 2 g proline, 2 g serine, 2 g threonine, 2 g tryptophan, 2 g tyrosine, and 2 g valine (Fisher). Drop-out (DO) powder mixture is a combination of the above ingredients minus the appropriate supplement. Two grams of the DO powder mixture is used per liter of medium.

6. *Amino acids supplement for sporulation medium*: Contains 2 g histidine, 10 g leucine, 2 g lysine, 2 g uracil; 0.1 g of the amino acid supplements powder mixture is used per liter of sporulation medium.

7. *Glucose (Dextrose, Fisher)*: Prepare 40% solution, autoclave, and store at room temperature.

8. *YEPD*: Add 120 mg adenine (Sigma), 10 g yeast extract, 20 g peptone, 20 g bacto agar (BD Difco) to 950 ml water in a 2 l flask. After autoclaving, add 50 ml of 40% glucose solution, mix thoroughly, cool to ∼65 °C and pour plates.

9. *YEPD + G418*: Cool YEPD medium to ∼65 °C, add 1 ml of G418 stock solution (final concentration 200 mg/l), mix thoroughly, and pour plates.

10. *YEPD + clonNAT*: Cool YEPD medium to ∼65 °C, add 1 ml of clonNAT stock solution (final concentration 100 mg/l), mix thoroughly, and pour plates.

11. *YEPD + G418/clonNAT*: Cool YEPD medium to ∼65 °C, add 1 ml of G418 (final concentration 200 mg/l), and 1 ml of clonNAT (final concentration 100 mg/l) stock solutions, mix thoroughly, and pour plates.

12. *Enriched sporulation*: Add 10 g potassium acetate (Fisher), 1 g yeast extract, 0.5 g glucose, 0.1 g amino acids supplement powder mixture for sporulation, 20 g bacto agar to 1 l water in a 2 l flask. After autoclaving, cool medium to ∼65 °C, add 250 μl of G418 stock solution (final concentration 50 mg/l), mix thoroughly, and pour plates.

13. *(SD/MSG) − His/Arg/Lys + canavanine/thialysine/G418*: Add 1.7 g yeast nitrogen base w/o amino acids or ammonium sulfate (BD Difco), 1 g MSG (L-glutamic acid sodium salt hydrate, Sigma), 2 g amino acids supplement powder mixture (DO − His/Arg/Lys), 100 ml water in a 250 ml flask. Add 20 g bacto agar to 850 ml water in a 2 l flask. Autoclave separately. Combine autoclaved solutions, add 50 ml 40%

glucose, cool medium to ~65 °C, add 0.5 ml canavanine (50 mg/l), 0.5 ml thialysine (50 mg/l), and 1 ml G418 (200 mg/l) stock solutions, mix thoroughly, and pour plates. Ammonium sulfate impedes the function of G418 and clonNAT. Hence, synthetic medium containing these antibiotics is made with monosodium glutamic acid as a nitrogen source (Cheng *et al.*, 2000).

14. *(SD/MSG) − His/Arg/Lys + canavanine/thialysine/clonNAT*:   Add 1.7 g yeast nitrogen base w/o amino acids or ammonium sulfate, 1 g MSG, 2 g amino acids supplement powder mixture (DO − His/Arg/Lys), 100 ml water in a 250 ml flask. Add 20 g bacto agar to 850 ml water in a 2 l flask. Autoclave separately. Combine autoclaved solutions, add 50 ml 40% glucose, cool medium to ~65 °C, add 0.5 ml canavanine (50 mg/l), 0.5 ml thialysine (50 mg/l), and 1 ml clonNAT (100 mg/l) stock solutions, mix thoroughly, and pour plates.

15. *(SD/MSG) − His/Arg/Lys + canavanine/thialysine/G418/clonNAT*: Add 1.7 g yeast nitrogen base w/o amino acids or ammonium sulfate, 1 g MSG, 2 g amino acids supplement powder mixture (DO − His/Arg/Lys), 100 ml water in a 250 ml flask. Add 20 g bacto agar to 850 ml water in a 2 l flask. Autoclave separately. Combine autoclaved solutions, add 50 ml 40% glucose, cool medium to ~65 °C, add 0.5 ml Canavanine (50 mg/l), 0.5 ml thialysine (50 mg/l), 1 ml G418 (200 mg/l) and 1 ml clonNAT (100 mg/l) stock solutions, mix thoroughly, and pour plates.

16. *(SD/MSG) Complete*: Add 1.7 g yeast nitrogen base w/o amino acids or ammonium sulfate, 1 g MSG, 2 g amino acids supplement powder mixture (complete), 100 ml water in a 250 ml flask. Add 20 g bacto agar to 850 ml water in a 2 l flask. Autoclave separately. Combine autoclaved solutions, add 50 ml of 40% glucose, mix thoroughly, cool medium to ~65 °C and pour plates.

17. *SD − His/Arg/Lys + canavanine/thialysine*: Add 6.7 g yeast nitrogen base w/o amino acids (BD Difco), 2 g amino acids supplement powder mixture (DO − His/Arg/Lys), 100 ml water in a 250 ml flask. Add 20 g bacto agar to 850 ml water in a 2 l flask. Autoclave separately. Combine autoclaved solutions, add 50 ml 40% glucose, cool medium to ~65 °C, add 0.5 ml canavanine (50 mg/l) and 0.5 ml thialysine (50 mg/l) stock solutions, mix thoroughly, and pour plates. This medium does not contain any antibiotics such as G418 or clonNAT and therefore ammonium sulfate is used as the nitrogen source.

18. *SD − Leu/Arg/Lys + canavanine/thialysine*: Add 6.7 g yeast nitrogen base w/o amino acids, 2 g amino acids supplement powder mixture (DO − Leu/Arg/Lys), 100 ml water in a 250 ml flask. Add 20 g bacto agar to 850 ml water in a 2 l flask. Autoclave separately. Combine autoclaved solutions, add 50 ml 40% glucose, cool medium to ~65 °C, add 0.5 ml canavanine (50 mg/l) and 0.5 ml thialysine (50 mg/l) stock solutions, mix thoroughly, and pour plates.

3.2. *Sp*SGA media and stock solutions

1. *250 mg/ml of G418 solution.* Dissolve 2.5 g G418 in 10 ml of water. Filter sterilize with a 0.22 µm screw cap filter.
2. *100 mg/ml of clonNat solution.* Dissolve 1 g clonNat in 10 ml of water. Filter sterilize with a 0.22 µm screw cap filter.
3. *SPA-agar plates (SPA plates).* Add 30 g agar, 10 g dextrose, 1 g of KH$_2$PO$_4$, and 1 ml of 1000× vitamin stock to 1 l of water. Autoclave and then pour 50 ml per PlusPlate dish. Allow to solidify and either use right away or store at 4 °C until needed.
4. *1000× vitamin stock.* Dissolve 1 g pantothenic acid, 10 g nicotinic acid, 10 g inositol, and 10 mg biotin in 1 l of water.
5. *YES media and plates.* Add 30 g glucose, 5 g yeast extract, 225 mg adenine, 225 mg L-histidine, 225 mg leucine, 225 mg uracil, and 225 mg lysine to 1 l of water. Autoclave and then cool to 55 °C before adding appropriate antibiotics (1:1000 dilution). To make solid media, add 30 g agar prior to autoclaving and pour 50 ml per PlusPlate.

4. Applications of SGA Methodology

Most large-scale studies have focused on fitness as the primary phenotype to identify genetic interactions (St Onge *et al.*, 2007; Tong *et al.*, 2004). In theory, all phenotypes are measurable and amenable to genetic interaction analyses. SGA methodology provides an efficient and systematic means for combining mutations and can be readily applied to identify additional genetic interactions that do not result in overt fitness defects. For example, reporter-gene constructs can be incorporated into the SGA methodology to monitor specific transcriptional responses in the ∼5000 deletion mutant backgrounds (Costanzo *et al.*, 2004; Fillingham *et al.*, 2009) and used as an alternative to fitness for uncovering genetic interactions (Jonikas *et al.*, 2009). Integrating SGA technology with methodologies for measuring a diverse set of phenotypes generates networks that provide comprehensive genome coverage and accurately reflect global cellular functions.

4.1. Integrating SGA and high-content screening

Combining SGA technology with different cytological reporters and high content screening (HCS) methodologies also enables identification of mutant combinations that lack obvious growth defects but elicit subtle yet unexpected cell biological phenotypes (Vizeacoumar *et al.*, 2009, 2010). A HCS platform gathers cell biological information for genome-wide arrays of mutants by first acquiring cell images and then quantifying specific morphological phenotypes following image processing (Fig. 7.7). The following section describes the steps involved in this procedure and can be applied to monitor virtually any subcellular event within the cell. The media required for various selections are described in detail in Section 3.

1. Cross the SGA query strain expressing the cell biological reporter to the deletion array as described in Section 2.4.
2. Once the *MATa-kanMX4* meiotic progeny is selected on agar plates as described in Section 2.4, transfer the cultures to liquid. Select for *MATa-kanMX4-natMX4* meiotic progeny by inoculating the cells in (SD/MSG) − His/Arg/Lys + canavanine/thialysine/G418/clonNAT liquid media in 96-well format plates. Note to include the auxotrophic selection of the reporter throughout the process.
3. Grow liquid cultures overnight at the desired temperature in 96-well plates containing 3 mm glass beads (Fisher Scientific). Seal the plates with breathable adhesive membranes (Corning).
4. Dispense appropriate volume of samples based on the optical density of each sample to 96-well filter plates (Whatmann) using a liquid handling robot (Biomek FX), to ensure uniform and optimal cell densities for subsequent image analysis.
5. Vacuum the filter plates (NucleoVac96 vacuum manifold, Macherey-Nagel) and wash the cells in sterile distilled water using plate washer (Multidrop384 Liquid Dispenser System, Thermo Electron Corporation). Vacuum the plates again so that only the washed cells remain in the filter plates.
6. Add low-fluorescence medium (Sheff and Thorn, 2004) with appropriate drug selection to the filter plates to resuspend the washed cells and transfer cells from the filter plates to 96-well optical plates (Matrical), using the liquid handler (Biomek FX) and let the cells settle down. Seal the plates with aluminum foil (VWR) to prevent drying of the samples.
7. Load the plates into an automated incubator (Cytomat, Thermo Fisher Scientific, Inc.) and store the plates at 4 °C prior to imaging to avoid over growth. Automate the platform such that each plate is incubated at 30 °C for 30–45 min prior to imaging. Using a robotic arm (CRS Catalyst Express, Thermo Electron Corporation) transfer the plates to the wide-field HCS imager (ImageXpress5000A, MDS Analytical, Inc.).
8. Image plates using a 60× air-objective collecting at least 4–6 images per well averaging to as many as 200 cells per mutant. Ensure proper storage of images in a database to readily access them for automated image analysis.
9. Analyze the images using an image-analysis software with throughput capabilities (MetaXpress software v1.6, MDS Analytical, Inc.) and extract morphometric features to generate a unique profile for each mutant. Open source software developed by academic labs, such as

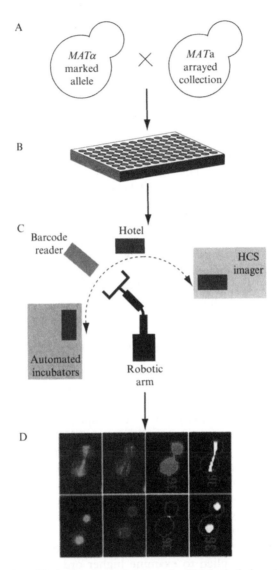

Figure 7.7   SGA-HCS pipeline for evaluation of cell biological phenotypes. (A) Using the SGA methodology, a fluorescent marker can be introduced into the arrayed deletion collection . (B) Deletion mutant colonies expressing a fluorescent marker are transferred to optical plates containing liquid selection media for imaging. (C) A robotic arm is used to move optical plates between incubators and the HCS imaging system. (D) Automated image analysis software, such as MetaXpress, is used to detect fluorescent signal and measure morphological features.

CellProfiler, may also be used for this purpose (Carpenter, 2007). Briefly, after shade correction and background subtraction, apply a segmentation technique such as thresholding, so that the range of signal intensity that pertains to cellular objects is selected separating bright cellular objects from the dark image background. Following object identification, quantitative measurements of each cell can be used to extract numerous morphometric parameters, a feature available in most image-analysis software.

Statistical analysis and data mining for multiplexed high content analysis (HCA) is still in its infancy. Hence, most HCA are custom designed and also require manual inspection of images (Bakal *et al.*, 2007; Corcoran *et al.*, 2004; Eggert *et al.*, 2004; Gururaja *et al.*, 2006; Loo *et al.*, 2007; Narayanaswamy *et al.*, 2006; Tanaka *et al.*, 2005). However, a more straightforward approach is to compare each mutant population read out against that of the wild-type cell population, and identify statistically deviant mutants for any desired feature. Methods to standardize HCA are still in progress and the continued development of these procedures to process HCS data without sacrificing information accuracy will be of paramount significance.

4.2. Essential gene and higher order genetic interactions

This chapter focuses on the application of SGA analysis to generate double mutant strains and identify genetic interactions among nonessential deletion mutants. However, SGA can also be applied to examine synthetic genetic interactions involving essential genes. For example, an SGA query strain can be crossed to the Tet-promoter collection (yTHC, Open Biosystems), double mutants can be selected and scored for growth defects in the presence of doxycycline, which downregulates the expression of the essential genes (Davierwala *et al.*, 2005). Other essential gene mutant collections (Ben-Aroya *et al.*, 2008; Schuldiner *et al.*, 2005) are now available in arrayed formats and are amenable to genetic interaction analyses using SGA technology (Costanzo *et al.*, in press).

While digenic interactions are more commonly studied, SGA methodology can be easily applied to examine higher order genetic interactions involving more than two genes (Tong *et al.*, 2004).

4.3. Combining SGA and gene overexpression libraries

In addition to loss-of-function mutations, SGA can be easily adapted to examine different forms of genetic interactions involving high-copy plasmid or regulatory expression of yeast or heterologous genes. Several gene overexpression libraries have been constructed, in recent years (Gelperin *et al.*, 2005; Hu *et al.*, 2007; Jones *et al.*, 2008; Zhu *et al.*, 2001). One of these libraries was used to assemble a Yeast Overexpression Array, containing ~6000 ORFs (Yeast GST-Tagged Collection, Open Biosystems), and was combined with SGA to screen for synthetic dosage lethality and suppression (Sopko et al., 2006). This proved to be a successful strategy for identifying downstream targets regulated by specific signaling pathways (Sopko et al., 2006, 2007). The development of new plasmid libraries carrying genes under inducible expression will expand the potential for dosage lethality (Gelperin et al., 2005; Hu et al., 2007), whereas the development of libraries in which each gene is under the control of its own promoter (Ho et al., 2009; Jones et al., 2008) offers the potential for building other overexpression arrays that may be particularly useful for dosage suppression experiments.

4.4. Applying SGA as a method for high-resolution genetic mapping (SGAM)

Because double mutants are created by meiotic recombination, a set of gene deletions that is linked to the query gene, which we refer to as the "linkage group" form double mutants at a reduced frequency, thus, appearing synthetic lethal/sick with the query mutation. Since the gene deletions represent mapping markers covering all chromosomes in the yeast genome, SGA mapping (SGAM) has been shown as an effective method for high-resolution genetic mapping (Chang et al., 2005; Jorgensen et al., 2002). In addition to mapping of recessive alleles, SGAM is particularly useful for rapid mapping of dominant mutations, which are challenging to clone using standard techniques (Menne et al., 2007).

4.5. Chemical genomics

Because chemical perturbations mimic genetic perturbations, genetic networks also provide a key for predicting the targets of inhibitory bioactive molecules (Parsons et al., 2004) (Fig. 7.8). If a compound inhibits a specific target protein, then the chemical-genetic profile, the set of mutants that are hypersensitive to the compound, should overlap with the genetic interaction profile of the target gene. As a result, the compound and its target should cocluster together and with other genes and compounds involved in the same biological process (Fig. 7.8).

The integration of ~1700 genetic interaction profiles with ~400 drug sensitivity profiles (Costanzo et al., in press) proved that compounds with known functions, such as hydroxyurea, which inhibits DNA synthesis, or tunicamycin, which inhibits glycosylation, cluster to their expected biological process (Giaever et al., 1999). Furthermore, the target of a novel drug, now named Erodoxin, was identified via inspection of a combined chemical-genetic correlation network. Thus, this chemical-genetic approach to

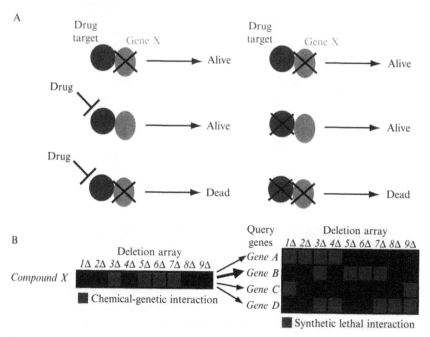

Figure 7.8 Chemical-genetic interactions can be modeled by synthetic genetic interactions. (A) In a chemical-genetic interaction (at left), a deletion mutant, lacking the product of the deleted gene (represented by a black X), is hypersensitive to a normally sublethal concentration of a growth-inhibitory compound. In a synthetic lethal genetic interaction (right), two single deletions lead to viable mutants but are inviable in a double-mutant combination. Gene deletion alleles that show chemical-genetic interactions with a particular compound should also be synthetically lethal or sick with a mutation in the compound target gene. (B) Comparison of a chemical-genetic profile to a compendium of genetic interaction (synthetic lethal) profiles should identify the pathways and targets inhibited by drug treatment. In this hypothetical figure, chemical-genetic and genetic interactions are both designated by red squares. For example, deletion mutants 3, 5, 6, and 7 are hypersensitive to compound X and a mutation in query gene A leads to a fitness defect when combined with deletion alleles 1, 2, 3, and 4. Here, the chemical-genetic profile of compound X resembles the genetic profile of gene B, thereby identifying the product of gene B as a putative target of compound X.

mode-of-action analysis complements haploinsufficiency profiling, which focuses on identifying the drug target directly (Giaever *et al.*, 1999; Hillenmeyer *et al.*, 2008).

REFERENCES

Aravind, L., *et al.* (2000). Lineage-specific loss and divergence of functionally linked genes in eukaryotes. *Proc. Natl. Acad. Sci. USA* **97,** 11319–11324.

Bader, J. S., *et al.* (2004). Gaining confidence in high-throughput protein interaction networks. *Nat. Biotechnol.* **22,** 78–85.

Bakal, C., *et al.* (2007). Quantitative morphological signatures define local signaling networks regulating cell morphology. *Science* **316**, 1753–1756.

Bandyopadhyay, S., *et al.* (2008). Functional maps of protein complexes from quantitative genetic interaction data. *PLoS Comput. Biol.* **4**, e1000065.

Bateson, W., *et al.* (1905). Reports to the Evolution Committee of the Royal Society. Report II. Harrison and Sons, London.

Ben-Aroya, S., *et al.* (2008). Toward a comprehensive temperature-sensitive mutant repository of the essential genes of *Saccharomyces cerevisiae*. *Mol. Cell* **30**, 248–258.

Boone, C., *et al.* (2007). Exploring genetic interactions and networks with yeast. *Nat. Rev. Genet.* **8**, 437–449.

Carpenter, A. E. (2007). Image-based chemical screening. *Nat. Chem. Biol.* **3**, 461–465.

Chang, M., *et al.* (2005). RMI1/NCE4, a suppressor of genome instability, encodes a member of the RecQ helicase/Topo III complex. *EMBO J.* **24**, 2024–2033.

Cheng, T. H., *et al.* (2000). Controlling gene expression in yeast by inducible site-specific recombination. *Nucleic Acids Res.* **28**, E108.

Collins, S. R., *et al.* (2006). A strategy for extracting and analyzing large-scale quantitative epistatic interaction data. *Genome Biol.* **7**, R63.

Collins, S. R., *et al.* (2007). Functional dissection of protein complexes involved in yeast chromosome biology using a genetic interaction map. *Nature* **446**, 806–810.

Corcoran, L. J., *et al.* (2004). A novel action of histone deacetylase inhibitors in a protein aggresome disease model. *Curr. Biol.* **14**, 488–492.

Costanzo, M., *et al.* (2004). CDK activity antagonizes Whi5, an inhibitor of G1/S transcription in yeast. *Cell* **117**, 899–913.

Davierwala, A. P., *et al.* (2005). The synthetic genetic interaction spectrum of essential genes. *Nat. Genet.* **37**, 1147–1152.

Decourty, L., *et al.* (2008). Linking functionally related genes by sensitive and quantitative characterization of genetic interaction profiles. *Proc. Natl. Acad. Sci. USA* **105**, 5821–5826.

Dixon, S. J., *et al.* (2008). Significant conservation of synthetic lethal genetic interaction networks between distantly related eukaryotes. *Proc. Natl. Acad. Sci. USA* **105**, 16653–16658.

Dixon, S. J., *et al.* (2009). Systematic mapping of genetic interaction networks. *Annu. Rev. Genet.* **43**, 601–625.

Eggert, U. S., *et al.* (2004). Parallel chemical genetic and genome-wide RNAi screens identify cytokinesis inhibitors and targets. *PLoS Biol.* **2**, e379.

Elena, S. F., and Lenski, R. E. (1997). Test of synergistic interactions among deleterious mutations in bacteria. *Nature* **390**, 395–398.

Fay, D. S., *et al.* (2002). fzr-1 and lin-35/Rb function redundantly to control cell proliferation in *C. elegans* as revealed by a nonbiased synthetic screen. *Genes Dev.* **16**, 503–517.

Fillingham, J., *et al.* (2009). Two-color cell array screen reveals interdependent roles for histone chaperones and a chromatin boundary regulator in histone gene repression. *Mol. Cell* **35**, 340–351.

Gelperin, D. M., *et al.* (2005). Biochemical and genetic analysis of the yeast proteome with a movable ORF collection. *Genes Dev.* **19**, 2816–2826.

Giaever, G., *et al.* (1999). Genomic profiling of drug sensitivities via induced haploinsufficiency. *Nat. Genet.* **21**, 278–283.

Giaever, G., *et al.* (2002). Functional profiling of the *Saccharomyces cerevisiae* genome. *Nature* **418**, 387–391.

Goldstein, A. L., and McCusker, J. H. (1999). Three new dominant drug resistance cassettes for gene disruption in *Saccharomyces cerevisiae*. *Yeast* **15**, 1541–1553.

Gururaja, T. L., *et al.* (2006). R-253 disrupts microtubule networks in multiple tumor cell lines. *Clin. Cancer Res.* **12**, 3831–3842.

Hartman, J. L., *et al.* (2001). Principles for the buffering of genetic variation. *Science* **291**, 1001–1004.

Hillenmeyer, M. E., *et al.* (2008). The chemical genomic portrait of yeast: Uncovering a phenotype for all genes. *Science* **320**, 362–365.

Ho, C. H., *et al.* (2009). A molecular barcoded yeast ORF library enables mode-of-action analysis of bioactive compounds. *Nat. Biotechnol.* **27**, 369–377.

Hu, Y., *et al.* (2007). Approaching a complete repository of sequence-verified protein-encoding clones for *Saccharomyces cerevisiae*. *Genome Res.* **17**, 536–543.

Jones, G. M., *et al.* (2008). A systematic library for comprehensive overexpression screens in *Saccharomyces cerevisiae*. *Nat. Methods* **5**, 239–241.

Jonikas, M. C., *et al.* (2009). Comprehensive characterization of genes required for protein folding in the endoplasmic reticulum. *Science* **323**, 1693–1697.

Jorgensen, P., *et al.* (2002). High-resolution genetic mapping with ordered arrays of *Saccharomyces cerevisiae* deletion mutants. *Genetics* **162**, 1091–1099.

Kelley, R., and Ideker, T. (2005). Systematic interpretation of genetic interactions using protein networks. *Nat. Biotechnol.* **23**, 561–566.

Kim, S. K., *et al.* (2001). A gene expression map for *Caenorhabditis elegans*. *Science* **293**, 2087–2092.

Koh, *et al.* (2010). *Nucelic Acids Res.* **38**, D502–D507.

Levy, S., *et al.* (2007). The diploid genome sequence of an individual human. *PLoS Biol.* **5**, e254.

Loo, L. H., *et al.* (2007). Image-based multivariate profiling of drug responses from single cells. *Nat. Methods* **4**, 445–453.

Lucchesi, J. C. (1968). Synthetic lethality and semi-lethality among functionally related mutants of *Drosophila melanogaster*. *Genetics* **59**, 37–44.

Mani, R., *et al.* (2008). Defining genetic interaction. *Proc. Natl. Acad. Sci. USA* **105**, 3461–3466.

Menne, T. F., *et al.* (2007). The Shwachman–Bodian–Diamond syndrome protein mediates translational activation of ribosomes in yeast. *Nat. Genet.* **39**, 486–495.

Mnaimneh, S., *et al.* (2004). Exploration of essential gene functions via titratable promoter alleles. *Cell* **118**, 31–44.

Myers, C. L., *et al.* (2006). Finding function: Evaluation methods for functional genomic data. *BMC Genomics* **7**, 187.

Narayanaswamy, R., *et al.* (2006). Systematic profiling of cellular phenotypes with spotted cell microarrays reveals mating-pheromone response genes. *Genome Biol.* **7**, R6.

Pan, X., *et al.* (2004). A robust toolkit for functional profiling of the yeast genome. *Mol. Cell* **16**, 487–496.

Pan, X., *et al.* (2006). A DNA integrity network in the yeast *Saccharomyces cerevisiae*. *Cell* **124**, 1069–1081.

Parsons, A. B., *et al.* (2004). Integration of chemical-genetic and genetic interaction data links bioactive compounds to cellular target pathways. *Nat. Biotechnol.* **22**, 62–69.

Roguev, A., *et al.* (2007). High-throughput genetic interaction mapping in the fission yeast *Schizosaccharomyces pombe*. *Nat. Methods* **4**, 861–866.

Schuldiner, M., *et al.* (2005). Exploration of the function and organization of the yeast early secretory pathway through an epistatic miniarray profile. *Cell* **123**, 507–519.

Segre, D., *et al.* (2005). Modular epistasis in yeast metabolism. *Nat. Genet.* **37**, 77–83.

Shannon, P., *et al.* (2003). Cytoscape: A software environment for integrated models of biomolecular interaction networks. *Genome Res.* **13**, 2498–2504.

Sheff, M. A., and Thorn, K. S. (2004). Optimized cassettes for fluorescent protein tagging in *Saccharomyces cerevisiae*. *Yeast* **21**, 661–670.

Sopko, R., *et al.* (2006). Mapping pathways and phenotypes by systematic gene overexpression. *Mol. Cell* **21**, 319–330.

Sopko, R., *et al.* (2007). Activation of the Cdc42p GTPase by cyclin-dependent protein kinases in budding yeast. *EMBO J.* **26,** 4487–4500.

St Onge, R. P., *et al.* (2007). Systematic pathway analysis using high-resolution fitness profiling of combinatorial gene deletions. *Nat. Genet.* **39,** 199–206.

Tanaka, M., *et al.* (2005). An unbiased cell morphology-based screen for new, biologically active small molecules. *PLoS Biol.* **3,** e128.

Tong, A. H., *et al.* (2001). Systematic genetic analysis with ordered arrays of yeast deletion mutants. *Science* **294,** 2364–2368.

Tong, A. H., *et al.* (2004). Global mapping of the yeast genetic interaction network. *Science* **303,** 808–813.

Tong, A. H., and Boone, C. (2006). Synthetic genetic array analysis in *Saccharomyces cerevisiae*. *Methods Mol. Biol.* **313,** 171–192.

Vizeacoumar, F. J., *et al.* (2009). A picture is worth a thousand words: Genomics to phenomics in the yeast *Saccharomyces cerevisiae*. *FEBS Lett.* **583,** 1656–1661.

Vizeacoumar, F., *et al.* (2010). Integrating high-throughput genetic interaction mapping and high-content screening to explore yeast spindle morphogenesis. *J. Cell Biol.* **188,** 69–81.

Waddington, C. H. (1957). *The Strategy of the Gene*. Allen and Unwin, London.

White, S. A., and Allshire, R. C. (2008). RNAi-mediated chromatin silencing in fission yeast. *Curr. Top. Microbiol. Immunol.* **320,** 157–183.

Winzeler, E. A., *et al.* (1999). Functional characterization of the *S. cerevisiae* genome by gene deletion and parallel analysis. *Science* **285,** 901–906.

Wood, V., *et al.* (2002). The genome sequence of *Schizosaccharomyces pombe*. *Nature* **415,** 871–880.

Ye, P., *et al.* (2005). Gene function prediction from congruent synthetic lethal interactions in yeast. *Mol. Syst. Biol.* **1,** 2005.0026.

Zhu, H., *et al.* (2001). Global analysis of protein activities using proteome chips. *Science* **293,** 2101–2105.

CHAPTER EIGHT

MAKING TEMPERATURE-SENSITIVE MUTANTS

Shay Ben-Aroya,* Xuewen Pan,† Jef D. Boeke,‡ *and* Philip Hieter*

Contents

Abstract

The study of temperature-sensitive (Ts) mutant phenotypes is fundamental to gene identification and for dissecting essential gene function. In this chapter, we describe two "shuffling" methods for producing Ts mutants using a combination of PCR, *in vivo* recombination, and transformation of diploid strains heterozygous for a knockout of the desired mutation. The main difference between the two methods is the type of strain produced. In the "plasmid" version, the product is a knockout mutant carrying a centromeric plasmid carrying the Ts mutant. In the "chromosomal" version, The Ts alleles are integrated directly into the endogenous locus, albeit not in an entirely native configuration. Both variations have their strengths and weaknesses, which are discussed here.

* Michael Smith Laboratories, Department of Medical Genetics, University of British Columbia, Vancouver, British Columbia, Canada
† Verna and Marrs McLean Department of Biochemistry and Molecular Biology, Department of Molecular and Human Genetics, Baylor College of Medicine, Houston, Texas, USA.
‡ High Throughput Biology Center, Johns Hopkins University school of Medicine, Baltimore, Maryland, USA

Methods in Enzymology, Volume 470
ISSN 0076-6879, DOI: 10.1016/S0076-6879(10)70008-2

1. INTRODUCTION

The study of temperature-sensitive (Ts) mutant phenotypes has proven to be a fundamental approach both for the identification of gene sets essential for various aspects of biology and for obtaining a detailed understanding of essential gene function. While the observation that temperature-sensitive mutations represent a general class of mutation was recognized in the 1950s (Horowitz, 1950), the first targeted screen, isolation, and analysis of Ts mutants (382 mutations located in 37 genes scattered widely over the bacteriophage T4 genome) were made by Edgar and Lielausis in 1963 (Edgar and Lielausis, 1964). Hartwell (1967) reported the isolation of 400 Ts mutations in *Saccharomyces cerevisiae*, which caused defects in essential processes including cell division, and protein, RNA, and DNA synthesis. Over the past 40 years, the isolation and analysis of Ts mutations in essential genes has been a linchpin technology for investigating the genetics and molecular biology of essential processes in all experimental organisms.

Ts mutations are typically missense mutations, which retain the function of a specific essential gene at standard (permissive) low temperature, lack that function at a defined high (nonpermissive) temperature, and exhibit partial (hypomorphic) function at an intermediate (semipermissive) temperature. Such mutants make possible the analysis of physiologic changes that follow controlled inactivation of a gene or gene product by shifting cells to a nonpermissive temperature, offering a powerful approach to the analysis of gene function.

Essential genes, by definition, encode critical cellular functions that are not buffered by redundant functions or pathways (Hartman *et al.*, 2001). Essential genes have been shown to be highly dense hubs within genetic interaction networks and are involved in all aspects of basic cellular function (Jeong *et al.*, 2001). Furthermore, essential genes tend to be more highly conserved in evolution; 38% of essential yeast proteins have easily identifiable counterparts in humans, versus 20% for nonessential genes (Hughes, 2002).

Despite their importance, the functions of many essential yeast proteins have not been studied. In part, this is due to the absence of essential gene representation in the genome-wide haploid mutant collections, which cover all of the ~ 5000 nonessential yeast genes. Thus, no comparable systematic haploid mutant collection currently exists for the ~ 1000 essential genes in *S. cerevisiae*. The frequency of sites mutable to a reduced or conditional function is highly gene-specific; for example, for highly conserved proteins, random single missense mutation would be expected, for the vast majority of positions within the protein, to cause complete loss of function when mutated. Therefore, genetic screens using a random mutagenesis approach rarely reach saturation because "mutability" varies widely among genes.

Here, we report detailed protocols for two methodologies that allow the systematic isolation of Ts alleles in essential genes of interest. The first method is plasmid-based, and the second is genome integration-based, and each has its specific advantages depending on the application. Both methods exploit features of the "haploid-convertible" heterozygous diploid collection, which allows introduction of the library of mutagenized essential gene copies into the heterozygous diploid and subsequent direct selection of haploids which are deleted for the target essential gene and that carry individual members of the mutagenized essential gene library, using the "diploid shuffle" technique (see below). Several other useful corollary methods for transferring extant Ts alleles or specific gene constructs (e.g., fusion proteins) are also presented. Finally, we note that our laboratories are in the process of generating a complete set of Ts alleles for each of the essential genes in S. cerevisiae, which will be distributed as a resource to the scientific community when completed (see Ben-Aroya et al., 2008 for details). For specific essential genes under study in individual laboratories, however, it may be useful, using the methods described here, to generate an additional series of independent Ts alleles for detailed functional analysis. Furthermore, mutagenized libraries of specific essential or nonessential genes can be screened for conditional viability under a variety of conditions that are normally sublethal in the wild-type strain (e.g., sublethal doses of drugs) using the methods described here.

2. DIPLOID SHUFFLE—PLASMID METHOD

2.1. General description of the diploid shuffle—plasmid method

Much like traditional plasmid-shuffling methods, mutants generated by this version of the diploid shuffle are plasmid-borne alleles that can be very easily transferred to and tested in different strain backgrounds. The experimental procedure is outlined in Fig. 8.1. First, the endogenous promoter (including the 5′-untranslated region, 5′-UTR, ~500 bp) and the terminator (3′-UTR, ~500 bp) of a gene of interest (or your favorite gene, YFG) are PCR-amplified and cloned in tandem onto a centromere-based yeast–Escherichia coli shuttle vector, which contains URA3 as the selectable marker in yeast cells, such as pRS416. The resultant promoter/terminator clone is subsequently linearized with an endonuclease, typically NotI, which cuts at a site preengineered between the promoter and terminator. Simultaneously, the sequence of YFG, including the whole open reading frame (ORF),

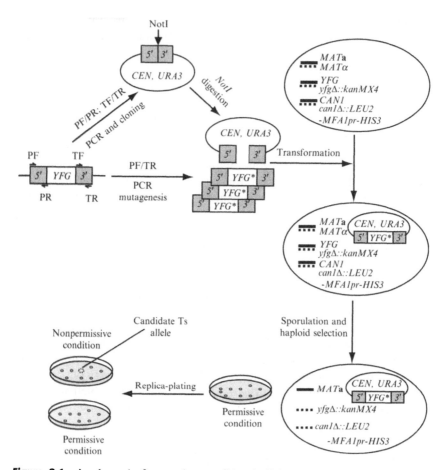

Figure 8.1 A schematic for creating conditional alleles using plasmid–chromosome shuffle. The promoter (5′) and terminator (3′) of *YFG* are separately PCR-amplified with primer pairs PF/PR and TF/TR, respectively, and cloned together onto a centromeric (CEN) yeast–*E. coli* shuttling vector. Here, PF stands for promoter forward, PR for promoter reverse, TF for terminator forward, and TR for terminator reverse. A *Not*I recognition site is engineered between the promoter and terminator. The resultant promoter/terminator clone is linearized with *Not*I digestion. In the meanwhile, the entire sequence of *YFG* gene, including the coding region and the promoter and terminator sequences, is mutagenized using error-prone PCR with the primer pair PF/TR. The mutagenesis PCR products and the linearized promoter/terminator plasmid DNA are mixed and transformed together into a haploid-convertible heterozygous diploid knockout mutant of the same gene (*MAT*a/α *YFG/yfg*Δ::*kanMX4 CAN1/can1*Δ::*LEU2-MFA1pr-HIS3*). The linearized promoter/terminator clones are repaired inside yeast cells mostly via homologous recombination using the cotransformed mutagenesis products (or YFG★ alleles) as the templates. Due to the extensive homology between the ends of the PCR product and the vector, >10⁵ recombinant clones can be easily generated. This pool of recombinants is then sporulated. Haploid *MAT*a G418ᴿ Ura⁺ cells are selected under a permissive condition on solid SC–Ura–Leu–His–Arg+G418+Can medium as single colonies, which are replica-plated onto two fresh plates and incubated under both permissive and nonpermissive conditions. Candidate alleles will grow under the permissive condition but not under the nonpermissive condition.

complete with the promoter and terminator regions, is randomly mutagenized with error-prone PCR. The linearized promoter/terminator plasmid and PCR products are then combined and cotransformed into a haploid-convertible heterozygous diploid deletion mutant ($YFG/yfg\Delta::kanMX4$) in the same gene being mutagenized. The mutagenized PCR product is thereby cloned into the $URA3$ plasmid via recombination mediated by the terminal homologous DNA sequences of both the PCR products and the linearized vector. The Ura$^+$ transformants are subsequently cultured in a sporulation medium and converted into haploid cells by growing on a medium that allow growth of only haploid $MATa$ G418^R Ura$^+$ cells. In these cells, the chromosomal wild-type copy of YFG is deleted, allowing direct observation of any phenotypes of the plasmid-borne alleles. To screen for conditional alleles, such haploid cells are first grown under a permissive condition such as low temperature. Colonies formed are subsequently replica-plated to fresh plates at permissive and nonpermissive conditions. Conditional alleles are identified as those grow under the permissive but not the nonpermissive condition and subsequently verified. This method has been used to create thermosensitive (Ts) alleles of multiple essential genes as well as a large collection of methyl methanesulfonate (MMS) hypersensitive alleles of $POL30$ (Huang et al., 2008; Lin et al., 2008). Here, we will outline the detailed methods to generating and verifying Ts alleles of an essential gene.

2.2. Materials

2.2.1. Media

Haploid selection synthetic medium SC–Ura–Leu–His–Arg+G418+Can: dextrose, 20 g/l; yeast nitrogen base without amino acids and ammonium sulfate, 1.7 g/l; SC–Ura–Leu–His–A dropout mix, 2 g/l; sodium glutamate, 1 g/l; G418, 200 mg/l; L-canavanine (Sigma, Cat# C1625), 60 mg/l; Agar, 2%. The sodium glutamate is substituted for ammonium sulfate as the nitrogen source and makes the G418 selection more reliable on the minimal medium.

SC–Ura: dextrose, 20 g/l; yeast nitrogen base without amino acids and ammonium sulfate, 1.7 g/l; SC–Ura dropout mix, 2 g/l; ammonium sulfate, 5 g/l; Agar, 2%.

Liquid YPD: yeast extract, 10 g/l; peptone, 20 g/l; dextrose, 20 g/l.

Solid and liquid sporulation medium: potassium acetate, 10 g/l; zinc acetate 0.05 g/l, with or without 2% agar, respectively.

Solid Luria Broth (LB) plus carbenicillin: yeast extract, 10 g/l; Tryptone, 5 g/l; sodium chrolide, 10 g/l; carbenicillin, 50 mg/l; Agar, 2%.

Liquid LB plus ampicillin: yeast extract, 10 g/l; Tryptone, 5 g/l; sodium chrolide, 10 g/l; ampicillin, 50 mg/l.

2.2.2. Yeast strains

The haploid-convertible heterozygous diploid knockout mutants (*MAT* **a**/α *ura3Δ0/ura3Δ0 leu2Δ0/leu2Δ0 his3Δ1/his3Δ1 lys2Δ0/LYS2 met15Δ0/ MET15 can1Δ::LEU2-MFA1pr::His3/CAN1 YFG/yfgΔ::KanMX*; Open-Biosystems Cat# YSC4428) (Pan *et al.*, 2006) are used to screen for Ts alleles. Chemically competent DH5α cells prepared as described (Inoue *et al.*, 1990) are used for cloning and plasmid recovering from yeast.

2.2.3. Plasmids

It is essential for this method to use a plasmid vector that contains the *YFG* promoter and terminator separated by a unique endonuclease (typically *Not*I) recognition site. Due to the limited auxotrophic markers available in the haploid-convertible heterozygous diploid knockout mutants, we normally use plasmids containing *URA3* as the selectable marker such as pRS416 (Brachmann *et al.*, 1998; Sikorski and Hieter, 1989) and YCplac33 (Gietz and Sugino, 1988). Other *URA3 CEN* vectors should also work.

2.2.4. Yeast genomic DNA

A genomic DNA sample isolated from the wild-type yeast strain BY4743 *MAT***a**/α (Brachmann *et al.*, 1998) was used as the template for cloning the promoter and terminator of *YFG* and for mutagenizing its entire sequence with PCR.

2.3. Methods

2.3.1. Constructing the promoter/terminator clone

In the past, we mostly used endonuclease restriction enzyme digestion and ligation to construct the promoter/terminator clone. First, the promoter and terminator of *YFG* are separately PCR-amplified using primers that contain endonuclease recognition sites, for example, *Hind*III/*Not*I for the promoter and *Not*I/*Bam*HI for the terminator. The PCR products are then digested with *Hind*III/*Not*I and *Not*I/*Bam*HI, respectively, and ligated to pRS416 (or YCplac33) digested with *Hind*III/*Bam*HI in a 3-piece ligation reaction. The ligation products are transformed into DH5α competent cells and candidate clones are selected on solid LB plus carbenicillin. More recently, we have adopted a modified version of the sequence and ligation independent cloning (SLIC) procedure (Li and Elledge, 2007) (Fig. 8.2). This method does not require endonuclease digestion of the inserts and thus greatly simplifies primer designs and experimental procedures, especially when a large number of genes are processed simultaneously. This is also a relatively new cloning technique and is thus described below in greater detail.

1. Design four PCR primers: promoter forward (PF), promoter reverse (PR), terminator forward (TF), and terminator reverse (TR). In addition to gene-specific sequences on their 3′ termini, the PF and TR

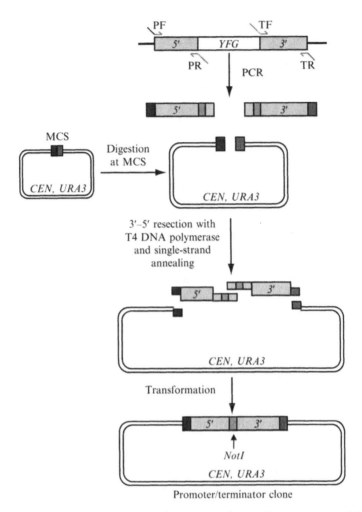

Figure 8.2 Constructing a promoter/terminator clone using sequence and ligation independent cloning (SLIC). The promoter (5′) and terminator (3′) of *YFG* are separately PCR-amplified from yeast genomic DNA with primer pairs PF/PR and TF/TR, respectively. In the meanwhile, a centromeric yeast–*E. coli* shuttling vector is linearized with endonuclease digestion at the multicloning site (MCS). The PCR products and the linear vector plasmid are mixed together and processed with T4 DNA polymerase to create 5′ single-stranded overhangs. The PCR primers are designed in such a way that the PCR products and the vector can be assembled via a homology-mediated single-strand annealing process. A *Not*I site is engineered between the cloned promoter and terminator. An aliquot of the annealing reaction is transformed into *E. coli* competent cells. This is a modified version of the SLIC procedure originally described by Li and Elledge (2007).

primers each contains a 30 bp sequence at the 5′ end that is either identical or complimentary to the ends of pRS416 (or YCplac33) linearized at the multicloning site (Fig. 8.2). The PR and TF primers are completely complementary and both contain three parts: ~20 bp promoter- or terminator-specific sequences on both ends and a *Not*I recognition site (8 bp) in the middle (Fig. 8.2).

2a. PCR amplify the promoter and terminator according to the conditions listed in Table 8.1. Here genomic DNA from BY4743**a**/α is used as the PCR template. Platinum Pfx DNA polymerase (Invitrogen, Cat# 11708) is used due to its relative robustness and high fidelity. Other enzymes with similar features such as the Phusion (New England Biolabs, NEB, Cat# F-540) and KOD (Novagen, Cat# 71085-3) DNA polymerases can also be used.

2b. Digest ~1 μg of pRS416, YCplac33, or any other *URA3* plasmid with an endonuclease at the multicloning site to generate ends that are competent in cloning the PCR products via SLIC.

3. Gel-purify the PCR products and the digested vector DNA using a gel extraction kit (Qiagen, Cat# 28706 or equivalent) by following the manufacturer's instruction. Here the PCR products of both the promoter and terminator are combined during purification. Elute each purified sample in 25 μl of provided elution buffer.

4. Set up a reaction according to Table 8.2. Here DNA resection in the 3′–5′ direction (by T4 DNA polymerase, NEB, Cat# M0203) and annealing between complementary single-strand overhangs occurs in the same reaction.

5. Immediately use 2 μl of the above reaction to transform 20 μl of chemically competent DH5α cells by following a standard protocol. This includes incubating the cell/DNA mixture sequentially on ice for 30 min, at 42 °C for 90 s, back on ice for 2 min, and at 37 °C for 30 min (in the presence of 200 μl of liquid LB medium).

Table 8.1 PCR amplification of promoters and terminators

Component	Volume/reaction (μl)
10× Platinum Pfx DNA Polymerase buffer	2.5
dNTP (2.5 mM each)	2
Primer mix (5 μM each)	2
Yeast genomic DNA (~200 ng/μl)	1
ddH$_2$O	17.25
Platinum Pfx DNA Polymerase (2.5 U/μl)	0.25
Total	25

PCR conditions: 94 °C 4 min; 30× (94 °C for 30 s, 55 °C for 30 s, 72 °C for 45 s); 72 °C for 7 min; hold at 4 °C.

Table 8.2 A T4 DNA polymerase resection reaction

Component	Volume (μl)
10× NEB *Bam*HI Buffer	2
10× BSA	2
ddH$_2$O	3.7
T4 DNA polymerase (3 U/μl)	0.3
Vector (100 ng/μl)	2
PCR product (50 ng/μl)	10
Total	20

Upon set up, the reaction is incubated at 25 °C for 30 min. Then, place it on ice until *E. coli* transformation.

6. Plate 100 μl of the transformation mixture on solid LB plus carbenicillin to select for single colony transformants.
7. Screen for positive clones using colony-PCR with the PF and PR primers according to Table 8.1 but with subtle modifications. Here cells from single colonies instead of yeast genomic DNA is used to provide PCR templates.
8. Prepare plasmid DNA samples from two to three positive clones using a mini-spin kit (Qiagen, Cat# 27106 or equivalent) by following the manufacturer's instructions.
9. Verify each plasmid with DNA sequencing by using two primers that read toward the inserts (promoter and terminator) in both directions from the vector backbone.

2.3.2. Mutagenizing YFG with error-prone PCR

Mutagenesis of *YFG* using error-prone PCR is performed essentially as described previously (Leung *et al.*, 1989). Again, a genomic DNA sample of the wild-type yeast strain BY4743**a**/α is used as the DNA template. The PF and TR primers described above are used to amplify the full-length gene and ~500 bp flanking sequences. TaKaRa Ex Taq (Cat# RR001A) or LA Taq (Cat# RR002B), which are ~4 times more accurate than the normal Taq polymerase, are used here due to their robustness. Induction of mutation rates is achieved by adding Mn^{2+} in the PCR at a final concentration of 10–150 μM that are arbitrarily defined, with higher concentrations for smaller genes (genes' sizes range from 0.5 to 5 kb).

1. Set up four independent reactions for each gene according to Table 8.3 to reduce potential founder effects.
2. After PCR, pool the samples.
3. Examine the PCR products by agarose gel electrophoresis by using a 1–2 μl sample to ensure successful PCR amplification.

Table 8.3 Error-prone PCR conditions

Component	Volume/reaction (μl)
$10\times$ Ex Taq buffer	5
dNTP (2.5 mM each)	4
Primer mixture (5 μM each)	4
Yeast genomic DNA ($\sim$200 ng/μl)	2
MnCl$_2$ (1–15 mM)	0.5
ddH$_2$O	34.25
Ex Taq DNA polymerase (5 U/ml)	0.25
Total	50

PCR conditions: 94 °C 4 min; 30 $\times$ (94 °C for 30 s, 55 °C for 30 s, 72 °C for 1 min/kb); 72 °C for 7 min; hold at 4 °C. Lower MnCl$_2$ concentrations are used for larger genes and higher concentrations for smaller genes.

2.3.3. Linearizing plasmid DNA of the promoter/terminator clone

Two micrograms of plasmid DNA of the promoter/terminator clone is digested with *Not*I (NEB, Cat# R0189) in NEB buffer 3 in a 20 μl reaction. The reaction is incubated at 37 °C for an overnight and subsequently at 65 °C for 20 min to inactivate *Not*I. A small aliquot of the digestion product is examined by agarose gel electrophoresis to ensure complete digestion of the plasmid.

2.3.4. Combining and concentrating the PCR products and digested vector

The mutagenized PCR products (approximately 10–20 μg in 200 μl) and linearized plasmid DNA of the promoter/terminator clone ($\sim$2 μg) are next combined and concentrated by ethanol precipitation.

1. Transfer both the mutagenesis PCR products and *Not*I-digested promoter/terminator plasmid DNA into a 1.7 ml microcentrifuge tube and adjust volume to $\sim$200 μl by adding ddH$_2$O.
2. Add 5 μl of 4 M ammonium acetate (pH 7.0) and 500 μl of 100% ethanol to the DNA sample and mix well by briefly vortexing. Place on ice for 10 min.
3. Precipitate DNA by spinning at >12,000 rpm in a microcentrifuge for 7 min. There should be a tiny whitish DNA pellet at the bottom of the tube.
4. Carefully aspirate the liquid and wash the DNA pellet once with 300 μl of 70% ethanol.
5. Spin at >12,000 rpm in a microcentrifuge for 3 min and carefully aspirate ethanol.
6. Dry the DNA pellet in a speed-vac.
7. Resuspend DNA in 28 μl of sterile ddH$_2$O.

2.3.5. Transforming yeast cells

The concentrated DNA sample of PCR products and the linearized vector is next transformed into the corresponding haploid-convertible heterozygous diploid yeast knockout mutant to create a mutagenized library of YFG.

1. Inoculate the haploid-convertible $YFG/yfg\Delta::kanMX4$ heterozygous diploid mutant into 5 ml liquid YPD and incubate at 30 °C for overnight in a roller drum.

2. Transfer an aliquot of the overnight culture into 50 ml YPD liquid (starting at 0.125 OD_{600nm}/ml in a 250 ml Erlenmeyer flask) and incubate at 30 °C with shaking (at 200 rpm) until a cell density of ~0.5 OD_{600nm}/ml is obtained.

3. Harvest the culture in a 50 ml conical tube by spinning at 4000 rpm for 3 min in a bench top centrifuge and discard the medium.

4. Resuspend cells in 10 ml of sterile ddH_2O, centrifuge as described in the previous step, and discard the supernatant.

5. Resuspend cells in 10 ml of 0.1 M lithium acetate (LiOAc), centrifuge, and discard the supernatant as before.

6. Resuspend cells in residual 0.1 M LiOAc in a total volume of 100 μl in a 1.7 ml microcentrifuge tube.

7. Make a transformation mixture in this order and mix well: 480 μl of 50% polyethylene glycol (PEG-3350, JTBaker, Cat# JTU221-9), 72 μl of 1 M lithium acetate, 40 μl of heat-denatured herring sperm DNA (10 mg/ml, Sigma, Cat# D6898) and 28 μl of DNA sample to be transformed (error-prone PCR products and linearized promoter/terminator clone).

8. Add the transformation mixture into the yeast competent cells prepared in step 6 and immediately mixed well by pipetting with a P1000 pipettor followed by vortexing (VMR, Vortexer 2) at top speed for 5–10 s.

9. Incubate the transformation reaction in a 30 °C incubator for 30 min.

10. Add 72 μl of dimethyl sulfoxide (DMSO; Qbiogene DMSO0001, Molecular Biology Grade) to the transformation reaction and immediately mixed thoroughly by vortexing at top speed for 5–10 s. DMSO is intrinsically sterile and no further sterilization is needed.

11. Incubate the transformation reaction in a 42 °C water bath for 13 min.

12. Spin down cells at 3600 rpm in a microcentrifuge for 30 s to pellet cells.

13. Aspirate the supernatant and resuspend cells in 1 ml of sterile ddH_2O.

14. Take 0.2 μl (1/5000) of the resuspended cells and plate on a SC–Ura plate to determine transformation yield. A successful reaction typically yields a library of 10^5–10^6 independent Ura$^+$ transformants, with >90% being the recombinants between the PCR products and the linearized plasmid DNA of the promoter/terminator clone. The remainder is primarily empty vector molecules that were never cut in the first place or rejoined by nonhomologous end joining.

2.3.6. Sporulation

The rest of the yeast transformation reaction is either incubated in 50 ml of fresh liquid SC–Ura at 30 °C for 2 days to allow propagation of the library or can be sporulated immediately to convert the transformants into a library of haploid spores that harbor mutant alleles of *YFG* on a plasmid (see below).

1. Grow the yeast transformants nonselectively in 50 ml of liquid YPD by incubating at 30 °C with shaking (180–200 rpm) for 3 h to refresh cells.
2. Harvest cells in a 50 ml conical tube by spinning at 4000 rpm for 3 min in a bench top centrifuge and discard the medium.
3. Resuspend cells in 40 ml of sterile ddH$_2$O and spin at 4000 rpm for 3 min.
4. Discard the supernatant and resuspend cells in 50 ml of liquid sporulation medium.
5. Incubate the sporulation culture in a 250 ml flask at 25 °C for 4–6 days with shaking (180–200 rpm). This will typically give rise to 20–40% of sporulation efficiency when checked under a microscope.
6. Repeat Steps 2 and 3.
7. Discard the supernatant and resuspend cells in 10 ml of sterile ddH$_2$O.
8. Spread aliquots (200 μl) of 10× serial dilutions of the sporulation culture onto individual plates of the haploid selection medium SC–Ura–Leu–His–Arg+G418 +Can and incubate at 25 °C for 2–3 days to determine the efficiency of producing *MA Ta* G418^R Ura$^+$ haploid cells.
9. Store the rest of the spores in ddH$_2$O at 4 °C for later use. The viability of spores can be maintained for a few weeks in this way.

2.3.7. Screening for Ts alleles

After the titer of *MA Ta* G418^R Ura$^+$ haploid cells is determined, the library is screened for potential Ts mutants. We typically screen $\sim$4000 clones for each gene.

1. Spread the spores on solid SC–Ura–Leu–His–Arg+G418+Can, aiming for a density of $\sim$400 *MA Ta* G418^R Ura$^+$ haploid colonies formed on each of 10 plates. Store the rest of the spores at 4 °C as a backup.
2. Incubate the plates at 25 °C for 3 days to allow formation of colonies of $\sim$2 mm in diameter.
3. Replica-plate the colonies from each plate to two fresh plates of the same haploid selection medium and mark orientations of the plates. It is best to prewarm the "nonpermissive" plate to 37 °C prior to replica plating.
4. Incubate one of the daughter plates at 25 °C and the other at 37 °C for 1 day.
5. Compare growth of each colony on both plates to assess potential Ts phenotype and select alleles that form relatively robust colonies at 25 °C

but ghost colonies at 37 °C as candidate Ts mutants. If no Ts colony is detected, one may need to screen more clones by using the backup spores stored at 4 °C, by using 38 °C as the restrictive temperature, or both.

2.3.8. Confirming Ts mutants

Candidate mutants are picked and restreaked onto the same haploid selection media and retested for the Ts phenotypes by incubating at 25 and 37 °C. The plasmids are next recovered from those confirmed to be Ts in this initial assay and reintroduced individually into the same haploid-convertible heterozygous diploid mutant to test whether the Ts phenotype is linked to the plasmids.

1. Grow each candidate Ts mutant in 1.5 ml of liquid SC–Ura at 25 °C until saturated, typically takes 1–2 days, depending on the particular alleles.
2. Harvest cells of each strain in a microcentrifuge tube by spinning at 4000 rpm for 1 min and discard the medium.
3. Resuspend cells in 500–1000 μl of sterile ddH$_2$O and repeat Step 2.
4. Resuspend cells in 40 μl Lysis Buffer (50 mM Tris–HCl 7.5, 10 mM EDTA) containing 5 mg/ml of Zymolyase 100T (MP Biomedicals, Cat# 320931) and incubate at 37 °C for 1 h with shaking at 15-min intervals.
5. Add 40 μl of 10% SDS and mix well by vortexing or pipetting.
6. Add 160 μl of 7.5 M ammonium acetate and mix well by vortexing or pipetting.
7. Incubate the sample at −80 °C for 15 min.
8. Centrifuge at >12,000 rpm for 5 min at 4 °C.
9. Carefully transfer 100 μl of clear supernatant to a new tube that contains 75 μl of isopropanol and mix well by inverting the tube several times.
10. Centrifuge at >12,000 rpm for 7 min at room temperature to precipitate DNA.
11. Wash the DNA pellet once with 100 μl 70% ethanol.
12. Carefully aspirate ethanol and dry the DNA pellet in a speed-vac.
13. Resuspend DNA pellet in 20 μl of sterile ddH$_2$O.
14. Use 2 μl of the DNA sample to transform chemically competent DH5α cells as mentioned in the "*constructing the promoter/terminator clone*" section.
15. Purify plasmid DNA from a representative bacterial transformant using a kit (Qiagen, Cat# 27106 or equivalent).
16. Individually transform each plasmid into the haploid-convertible heterozygous diploid knockout mutant as described previously (Gietz *et al.*, 1995) and select for Ura[+] transformants.

17. Patch two representative Ura$^+$ transformants for each plasmid on solid SC–Ura and incubate at 30 °C for overnight.
18. Transfer cells to a sporulation plates and incubate at room temperature for 4–6 days.
19. Resuspend cells from each sporulated culture in sterile ddH$_2$O.
20. Spot aliquots of 10× serial dilutions onto two SC–Ura–Leu–His–Arg+G418+Can plates.
21. Incubate one plate at 25 °C and the other at 37 °C for 2–3 days.
22. Compare the growth of *MAT*a G418^R Ura$^+$ cells at both temperatures. True Ts alleles will form colonies at 25 °C but not at 37 °C.

3. Diploid Shuffle—Chromosome Method

3.1. A general description

The "diploid shuffle" chromosome method has now been used to systematically screen for missense mutations that result in temperature-sensitive (Ts) alleles of hundreds of essential genes, with each allele directly integrated at its endogenous chromosomal location and flanked with the "barcodes" of the corresponding yeast knockout mutant (Fig. 8.3; Ben-Aroya *et al.*, 2008, unpublished data). First, *YFG*, including its promoter and terminator regions, is mutagenized with error-prone PCR (Fig. 8.3A). The mutagenized PCR product is next cloned into SB221+Topo-TA (Fig. 8.3B). This plasmid contains the *URA3* gene flanked by the 5′ and 3′ regions of *KanMX*. The Topo-TA cloning site (Invitrogen) has been inserted in between the *KanMX* 5′ end and the *URA3* gene. This site allows direct cloning of each of the PCR products, without the need for any further modifications. The result of the cloning step is a library of mutagenized *YFG*, which is then transformed into *E. coli*, and digested to release linear fragments (following DNA purification) (Fig. 8.3B). The linear fragments are directly transformed into the corresponding strain from the haploid-convertible heterozygous *YFG/yfg*Δ*::kanMX* diploid YKO collection by selecting for Ura$^+$ transformants (Fig. 8.3C). The ∼700 bp *KanMX5′* and *KanMX3′* fragments direct the mutagenized *YFG* library into the *yfg*Δ*::KanMX* genomic locus via homologous recombination, with retention of the original bar codes flanking each gene (Fig. 8.3C). Pools of Ura$^+$ cells containing the mutant alleles are sporulated (Fig. 8.3E). Spores thus formed are spread on a haploid selective medium and incubated at 25 °C for colony formation. Only haploid *MAT*a Ura$^+$ spores containing the integrated mutant allele of *YFG* can grow on this medium. Colonies formed on selective medium are replica-plated and incubated at 37 °C (Fig. 8.3F). Colonies growing at 25 °C but not at 37 °C are selected as potential Ts

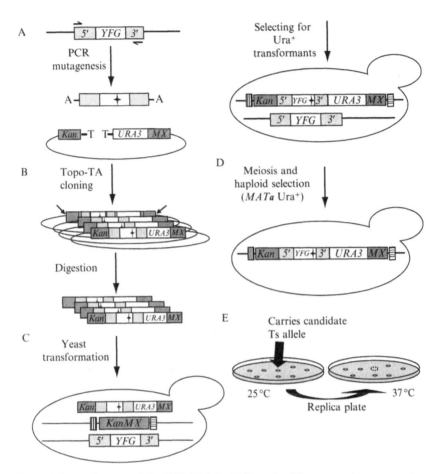

Figure 8.3 A diagram of the "Diploid shuffle" method for generating temperature-sensitive alleles. (A) Genomic DNA containing *YFG* and its 5′ and 3′ regions is used as the template for PCR mutagenesis. Two black horizontal arrows represent the gene-specific primers used. The mutagenized PCR products are cloned into the vector SB221+Topo-TA (mutations are represented by black stars). The Topo-TA cloning site is represented by a black T, the A overhang protruding from the PCR product is represented by a black A. Left gray bar represents the 5′ half of the *KanMX* selectable marker (*Kan*), the right gray bar represents the other half of the *KanMX* selectable marker (MX). The *NotI* restriction sites are indicated by two diagonal black arrows. (B) The product of the cloning step is a library of a mutagenized YFEG. The library is then transformed into *E. coli*, and digested with *NotI* to release linear fragments (following DNA purification). (C) The linearized library is transformed into the corresponding heterozygous diploid strain. Bars that flank the *KanMX* knockout represent the two barcodes. (D) Heterozygous diploid transformants are sporulated (following meiosis), and *MAT*a Ura⁺ haploids spores are selected on haploid selective medium at 25 °C. (E) Selection of temperature-sensitive candidates following the replica plating and incubating at 25 and 37 °C. Back arrows identify a potential Ts allele.

alleles, and retested. In summary, the final product of the diploid shuffle approach is a confirmed *MAT***a** strain from the YKO collection genetic background containing a *URA3* marked Ts allele of a specific gene integrated into its endogenous locus and flanked by both barcodes. In addition, each strain contains a *LEU2-MFA1pr-HIS3* reporter integrated at the *CAN1* locus.

In addition to creating Ts alleles, the diploid shuffle-chromosomal method can also be used to transfer existing alleles into the knockout strain background from other strain backgrounds and vice versa. Using the Topo-TA plasmid and protocol described in Fig. 8.1, any extant Ts allele can be easily transferred to the deletion collection genetic background (referred to as "allele transfer-in"). The result is an integrated allele, marked by *URA3*, and flanked by the appropriate barcodes. Moreover, using the Topo-TA plasmid, any PCR product (mutant allele, fusion protein, heterologous gene expression cassette, etc.) can be introduced at any of the 6000 genomic sites carrying a *KanMX* replacement cassette as the integration site, depending on the specific deletion mutant chosen as the recipient strain. By using primers that are external to the primers used for the original mutagenesis, the *URA3* marked Ts-allele or gene construct can be easily transferred from the deletion set genetic background to any other strain of interest, and replace the wild-type copy in the recipient strain by homologous recombination (referred to as "allele transfer-out," see Fig. 8.4). Thus, each Ts mutation or gene construct can be analyzed for more specific phenotypes of interest in a variety of genetic contexts.

3.2. Materials

3.2.1. Media

Haploid selection medium SC–Ura–Leu–His–Arg+Can: dextrose, 20 g/l; yeast nitrogen base without amino acids and ammonium sulfate, 1.7 g/l; SC–Ura–Leu–His–Arg dropout mix, 2 g/l; sodium glutamate, 1 g/l; L-canavanine, 60 mg/l; Agar, 2%.

YPD: yeast extract, 10 g/l; peptone, 20 g/l; dextrose, 20 g/l; Agar, 2%.

YPD+G418: yeast extract, 10 g/l; peptone, 20 g/l; dextrose, 20 g/l; Agar, 2%; G418, 200 mg/l.

Diploid selection medium SC–Ura+ClonNAT: dextrose, 20 g/l; yeast nitrogen base without amino acids and ammonium sulfate, 1.7 g/l; SC–Ura dropout mix, 2 g/l; sodium glutamate, 1 g/l; ClonNAT, 200 mg/l; Agar, 2%. The sodium glutamate is substituted for ammonium sulfate as the nitrogen source and makes the ClonNAT selection more reliable on the minimal medium.

Others are the same as described in Section 2.

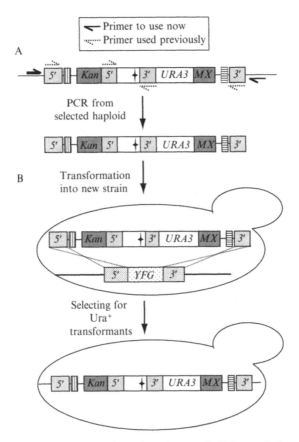

Figure 8.4 Allele transfer-out. Unless otherwise stated, all the symbols are as in Fig. 3. (A) In this specific example, genomic DNA containing a Ts allele that was generated by the diploid shuffle method PCR-amplified is used. Two black arrows represent the primers used for amplification. These primers are specific to the 5′ and 3′ regions of YFG in the recipient strain. They are also external to the two primers originally used to generate the Ts allele in the donor strain (represented by two broken arrows). (B) The PCR product is transformed to the strain of interest. The specific example shows allele transfer to a haploid strain. However, it may be desirable to transfer the allele to a diploid strain first, followed by sporulation and tetrad dissection, if the Ts allele being transferred is inviable in the specific genetic context of interest. The Ts allele replaces the wild-type YFG by homologous recombination (represented by dashed lines) and give rise to Ura⁺ transformants (indicated by the Ts phenotype).

3.2.2. Strains

The genotype of haploid-convertible heterozygous diploid strains is similar to those described in Section 2. Haploid Ts strains have the following genotype: *MAT*a *ura3Δ0 leu2Δ0 his3Δ1 lys2Δ0* (or *LYS2*) *met15Δ0* (or *MET15) can1Δ::LEU2-MFA1pr::HIS3 yfg-ts::URA3*.

BY4742-*ade2101-NatMX* is a *MATα* wild-type haploid strain in which the *NatMX* gene is linked to the *ade2-101* ochre mutation. This strain is used to confirm the *MATa* Ura$^+$ Ts candidate: *MATα his3Δ1 leu2Δ0 lys2Δ0 ura3Δ0 ade2-101-NatMX*.

OneShot® TOP 10 Electrocomp Cells (Invitrogen Cat# C4040-52).

3.2.3. Plasmids

SB221 was derived from M4758 (Voth *et al.*, 2003). In M4758, the *Bgl*II/ *Kpn*I and *Sph*I/*Eco*RI fragments contain the TEF promoter (388bp) and terminator (262bp), respectively, of *Ashbya gossypii*. In SB221 these fragments were replaced by a *Bam*HI/*Kpn*I PCR fragment (731bp) containing the TEF promoter plus half of the *KanMX* gene, and a *Sph*I/*Eco*RI fragment (751bp) containing the other half of the *KanMX* gene and the TEF terminator. In both cases, the template for PCR products was the *KanMX* gene used for constructing the heterozygous diploid collection. Finally, the *Bam*HI site of SB221 was adjusted with a Topo-TA site (invitrogen) to create SB221-Topo-TA.

3.3. Methods

3.3.1. PCR mutagenesis

This can be carried out as described in Section 2. However, we have successfully used a slightly different condition for all the experiments using this diploid shuffle-chromosomal method. Here, we have exclusively used LA Taq DNA polymerase (Cat# RR002B) and 150 μM MnCl$_2$. One PCR of 50 μl, instead of four, is normally set up for each gene. Two primers, which allow amplification of the entire coding region, 250–300 bp of the 5'-UTR, and 150–200 bp of the 3'-UTR of each gene, are used for PCR.

3.3.2. Cloning the PCR products

The mutagenized PCR products are purified and cloned into *E. coli* cells via electroporation.

1. Purify the PCR product using the ChargeSwitch® Clean-up Kit (Invitrogen Cat# CS12000).
2. Set up a ligation reaction that includes the following components: 0.5 μl SB221 Topo-TA, 1.0 μl ligation Buffer (300 mM NaCl, 15 mM MgCl$_2$), purified PCR product (100 ng/1 kb), and ddH$_2$O to a total volume of 6 μl.
3. Incubate the reaction at room temperature overnight.
4. Store the reaction at 4 °C (if using soon) or at (−20 °C).
5. Take 1.5 μl of the ligation products and transform into an aliquot of 50 μl OneShot® TOP 10 Electrocomp Cells (Invitrogen Cat# C4040-52) via electroporation.

6. Add 200 μl of LB into the transformation and incubate at 37 °C for 1 h.
7. Plate a 1 μl aliquot onto a LB plus ampicillin (LB-amp) plate to estimate the transformation efficiency.
8. Transfer 100 μl of the transformation suspension into a flask containing 50 ml of LB-amp liquid media and incubate at 37 °C for ~12 h to amplify the plasmid library. If necessary, store the rest of the transformation suspension (approximately 150 μl) in 20% glycerol at −70 °C for later use.
9. Count the number of colonies on the 1 μl plate and multiply by the microliters inoculated into the LB-amp liquid media to estimate the complexity of the library. (The total number of colonies should be ~150,000–200,000).
10. Purify the plasmid DNA library using a Plasmid Midi prep kit (Qiagen Cat# 12745).
11. Digest 2 μl of 10× diluted library DNA sample with NotI (10 U/μl, NEB, Cat# R018S) in a 10 μl reaction and examine the digests with agarose gel electrophoresis. This will allow estimation of the overall ligation efficiency. 2.5 kb+2.7 kb fragments represent the empty vector (Fig. 8.5A), while a successful cloning reaction is indicated by a band shift of the 2.7 kb fragment (in accordance with the insert size) (Fig. 8.5B).

3.3.3. Yeast transformation

Purified plasmid DNA sample of the random mutagenesis library is next digested with NotI or EcoRI (NEB, Cat# R0101T, 100 U/μl) and transformed into the corresponding haploid-convertible heterozygous knockout diploid mutant. The mutant alleles will be integrated into the endogenous locus via homologous recombination.

1. Digest 40 μg of library plasmid DNA (over night at 37 °C) in a total volume of 400 μl using either of the two restriction enzymes mentioned above.
2. Reduce the reaction volume to 100 μl using a speed-vac.
3. Use the digested library DNA to transform the corresponding heterozygous diploid strain. Split the digested DNA to two 50 μl aliquots, and use each of them to prepare a separate transformation suspension as already described (Pan et al., 2004).
4. Combine the two transformation suspensions (~1284 μl) and plate 1 μl onto a SC–Ura plate to estimate the transformation yield.
5. Transfer the rest of the transformation suspension to 50 ml liquid SC–Ura and incubate at 30 °C for 2 days to amplify the yeast library.
6. Count the number of colonies on the 1 μl plate and multiply by the microliters inoculated into the SC–Ura to estimate the total transformation yield. This number should be no less than 15,000. Otherwise, repeat yeast transformation procedure.

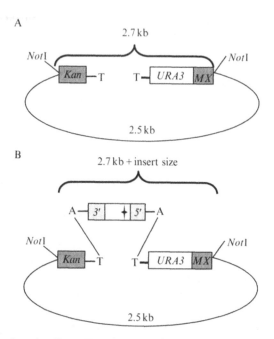

Figure 8.5 Testing the Topo-TA cloning efficiency in SB221. Unless otherwise stated, all the symbols are as in Fig. 3. (A) The SB221 vector is made up of the *URA3* gene flanked by a fragment containing the TEF promoter plus half of the *KanMX* gene (*Kan*), and another fragment containing the other half of the *KanMX* gene and the TEF terminator (*Mx*). This 2.7 kb fragment can be excised from the 2.5 kb backbone by restriction digestion with *Not*I. (B) Restriction digest with *Not*I following a successful Topo-TA cloning reaction is indicated by the 2.5 kb backbone and a band shift of the 2.7 kb fragment, in accordance with the insert size.

3.3.4. Yeast sporulation

Amplified yeast library is then sporulated to convert the heterozygous diploid into haploid spores similarly as described in Section 2. However, a slightly different haploid selection medium, SC–Ura–Leu–His–Arg+Can, is used to determine the efficiency of producing haploid *MA T***a** Ura$^+$ cells from the sporulation culture.

3.3.5. Screening for Ts alleles

After the plating efficiency is determined, typically ∼6000 haploid *MA T***a** Ura$^+$ colonies are screened for candidate Ts alleles for each gene.

1. Spread the sporulation culture on 15 plates of SC–Ura–Leu–His–Arg+Can at ∼400 colonies per plate.
2. Incubate these plates at 25 °C for 3–4 days.
3. Replica-plate colonies on each plate to a fresh plate of the same haploid selection media and mark the orientation of both the mother and the daughter plates.

4. Incubate the mother plates at 25 °C and the daughter plates at 37 °C.
5. Assess the Ts phenotype of each clone by comparing its growth on both mother and daughter plates on the next day.

3.3.6. Confirming Ts phenotypes

Restreak the Candidate Ts mutants onto two haploid selection media plates, and incubate at 25 and 37 °C. Once the Ts phenotype is confirmed, backcross the *MAT***a** Ura$^+$ Ts candidate to a wild-type BY4742-*ade2101-NatMX MAT*α strain. In this strain, the *NatMX* gene (which provides resistance to the drug ClonNAT) is linked to the *ade2-101* ochre mutation. Select the diploid cells by streaking onto SC–Ura+ClonNAT media, and then subjected to sporulation and dissection. Replicate the dissected tetrads onto YPD (25 and 37 °C), SC–Ura (25 °C), and YPD supplemented with G418 (25 °C). This confirms that: (1) the temperature sensitivity segregates in a Mendelian manner (2:2), and indicates that the Ts phenotype depends on a single mutated gene, (2) the Ts phenotype is linked to the *URA3* gene, and therefore cosegregates with the mutated PCR product, (3) the mutagenized PCR product was integrated at the correct genomic locus rendering the cells G418 sensitive (5′*kanMX::yfg-ts-URA3::*3′*kanMX*).

3.3.7. Allele transfer-in

This is carried out similarly as described above but with subtle modifications. A preexisting Ts allele (or other gene construct) is first PCR-amplified with the appropriate plasmid or genomic DNA as the template using a proofreading-competent polymerase (e.g., LA Taq DNA polymerase) that generate an "A" overhang on each 3′ end of the PCR product. The PCR products are cloned into SB221 using Topo-TA cloning. The cloned PCR products are then released together with the *URA3* marker and the *KanMX*5′ and *KanMX*3′ fragments from the vector backbone and transformed into the corresponding haploid-convertible heterozygous diploid mutant. Ura$^+$ yeast transformants are sporulated as a population and plated on SC–Ura–Leu–His–Arg+Can to select for haploid *MAT***a** Ura$^+$ cells at an appropriate colony density. Single colonies are then tested for the phenotype of interest (e.g., temperature sensitivity). Candidate clones are then backcrossed to a wild-type *MAT*α strain and analyzed with tetrad dissection to further confirm the phenotype.

3.3.8. Allele transfer-out

Here, we describe how to transfer a Ts allele generated by the diploid shuffle method to any other *ura3* strain background. Unless otherwise stated, methodologies are as mentioned above.

1. Design PCR primers that will amplify the whole 5′*kanMX::yfg-ts-URA3::*3′*kanMX* cassette. These primers are specific to the 5′ and 3′

regions of *YFG* in the recipient strain. They should also be external to the two primers originally used to generate the Ts allele in the donor strain (Fig. 8.4A).

2. Set up a PCR using genomic DNA of the Ts mutant as DNA template.
3. Transform the PCR product into your strain of interest and select for Ura$^+$ transformants.
4. If the recipient strain is haploid, screen for colonies with a Ts phenotype.
5. Backcross to a wild-type strain of the opposite mating type.
6. Verify that the Ts phenotype is linked to the Ura$^+$ by tetrad analysis.

The recipient strain can also be diploid. In this case, Ura$^+$ transformants are selected after Step 3, sporulated, and characterized by tetrad analysis to ensure that the Ts and Ura$^+$ phenotypes cosegregate. This will also allow testing whether the Ts allele is viable in the particular genetic context of the recipient strain. It is also possible that this allele is no longer Ts in this strain background. If so, representative Ura$^+$ transformants will need to be characterized with diagnostic PCR or sequencing to ensure that the 5′*kanMX*::*yfg-ts-URA3*::3′*kanMX* cassette is indeed integrated at the right locus.

4. PERSPECTIVES

The two variations of the "diploid shuffle" are both highly efficient methods for making Ts mutants. It is essentially always possible, by screening enough mutants using the methods outlined here, to find such alleles. An adaptation of the methods outlined here will be required for making Ts mutants in very large essential genes (in this case, the gene would be mutagenized in sections). The relative advantages of the methods described here are summarized in Table 8.4.

We have chosen to move forward with the chromosomal method for the generation of a genome-wide collection of Ts mutants as a community resource because it was felt that most users would prefer integrated copies that would not fluctuate or be lost from a subpopulation of cells at each division due to their being on an episome.

Table 8.4 Advantages of the two methods

Plasmid	Chromosomal	Both
Faster/easier	Mutant is single copy	Works for any essential gene
No special reagents needed	Mutant is stable (not episomal)	Mutants are tagged with molecular barcodes assigned to original knockout

The ability to generate a genome-wide collection compares favorably with other attempts to generate genome-wide resources for the study of essential genes, such as Tet-regulated alleles (Hartman *et al.*, 2001; Mnaimneh *et al.*, 2004) and dAMP (Schuldiner *et al.*, 2005). Both of those approaches, while having their distinct advantages, only produced well-behaved alleles in about 30% of the cases. Disadvantages of Ts mutants include the fact that they are not uniform with regard to nonpermissive temperature and leakiness, and the fact that the needed temperature shifts may induce heat shocks or other side effects that could potentially cloud phenotypic analyses. Nevertheless these alleles have been the bastion of traditional genetic analyses of essential genes. The Ts alleles we have sequenced include a mix of single amino acid substitutions and multi amino acid substitutions; however, we have not sequenced enough of these to develop extensive statistics on this. Studies of collections of Ts mutants in a variety of genes have allowed the empirical determination of their typical characteristics. On this basis, a scheme for predicting Ts mutants has been developed (P. Ye, J. Dymond, X. Shi, Y.-Y. Lin, X. Pan, J. D. Boeke, and J. S. Bader, submitted for publication). This is likely to prove very useful for designing Ts mutants in organisms in which extensive screening is impractical.

REFERENCES

Ben-Aroya, S., *et al.* (2008). Toward a comprehensive temperature-sensitive mutant repository of the essential genes of *Saccharomyces cerevisiae*. *Mol. Cell* **30**, 248–258.

Brachmann, C. B., *et al.* (1998). Designer deletion strains derived from *Saccharomyces cerevisiae* S288C: A useful set of strains and plasmids for PCR-mediated gene disruption and other applications. *Yeast* **14**, 115–132.

Edgar, R. S., and Lielausis, I. (1964). Temperature-sensitive mutants of bacteriophage T4d: Their isolation and genetic characterization. *Genetics* **49**, 649–662.

Gietz, R. D., and Sugino, A. (1988). New yeast–*Escherichia coli* shuttle vectors constructed with *in vitro* mutagenized yeast genes lacking six-base pair restriction sites. *Gene* **74**, 527–534.

Gietz, R. D., *et al.* (1995). Studies on the transformation of intact yeast cells by the LiAc/SS-DNA/PEG procedure. *Yeast* **11**, 355–360.

Hartman, J. L.T, *et al.* (2001). Principles for the buffering of genetic variation. *Science* **291**, 1001–1004.

Hartwell, L. H. (1967). Macromolecule synthesis in temperature-sensitive mutants of yeast. *J. Bacteriol.* **93**, 1662–1670.

Horowitz, N. H. (1950). Biochemical genetics of *Neurospora*. *Adv. Genet.* **3**, 33–71.

Huang, Z., *et al.* (2008). Plasmid–chromosome shuffling for non-deletion alleles in yeast. *Nat. Methods* **5**, 167–169.

Hughes, T. R. (2002). Yeast and drug discovery. *Funct. Integr. Genomics* **2**, 199–211.

Inoue, H., *et al.* (1990). High efficiency transformation of *Escherichia coli* with plasmids. *Gene* **96**, 23–28.

Jeong, H., *et al.* (2001). Lethality and centrality in protein networks. *Nature* **411**, 41–42.

Leung, D. W., Chen, E., and Goeddel, D. V. (1989). A method for random mutagenesis of a defined DNA segment using a modified polymerase chain reaction. *J. Cell. Mol. Biol.* **1**, 11–15.

Li, M. Z., and Elledge, S. J. (2007). Harnessing homologous recombination *in vitro* to generate recombinant DNA via SLIC. *Nat. Methods* **4**, 251–256.

Lin, Y. Y., *et al.* (2008). A comprehensive synthetic genetic interaction network governing yeast histone acetylation and deacetylation. *Genes Dev.* **22**, 2062–2074.

Mnaimneh, S., *et al.* (2004). Exploration of essential gene functions via titratable promoter alleles. *Cell* **118**, 31–44.

Pan, X., *et al.* (2004). A robust toolkit for functional profiling of the yeast genome. *Mol. Cell* **16**, 487–496.

Pan, X., *et al.* (2006). A DNA integrity network in the yeast *Saccharomyces cerevisiae*. *Cell* **124**, 1069–1081.

Schuldiner, M., *et al.* (2005). Exploration of the function and organization of the yeast early secretory pathway through an epistatic miniarray profile. *Cell* **123**, 507–519.

Sikorski, R. S., and Hieter, P. (1989). A system of shuttle vectors and yeast host strains designed for efficient manipulation of DNA in *Saccharomyces cerevisiae*. *Genetics* **122**, 19–27.

Voth, W. P., *et al.* (2003). New 'marker swap' plasmids for converting selectable markers on budding yeast gene disruptions and plasmids. *Yeast* **20**, 985–993.

CHAPTER NINE

QUANTITATIVE GENETIC INTERACTION MAPPING USING THE E-MAP APPROACH

Sean R. Collins,* Assen Roguev,† *and* Nevan J. Krogan†

Contents

Abstract

Genetic interactions represent the degree to which the presence of one mutation modulates the phenotype of a second mutation. In recent years, approaches for measuring genetic interactions systematically and quantitatively have proven to be effective tools for unbiased characterization of gene

* Department of Chemical and Systems Biology, Stanford University School of Medicine, Stanford, California, USA
† Department of Cellular and Molecular Pharmacology, California Institute for Quantitative Biomedical Research, University of California, San Francisco, California, USA

Methods in Enzymology, Volume 470
ISSN 0076-6879, DOI: 10.1016/S0076-6879(10)70009-4

function and have provided valuable data for analyses of evolution. Here, we present protocols for systematic measurement of genetic interactions with respect to organismal growth rate for two yeast species.

1. INTRODUCTION

Genetic interactions, which represent the modulation of the phenotype of one mutation by the presence of a second mutation, have long been used as a tool to dissect the functional relationships among sets of genes (Guarente, 1993; Kaiser and Schekman, 1990). Classically, researchers have looked for strong qualitative differences between observed phenotypes of double mutants and the phenotypes of the two related single mutants. For example, a relationship referred to as synthetic lethality is observed when two mutations are not lethal when present individually but, when combined, result in an inviable organism. Synthetic sick/lethal, or negative, interactions have been used as evidence that two genes act in independent but complementary pathways. For example, strong negative genetic interactions exist between two key machines working in parallel pathways involved in chromatin assembly, the HIR complex (Hir1, Hir2, Hir3, and Hpc2) and the CAF complex (Msi1, Cac2, and Rlf2) (Collins *et al.*, 2007a; Loyola and Almouzni, 2004) (Fig. 9.1B). On the other hand, if a normally deleterious mutation has no effect in the context of a second mutation, this is referred to as a positive genetic interaction, and often it identifies genes that act in the same pathway. Indeed, positive genetic interactions exist between members of the HIR complex as well as between the components that comprise the CAF complex (Collins *et al.*, 2007a) (Fig. 9.1B). These two classes of interaction have been extremely useful for deciphering the organization of molecular pathways in model organisms, but in fact they represent only two special cases within a much larger spectrum of interactions (Fig. 9.1A).

Recent advances in technology now make it possible to measure large numbers of genetic interactions systematically and in parallel in yeast (Pan *et al.*, 2004; Roguev *et al.*, 2007; Tong *et al.*, 2001), and it is possible to make these measurements quantitatively (Decourty *et al.*, 2008; Roguev *et al.*, 2008; Schuldiner *et al.*, 2005). Synthetic growth defects of varying magnitudes as well as a broad range of suppressive and masking interactions can be detected. Importantly, the ability to make these measurements in a high-throughput fashion also provides a novel and valuable context for understanding quantitative genetic interactions. Work has demonstrated that the pattern of interactions (represented as a mathematical vector) for a given mutation can be used as a multidimensional phenotype. These patterns can be compared, and sets of genes producing similar patterns (and hence sets that are functionally closely related) can be accurately identified (Schuldiner *et al.*, 2005; Tong *et al.*, 2004;

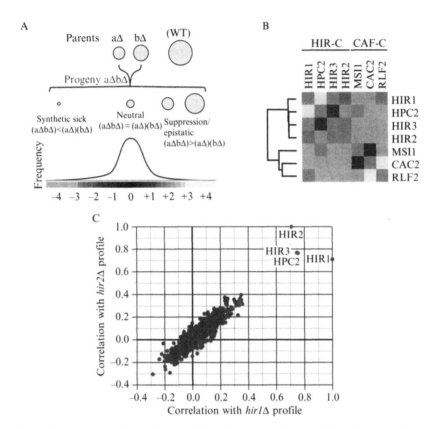

Figure 9.1 Epistatic interactions within and between chromatin assembly complexes. (A) The entire spectrum of genetic interactions. Quantitative genetic analysis can identify negative ((aΔbΔ) < (aΔ)(bΔ)), positive ((aΔbΔ) > (aΔ)(bΔ)), and neutral ((aΔbΔ) = (aΔ)(bΔ)) genetic interactions. (B) Genetic interactions between and within the HIR and CAF chromatin assembly complexes. Using the E-MAP approach (Collins *et al.*, 2007a), strong negative interactions were detected between components of the HIR-C (*HIR1, HPC2, HIR3,* and *HIR2*) and the CAF-C (*MSI1, CAC2,* and *RLF2*), which are known to function in parallel pathways to ensure efficient chromatin assembly. Conversely, positive genetic interactions were observed between components within each complex. Blue and yellow interactions correspond to negative and positive genetic interactions, respectively. (C) Plot of correlation coefficients generated from comparison of the genetic profiles from *hir1Δ* and *hir2Δ* to all other ~750 profiles from the chromosome biology E-MAP (Collins *et al.*, 2007a). Note the high pairwise correlations with *HIR1, HIR2, HIR3,* and *HPC2*. (See Color Insert.)

Ye *et al.*, 2005). For example, the patterns of interactions for HIR complex gene deletions are more strongly correlated to each other than to the patterns for other genes (Collins *et al.*, 2007a) (Fig. 9.1C).

We designed the epistatic miniarray profile (E-MAP) approach focusing on two key strategies to maximize the value of high-throughput genetic interaction measurements. In this approach, quantitative genetic interactions are measured (using a simple growth phenotype) systematically between all pairwise combinations of 400–800 rationally chosen mutations. The first strategy is measuring interactions quantitatively, which allows the detection and analysis of the complete spectrum of interaction strengths. We have found that this both improves our ability to use patterns of genetic interactions to identify sets of genes acting in a common pathway (Collins *et al.*, 2006), and that positive interactions (i.e., where a double mutant is fitter than expected) are particularly useful for making conjectures about gene function. Second, we aim to measure all pairwise interactions among a rationally chosen set of genes. This approach has several advantages. It increases the signal-to-noise ratio because the frequency of genetic interactions is higher between genes acting in related pathways (thus signal increases while noise is constant). It also provides a richer set of patterns for analysis. For example, having data for the components of all known DNA damage repair genes in an organism allows a researcher not only to classify a newly identified gene as a damage repair component, but also to assess which (if any) of the known repair pathways the new gene is most likely closely involved in. Importantly, with such a strategy, expansion is easy. New mutations of interest can rapidly be screened, and the results can be readily compared to and merged with the existing dataset.

Genetic interactions are measured not only in terms of growth or viability, but also can be (and have been) derived from other phenotypes (Jonikas *et al.*, 2009). In general, we can define a genetic interaction (ε_{AB}) between mutations A and B in terms of any quantitative phenotype P as the difference between the observed phenotype of the double mutant ($P_{AB,observed}$) and the expected phenotype of the double mutant ($P_{AB,expected}$) if no interaction exists between the two mutations:

$$\varepsilon_{AB} = P_{AB}^{observed} - P_{AB}^{expected}$$

This formulation clearly depends on our ability to compute $P_{AB,expected}$ as a function of $P_{A,observed}$ and $P_{B,observed}$. A theoretically derivable form for $P_{AB,expected}$ does not necessarily exist. However, a practically useful rule is that $P_{AB,expected}$ should account for the typical combined effect of two individual mutations with phenotypes $P_{A,observed}$ and $P_{B,observed}$. Since strong genetic interactions are rare (Pan *et al.*, 2004; Schuldiner *et al.*, 2005; Tong *et al.*, 2004), they manifest themselves as outliers that deviate from the surface that describes the broad trends in the majority of double-mutant data (Fig. 9.2). With this motivation, we define $P_{AB,expected}$ empirically to represent this typical combined effect. Additionally, if a simple functional form (e.g., the product $P_{A,observed} \times P_{B,observed}$) captures accurately the empirical relationship, this can be an extremely useful simplification.

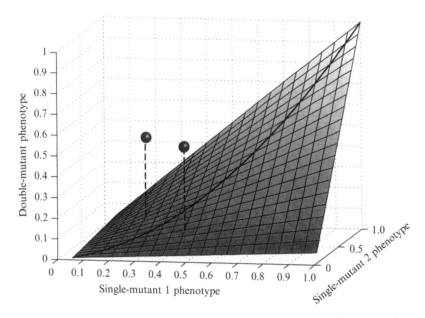

Figure 9.2   Genetic interactions as deviations from expected double-mutant pheno-types. An idealized smooth surface is shown to represent the expected combined effects of independent mutations. The surface shown is not based on real data, but is intended to serve as an abstract example. The *x*- and *y*-axes represent single-mutant phenotypes, scaled between 0 and 1. The height of the surface (along the *z*-axis) represents the corresponding expected double-mutant phenotype. This surface should accurately describe the empirical typical double-mutant phenotypes, and it should be symmetric about the line $y = x$. The grey spheres represent observations for specific double mutants. The quantitative interaction is represented by the vertical distance from the point to the surface. As both of these points lie above the surface, they represent positive interactions.

In this chapter, we describe in detail a method for measuring quantitative genetic interactions based on the area of yeast colonies. We have used this strategy (with some technical differences) for both *Saccharomyces cerevisiae* and *Schizosaccharomyces pombe* (Roguev *et al.*, 2007; Schuldiner *et al.*, 2006), and we provide here the details for each protocol. We also give a brief description of how analysis of the resulting data can be used to generate specific biological hypotheses.

 ## 2. SELECTION OF MUTATIONS FOR GENETIC ANALYSIS

Ultimately, comprehensive genetic interaction maps in both budding and fission yeasts may be generated where every possible pairwise double mutant is created and phenotypically assessed. Completion of such comprehensive genome-wide maps will represent a major accomplishment; however, it will still be some time until this is achieved. Additionally, our observations so far indicate that such maps will consist primarily of neutral genetic interactions. As mentioned above, the frequency of strong negative or positive genetic interactions between randomly chosen gene pairs has been estimated to be as little as 0.5%, but interactions are much more frequent between functionally related pairs of genes (Schuldiner *et al.*, 2005; Tong *et al.*, 2004). The E-MAP approach was devised to take advantage of this fact by specifically targeting genes that are likely to be functionally related. There are various ways by which sets of mutations can be selected for high-density, quantitative genetic interaction screening, and which is best strongly depends on the biological process that one wishes to interrogate, and the types of answers one would like to uncover. Genes can be chosen with the goal of identifying missing links within and between known pathways (e.g., what genes control the deposition of variant histones?), and they can also be chosen to address broader questions (to what extent are genetic interactions conserved during evolution?). We provide here an overview of strategies that have been used in the past.

The first E-MAP focused on approximately 400 genes likely to be involved in the early secretory pathway. In this case, gene selection was based largely on the localization of the corresponding proteins to either the Golgi apparatus or the endoplasmic reticulum (Schuldiner *et al.*, 2005). The rationale was that proteins residing in a common subcellular compartment are more likely to be functionally related. Of course, gene selection solely based on localization would miss important factors acting in an indirect fashion. For example, signaling proteins such as kinases may mediate strong effects on spatially distant processes.

Another recently generated E-MAP focused on factors involved in various aspects of chromosome function, including transcription, DNA repair/replication, and chromosome segregation (Collins *et al.*, 2007a). For this study, protein–protein interaction datasets were used as a primary source for gene selection. Using information from systematic affinity tag/purification mass spectrometry experiments (Collins *et al.*, 2007b; Gavin *et al.*, 2006; Krogan *et al.*, 2006), the genes whose corresponding proteins were contained within one or more complexes involved in the chromatin-related functions were targeted. Once again, this method alone would miss factors indirectly impinging on the processes.

Sets of genes have also been targeted based on the presence of character-ized domains (e.g., kinase domain, SET domain, etc.) or their likely molec-ular activity. For instance, an E-MAP was generated that comprised all kinases and phosphatases (both protein and nonprotein), regulatory subunits for these enzymes, phospho-binding proteins, and many factors known to be phosphorylated (Fiedler *et al.*, 2009). The genes in this E-MAP impinge on all processes in the cell and therefore this approach allowed for a global view of the genetic architecture of the signaling apparatus instead of high-density information on one specific biological process. This work revealed an enrichment of positive genetic interactions between kinases, phosphatases, and their corresponding substrates (Fiedler *et al.*, 2009) which would have been difficult to assess without a broad survey of these protein classes.

Finally, genes can be selected based on global, unbiased phenotypic screens. A recent study began with a genome-wide screen for mutations that affect the activity of the unfolded protein response (UPR) pathway (Jonikas *et al.*, 2009). Then, approximately 400 mutations that modulate basal UPR activity were selected and further characterized using systematic double-mutant analysis. Of course, gene selection in this fashion requires single mutants to have a measurable effect on the process in question, whereas multiple mutations may be required to see an effect when compensatory pathways exist.

In general, a combination of the approaches described above is likely to be most effective. For example, in the final chromosome function E-MAP (Collins *et al.*, 2007a), selection was not only based on composition of protein complexes but also on genome-wide genetic interaction screens using mutations of genes known to be integrally involved in processes of interest. In this way, a number of genes not associated with a known chromatin remodeling complex and not yet functionally annotated could be included in the analysis. Using this approach, we included the previously uncharacterized protein Rtt109 in the genetic analysis. Based on the genetic interaction profiling, we were able to link its function to the chromatin assembly protein Asf1 and identify it as the founding member of a new family of histone acetyltransferases responsible for K56 histone H3 acetylation (Collins *et al.*, 2007a; Driscoll *et al.*, 2007; Han *et al.*, 2007).

Recently, this analysis was extended to another, evolutionarily distant organism (*S. pombe*). Selection of genes for analysis was guided by the same criteria. Additionally, mapping of direct orthologs between the two organisms made it possible to observe some of the general trends in genetic interaction network evolution (Roguev *et al.*, 2007, 2008).

3. GENERATION AND MEASUREMENT OF DOUBLE MUTANT STRAINS

Here, we describe protocols for screening in both *S. cerevisiae* (SGA, synthetic genetic array) (Collins *et al.*, 2006; Schuldiner *et al.*, 2006; Tong and Boone, 2006) and *S. pombe* (PEM, pombe epistatic mapper) (Roguev *et al.*, 2007). A flowchart comparing the two protocols is presented in Fig. 9.3. Both methodologies are, in essence, high-throughput procedures for random spore analysis where the growth of double mutant cell pools is

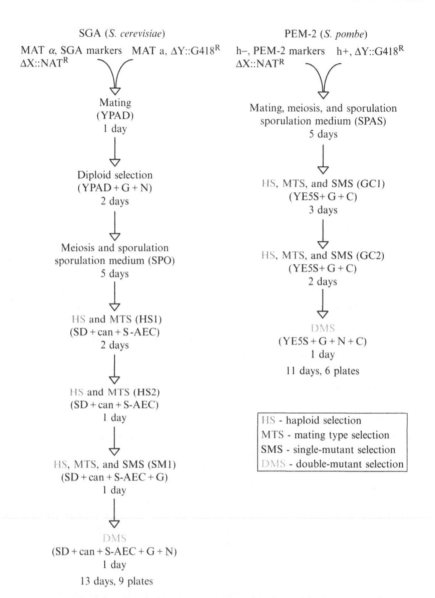

Figure 9.3   Overview of the experimental protocol. Flow charts outlining the series of selections used in *S. cerevisiae* E-MAP screens (left) and *S. pombe* PEM screens (right) are presented.

monitored on agar using high-density cell arrays. The same logic is followed in both model organisms. In each screen, a query strain containing one NAT-marked mutation is crossed to an array of strains carrying G418-marked mutations. An array of diploid strains is generated by mating and a change of growth media is used to induce meiosis and sporulation. A series of selections are then used to select for haploid double-mutant cells of a particular mating type carrying both mutations. The growth phenotypes of the resulting double mutants are assessed by measuring colony area after a defined period of time.

The two protocols differ in several important technical details due to biological differences between the two organisms. In *S. cerevisiae*, the mating step occurs on rich medium and nitrogen starvation is used to induce meiosis and sporulation. However, in *S. pombe*, the entire sequence (mating, meiosis, and sporulation) is induced by limited nitrogen, allowing the whole process to be carried out in a single step. Unlike in *S. pombe*, the diploid phase in *S. cerevisiae* is very stable, so an additional step enriching for diploids (diploid selection) is usually included in the SGA screen. In both systems, after the spores are germinated, the remaining diploid cells are killed (during haploid selection steps, HS), and only haploid cells of one mating type are allowed to grow (mating type selection, MTS). Importantly, one marker used in this selection comes from the parent strain of the opposite mating type, thus requiring that only haploid progeny from the initial mating can pass the selection. HS in both systems and MTS in *S. pombe* take advantage of recessive selectable markers. In *S. cerevisiae*, canavanine and S-AEC are used to select against the parent diploid cells, which are heterozygous for the *CAN1* and *LYP1* genes encoding transporters needed for import of these toxic compounds. A mating type-specific promoter driving the transcription of a conditionally essential metabolic gene (HIS3) is used for MTS. In *S. pombe*, a recessive allele providing resistance to cycloheximide has been engineered to perform both selections using a single marker (Roguev *et al.*, 2007). In both the SGA and PEM protocols, two rounds of HS and MTS are carried out. In *S. cerevisiae*, an additional single mutant selection (SMS) step enriching for single and double mutant haploids is performed. In the PEM protocol, the same media is used for HS, MTS, and SMS. Finally, a double-mutant selection (DMS) is performed, the arrays are photographed and the images analyzed.

All selections before the final DMS step (HS, MTS, and SMS) let single and double mutants compete within the same cell mixture. These competition steps greatly improve the sensitivity and dynamic range of the method by allowing the detection of subtle synthetic growth defects. The final colony sizes measured after growth on the DMS plates reflect both the growth rate of the double mutant cells during this final stage, as well as the fraction of cells deposited on the final plate which has the right genotype (both NAT- and KAN-marked mutations). This latter quantity is largely determined by competitive growth during earlier steps.

In planning a set of screens, one must decide how many replicate measurements will be sufficient to generate a high-quality dataset. By reanalyzing our earlier published data (Schuldiner *et al.*, 2005), we find that the first three independent replicate measurements give substantial improvements in the precision of the measured average colony size (Fig. 9.4B). Additional replicates beyond three improve data quality, but it may be worth sacrificing these marginal gains in exchange for the ability to complete more screens at lower cost.

Detailed SGA and PEM protocols using a Singer RoToR pinning station are given below. The protocols below may also be executed with other types of pinning devices such as hand pinning tools or alternate robotic pinners (Schuldiner *et al.*, 2006). If hand-pinning tools are used, a pinner with small diameter pins (e.g., VP384FP4 from V&P Scientific, San Diego, CA) should be used for the last step. The length of incubation times may need to be adjusted slightly for other pinning systems. We have used two days rather than one for the diploid selection and for the SMS steps for *S. cerevisiae* screens in 768 colony format with hand pinning tools. In our experience, results obtained using the Singer RoToR have better signal-to-noise than those obtained with hand-pinning devices. However, satisfactory results can be obtained with either method.

3.1. Basic SGA protocol

3.1.1. Genotypes

Query: MATα; his3Δ1; leu2Δ0; ura3Δ0; LYS2+; can1::STE2pr-SpHIS5 (SpHIS5 is the *S. pombe* HIS5 gene); lyp1Δ::STE3pr-LEU2; XXX:: NatMX
Library: MATa; his3Δ1; leu2Δ0; ura3Δ0; met15Δ0; LYS2+; CAN1+; LYP1+; YYY::KanMX

3.1.2. Growth media (all recipes for 1 l of media)

YPAD (YEPD + adenine)—Mix 10 g yeast extract, 20 g peptone, 120 mg adenine, 20 g Difco Agar, and DDW up to a final volume of 1 l. Autoclave. Add 50 ml of sterile 40% glucose.
SPO (sporulation media: NGS Agar)—Mix 20 g Difco Agar and 820 ml DDW in one flask. Mix 0.5 g –ura-trp amino acid powder mix (Sunrise Science #1010-100), 2.5 ml 20 m*M* uracil, 2.5 ml 20 m*M* tryptophan, and 163 ml DDW in a second flask. Autoclave each flask separately. Mix the two flasks and add 20 ml of 500 mg/ml *filter sterilized* potassium acetate. *Note*: do not autoclave the potassium acetate!
SD–HIS–LYS–ARG (for haploid selections)—Mix 20 g Difco Agar and 850 ml DDW in one flask. Mix 6.7 g yeast nitrogen base without amino acids, 2 g amino acid drop out mix (recipe below), and 100 ml DDW in a second flask. Autoclave both flasks. Mix the flasks and add 50 ml of 40% glucose.
SD(MSG)–HIS–LYS–ARG (for single- and double-mutant selections)— Mix 20 g Difco Agar and 850 ml DDW in one flask. Mix 1.7 g yeast

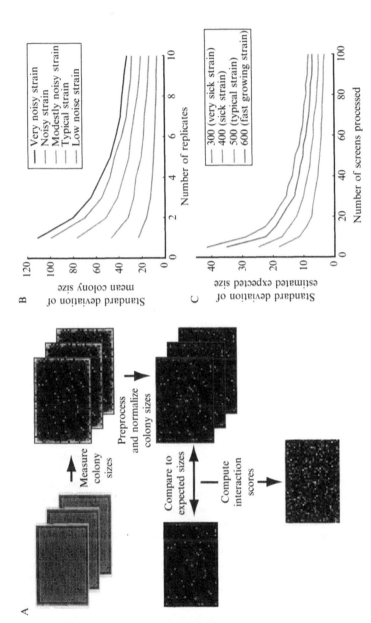

Figure 9.4 Overview of the data processing procedure. (A) A flowchart describing the data processing procedure for a single screen is shown. The first images are digital photographs of arrays of yeast colonies. In the following images heatmaps of either measured colony sizes or genetic interaction scores are shown. In colony size heatmaps, blue represents small colonies, black represents average-sized colonies, and yellow represents large colonies. In the genetic interaction heatmap, blue represents negative interactions, black represents neutral interactions, yellow represents positive interactions, and gray represents missing data (or data filtered out during quality control).

nitrogen base without amino acids and without ammonium sulfate (Becton, Dickinson and Company #233520), 2 g amino acid drop out mix (recipe below), 1 g monosodium glutamatic acid, and 100 ml DDW in a second flask. Autoclave the first flask and filter sterilize the second. Mix the two flasks together and add 50 ml 40% glucose.

Amino acid drop out mix: Mix 3 g adenine, 10 g leucine, 0.2 g para-aminobenzoic acid, and 2 g each of alanine, asparagine, aspartic acid, cysteine, glutamine, glutamic acid, glycine, inositol, isoleucine, methionine, phenylalanine, proline, serine, threonine, tryptophan, tyrosine, uracil, and valine.

Note: Be sure to pour level plates, as this is very important for the effectiveness of the robotic pinning steps.

3.1.3. Drug concentrations

NAT (N)—100 mg/l; G418 (G)—100 mg/l; S-AEC (S)—50 mg/l; canavanine (c)—50 mg/l

Note: for adding labile compounds (NAT, G418, S-AEC, and canavanine), the media should be cooled first until is hot, but no longer painful to touch the container.

Note: previous protocols used 200 mg/l NAT and G418, but we have found 100 mg/l to be sufficient.

3.1.4. Plates nomenclature

YPAD = YPAD; SPO = SPO; Diploid = YPAD + N + G; NAT = YPAD + N; G418 = YPAD + G; HS = SD–HIS–LYS–ARG + S + c; SM = SD–HIS–LYS–ARG + S + c + G; DM = SD–HIS–LYS–ARG + S + c + G + N

(B) The variability in measured growth phenotype (mean colony size over the replicate measurements) is shown as a function of number of experimental replicates. The curves shown were generated using data from 36 replicates of a control screen run while generating the early secretory pathway E-MAP (Schuldiner *et al.*, 2005). On a given curve, the point corresponding to *N* replicates was generated by randomly drawing *N* of the 36 replicates for a particular strain and computing the mean normalized colony size. This process was repeated 1000 times, and the standard deviation over these 1000 repeats was plotted. Each curve represents data for a different strain. Five different representative strains with different levels of measurement variability were chosen for analysis. (C) The variability of the empirically determined expected double-mutant phenotype was estimated as a function of the number of screens analyzed in parallel. In this case, sets of screens of the indicated size were drawn at random from a set of 329 screens completed at approximately the same time from the early secretory pathway E-MAP (Schuldiner *et al.*, 2005). For each point on each curve, 1000 random draws were completed, and each time expected colony size values were computed. The standard deviation of the expected colony sizes (over the 1000 random draws) was plotted for four representative strains with different single mutant phenotypes. (See Color Insert.)

HS1 and HS2 are HS plates used in two consecutive steps of the protocol.

3.1.5. Procedure

3.1.5.1. Preparation of query lawn

A lawn of cells of the query strain is prepared for use in the mating step.
Inoculate a liquid culture in YEPD from a single colony of the query strain (from a YEPD + NAT plate) and grow to saturation overnight.
Prepare a lawn of cells—Spread up to 500 μl of thick culture onto a NAT plate using glass beads and incubate at 30 °C for 2 days.

3.1.5.2. Preparation of library arrays (or "T-arrays" for "Target arrays") in 1536 format

The library array is replicated for use in the mating step.
Program: Agar-Agar. Replicate. Replicate Many. 1536
Parameters: Source plate: G418; Target plate: G418; Source pressure: 100%; Target pressure: 100%; Offset: Manual; Number of replicas: 3; Economy: ON; Revisit source: ON
Note: More than three replicas/source at this density may not be consistent and slow growing mutants (e.g., small colonies) may be lost. Incubate at 30 °C for 1 day.

3.1.5.3. Mating

Query and array cells are pinned on top of each other on a fresh plate for mating. First pin the T-array twice and then pin the lawn twice on top of it.
Program: Agar-Agar. Replicate. 1536
Parameters: Source plate: T-array and lawn; Target plate: YPAD; Source pressure: 100%; Target pressure: 100%; Offset: Manual; Economy: OFF
Incubate at 30 °C for 1 day.

3.1.5.4. Diploid selection

Cells are pinned from the mating plate and diploids are selected by using NAT and G418.
Program: Agar-Agar. Replicate. 1536
Parameters: Source plate: mating; Target plate: Diploid; Source pressure: 100%; Target pressure: 100%; Offset: Manual; Economy: OFF
Incubate at 30 °C for 1 day.

3.1.5.5. Sporulation

Replicate the arrays from Diploid onto SPO plates using 384 pads.
Do 2 pins per array onto the same target GC plate. This is the most critical stage of the protocol and transferring enough cells on the target plate is very important.

Program: Agar–Agar. Replicate. Replicate One. 384
Parameters: Source plate: Diploid; Target plate: SPO; Source pressure: 100%; Target pressure: 100%; Offset: OFF; Economy: OFF
Incubate for 5 days at room temperature in a humid environment.
Note: When loading the pads click "Modify" to change to 384 pads.
Note: The plates should be packed in plastic bags and kept in a humid environment to prevent drying.

3.1.5.6. HS1

Replicate the arrays from the SPO plates onto HS plates using 384 pads. Do 2 pins per array. The cells do not divide on the SPO media, so maximizing cell transfer at this step and at the previous step is critical.
Program: Agar–Agar. Replicate. Replicate One. 384
Parameters: Source plate: SPO; Target plate: HS (HS1); Source pressure: 100%; Target pressure: 100%; Offset: Automatic
Note: Incubate for 2 days at 30 °C.

3.1.5.7. HS2

Replicate the arrays from the HS1 plates onto HS2 plates using 1536 pads. Do 1 pin per array.
Program: Agar–Agar. Replicate. Replicate One. 1536
Parameters: Source plate: HS1; Target plate: HS (HS2); Source pressure: 100%; Target pressure: 100%; Offset: Automatic
Note: Incubate for 1 day at 30 °C.

3.1.5.8. SM

Replicate the arrays from the HS2 plates onto SM plates using 1536 pads. Do 1 pin per array.
Program: Agar–Agar. Replicate. Replicate One. 1536
Parameters: Source plate: HS1; Target plate: HS (HS2); Source pressure: 100%; Target pressure: 100%; Offset: Automatic
Note: Incubate for 1 day at 30 °C.

3.1.5.9. DM

Replicate the arrays from the SM plates onto DM plates using 1536 pads. Do 1 pin per array.
Program: Agar–Agar. Replicate One. 1536
Parameters: Source plate: SM; Target plate: DM; Source pressure: 100%; Target pressure: 100%; Offset: Automatic
Incubate at 30 °C.
Take pictures of the DM after 48 h.
Store the final plates in coldroom.

3.2. Basic PEM protocol

3.2.1. Genotypes

Query: h–; ade6-M210; leu1-32; ura4-D18; mat1_m-cyhS, smt0; rpl42::
cyhR (sP56Q); XXX::NatMX6
Library: h+; ade-M210 (or M216); ura4-D18; leu1-32; YYY::KanMX6

3.2.2. Growth media

YE5S (general purpose rich media)—5 g/l yeast extract; 30 g/l glucose; 225
mg/l of each adenine, leucine, histidine, uracil, and lysine; 20 g/l Difco
Agar
SPAS (mating and sporulation media)—10 g/l glucose; 1 g/l KH_2PO_4
45 mg/l of each adenine, histidine, leucine, uracil, and lysine hydrochloride; 1 ml 1000× vitamin stock (1 g/l pantotenic acid; 10 g/l nicotinic acid;
10 g/l inositol; 10 mg/l biotin)

3.2.3. Drug concentrations

NAT (N)—100 mg/l; G418 (G)—100 mg/l; CYH (C)—100 mg/l

3.2.4. Plates nomenclature

YE5S = YE5S; SPAS = SPAS; NAT = YE5S + N; G418 = YE5S + G;
GC = YE5S + G + C; GNC = YE5S + G + N + C

GC1 and GC2 are GC plates used in two consecutive steps of the
protocol.

3.2.5. Procedure

3.2.5.1. Preparation of Query arrays (Q-arrays) in 1536 format

Prepare a lawn of cells—Spread up to 500 μl of thick culture onto a NAT
plate using glass beads and incubate at 30 °C for 2–3 days.
Program: Agar-Agar. Replicate. Replicate One. 1536
Parameters: Source plate: NAT; Target plate: NAT
Source pressure: 100%; Target pressure: 100%; Offset: Manual; Offset
radius: 1 mm
Note: Do 2 pins per plate picking cells from different parts of the lawn plate.
Incubate at 30 °C for 2–3 days.

3.2.5.2. Preparation of library arrays (T-arrays) in 1536 format

Program: Agar-Agar. Replicate. Replicate Many. 1536
Parameters: Source plate: G418; Target plate: G418; Source pressure: 100%;
Target pressure: 100%; Offset: Manual; Number of replicas: 3; Economy:
ON; Revisit source: ON

Note: More than three replicas/source at this density may not be consistent and slow growing mutants (e.g., small colonies) may be lost. Incubate at 30 °C for 2–3 days.

3.2.5.3. Mating

Combine the T–array and the Q–array onto a SPAS plate generating a 1536 density mating array. First pin the T–array twice (2×) and then pin the Q-array twice on top of it with agar mixing.
Program: Agar-Agar. Replicate. 1536
Parameters: Source plate: T-array and Q-array; Target plate: SPAS; Source pressure: 100%; Target pressure: 100%; Offset: Manual; Economy: OFF
Note: Incubate for 5–6 days at room temperature packing the plates in plastic bags to prevent drying.

3.2.5.4. SPAS-GC1

Replicate the mating arrays from SPAS onto GC plates using 384 pads.
Do 2 pins per array onto the same target GC plate. You will need eight (Schuldiner *et al.*, 2005) 384 pads per array. This is the most critical stage of the protocol and transferring enough cells on the target plate is very important.
Program: Agar-Agar. Replicate. Replicate One.1536
Parameters: Source plate: SPAS; Target plate: GC; Source pressure: 100%; Target pressure: 100%; Offset: OFF; Economy: OFF
Note: When loading the pads click "Modify" to change to 384 pads. Incubate for 3 days at 30 °C.

3.2.5.5. GC1-GC2

Replicate the arrays from the GC1 plates onto GC2 plates using 1536 pads.
Do 1 pin per array.
Program: Agar-Agar. Replicate. Replicate One.1536
Parameters: Source plate: GC (GC1); Target plate: GC (GC2); Source pressure: 100%; Target pressure: 100%; Offset: Automatic
Note: Incubate for 2 days at 30 °C.

3.2.5.6. GC2-GNC

Replicate the arrays from the GC2 plates onto GNC plates using 1536 pads.
Do 1 pin per array.
Program: Agar-Agar. Replicate. Replicate One.1536
Parameters: Source plate: GC (GC2); Target plate: GNC; Source pressure: 100%; Target pressure: 100%; Offset: Automatic
Note: Incubate at 30 °C. Take pictures of the GNC at 24, 36, and 48 h. Store the final plates in coldroom.

3.3. Digital photography

We take color digital photographs of the final plates using a Canon PowerShot S3 IS camera (6.0 megapixels) at a resolution of 180 dpi, focal length $= 18.2$ and F-number $= 8$. Images are taken at a distance of 60 cm (24 in.) by mounting the camera on a KAISER camera stand (Germany). The position of the plate is fixed using a custom-made metal plate holder permanently bolted to the camera stand. The base of the camera stand is covered with black velvet to create a uniform black background for the images. Illumination is provided by two fixed luminescent lamps (25–30 W) outside of a 20 × 20 in. nylon soft light tent which serves to even the illumination and prevent reflections.

4. Data Processing and Computation of Scores

The data from the screens is collected as digital photographs of arrays of yeast colonies. These images are converted into measures of interactions between mutations through a multistep computational process (Collins *et al.*, 2006). Colony areas are measured digitally using the HT Colony Grid Analyzer Java program (Collins *et al.*, 2006). The resulting sizes are then processed using a software toolbox we have developed for use with MATLAB. Both pieces of software are freely available for download (http://sourceforge.net/projects/ht-col-measurer/ and http://sourceforge.net/projects/emap-toolbox/). The computational steps for converting colony area measurements into genetic interaction scores is depicted in Fig. 9.4A and outlined in the following steps:

4.1. Preprocessing and normalization

A preprocessing and normalization step is used to correct for systematic artifacts (uneven image lighting, artifacts due to physical curvature of the agar surface on which the colonies grew, differences in growth time, etc.), and also to account for the growth phenotype of the query strain.

Several types of systematic artifacts can arise in the data collection process that give rise to spatial patterning of measured colony sizes which is independent of the growth properties of the yeast strains in the experiment. For instance, uneven lighting can result in apparently larger colonies in areas with brighter light. Additionally, an uneven agar surface can result in deposition of a larger number of cells in vertically elevated areas of the agar surface, and deposition of fewer cells in lower areas. In our experience, this agar curvature artifact is much more pronounced using the Singer robot plastic pad-based cell deposition method rather than a floating pin-based method.

We correct for these artifacts using a spatial flattening of the colony sizes. Specifically, the colony size measurement at each position on one agar plate is compared to the median size at that position over all plates in the dataset to compute a log-ratio indicating whether that colony is larger or smaller than is typical. The resulting set of log-ratios is fit using robust linear regression (using MATLAB's robustfit function) to a second order surface (i.e., one of the form: $z = Ax^2 + By^2 + Cxy + Dx + Ey + F$). The resulting surface is then subtracted from the log-ratios to remove spatial artifacts, and the corrected colony sizes are recovered by exponentiating the result and multiplying by the original median size at the corresponding position. This correction requires only six parameters and is calculated using 384, 768, or 1536 measurements, depending on the number of colonies on the plate, so it is unlikely to be strongly affected by real genetic interactions. Additionally, robust regression, rather than standard linear regression, is used to minimize the impact of individual real interactions on the calculated correction. Colony sizes of zero and the colonies on the edges of the plate are excluded from the correction calculation. In our experience, this correction effectively removes spatial patterning, without compromising detection of interactions. The MATLAB toolbox contains a graphical user interface which displays heatmaps (similar to those seen in Fig. 9.4A) before and after the spatial flattening so that the success of this step can be assessed.

A separate correction is applied for the colonies on the edges of the plate. For each edge row or column, the colony sizes are scaled such that the median size in that row or column is equal to the median size in the interior of the plate. This correction is applied because we have found that edge rows and columns can be systematically larger or smaller (usually larger) than other colonies on the plate due to proximity or distance from the physical edge of the agar.

Finally, colony sizes are normalized to account for differences in the growth phenotype of the query mutation which is present in all double-mutant colonies on the same plate. We apply this normalization in addition to the above-described spatial flattening to account for the possibility that the query mutation may have more synthetic interactions (where double mutants grow more slowly than expected) than positive interactions (where double mutants grow faster), or vice versa. We do assume that most mutations in the array will have little or no growth defect, as is the case for gene deletions in yeast (Breslow *et al.*, 2008; Giaever *et al.*, 2002), and that most mutation pairs will be noninteracting. We, therefore, normalize the colony sizes according to the peak of the histogram of colony sizes on a given plate (this is the "Parzen" setting in the toolbox menu). Heatmap images showing the spatial pattern of colony size measurements at different stages of the normalization process can be seen in Fig. 9.4A. All of these preprocessing and normalization steps are implemented in the MATLAB toolbox.

4.2. Computing expected colony sizes

After normalization, the growth phenotype of the query mutation has been accounted for, and we assume that differences in normalized colony size now result from either the growth phenotype of the array mutation or a genetic interaction. For a given array mutation (which is always present in the exact same spatial position within the array), we then empirically estimate the expected normalized colony size as the typical normalized colony size over a set of screens. If the number of screens is large (generally 50 or more), we prefer to use the peak of the histogram of normalized sizes (the "Parzen" setting in the toolbox), similar to our normalization procedure. However, if the number of screens analyzed is smaller we have found that the median normalized size is a more robust estimate.

We sometimes observe batch-to-batch variability, where the typical colony size estimated for one group of screens completed at approximately the same time using the same preparation of media differ from the values estimated for another group of screens completed at a different time (perhaps weeks or months apart). If such batch-to-batch variability is apparent, it is preferable to compute the expected colony sizes independently for each batch. This can be done easily with options included in the MATLAB toolbox (web address provided above).

A natural question is then how many different screens need to be included in a batch, such that the estimated expected colony size values will be reliable? We have investigated this question empirically, using measurements from the early secretory pathway E-MAP (Schuldiner *et al.*, 2005). As increasingly many screens are completed in the same batch, the error in estimates of expected colony sizes decreases (Fig. 9.4C). There can be substantial error if a batch includes fewer than 20 screens. On the other hand, each additional screen beyond about the 40th gives only marginal improvement.

4.3. Computing genetic interaction scores

We compute genetic interaction scores as we have previously described (Collins *et al.*, 2006) as S-scores, which are closely related to *t*-values and account for both the magnitude of the interaction effect, as well as our confidence that the measurement constitutes a real genetic interaction. The S-score differs from a standard *t*-value in several ways. Rather than comparing experimental observations explicitly to control screens, we use the expected colony size estimates described above. Additionally, a standard *t*-value can be very sensitive to the standard deviation of a small number of experimental replicate measurements. In particular, if the replicates are unusually similar, resulting in an unusually small standard deviation, this can result in a large score even if the magnitude of the interaction is small.

We have found that the reproducibility of S-scores is substantially improved by implementing a lower bound on the standard deviation measurement (i.e., if the measured standard deviation is below the lower bound, we use the lower bound instead) (Collins *et al.*, 2006). This lower bound is an estimate of the typical standard deviation for double-mutant measurements with similar parent query and array strain phenotypes.

4.4. Quality control

Careful quality control is an essential part of the screening process. In our experience, it is not uncommon for ∼10% of all strains (both array and query) to be incorrect. These incorrect strains need to be systematically identified, and removed from the analysis.

The most effective tool for identifying incorrect strains is analysis of the apparent interaction scores between mutations at chromosomally linked loci (Collins *et al.*, 2006; Roguev *et al.*, 2008). The absence of apparent negative interactions between a mutation and mutations at loci within ∼100 kb in *S. cerevisiae* or ∼200 kb in *S. pombe* is strong evidence that a strain is incorrect. The MATLAB toolbox contains a graphical user interface for browsing the linkage data which facilitates the identification of incorrect strains. Additionally, cases where the data for an array strain is completely uncorrelated to data for a corresponding query strain should be identified systematically, and if found, the corresponding strains should be checked by PCR.

5. EXTRACTION OF BIOLOGICAL HYPOTHESES

Completion of a genetic interaction map creates a huge quantity of data. These data have proven useful in numerous instances for generating new hypotheses and helping guide ongoing work (Collins *et al.*, 2007a; Fiedler *et al.*, 2009; Keogh *et al.*, 2005; Kornmann *et al.*, 2009; Kress *et al.*, 2008; Laribee *et al.*, 2007; Morrison *et al.*, 2007; Nagai *et al.*, 2008; Schuldiner *et al.*, 2005; Wilmes *et al.*, 2008). However, determining the best way to navigate these maps and generate hypotheses from them remain areas of active work. We expect that we and others will continue to find new and better ways to use the data, but we also describe here several general approaches that have proven useful in the past.

5.1. Identifying genes acting in the same pathway using patterns of interactions

One of the simplest and most useful applications of high-throughput genetic interaction data is the identification of sets of genes whose products work together very closely in a common biochemical pathway. For each pair of mutations in a quantitative genetic interaction map, two distinct measures of their relationship can be recognized. First, the genetic interaction score (S-score) represents the degree of synergizing or mitigating effects of the two mutations in combination; this can be neutral (e.g., no interaction), positive (e.g., suppression), or negative (e.g., synthetic lethality). Second, the similarity (typically measured as a correlation) of their genetic interaction profiles represents the congruency of the phenotypes of the two mutations across a wide variety of genetic backgrounds. One would logically expect both measures to be indicative of whether two genes act in the same pathway. Indeed, pairs of genes exhibiting positive genetic interactions and highly correlated genetic interaction profiles very frequently encode proteins that are physically associated (Collins *et al.*, 2007a). Furthermore, in cases where the proteins do not physically associate, they tend to act coherently in a biochemical pathway. This latter case is particularly informative as these are very close functional relationships which could be difficult to detect by other methods. The combined signature of a positive genetic interaction and highly correlated interaction patterns can be formalized into a score (the COP score (Collins *et al.*, 2006, 2007a; Schuldiner *et al.*, 2005), and the sets of genes with this signature are also often readily apparent after hierarchical clustering of genetic interaction profiles.

In general, the similarity of interaction profiles is much stronger evidence for membership in the same pathway than an individual positive interaction. However, the direct interaction score sometimes provides the critical distinction. For example, deletions of *DOA1* and *UBP6* give very similar patterns of genetic interactions (Fig. 9.5B), which likely reflects the fact that each deletion leads to depletion of ubiquitin (Collins *et al.*, 2007a). However, the double deletion results in a strong negative genetic interaction, presumably because they largely function independently of each other to maintain ubiquitin levels.

It should be noted that interpretation of genetic interaction data derived from hypomorphic alleles of essential genes may differ from the data obtained from deletions of nonessential genes. For example, two hypomorphic alleles may, in fact, give rise to a negative genetic interaction, if each mutation partially cripples the same pathway. However, a positive genetic interaction may still be observed if the two encoded proteins form a tight heterodimer, where the minimum of the two concentrations determines the cellular phenotype.

5.2. Using individual interactions to predict enzyme–substrate relationships

In addition to identifying the core components of coherent pathways, one would like to have efficient strategies to suggest points of integration or cross-talk between pathways. One important class of such connections

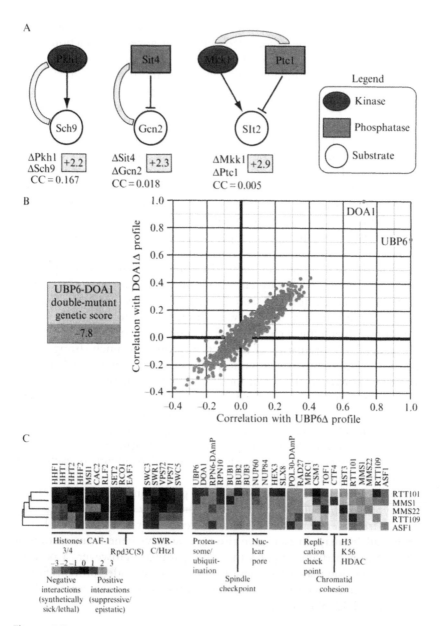

Figure 9.5 Quantitative genetic data reveals insight into functional pathways. (A) Individual genetic interactions identify enzyme–substrate relationships. An E-MAP focused on the regulation of phosphorylation in budding yeast (Fiedler *et al.*, 2009) revealed positive genetic interactions between the kinase Pkh1 and it substrate, Sch9 (+2.2), between the phosphatase Sit4 and its substrate, Gcn2 (+2.3), and between a kinase (Mkk1) and a phosphatase (Ptc1) (+2.9) that acts on Slt2. The correlation of genetic interaction profiles between these pairs of genes is below the

between pathways includes protein-modifying enzymes and their cognate substrates. We have found that kinase–substrate and phosphatase–substrate pairs are enriched for positive genetic interactions (Fiedler *et al.*, 2009). Pkh1-Sch9 and Sit4-Gcn2 correspond to kinase–substrate and phosphatase–substrate relationships, respectively, and in both cases, the double deletions result in positive genetic interactions (Fig. 9.5A) (Fiedler *et al.*, 2009). However, in neither case is the correlation coefficient between the pairs notable. Thus, when looking for a critical substrate of a kinase, phosphatase, or other protein-targeting enzyme, the genes with the highest scoring positive interactions represent excellent candidates, even if the correlation between the interaction profiles is weak. It should be noted though, that kinases and phosphatases may have multiple substrates. In these cases, only enzyme–substrate pairs corresponding to modifications that significantly affect the phenotype being measured are likely to be detectable.

5.3. Using individual interactions to predict opposing enzyme relationships

Similarly, we have found that pairs of opposing kinases and phosphatases which share a common substrate are highly enriched for positive genetic interactions. Since these enzymes have opposing effects, they will also tend not to have correlated interaction patterns. An example is the mitogen-activated protein kinase kinase Mkk1 and the phosphatase Ptc1. These enzymes have opposing roles regulating the activity of the downstream kinase Slt2. The two deletion mutations have a strong positive genetic interaction, but the interaction profiles are not correlated (Fiedler *et al.*, 2009) (Fig. 9.5A). Such positive interactions can be key clues for deciphering the role of uncharacterized genes and pathways. Indeed, strong positive interactions with the H3 K56 histone deacetylase Hst3 was an important piece of data suggesting that Rtt109 was the opposing acetylase (see below for further discussion of this pathway) (Collins *et al.*, 2007a).

---

individual genetic interactions and was derived from the kinase E-MAP (Fiedler *et al.*, 2009). (B) Functional connection between the deubiquitinase enzyme, Ubp6 and the ubiquitin chaperone, Doa1. A strain containing deletions in both *UBP6* and *DOA1* results in a strong negative genetic interaction (-7.8) and the genetic profiles generated from these deletions are highly correlated. (C) Using genetic interaction profiles to identify a pathway involved in genome integrity. Subsets of interactions (both negative and positive) for *rtt101Δ*, *mms1Δ*, *mms22Δ*, *rtt109Δ*, and *asf1Δ* are displayed. Some interactions are observed for all five deletions, interactions with the SWR complex are seen with only *asf1Δ* and *rtt109Δ* whereas only deletion of *ASF1* result in negative interactions with histones H3/H4, the CAF complex, and factors involved in the Rpd3C(S) pathway (Set2, Eaf3, and Rco1).

5.4. Dissecting multiple roles of a single gene by detailed comparison of interaction patterns

A major challenge in interpreting genetic interaction data is that many genes encode multifunctional proteins acting in multiple pathways. While it would be useful to be able to extract pathway-specific information, genetic interactions for these genes may arise from a role in one pathway or another, or the interactions may represent a mixture of effects from more than one pathway. We have found that in some cases, comparison of the interaction profile for a multifunctional gene with the profiles for other related genes can give important pathway-specific insights. For example, we identified a pathway involving the chromatin assembly protein Asf1, a then-uncharacterized protein Rtt109, and a putative ubiquitin ligase complex containing Rtt101, Mms1, and Mms22 (Collins *et al.*, 2007a). Asf1 was the best characterized member of the group, but it had also been implicated in multiple cellular roles (Loyola and Almouzni, 2004). Comparison of the genetic interaction profiles for this group of genes suggested that several of these roles were Asf1-specific and that all of these factors work together in a pathway intimately related to histone H3 K56 acetylation and maintenance of genomic integrity during DNA replication. Further experiments identified Rtt109 as the acetylase (Collins *et al.*, 2007a; Driscoll *et al.*, 2007; Han *et al.*, 2007).

In the above case, a critical observation was that a subset of *asf1Δ*'s interactions was unique to *asf1Δ*, and another large subset is shared uniformly with the rest of the group. All members of this pathway display positive genetic interactions with one another, positive interactions with replication checkpoint genes (*MRC1*, *TOF1*, and *CSM3*), and negative genetic interactions with genes involved in DNA replication (*POL30*, *ELG1*, *RAD27*), the spindle checkpoint (*BUB1*, *BUB2*, and *BUB3*), DNA Repair (*NUP60*, *NUP84*, *HEX3*, *SLX8*) and ubiquitin regulation (*UBP6*, *DOA1*, *RPN6*, *RPN10*). Asf1 was also known to be required for histone H3 K56 acetylation, and the shared positive interaction between all members of this pathway and the gene encoding a K56 deacetylase (*HST3*) suggested that this role of Asf1 was likely to be central to the pathway (Celic *et al.*, 2006; Maas *et al.*, 2006).

On the other hand, only *asf1Δ* displays negative interactions with factors involved in general chromatin assembly: the histone genes *HHF1*, *HHF2*, *HHT1*, *HHT2* and components of the CAF complex (*MSI1*, *CAC2*, and *RLF2*), arguing that Asf1's general role in chromatin assembly is independent of its function with Rtt109. Similarly, only deletion of *ASF1* results in negative interactions when combined with deletions of the Rpd3C(S) histone deacteylation complex (*EAF3* and *RCO1*) and *SET2*, which codes for a histone methyltransferase enzyme. Eaf3, Rco1, and Set2 function together in a histone deacetylation/methylation pathway that is required for maintaining chromatin integrity during transcription and suppressing cryptic initiation by RNAPII (Carrozza *et al.*, 2005; Joshi and Struhl, 2005; Keogh *et al.*, 2005), again suggesting that Asf1 alone impinges on this process. Indeed, deletion of *ASF1* and not the other factors functioning in the K56 acetylation pathway results in spurious transcription rising from the inability to suppress cryptic initiation (Schwabish and Struhl, 2006).

6. Perspective

The term "epistatic" was originally used in 1907 by Bateson to describe a masking effect whereby a variant or allele at one locus prevents the variant at another locus from manifesting its effect (Bateson, 1907). Over the past one hundred years, geneticists have uncovered important biological insight from "epistatic" interactions, first qualitatively and then much more efficiently through quantitative analysis. The vast majority of genetic interaction data has been collected from simpler systems like yeast and bacteria using basic read-outs like colony size or growth rates. We are now in a position to collect this type of data in multicellular systems and the phenotypic read-outs, which are expanding exponentially, can occur at the organismal level, providing invaluable information about not only functional pathways that comprise key biological processes but also about evolution and behavior.

REFERENCES

Bateson, W. (1907). Facts limiting the theory of heredity. Lipid-modifying therapy and attainment of cholesterol goals in Hungary: the return on expenditure achieved for lipid therapy (REALITY) study. *Science* **26,** 647–660.

Breslow, D. K., Cameron, D. M., Collins, S. R., Schuldiner, M., Stewart-Ornstein, J., Newman, H. W., Braun, S., Madhani, H. D., Krogan, N. J., and Weissman, J. S. (2008). A comprehensive strategy enabling high-resolution functional analysis of the yeast genome. *Nat. Methods* **5,** 711–718.

Carrozza, M. J., Li, B., Florens, L., Suganuma, T., Swanson, S. K., Lee, K. K., Shia, W. J., Anderson, S., Yates, J., Washburn, M. P., and Workman, J. L. (2005). Histone H3 methylation by Set2 directs deacetylation of coding regions by Rpd3S to suppress spurious intragenic transcription. *Cell* **123,** 581–592.

Celic, I., Masumoto, H., Griffith, W. P., Meluh, P., Cotter, R. J., Boeke, J. D., and Verreault, A. (2006). The sirtuins hst3 and Hst4p preserve genome integrity by controlling histone h3 lysine 56 deacetylation. *Curr. Biol.* **16,** 1280–1289.

Collins, S. R., Schuldiner, M., Krogan, N. J., and Weissman, J. S. (2006). A strategy for extracting and analyzing large-scale quantitative epistatic interaction data. *Genome Biol.* **7,** R63.

Collins, S. R., Miller, K. M., Maas, N. L., Roguev, A., Fillingham, J., Chu, C. S., Schuldiner, M., Gebbia, M., Recht, J., Shales, M., Ding, H., Xu, H., *et al.* (2007a).

Functional dissection of protein complexes involved in yeast chromosome biology using a genetic interaction map. *Nature* **446,** 806–810.

Collins, S. R., Kemmeren, P., Zhao, X. C., Greenblatt, J. F., Spencer, F., Holstege, F. C., Weissman, J. S., and Krogan, N. J. (2007b). Toward a comprehensive atlas of the physical interactome of *Saccharomyces cerevisiae. Mol. Cell. Proteomics* **6,** 439–450.

Decourty, L., Saveanu, C., Zemam, K., Hantraye, F., Frachon, E., Rousselle, J. C., Fromont-Racine, M., and Jacquier, A. (2008). Linking functionally related genes by sensitive and quantitative characterization of genetic interaction profiles. *Proc. Natl. Acad. Sci. USA* **105,** 5821–5826.

Driscoll, R., Hudson, A., and Jackson, S. P. (2007). Yeast Rtt109 promotes genome stability by acetylating histone H3 on lysine 56. *Science* **315,** 649–652.

Fiedler, D., Braberg, H., Mehta, M., Chechik, G., Cagney, G., Mukherjee, P., Silva, A. C., Shales, M., Collins, S. R., van Wageningen, S., Kemmeren, P., Holstege, F. C., *et al.* (2009). Functional organization of the S. cerevisiae phosphorylation network. *Cell* **136,** 952–963.

Gavin, A. C., Aloy, P., Grandi, P., Krause, R., Boesche, M., Marzioch, M., Rau, C., Jensen, L. J., Bastuck, S., Dumpelfeld, B., Edelmann, A., Heurtier, M. A., *et al.* (2006). Proteome survey reveals modularity of the yeast cell machinery. *Nature* **440,** 631–636.

Giaever, G., Chu, A. M., Ni, L., Connelly, C., Riles, L., Veronneau, S., Dow, S., Lucau-Danila, A., Anderson, K., Andre, B., Arkin, A. P., Astromoff, A., *et al.* (2002). Functional profiling of the *Saccharomyces cerevisiae* genome. *Nature* **418,** 387–391.

Guarente, L. (1993). Synthetic enhancement in gene interaction: a genetic tool come of age. *Trends Genet* **9,** 362–366.

Han, J., Zhou, H., Horazdovsky, B., Zhang, K., Xu, R. M., and Zhang, Z. (2007). Rtt109 acetylates histone H3 lysine 56 and functions in DNA replication. *Science* **315,** 653–655.

Jonikas, M. C., Collins, S. R., Denic, V., Oh, E., Quan, E. M., Schmid, V., Weibezahn, J., Schwappach, B., Walter, P., Weissman, J. S., and Schuldiner, M. (2009). Comprehensive characterization of genes required for protein folding in the endoplasmic reticulum. *Science* **323,** 1693–1697.

Joshi, A. A., and Struhl, K. (2005). Eaf3 chromodomain interaction with methylated H3-K36 links histone deacetylation to Pol II elongation. *Mol. Cell* **20,** 971–978.

Kaiser, C. A., and Schekman, R. (1990). Distinct sets of SEC genes govern transport vesicle formation and fusion early in the secretory pathway. *Cell* **61,** 723–733.

Keogh, M. C., Kurdistani, S. K., Morris, S. A., Ahn, S. H., Podolny, V., Collins, S. R., Schuldiner, M., Chin, K., Punna, T., Thompson, N. J., Boone, C., Emili, A., *et al.* (2005). Cotranscriptional set2 methylation of histone H3 lysine 36 recruits a repressive Rpd3 complex. *Cell* **123,** 593–605.

Kornmann, B., Currie, E., Collins, S. R., Schuldiner, M., Nunnari, J., Weissman, J. S., and Walter, P. (2009). An ER-mitochondria tethering complex revealed by a synthetic biology screen. *Science* **325,** 477–481.

Kress, T. L., Krogan, N. J., and Guthrie, C. (2008). A single SR-like protein, Npl3, promotes pre-mRNA splicing in budding yeast. *Mol. Cell* **32,** 727–734.

Krogan, N. J., Cagney, G., Yu, H., Zhong, G., Guo, X., Ignatchenko, A., Li, J., Pu, S., Datta, N., Tikuisis, A. P., Punna, T., Peregrin-Alvarez, J. M., *et al.* (2006). Global landscape of protein complexes in the yeast *Saccharomyces cerevisiae. Nature* **440,** 637–643.

Laribee, R. N., Shibata, Y., Mersman, D. P., Collins, S. R., Kemmeren, P., Roguev, A., Weissman, J. S., Briggs, S. D., Krogan, N. J., and Strahl, B. D. (2007). CCR4/NOT complex associates with the proteasome and regulates histone methylation. *Proc. Natl. Acad. Sci. USA* **104,** 5836–5841.

Loyola, A., and Almouzni, G. (2004). Histone chaperones, a supporting role in the limelight. *Biochim. Biophys. Acta* **1677,** 3–11.

Maas, N. L., Miller, K. M., DeFazio, L. G., and Toczyski, D. P. (2006). Cell cycle and checkpoint regulation of histone H3 K56 acetylation by Hst3 and Hst4. *Mol. Cell* **23**, 109–119.

Morrison, A. J., Kim, J. A., Person, M. D., Highland, J., Xiao, J., Wehr, T. S., Hensley, S., Bao, Y., Shen, J., Collins, S. R., Weissman, J. S., Delrow, J., *et al.* (2007). Mec1/Tel1 phosphorylation of the INO80 chromatin remodeling complex influences DNA damage checkpoint responses. *Cell* **130**, 499–511.

Nagai, S., Dubrana, K., Tsai-Pflugfelder, M., Davidson, M. B., Roberts, T. M., Brown, G. W., Varela, E., Hediger, F., Gasser, S. M., and Krogan, N. J. (2008). Functional targeting of DNA damage to a nuclear pore-associated SUMO-dependent ubiquitin ligase. *Science* **322**, 597–602.

Pan, X., Yuan, D. S., Xiang, D., Wang, X., Sookhai-Mahadeo, S., Bader, J. S., Hieter, P., Spencer, F., and Boeke, J. D. (2004). A robust toolkit for functional profiling of the yeast genome. *Mol. Cell* **16**, 487–496.

Roguev, A., Wiren, M., Weissman, J. S., and Krogan, N. J. (2007). High-throughput genetic interaction mapping in the fission yeast Schizosaccharomyces pombe. *Nat. Methods* **4**, 861–866.

Roguev, A., Bandyopadhyay, S., Zofall, M., Zhang, K., Fischer, T., Collins, S. R., Qu, H., Shales, M., Park, H. O., Hayles, J., Hoe, K. L., Kim, D. U., *et al.* (2008). Conservation and rewiring of functional modules revealed by an epistasis map in fission yeast. *Science* **322**, 405–410.

Schuldiner, M., Collins, S. R., Thompson, N. J., Denic, V., Bhamidipati, A., Punna, T., Ihmels, J., Andrews, B., Boone, C., Greenblatt, J. F., Weissman, J. S., and Krogan, N. J. (2005). Exploration of the function and organization of the yeast early secretory pathway through an epistatic miniarray profile. *Cell* **123**, 507–519.

Schuldiner, M., Collins, S. R., Weissman, J. S., and Krogan, N. J. (2006). Quantitative genetic analysis in *Saccharomyces cerevisiae* using epistatic miniarray profiles (E-MAPs) and its application to chromatin functions. *Methods* **40**, 344–352.

Schwabish, M. A., and Struhl, K. (2006). Asf1 mediates histone eviction and deposition during elongation by RNA polymerase II. *Mol. Cell* **22**, 415–422.

Tong, A. H., and Boone, C. (2006). Synthetic genetic array analysis in *Saccharomyces cerevisiae*. *Methods Mol. Biol.* **313**, 171–192.

Tong, A. H., Evangelista, M., Parsons, A. B., Xu, H., Bader, G. D., Page, N., Robinson, M., Raghibizadeh, S., Hogue, C. W., Bussey, H., Andrews, B., Tyers, M., *et al.* (2001). Systematic genetic analysis with ordered arrays of yeast deletion mutants. *Science* **294**, 2364–2368.

Tong, A. H., Lesage, G., Bader, G. D., Ding, H., Xu, H., Xin, X., Young, J., Berriz, G. F., Brost, R. L., Chang, M., Chen, Y., Cheng, X., *et al.* (2004). Global mapping of the yeast genetic interaction network. *Science* **303**, 808–813.

Wilmes, G. M., Bergkessel, M., Bandyopadhyay, S., Shales, M., Braberg, H., Cagney, G., Collins, S. R., Whitworth, G. B., Kress, T. L., Weissman, J. S., Ideker, T., Guthrie, C., *et al.* (2008). A genetic interaction map of RNA-processing factors reveals links between Sem1/Dss1-containing complexes and mRNA export and splicing. *Mol. Cell* **32**, 735–746.

Ye, P., Peyser, B. D., Pan, X., Boeke, J. D., Spencer, F. A., and Bader, J. S. (2005). Gene function prediction from congruent synthetic lethal interactions in yeast. *Mol. Syst. Biol.* **1**, 2005.0026.

CHAPTER TEN

Exploring Gene Function and Drug Action Using Chemogenomic Dosage Assays

Elke Ericson,[*,†,‡] Shawn Hoon,[§,¶] Robert P. St.Onge,[§] Guri Giaever,[*,†,‖] *and* Corey Nislow[†,‖,**]

Contents

Abstract

In this chapter, we describe a series of genome-wide, cell-based assays that provide a solid basis for understanding drug–gene interactions, gene function, and for defining the mechanism of action of small molecules. Each of these assays takes advantage of the ability to grow complex pools competitively and to use high-density microarrays that report the results of such screens.

[*] Department of Pharmaceutical Sciences, University of Toronto, Toronto, Ontario, Canada
[†] Donnelly Centre for Cellular and Biomolecular Research, University of Toronto, Toronto, Ontario, Canada
[‡] Department of Bioscience, AstraZeneca R&D Mölndal, Mölndal, Sweden
[§] Stanford Genome Technology Center, Stanford University, Palo Alto, California, USA
[¶] Molecular Engineering Lab, Science and Engineering Institutes, Agency for Science Technology and Research, Singapore
[‖] Department of Molecular Genetics, University of Toronto, Toronto, Ontario, Canada
[**] Banting and Best Department of Medical Research, University of Toronto, Toronto, Ontario, Canada

Methods in Enzymology, Volume 470
ISSN 0076-6879, DOI: 10.1016/S0076-6879(10)70010-0

The assays described here take advantage of alterations in gene dosage of *Saccharomyces cerevisiae*, and include HIP (haploinsufficiency profiling), HOP (homozygous profiling), and MSP (multicopy suppression profiling) as genetic tools to understand gene function and drug mechanism. The common experimental theme is that, in each assay, strains are pooled and screened in parallel to investigate the relative contribution of each gene product to sensitivity or resistance to a drug or environmental perturbation across the genome in a single assay. Further, the compendium of results from these screens can inform large-scale network analysis of genetic function, gene–gene interactions, and mechanism of drug action.

1. Introduction

Our initial pooled screening platform was designed to interrogate the yeast deletion collection, a genome-wide set of arrayed single-gene deletion strains in *Saccharomyces cerevisiae* (Giaever *et al.*, 2002) that can be used in place of (or as a complement to) random-mutant libraries and individually constructed strains. One advantage of this collection over others is that it includes all genes and therefore does not suffer from biases. The collection has enabled the development of methods for studying all $\sim$6000 deletion strains in a single culture in parallel, in a single assay (Giaever *et al.*, 2002; Shoemaker *et al.*, 1996; Winzeler *et al.*, 1999). Specifically, unique 20 bp DNA "barcodes" or "tags" incorporated into each strain enable relative strain abundance to be determined by amplifying the barcodes using common flanking primers and hybridizing the amplicons to a microarray that carry the tag complements (Fig. 10.1). For example, pooled analysis of all $\sim$6000 heterozygous deletion strains can be used to identify novel drug targets (Giaever *et al.*, 1999, 2004; Lum *et al.*, 2004) based on their relative sensitivity in the *haplo*insufficiency profiling (HIP) assay. The rationale behind this technique is as follows: if a locus encodes the target of a drug or small molecule and this target is important for growth, then the decreased gene dosage in the heterozygous deletion strain, combined with further reduction in gene function due to drug binding, will confer increased sensitivity to the drug. Therefore, in theory, the most sensitive strain in the pool will be heterozygous for the drug target (Ghose *et al.*, 1999; Giaever *et al.*, 2004; Lum *et al.*, 2004). In a similar manner, analysis of competitive assays using the collection of $\sim$4700 nonessential genes with the *homo*zygous profiling (HOP) assay can reveal information about the drug target pathway, such as buffering interactions. In this case, the assay mimics a double deletion mutant, because one gene is completely absent, and the second is diminished in function by the action of the compound (Hillenmeyer *et al.*, 2008; Parsons *et al.*, 2004, 2006). Thousands of

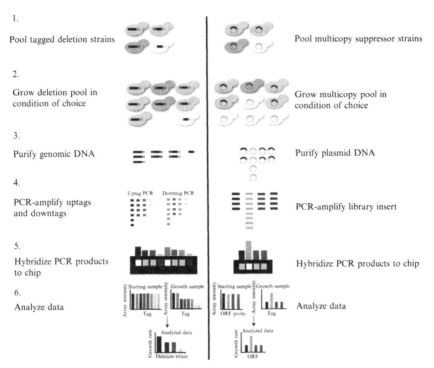

Figure 10.1 Illustration of the chemogenomic platform that interrogates yeast deletion and overexpression pools with a single TAG4 array. Fitness profiling of pooled deletion strains involves six main steps: (1) Strains (or multicopy transformants) are pooled at approximately equal abundance. (2) The pool is grown competitively in the condition of choice. If a gene is required for growth under this condition, the strain carrying this deletion will grow more slowly and become underrepresented in the culture (light-coloured strain). Conversely, the strain carrying a plasmid that confers resistance to compound will become overrepresented in the culture (light-coloured strain). (3) In deletion profiling, genomic DNA is isolated from cells harvested after a predefined number of generations. For MSP, plasmid DNA is isolated when the experiment is complete. (4) Barcodes are amplified from the genomic DNA with universal primers in either one (MSP) or two (deletion profiling) PCR reactions. (5) PCR products are hybridized to an array with complementary probes. (6) Tag intensities for the treatment sample are compared to tag intensities for a control sample to determine the relative fitness of each strain. Genes that confer both resistance when overexpressed and sensitivity when deleted are more likely to be directly related to the drug's mechanism of action.

genome-wide screens performed to date, including those described above, have provided a wealth of functional information on the yeast genome (Alberts, 1998; Birrell *et al.*, 2002; Deutschbauer *et al.*, 2005; Giaever *et al.*, 1999, 2002, 2004; Kastenmayer *et al.*, 2006; Lee *et al.*, 2005; Lum *et al.*, 2004; Ooi *et al.*, 2001; Parsons *et al.*, 2004, 2006; Shoemaker *et al.*, 1996; Steinmetz *et al.*, 2002; Winzeler *et al.*, 1999) and have helped make yeast the most well-characterized organism to date.

Multicopy suppression profiling (MSP) is essentially the reverse of the HIP and HOP assays, focusing on increased gene dose versus decreased gene dose. While the concept is not new, current parallel screening technologies provide results on a genome-wide scale at a far greater level of resolution. DNA clone libraries overexpressing gene products are screened competitively to identify those that confer resistance to compounds (Fig. 10.1). Traditional multicopy suppressor screens involve cumbersome plating techniques (requiring large amounts of compound) and necessitate sequencing of individual clones (Rine et al., 1983). Moreover, if the wrong time point is assayed, or the wrong drug concentration is used, the result can be dominated by a gene product unrelated to the drug target, for example, a drug pump. MSP is a variation of the overexpression resistance concept and uses a high copy, random genomic library (Hoon et al., 2008) or an inducible ORF library (Butcher and Schreiber, 2006). MSP easily allows for collection of several time points to avoid domination of the culture by any one strain. Subsequent analysis allows assessment of the relative resistance of all clones, and an amplification step followed by hybridization to an array (TAG4) ranks the most predominant clones based on their relative abundance. More recently, systematic collections of yeast overexpression clones have become available (Ho et al., 2009; Jones et al., 2008). Because these collections are complete, not subject to library biases, and (in the case of Ho et al., 2009) barcoded, these libraries will clearly improve and expand the results that can be obtained using the MSP assay.

It is worth noting that both the HIP/HOP (loss-of-function) screens and MSP (gain-of-function) screens are each valuable on a screen-by-screen basis. However, combining results from all assays allows one to examine the effect of both increasing and decreasing gene dosage which can be particularly useful for distinguishing the bona fide target from a longer list of potential candidates. Table 10.1 provides a summary of the uses and complementary information gleaned by using the three screening methods. For example, sensitivity in the HIP assay and resistance in the MSP assay does not guarantee that the affected gene product is the bona fide drug target as the gene could be a drug pump. However, the HOP assay can eliminate such a pump as a possible target because if the corresponding deletion strain displays increased sensitivity in the HOP assay it is unlikely to be a direct target of the drug.

In addition to generating information about single genes and individual drugs, large sets of experiments can reveal novel relationships between genes and drugs. For example, for a given gene pair, a greater correlation of fitness values (cofitness) across conditions suggests functional relatedness of the two genes. Similarly, for a given drug pair, the correlation of fitness values across the strains is a measure of the similarity of their mechanism of action, and often, their structure (Hillenmeyer et al., 2008). This type of analysis allows characterization of both genes and compounds with previously unknown function, and is one of the primary aims of functional genomics.

Table 10.1 Interpreting results from HIP, HOP, and MSP

Assay results	HIP[a]	HOP[b]	MSP[c]
Direct molecular target of drug, essential gene	S	NA	R
Direct molecular target of drug, nonessential gene	S	R	R
Essential genes encoding proteins involved in target pathway	S	NA	not S or R
Genes involved in drug detoxification, essential genes	S	NA	R
Genes involved in drug detoxification, nonessential genes	S	S	R

[a] genes essential for survival
[b] genes not essential for survival
[c] All genes
S, sensitive; R, resistant; NA, not applicable.
Because assay readout is growth, most targets are necessarily essential for survival except those nonessential deletions that have a growth phenotype in the absence of compound.

2. METHODOLOGY

The pooled fitness assay involves seven main steps, described in detail below.

- Yeast deletion strain pool and MSP pool construction
- Determining the drug dosage
- Experimental pool growth
- Purification and amplification
- Hybridization
- Analysis of results
- Confirmation of microarray data

2.1. Yeast deletion strain pool and MSP pool construction

Allow 1 week to generate pooled aliquots of cells. Pooling is performed infrequently and cells can be stored indefinitely at $-80\,^{\circ}\text{C}$.

2.1.1. Yeast deletion strain pool construction

1. Thaw the frozen glycerol stocks completely for the strains of interest (such as the genome-wide homozygous deletion collection) but avoid exposing thawed cells to room temperature for more than 2 h.

2. To sterilize a 96-well pin tool, dip the pin tool in water to rinse away any remaining cells, followed by 2 dips in 70% ethanol baths (pipette tip box lids work well), flame the pin tool, and cool for 1 min. The level of the ethanol baths should exceed the level in the water bath to ensure all carry-over cells are flamed and removed. Replace water every 4–6 pinnings.

3. Insert the sterile 96-well pin tool into a thawed 96-well plate, swirl gently and transfer it to a Nunc Omni Tray containing YPD-agar including the appropriate antibiotic. Following the same procedure, transfer the cells from all remaining 96-well plates to agar plates. Grow colonies until they reach maximal size at 30 °C (2–3 days).

4. After colonies have reached full size, make note of any missing or slow growing strains. These should be individually repinned and added at twice the amount as the rest of the collection.

5. Working in a sterile environment, flood plate with 5–10 ml media, soak for 5 min, and resuspend colonies with a cell spreader. Pour the liquid plus cells into a 50 ml conical centrifuge tube and add glycerol to 15% or DMSO to 7% (vol/vol).

6. Measure the OD_{600} of the pool and adjust (by dilution or centrifugation and resuspension) to a final concentration of 50 OD_{600}/ml with media containing 15% glycerol or 7% DMSO.

2.1.2. MSP pool construction

1. Take a suitable *S. cerevisiae* random genomic library (we typically use a library constructed in a high-copy 2 μm expression vector (YEplac195)) and transform into yeast (BY4743) (Brachmann *et al.*, 1998; Winzeler *et al.*, 1999) by a standard lithium acetate method (Gietz and Schiestl, 2007) and select on medium lacking uracil (ura–). The MOB-ORF collection (Ho *et al.*, 2009) should work quite well for MSP.

2. After 3 days of growth, make sure you have at least 10^6 transformant colonies.

3. Resuspend colonies by flooding the plates as described above and pool into ura-medium containing 7% DMSO, aliquot as 10–25 μl samples of pool into individually capped PCR tubes, and store at − 80 °C aliquot until use.

2.2. Determining the drug dosage

Successful genome-wide assays require a drug/compound dosage that affects the rate of pooled growth. To determine this dose for HIP/HOP and MSP, we prescreen compounds using wild-type cells. For the deletion (loss-of-function) screens, we aim for treatment doses that cause a 10–20% decreased growth rate on wild type (Fig. 10.2). At 10–20% inhibition, optimal results for the heterozygous collection are usually obtained at the 20 generation time point, and optimal results for the homozygous collection

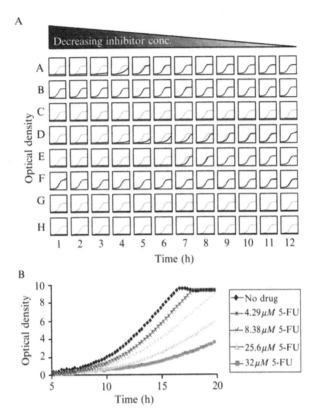

Figure 10.2 Prescreening compounds against wild-type yeast to determine an appropriate dose for genome-wide screening. (A) A 96-well flat bottom plate is filled with 100 μl of cell suspension at an OD of 0.062. Two microliters of compound (typically dissolved in DMSO) is added using a slotted pin tool or multichannel pipette. Cells are grown with constant shaking for 16–20 h at 30 °C. The final concentration of DMSO should not exceed 2%. In this example, column 12 contains vehicle (DMSO) control, and columns 1–11 contain decreasing amounts of test compound. In each well of the plate, the growth curve in test compound is plotted in black against a plot of the control growth curve in grey. (B) Higher resolution image of several prescreens obtained with 5-FU overlaid on top of one another. In this titration series, an IC_{10-15} is obtained with 4.29 μM 5-FU and an IC_{70} is obtained with 32 μM 5-FU. The former dose would be appropriate for deletion profiling (HIP and HOP) whereas the latter dose would be appropriate for resistance profiling (MSP). Due to the nonlinearity at higher optical densities, Tecan ODs were converted to "traditional" 1 mm cuvette ODs using the calibration function "real OD"=−1.0543 + 12.2716 × measured OD.

are usually obtained at the 5 generation time point. Heterozygous strains have more subtle differences in growth rate (typically <5%) and therefore require a longer time to resolve growth differences. For the MSP (gain-of-functions) screens, we aim for a compound dose that inhibits wild-type growth by 70–90%. In this case, cells are collected when the OD of the culture reaches 2.0, regardless of the number of generations required to achieve this culture density. These generation times and doses have been empirically determined over the last several years and provide a well-informed starting point; nonetheless variations in generation times and degree of inhibition may improve results.

2.2.1. Materials for yeast deletion strain pool and MSP pool construction and determining the drug dosage

1. Frozen glycerol stocks of the yeast deletion collection in 96-well microtiter plates (OpenBiosystems, Part Nos. YSC1056 and YSC1055).
2. Nunc Omni Trays (VWR, Catalog No. 62409-600).
3. 96-well pin tool (V&P Scientific, Catalog No. VP407A).
4. 30 °C incubator for growing plated yeast.
5. Spectrophotometer.
6. G418, Nourseothricin (Agri-Bio, Catalog No. 3000, Werner Bio-Agents, Catalog No. 500100).
7. Cell spreader (VWR, Catalog No. 89042-020).

2.3. Experimental pool growth for deletion and overexpression collections

The laboratory procedure starting from thawing the frozen aliquots of pooled cells is described below and visualized in Fig. 10.3. Examples of results using the HIP, HOP, or MSP assay can be seen in Fig. 10.4.

2.3.1. Deletion profiling (HIP/HOP)

1. Thaw a frozen aliquot of pooled cells on ice.
2. Immediately dilute the pool into media with drug or condition of choice, in parallel with the appropriate solvent controls. Inoculate cultures using either option a or b.
   (a) *Automated robotic cell growth.* Inoculate cells in medium with drug or vehicle at OD_{600} of 0.0625 in a total volume of 700 μl in a 48-well plate, and seal with a plastic plate seal. Similarly, prepare wells as above but without cells to be used by the robot for the inoculation at 5-generation intervals. If the condition requires optimal aerobic growth (e.g., nonfermentable carbon sources) poke a small hole in the membrane seal toward the side of each well. Grow in a spectrophotometer at 30 °C with at an experimentally determined optimal shaking regimen. Cells can normally be grown to a final OD_{600} of 2.0 which equates 5 generations of growth. Part of the cell suspension can be saved on a cold plate on the robotic deck at user defined generation times (http://med.stanford.edu/sgtc/technology/access. html, for details contact C. Nislow or G. Giaever).

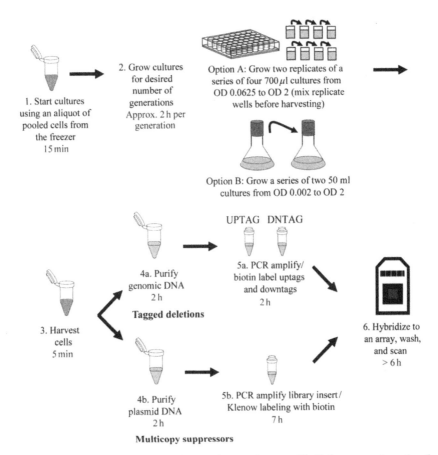

Figure 10.3 Timeline for pooled growth experiments. (1) Cultures are inoculated using thawed aliquots of pooled cells. (2) Cells are grown for the desired number of generations (typically 5–20 generations, 16–72 h). The specific amount of time needed for growth will vary depending on the number of generations and the level of inhibition of the treatment. (3) Cells are harvested by centrifugation. (4) Genomic DNA or plasmid DNA is purified from the cell pellets using standard column-based purification kits. (5) Tags (or genomic DNA fragments) are PCR-amplified. (6) PCR products are hybridized to an array, which is then washed and scanned.

(b) *Manual cell growth.* Inoculate a 50 ml culture at a starting OD_{600} of 0.0020 in a 250 ml culture flask. Grow in a shaking incubator at 30 °C at 250 rpm. After cells reach a final OD_{600} of 2.0 they will have undergone 10 generations of growth.

3. Collect 1–2 OD_{600} of cells in a 1.5 ml microfuge tube after growth to the desired number of generations. Pellet cells, and remove the media. If option (b) above has been used, normalize the ODs between all samples.

4. If not proceeding directly to the next step, cell pellets can be stored at −20 °C.

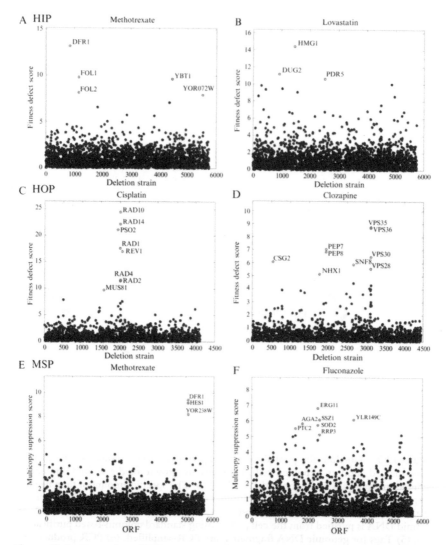

Figure 10.4 Examples of HIP–HOP and MSP results. Panels (A–D) show the results of HIP or HOP screens in which the heterozygote deletions strains were grown for 20 generations and the homozygous strains for 5 generations. For methotrexate (A) and lovastatin (B), the deletion screens identify the drug targets (DFR1 and HMG1, respectively) as the most sensitive heterozygotes. For cisplatin (C), several homozygotes deleted for components of the DNA damage response appear as sensitive (Lee *et al.*, 2005). For clozapine (D), an atypical antipsychotic compound for which no known target exists in yeast, a number of nonessential genes involved in vesicle traffic and protein sorting appear as significantly sensitive (Ericson *et al.*, 2008). Panels (E–F) show the results of two MSP screens. In the case of methotrexate and fluconazole, the known targets (DFR1 and ERG11, respectively) appear as the most resistant clones (Hoon *et al.*, 2008).

2.3.2. MSP profiling

MSP assays are performed as described above, except that cells are grown in medium lacking uracil, in the presence of high doses of compound, and that cells are collected when the OD of the culture reaches 2.0. Because it is impossible to predict when a resistant clone (or clones) will "overtake" the culture, we typically allow 2–4 days for a typical MSP growth assay.

Note: A starting cell sample (i.e., a "T0 time point") is required to assess initial strain representation in the newly created pool. To prepare such a sample, add 1–2 OD_{600} of pool directly from the freezer aliquots to a 1.5 ml tube and process as described above.

When choosing between option (a) or (b) for the pool growth, consider the following, option (a) features a higher throughput, smaller culture volume and compound cost, and automated inoculation and cell suspension steps compared to option (b). However, option (a) has a higher sampling error due to small culture volume. The sampling error during cell growth is currently the largest source of error when growing in small cultures. Option (b), allows cells to be collected at ∼1000 cells/strain which will increase accuracy. However, manual growth is limited in the ability to collect cells at exact generation times which is important for the downstream analysis. For both options (a) and (b), the condition and control samples need to be collected at the same number of generations.

2.3.3. Materials

1. Temperature controlled shaker for 250 ml flasks or shaking spectrophotometer such as Infinite F200 (Tecan; www.tecan.com). *Note*: many spectrophotometers do not shake sufficiently hard for growing yeast in suspension).
2. 48-well plates (Greiner, Part No. 677102).
3. Adhesive plate seals (ABgene, Catalog No. AB-0580).
4. 250 ml culture flasks (if growing cultures in flasks).

2.4. Purification and amplification of barcodes and ORFs

2.4.1. Deletion profiling (HIP/HOP)

1. Purify genomic DNA from ∼2 OD_{600} of cells with the Zymo Research YeaStar kit using Protocol I (included with the kit), or another suitable method for purifying yeast genomic DNA. Genomic DNA can be stored indefinitely at − 20 °C.
2. If desired, quantify genomic DNA using a gel or a UV spectrophotometer.
3. Set up two 60 μl PCR reactions for each sample, one for the uptags and one for the downtags (33 μl dH$_2$O, 6 μl 10× PCR buffer without MgCl$_2$, 3 μl 50 mM MgCl$_2$, 1.2 μl 10 mM dNTPs, 1.2 μl 50 μM Up

or Down primer mix (see below), 0.6 μl 5 U/μl Taq polymerase, $\sim$0.1 μg genomic DNA in 15 μl).

4. Cycle as follows in a thermocycler with a heated lid: 94 °C 3 min; repeat 30×: 94 °C 30 s, 55 °C 30 s, 72 °C 30 s; 72 °C 3 min, hold at 4 °C. PCR products can be stored at −20 °C.
5. Check the resulting PCR products on a gel, expecting a $\sim$60 bp product (see note 4 under Section 3).
6. If not proceeding to the next step immediately, store PCR products at −20 °C in a nonfrost free freezer.

2.4.1.1. Materials

1. YeaStar Genomic DNA Kit (Zymo Research, Catalog No. D2002).
2. Taq DNA polymerase (Invitrogen, Catalog No. 10342).
3. dNTPs (Invitrogen, Catalog No. 10297).
4. 10× PCR reaction buffer (MgCl$_2$) (Invitrogen, Part No. 52724).
5. 50 mM MgCl$_2$ (Invitrogen, Part No. 52723).
6. *Up primer mix*: Dissolve Uptag (5′-GAT GTC CAC GAG GTC TCT-3′) and Buptagkanmx4 (5′ biotin-GTC GAC CTG CAG CGT ACG-3′) each in dH$_2$O at 100 pmol/μl, then mix in a 1:1 ratio for a final concentration of 50 pmol/μl each. Store at −20 °C.
7. *Down primer mix*: Dissolve Dntag (5′-CGG TGT CGG TCT CGT AG-3′) and Bdntagkanmx4 (5′ biotin-GAA AAC GAG CTC GAA TTC ATC G-3′) each in dH$_2$O at 100 pmol/μl ($=$100 μM), then mix in a 1:1 ratio for a final concentration of 50 pmol/μl each. Store at −20 °C.
8. Thermocycler with heated lid.

2.4.2. MSP

1. Isolate plasmids using the Zymoprep II plasmid isolation kit.
2. Amplify inserts by PCR using common M13 primers in amplification conditions described above (M13 forward primer: 5′-GTT GTA AAA CGA CGG CCA GT-3′; M13 reverse primer: 5′-CAG GAA ACA GCT ATG ACC-3′).
3. Purify PCR products using QIAquick PCR purification kit.
4. Label inserts with biotin using the BioPrime labeling kit.
5. Hybridize labeled inserts to Affymetrix TAG4 arrays using the same protocols as described for TAG hybridization below.

2.4.2.1. Materials

1. Zymoprep II plasmid isolation kit (Zymo Research; Catalog No. D2004)
2. FailSafeTM PCR System (EPICENTRE Biotechnologies)
3. QIAquick PCR purification kit (Qiagen; Catalog No. 28104)

4. Bioprime labeling kit (Invitrogen; Catalog No. 18094-011)
5. M13 forward primer: 5'-GTT GTA AAA CGA CGG CCA GT-3'
6. M13 reverse primer: 5'-CAG GAA ACA GCT ATG ACC-3'

2.5. Hybridization

1. Set up a boiling water bath with a floating rack for 1.5 ml tubes and a slushy ice bucket. Set hybridization oven temperature to 42 °C.
2. Fill the arrays with 1× hybridization buffer.
3. Prewet the array at 42 °C and 20 rpm for at least 10 min in the hybridization oven.
4. Immediately before use, prepare 90 μl hybridization mix per sample (75 μl 2× hybridization buffer, 0.5 μl B213 control oligonucleotide (0.2 fm/μl), 12 μl mixed oligonucleotides (12.5 pm/μl), 3 μl 50× Denhardt's solution) in 1.5 ml tubes suitable for boiling. MSP hybridization conditions are identical except MSP does not employ any blocking oligos.
5. While arrays are equilibrating, add 30 μl of uptag PCR and 30 μl of downtag PCR to 90 μl of hybridization mix for a total volume of 150 μl. (For MSP add the entire labeling reaction.)
6. Boil for 2 min and set on ice for at least 2 min.
7. Spin tubes briefly.
8. Remove hybridization buffer from the arrays and replace with the hybe mix (kept on ice).
9. Place a Tough-Spot (or other adhesive tape) over each of the two gaskets to prevent evaporation and hybridize for 16 h at 42 °C and 20 rpm.
10. Immediately before use, prepare 600 μl biotin labeling mix per sample (180 μl 20× SSPE, 12 μl 50× Denhardt's solution, 6 μl 1% Tween 20 (vol/vol), 1 μl 1 mg/ml streptavidin-phycoerythrin, 401 μl dH2O).
11. Aliquot 600 μl biotin labelling mix per chip into tubes. Remove Tough-Spots from chips.
12. Remove hybridization mix from the microarrays and save it in −20 °C (can be reused if needed). Fill chips with Wash A.
13. Wash the microarrays using the Affymetrix fluidics station according to the manufacturer's instructions. After priming the station, use the "Gene-Flex_Sv3_450" protocol with the following modifications: 1 extra step with Wash A (1 cycle, 2 mixes) before staining, Wash B temperature 42 °C instead of 40 °C, stain at 42 °C instead of 25 °C. This same protocol is used for processing MSP samples. *Note*: it is possible to perform the post hybridization wash, the biotin staining and the post staining wash manually, see page 396 in C. Nislow and G. Giaever, 2007.

14. Verify the absence of air bubbles in the microarrays. If bubbles are present, insert the chip again, and it will automatically be refilled with Wash A. If there are any marks or smudges on the array surface, clean the glass window on each array with isopropanol and a cotton swab or lint-free wipe. Put Tough-Spots on the gaskets to prevent evaporation and put arrays in scanner.
15. Scan at an emission wavelength of 560 nm.
16. When all fluidics operations are complete, run the fluidics station "SHUTDOWN_450" protocol.

2.5.1. Materials

1. Genflex Tag 16K Array v2 (Affymetrix, Part No. 511331).
2. Hybridization Oven 640 (Affymetrix, Part No. 800138).
3. GeneChip Fluidic Station 450 (Affymetrix, Part No. 00-0079).
4. GeneArray Scanner 3000 (Affymetrix, Part No. 00-0212).
5. 1.5 ml Microfuge tubes (suitable for boiling).
6. Boiling water bath with floating rack for 0.5 ml tubes.
7. Teeny Tough-Spots (Diversified Biotech, Catalog No. TS-TNY).
8. Denhardt's Solution, 50× concentrate (e.g., Sigma, Catalog No. D-2532).
9. Streptavidin, R-phycoerythrin conjugate (SAPE) (Invitrogen, Catalog No. S-866). Store at 4 °C. *Do not freeze.*
10. B213 oligonucleotide: (5′ biotin-CTGAACGGTAGCATCTTGAC-3′).
11. Mixed oligonucleotides: Dissolve each of the following eight oligos in dH$_2$O at 100 pmol/μl: Uptag (5′-GAT GTC CAC GAG GTC TCT-3′), Dntag (5′-CGG TGT CGG TCT CGT AG-3′), Uptagkanmx (5′-GTC GAC CTG CAG CGT ACG-3′), Dntagkanmx (5′-GAA AAC GAG CTC GAA TTC ATC G-3′), Uptagcomp (5′-AGA GAC CTC GTG GAC ATC-3′), Dntagcomp (5′-CTA CGA GAC CGA CAC CG-3′), Upkancomp (5′-CGT ACG CTG CAG GTC GAC-3′), Dnkancomp (5′-CGA TGA ATT CGA GCT CGT TTT C-3′).
12. 0.5 *M* EDTA (BioRad, Catalog No. 161-0729).
13. 10% Tween: (Sigma, Catalog No. T2700).
14. 12× MES stock: For 10 ml, dissolve 0.7 g MES free acid monohydrate (Sigma, Catalog No. M5287) and 1.93 g MES sodium salt (Sigma, Catalog No. M5057) in 8 ml Molecular Biology Grade water. After mixing well, check pH and adjust if needed to a pH between 6.5 and 6.7. Add Molecular Biology Grade water to a total volume of 10 ml. Filter through a 0.2 μM filter, shield from light by wrapping tube in foil, and store at 4 °C. Replace if solution becomes visibly yellow or after 12 months, whichever comes first.
15. 2× Hybridization buffer: For 50 ml, mix 8.3 ml of 12× MES stock (prepared as above), 17.7 ml of 5 *M* NaCl (J. T. Baker, Catalog No.

3624-01), 4.0 ml of 0.5 M EDTA, 0.1 ml of 10% Tween 20 (vol/vol), 19.9 ml filtered dH_2O. Filter through a 0.2 μM filter.

16. Wash A: Mix 300 ml 20× SSPE (Sigma, Catalog No. S2015), 1 ml 10% Tween (vol/vol), 699 ml filtered dH_2O. Filter through a 0.2 μM filter.

17. Wash B: Mix 150 ml 20× SSPE (Sigma, Catalog No. S2015), 1 ml 10% Tween (vol/vol), 849 ml dH_2O. Filter through a 0.2 μM filter.

2.6. Analysis of results

2.6.1. Outlier masking (for HIP and HOP)

The Affymetrix TAG4 array contains at least five replicate features for each tag probe, dispersed across the array so that outlier features can be identified and discarded before calculating an average intensity value for each tag.

1. For each array feature, examine the surrounding 5 features × 5 features. If 13 (or more) of the 25 probes in this region differ from their trimmed replicate mean (the mean of the three middle replicates, excluding the highest and lowest replicates) by more than 10%, this probe is not suitable for data analysis.

2. Once these outlier-dense regions have been identified, pad them by including all probes within a 5-probe radius, as defined by $((x_1 - x_2)^2 + (y_1 - y_2)^2)^{1/2} < 6$ where x_1, x_2, y_1, and y_2 are the x and y coordinates for the two features.

3. Discard features for which (standard deviation of feature pixels/mean feature pixels) > 0.3. The standard deviation is included in the .cel file for Affymetrix arrays.

4. After identification and removal of outliers, calculate intensity values for each tag by averaging all unmasked replicates.

2.6.2. Saturation correction (for HIP, HOP, and MSP)

The signal on the TAG4 array is not linearly related to tag concentration because of the phenomenon of feature saturation (Pierce et $al.$, 2006). If uncorrected, this saturation will cause the degree of sensitivity or resistance to be underestimated for strains with brighter tags. Saturation is corrected by comparing uptag and downtag ratios, specifically:

1. Using a pair of arrays that are not sample replicates, calculate $\ln(i_c - bg)/(i_t - bg)$ for each tag, where i_c is the control intensity, i_t is the treatment intensity, and bg is the background as estimated by taking the mean intensity of the unassigned tag probes.

2. Mark ratios for any tags with minimum values less than 3× background as unusable.

3. Pair uptag and downtag ratios by strain. Ignore ratios for any strains with less than two usable tags.

4. For each strain, calculate the difference in average intensity for the two tags: $(i_{tu} + i_{cu})/2 - (i_{td} + i_{cd})/2$, where the subscript u indicates the uptag, and the subscript d indicates the downtag.

5. Sort ratios by the difference in average intensity. Take a sliding window of 600 ratio pairs, sliding 100 pairs at a time. For each window fit a line to the uptag ratios (x-axis) versus the downtag ratios (y-axis) using least-squares and take the slope. Calculate the mean of the differences in average intensity for the window.

6. Fit a least-squares line to the intensity differences for all windows (x-axis), versus slopes for all windows (y-axis) and take the slope of this line.

7. Repeat using the reverse intensity difference: $(i_{td} + i_{cd})/2 - (i_{tu} + i_{cu})/2$, and taking the slope with the axes reversed: the downtag ratios on the x-axis, and the uptag ratios on the y-axis.

8. Average the slope calculated in step 6 with the slope in step 7. This is the saturation correction factor S. A typical range for S is 0.0001–0.0005.

9. Adjust the raw intensity data using the following transformation: $f(i) = ie^{iS}$.

10. To correct more than two arrays, calculate S for all possible pairs of arrays, then use the median as the correction factor for all arrays in the set. Using a larger group of arrays will improve the accuracy of S.

2.6.3. Array normalization

The uptags and downtags should be normalized separately, because they are amplified in separate PCR reactions, and the intensities of the individual PCR reactions will affect their array intensities. For MSP ORF probes, quantile or mean normalization is performed without distinguishing between probes because the MSP PCR is performed in a single reaction. Normalize using either quantile normalization or mean normalization.

To quantile normalize a set of arrays: rank values obtained from each array for each set of tags (up and down) in order of increasing intensity. For each rank, assign the tag at that rank for each array to the median of all values at that rank.

To mean normalize a set of arrays: for each set of tags (up and down), divide by the mean. Multiply each tag set by the mean across all arrays (this is for convenience only; it returns the tag intensities to approximately their original range).

2.6.4. Removing unusable tags

Tags with low-intensity values in control samples will give poor-quality results. An intensity value threshold for excluding these tags can be chosen by comparing the correlation of uptag and downtag ratios as a function of tag intensity. Specifically:

1. Using any treatment-control pair, calculate $\log 2(i_c - bg)/(i_t - bg)$ for each tag, where i_c is the control intensity, i_t is the treatment intensity, and bg is the mean intensity of the unassigned tag probes.
2. Pair uptag and downtag ratios by strain and for each tag pair, take the minimum intensity for the two tags in the two samples. Sort the ratio pairs by this minimum intensity.
3. Use a sliding window of size 50 on the ranked ratio pairs, starting with the lowest intensity pairs. Calculate the correlation of uptag and downtag ratios for pairs within the window. Also calculate the average of the minimum intensities calculated in the previous step.
4. Slide the window by 25 pairs, and repeat the previous step until all pairs have been traversed.
5. Plot the average minimum intensity versus the uptag–downtag correlation for all windows.
6. Chose an intensity threshold, for example, we find the intensity value where the correlation first reaches 80% of its maximum level is a good cutoff.
7. Mark any tags that are below this cutoff in either of the samples as unusable for subsequent analysis.

2.6.5. Calculating log2 ratios for deletion pool screens

1. For each tag, calculate $\log 2(\mu_c - bg)/(\mu_t - bg)$ where μ_c is the mean intensity for the control samples, μ_t is the mean intensity for the treatment samples, and bg is the mean intensity of the unassigned probes.
2. For each strain, average the log2 ratios for all usable tags to obtain a final sensitivity score. Strains that are sensitive will have positive scores, while strains that are resistant will have negative scores. This score is proportional to the log2 ratio of cells present in the control sample versus the treatment sample.

Note: Whereas p values describe the level of confidence for calling strain sensitivity, the log2 ratios give the best estimate of the actual level of sensitivity for each strain.

For MSP, as we are looking for gain of signal, it does not affect the results significantly if low-intensity tags are present in the analysis.

2.6.6. Calculating MSP ratios

For MSP, each ORF is represented by at least two probes and the log2 ratios of each probe are averaged to generate a single score for each gene. To identify each suppressor locus, order the log2 ratio of intensities by each ORF's genomic location and perform the analysis using a sliding window to identify loci that have at least two adjacent ORFs with log2 ratios ≥ 1.6.

All scripts for performing these analyses on the array data can be found at these two URLs, the first describes the analyses, the second contains the actual PERL scripts.

http://chemogenomics.stanford.edu/supplements/04tag/analysis

http://chemogenomics.stanford.edu/supplements/04tag/download. html#scripts

2.7. Confirmation of microarray data

After performing a pooled fitness assay, several strains are often candidates for interacting with the test compound, based on their sensitivity or resistance. The choice of which strains to confirm after performing a pooled assay is somewhat arbitrary, but as a general guideline, the sensitivity scored using the pooled fitness approach tends to be confirmed for the majority of deletion strains with z-scores>3 (Lee *et al.*, 2005). Requiring a fold change of at least 2 (log2=1) avoids testing strains associated with z-scores that are high only because they have a very reproducible abundance pattern in the control condition.

While individual confirmation assays can be conducted in any desired growth format, we perform them in 100 μl in 96-well plates, using the same drug (and vehicle) concentration that was used in the genome-wide screen. The 96-well plate format allows for the testing of several strains in replicate in parallel, and limits the amount of drug required. Because some drugs loose efficacy over time, it is crucial to include a wild-type strain, preferably on the same growth plate. Ideally, all strains should be tested in triplicate in both drug and vehicle, for robust statistical analysis. In most cases, a deletion strain is confirmed as sensitive to the drug when its growth (compared to wild type) is not affected in the control condition, but is inhibited by drug (see *ypt7*Δ in Fig. 10.5). The vehicle control accounts for deletion strains that exhibit a growth defect relative to wild type in the absence of compound (e.g., *erg4*Δ in Fig. 10.5). In some cases, a growth defect may also be observed for both vehicle and drug, but the strain is still considered significantly sensitive because the growth defect in drug is greater than the growth defect in vehicle (see *vps3*Δ in Fig. 10.5). Similarly, MSP assay results can be confirmed using these growth assays by comparing the growth of over-expression strains to that of the control in the conditions of interest.

Occasionally, strains fail to confirm. This may be due to biological or technical reasons. For example, cross-contamination can occur between wells on the agar plate where the deletion collection is stored, or on the plate used for the actual confirmation screen, arguing for careful handling of the strains. To test for cross-contamination and to verify a strain's identity, deletion strains can be tested using PCR after streaking for single colonies (http://www-sequence. stanford.edu/group/yeast_deletion_project/deletions3.html). For homozygous deletion strains, it is necessary to test for both the presence of the deletion cassette at the intended locus as well as for the absence of a wild-type copy of the ORF.

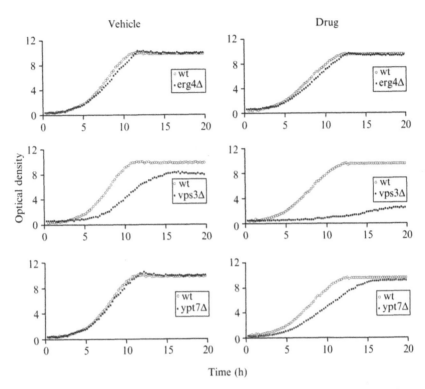

Figure 10.5 Confirmation of microarray-detected growth sensitivity. Three deletion strains (*erg4Δ*, *vps3Δ*, and *ypt7Δ*) with microarray-inferred sensitivity to drug were tested by growing the strains in isolation. We used a Tecan Genios reader, with optical density readings taken every 15 min over 20 h. Each strain was grown in triplicate and compared to the wild-type growth both in vehicle (left panel) and in drug (right panel). Growth defects can be either severe (exemplified by *vps3Δ*) or mild (exemplified by *erg4Δ*). In this example, the sensitivity scored by microarray could be confirmed for *vps3Δ* and *ypt7Δ*. Wild-type growth is represented by open circles and mutant growth by filled squares. Representative growth curves for the replicates are shown. Tecan ODs were converted to traditional "cuvette" ODs using the calibration function "real OD"=−1.0543 + 12.2716 × measured OD.

In cases where the observed growth defects for deletion strains are minor or difficult to discern between drug and control, statistical tests can help to determine if the phenotype in question is significant or not. To perform these tests, three fitness scores (*W*) are derived for each of the mutants and conditions tested, using the three replicates of wild type and mutant (shown here for the first replicate only):

$W_{1vehicle}$ = doubling time of $wt_{1vehicle}$/doubling time of $mutant_{1vehicle}$, where 1 = strain replicate 1.

W_{1drug} = doubling time of wt_{1drug}/doubling time of $mutant_{1drug}$, where 1 = strain replicate 1.

To obtain a p value for whether the difference between W observed in vehicle and drug is significant, a student's t-test can be performed on the two populations of three W values (using a two-tailed distribution, two-sample equal variance). The choice of p value is arbitrary, but as a guideline, we typically require a p value of < 0.05 for the observed growth defect to be deemed significant. For example, performing these calculations, $erg4\Delta$ in Fig. 10.5 is not significantly sensitive to the drug tested, but $vps3\Delta$ and $ypt7\Delta$ are.

A W value below 1 indicates sensitivity of the mutant compared to wild type. To account for any growth defects due to the vehicle, a normalized fitness (NW) can be calculated as follows:

$$NW = \text{average } W_{\text{drug}}/\text{average } W_{\text{vehicle}}$$

A NW value below 1 indicates that the strain has a drug-induced growth defect. This NW value, along with its p value as calculated above, allows for easy comparisons of phenotypes between drugs and strains, and can be reported as the "normalized fitness value" for the individual growth assays (Fig. 10.5).

3. EXPERIMENTAL CONSIDERATIONS

1. *Choosing the appropriate culture volume and starting OD* Choosing an appropriate starting culture volume and cell concentration is critical for obtaining good results and using at least 300 cells per strain is recommended. If the culture vessel will not accommodate the volume needed to reach the desired cell numbers, multiple cultures can be grown in parallel and pooled at the end of the experiment.
2. *Replicates* Each experiment requires at least two arrays—a control sample, and a treatment sample. We often use a large (more than 10) set of control arrays for analyzing many different experimental arrays, each with only one replicate. This control set can be used to calculate the statistical significance of the final results, and helps minimize the total number of experimental arrays needed.
3. *Comparing OD values between plates and cuvettes* If cells are grown in a shaking spectrophotometer, note that the OD measured for each well in the plate will differ from the OD of the same culture when measured in a cuvette due to differences in path length. Similarly, OD readings in a shaking spectrophotometer will also vary with differences in culture volume. All OD values reported here refer to those measured in a 1 cm path-length cuvette.
4. *Checking for successful tag amplification* Tag PCR products should be evaluated by agarose gel electrophoresis. The desired product is ~ 60 bp. A second band is often seen because noncomplementary tags can

hybridize at their common primer regions, forming a partially single-stranded structure that migrates faster than the fully double-stranded tag products. It is also common to observe amplification in no-template control reactions, nonetheless these spurious bands do not adversely affect the results.

5. *Barcode sequences* A full list of sequences is available (Smith *et al.*, 2009).
6. *Array options* These protocols use Affymetrix arrays; however, they can be easily adapted to the array platform of your choice. For alternative array examples, see Pan *et al.* (2004) and Yuan *et al.* (2005).
7. *Evaluating data quality and replicate sample agreement* The most effective way to measure the quality of technical replicate samples is to measure the correlation of the log-transformed, normalized, and saturation-corrected intensity values. The correlation of these "processed raw values" should be at least 0.90 for replicates. For biological replicates grown and prepared separated in time, we typically observe correlations of about 0.7 (Ericson *et al.*, 2008; Hillenmeyer *et al.*, 2008).

4. PERSPECTIVES

Genome-wide chemogenomic assays have proven extremely valuable in determining gene function and the mechanism of action of drugs and small molecule probes. To date, the majority of such gene-dose assays have relied on homozygous and heterozygous deletions in yeast to probe loss-of-function effects, and a genomic library of clones to investigate gain-of-function effects. Recently the palette of available resources for such screens has expanded considerably, offering the promise of higher resolution gene-dose assays. These collections include; but are not limited to:

1. DAmP alleles (Breslow *et al.*, 2008; Yan *et al.*, 2008)
2. Ts alleles (Ben-Aroya *et al.*, 2008)
3. Systematic clone banks of ORFs (Ho *et al.*, 2009) and genome fragments (Jones *et al.*, 2008)

In addition, the recently introduced barcoder technology (Yan *et al.*, 2008) will allow rapid barcoding of any strain collection for parallel analysis. The experimental rationale that underlies these screens is not limited to *S. cerevisiae*, indeed work from several groups has produced genome-wide collections for *Escherichia coli* (Baba *et al.*, 2006; Kitagawa *et al.*, 2005), *Schizosaccharomyces pombe* (http://pombe.bioneer.co.kr/), and human cells using interfering RNAs (Luo *et al.*, 2008; Moffat *et al.*, 2006). Each of the collections can be screened in a manner analogous to what we describe in this chapter. Experiments that use these collections will certainly expand our perspective on cellular physiology at a systems level.

Finally, it is important to note that the screening platform described here relies on high-density microarrays. Recent advances in high-throughput, "Next-generation" sequencing technologies promise to eventually displace microarrays in this regard. Toward this end, we have directly compared results obtained using a traditional microarray readout with next-generation sequencing (Smith *et al.*, 2009) and have found that the performance of next-generation sequencing, is comparable, and in some aspects, superior to microarrays.

ACKNOWLEDGMENTS

We thank all members of the HIP–HOP laboratories at Stanford University and the University of Toronto. Ron Davis provided advice throughout the development of these assays. EE is supported by an Ontario Postdoctoral Fellowship. Research in the Nislow and Giaever laboratories is supported by the National Human Genome Research Institute of the NIH and by the CIHR (MOP-81340 to GG and MOP-84305 to CN).

REFERENCES

Alberts, B. (1998). The cell as a collection of protein machines: Preparing the next generation of molecular biologists. *Cell* **92,** 291–294.

Baba, T., *et al.* (2006). Construction of *Escherichia coli* K-12 in-frame, single-gene knockout mutants: The Keio collection. *Mol. Syst. Biol.* **2,** 2006.0008.

Ben-Aroya, S., *et al.* (2008). Toward a comprehensive temperature-sensitive mutant repository of the essential genes of *Saccharomyces cerevisiae. Mol. Cell* **30,** 248–258.

Birrell, G. W., *et al.* (2002). Transcriptional response of *Saccharomyces cerevisiae* to DNA-damaging agents does not identify the genes that protect against these agents. *Proc. Natl. Acad. Sci. USA* **99,** 8778–8783.

Brachmann, C. B., *et al.* (1998). Designer deletion strains derived from *Saccharomyces cerevisiae* S288C: A useful set of strains and plasmids for PCR-mediated gene disruption and other applications. *Yeast* **14,** 115–132.

Breslow, D. K., *et al.* (2008). A comprehensive strategy enabling high-resolution functional analysis of the yeast genome. *Nat. Methods* **5,** 711–718.

Butcher, R. A. and Schreiber, S. L. (2006). A microarray-based protocol for monitoring the growth of yeast overexpression strains. *Nat. Protoc.* **1,** 569–576.

Deutschbauer, A. M., *et al.* (2005). Mechanisms of haploinsufficiency revealed by genome-wide profiling in yeast. *Genetics* **169,** 1915–1925.

Ericson, E., *et al.* (2008). Off-target effects of psychoactive drugs revealed by genome-wide assays in yeast. *PLoS Genet.* **4,** e1000151.

Ghose, A. K., *et al.* (1999). A knowledge-based approach in designing combinatorial or medicinal chemistry libraries for drug discovery. 1. A qualitative and quantitative characterization of known drug databases. *J. Comb. Chem.* **1,** 55–68.

Giaever, G., *et al.* (1999). Genomic profiling of drug sensitivities via induced haploinsufficiency. *Nat. Genet.* **21,** 278–283.

Giaever, G., *et al.* (2002). Functional profiling of the *Saccharomyces cerevisiae* genome. *Nature* **418,** 387–391.

Giaever, G., *et al*. (2004). Chemogenomic profiling: Identifying the functional interactions of small molecules in yeast. *Proc. Natl. Acad. Sci. USA* **101**, 793–798.

Gietz, R. D., and Schiestl, R. H. (2007). Large-scale high-efficiency yeast transformation using the LiAc/SS carrier DNA/PEG method. *Nat. Protoc.* **2**, 38–41.

Hillenmeyer, M. E., *et al*. (2008). The chemical genomic portrait of yeast: Uncovering a phenotype for all genes. *Science* **320**, 362–365.

Ho, C. H., *et al*. (2009). A molecular barcoded yeast ORF library enables mode-of-action analysis of bioactive compounds. *Nat. Biotechnol.* **27**, 369–377.

Hoon, S., *et al*. (2008). An integrated platform of genomic assays reveals small-molecule bioactivities. *Nat. Chem. Biol.* **4**, 498–506.

Jones, G. M., *et al*. (2008). A systematic library for comprehensive overexpression screens in *Saccharomyces cerevisiae*. *Nat. Methods* **5**, 239–241.

Kastenmayer, J. P., *et al*. (2006). Functional genomics of genes with small open reading frames (sORFs) in *S. cerevisiae*. *Genome Res.* **16**, 365–373.

Kitagawa, M., *et al*. (2005). Complete set of ORF clones of *Escherichia coli* ASKA library (a complete set of *E. coli* K-12 ORF archive): Unique resources for biological research. *DNA Res.* **12**, 291–299.

Lee, W., *et al*. (2005). Genome-wide requirements for resistance to functionally distinct DNA-damaging agents. *PLoS Genet.* **1**, e24.

Lum, P. Y., *et al*. (2004). Discovering modes of action for therapeutic compounds using a genome-wide screen of yeast heterozygotes. *Cell* **116**, 121–137.

Luo, B., *et al*. (2008). Highly parallel identification of essential genes in cancer cells. *Proc. Natl. Acad. Sci. USA* **105**, 20380–20385.

Moffat, J., *et al*. (2006). A lentiviral RNAi library for human and mouse genes applied to an arrayed viral high-content screen. *Cell* **124**, 1283–1298.

Nislow, C. and Giaever, G. (2007). Chemical Genomic Tools for Understanding Gene Function and Drug Action. *Methods in Microbiology* **36**, 387–414, 708–709.

Ooi, S. L., *et al*. (2001). A DNA microarray-based genetic screen for nonhomologous end-joining mutants in *Saccharomyces cerevisiae*. *Science* **294**, 2552–2556.

Pan, X., *et al*. (2004). A robust toolkit for functional profiling of the yeast genome. *Mol. Cell* **16**, 487–496.

Parsons, A. B., *et al*. (2004). Integration of chemical-genetic and genetic interaction data links bioactive compounds to cellular target pathways. *Nat. Biotechnol.* **22**, 62–69.

Parsons, A. B., *et al*. (2006). Exploring the mode-of-action of bioactive compounds by chemical-genetic profiling in yeast. *Cell* **126**, 611–625.

Pierce, S. E., *et al*. (2006). A unique and universal molecular barcode array. *Nat. Methods* **3**, 601–603.

Rine, J., *et al*. (1983). Targeted selection of recombinant clones through gene dosage effects. *Proc. Natl. Acad. Sci. USA* **80**, 6750–6754.

Shoemaker, D. D., *et al*. (1996). Quantitative phenotypic analysis of yeast deletion mutants using a highly parallel molecular bar-coding strategy. *Nat. Genet.* **14**, 450–456.

Smith, A. M., *et al*. (2009). Quantitative phenotyping via deep barcode sequencing. *Genome Res.* **19**, 1836–1842.

Steinmetz, L. M., *et al*. (2002). Systematic screen for human disease genes in yeast. *Nat. Genet.* **31**, 400–404.

Winzeler, E. A., *et al*. (1999). Functional characterization of the *S. cerevisiae* genome by gene deletion and parallel analysis. *Science* **285**, 901–906.

Yan, Z., *et al*. (2008). Yeast Barcoders: A chemogenomic application of a universal donor-strain collection carrying bar-code identifiers. *Nat. Methods* **5**, 719–725.

Yuan, D. S., *et al*. (2005). Improved microarray methods for profiling the Yeast Knockout strain collection. *Nucleic Acids Res.* **33**, e103.

PROTEOMICS

CHAPTER ELEVEN

YEAST EXPRESSION PROTEOMICS BY HIGH-RESOLUTION MASS SPECTROMETRY

Tobias C. Walther,* Jesper V. Olsen,[†] *and* Matthias Mann[‡]

Contents

Abstract

Comprehensive analysis of yeast as a model system requires to reliably determine its composition. Systematic approaches to globally determine the abundance of RNAs have existed for more than a decade and measurements of mRNAs are widely used as proxies for detecting changes in protein abundance. In contrast, methodologies to globally quantitate proteins are only recently becoming available. Such experiments are essential as proteins mediate the majority of biological processes and their abundance does not always correlate

* Organelle Architecture and Dynamics, Max Planck Institute of Biochemistry, Martinsried, Germany
† Novo Nordisk Foundation Center for Protein Research, Faculty of Health Sciences, University of Copenhagen, Denmark
‡ Proteomics and Signal Transduction, Max Planck Institute of Biochemistry, Martinsried, Germany

Methods in Enzymology, Volume 470
ISSN 0076-6879, DOI: 10.1016/S0076-6879(10)70011-2

well with changes in gene expression. Particularly translational and post-translational controls contribute majorly to regulation of protein abundance, for example in heat shock stress response. The development of new sample preparation methods, high-resolution mass spectrometry and novel bioinfomatic tools close this gap and allow the global quantitation of the yeast proteome under different conditions. Here, we provide background information on proteomics by mass-spectrometry and describe the practice of a comprehensive yeast proteome analysis.

1. INTRODUCTION

A major goal in analyzing yeast as a eukaryotic model is to understand how components of the system interact dynamically and to determine the "wiring" of the interacting parts. A prerequisite for such analysis is the ability to reliably and globally determine the composition of yeast cells under different conditions. While methods to determine RNA quantitatively and comprehensively, such as microarrays, have existed for more than a decade, the technology to globally determine changes of protein abundance is only recently becoming available. Proteins, however, constitute the majority of biologically active agents and information on their relative abundance is thus often crucial. Since changes in the amount of a protein are not always reflected in corresponding mRNA level changes (e.g., see Bonaldi *et al.*, 2008), it is essential for many experiments to measure them directly. This is particularly evident for regulatory processes that are mediated by posttranscriptional regulation affecting, for example, protein stability or production, such as heat stress.

For these reasons, many techniques to determine the relative abundance of proteins have been developed: Most notably, Western blot or fluorescence measurements of tagged proteins are routinely used and comprehensive libraries containing most yeast open reading frames fused to GFP or the TAP tag have been developed (for a global analysis using these resources see, e.g., Ghaemmaghami *et al.*, 2003; Huh *et al.*, 2003). However, in some cases, these tagging methods may interfere with protein function as they rely on altering the protein sequence by introducing tags. For example, C-terminally modified (e.g., by lipidation on a CaaX box) or tail-anchored proteins are not functional in these libraries as the tags are usually introduced at the C-terminus of proteins. In addition, assays relying on libraries of tagged proteins are not easily applied to global experiments, especially when the goal is to compare multiple experimental conditions, because one experiment for each gene or about 6000 individual experiments for all genes have to be performed for each condition.

In addition to these techniques, mass spectrometry (MS) is used to identify single proteins, for example, purified and resolved by denaturing

SDS polyacrylamide gel electrophoresis and this has become a standard tool in biochemistry. During the last 5 years, accelerating advances in MS technology, sample preparation, and computational proteomics led to the development of capabilities to comprehensively determine the relative amount of proteins in complex mixtures. Particularly the advent of precision proteomics due to new instrumentation, such as hybrid linear ion trap Fourier-transform mass spectrometers (e.g., the LTQ-Orbitrap) led to high mass resolving power (Mann and Kelleher, 2008). Resulting from these advances and the concomitant development of computational tools, "shotgun" approaches to sequence an increasingly high number of peptides from complex samples are now available (Cravatt et al., 2007; Domon and Aebersold, 2006). The high mass resolution obtained in these experiments reduces the number of potentially false-positive assignments and increases the number of identifications from a single chromatographic run (Cox and Mann, 2008). Together these advances enabled us in 2008 to report the first determination of a complete eukaryotic proteome from Saccharomyces cerevisiae (de Godoy et al., 2008).

To perform such global proteome quantitation, we used a state-of-the-art proteomics analysis setup consisting of sample preparation to separate peptides by isoelectric focusing (IEF), separation of peptides by liquid chromatography (LC), and online-injection of eluting peptides to the MS by electrospray ionization (ESI). In this chapter, we provide background on MS methods and describe the principles and practices of such a proteomic experiment. We also discuss how measurement time will be reduced drastically from our original report by new developments in sample preparation, novel MS instrumentation, and advanced computational methods. We expect that these developments together will in the future make expression proteomics of yeast a standard experiment for quantitative, comprehensive approaches.

2. Background, Methods, and Applications

2.1. The challenge

The principle challenge of proteomics is to reliably identify and quantitate proteins in dauntingly complex mixtures, for example, in a protein extract. In the case of yeast, at least 4200 proteins are expressed under normal growth conditions (de Godoy et al., 2008; Ghaemmaghami et al., 2003; Huh et al., 2003). Some fraction of these proteins is posttranslationally modified, for example, by phosphorylation, acetylation, or glycosylation, and this further increases the chemical complexity of the polypeptide mixture in an extract.

At the moment not every protein with all modifications can be quantitated from a single experiment. However, several methods are successfully employed to approach the problem. Initial experiments resolved proteins by sequential electrophoresis in two dimensions (2D-gel analysis), with each dimension fractionating the proteins based on a different principal characteristic of proteins (O'Farrell, 1975). Most commonly, IEF in one dimension and separation by size in denaturing SDS gels in the second dimension is used. The first "proteome-scale" experiments using this technology identified roughly 150 yeast proteins (Shevchenko *et al.*, 1996). However, this method often leads to multiple spots for the same protein, different, for example, in their modification(s) (Fountoulakis *et al.*, 2004). In addition, 2D gels are strongly biased toward only the most abundant proteins. Thus identification of more than a few hundred different proteins is generally not feasible. Moreover, accurate quantitation is often not possible because of overlaying spots (Campostrini *et al.*, 2005). A further complication is that the technology by itself does not allow the identification of proteins observed and therefore it is usually is combined with another analytical method, such as MS or Western blotting. For these combined reasons, 2D gels have not developed into a comprehensive proteomics technology.

In contrast, MS-based proteomics can unambiguously identify proteins in a very complex mixture with minimal prior separation. Because MS is a versatile tool that combines several unique capabilities; such as quantification of proteins from a cell or organism and characterization of important posttranslational modifications (PTMs) of proteins (e.g., phosphorylation) in addition to identification of individual proteins in a complex mixture, it has become the most important technology in proteomics today (Aebersold and Mann, 2003).

In an MS-based proteomic experiment, complex protein mixtures are usually digested by a protease, yielding a mixture of tens of thousands of peptides with a range of abundance over more than four orders of magnitude. Until recently, this complexity has limited the feasibility of "shotgun" approaches that aim to directly identify the peptides in the mixture. To deal with this problem, several methods to reduce the complexity of peptide mixtures for analysis were introduced. Traditionally, the starting material, that is, the yeast extract, is further fractionated using, for example, subcellular fractionation or denaturing SDS gel electrophoresis. Peptides resulting from the digest of such fractions are then separated by reversed-phase LC right before being analyzed "online" in the MS. Our recent experience, however, has been that extensive fractionation and separation at the protein level leads to rapidly diminishing returns in terms of the number of identified proteins (Bonaldi *et al.*, 2008; de Godoy *et al.*, 2008). Instead, several strategies for the fractionation of peptides after proteolytic digestion of protein mixtures have been designed. In one variation of this principle, termed *multi*dimensional *p*rotein *i*dentification *t*echnology or "MudPIT," the resulting peptides are separated by strong cation exchange chromatography (Washburn *et al.*, 2001). In addition, some variant techniques with mixed-anion/cation beds also exist (e.g., see Motoyama *et al.*, 2007). Recently, an alternative method employing IEF of peptides in a combined stationary and liquid phase was developed and this was an important contribution to directly asses the protein composition of total yeast extract (de Godoy *et al.*, 2008; Hubner *et al.*, 2008). This protocol, which is described in detail below, uses immobilized p*I* strips to separate peptides by IEF and further resolves the peptides in the resulting fractions by LC to directly analyze them by MS.

2.2. Background on MS instrumentation for "shotgun" proteomics

MS is essentially a technique for weighing molecules, but the measurements are not performed with a conventional balance or scale. Instead, in MS gas phase ions of peptides are separated or filtered according to their mass-to-charge (m/z) ratio in a magnetic or electrostatic field and finally recorded by a detector. The resulting mass spectrum is a plot of the relative abundances of the produced ions as a function of their m/z ratio (see Fig. 11.2).

Because every peptide molecule and modification has a characteristic mass, MS is a very powerful and nearly universal tool in proteomics and can result in determination of the chemical composition when mass accuracy is sufficiently high. Peptides furthermore have distinct fragmentation patterns that provide structural information to identify their amino acid sequences and modifications.

MS instrumentation developments have greatly contributed to recent breakthroughs in proteomic research. Several types of MS are currently employed in proteomics. They are distinguished by the ionization method used to charge peptides and by the type of mass analyzer used to determine the mass-to-charge (m/z) ratio of the resulting ions.

As traditional ionization methods such as chemical ionization are often too harsh, "soft" methods that allow ionization of intact biomolecules are necessary for MS-based proteomics. The two ionization methods employed in proteomics are *m*atrix *a*ssisted *l*aser *d*esorption *i*onization (MALDI) and ESI. ESI, which is used most commonly, allows large, nonvolatile molecules such as peptides and proteins to be ionized nondestructively directly from a liquid phase, usually consisting of a mixture of volatile organic solvent and acidified water. In electrospray, a liquid is passed through a nozzle to which a high voltage is applied (Fenn *et al.*, 1989; Whitehouse *et al.*, 1985). The charged liquid becomes unstable as it is forced to hold more and more charges. Soon the liquid reaches a critical point and near the tip of the nozzle it blows apart into a cloud of tiny, highly charged droplets. These droplets rapidly shrink as solvent molecules evaporate from their surface increasing the electric field at the droplet surface. By a process of "ion evaporation" (Iribarne and Thomson model) or simple solvent evaporation (charged residue model), the "naked" biomolecule becomes a gas-phase ion (Iribane and Thomson, 1976).

The other "soft" ionization technique, MALDI, was also developed in the late 1980s (Hillenkamp and Karas, 1990; Hillenkamp *et al.*, 1991; Karas and Hillenkamp, 1988). In this technique, analyte molecules are cocrystallized with an UV- or IR-absorbing substance—termed the matrix—which is usually an organic carboxylic acid such as 2,5-dihydroxybenzoic acid (UV-absorbing) or succinic acid (infrared absorbing). The analytes are desorbed and ionized by a laser beam (pulsed laser irradiation) from the solid or liquid surface containing the organic matrix compound in approximately 1000-fold excess. A widely accepted view how the matrix assists the ionization is that neutral sample molecules are ionized by acid–base proton transfer reactions with the protonated carboxylic acid matrix ions in a dense phase just above the surface of the matrix.

Like ESI, MALDI is capable of efficiently ionizing large biomolecules such as peptides and proteins and is often used with *time-of-flight* (TOF) MS (see below) due to the vacuum-compatibility and pulsing nature of the technique (the laser pulse frequency can easily be synchronized with the TOF extraction pulse). Both ESI and MALDI ionization allow introduction of biological molecules exceeding one million Daltons into MS, but they are by far most often used for analysis of peptides. For this purpose, the main difference between the two methods is that ESI predominately produces multiply charged ions, MH_n^{n+}. In contrast, MALDI almost exclusively generates singly charged peptide ions, MH^+, which can be difficult to sequence by the low-energy dissociation methods available on most proteomic mass analyzers; because when the single proton is fixed on the side chain of an arginine or lysine residue then there is no mobile proton available to induce peptide–amide bond fragmentations. 2D-gel-based proteomics is almost exclusively coupled to MALDI-TOF MS analysis, whereas most other areas of proteomics are increasingly based on ESI, since it is possible to integrate ESI with online LC-MS/MS.

After ionization, the mass of peptide ions is determined. The earliest analyzers used in MS-based proteomics combined a series of quadrupoles, each capable of selecting specific ions that can pass an applied electric field that deflects ions with other m/z ratios: In the first quadrupole, peptides are filtered (MS spectrum). In the second quadrupole, one filtered peptide at a time is fragmented at the peptide bond by collision with noble gases such as Argon or Helium and this is commonly called *collision-induced dissociation* ("CID"). Subsequently to the collisions, the resulting spectrum of the fragmented ions is filtered in the third quadrupole (MS/MS spectrum). Information about the sequence of the peptide analyzed is then contained in the mass difference between series of fragmented ions in this MS/MS spectrum.

As an alternative to CID, another fragmentation method, electron transfer dissociation ("ETD") has been employed in the last few years. In this method, electrons are transferred from radical anions to the positively multiply charged peptide ions, which then fragment adjacent to the amino group of the peptide bond (for review, see Mikesh et al., 2006). CID sometimes yields incomplete spectra, especially for very basic peptides. ETD often yields more uniformly fragmented peptides and in addition, the fragmentation is more specific for the peptide backbone and therefore PTMs, such as phosphorylations are less likely to be lost in the spectra of peptide-backbone-derived fragments. These different fragmentation methods can in principle be coupled with any type of MS instrument.

The classic setup consisting of triple quadrupoles described here is in principle very fast, but its mass accuracy (usually limited to 0.5 Da) and sensitivity are not high enough for experiments aimed at large-scale discovery of proteins in complex mixtures. Therefore, they are currently mainly employed in multiple reaction monitoring (MRM) experiments, where the fast MS/MS switching capability is used to detect a few preprogrammed fragmentation patterns for a number of predefined peptides, with the goal of accurate and targeted quantitation in complex mixtures (Anderson et al., 2009).

As an alternative instrument type, TOF MS measure the travel time of ions to the detector after they have all been accelerated to the same kinetic energy. This time is directly proportional to the mass-to-charge ratio (m/z), which can be quite accurately measured. Drawbacks of these instruments include that very high resolution and sensitivity are difficult to achieve at the same time.

A mass analyzer type currently used more commonly in proteomics is the ion trap. The basic principle of ion traps is similar to that of quadrupoles, with the exception that selected ions are trapped in the electric field and can be accumulated over time. This makes this instrument highly sensitive, but similar to quadrupole analyzers, its resolution is quite limited, sometimes leading to a mass uncertainty of several daltons. To overcome some of these drawbacks of conventional 3D-ion traps, a new generation of ion traps with superior ion capacity, dynamic range, scan speed, and sensitivity has been introduced. These are the linear ion traps (or 2D-ion traps)—essentially segmented quadrupole mass filters—capable of trapping and detecting a factor hundred more ions than traditional 3D-ion traps (Hager and Yves Le Blanc, 2003; Schwartz et al., 2002).

A major breakthrough in proteomics was the introduction of a "hybrid" MS, the LTQ-Orbitrap consisting of a linear ion trap and an orbitrap (Scigelova and Makarov, 2006; Fig. 11.1). The orbitrap is the first fundamentally new mass analyzer in more than 20 years. The instrument contains three components. It has a linear quadrupole ion trap (LTQ), in which it is possible to control and manipulate (e.g., accumulate and collisionally activate) ions in the subsecond time-scale. Detection can be achieved in two

A   Configuration of a tandem mass spectrometer

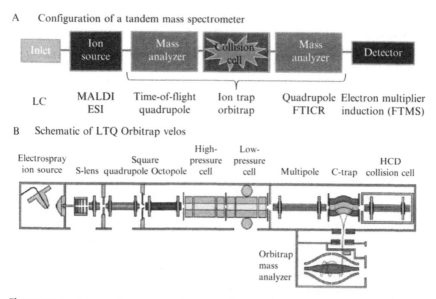

B   Schematic of LTQ Orbitrap velos

Figure 11.1 Schematic representation of typical tandem mass spectrometers for proteomics experiments. (A) Principal setup of a tandem mass spectrometer. Peptides are typically separated by liquid chromatography (LC) up-front and transferred to the gas phase in the ion source (by either MALDI or ESI). Peptide ions of interest can be separated or isolated in the first mass analyzer (either an ion trap or a quadrupole) and injected into the collision cell. The resulting fragmentation ions are analyzed in a second mass analyzer, for example, an ion trap or a TOF and recorded by electron multiplier detectors or by induction in Fourier-transform instruments. (B) Schematic overview of the LTQ-Orbitrap Velos. The front end of the instrument is a dual linear ion trap mass spectrometer capable of efficient ion accumulation, isolation, fragmentation, and detection of MS or MS^n ions. Accumulated ion populations are moved into the C-trap via an octapole ion guide, or for higher energy dissociation accelerates into the HCD collision cell and resulting fragments are subsequently moved back to the C-trap. In the C-trap, the motion of the ion population is damped by a residual pressure of nitrogen. Ions are then injected into the orbitrap in a short pulse and begin to circle the central electrode. The ion signals (peptide m/z values) are detected via a differential amplifier between the two halves of the outer orbitrap electrodes.

ways. In the linear ion trap, ions can be ejected radially through slits in the quadrupole rods and detected by two electron multiplier detectors. Alternatively, ions are ejected axially from the ion trap and transferred via octopole-ion guides into another ion trap (the C-trap) where they are collisionally cooled and focused, before they are orthogonally injected into the third component of the instrument, the Orbitrap mass analyzer, which operates in very high vacuum. The LTQ-Orbitrap instrument is particularly suitable for both qualitative and quantitative analysis of complex peptide mixtures, because of its high sensitivity, dynamic range, mass accuracy, and sequencing speed. This allows for sequencing of thousands of peptides by high-resolution tandem MS in less than 1 hour of LC–MS/MS analysis time. Due to these advantages that allow collection of spectra with very high resolution (60,000) and routinely with low parts-per-million mass accuracy this LTQ-Orbitrap instrument is used for all MS analyses described below.

2.3. Quantitative proteomics

In many proteomic experiments, the goal is not only to test the presence of a specific protein, but to quantitate its abundance as well. To date, most approaches rely on relative quantitation between different conditions. In a "label-free" approach, the integrated intensity of peptide peaks is compared between different experiments and used as a measure of protein abundance. However at this point, methods that compare intensities of peptides in the same LC–MS run are more reliable. Several methods exist that specifically label peptides of one condition during the proteolytic cleavage, for example, by use of water containing "heavy" ^{18}O, which is incorporated at the C-terminus of each peptide (Yao et al., 2001). Alternatively, chemical labeling of protein mixtures can be employed. Affinity tags with different mass are sometimes used and this technique was termed "Isotope-coded affinity tag" assay ("ICAT", Gygi et al., 1999). Most commonly, thiol-reactive reagents are used to covalently link an isotope labeled affinity tag (e.g., biotin) to cysteine containing peptides. These are subsequently affinity purified and analyzed by MS. When samples are mixed after cross-linking with differently labeled reagents, they can be distinguished afterwards by their mass difference. Disadvantages of this method are that they introduce an additional processing step that may have different efficiency in different samples and that comparatively few tryptic peptides contain a cysteine that can be modified. For these reasons, ICAT is not generally used today. Instead, iTRAQ labeling of amine groups has become popular (Ross et al., 2004). In this technique, peptides are labeled with up to eight different isobaric tandem mass tags, mixed and analyzed by LC–MS/MS. Each tag consists of a reporter and balance group, which has been designed such that it is very prone to fragmentation under CID. The technique is based on chemically tagging the free N-terminus and lysine-ε-amino group on peptides generated from protein digests that have been isolated from different cell states. The tagged samples are then combined and in the full scan spectra, peptides from the different conditions appear at the same m/z ratio. Upon fragmentation, however, different reporter ions are released from the different iTRAQ tags and their relative amount is then quantitated to calculate the relative contribution of peptides from each condition to the intensity signal of the protein.

For yeast cells, metabolic labeling is the preferred method for comparative proteomics since it is easily done with amino acids containing stable nonradioactive isotopes incorporated in a cell culture and results in highly uniform and efficient labeling, ("SILAC," Ong et al., 2002). Conveniently, amino acids are chosen that ensure that one and only one labeled residue is present in each peptide. Trypsin is a commonly used protease to digest protein mixtures and it cleaves at the carboxyl side of arginine and lysines. Thus, when proteins are labeled with these two amino acids (with [13C6/ 15N2] L-lysine and [13C6/15N4] L-arginine, respectively) and digested with trypsin, each resulting peptide contains one labeled amino acid. Analogously, the endoproteinase LysC that cleaves after lysines is used on proteins labeled only with lysine. For comparative proteomics, both the labeled and unlabelled proteins are treated the same way; in fact most easily, they are mixed together in equal ratio of total protein and processed together. The resulting peptides are easily distinguished in the MS by their characteristic mass difference and computational proteomics software automatically identifies and quantifies corresponding SILAC pairs, giving an accurate ratio of light and heavy peptides, which can then be averaged for protein ratios. This analysis is not limited to two different isotope forms of peptides and a third label is often used. However, each labeling increases the complexity of the mixture and therefore complicates the comprehensive analysis of all peptides contained in a mixture.

A complication of metabolic labeling of yeast by heavy isotope containing amino acids is potential conversion between each other due to coupled amino acid synthesis pathways. In practice, only conversion from arginine to proline is common (Ong *et al.*, 2003). Depending on the cell type and sequence of each peptide (the number of prolines contained) this can result in variable and complex patterns of SILAC peptide peaks, preventing accurate automated analysis. To avoid this problem, one possibility is to reduce the amount of labeled arginine added to the synthetic medium and another one is to omit arginine labeling altogether and to use LysC instead of trypsin for the digest. This latter approach has proven very powerful especially in yeast. It results in peptides of increased average length, but concomitantly reduces the complexity of the mixture. In addition, lysine labeling of yeast cells is easily achieved since many strain backgrounds are lysine auxotrophs. For example, the commonly used BY4739 strain used to derive the MAT alpha gene deletion collection bears the *lys2Δ0* allele, is lysine auxotroph and can therefore be directly labeled. (Giaever *et al.*, 2002; Winzeler *et al.*, 1999). Other commonly used lab strains, such as W303, can easily be made auxotroph by deleting *LYS2* and *ARG4* when double labeling with arginine and cleavage with trypsin is desired.

This *in vivo* labeling approach can be used to compare different conditions and Fig. 11.2 shows a representative workflow for such an experiment. In this way, we compared the complete proteome of haploid and diploid yeast cells. Of course the same logic can be applied to cells grown under different conditions, cells with a different genotype (e.g., harboring a deletion) or different biochemical fractions. For example, in a variation of this protocol, protein complexes purified from heavy lysine labeled cells by affinity chromatography are directly compared to background resulting

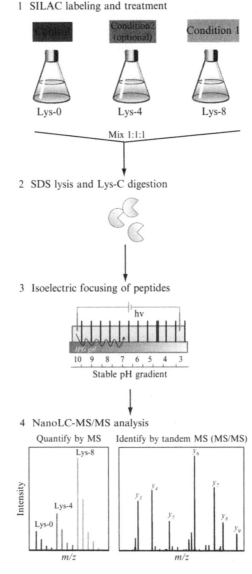

1 SILAC labeling and treatment

Control   Condition2 (optional)   Condition 1

Lys-0   Lys-4   Lys-8

Mix 1:1:1

2 SDS lysis and Lys-C digestion

3 Isoelectric focusing of peptides

hv

IPG gel

10 9   8   7   6   5   4   3

Stable pH gradient

4 NanoLC-MS/MS analysis

Quantify by MS   Identify by tandem MS (MS/MS)

Lys-8

Lys-4

Lys-0

Intensity

m/z m/z

y_6 y_4 y_3 y_7 y_5 y_8 y_9

Figure 11.2   Workflow of a SILAC experiment in yeast. 1. Yeast cultures are grown in either light or heavy amino acid containing medium. 2. After harvesting, proteins are digested with a protease to yield peptides. 3. These are subsequently separated by an analytic method that is orthogonal to reversed-phase chromatography (such as isoelectric focusing or ion exchange chromatography methods) prior to LC-MS analysis. Here, isoelectric focusing on immobilized p*I* strips is shown as an example. 4. The fractions of this separation are then analyzed by LC-MS/MS. In full scan spectra, the same peptide from different cell populations is quantified by the relative difference in intensities between the SILAC labels. Peptides are subsequently sequenced and identified by MS/MS.

from unspecific binding of proteins from light labeled cells to a control matrix (Vermeulen *et al.*, 2008). Similarly, many other variations of this general principle are possible. Examples also include the measurement of turnover of proteins (Doherty *et al.*, 2009). For such an experiment, cells are switched from medium containing "light" amino acids to medium containing "heavy" amino acids for defined times. The change of ratio over time will then indicate the time course of protein turnover, measured individually for each protein.

2.4. Computational proteomics and data analysis

The streamlining of acquisition and the large amount of data necessitate efficient and automated evaluation of the resulting spectra. For example, in a typical proteomic experiment, at least 12 fractions from the IEF of peptides are analyzed, each by an LC run of at least 120 min, collecting a spectrum every second. Together, these runs result in at least 86,400 MS spectra of high mass accuracy that need to be evaluated. In addition, the most abundant peaks (usually the top five in abundance) of every MS scan are fragmented and the MS/MS spectra are collected, adding further data to be evaluated. A major breakthrough in computational proteomics was recently achieved in the MaxQuant software suite (Cox and Mann, 2008). The algorithms use correlation analysis and graph theory to detect peaks and isotope clusters in the MS, using the m/z ratio, intensity, and LC elution time as parameters. Figure 11.3 shows a plot with every detected peptide plotted as its m/z versus its elution time in green. Successful identifications are shown in purple. As can be easily appreciated, the identification rates are very high, usually resulting in identification of the majority of peptides. This information is automatically submitted to a commercial search engine (Mascot). An important consideration in computational proteomics is the control of the accuracy of peptide identification. Today, 99% certainty of protein identification is usually desired and this is controlled by monitoring the number of identifications in a "nonsense" or decoy database consisting of reversed protein sequences (Elias and Gygi, 2007). This search results in automatic identification of peptides in the mixture with high confidence. In the next step, peak intensities for each member of a SILAC pair are calculated from the isotope pattern. Multiple measurements for each peptide are integrated and statistically evaluated, resulting in a measurement of the abundance ratio of the proteins in each sample and a confidence estimate for that measurement.

2.5. Perspective and outlook

While total proteome quantitation is still a time-consuming experiment requiring advanced instrumentation and specialized know-how, several trends will make such experiments much more routine and available to

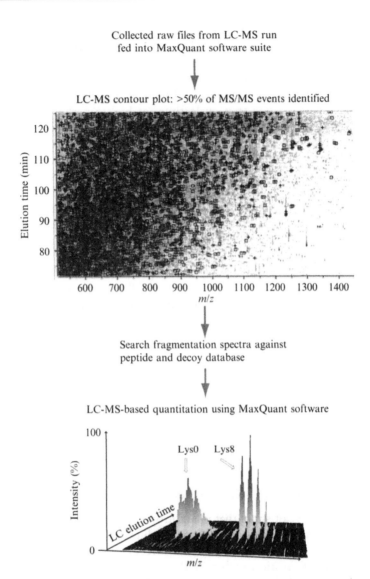

Collected raw files from LC-MS run
fed into MaxQuant software suite

LC-MS contour plot: >50% of MS/MS events identified

Search fragmentation spectra against
peptide and decoy database

LC-MS-based quantitation using MaxQuant software

Figure 11.3 Workflow of the analysis of a proteomic experiment. Complex LC-MS Raw files containing MS and MS/MS spectra collected from the LTQ–Orbitrap is fed into the MaxQuant software package that automatically searches fragmentation patterns against a target/decoy protein database. Currently, identified peptide sequences are filtered based on their database score and accepted at a false discovery rate (FDR) of less than 1%. MaxQuant then automatically determines the peptide ratios of SILAC pairs and calculates the significance of regulation for all identified proteins.

molecular biology laboratories in the future. Among them, the development of a new generation of hybrid MS instruments named "LTQ-Orbitrap Velos," which offers significantly higher sensitivity due to new ion optics systems that enhances the transfer of ions from the source to the MS by an order of magnitude (Olsen *et al.*, 2009). Similarly, this instrument enables faster scan cycle times at higher performance with a new dual pressure linear ion trap. Together this improves the scan speed more than twofold, which in effect will reduce MS measurement time requirements for yeast expression proteomics. In addition to developments in MS instrumentation, novel preparation methods (such as "filter aided sample preparation"; Wisniewski *et al.*, 2009) result in more uniform samples between experiments and therefore even higher reproducibility and identification rates. Finally, streamlined versions of MaxQuant and novel bioinformatic tools will result in faster evaluation of the data, a process that is still computationally intensive and time consuming.

With these enhancements, expression proteomics by MS is likely to continue to become more widespread and will soon be a common technique to comprehensively analyze the composition of yeast cells.

3. PROTOCOLS

In the following, we describe in detail how to grow SILAC labeled yeast for a proteomic experiment, how to process the samples for LC-MS, how to perform the measurements and how to analyze the resulting data.

3.1. Yeast strains for SILAC proteomics experiments

In principle, any yeast strain that is auxotroph for lysine and/or argenine is suitable for SILAC labeling, depending on which label is used. If a particular stain needs to be used, either *LYS2* or *ARG4* can be introduced either directly by transformation with a PCR product (Janke *et al.*, 2004) or by crossing with a deletion strain such as BY4739. Primers to delete *LYS2* using the system described by Janke *et al.* (2004) are Lys2-S1 (5′-atttcagtga aaaactgcta atagagagat atcacagagt tactcactaa tgcgtacgct gcaggtcgac-3′) and Lys2-S2 (5′-ctaattcat atttaattat tgtacatgga catatcatac gtaatgctca accttaatcg atgaattcga gctcg-3′); conversely, for *ARG4*, Arg4-S1 (5′-cgcaattgaa gagct-caaaa gcaggtaact atataacaag actaaggcaa acatgcgtac gctgcaggtc gac-3′) and Arg4-S2 (5′-gtcctagaag taccagacct gatgaaattc ttgcgcataa cgtcgccatc tgctaatcga tgaattcgag ctcg-3′) primers are used.

3.2. Media for SILAC labeling

Stock medium (any other drop out mix lacking arginine and lysine will work):

XYNB without amino acids	6.7	g/l
XGlucose	20	g/l

	Final concentration [mg/l]	Stock per 100 ml	Amount of stock [ml] for 1 l
Adenine sulfate	20	200 mg	10
Uracil	20	200 mg	10
L-Tryptophan	20	1 g	2
L-Histidine–HCl	20	1 g	2
L-Arginine–HCl[a]	20	1 g	2
L-Methionine	20	1 g	2
L-Tyrosine	30	200 mg	15
L-Leucine	60	1 g	6
L-Isoleucine	30	1 g	3
L-Phenylalanine	50	1 g	5
L-Glutamic acid	100	1 g	10
L-Aspartic Acid	100	1 g	110
L-Valine	150	3 g	5
L-Threonine	200	4 g	5
L-Serine	400	8 g	5

[a] ONLY for lysine labeling alone.

To prepare media ready to use (use light/medium/heavy amino acids as desired):

L-Lysine	30	mg/l
L-Argenine[a]	5	mg/l

[a] ONLY for double Lys/Arg labeling; concentration of arginine can be varied up to 20 mg/l; this low amount of 5 mg/l minimizes arginine to proline conversion.

SILAC amino acids (Isotec-Sigma):

Lys 4:	L-Lysine—4,4,5,5-d_4·Cl	Cat.# 616192
Lys 8:	L-Lysine—$^{13}C6$, $^{15}N_2$·Cl	Cat.# 608041
Arg 6:	L-Arginine—$^{13}C_6$·Cl	Cat.# 643440
Arg 10:	L-Arginine—$^{13}C_6$, $^{15}N_4$·Cl	Cat.# 608033

3.3. Growing yeast cultures for SILAC labeling

1. Grow preculture of a lysine auxotroph strain over night in SC medium
2. Inoculate SILAC medium (either heavy or light) with a 1:10,000 dilution of the preculture
3. Grow overnight at 30 °C
4. Check OD next morning (cells should reach OD = 0.5–0.7)
5. To harvest, spin down cells 10 min 4000 rpm 4 °C

3.4. Extract Preparation for SILAC experiments

1. Resuspend cells in buffer S (150 mM K acetate, 2 mM Mg acetate, 1× protease inhibitor cocktail (Roche), 20 mM HEPES, pH 7.4) at a density of 50 OD/ml (minimum buffer volume for the bead-mill (3 ml))
2. Slowly drop cell suspension in N2(l)
3. Grind cells in MM301 Ball Mill (Retsch), three cycles of 3 min at 10 Hz.
4. Thaw grinded cells
5. Detergent extraction (30 min; 1% TritonX 100, 4 °C rotating)
6. Spin down 10 min 1000 rpm 4 °C
7. Collect supernatant
8. Measure protein concentration by Bradford assay
9. Mix extracts from strains that are used for the comparison in a 1:1 ratio of protein amounts

3.5. In-solution digest of proteins for MS

1. Reduce proteins for 20 min at RT in 1 mM dithiothreitol (DTT)
2. Alkylate proteins for 15 min using 5.5 mM iodoacetamide (IAA) at RT in the dark
3. Digest with the endoproteinase Lys-C (Wako) using 1:50 w/w over night at RT (arginine and lysine labeled yeast proteins are digested with Lys-C in a similar manner)
4. Dilute the resulting peptide mixtures with Millipore water to achieve a final urea concentration below 2 M
5. *For argenine labeled cells*, add trypsin (modified sequencing grade, Promega) 1:50 w/w and digest overnight
6. Trypsin and Lys-C activity are quenched by acidification of the reaction mixtures with TFA to pH ∼2.

3.6. Test of label incorporation

It is straightforward to calculate the SILAC incorporation efficiency of a peptide by high-resolution MS. To this end, a small aliquot of yeast cells grown in the presence of either [$^{13}C_6$]arginine or [$^{13}C_6$]lysine for several generations is analyzed by MS. Proteins are extracted from the sample as described above and separated by one-dimensional gel electrophoresis (SDS-PAGE). The gel is Coomassie stained and a slice corresponding to 30–40 kDa protein size range is excised and digested *in situ* with LysC (or alternative trypsin for arginine labeling). The resulting peptide mixture is desalted and analyzed by nanoflow LC-tandem MS. The raw MS data can be analyzed in the MaxQuant software suite as described below. As an output, MaxQuant will generate a list of identified and quantified proteins. The median (H/L) peptide and protein ratio for all proteins directly reflects the SILAC amino acid incorporation rate. In order to allow accurate quantitation of ratios in the subsequent proteomic experiment, the ratio indicating labeling efficiency should be at least 99%.

3.7. Peptide IEF (Optional)

Alternative to the peptide isoelectric focusing described here, other peptide fractionation procedures yield good results. For example, anion exchange fractionation can be used as described in Wiśniewski *et al.* (2009).

1. To separate peptides according to their isoelectric point, 75 µg of in-solution digested peptides are fractionated using the Agilent 3100 OFFGEL fractionator (Agilent, G3100AA).
2. Set up the system according to the manual of the High Res Kit, pH 3–10 (Agilent, 5188-6424) but exchange strips by 24 cm Immobiline Dry-Strip, pH 3–10 (GE Healthcare, 17-6002-44) and ampholytes by IPG Buffer, pH 3–10 (GE Healthcare, 17-6000-87).
3. Rehydrate strips for 20 min with 20 µl rehydration buffer per well containing 5% glycerol and ampholytes diluted 1:50.
4. Prepare 150 µl of peptide solution containing 3.125 µg yeast digest, 5% glycerol, and ampholytes diluted 1:50.
5. Apply mixture to each well of the OFFGEL device.
6. Close wells with a silicon cover seal to prevent evaporation of liquid.
7. Focus peptides for 50 kVh at maximum current of 50 µA, maximum voltage of 8000 V and maximum power of 200 mW into 24 fractions.

Average run time is approximately 30 h.

8. Add 3% acetonitrile, 1% trifluoroacetic acid, and 0.5% acetic acid to acidify each peptide fraction.
9. Desalt and concentrate fraction on a reversed-phase C18 StageTip (Rappsilber *et al.*, 2007).

3.8. MS analysis

3.8.1. Equipment

We perform all MS experiments on a nanoflow high-performance liquid chromatography (HPLC) system (Agilent Technologies 1200, Waldbronn, Germany) connected to a hybrid LTQ-Orbitrap classic or XL (Thermo Fisher Scientific, Bremen, Germany) equipped with a nanoelectrospray ion source (Proxeon Biosystems, Odense, Denmark). The HPLC system consists of a solvent degasser nanoflow pump and a temperature controlled microautosampler kept constantly at 4 °C in order to reduce sample evaporation. The peptide mixtures are loaded onto a 15 cm analytical column (75 μm inner diameter) packed in-house with a methanol slurry of 3 μm reverse-phased, fully end-capped C18 beads (Reprosil-AQ Pur, Dr. Maisch) using a pressurized "packing bomb" operated at 50–60 bar. Mobile phases for HPLC consist of (A) 99.5% Milli-Q water and 0.5% acetic acid (v/v); (B) 19.5% Milli-Q water, 80% acetonitrile, and 0.5% acetic acid (v/v).

3.8.2. Procedure of sample preparation/injection

1. Prior to MS analysis elute all samples of the C18 StageTip directly into a 96 sample well plate (Abgene, UK) using two times 20 μl buffer B (80% acetonitrile, 0.5% acetic acid).
2. Concentrate samples in a "speed-vac" for 12 min in order to remove all organic solvent.
3. Adjust sample volume to approximately 8 μl by adding an appropriate volume of buffer A (0.5% acetic acid).
4. Load 5 μl of prepared peptide mixture onto the analytical column for 20 min in 2% buffer B at a flow rate of 500 nl/min followed by reverse-phased separation through a 90 min gradient ranging from 5% to 40% acetonitrile in 0.5% acetic acid.
5. Wash the column for 10 min with high concentration of organic solvent (90% buffer B) and equilibrate it for another10 min with buffer A (0.5% acetic acid) prior to loading of the next sample.
6. The eluted peptides from the HPLC column are directly electrosprayed into the MS for detection.

3.8.3. Mass spectrometry

We operate the MS instrument in data-dependent mode by automatically switching between full survey scan MS and consecutive MS/MS acquisition. Survey full scan MS spectra (mass range m/z 300–2000) are acquired in the orbitrap section of the instrument with a resolution of $R = 60,000$ at m/z 400 (after accumulation to a "target value" of 1,000,000 in the linear ion trap). The 10 most intense peptide ions in each survey scan with an ion intensity above 500 counts and a charge state ≥ 2 are sequentially isolated to a target value of 5000 and fragmented in the linear ion trap by collisionally induced dissociation (CID/CAD).

All peaks selected for fragmentation are automatically put on an exclusion list for 90 s, which ensures that the same ion would not be selected for fragmentation more than once. For optimal duty cycle, the fragment ion spectra are recorded in the LTQ-MS "in parallel" with the orbitrap full scan detection. For all survey scan measurements with the orbitrap detector, a lock-mass ion from ambient air (m/z 391.284286, 429.08875, and 445.120025) is used for internal calibration as described ensuring an overall sub-ppm mass accuracy for all detected peptides (Olsen et $al.$, 2005).

For all MS experiments, data is saved in RAW file format (Thermo Scientific, Bremen, Germany) using the Xcalibur 2.0 with Tune 2.2 or 2.4. All data was loaded into the in-house written software MaxQuant and analyzed as described below.

3.9. Identification and quantitation of peptides and proteins

The data analysis is performed with the MaxQuant software as described in Cox and Mann (2008) supported by Mascot as the database search engine for peptide identifications. In addition, a step by step protocol for the analysis can be found in Cox et $al.$ (2009). In short, peaks in MS scans are determined as three-dimensional hills in the mass-retention time plane. They are then assembled to isotope patterns and SILAC pairs by graph theoretical methods. MS/MS peak lists are filtered to contain at most six peaks per 100 Da intervals and searched by Mascot (www.matrixscience.com) against a concatenated forward and reversed version of the yeast ORF database ($Saccharomyces$ Genome Database SGDTM at Stanford University—www.yeastgenome.org). Protein sequences of common contaminants, for example, keratins, were added to the database. The initial mass tolerance in MS mode was set to 7 ppm and MS/MS mass tolerance was 0.5 Da. Cysteine carbamidomethylation is searched as a fixed modification, whereas N-acetyl protein, N-pyroglutamine, and oxidized methionine are searched as variable modifications. Labeled arginine and lysine are specified as fixed or variable modifications, depending on the prior knowledge about the parent ion. The resulting Mascot.dat files are loaded into the MaxQuant software together with the raw data for further analysis. SILAC peptide and protein quantitation is performed automatically with MaxQuant using default settings for parameters. Here, for each SILAC pair, the ratio is determined by a robust regression model fitted to all isotopic peaks and all scans that the pair elutes in. SILAC protein ratios are determined as the median of all peptide ratios assigned to the protein. Absolute protein quantitation is based on extracted ion chromatograms (XICs) of contained peptides. To minimize false identifications all top-scoring peptide assignments made by Mascot are filtered based on prior knowledge of individual peptide mass error, SILAC state, and the correct number of lysine and arginine residues specified by the mass difference observed in the full scan between the SILAC partners. Furthermore peptide assignments are statistically evaluated in a Bayesian model based on sequence length and Mascot score. We accept peptides and proteins with a false discovery rate (FDR) of less than 1%, estimated based on the number of accepted reverse hits.

ACKNOWLEDGMENTS

We thank Florian Fröhlich for critical reading of the manuscript. This study was supported by the Max Planck Society (Tobias C. Walther and Matthias Mann) and the EU Council's 7th Framework "Prospect" grant (to Matthias Mann). Tobias C. Walther was supported by the German Academic Research Council (DFG; WA 1669/2-1). Jesper V. Olsen and the Center for Protein Research are supported by a generous grant from the Novo Nordisk Foundation.

REFERENCES

Aebersold, R., and Mann, M. (2003). Mass spectrometry-based proteomics. *Nature* **422,** 198–207.

Anderson, N. L., Anderson, N. G., Pearson, T. W., Borchers, C. H., Paulovich, A. G., Patterson, S. D., Gillette, M., Aebersold, R., and Carr, S. A. (2009). A human proteome detection and quantitation project. *Mol. Cell. Proteomics* **8,** 883–886.

Bonaldi, T., Straub, T., Cox, J., Kumar, C., Becker, P. B., and Mann, M. (2008). Combined use of RNAi and quantitative proteomics to study gene function in *Drosophila. Mol. Cell* **31,** 762–772.

Campostrini, N., Areces, L. B., Rappsilber, J., Pietrogrande, M. C., Dondi, F., Pastorino, F., Ponzoni, M., and Righetti, P. G. (2005). Spot overlapping in two-dimensional maps: A serious problem ignored for much too long. *Proteomics* **5,** 2385–2395.

Cox, J., and Mann, M. (2008). MaxQuant enables high peptide identification rates, individualized p.p.b.-range mass accuracies and proteome-wide protein quantification. *Nat. Biotechnol.* **26,** 1367–1372.

Cox, J., Matic, I., Hilger, M., Nagaraj, N., Selbach, M., Olsen, J. V., and Mann, M. (2009). A practical guide to the MaxQuant computational platform for SILAC-based quantitative proteomics. *Nat. Protoc.* 4(5), 698–705.

Cravatt, B. F., Simon, G. M., and Yates, J. R. III (2007). The biological impact of mass-spectrometry-based proteomics. *Nature* **450,** 991–1000.

de Godoy, L. M., Olsen, J. V., Cox, J., Nielsen, M. L., Hubner, N. C., Frohlich, F., Walther, T. C., and Mann, M. (2008). Comprehensive mass-spectrometry-based proteome quantification of haploid versus diploid yeast. *Nature* **455,** 1251–1254.

Doherty, M. K., Hammond, D. E., Clague, M. J., Gaskell, S. J., and Beynon, R. J. (2009). Turnover of the human proteome: Determination of protein intracellular stability by dynamic SILAC. *J. Proteome Res.* **8,** 104–112.

Domon, B., and Aebersold, R. (2006). Mass spectrometry and protein analysis. *Science* **312,** 212–217.

Elias, J. E., and Gygi, S. P. (2007). Target-decoy search strategy for increased confidence in large-scale protein identifications by mass spectrometry. *Nat. Methods* **4,** 207–214.

Fenn, J. B., Mann, M., Meng, C. K., Wong, S. F., and Whitehouse, C. M. (1989). Electrospray ionization for mass spectrometry of large biomolecules. *Science* **246,** 64–71.

Fountoulakis, M., Tsangaris, G., Oh, J. E., Maris, A., and Lubec, G. (2004). Protein profile of the HeLa cell line. *J. Chromatogr. A* **1038,** 247–265.

Ghaemmaghami, S., Huh, W. K., Bower, K., Howson, R. W., Belle, A., Dephoure, N., O'Shea, E. K., and Weissman, J. S. (2003). Global analysis of protein expression in yeast. *Nature* **425,** 737–741.

Giaever, G., Chu, A. M., Ni, L., Connelly, C., Riles, L., Veronneau, S., Dow, S., Lucau-Danila, A., Anderson, K., Andre, B., Arkin, A. P., Astromoff, A., *et al.* (2002). Functional profiling of the *Saccharomyces cerevisiae* genome. *Nature* **418,** 387–391.

Gygi, S. P., Rist, B., Gerber, S. A., Turecek, F., Gelb, M. H., and Aebersold, R. (1999). Quantitative analysis of complex protein mixtures using isotope-coded affinity tags. *Nat. Biotechnol.* **17,** 994–999.

Hager, J. W., and Yves Le Blanc, J. C. (2003). Product ion scanning using a Q-q-Q linear ion trap (Q TRAP) mass spectrometer. *Rapid Commun. Mass Spectrom.* **17,** 1056–1064.

Hillenkamp, F., and Karas, M. (1990). Mass spectrometry of peptides and proteins by matrix-assisted ultraviolet laser desorption/ionization. *Methods Enzymol.* **193,** 280–295.

Hillenkamp, F., Karas, M., Beavis, R. C., and Chait, B. T. (1991). Matrix-assisted laser desorption/ionization mass spectrometry of biopolymers. *Anal. Chem.* **63,** 1193A–1203A.

Hubner, N. C., Ren, S., and Mann, M. (2008). Peptide separation with immobilized pI strips is an attractive alternative to in-gel protein digestion for proteome analysis. *Proteomics* **8,** 4862–4872.

Huh, W. K., Falvo, J. V., Gerke, L. C., Carroll, A. S., Howson, R. W., Weissman, J. S., and O'Shea, E. K. (2003). Global analysis of protein localization in budding yeast. *Nature* **425,** 686–691.

Iribane, J. V., and Thomson, B. A. (1976). On the evaporation of small ions from charged droplets. *J. Chem. Phys.* **64,** 2287–2294.

Janke, C., Magiera, M. M., Rathfelder, N., Taxis, C., Reber, S., Maekawa, H., Moreno-Borchart, A., Doenges, G., Schwob, E., Schiebel, E., and Knop, M. (2004). A versatile toolbox for PCR-based tagging of yeast genes: New fluorescent proteins, more markers and promoter substitution cassettes. *Yeast* **21,** 947–962.

Karas, M., and Hillenkamp, F. (1988). Laser desorption ionization of proteins with molecular masses exceeding 10,000 daltons. *Anal. Chem.* **60,** 2299–2301.

Mann, M., and Kelleher, N. L. (2008). Precision proteomics: The case for high resolution and high mass accuracy. *Proc. Natl. Acad. Sci. USA* **105,** 18132–18138.

Mikesh, L. M., Ueberheide, B., Chi, A., Coon, J. J., Syka, J. E., Shabanowitz, J., and Hunt, D. F. (2006). The utility of ETD mass spectrometry in proteomic analysis. *Biochim. Biophys. Acta* **1764,** 1811–1822.

Motoyama, A., Xu, T., Ruse, C. I., Wohlschlegel, J. A., and Yates, J. R. III (2007). Anion and cation mixed-bed ion exchange for enhanced multidimensional separations of peptides and phosphopeptides. *Anal. Chem.* **79,** 3623–3634.

O'Farrell, P. H. (1975). High resolution two-dimensional electrophoresis of proteins. *J. Biol. Chem.* **250,** 4007–4021.

Olsen, J. V., de Godoy, L. M., Li, G., Macek, B., Mortensen, P., Pesch, R., Makarov, A., Lange, O., Horning, S., and Mann, M. (2005). Parts per million mass accuracy on an Orbitrap mass spectrometer via lock mass injection into a C-trap. *Mol. Cell. Proteomics* **4,** 2010–2021.

Olsen, J. V., Schwartz, J.C., Griep-Raming, J., Nielsen, M. L., Damoc, E., Denisov, E., Lange, O., Remes, P., Taylor, D., Splendore, M., Wouters, E. R., Senko, M., *et al.* (2009). A dual pressure linear ion trap orbitrap instrument with very high sequencing speed. *Mol. Cell Proteomics.* **8**(12), 2759–2769. [Epub 2009 Oct 14].

Ong, S. E., Blagoev, B., Kratchmarova, I., Kristensen, D. B., Steen, H., Pandey, A., and Mann, M. (2002). Stable isotope labeling by amino acids in cell culture, SILAC, as a simple and accurate approach to expression proteomics. *Mol. Cell. Proteomics* **1**, 376–386.

Ong, S. E., Kratchmarova, I., and Mann, M. (2003). Properties of 13C-substituted arginine in stable isotope labeling by amino acids in cell culture (SILAC). *J. Proteome Res.* **2**, 173–181.

Rappsilber, J., Mann, M., and Ishihama, Y. (2007). Protocol for micro-purification, enrichment, pre-fractionation and storage of peptides for proteomics using StageTips. *Nat. Protoc.* **2**, 1896–1906.

Ross, P. L., Huang, Y. N., Marchese, J. N., Williamson, B., Parker, K., Hattan, S., Khainovski, N., Pillai, S., Dey, S., Daniels, S., Purkayastha, S., Juhasz, P., *et al.* (2004). Multiplexed protein quantitation in *Saccharomyces cerevisiae* using amine-reactive isobaric tagging reagents. *Mol. Cell. Proteomics* **3**, 1154–1169.

Schwartz, J. C., Senko, M. W., and Syka, J. E. (2002). A two-dimensional quadrupole ion trap mass spectrometer. *J. Am. Soc. Mass Spectrom.* **13**, 659–669.

Scigelova, M., and Makarov, A. (2006). Orbitrap mass analyzer—overview and applications in proteomics. *Proteomics* **6**(Suppl. 2), 16–21.

Shevchenko, A., Jensen, O. N., Podtelejnikov, A. V., Sagliocco, F., Wilm, M., Vorm, O., Mortensen, P., Shevchenko, A., Boucherie, H., and Mann, M. (1996). Linking genome and proteome by mass spectrometry: Large-scale identification of yeast proteins from two dimensional gels. *Proc. Natl. Acad. Sci. USA* **93**, 14440–14445.

Vermeulen, M., Hubner, N. C., and Mann, M. (2008). High confidence determination of specific protein–protein interactions using quantitative mass spectrometry. *Curr. Opin. Biotechnol.* **19**, 331–337.

Washburn, M. P., Wolters, D., and Yates, J. R. III (2001). Large-scale analysis of the yeast proteome by multidimensional protein identification technology. *Nat. Biotechnol.* **19**, 242–247.

Whitehouse, C. M., Dreyer, R. N., Yamashita, M., and Fenn, J. B. (1985). Electrospray interface for liquid chromatographs and mass spectrometers. *Anal. Chem.* **57**, 675–679.

Winzeler, E. A., Shoemaker, D. D., Astromoff, A., Liang, H., Anderson, K., Andre, B., Bangham, R., Benito, R., Boeke, J. D., Bussey, H., Chu, A. M., Connelly, C., *et al.* (1999). Functional characterization of the *S. cerevisiae* genome by gene deletion and parallel analysis. *Science* **285**, 901–906.

Wisniewski, J. R., Zougman, A., Nagaraj, N., and Mann, M. (2009). Universal sample preparation method for proteome analysis. *Nat. Methods* **6**, 359–362.

Wiśniewski, J. R., Zougman, A., Mann, M. (2009). Combination of FASP and StageTip-based fractionation allows in-depth analysis of the hippocampal membrane proteome. *J. Proteome Res.* **8**(12), 5674–5678.

Yao, X., Freas, A., Ramirez, J., Demirev, P. A., and Fenselau, C. (2001). Proteolytic 18O labeling for comparative proteomics: Model studies with two serotypes of adenovirus. *Anal. Chem.* **73**, 2836–2842.

CHAPTER TWELVE

HIGH-QUALITY BINARY INTERACTOME MAPPING

Matija Dreze,[*,†,‡,§] Dario Monachello,[*,†,¶] Claire Lurin,[¶] Michael E. Cusick,[*,†,‡] David E. Hill,[*,†,‡] Marc Vidal,[*,†,‡] *and* Pascal Braun[*,†,‡]

Contents

Abstract

Physical interactions mediated by proteins are critical for most cellular functions and altogether form a complex macromolecular "interactome" network.

* Center for Cancer Systems Biology (CCSB), Dana–Farber Cancer Institute, Boston, Massachusetts, USA
† Department of Cancer Biology, Dana–Farber Cancer Institute, Boston, Massachusetts, USA
‡ Department of Genetics, Harvard Medical School, Boston, Massachusetts, USA
§ Facultés Universitaires Notre-Dame de la Paix, Namur, Belgium
¶ Unité de Recherche en Génomique Végétale (URGV), Evry Cedex, France

Methods in Enzymology, Volume 470
ISSN 0076-6879, DOI: 10.1016/S0076-6879(10)70012-4

Systematic mapping of protein–protein, protein–DNA, protein–RNA, and protein–metabolite interactions at the scale of the whole proteome can advance understanding of interactome networks with applications ranging from single protein functional characterization to discoveries on local and global systems properties. Since the early efforts at mapping protein–protein interactome networks a decade ago, the field has progressed rapidly giving rise to a growing number of interactome maps produced using high-throughput implementations of either binary protein–protein interaction assays or co-complex protein association methods. Although high-throughput methods are often thought to necessarily produce lower quality information than low-throughput experiments, we have recently demonstrated that proteome-scale interactome datasets can be produced with equal or superior quality than that observed in literature-curated datasets derived from large numbers of small-scale experiments. In addition to performing all experimental steps thoroughly and including all necessary controls and quality standards, careful verification of all interacting pairs and validation tests using independent, orthogonal assays are crucial to ensure the release of interactome maps of the highest possible quality. This chapter describes a high-quality, high-throughput binary protein–protein interactome mapping pipeline that includes these features.

1. INTRODUCTION

Interactions mediated by proteins and the complex "interactome" networks resulting from these interactions are essential for biological systems. Mapping protein–protein, protein–DNA, protein–RNA, and protein–metabolite interactions that form "interactome" networks is a major goal of functional genomics, proteomics, and systems biology (Vidal, 2005). Information obtained from large-scale efforts to identify protein interaction partners yields crucial biological insights throughout a range of applications. At the single protein level, interactome maps have helped assign functions to both uncharacterized and well-studied gene products (Oliver, 2000). At the systems level, interactome maps have enabled investigations of how regulatory circuits and global cellular network properties relate to biological functions (Han *et al.*, 2004; Jeong *et al.*, 2001; Milo *et al.*, 2002; Yu *et al.*, 2008).

The two major high-throughput strategies used so far to delineate protein–protein interactome networks are: (i) binary protein–protein interaction assays, which detect direct pairwise interactions, and (ii) affinity purification followed by mass spectrometry (AP–MS) approaches, which detect biochemically stable, copurifying protein complexes containing both direct and indirect protein associations. Classically, binary interaction assays have been based on the yeast two-hybrid (Y2H) system developed 20 years ago (Fields and Song, 1989), and which has been improved over time to increase efficiency and quality (Durfee *et al.*, 1993; Gyuris *et al.*, 1993; Vidal *et al.*, 1996). Of late, alternative approaches have been developed to detect binary interactions, such as protein arrays, protein complementation assays, and the split ubiquitin method (Miller *et al.*, 2005; Tarassov *et al.*, 2008; Zhu *et al.*, 2001).

Until recently high-throughput methods were regarded as more likely to produce lower quality information than low-throughput experiments. It has now been shown that highly reliable interactome datasets can be obtained at the scale of the whole proteome (Braun *et al.*, 2009; Cusick *et al.*, 2009; Simonis *et al.*, 2009; Venkatesan *et al.*, 2009) provided that all experimental steps are thorough and all necessary controls and quality standards are included. Lastly, careful verification of all candidate interactions and experimental validation using independent interaction assays are necessary to ensure the release of interactome maps of the highest possible quality.

Even when highly reliable, interactome maps should be considered as network models of interactions that *can* happen between all proteins encoded by the genome of an organism of interest. As such, they correspond to static representations of collapsed time-, space-, and condition-dependent interactions that dynamically regulate the behavior and developmental fate of diverse tissues. Thus, interactome maps should be used as static scaffold-like information from which the dynamic features of biologically relevant interactions, that is, those that *do* happen *in vivo*, can be modeled by integrating additional functional information such as transcriptional and phenotypic profiling data (Ge *et al.*, 2001, 2003; Gunsalus *et al.*, 2005; Vidal, 2001). Ultimately, novel potentially insightful interactions need to be evaluated for their biological significance using genetic experiments, where specific *cis*-acting interaction-defective alleles (IDAs) of one or both proteins or *trans*-acting disruptors are tested functionally (Dreze *et al.*, 2009; Endoh *et al.*, 2002; Vidal and Endoh, 1999; Vidal *et al.*, 1996; Zhong *et al.*, 2009).

2. HIGH-QUALITY BINARY INTERACTOME MAPPING

The quality of any dataset can be affected by a high rate of "false positives" and need to be addressed in two fundamentally different contexts. One relates to avoidable experimental errors leading to wrong information, and the other relates to as yet undiscovered fundamental properties of proteins (Fig. 12.1). Our binary interactome mapping strategy is designed to differentiate between these two classes of issues designated "technical" and "biological" false positives.

- *Technical false positives*

All techniques used to map protein interactions can give rise to artifacts. It goes without saying that artifacts or technical false positives should be

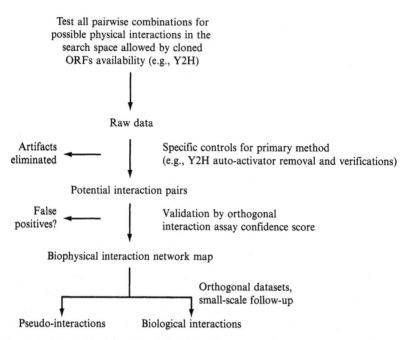

Test all pairwise combinations for
possible physical interactions in the
search space allowed by cloned
ORFs availability (e.g., Y2H)

Raw data

Artifacts eliminated ←——— Specific controls for primary method
(e.g., Y2H auto-activator removal and verifications)

Potential interaction pairs

False positives? ←——— Validation by orthogonal
interaction assay confidence score

Biophysical interaction network map

Orthogonal datasets,
small-scale follow-up

Pseudo-interactions   Biological interactions

Figure 12.1 General strategy to map binary interactome networks. All possible pairs of a search space are tested using a large-scale binary interaction detection assay such the yeast two-hybrid (Y2H) system. First-round positives constitute the raw dataset in which artifacts need to be identified and eliminated. The resulting set of putative interactions is then validated using alternative binary interaction detection assays. This step allows determination of the dataset precision or experimentally determined confidence scores for all individual interactions. Overlap of biophysical interactions with other types of datasets, such as coexpression or phenotypic profiles, or small-scale experimental follow-up, allows the identification of biologically relevant binary interactions.

identified and removed as much as possible with appropriately designed experimental conditions and controls. Potential artifacts are different for every method and can arise systematically or sporadically. Often it takes several years of collective use, after the original description of a method, for systematic artifacts to be understood and thus become avoidable.

In biochemical AP–MS experiments, or in the design and use of antibodies, nonspecific binding by abundant proteins or contaminant proteins introduced while carrying out experiments represent technical false positives that can and should be removed. Y2H is based on a set of growth selections designed to identify the reconstitution of a transcription factor mediated by two hybrid proteins. Although powerful, the system needs to be carefully controlled because unrelated spontaneous genetic suppressors can appear during these growth selections. Such artifacts can reliably be removed by thorough implementation of the methods described below.

However careful the execution of Y2H mapping experiments or any other high-throughput methodology is, the precision of the obtained dataset (i.e., the inverse of the false discovery rate (FDR)) still needs to be determined to estimate both systematic and sporadic technical false positives that might remain undetected (Fig. 12.1). We advocate below further rigorous experimental *verifications* of all interacting pairs using the Y2H version used to produce a dataset, followed by careful *validation* using orthogonal protein interaction assays to determine overall quality. Once these steps have been implemented the result is a set of well-demonstrated interactions, proven to physically interact. We refer to such protein pairs as "biophysical interactors".

- *Biological false positives*

While it is plausible that most biophysical interactions are biologically relevant, their relevance, and the mechanism by which biophysically demonstrated protein interactions affect the physiology of an organism, remains to be demonstrated in subsequent, often laborious experiments (Fig. 12.1). It is theoretically possible that a subset of biophysical interactions might be biologically inconsequential because, among other possibilities, they remain either spatially or temporally separated throughout the lifetime of an organism. Such "pseudo-interactions" can be viewed as biological false positives that need to be eliminated or, alternatively, might represent interesting evolutionary remnants similar to the existence of pseudo-genes in many organisms (Venkatesan *et al.*, 2009).

2.1. Production and verification of Y2H datasets

Fields and Song (1989) first described the Y2H system as the reconstitution of a transcription factor through expression of two hybrid proteins, one fusing the DNA-binding (DB) domain to a protein X (DB-X) and the other fusing an activation domain (AD) to a protein Y (AD-Y). In the last 20 years much has been learned about possible artifacts and appropriate controls, so that today Y2H can be considered not only one of the most efficient, but also one of the most reliable binary interaction assays available for small-, medium-, and large-scale interaction mapping. We next discuss specific artifacts of the Y2H system and the appropriate controls developed to detect and remove them.

2.1.1. Autoactivators

A common artifact of the Y2H system is autoactivation of Y2H-inducible reporter genes. This occurs when DB-X (where X is a full-length protein or a protein fragment) activates transcription of Y2H reporter genes irrespective of the presence of any AD-Y. Three classes of autoactivators need to considered: (i) genuine transcription factors that contain a *bona fide* AD and consequently will likely score as autoactivators when fused to DB, (ii) proteins that are not transcription factors in their natural context but can behave as autoactivators because they contain a cryptic AD (cognate autoactivators), and (iii) nontranscription factor proteins that contain one or more cryptic ADs that are only functional as truncated fragments and not when expressed in the context of full-length proteins (*de novo* autoactivators).

Both genuine transcription factors and cognate DB-X autoactivators can be identified and removed by performing prescreens for reporter gene activation either with AD expressed alone (i.e., in the absence of any Y fused protein) or even with no AD at all.

De novo autoactivators are more difficult to detect than transcription factors and cognate autoactivators. The Y2H system is based on positive growth selections for potentially rare events, such as the finding of a single cDNA out of a complex library. The Y2H system can just as rigorously select for mutations that occur during the course of a screen and which convert a nonactivator protein into a *de novo* autoactivator. Such events are relatively frequent and some early Y2H datasets may have been inadvertently overpopulated by spontaneous autoactivators (Ito *et al.*, 2001; Yu *et al.*, 2008). A method to systematically remove these artifacts (Walhout and Vidal, 1999) employs a counter-selectable marker *CYH2* present on the AD-Y coding plasmid together with control plates that contain cycloheximide (CHX). At every stage of the interactome mapping pipeline reporter gene activity is evaluated in parallel both on regular selective plates and on selective plates that contain CHX. The *CYH2* marker allows the selection of yeast cells that do not contain any AD-Y and thus the convenient identification of DB-X autoactivators.

2.1.2. Retesting to verify candidate interactions

In addition to autoactivating mutations in the DB-X protein, other genetic changes can occur during a screen. Mutations of the full-length DB-X or AD-Y protein might permit interactions that are otherwise undetectable or inhibited. Other mutations, such as *cis*-acting mutations in reporter genes and *trans*-acting mutations at unlinked genetic loci, could lead to reporter gene activation in the absence of any physical interaction between DB-X and AD-Y. To identify and remove such artifacts, all interaction candidates are systematically verified using yeast transformants freshly thawed from DB-X and AD-Y archival stocks. Haploid yeast cells of opposite mating-type, each containing DB-X or AD-Y expression plasmids, are mated according to the interacting pairs identified in the original screens and are tested for reproducible Y2H phenotypes to confirm reporter gene activation. Usually ~50% of interaction candidates can be successfully verified, which suggest that perhaps half of all primary Y2H positives belong to the classes of artifacts described above.

2.1.3. A high-quality Y2H implementation

Besides the precautions already mentioned, the Y2H version we have developed presents the following features that ensure high data quality.

Low DB-X and AD-Y hybrid protein expression   The use of low copy number yeast expression vectors together with the presence of weak promoters expressing DB-X and AD-Y hybrid proteins leads to low expression, which minimizes artifactual interactions driven by mass action. Use of high copy number vectors can increase DB-X and AD-Y protein expression and increase the sensitivity of the assay. This comes at the cost of increasing the detection of unspecific interactions (Braun *et al.*, 2009). The use of high copy number vectors should be accompanied by rigorous quality control and validation of every individual interaction with multiple assays.

Yeast strains   We have used two different Y2H strain backgrounds over the years (Vidal *et al.*, 1996; Yu *et al.*, 2008). The protocols described are applicable to Y8800 and Y8930, *MAT***a** and *MAT*α, respectively, two strains derived from PJ69-4 (James *et al.*, 1996) which harbor the following genotype: *leu2-3,112 trp1-901 his3-200 ura3-52 gal4Δ gal80Δ GAL2-ADE2 LYS2::GAL1-HIS3 MET2::GAL7-lacZ cyh2^R*. The availability of two haploid strains of opposite mating types enables the use of mating to efficiently combine large collections of DB-X and AD-Y constructs. By convention the Y8800 *MAT***a** and Y8930 *MAT*α strains are transformed with AD-Y and DB-X constructs, respectively.

Y2H-inducible reporter genes   The reporter genes *GAL2-ADE2* and *LYS2::GAL1-HIS3* are integrated into the yeast genome. Expression of the *GAL1-HIS3* reporter gene should be tested with 1 mM 3AT (3-amino-1,2,4-triazole, a competitive inhibitor of the *HIS3* gene product). When dealing with DB-X autoactivators, higher 3AT concentrations can be used to circumvent autoactivator-dependent activity of *GAL1-HIS3*. Interactions identified at higher 3AT concentrations should be accompanied by rigorous quality control and validation of every individual interaction using multiple assays.

Y2H controls   Y2H-inducible reporter gene expression levels can vary from weak to very strong, although these levels may not reflect the actual affinity of protein–protein interactions as they take place in their native environment. To help determine which candidate clones likely represent genuine biophysical interactors, six controls are added systematically to the master plates of Y2H experiments (Walhout and Vidal, 2001). This collection of diploid control strains contains plasmid pairs expressing DB-X and AD-Y hybrid proteins across a wide spectrum of interaction read-outs. For each control strain, a short description of plasmids and DB-X and AD-Y hybrid proteins are provided in Table 12.1 and expected phenotypes are shown in Fig. 12.2.

Please email "pascal_braun@dfci.harvard.edu" to request strains, plasmids, and controls.

2.2. Validation of Y2H datasets to produce reliable binary interactome maps

Despite the rigorous implementation of controls for identification of technical artifacts, a fraction of technical false positives can still be recovered in large-scale datasets. Well-described artifacts might have escaped detection, or it is possible that certain classes of artifacts have not been identified yet and consequently no controls are available to detect and remove them. Therefore, the quality of any dataset must be further assessed before it can be used as a reliable interactome map.

In earlier attempts at addressing this question for Y2H, protein pairs that activated two or more distinct reporters or pairs that were detected in two or more configurations (e.g., DB-X/AD-Y and DB-Y/AD-X) were

Table 12.1   Y2H controls

	Plasmid pairs	Protein	Interaction strength
Control 1	pDEST-AD	No insert	None, background
	pDEST-DB	No insert	
Control 2	pDEST-AD-E2F1	Human E2F1 aa 342–437	Weak (control for CHX control plates)
	pDEST-DB- CYH2-pRB	Human pRB aa 302–928	
Control 3	pDEST-AD-Jun	Mouse Jun aa 250–325	Moderately strong
	pDEST-DB-Fos	Rat Fos aa 132–211	
Control 4	pDEST-AD	No insert	Very strong
	pDEST-DB-Gal4	Yeast Gal4 aa 1–881	
Control 5	pDEST-AD-dE2F1	*Drosophila* E2F aa 225–433	Strong
	pDEST-DB-dDP	*Drosophila* DP aa 1–377	
Control 6	pDEST-AD- CYH2-dE2F1	*Drosophila* E2F aa 225–433	Strong (control for CHX plates)
	pDEST-DB-dDP	*Drosophila* DP aa 1–377	

Identities and description of expected phenotypes for the six controls used in every Y2H experiment.

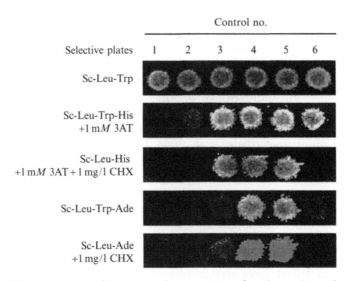

Figure 12.2 Phenotypes of Y2H controls. Six strains referred to as Control 1–6, each containing a different pair of DB-X and AD-Y hybrid proteins, are spotted on media selecting for the presence of both plasmids (top row) and, after an overnight incubation, replica-plated onto media selecting for Y2H-dependent reporter activation (rows 2 and 4). The six strains express DB-X/AD-Y pairs that result in reporter gene activation at various intensities. DB-X autoactivation is tested on plates that select for the loss of the AD-Y plasmid (rows 3 and 5).

considered to be of "higher quality", that is, more likely to be real biophysical interactors, than those pairs that activated only one reporter or were only found in a single orientation. Historically, and especially with cDNA screens, these criteria did indeed offer limited protection against artifacts, and enabled identification of more likely "true" interactions (Vidalain *et al.*, 2003). Today, however, such artifacts can be removed more systematically and more reliably by the controls described in Sections 2.1.1 (CHX control) and 2.1.2 (verification). All interactions that pass these controls are considered high-quality Y2H interactions, irrespective of whether or not they are detected in only one orientation or if they activate only one reporter.

Many "true" interaction pairs activate only one Y2H reporter or are detected in only one configuration. This is due to effects that are unrelated to the interaction capacity of the two examined proteins. The genomic context of the different reporters or use of promoters that require different levels of reconstituted transcription factor can lead to differential reporter activation. Similarly, the use of hybrid proteins imposes steric constraints on proteins that can interfere with detection of many interactions in at least one configuration. This was shown by testing a set of well documented positive control interaction pairs in Y2H and four other binary interaction assays. Consistently, only half of the positive scoring controls were detected in both orientations in any of the five assays (Braun *et al.*, 2009). Thus, while activation of multiple reporters and detection of interactions in multiple configurations can be comforting, these attributes are neither necessary nor sufficient requirements for high quality interactions.

Various experimental methods and computational approaches have been described to evaluate the quality of large-scale interactome datasets. Most computational methods estimate the correlation between physical interaction data and secondary data, such as expression profiling or types of functional annotations (Bader *et al.*, 2004; Deane *et al.*, 2002). Determination of data quality using this approach can effectively lead to filtered datasets that might be biased for particular classes of interactors, such as those with strong coexpression correlation. Such correlative data evaluation approaches make implicit assumptions about the nature of protein–protein interactions, which can potentially lead to erroneous conclusions (Yu *et al.*, 2008). Interactome maps can be productively integrated with orthogonal datasets to gain novel insights into biology (Pujana *et al.*, 2007; Vidal, 2001). If interaction datasets have been prefiltered using orthogonal data then such higher level analysis becomes less informative.

Another approach is to overlap the information from different interaction datasets to assess the FDR. In these analyses crucial details of the underlying experiments used in the respective screens are often ignored. Four critical parameters have to be taken into consideration (Venkatesan *et al.*, 2009): (i) the number and identity of ORFs used in each screen (*completeness*), (ii) the detection limitations of the assays used (*assay sensitivity*) (affected by many parameters like strains, location of protein tags, detection methods), (iii) the extent of incomplete sampling in each search space (*sampling sensitivity*), and (iv) the potential presence of technical false positives (*precision*). Without knowledge of these parameters for each dataset, any conclusion about data quality based on their overlap is meaningless. Thus, given the inherent limitations of computational approaches for quality control, experimental methods involving alternative protein interaction assays are strongly preferred.

2.2.1. Quality control I: Experimental assessment of dataset *precision*

One experimental approach to validate dataset quality consists in testing a representative sample of potential interactions from a given dataset with an orthogonal interaction assay. Since there is apparently not a single assay capable of detecting all protein–protein interactions tested, and considering that the subset of interaction pairs scoring positive in any two assays is rarely identical (Braun *et al.*, 2009), it is to be expected that only a fraction of interactions from a particular dataset will be detected by a validation assay. This is a consequence of the nature of interaction assays and the biochemical diversity of interactions, and not *per se* an indication of the quality of the original dataset. Characterizing the validation assay of choice using a positive reference set (PRS) and random reference set (RRS) of well-documented and random protein–protein interactions, respectively (Braun *et al.*, 2009; Cusick *et al.*, 2009; Venkatesan *et al.*, 2009; Yu *et al.*, 2008) provides an estimate of the *assay sensitivity* and the background of the validation assay. If desired, the stringency of the validation assay can be adjusted to decrease the background (and *assay sensitivity*) or to increase the *assay sensitivity* (and background). The validation assay results of the dataset sample relative to the PRS/RRS benchmark data enables estimation of the dataset *precision* (Braun *et al.*, 2009; Venkatesan *et al.*, 2009; Yu *et al.*, 2008).

2.2.2. Quality control II: Experimental confidence scores for individual interactions

When dataset precision is determined using a single assay, validation rates between 20% and 40% can be expected for both PRS and high-quality datasets under conditions in which the RRS detection rate is below 5% (Braun *et al.*, 2009). In the long term, it will be highly desirable to not only estimate the overall precision of a dataset but to validate all protein–protein interactions individually. Validation for individual interactions in a dataset can be made stronger if multiple complementary assays are used to test the interactions (Braun *et al.*, 2009). The concept of calibrating and bench-marking assay performance with the PRS/RRS can be applied to multiple assays and can be used to calculate a confidence score for individual bio-physical protein–protein interactions. Multiple interaction assays are first benchmarked against common PRS and RRS reference sets to obtain comparable calibrations of *assay sensitivity* and background. Then, all inter-actions identified in a large-scale interactome screen are characterized using the same assay implementations. After the results from all assays have been collected for any interaction pair, a confidence score can be calculated based on prior PRS/RRS calibration of the assays and the validation results of the respective interaction (Braun *et al.*, 2009). PRS/RRS clones for several organisms are available upon request (Braun *et al.*, 2009; Simonis *et al.*, 2009; Venkatesan *et al.*, 2009; Yu *et al.*, 2008).

2.3. Biological evaluation of binary interactome maps

The identification of high-confidence biophysical interactions is an important first step towards answering many biological questions both at small and large scale. However, even robustly demonstrated biophysical interactions might be biological false positives, or pseudo-interactions, that never occur *in vivo*.

Biological relevance of protein–protein interactions has been inferred from network analyses or by combining interactome information with systematic genetic data (Collins *et al.*, 2007a,b; Pujana *et al.*, 2007). Despite some success, these approaches remain constrained by the availability of high-quality datasets, and are limited as they are predictions.

Until demonstrated by thorough mechanistic studies of all proteins involved, the biological role of protein–protein interactions remains elusive. Such mechanistic studies are typically carried out at small scale, so this approach is unsustainable and cost prohibitive for characterizing hundreds of thousands of soon to be discovered human protein interactions (Venkatesan et al., 2009).

Biological relevance of a biophysical protein–protein interaction may be derived from an observed phenotype following genetic disruption of this specific interaction in vivo. Such IDAs can occur naturally, as has been found for some inherited Mendelian diseases (De Nicolo et al., 2009; Zhong et al., 2009). For these alleles the causal link between disruption of the biophysical interaction and the observable phenotype must be demonstrated. Alternatively, IDAs can be generated experimentally using a reverse two-hybrid approach (Dreze et al., 2009; Vidal et al., 1996). For defining a biological function of a biophysical interaction using such experimentally generated IDAs, a critical step is the identification of a phenotype and subsequent demonstration of causality.

Certain interactions may have subtle or modifying roles in the regulation of cellular functions. Disruption of such interactions individually may lead to subtler, easy to overlook phenotypes. Disruption of such interactions in the presence of other genetic or environmental perturbations may produce more observable systems alterations. For those, quantitative mathematical modeling may be useful for analyzing small or synergistic phenotypic consequences.

3. High-Throughput Y2H Pipeline

3.1. Assembly of DB-X and AD-Y expression plasmids

The first step towards binary interactome mapping is the generation of expression plasmids. For high-throughput experiments it is preferable to use sequence independent recombinational subcloning technologies such as Gateway cloning (Walhout et al., 2000). Large resources containing thousands of distinct ORFs in Gateway entry vectors are available for a few organisms (Lamesch et al., 2004, 2007; Reboul et al., 2003; Rual et al., 2004). These ORFs can be transferred into Gateway-compatible expression vectors in a simple single-step reaction (Fig. 12.3). Albeit not mandatory, linearizing the destination vectors by restriction digestion improves recombination efficiency and decreases background as well as chances of obtaining incorrect LR recombination clones. The restriction enzyme should be chosen so that the destination vector is digested only once between the two Gateway recombination sites. The Gateway LR reaction, carried out using enzyme and buffer concentrations optimized by titration, gives best yields at 25 °C for 18 h but can also be carried out for ~2 h at room

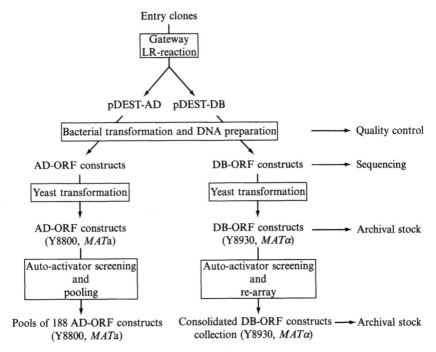

Figure 12.3 Pipeline for preparation of Y2H reagents. The pipeline from producing Gateway entry clones to transformation and quality control of yeast strains used in Y2H screens. Protein-encoding ORFs are first transferred by Gateway LR reactions into pDEST-AD and pDEST-DB, and amplified in bacteria. DNA is then extracted for yeast transformations. After transformation DB-X hybrid proteins are tested for autoactivator phenotypes and then rearrayed before screening. AD-Y hybrid proteins are combined into minipools of 188 different clones per pool.

temperature. Completed recombination reactions are transformed into *Escherichia coli*, grown for 18 h, and plasmids are isolated. This step can be done manually or by using liquid handling robots. Because all steps are carried out in 96-well microtiter plates, protocols are provided for the equivalent of one 96-well plate.

Protocol 1: Restriction digestion of Y2H destination vectors

1. Combine in one tube:
    – 11 μg destination vector (pDEST-AD or pDEST-DB).
    – 11 μl of 10× restriction enzyme buffer.
    – 2.5 μl of *Sma*I restriction enzyme (50 units).
    – 85.5 μl filter-sterilized water.
2. Mix well by pipetting up and down several times.
3. Incubate at 25 °C for 12–16 h.
4. Incubate at 65 °C for 20 min to heat-inactivate the restriction enzyme.

Display 500 ng of digested destination vector on a 1% agarose gel alongside 500 ng of undigested destination vector to confirm complete digestion. The heat-inactivated reaction mix can be used for Gateway LR reactions without further purification.

Protocol 2: High-throughput Gateway LR recombinational cloning

1. Combine in one tube:
   - 110 μl of SmaI digested destination vector (11 μg).
   - 110 μl of LR clonase buffer 5×.
   - 55 μl of TE 1×.
   - 55 μl of LR clonase enzyme mix (Invitrogen) (keep this mix on ice).
2. Homogenize by gently pipetting up and down.
3. With a multichannel pipette, distribute 3 μl of this solution into every well of a 96-well microtiter plate.
4. Add 2 μl of entry clone per well.
5. Centrifuge briefly.
6. Incubate at 25 °C for 18 h.

Protocol 3: Bacterial transformation
The following protocol is used to transform, amplify, and isolate Gateway LR reaction products:

1. Thaw 1 ml of competent DH5α-T1^R (Invitrogen) cells on ice (with a transformation efficiency greater than 5×10^7 antibiotic resistant colonies per μg of input DNA).
2. Add 10 μl of competent cells per well directly into a 96-well plate containing 5 μl Gateway LR reaction mix in each well.
3. Seal the plate with adhesive foil.
4. Incubate on ice for 20 min.
5. Heat shock at 42 °C in a standard thermocycler for 1 min.
6. Incubate on ice for 2 min.
7. Add 100 μl of prewarmed (37 °C) SOC media per well. Seal the plate with adhesive foil to avoid contamination.
8. Incubate at 37 °C for 1 h.
9. Transfer the transformation mix into a 96-well deep-well plate containing 1 ml of LB media with 100 μg/ml of ampicillin.
10. Incubate on a 96-well plate shaker at 37 °C for 20 h.
11. Remove 5 μl for subsequent analysis by PCR (Protocol 4).
12. Remove 80 μl of the overnight culture, mix with 80 μl of 40% (w/v) autoclaved glycerol and store at −80°C.
13. Use the remainder of the overnight culture for plasmid isolation.

SOC medium 0.5% yeast extract, 2% tryptone, 10 mM NaCl, 2.5 mM KCl, 10 mM MgCl$_2$, 10 mM MgSO$_4$, 20 mM glucose. Add glucose after autoclaving the solution with the remaining ingredients, and let cool down. Sterilize the final solution by passing it through a 0.2 μm filter. SOC medium can be stored at room temperature.

Transformation controls It is good practice to systematically control for media contamination (no cells), competent cells contamination (cells only), Gateway LR reaction contamination (negative control of LR reaction), and transformation efficiency (10 pg of pUC19). If the four controls indicate clean and successful transformation, proceed to the next quality control step.

Recombination control To confirm that Gateway LR reactions occurred properly, analyze recombination products by bacterial culture PCR using destination vector specific primers (Protocol 4). For each transformation plate select one row for PCR.

Protocol 4: Bacterial culture PCR
Dilute 5 μl of bacterial culture into 95 μl of sterile water and mix by pipetting up and down. Keep bacterial cultures at 4 °C until PCR results are determined.
   For one 96-well plate of PCR, prepare in a tube on ice:

- 330 μl of HiFi Platinum Taq polymerase buffer 10× (Invitrogen).
- 120 μl of 50 mM MgSO$_4$ (final concentration 1.8 mM).
- 33 μl of 40 μM dNTPs (final concentration 400 nM).
- 3.3 μl of 200 μM AD or DB forward primer (final concentration 180 nM).
- 3.3 μl of 200 μM Term reverse primer (final concentration 180 nM).
- 20 μl of HiFi Platinum Taq polymerase (Invitrogen).
- 2.5 ml of filter-sterilized water.

Aliquot 30 μl of the reaction mix into every well of a soft shell, V-bottom 96-well microtiter plate. Keep on ice. Add 3 μl of the diluted bacterial culture per well as DNA template. Wells G12 and H12 are used as negative control (water as template) and positive control (10 ng of empty pDEST-AD or pDEST-DB), respectively.
   Place the PCR plate on a thermocycler and run the following program:

Step 1: Denaturation at 94 °C for 4 min.
Step 2: Denaturation at 94 °C for 30 s.
Step 3: Annealing at 58 °C for 30 s.
Step 4: Elongation at 68 °C for 3 min.
        Repeat Steps 2-3-4, 34 times.
Step 5: Final elongation at 68 °C for 10 min.
Step 6: Hold at 10 °C.

Primer sequences The primers are designed such that the 5′-primer confers AD and DB vector specificity, respectively, whereas the Term 3′-primer is identical for both vectors.

AD: 5′-CGCGTTTGGAATCACTACAGGG-3′
DB: 5′-GGCTTCAGTGGAGACTGATATGCCTC-3′
Term: 5′-GGAGACTTGACCAAACCTCTGGCG-3′

Once PCR reactions are completed, analyze 5 μl of PCR product on a 1% agarose gel by comparing sizes to that of the control from well H12 (the PCR amplicon from a destination vector containing the Gateway cassette has an expected size $\sim$1.9 kb). The H12 control serves simultaneously as a positive control for the PCR and as a negative control for the LR recombination reaction. PCR failure is indicated by the absence of the H12 product, and failure of the LR reaction may be indicated by a dominant band of 1.9 kb across all wells. Successful LR reactions will give rise to the size distribution of the original ORFs to which $\sim$280 bp of vector sequences are added due to the AD, DB, and Term primer positions. If the PCR results indicate successful LR recombinations, prepare archival stocks by combining 80 μl of bacterial cultures with 80 μl of 40% (w/v) glycerol in a round-bottom 96-well microtiter plate. The rest of the cultures are used for plasmid isolation. Afterwards, ensure successful plasmid isolation by analyzing 2 μl of the DNA preparation on a 1% agarose gel.

To ensure the absence of plate orientation mistakes when processing multiple plates, sequence verify PCR products amplified from one column of each 96-well miniprep plate. Use 1 μl of the DNA preparation as template for PCR. The primers, recipes, and PCR conditions are identical to those presented in Protocol 4. BLASTn of the acquired sequences against a reference database identifies clones and allows verification of their correct locations.

3.2. Yeast transformation

DB-X and AD-Y expression plasmid constructs are individually transformed into competent Y8930 (*MAT*α) and Y8800 (*MAT*a) strains, respectively.

Protocol 5: Yeast transformation
This protocol requires two solutions that need to be freshly prepared from stock solutions in order to obtain maximum transformation efficiencies. Tris–EDTA–lithium acetate solution (TE/LiAc) is prepared by 10-fold dilution of 10× TE and 1 M LiAc stocks to give 10 mM Tris–HCl (pH 8.0), 1 mM EDTA (pH 8.0), and 100 mM LiAc final concentration. TE/LiAc polyethyleneglycol (PEG) solution is prepared by combining 8 volumes of 44% (w/v) PEG 3350 with 1 volume of 10× TE and 1 volume

of 1 M LiAc. The following volumes and quantities are given for carrying out one 96-well plate of transformations.

1. Streak Y8800 and Y8930 on separate YEPD plates and incubate at 30 °C for 48–72 h to obtain isolated colonies.
2. For each strain, inoculate 20 ml of YEPD with 10 isolated colonies. Incubate at 30 °C on a shaker for 14–18 h.
3. Measure and record the OD_{600}, which should be between 4.0 and 6.0. Dilute cells into YEPD media to obtain a final $OD_{600} = 0.1$. Use 100 ml of YEPD media per 96-well plate of transformations.
4. Incubate at 30 °C on a shaker until OD_{600} reaches 0.6–0.8 (4–6 h).
5. Boil carrier DNA (salmon sperm DNA, Sigma-D9156) for 5 min then place on ice until needed.
6. Harvest cells by centrifugation at $800 \times g$ for 5 min. Discard the supernatant and resuspend cells gently in 10 ml of sterile water.
7. Centrifuge as described in step 6 and discard the supernatant.
8. Resuspend cells in 10 ml of TE/LiAc solution, centrifuge, and discard the supernatant.
9. Resuspend cells in 2 ml of TE/LiAc solution, then add 10 ml of TE/LiAc/PEG solution supplemented with 200 μl of boiled carrier DNA. Mix the solution by inversion.
10. Dispense 120 μl of this mix into each well of a round-bottom 96-well microtiter plate.
11. Add 10 μl of plasmid DNA to the competent yeast and mix by pipetting up and down several times. Use liquid handling robots to transfer and mix 96 samples at a time. Seal the plate with adhesive foil.
12. Incubate at 30 °C for 30 min.
13. Subject to heat shock in a 42 °C water bath for 15 min.
14. Centrifuge the 96-well plate for 5 min at $800 \times g$. Carefully remove the supernatant using a multichannel pipette.
15. To each well add 100 μl of sterile water and resuspend cell pellets by pipetting up and down.
16. Centrifuge the 96-well plate for 5 min at $800 \times g$, then carefully remove 90 μl of water from each well using a multichannel pipette.
17. Resuspend cell pellets by vortexing the 96-well plate on a shaker.
18. Spot 5 μl/well of cell suspension onto an appropriate selective plate (Sc-Trp for AD-Y, Sc-Leu for DB-X). For a consistent footprint, use of a liquid handling robot is recommended.
19. Incubate at 30 °C for 72 h.
20. Using sterile flat-end toothpicks, pick transformed yeast colonies into individual wells of a 96-well round-bottom plate containing 160 μl of selective media (Sc-Trp for AD-Y, Sc-Leu for DB-X).
21. Incubate on a shaker at 30 °C for 72 h.

22. Prepare archival stocks by combining 80 μl of the yeast culture with 80 μl of 40% (w/v) autoclaved glycerol in a round-bottom 96-well plate. Store at $-80\,^{\circ}C$.

3.3. Autoactivator removal and AD-Y pooling

3.3.1. Autoactivator identification and removal

To be as close as possible to the physiology of the cells in which interactions are detected, the identification of autoactivators is achieved in diploid yeast strains obtained by mating DB-X yeast strains with the Y8800 yeast strain transformed with the AD encoding plasmid containing no insert (empty pDEST-AD). All diploid yeast strains showing a growth phenotype stronger than the "no interaction" Y2H control (control 1) are considered autoactivators. Because activation of the *GAL1::HIS3* reporter gene is easier to achieve than that of *GAL7::ADE2*, it is used for autoactivator identification.

Protocol 6: Identification of autoactivating DB-X hybrid proteins

Before starting the experiment, prepare one YEPD plate, one Sc-Leu-Trp plate and one Sc-Leu-Trp-His + 1 m*M* 3AT plate for each 96-well plate of DB-X yeast strains to be tested. The YEPD plates should be prepared at least 1 week in advance to allow them to dry. This allows fast penetration of liquid in the mating step and prevents merging of adjacent spots due to excess liquid.

1. Add 160 μl of fresh liquid Sc-Leu media to each well of a round-bottom 96-well microtiter plate followed by 5 μl from individual glycerol stocks of each DB-X yeast strain to each well.
2. For every plate of DB-X yeast strains to be tested, inoculate a test tube containing 0.55 ml of Sc-Trp media with Y8800 transformed with empty pDEST-AD.
3. Incubate at 30 °C for 72 h on a shaker.
4. Spot 5 μl of DB-X liquid cultures on a YEPD plate using a liquid handling robot.
5. Allow the spots to dry for 30–60 min.
6. Aliquot the pDEST-AD transformed Y8800 yeast culture into a round-bottom 96-well plate.
7. Spot 5 μl of pDEST-AD transformed Y8800 on top of the DB-X spots.
8. Spot Y2H controls at the bottom of the plate.
9. Incubate mating plates at 30 °C for 14–18 h.
10. Replica-plate onto Sc-Leu-Trp media to select for diploid cells.
11. Incubate at 30 °C for 14–18 h.
12. Replica-plate from the Sc-Leu-Trp media onto Sc-Leu-Trp-His + 1 m*M* 3AT media. Nonautoactivating yeast cells are not able to activate the *GAL1::HIS3* reporter gene hence should not grow on this media.

13. Incubate at 30 °C for 14–18 h.
14. "Replica-clean" Sc-Leu-Trp-His + 1 mM 3AT plates by placing each plate on a piece of velvet stretched over a replica-plating block and pushing evenly on the plate to remove excess yeast. Replace the cloth and move to process the next plate until all plates have been cleaned.
15. Incubate at 30 °C for 72 h.
16. Score growth phenotypes.

Growth phenotypes are scored by comparison to the "no interaction" Y2H control (control 1). All yeast strains showing a stronger growth phenotype than control 1 are considered autoactivators. To reliably identify autoactivators it is best to score growth twice independently. If a yeast clone is given two different scores, accepting the most stringent one ensures high quality of the starting material for subsequent interactome mapping.

Autoactivators are physically removed from the collection of DB-X transformed yeast clones by robotic rearraying of nonautoactivator yeast clones into new plates. During the rearray step plate positions G12 and H12 are left empty for control purposes. New glycerol stocks are prepared from this consolidated collection of nonautoactivating yeast strains and used for subsequent Y2H screens.

1. From archival glycerol stocks containing all of the individual DB-X yeast clones, cherry pick nonautoactivating DB-X yeast clones into plates containing 160 μl Sc-Leu (DB-X) liquid media.
2. Incubate at 30 °C for 72 h on a 96-well plate shaker.
3. Prepare an archival stock by combining 80 μl of the yeast culture with 80 μl of 40% (w/v) autoclaved glycerol in a round-bottom 96-well plate.

Albeit much less frequent, autoactivating AD-Y can also occur. The previous protocol can easily be adapted for AD-Y autoactivator identification by use of AD specific reagents wherever appropriate. As an alternative, to reduce time and cost, it is possible to test AD-Y autoactivation using pools. For this, each AD pool, described in Protocol 7, is mated with Y8930 transformed with the DB encoding plasmid containing no insert (empty pDEST-DB) then processed (Protocol 6). If a diploid strain shows growth on autoactivator detection plates, the responsible AD-Y yeast clone can be identified by deconvoluting the AD-Y pool. This step is achieved by testing all 188 AD-Y yeast clones constituting the pool for autoactivation. Once identified, the autoactivating AD-Y yeast clone is removed and the affected pool is reassembled without it (Protocol 7).

3.3.2. Efficient screening by AD pooling

The pools used in the Y2H pipeline combine 188 different AD-Y expressing yeast clones. This experimentally defined pool-size provides an optimal compromise between screening efficiency (number of plates to be processed) and screen sensitivity (number of interactors identified).

Protocol 7: Construction of AD pools

This protocol describes the construction of one pool of 188 different AD-Y hybrid constructs transformed into Y8800, starting from two 96-well plates of AD-Y yeast strains.

1. For each of the two plates of 94 AD-Y constructs: inoculate 500 μl per well of Sc-Trp media with 5 μl per well of AD-Y yeast strains.
2. Grow on a shaker at 30 °C for 4 days.
3. Resuspend yeast cell cultures by thoroughly vortexing the culture plates.
4. Measure the OD_{600} to ensure that growth is homogenous throughout each plate, hence that each AD-Y yeast strains will be represented in the same proportion.
5. Transfer the contents of the two culture plates into a sterile trough.
6. Mix thoroughly to ensure equal representation of all AD-Y yeast strains in the pool.
7. On a liquid handling platform, prepare archival stocks by combining 80 μl of the pooled yeast cultures with 80 μl of 40% (w/v) autoclaved glycerol in round-bottom 96-well microtiter plates.

If additional copies of the AD-Y pools are required, these should be made according to the protocol above and not by amplification of existing pools, as amplification can lead to loss of representation within the pool.

Protocol 8: Assessing equal representation of AD-Y clones in pools

Before the AD pools are used for Y2H experiments, equal representation of each of the 188 AD-Y clones in the pools should be confirmed. Biased pools and low representation of some AD-Y yeast cells will decrease if not eliminate the ability to detect protein interactions involving the underrepresented hybrid proteins.

1. For each plate of AD-Y pools streak 5 μl of glycerol stock from two randomly selected wells onto Sc-Trp plates.
2. Grow at 30 °C for 72 h.
3. From each Sc-Trp plate, pick 96 isolated colonies and lyse yeast cells according to Protocol 9.
4. Add 3 μl of the yeast cell lysate as PCR template.
5. Carry out PCR according to Protocol 10.
6. Run 5 μl of PCR product on a 1% agarose gel. If PCR products can be detected in most reactions, proceed to analyze the corresponding plate by end-read sequencing.
7. Identify the obtained sequences by BLASTn. If the created pools are not biased, the frequency at which yeast cells containing identical AD-Y plasmids were picked (and hence the sequence identifications) should follow a normal distribution.

Protocol 9: Yeast cell lysis

1. Prepare lysis buffer by dissolving 2.5 mg/ml zymolase 20T (21,100 U/g, Seikagaku Corp.) in 0.1 M sodium phosphate buffer (pH 7.4). Keep on ice.
2. Aliquot 15 μl of lysis buffer into the wells of a 96-well PCR plate. Keep on ice.
3. Pick a small amount of yeast cells (not more than fits on the very end of a standard 200 μl tip) and resuspend in the lysis buffer in the PCR plate.
4. Place the PCR plate on a thermocycler and run the following program:
   Step 1: 37 °C for 15 min
   Step 2: 95 °C for 5 min
   Step 3: Hold at 10 °C
5. Add 100 μl of filter-sterilized water to each well.
6. Centrifuge 10 min at 800×g.
7. Store at −20 °C.

Protocol 10: Yeast lysate PCR

For each 96-well plate of PCR reactions, prepare the following reaction mix on ice:

− 330 μl of HiFi Platinum Taq (Invitrogen) polymerase buffer 10×.
− 120 μl of 50 mM MgSO$_4$ (final concentration 1.8 mM).
− 33 μl of 40 μM dNTPs (final concentration 400 nM).
− 3.3 μl of 200 μM AD primer (final concentration 180 nM).
− 3.3 μl of 200 μM Term primer (final concentration 180 nM).
− 20 μl of HiFi Platinum Taq polymerase (Invitrogen).
− 2.5 ml of filter-sterilized water.

Aliquot 30 μl into every well of a 96-well PCR plate. Keep on ice. To each well, add 3 μl of the yeast cell lysate (Protocol 9) as DNA template. Seal plate with adhesive aluminum foil.

Place the PCR plate on a thermocycler and run the following program:

Step 1: Denaturation at 94 °C for 4 min.
Step 2: Denaturation at 94 °C for 30 s.
Step 3: Annealing at 58 °C for 30 s.
Step 4: Elongation at 68 °C for 3 min.
Repeat Step 2-3-4, 34 times.
Step 5: 68 °C for 10 min.
Step 6: Hold at 10 °C.

3.4. Screening and phenotyping

The Y2H pipeline consists of three essential stages, which together yield highly reliable interactions: primary screening, secondary phenotyping, and verification (Fig. 12.4). The high-throughput Y2H pipeline presented here

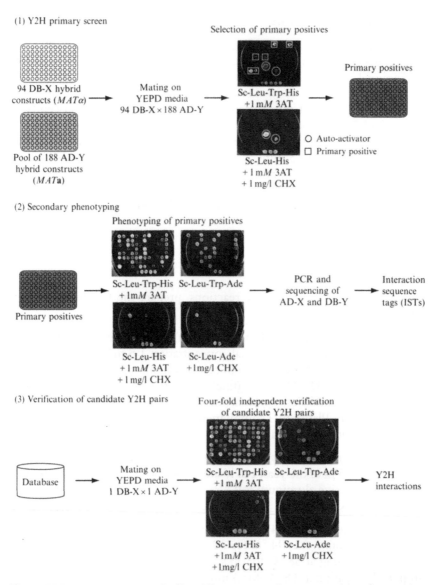

Figure 12.4 Y2H screening pipeline. Three steps, primary screening, phenotyping, and retesting, ensure high-throughput and reliable removal of artifacts. For primary screens, 94 distinct DB-X constructs are mated against a minipool containing 188 AD-Y hybrids. Positive colonies are picked from selective plates and in "secondary phenotyping" are evaluated on two types of selective plates and respective autoactivation control plates. Protein pairs considered as "candidate Y2H interactions" are identified by DNA sequencing of PCR products amplified from positive colonies. All identified pairs are verified using fresh archival yeast stocks. DB-X/AD-Y pairs that score positive on at least three out of four independent plate sets are considered high-quality Y2H interactions (see text for details).

has been used to produce several high-quality proteome-scale binary inter-actome maps (Rual *et al.*, 2005; Simonis *et al.*, 2009; Yu *et al.*, 2008).

Protocol 11: Y2H primary screening

Pour all required agar plates at least 1 week before starting the experiments and store them without wrapping at room temperature. Storage ensures that the plates are sufficiently dry, which in turn will prevent merging of spotted yeast cultures in the mating step which can otherwise occur due to excess liquid and slow absorption into the agar.

[Day 0: Inoculation]

1. Thaw glycerol stocks of the DB-X yeast strains and AD pools to be tested. One person can easily handle a batch of 100 mating plates, for example, 10 96-well plates of DB-X yeast clones tested against 10 96-well plates with AD-Y pools.
2. Inoculate 96-well plates that contain 160 μl selective media in every well (Sc-Leu for DB plates, Sc-Trp media for AD pool plates), with 5 μl/well of the thawed glycerol stock plates.
3. Seal all plates with adhesive tape and return glycerol stocks to $-80°C$.
4. Incubate the inoculated cultures at 30 °C on a shaker for 72 h.

[Day 3: Mating]

1. For each combination [AD-Y pool plate$\times$96 DB-X plate] spot 5 μl/well of the respective AD-Y pool liquid culture onto a mating plate (YEPD) using a liquid handling robot.
2. Allow spots to dry for 30–60 min.
3. Spot 5 μl/well of each DB-X on top of the AD pool spots.
4. Spot Y2H controls onto every plate.
5. Incubate mating plates at 30 °C for 14–18 h.

[Day 4: Replica-plating]

1. Replica-plate mated yeast cells from mating plates onto screening plates (Sc-Leu-Trp-His + 1 mM 3AT).
2. To detect *de novo* autoactivators, for each distinct plate of DB-X yeast clones, replica-plate yeast from three mating plates (with three different AD pools) onto Sc-Leu-His + 1 mM 3AT + 1 mg/l CHX plates.
3. Incubate at 30 °C for 14–18 h.

[Day 5: Replica-clean]

1. Replica-clean all plates by placing each plate on a piece of velvet stretched over a replica-plating block and pushing evenly on the plate to remove excess yeast cells. Replace the cloth and move to process the next plate until all plates have been cleaned. The plates need to be cleaned enough to reduce background, but excessive cleaning can also lead to accidental removal of positives.

2. Incubate at 30 °C for 5 days.

[Day 10: Score and pick colonies]
Pick primary positive colonies from screening plates and resuspend in a 96-well plate containing liquid media (Sc-Leu-Trp). Only consider colonies that grew better than background as indicated by control 1 of the six Y2H controls (Fig. 12.2). Only pick primary positives where the corresponding spots on the CHX plates are negative. Consider all three CHX plates as controls. Since every individual DB-X construct is mated against a pool of 188 AD-Y constructs, it is possible to obtain multiple interactions per spot. To account for this infrequent yet possible event we pick at most three colonies per spot.

1. Pick positive yeast colonies into a 96-well plate containing 160 µl/well Sc-Leu-Trp media. Leave positions G12 and H12 empty for subsequent controls.
2. Incubate the culture plate at 30 °C for 72 h.
3. The cultures can be used directly for phenotyping (Protocol 12—start at Step 2). It is also recommended to prepare an archival glycerol stock by combining 80 µl of the yeast culture with 80 µl of 40% (w/v) autoclaved glycerol in a 96-well plate, sealing the plates with adhesive tape and storing at −80°C.

Protocol 12: Phenotyping
[Day 0: Inoculation]

1. Thaw glycerol stocks of primary positives.
2. Spot 5 µl/well onto Sc-Leu-Trp plates using a 96-well liquid handling robot.
3. Seal all glycerol stock plates with adhesive tape and return to −80°C.
4. Add Y2H controls.
5. Incubate the Sc-Leu-Trp plates at 30 °C for 48 h.

[Day 2: Replica-plating]

1. Replica-plate from Sc-Leu-Trp plates onto four phenotyping plates:
   − Sc-Leu-Trp-His + 1 mM 3AT
   − Sc-Leu-Trp-Ade
   − Sc-Leu-His + 1 mM 3AT + CHX (1 mg/l)
   − Sc-Leu-Ade + CHX (1 mg/l)

The first two plates are used to assess Y2H reporter activity; the two CHX plates enable detection of autoactivators.

2. Clean the plates immediately after replica-plating. This step will minimize background growth.
3. Incubate the phenotyping plates at 30 °C for 72 h.

[Day 5: Scoring]

1. Identify autoactivators by inspecting CHX plates. Any yeast spot showing growth on these plates should not be considered for further processing.
2. Identify candidate interactions (secondary positives). It is useful to differentiate positives activating one reporter gene (most often *GAL1::HIS3*) from those activating both reporter genes. An example of a Sc-Leu-Trp plate and the four assay plates along with proper scoring are shown (Fig. 12.5).
3. Patch all secondary positives on fresh Sc-Leu-Trp plates.
4. Incubate the Sc-Leu-Trp plates at 30 °C for 48 h.
5. Lyse cells according to Protocol 9.
6. Amplify the inserts of the DB-X and AD-Y inserts of positive colonies by yeast colony PCR according to Protocol 10 for subsequent ORF identification by end-read sequencing. At this stage the matched PCR products coding for putatively interacting proteins are physically separated. It is critical to track the matching AD-Y and DB-X PCR products so that interacting pairs can be identified after sequencing.

Once sequencing data have been received and the candidate protein pairs have been identified, a list of unique candidate interaction pairs can be compiled.

3.5. Verification

Protocol 13: Verification of candidate Y2H interaction pairs

While the CHX control at every step identifies spontaneous autoactivators arising from mutations in DB-X, this last verification step protects against other potential artifacts, for example, from mutations elsewhere in the yeast genome, and ensures robust high data quality. To reach maximum reproducibility, robustness, and reliability of Y2H interactions, this critical step is carried out a total of four times independently (16 plates corresponding to four sets of four assay plates), ideally by four different experimenters. Only interactions that score positive at least three out of four plate sets, and do not once score as autoactivators, are accepted as verified Y2H interactions.

Before the verification experiment can be done, it is necessary to rearray yeast clones corresponding to candidate Y2H interacting pairs into new plates. During the rearray step, plate positions G12 and H12 should be left empty for subsequent controls.

1. From archival glycerol stocks of the individual AD and DB transformed yeast clones, rearray the (candidate) interaction partner clones into matching positions of plates containing 160 μl Sc-Trp (AD-Y) and Sc-Leu (DB-X) liquid media.
2. Incubate at 30 °C on a 96-well plate shaker for 72 h.
3. Prepare an archival stock by combining 80 μl of the yeast culture with 80 μl of 40% (v/v) autoclaved glycerol in a round-bottom 96-well plate.

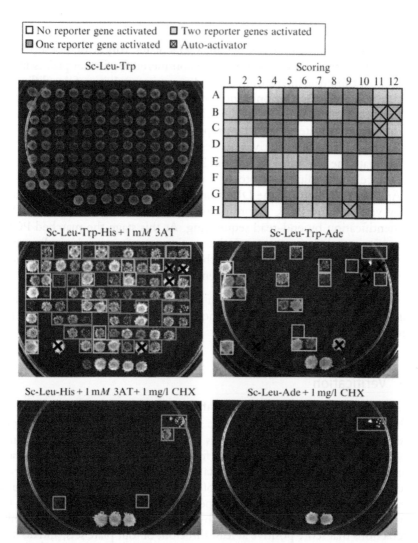

Figure 12.5 Phenotyping plates and scoring. First, autoactivators are identified and crossed out. The stringency of autoactivator detection is high such that even slight growth on the CHX control plates leads to elimination of the respective candidate. Subsequently, growth is evaluated on the selective –His and –Ade plates using the six controls (Fig. 12.1) as reference. (See Color Insert.)

[Day 0: Inoculation]

1. Thaw glycerol stocks of rearrayed Y2H candidate pairs completely.
2. With 5 μl of glycerol stock, inoculate 160 μl of fresh Sc-Leu (DB-X) and Sc-Trp (AD-Y) liquid media dispensed in round-bottom 96-well culture plates.
3. Incubate at 30 °C for 72 h.

[Day 3: Mating]

1. Dispense 5 μl/well of AD-Y liquid culture onto a YEPD mating plate.
2. From the matching DB-X plate, dispense 5 μl/well of DB-X on top of the AD-Y spots.
3. Add Y2H controls.
4. Incubate at 30 °C for 14–18 h.

[Day 4: Selection of diploids]

1. Replica-plate mated yeast cells onto Sc-Leu-Trp diploid selection plates.
2. Incubate at 30 °C for 14–18 h.

[Day 5: Phenotyping of diploids]

1. Replica-plate diploid yeast cells onto the four phenotyping plates and autoactivator identification plates.
2. Immediately after, replica-clean all plates thoroughly by placing each plate on a piece of velvet stretched over a replica-plating block and pushing evenly on the plate to remove excess yeast. Replace the cloth and move to process the next plate until all plates have been cleaned.
3. Incubate at 30 °C for 3 days.

[Day 10: Scoring]
The scoring of each of the four plate sets is done independently in the same way as for secondary phenotyping. We consider as verified only those Y2H pairs that scored positive in at least three out of four plate sets and are never scored as an autoactivator.

3.6. Media and plates

3.6.1. Nonselective rich yeast medium (YEPD)
The Y8800 and Y8930 yeast strains are propagated on solid agar YEPD plates or in liquid YEPD medium.

YEPD media

1. Mix 20 g of yeast extract, 40 g of bacto-peptone and 1900 ml of water.
2. Autoclave for 45 min.
3. Store at room temperature.
4. Before use add 50 ml of 40% (w/v) autoclaved glucose and 15 ml of 65 mM adenine solution per liter of media.

YEPD agar plates

1. Mix 20 g of yeast extract and 40 g of bacto-peptone with 950 ml of water in a 2 l flask.

2. Add a stir bar.
3. Mix 40 g of agar and 950 ml of water in a second 2 l flask and shake well.
4. Autoclave the two flasks for 45 min.
5. Transfer the contents of each agar flask to one media flask and mix well.
6. Cool to 55 °C and keep in a water bath until ready to pour.
7. Before pouring, add 100 ml of autoclaved 40% (w/v) glucose and 30 ml of 65 mM adenine solution.
8. Pour 15 cm agar plates.
9. Dry for 5–7 days at room temperature and store at room temperature. If the plates need to be used earlier, they can be dried for 30 min in a sterile hood with the ventilation on.

3.6.2. Selective yeast media

Selective media are used for maintaining the AD-Y and DB-X plasmids and detection of reporter activity. Prototrophic markers are used for selection on plates lacking the appropriate amino acid or nucleotide. In our system the DB-expressing plasmid contains the selectable marker *LEU2* which enables growth of the Y8800/Y8930 yeast strains on plates lacking leucine (-Leu), while the AD-expressing plasmid contains the *TRP1* marker which enables growth on plates lacking tryptophan (-Trp). The other two proto-trophic markers (*HIS3* and *ADE2*) are used as reporter genes in our experiments. Expression of these markers is selected on plates lacking histidine (-His) (supplemented with 1 mM 3-amino-1,2,4-triazole, 3AT) or lacking adenine (-Ade). Autoactivator detection plates are supplemented with 1 mg/l of CHX and contain tryptophan to allow growth of yeast cells without the AD-Y plasmid.

Synthetic complete (Sc) media The different selective media are based on the same Sc drop-out media recipe, but then supplemented with different amino acids to prepare the media appropriate for the various applications.

• *Sc media*

1. Mix 5.2 g of amino acid powder lacking leucine, tryptophan, histidine, and adenine, 6.8 g of yeast nitrogen base (without ammonium sulfate and amino acids), and 20 g of ammonium sulfate.
2. Dissolve in 1900 ml water and add a stir bar.
3. Adjust the pH to 5.9 by adding a few drops of 10 M NaOH.
4. Autoclave the flasks for 45 min.
5. Add 8 ml of each stock solution as needed. Store at room temperature.

• *Sc agar plates*

For a 4 l preparation of 15 cm agar Petri plates containing Sc medium lacking particular amino acids or nucleotides:

1. Place a magnetic stir bar into two 2 l flasks and label as the "media flasks."
2. Mix 5.2 g of amino acid powder lacking leucine, tryptophan, histidine, and adenine, 6.8 g of yeast nitrogen base (without ammonium sulfate and amino acids), and 20 g of ammonium sulfate.
3. Dissolve in 1900 ml water and add a stir bar.
4. Adjust the pH to 5.9 by adding a few drops of 10 M NaOH.
5. Add 40 g of agar and 900 ml of water to two 2 l flasks.
6. Autoclave the four flasks for 45 min.
7. Transfer the contents of each agar flask to one media flask and mix well.
8. Cool to 55 °C and keep in a water bath until ready to pour.
9. Add 100 ml of autoclaved 40% glucose (w/v).
10. Add the required concentrated stock solutions, and 3AT or CHX as needed.
11. Pour approximately 100 ml in 15-cm sterile Petri plates.
12. Dry for 5–7 days at room temperature then store indefinitely at 4 °C. If the plates need to be used earlier, they can be dried for 30 min in a sterile hood with ventilation on.

Amino acid powder mix and stock solutions All amino acids that are never used as prototrophic markers are combined in a amino acid mix that is added to all Sc plates.

To prepare the amino acid powders:

1. Mix 6 g of each of the following amino acids: alanine, arginine, aspartic acid, asparagine, cysteine, glutamic acid, glutamine, glycine, isoleucine, lysine, methionine, phenylalanine, proline, serine, threonine, tyrosine, and valine.
2. For the amino acid powder containing adenine, add 6 g of adenine sulfate.

Tryptophan, histidine, leucine, uracil, and adenine are omitted so they can be added to batches of plates as needed. The concentrated stock solutions are used at 8 ml/l of media, except for adenine which is used at 15 ml/l of media. The different stock solutions are prepared at the following concentrations: 100 mM histidine–HCl (store light protected), 100 mM leucine, 65 mM adenine sulfate, and 40 mM tryptophan. These stock solutions are stored at room temperature, except for tryptophan, which should be stored in the dark at 4 °C.

4. VALIDATION USING ORTHOGONAL BINARY INTERACTION ASSAYS

Complementary assays are essential to assess the precision of a dataset against PRS and RRS (see 2.2.1.). The following complementary assays can be used to determine the precision of a dataset by testing a random sample, and as part of an interaction assay tool-kit for confidence scoring of individual interactions (Braun *et al.*, 2009). We describe the yellow fluorescent protein (YFP) based protein complementation assay (Nyfeler *et al.*, 2005) and the sandwich ELISA-like well-NAPPA protein interaction assay (Braun *et al.*, 2009). All expression constructs for these methods can be assembled using Gateway recombinational cloning or other high-throughput cloning methods.

Protocol 14: Yellow fluorescent protein complementation assay (YFP-PCA)

In YFP–PCA, two nonfluorescent fragments of YFP (F1 and F2) are genetically attached to ORFs coding for the two proteins that are to be tested in this assay. If the two proteins interact functional YFP can be reconstituted and detected by fluorescence-activated cell sorting (FACS). In this protocol a cyan fluorescent protein (CFP) coding plasmid is cotransfected as a transfection control.

[Day 0: Seed cells, measure DNA]

1. In a 96-well tissue culture plate, seed CHO-K1 cells at 6×10^4 cells/well in 100 μl Ham's F12 media + 10% fetal calf serum. After 24 h, confluence should reach 70%.
2. Determine the concentration of the expression plasmids with PicoGreen assay (Invitrogen) or related assay.

[Day 1: Transfection]

1. Replace growth media on cells with 100 μl Opti-MEM media (Invitrogen) equilibrated to 37 °C.
2. Combine 30 ng of each PCA construct with 140 ng CFP plasmid for a total of 200 ng DNA in 25 μl Opti-MEM media per well to obtain the DNA mix.
3. Combine 0.5 μl Lipofectamine 2000 reagent (Invitrogen) with 25 μl Opti-MEM media per well to obtain the transfection reagent mix.
4. Incubate 5–25 min at room temperature.
5. Combine the DNA and transfection reagent mixes to yield 50 μl transfection mix.
6. Incubate for at least 20 min (not longer than 6 h).
7. Add transfection mix to the cells.
8. Incubate for 18 h.

[Day 3: FACS Analysis]

1. Wash cells three times gently with phosphate-buffered saline (PBS).
2. Add 20 μl trypsin.
3. Incubate $\sim$10 min at room temperature until cells are detached.
4. Resuspend in 100 μl PBS.
5. Analyze cells by FACS.

Count a minimum of 10,000 events. Gate for CFP positive cells and analyze YFP fluorescence only in this subpopulation. Discard any result that is supported by less than 200 cells or if the CFP transfection rate is unacceptably low (<5%). On every FACS instrument the voltages and gates need to be calibrated using YFP and CFP controls. The best criteria for scoring positive interactions should be identified using a large enough set of controls (at least one plate worth of each PRS and RRS). After such a calibration, score a pair positive if at least 30% of CFP positive cells are YFP positive and if the average YFP signal is above background and if the YFP/CFP ratio was at least twice as high as the ratio of the average YFP signal over the average CFP ratio on that plate. Calibrate gating of the instrument by using full-length YFP and CFP constructs. Scoring parameters must be recalibrated on PRS/RRS data for each implementation.

Protocol 15: Well-nucleic acid programmable protein array (wNAPPA)

In well-NAPPA, the two proteins are genetically fused to a glutathione-S-transferase tag and to an HA epitope tag respectively and expressed in a coupled transcription/translation reticulocyte lysate. The GST-tagged protein (GST-X) is captured using an anti-GST antibody that is immobilized at the bottom of a 96-well microtiter plate. If the two proteins are interacting, this interaction can be detected with an anti-HA antibody. Like all assays, this biochemical pull-down assay from *in vitro* coupled transcription–translation needs to be calibrated against PRS and RRS datasets to evaluate performance.

[Day 0: Blocking]

1. Add 200 μl/well blocking buffer (5% (w/v) fat-free dry milk powder dissolved in PBS prepared according to standard protocols) to a microtiter plated coated with rabbit anti-GST antibody (GST 96-well Detection Module, GE Healthcare).
2. Block at 4 °C for 14–24 h.

[Day 1: wNAPPA assay]

1. Determine the DNA concentration of expression plasmids using PicoGreen or a similar assay.
2. Add 0.5–1 μg of each of the two plasmids to complete reticulocyte lysate reaction mix (25 μl) (TnT Coupled Transcription/Translation System, Promega).
3. Incubate for 1.5 h at 30 °C on a shaker.
4. Dilute the reaction mix with 100 μl/well blocking solution.
5. Transfer the diluted reaction mix to the prepared anti-GST coated plate.

6. Incubate at 15 °C on a shaker for 2 h.
7. Discard reaction mix and wash three times with 200 μl blocking buffer for 5 min.
8. Add 150 μl anti-HA monoclonal antibody (Cell Signaling Technologies) 1:5000 in blocking buffer.
9. Wash three times with 150 μl with blocking buffer for 5 min each.
10. Add horseradish peroxidase (HRP) coupled goat anti-mouse antibody (Amersham) 1:1000–1:2000 in blocking buffer.
11. Wash three times with 150 μl PBS for 5 min.
12. Develop with 100 μl enhanced chemiluminescence (ECL) reagent like Pierce PicoWest ECL reagent. Alternatively a colorimetric HRP substrate will give similar results.
13. Chemiluminescence is measured with a Biorad molecular imager gel doc system, but measurement could also be done with a 96-well plate spectrophotometer reader.

 ## 5. CONCLUSION

Information on interactome networks constitutes a critical element of systems biology. We have spelled out a general approach to high-quality interactome mapping in which a reliable high-throughput assay is used as a primary screening platform. Subsequently, alternative validation assays are used to demonstrate data quality in a way unprejudiced by preconceived ideas and biases about what protein interactions are supposed to look like. To produce high-quality data, appropriate controls need to be implemented at every stage of a binary interactome mapping pipeline, including thorough controls for technical artifacts and subsequent experimental determination of the quality of interactome network maps. Experimental validation of primary screening data ensures data quality unbiased by current scientific perceptions and hence of greatest utility for exploring biology.

Use of this general framework of interactome mapping, the main features of which are stringent removal of technical artifacts and experimental control of data quality, will enable production of high-quality datasets.

ACKNOWLEDGMENTS

We thank past and current members of the Vidal Lab and the Center for Cancer Systems Biology (CCSB) for their help and constructive discussions over the course of developing our binary interaction mapping strategies, framework, and protocols. This work was supported by National Human Genome Research Institute grants R01-HG001715 awarded to M.V. and D.E.H and P50-HG004233 awarded to M.V., grant DBI-0703905 from the National Science Foundation to M. V. and D. E. H., and by Institute Sponsored Research funds from the Dana-Farber Cancer Institute Strategic Initiative awarded to M. V. and CCSB. D. M. is supported by grant LSHG-CT-2006-037704 from the 6th Framework Program of the European Commission. M. V. is a "Chercheur Qualifié Honoraire" from the Fonds de la Recherche Scientifique (FRS-FNRS, French Community of Belgium).

REFERENCES

Bader, J. S., Chaudhuri, A., Rothberg, J. M., and Chant, J. (2004). Gaining confidence in high-throughput protein interaction networks. *Nat. Biotechnol.* **22,** 78–85.

Braun, P., Tasan, M., Dreze, M., Barrios-Rodiles, M., Lemmens, I., Yu, H., Sahalie, J. M., Murray, R. R., Roncari, L., De Smet, A.-S., Venkatesan, K., Rual, J.-F., *et al.* (2009). An experimentally derived confidence score for binary protein–protein interactions. *Nat. Methods* **6,** 91–97.

Collins, S. R., Kemmeren, P., Zhao, X. C., Greenblatt, J. F., Spencer, F., Holstege, F. C., Weissman, J. S., and Krogan, N. J. (2007a). Towards a comprehensive atlas of the physical interactome of *Saccharomyces cerevisiae*. *Mol. Cell Proteomics* **6,** 439–450.

Collins, S. R., Miller, K. M., Maas, N. L., Roguev, A., Fillingham, J., Chu, C. S., Schuldiner, M., Gebbia, M., Recht, J., Shales, M., Ding, H., Xu, H., *et al.* (2007b). Functional dissection of protein complexes involved in yeast chromosome biology using a genetic interaction map. *Nature* **446,** 806–810.

Cusick, M. E., Yu, H., Smolyar, A., Venkatesan, K., Carvunis, A. R., Simonis, N., Rual, J.-F., Borick, H., Braun, P., Dreze, M., Vandenhaute, J., Galli, M., *et al.* (2009). Literature-curated protein interaction datasets. *Nat. Methods* **6,** 39–46.

Deane, C. M., Salwinski, L., Xenarios, I., and Eisenberg, D. (2002). Protein interactions: Two methods for assessment of the reliability of high throughput observations. *Mol. Cell Proteomics* **1,** 349–356.

De Nicolo, A., Parisini, E., Zhong, Q., Dalla Palma, M., Stoeckert, K. A., Domchek, S. M., Nathanson, K. L., Caligo, M. A., Vidal, M., Cusick, M. E., and Garber, J. E. (2009). Multimodal assessment of protein functional deficiency supports pathogenicity of BRCA1 p.V1688del. *Cancer Res.* **69,** 7030–7037.

Dreze, M., Charloteaux, B., Milstein, S., Vidalain, P.-O., Yildirim, M. A., Zhong, Q., Svrzikapa, N., Romero, V., Laloux, G., Brasseur, R., Vandenhaute, J., Boxem, M., *et al.* (2009). 'Edgetic' perturbation of a *C. elegans* BCL-2 ortholog. *Nat. Methods* **7,** 843–849.

Durfee, T., Becherer, K., Chen, P. L., Yeh, S. H., Yang, Y., Kilburn, A. E., Lee, W. H., and Elledge, S. J. (1993). The retinoblastoma protein associates with the protein phosphatase type 1 catalytic subunit. *Genes Dev.* **7,** 555–569.

Endoh, H., Vincent, S., Jacob, Y., Real, E., Walhout, A. J., and Vidal, M. (2002). Integrated version of reverse two-hybrid system for the postproteomic era. *Methods Enzymol.* **350,** 525–545.

Fields, S., and Song, O. (1989). A novel genetic system to detect protein–protein interactions. *Nature* **340,** 245–246.

Ge, H., Liu, Z., Church, G. M., and Vidal, M. (2001). Correlation between transcriptome and interactome mapping data from *Saccharomyces cerevisiae*. *Nat. Genet.* **29,** 482–486.

Ge, H., Walhout, A. J., and Vidal, M. (2003). Integrating 'omic' information: A bridge between genomics and systems biology. *Trends Genet.* **19,** 551–560.

Gunsalus, K. C., Ge, H., Schetter, A. J., Goldberg, D. S., Han, J. D., Hao, T., Berriz, G. F., Bertin, N., Huang, J., Chuang, L. S., Li, N., Mani, R., *et al.* (2005). Predictive models of molecular machines involved in *Caenorhabditis elegans* early embryogenesis. *Nature* **436,** 861–865.

Gyuris, J., Golemis, E., Chertkov, H., and Brent, R. (1993). Cdi1, a human G1 and S phase protein phosphatase that associates with Cdk2. *Cell* **75,** 791–803.

Han, J. D., Bertin, N., Hao, T., Goldberg, D. S., Berriz, G. F., Zhang, L. V., Dupuy, D., Walhout, A. J., Cusick, M. E., Roth, F. P., and Vidal, M. (2004). Evidence for dynamically organized modularity in the yeast protein–protein interaction network. *Nature* **430,** 88–93.

Ito, T., Chiba, T., Ozawa, R., Yoshida, M., Hattori, M., and Sakaki, Y. (2001). A comprehensive two-hybrid analysis to explore the yeast protein interactome. *Proc. Natl. Acad. Sci. USA* **98,** 4569–4574.

James, P., Halladay, J., and Craig, E. A. (1996). Genomic libraries and a host strain designed for highly efficient two-hybrid selection in yeast. *Genetics* **144,** 1425–1436.

Jeong, H., Mason, S. P., Barabasi, A. L., and Oltvai, Z. N. (2001). Lethality and centrality in protein networks. *Nature* **411,** 41–42.

Lamesch, P., Milstein, S., Hao, T., Rosenberg, J., Li, N., Sequerra, R., Bosak, S., Doucette-Stamm, L., Vandenhaute, J., Hill, D., and Vidal, M. (2004). *C. elegans* ORFeome version 3.1: Increasing the coverage of ORFeome resources with improved gene predictions. *Genome Res.* **14,** 2064–2069.

Lamesch, P., Li, N., Milstein, S., Fan, C., Hao, T., Szabo, G., Hu, Z., Venkatesan, K., Bethel, G., Martin, P., Rogers, J., Lawlor, S., *et al.* (2007). hORFeome v3.1: A resource of human open reading frames representing over 10, 000 human genes. *Genomics* **89,** 307–315.

Miller, J. P., Lo, R. S., Ben-Hur, A., Desmarais, C., Stagljar, I., Noble, W. S., and Fields, S. (2005). Large-scale identification of yeast integral membrane protein interactions. *Proc. Natl. Acad. Sci. USA* **102,** 12123–12128.

Milo, R., Shen-Orr, S., Itzkovitz, S., Kashtan, N., Chklovskii, D., and Alon, U. (2002). Network motifs: Simple building blocks of complex networks. *Science* **298,** 824–827.

Nyfeler, B., Michnick, S. W., and Hauri, H. P. (2005). Capturing protein interactions in the secretory pathway of living cells. *Proc. Natl. Acad. Sci. USA* **102,** 6350–6355.

Oliver, S. (2000). Guilt-by-association goes global. *Nature* **403,** 601–603.

Pujana, M. A., Han, J. D., Starita, L. M., Stevens, K. N., Tewari, M., Ahn, J. S., Rennert, G., Moreno, V., Kirchhoff, T., Gold, B., Assmann, V., ElShamy, W. M., *et al.* (2007). Network modeling links breast cancer susceptibility and centrosome dysfunction. *Nat. Genet.* **39,** 1338–1349.

Reboul, J., Vaglio, P., Rual, J.-F., Lamesch, P., Martinez, M., Armstrong, C. M., Li, S., Jacotot, L., Bertin, N., Janky, R., Moore, T., Hudson, J. R. Jr., *et al.* (2003). *C. elegans* ORFeome version 1.1: Experimental verification of the genome annotation and resource for proteome-scale protein expression. *Nat. Genet.* **34,** 35–41.

Rual, J.-F., Hirozane-Kishikawa, T., Hao, T., Bertin, N., Li, S., Dricot, A., Li, N., Rosenberg, J., Lamesch, P., Vidalain, P. O., Clingingsmith, T. R., Hartley, J. L., *et al.* (2004). Human ORFeome version 1.1: A platform for reverse proteomics. *Genome Res.* **14,** 2128–2135.

Rual, J.-F., Venkatesan, K., Hao, T., Hirozane-Kishikawa, T., Dricot, A., Li, N., Berriz, G. F., Gibbons, F. D., Dreze, M., Ayivi-Guedehoussou, N., Klitgord, N., Simon, C., *et al.* (2005). Towards a proteome-scale map of the human protein–protein interaction network. *Nature* **437,** 1173–1178.

Simonis, N., Rual, J.-F., Carvunis, A.-R., Tasan, M., Lemmons, I., Hirozane-Kishikawa, T., Hao, T., Sahalie, J. M., Venkatesan, K., Gebreab, F., Cevik, S., Klitgord, N., *et al.* (2009). Empirically controlled mapping of the *Caenorhabditis elegans* protein–protein interactome network. *Nat. Methods* **6,** 47–54.

Tarassov, K., Messier, V., Landry, C. R., Radinovic, S., Molina, M. M., Shames, I., Malitskaya, Y., Vogel, J., Bussey, H., and Michnick, S. W. (2008). An *in vivo* map of the yeast protein interactome. *Science* **320,** 1465–1470.

Venkatesan, K., Rual, J.-F., Vazquez, A., Stelzl, U., Lemmens, I., Hirozane-Kishikawa, T., Hao, T., Zenkner, M., Xin, X., Goh, K. I., Yildirim, M. A., Simonis, N., et al. (2009). An empirical framework for binary interactome mapping. Nat. Methods 6, 83–90.

Vidal, M. (2001). A biological atlas of functional maps. Cell 104, 333–339.

Vidal, M. (2005). Interactome modeling. FEBS Lett. 579, 1834–1838.

Vidal, M., and Endoh, H. (1999). Prospects for drug screening using the reverse two-hybrid system. Trends Biotechnol. 17, 374–381.

Vidal, M., Brachmann, R. K., Fattaey, A., Harlow, E., and Boeke, J. D. (1996). Reverse two-hybrid and one-hybrid systems to detect dissociation of protein–protein and DNA-protein interactions. Proc. Natl. Acad. Sci. USA 93, 10315–10320.

Vidalain, P. O., Boxem, M., Ge, H., Li, S., and Vidal, M. (2004). Increasing specificity in high-throughput yeast two-hybrid screens. Methods 32, 363–370.

Walhout, A. J., and Vidal, M. (1999). A genetic strategy to eliminate self-activator baits prior to high-throughput yeast two-hybrid screens. Genome Res. 9, 1128–1134.

Walhout, A. J., and Vidal, M. (2001). High-throughput yeast two-hybrid assays for large-scale protein interaction mapping. Methods 24, 297–306.

Walhout, A. J., Temple, G. F., Brasch, M. A., Hartley, J. L., Lorson, M. A., van den Heuvel, S., and Vidal, M. (2000). GATEWAY recombinational cloning: Application to the cloning of large numbers of open reading frames or ORFeomes. Methods Enzymol. 328, 575–592.

Yu, H., Braun, P., Yildirim, M. A., Lemmens, I., Venkatesan, K., Sahalie, J., Hirozane-Kishikawa, T., Gebreab, F., Li, N., Simonis, N., Hao, T., Rual, J.-F., et al. (2008). High-quality binary protein interaction map of the yeast interactome network. Science 322, 104–110.

Zhong, Q., Simonis, N., Li, Q.-R., Charloteaux, B., Heuze, F., Klitgord, N., Tam, S., Yu, H., Venkatesan, K., Mou, D., Swearingen, V., Yildirim, M. A., et al. (2009). Edgetic perturbation models of human genetic disorders. Mol. Syst. Biol. 5, 321.

Zhu, H., Bilgin, M., Bangham, R., Hall, D., Casamayor, A., Bertone, P., Lan, N., Jansen, R., Bidlingmaier, S., Houfek, T., Mitchell, T., Miller, P., et al. (2001). Global analysis of protein activities using proteome chips. Science 293, 2101–2105.

CHAPTER THIRTEEN

QUANTITATIVE ANALYSIS OF PROTEIN PHOSPHORYLATION ON A SYSTEM-WIDE SCALE BY MASS SPECTROMETRY-BASED PROTEOMICS

Bernd Bodenmiller*,[1] *and* Ruedi Aebersold*,[†,‡]

Contents

Abstract

Systems biology at the molecular level is concerned with networks of interacting molecules, their structure, and dynamic response to perturbations that give rise to systems' properties that determine measurable, macroscopic phenotypes. At any time, in any cell, multiple types of molecular networks are concurrently active.

One of the most important known regulatory systems in eukaryotic cells is reversible protein phosphorylation catalyzed by protein kinases and phosphatases, respectively. Therefore, it is essential to understand and eventually model the protein phosphorylation-mediated informational fluxes in cells from sensors and signaling systems to effector molecules, to comprehensively analyze the dynamic system of kinases/phosphatases and their substrates and to determine the basic rules of information processing in cells. In this chapter, we describe the protocols necessary to comprehensively and quantitatively measure the phosphorylation-modulated informational networks in cells.

* Institute of Molecular Systems Biology, ETH Zurich, Zurich, Switzerland
† Institute for Systems Biology, Seattle, Washington, USA
‡ Faculty of Science, University of Zurich, Zurich, Switzerland
[1] Current address: Department of Microbiology and Immunology, Stanford University Stanford, California, USA

Methods in Enzymology, Volume 470
ISSN 0076-6879, DOI: 10.1016/S0076-6879(10)70013-6

The pipeline relies on the selective, quantitative isolation of phosphopeptides generated by the tryptic digestion of complex protein mixtures and their subsequent mass spectrometric and computational analysis.

We believe that the protocols and data processing tools described in this chapter will be a valuable resource for biologists interested in the analysis of protein phosphorylation-based signal transduction.

1. INTRODUCTION

Reversible phosphorylation of proteins, carried out by kinases and phosphatases, constitutes one of the most important regulatory mechanisms in eukaryotic cells. Kinases, phosphatases, and their substrates form a network that controls and processes the flow of information, from sensors via signaling relays to effector molecules, thereby regulating processes like cellular growth, cell division, or apoptosis (Hunter, 2000).

In yeast over 150 kinases and phosphatases are currently known, but our knowledge of their cellular roles, and especially their substrates, is still very sparse (Hunter and Plowman, 1997; Ptacek et al., 2005). This is illustrated by the fact that only for several hundreds of the $\sim$2000 phosphoproteins, with their over 10,000 phosphorylation sites, the upstream kinases or phosphatases are known in vivo (Ptacek et al., 2005).

To thoroughly comprehend cellular processes and their adaptation to stimuli, it is of crucial importance to investigate the connections between kinases, phosphatases, and their in vivo responders (proteins that change their phosphorylation status based on the activity of a given kinase or phosphatase). The elucidation of these connections constitutes one of the major biological questions which, if addressed, would open new avenues for biological and medical research (Hunter, 2000; Tan et al., 2009).

Over the last decade, several phosphoproteomic techniques that allow elucidating these connections have emerged. They are based on the reproducible and highly specific isolation of phosphopeptides from the tryptic digests of the proteome, their analyses using liquid chromatography–tandem mass spectrometry (LC–MS/MS), and the subsequent analysis of such data using computational tools (Aebersold and Goodlett, 2001; Aebersold and Mann, 2003; Gruhler et al., 2005; Olsen et al., 2006; Tao et al., 2005; Zhou et al., 2001).

However, such analyses are still far from being routine. First, signaling proteins are often of low abundance, and to complicate things only a fraction of a given protein may be phosphorylated at a given time, making the detection of the corresponding phosphorylation sites very challenging (Salih, 2004).

Second, it is often difficult to preserve phosphorylated amino acid residues as phosphorylations on amino groups and acyl-phosphorylations are extremely acid labile, while phosphorylations on hydroxyl groups are base sensitive. Therefore, special precautions need to be taken when processing the samples (Salih, 2004).

Third, changes in the state of the phosphoproteome are highly dynamic; therefore, it is essential to preserve the cellular phosphorylation state of interest (Gruhler et al., 2005; Olsen et al., 2006). We found, for example, that the widely applied washing of yeast cells with ice-cold, low molarity buffers alone triggers both the starvation and the osmotic shock responses, while centrifuging yeasts triggers their stress response, all changing the phosphoproteome. Fourth and finally, challenges associated to the MS-based analysis of protein phosphorylation still exists, among them the impaired fragmentation of phosphopeptides under collision-induced dissociation (CID) in ion trap instruments or their nonstandard chromatographic behavior (Macek et al., 2009).

We have developed an MS approach based on label-free quantification that allows to comprehensively, quantitatively, and reproducibly monitoring a significant fraction of the phosphoproteome in yeast ($> 10,000$ phosphorylation sites on > 2000 phosphoproteins). We have used this approach to successfully detect changes in the phosphoproteome upon the inhibition of the Tor kinase using a specific inhibitor, rapamycin (Huber et al., 2009), and to determine the first phosphorylation network of yeast, connecting most kinases, phosphatases with their responders in vivo (Bodenmiller et al., submitted).

The approach presented here (for an overview see Fig. 13.1), even though optimized for yeast, is with minor adaptations generally applicable to detect and quantify changes in the phosphoproteome of any organism, with high throughput and low effort, upon any stimulus of interest.

In the following, we use the comparison between yeast wild-type cells and yeast kinase gene deletion mutant cells to illustrate our workflow.

2. PROTOCOLS

2.1. Generation of peptide samples

For any phosphoproteomic approach it is essential to preserve the in vivo phosphorylation state of the cells. For that purpose we established a method in which trichloroacetic acid (TCA) is added directly into the growth medium to final 6% before the yeast harvest (Urban et al., 2007). TCA rapidly enters the cells and first, lowers the pH to ~ 1.5 and thereby protonates all proteins (including kinases and phosphatases) and second,

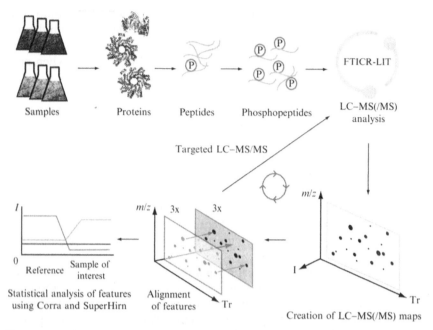

Figure 13.1 An experimental workflow to quantify changes in the phosphoproteome. It consists of the following steps: (i) triplicate cultures of the reference and cell state of interest are grown; (ii) phosphopeptides are highly selectively isolated from the corresponding proteome digests; (iii) for each phosphopeptide isolate, LC–MS(/MS) maps (also called phosphorylation patterns) are generated; (iv) the phosphoproteome patterns are compared and correlated using the algorithms *SuperHirn* (Mueller *et al.*, 2007) and *Corra* (Brusniak *et al.*, 2008). This analysis yields phosphopeptide features which are significantly regulated upon the stimulus of interest; (v) if significantly regulated phosphopeptide ions could not be annotated with an amino acid sequence in (iii), they are reanalyzed using targeted LC–MS/MS; (vi) a list of phosphopeptides displaying a significant abundance change between the reference cells and cells of interest is generated for further analyses.

denatures them, thus quenching all enzymatic activity and preserving the cellular phosphorylation state *in vivo* (Urban *et al.*, 2007).

To perform the subsequent phosphopeptide isolation, ~ 3 mg of cellular protein lysate are needed. Such amount can be easily isolated from ~ 50 ml of a yeast culture grown to OD_{600} 0.8–1.0. In case system-wide phospho-peptide quantification is performed, it is strongly recommended to grow biological triplicates, allowing to compute solid statistical significances (Brusniak *et al.*, 2008). Finally, we propose to use synthetic defined (SD) medium to culture yeast as, first, the exact composition of the medium can be controlled and, second, protein contaminations deriving from YPD medium can be avoided.

2.1.1. Medium and growth conditions

Saccharomyces cerevisiae strains are streaked out on appropriate plates and three replicates (from three single colonies), both of the wild-type and the gene deletion strain, are grown in a 1-ml preculture over night at 30 °C in SD medium (per liter: 1.7 g yeast nitrogen base without amino acids (Chemie Brunschwig, Basel, Switzerland), 5 g ammonium sulfate, 2% glucose (w/v), 0.03 g isoleucine, 0.15 g valine, 0.04 g adenine, 0.02 g arginine, 0.02 g histidine, 0.1 g leucine, 0.03 g lysine, 0.02 g methionine, 0.05 g phenylalanine, 0.2 g threonine, 0.04 g tryptophan, and 0.03 g tyrosine). The preculture is then used to inoculate 50 ml of SD medium in a 500-ml shaking flask to an OD_{600} of 0.05.

2.1.2. Harvest of the yeast cells

At an OD_{600} of 0.8, 100% (w/v) TCA is added to a final concentration of 6% directly into the yeast cultures (Urban *et al.*, 2007). The flasks containing the cultures are put for 10 min into ice water (all subsequent steps are performed at 4 °C). The cultures are then transferred into 50 ml tubes and the yeasts are pelleted by centrifugation at $1500 \times g$. The supernatant is discarded and the cells are resuspended in 10 ml ice-cold acetone (at this step the yeast form aggregates, resembling snowflakes). Then the cells are again pelleted at $1500 \times g$ and the supernatant is discarded. This step is repeated once and then the yeast pellet is transferred to a 2-ml safe-lock (important!) Eppendorf tube and can be stored at -80 °C until further processing.

2.1.3. Lysis of yeast cells and peptide generation

After removing residual acetone by pipetting, 800 μl of a solution consisting of 8 M urea, 50 mM ammonium bicarbonate and 5 mM EDTA is added to each yeast pellet (3 replicates $\times$ 2 strains). In addition, acid-washed glass beads (500 μm diameter) are added to a final volume of 1.5 ml. Then cells are lyzed by beat beating (five times beating for 1 min, with 1 min breaks on ice between consecutive beatings). After that step, the cell debris is pelleted by centrifugation at $16{,}000 \times g$ at room temperature and the supernatant is collected. Again 800 μl of the urea solution is added and the bead beating procedure described above is repeated. Then the protein concentration of the pooled supernatants of each replicate is determined using a BCA protein assay kit (Pierce, Rockford, IL, USA). For each replicate, 3 mg of protein is reduced for 30 min by 5 mM tris(2-carboxyethyl)phosphine (TCEP) and alkylated for 1 h using 10 mM iodoacetamide. After diluting the urea solution with 50 mM ammonium bicarbonate to < 1.5 M, trypsin is added in a ratio of 1:125 (w/w) and the digestion is performed for 8 h at 37 °C (each step (reduction, alkylation, and digestion) is very insensitive to the total volume and have been successfully performed in several microliters up to 100 ml).

2.1.4. Purification of peptides

Prior to the phosphopeptide isolation, peptides have to be cleaned using reverse phase chromatography. First, the pH of the digestion mixture is lowered to pH 2.5 using 10–100% trifluoroacetic acid (TFA). Of note: TFA (and also TCA) solutions > 1% must always be stored in glass containers but never in plastics as otherwise high amounts of polymers are generated which impede the mass spectrometry measurements. For the same reason TFA/ TCA solutions > 1% must always be added with a glass syringe. Second, an appropriate reverse phase column (e.g., C18 Sep-Pak with 500 mg resin, Waters, Milford, MA, USA) is wetted with acetonitrile and equilibrated using 0.1% TFA, 2% acetonitrile. After the peptides have been bound to the resin in the columns, they are washed with 0.1% TFA, 2% acetonitrile, and the peptides are eluted using 0.1% TFA, 40% acetonitrile. Finally, the eluent is dried in a vacuum concentrator to completion.

2.2. Phosphopeptide isolation

The methods commonly used for the selective isolation of phosphopeptides can be grouped into affinity based and chemical methods. In the following, three different protocols are described: the first one is based on immobilized metal affinity chromatography (IMAC) using Fe(III) metal ions (Andersson and Porath, 1986; Bodenmiller et al., 2007a; Ficarro et al., 2002), the second one, which is also based on affinity chromatography, uses a titanium dioxide (TiO_2) resin (Andersson and Porath, 1986; Bodenmiller et al., 2007b; Larsen et al., 2005; Pinkse et al., 2004) and the third one uses phosphoramidate chemistry (PAC) (Bodenmiller et al., 2007c; Tao et al., 2005; Zhou et al., 2001).

Each of the methods, if correctly used, is highly specific and reproducible for the isolation of phosphopeptides, and therefore suitable for label-free quantification, however, isolating distinct yet overlapping parts of the phosphoproteome. Therefore, it is advantageous to apply all of the methods for a given sample to increase the covered phosphoproteome (Bodenmiller et al., 2007b). In addition, the following specific strengths and weaknesses exist. First, for both affinity-based methods the under loading of the resin will result in low specificity of phosphopeptide isolation. We found that PAC is more tolerant against this event (Bodenmiller et al., 2007b). Second, both TiO_2 and PAC have a strong bias toward singly phosphorylated phosphopeptides, while in the case of IMAC the amount of peptide loaded per resin will influence if singly or multiply phosphorylated peptides are isolated (Bodenmiller et al., 2007b; Thingholm et al., 2008, 2009). Third, of the three methods PAC is the most reproducible while IMAC is the least one. Fourth and finally, in terms of practicability, TiO_2 is fast, straight forward, and tolerant to impurities and experimental errors. Therefore, we propose to start with that method. However, also IMAC and PAC are easy to master and only need half a day more to perform.

For all methods the correct pH at different steps in the protocol is crucial for the success and reproducibility of the isolations. In addition, for the affinity chromatography-based methods the optimum peptide to resin ratio is essential for their specificity (Bodenmiller et al., 2007b).

2.2.1. Phosphopeptide isolation using a TiO_2 resin

Three milligrams of dried peptides are reconstituted in 280 μl of a washing solution (WS) consisting of 80% acetonitrile, 3.5% TFA which is saturated with phthalic acid ($\sim$100 mg phthalic acid/ml, precipitate must be visible at the bottom of the vial after vigorous shaking). Then 1.25 mg TiO_2 (GL Science, Saitama, Japan) resin is placed into a 1-ml Mobicol spin column (MoBiTec, Göttingen, Germany) and is subsequently washed with 280 μl water, 280 μl methanol, and finally is equilibrated with 280 μl WS for at least 10 min (the liquid is always removed by centrifugation using 500×g).

After removal of the WS, the peptide solution is added to the equilibrated TiO_2 in the blocked Mobicol spin column and is incubated for >30 min with end-over-end rotation. After the incubation step the peptide solution is removed by centrifugation, and the resin is thoroughly washed two times each with 280 μl of the WS, with a 80% acetonitrile, 0.1% TFA solution, and finally with 0.1% TFA. In the final step, phosphopeptides are eluted from the TiO_2 resin using two times 150 μl of a 0.3-M NH_4OH solution (pH $\sim$ 10.5). After elution, the pH of the pooled eluents is rapidly adjusted to 2.7 using 10% TFA, and phosphopeptides are purified using an appropriate reverse phase column suitable for up to 20 μg peptide (e.g., C18 Micro Spin Column, Nest Group or Harvard Apparatus, MA, USA; see above for the detailed protocol).

2.2.2. Phosphopeptide isolation using IMAC

Three milligrams of peptides are reconstituted in 280 μl of a WS consisting of 250 mM acetic acid with 30% acetonitrile at pH 2.7. Then 60 μl of uniformly suspended PHOS-Select iron affinity gel (Sigma Aldrich), corresponding to $\sim$30 μl resin, is placed into a 1-ml Mobicol spin column (MoBiTec) using a cut pipette tip. The resin is washed and thereby equilibrated by using three times 280 μl of the WS (the liquid is always removed by centrifugation at 500×g). After removal of the WS the peptide solution is added to the equilibrated IMAC resin in the blocked Mobicol spin column. To obtain reproducible results it is crucial that the pH in all replicate samples is maintained at $\sim$2.5. The affinity gel is then incubated with the peptide solution for 120 min with end-over-end rotation. After the incubation, the liquid is removed by centrifugation and the resin is thoroughly washed two times with 280 μl of the WS, and once with ultrapure water. In the final step, phosphopeptides are eluted once using 150 μl of a 50-mM phosphate buffer (pH 8.9) and once using 150 μl of a 100-mM phosphate buffer (pH 8.9), each time incubating the resin <3 min with the elution buffer. Both elutes are pooled, the pH is rapidly adjusted to 2.7 using 10% TFA, and phosphopeptides are purified using an appropriate reverse phase column (C18 Micro Spin Column, Nest Group or Harvard Apparatus; see above for the detailed protocol).

2.2.3. Phosphopeptide isolation using phosphoramidate chemistry

In the first reaction step, the peptides are methyl esterified. Here it is essential to use water free reagents and to work under dry conditions, otherwise unwanted side reactions can take place (He *et al.*, 2004). In detail, 3 mg of dried peptide are reconstituted in 1.5 ml of methanolic HCl which was prepared by slowly adding 240 μl of acetyl chloride to 1.5 ml of anhydrous methanol (careful—strong heat development). The methyl esterification is then allowed to proceed at 12 °C for 120 min. The solvent is quickly removed in a cool vacuum concentrator and peptide methyl esters are dissolved in 120 μl methanol, 120 μl water, and 240 μl acetonitrile.

Using glass beads as a solid phase (see Fig. 13.2): 750 μl of a solution containing 50 mM N-(3-dimethylaminopropyl)-N'-ethylcarbodiimide (EDC), 100 mM imidazole (pH 5.6), 100 mM 2-(N-morpholino)ethanesulfonic acid (MES) (pH 5.6), and 2 M cystamine are added to the peptide solution (Bodenmiller *et al.*, 2007b).

Of note: for the success of the reaction the correct pH is crucial. It is, therefore, strongly recommended to check the pH of the final reaction solution containing the peptides using a micro-pH electrode and, if necessary, adjust the pH to 5.5–6.0 (Bodenmiller *et al.*, 2007c).

The reaction is allowed to proceed at room temperature with vigorous shaking for 8 h. The solution is then loaded onto an appropriate reverse phase column (C18 Sep-Pak with 500 mg resin, Waters) and the derivatized peptides are, first, washed with 5 ml 0.1% TFA, second, treated with 3 ml 10 mM TCEP (pH should be adjusted to ~3 prior to loading using sodium hydroxide (NaOH)) for 8 min, in order to produce free thiol groups, third, washed again with 5 ml 0.1% TFA to remove residual TCEP, and fourth and finally, the derivatized peptides are eluted with 1.6 ml 80% acetonitrile, 0.1% TFA, and the pH is adjusted to 6.0 using phosphate buffer pH 8.9.

The acetonitrile/water solution is partially removed in the vacuum concentrator to a final volume of ~500 μl and the derivatized phosphopeptides are incubated with 5 mg maleimide functionalized-glass beads for 1 h at pH 6.2 in a Mobicol column. (The beads are synthesized by dissolving 120 μmol hydroxybenzotriazole, 120 μmol of 3-maleimidopropionic acid, and 120 μmol diisopropylcarbodiimide in 1 ml of dry dimethylformamide

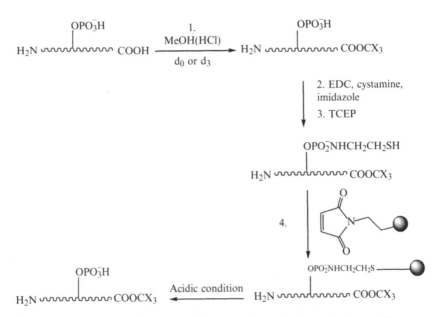

Figure 13.2 Schematic illustration of the procedure to isolate phosphopeptides using phosphoramidate chemistry. It consists of the following steps: First, the carboxylate groups of the peptides are protected using methyl esterification. Second, the phosphate groups are derivatized with cystamine. Third, the cystamine is reduced using TCEP, thereby generating free thiol groups. Forth, the derivatized phosphopeptides are coupled to a maleimide derivatized solid phase (e.g., amino propyl containing glass beads). Fifth and finally, nonphosphopeptides are removed by washing the solid phase and phosphopeptides are released from the solid phase using acidic conditions (from Bodenmiller et al., 2007c).

completely. After 30 min incubation, 100 mg CPG beads (Proligo Biochemie, Hamburg, Germany) corresponding to 40 μmol free amino groups are added for 90 min. After the reaction they are washed using dimethylformamide and dried using a vacuum concentrator. Beads are stored dry at 4 °C.)

Then the derivatized beads are washed two times sequentially with 300 μl 3 M NaCl, water, methanol and, finally, with 80% acetonitrile to remove nonspecifically bound peptides. In the last step, the beads are incubated with 300 μl 5% TFA, 30% acetonitrile for 1 h to recover the phosphopeptides. The recovered sample is dried in the vacuum concentrator and is reconstituted in an appropriate solvent for the LC–MS(/MS) analysis.

Using amino-derivatized dendrimer as a solid phase: For the phosphoramidate reaction, dissolve the peptide methyl esters in 600 μl of reaction solution (50 mM EDC, 100 mM imidazole, 100 mM MES, and 1 M

PAMAM dendrimer Generation 5 (Dendritech Inc., Midland, MI, USA), pH 5.5), and incubate at the room temperature with strong shaking for 18–24 h (Tao *et al.*, 2005).

To remove unspecifically bound peptides after the reaction, transfer the reaction solution into a membrane filter device (molecular mass cut-off of 5000) in which the membrane side is inward or perpendicular. Then wash the dendrimer three times with 2 M NaCl, 2 M NaCl in 50% methanol, and three times with 50% methanol (for all steps mix well and make sure that the dendrimer is dissolved) always using an appropriate amount of the solvents that allow for vigorous vortexing. Discard the flow through to remove nonspecifically attached/bound nonphosphopeptides, and incubate the dendrimer with 300 μl 5% TFA for 1 h to break the phosphoramidate bonds and thereby recovering the phosphopeptides.

Spin down and collect the eluent. Add 150–300 μl 50% methanol to the dendrimer, mix well, and spin down to pool the filtrates. The recovered sample is dried in the vacuum concentrator and is reconstituted in an appropriate solvent for the LC–MS(/MS) analysis.

2.3. Mass spectrometric analyses of the phosphopeptide isolates

As the phosphopeptide quantification described in this chapter is based on a label-free approach, it is important that the mass spectrometric measurements are performed on high mass accuracy, high-resolution instruments such as the hybrid LTQ-Orbitrap, LTQ-FT, or quadrupole time-of-flight (QTOF) and that the chromatographic retention time is highly reproducible.

Ideally, the phosphopeptides are separated using an LC system employing a nanoflow (200–300 nl/min) acetonitrile/water gradient using a reverse phase resin. As phosphopeptides are on average more hydrophilic than their nonphosphorylated counterparts, it is advisable to, first, use C18 beads with 3 μm diameter or less, as the retention and separation of phosphopeptides is increased, and second, to adopt the acetonitrile/water gradient normally used for the corresponding nonphosphopeptides, for example, by starting with 2% and ending with 24% acetonitrile over 60–90 min and by doubling the used formic acid concentration to 0.2%.

An example LC setup consists of an Eksigent nano LC system (Eksigent Technologies, Dublin, CA, USA), equipped with a 11-cm fused silica emitter, 75 μm inner diameter (BGB Analytik, Böckten, Switzerland), packed with Magic C18 AQ, 3 μm beads, loaded from a cooled (4 °C) Spark Holland auto sampler.

Example MS settings to analyze phosphopeptides on an LTQ-Orbitrap-MS are the following: each MS1 scan is acquired at 60,000 FWHM nominal resolution settings, with an overall cycle time of $\sim$1.2 s. For injection control, the automatic gain control (AGC) is set to 5×10^5 ions for a full Orbitrap-MS (1×10^4 ions for the linear ion trap MS/MS). The instrument is calibrated externally, according to the manufacturer's instructions. To ensure a high mass accuracy, the internal lock mass calibration at m/z 429.088735 and 445.120025 should be used.

Often peptides phosphorylated on a serine or threonine residue exhibit a loss of phosphoric acid in ion trap-based CID MS2 spectra, severely impairing the phosphopeptide backbone fragmentation, thereby reducing the ability of database searching algorithms to unambiguously identify the phosphopeptide (see Fig. 13.3). To address this problem several strategies have been developed. The first one is called MS/MS/MS (MS3) experiment (Beausoleil et al., 2004). Here the dominant neutral loss fragment ion from the MS2 measurement is resubjected to CID, and the resulting MS3 spectrum is used to identify the phosphopeptide (see Fig. 13.3). The second strategy is called multistage activation (MSA or "pseudo MS3") (Schroeder et al., 2004). Here the neutral loss ion is collisionally activated while the fragments from the precursor ion are still present in the ion trap. As a result, the spectrum contains ions from the precursor fragmentation and the neutral loss product (hence "pseudo MS3"). The main advantages of this method are that, first, a higher number of fragment ion spectra compared to the MS3 strategy can be recorded in the same time and, second, that the subsequent data processing is facilitated compared to the MS2/MS3 strategy (see below). Finally, the last method to fragment phosphopeptides in ion trap instruments was recently developed and is based on electron transfer dissociation (ETD) (Syka et al., 2004). Here radical anions are used to transfer electrons to multiply charged peptides stored in the ion trap. As a result, the peptide backbone is fragmented by a nonergodic process, yielding extensive fragmentation of the phosphopeptide backbone. Particularly, important for the analysis of phosphopeptides is the fact that the phosphate group remains attached to the phosphopeptide during the ETD fragmentation process, facilitating the unambiguous site assignment.

As the phosphopeptides identified by CID and ETD only partially overlap, it is also recommended to use both methods if available, to increase the number of identified phosphopeptides (Molina et al., 2007).

Example settings for the different fragmentation methods are as following: for all three of them, the 3–6 most intense phosphopeptide ions identified in the MS1 measurement are typically selected for fragmentation (importantly, the MS2 measurements should not delay the recording of the MS1 spectra). For the CID-based methods (MS2, MS3, and MSA), the isolation width of the parental ion is set to 2 m/z, the normalized collision energy to 35, the activation Q to 0.25 and the activation time to 30–100 ms. For the MS3 and MSA experiments, the neutral loss of -98 Da for singly, -49 Da for doubly, -32.7 Da for triply, and -24.5 Da for quadruply

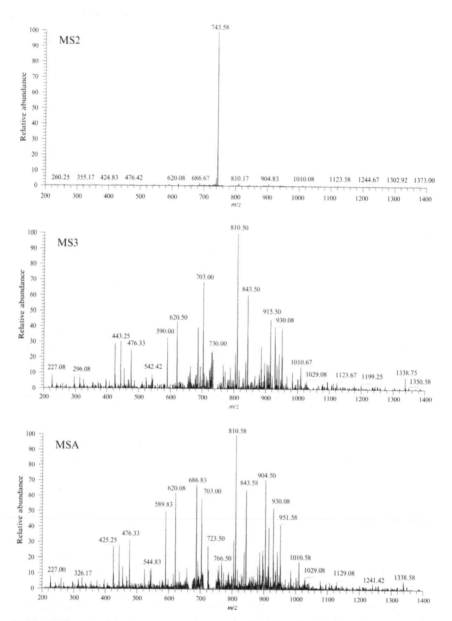

Figure 13.3 Phosphopeptide fragmentation spectra. The MS2, MS3, and MSA spectra of the phosphopeptide with the parental mass at *m/z* 767.89 are shown.

charged peptides must be defined to trigger the additional fragmentation step. Typical settings for an ETD experiment performed in an ion trap are analogous to the CID methods except that many more ions are collected in the ion trap prior to fragmentation (80,000 vs. 30,000).

2.4. Data analyses

2.4.1. Database search

To annotate the phosphopeptide fragmentation spectra with the corresponding peptide sequences, sequences database searches must be performed (Eng et al., 1994). Currently, a wide variety of algorithms exist, all with specific strengths and weaknesses, but it has been shown that the Sequest (Eng et al., 1994) and Mascot (Perkins et al., 1999) algorithms perform particularly well in case of phosphopeptides (Bakalarski et al., 2007).

After the database search has been completed, it is necessary to estimate the false positive rate of phosphopeptide identifications. For such analyses, either the statistical tool PeptideProphet (Keller et al., 2002, 2005) or a decoy database can be used (Elias and Gygi, 2007).

To perform the database search, the MS data must be converted to the centroid mzXML or mzML format (http://tools.proteomecenter.org/software.php). For yeast, the data is typically searched against a decoy version of the SGD nonredundant database (www.yeastgenome.org).

General settings for the database search are as follows: for the *in silico* digest of the SGD database, trypsin is defined as protease, cleaving after K and R (if followed by P, the cleavage is not allowed). As phosphorylation sites close to an R or K impair the tryptic cleavage, at least two missed cleavages should be allowed in addition to one nontryptic terminus. The peptide mass is set to 600–6000 Da (if ETD data is searched, the mass range is extended to 12,000). The precursor ion tolerance is set to <20 ppm (depending on the MS instrument used), and the fragment ion tolerance is set to 0.5–0.8 Da. In addition, cysteine is defined with a fixed carboxyamidomethylation modification (+57.0214 Da), and the phosphate group must be defined as a variable modification of serine/threonine (MS2, ETD, MSA +79.9663 Da; MS3 −18.01528) and tyrosine (MS2, MSA, MS3, ETD +79.9663 Da). Finally, y and b fragment ions are defined for all CID-based data, while c and z fragment ions are defined for ETD.

The results obtained from the database search are then subjected to statistical filtering using PeptideProphet (http://tools.proteomecenter.org/software.php), and a PeptideProphet cut-off of 0.9 is typically set in order to classify peptides as correctly or incorrectly identified. Alternatively, the false discovery rate (FDR) can be determined using the decoy entries of the decoy database (typically cut-offs between 1% and 5% are used).

2.4.2. Quantification

Two points should be kept in mind for the label-free quantification (Mueller et al., 2007, 2008; Rinner et al., 2007). First, an essential point throughout the entire experimental pipeline is reproducibility, in terms of the generation of the phosphopeptide isolates and their analyses using LC–MS (e.g., retention time and mass accuracy). Second, if subtle regulatory events shall be identified (e.g., 50% upregulation of a phosphorylation site), more than three biological triplicates should be processed and analyzed, to increase the sensitivity of the statistical analyses. If higher precision is required stable isotope labeling-based quantification methods might be preferably applied.

A typical data analysis workflow applied in label-free quantification consists of the following steps (Mueller *et al.*, 2007, 2008; Rinner *et al.*, 2007). First, the MS data is converted into the profile mzXML format (http://tools.proteomecenter.org/software.php). Second, phosphopeptide ion peaks are detected ("feature detection"). Third, the phosphopeptide ion peak areas are computed based on the LC–MS data. Fourth, the peaks are aligned over the analyzed LC–MS runs. Fifth, each peak is annotated with the phosphopeptide sequence and, sixth, the ratio between the phosphopeptide ions derived of the wild-type and kinase gene deletion mutant is computed.

A widely used and freely available tool to perform these tasks is called SuperHirn. Typical parameters used for the analyses of LTQ-Orbitrap data, their description and the download of the tool can be found under the link http://tools.proteomecenter.org/wiki/index.php?title=Software:Super Hirn.

Of note, the peptide identifications must be imported into the SuperHirn analysis using the output format (.pepXML) of the PeptideProphet.

2.4.3. Computation of statistical significances of observed regulations

Based on the SuperHirn output file (called MasterMap), which contains all detected phosphopeptide ions derived from the triplicate wild-type samples and the triplicate mutant samples along with the sequence annotation of the identified phosphopeptides, the significance of an observed abundance variation is computed. Before the significance analysis, the phosphopeptides in the MasterMap are separated into different statistical classes. First, a class consisting either of phosphopeptides for which the MS1 signal was detected in all replicates (three times wild-type and three times the mutant samples) or of phosphopeptides for which 1–5 signals are missing. In the category "3 signals missing" the phosphopeptides are further separated between (i) the signal is reproducibly present in either all wild-type or all mutant samples and (ii) the signal is spread over all wild-type and mutant samples.

Before statistical analysis, the missing data values are computed using the integrated background noise given by the LC–MS analysis and determined using SuperHirn as a baseline. These datasets are then further analyzed using the freely available software tool called Corra (Brusniak *et al.*, 2008) (http://tools.proteomecenter.org/Corra/corra.html). It wraps around the Limma software package (http://www.bioconductor.org/) and performs an Empirical Bayes in alternative to the Welsh *t*-test, yielding a *p*-value that is further adjusted for multiple comparisons, according to the procedure by

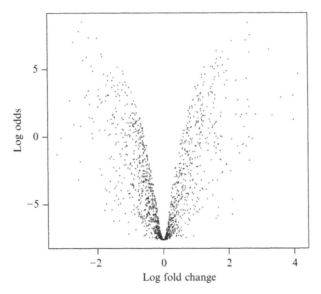

Figure 13.4 Example volcano plot as generated by the Corra analysis. Each dot corresponds to a measured phosphopeptide ion. The y-axis shows the Log Odds that the observed regulations are true and the x-axis the change of a phosphopeptide ion in the $\log_2$ scale.

Benjamini and Hochberg (1995) which controls the FDR. After this last analysis, the phosphopeptides which significantly change their abundance due to a changed activity of a kinase or phosphatase can be identified (see Fig. 13.4).

These data can then be used, for example, to infer in which biological processes a given kinase is involved or which signaling pathways are affected.

Overall, we found that such generated data is a strong and solid basis for follow up, in depth, experiments.

ACKNOWLEDGMENTS

Many past and present members of the Aebersold lab have contributed to the development of this protocol. We are especially grateful to Lukas Müller who programmed SuperHirn, Mi-Youn Brusniak who programmed Corra and Paola Picotti, and Alexander Schmidt who developed the targeted proteomics approach. We also want to thank Paola Picotti and Stephania Vaga for proof reading the chapter. This project has been funded in part by ETH Zurich, by Federal funds from the National Heart, Lung, and Blood Institute, National Institutes of Health, under contract no. N01-HV-28179, and by the PhosphoNetX project of SystemsX.ch, the Swiss initiative for systems biology. Bernd Bodenmiller was supported by the Boehringer Ingelheim Fellowship.

REFERENCES

Aebersold, R., and Goodlett, D. R. (2001). Mass spectrometry in proteomics. *Chem. Rev.* **101,** 269–295.

Aebersold, R., and Mann, M. (2003). Mass spectrometry-based proteomics. *Nature* **422,** 198–207.

Andersson, L., and Porath, J. (1986). Isolation of phosphoproteins by immobilized metal (Fe^{3+}) affinity-chromatography. *Anal. Biochem.* **154,** 250–254.

Bakalarski, C. E., Haas, W., Dephoure, N. E., and Gygi, S. P. (2007). The effects of mass accuracy, data acquisition speed, and search algorithm choice on peptide identification rates in phosphoproteomics. *Anal. Bioanal. Chem.* **389,** 1409–1419.

Beausoleil, S. A., Jedrychowski, M., Schwartz, D., Elias, J. E., Villen, J., Li, J. X., Cohn, M. A., Cantley, L. C., and Gygi, S. P. (2004). Large-scale characterization of HeLa cell nuclear phosphoproteins. *Proc. Natl. Acad. Sci. USA* **101,** 12130–12135.

Benjamini, Y., and Hochberg, Y. (1995). Controlling the false discovery rate—A practical and powerful approach to multiple testing. *J. R. Stat. Soc. Ser. B-Methodological* **57,** 289–300.

Bodenmiller, B., Malmstrom, J., Gerrits, B., Campbell, D., Lam, H., Schmidt, A., Rinner, O., Mueller, L. N., Shannon, P. T., Pedrioli, P. G., Panse, C., Lee, H. K., et al. (2007a). PhosphoPep–A phosphoproteome resource for systems biology research in *Drosophila* Kc167 cells. *Mol. Syst. Biol.* **3,** 139.

Bodenmiller, B., Mueller, L. N., Mueller, M., Domon, B., and Aebersold, R. (2007b). Reproducible isolation of distinct, overlapping segments of the phosphoproteome. *Nat. Methods* **4,** 231–237.

Bodenmiller, B., Mueller, L. N., Pedrioli, P. G., Pflieger, D., Junger, M. A., Eng, J. K., Aebersold, R., and Tao, W. A. (2007c). An integrated chemical, mass spectrometric and computational strategy for (quantitative) phosphoproteomics: Application to *Drosophila melanogaster* Kc167 cells. *Mol. Biosyst.* **3,** 275–286.

Bodenmiller, B., Wanka, S., Kraft, C., Urbano, J., Campbell, D., Gerrits, B., Picotti, P., Lam, H., Vitek, O., Brusniak, M. Y., Schlapbach, R., Shokat, K., et al. (submitted). An In Vivo Protein Phosphorylation Network of Budding Yeast.

Brusniak, M. Y., Bodenmiller, B., Campbell, D., Cooke, K., Eddes, J., Garbutt, A., Lau, H., Letarte, S., Mueller, L. N., Sharma, V., Vitek, O., Zhang, N., et al. (2008). Corra: Computational framework and tools for LC–MS discovery and targeted mass spectrometry-based proteomics. *BMC Bioinformatics* **9,** 542.

Elias, J. E., and Gygi, S. P. (2007). Target-decoy search strategy for increased confidence in large-scale protein identifications by mass spectrometry. *Nat. Methods* **4,** 207–214.

Eng, J. K., McCormack, A. L., and Yates, J. R. (1994). An approach to correlate tandem mass spectral data of peptides with amino acid sequences in a protein database. *J. Am. Soc. Mass Spectrom.* **5,** 976–989.

Ficarro, S. B., McCleland, M. L., Stukenberg, P. T., Burke, D. J., Ross, M. M., Shabanowitz, J., Hunt, D. F., and White, F. M. (2002). Phosphoproteome analysis by mass spectrometry and its application to *Saccharomyces cerevisiae*. *Nat. Biotechnol.* **20,** 301–305.

Gruhler, A., Olsen, J. V., Mohammed, S., Mortensen, P., Faergeman, N. J., Mann, M., and Jensen, O. N. (2005). Quantitative phosphoproteomics applied to the yeast pheromone signaling pathway. *Mol. Cell. Proteomics* **4,** 310–327.

He, T., Alving, K., Feild, B., Norton, J., Joseloff, E. G., Patterson, S. D., and Domon, B. (2004). Quantitation of phosphopeptides using affinity chromatography and stable isotope labeling. *J. Am. Soc. Mass Spectrom.* **15,** 363–373.

Huber, A., Bodenmiller, B., Uotila, A., Stahl, M., Wanka, S., Gerrits, B., Aebersold, R., and Loewith, R. (2009). Characterization of the rapamycin-sensitive phosphoproteome reveals that Sch9 is a central coordinator of protein synthesis. *Genes Dev.* **23**, 1929–1943.

Hunter, T. (2000). Signaling—2000 and beyond. *Cell* **100**, 113–127.

Hunter, T., and Plowman, G. D. (1997). The protein kinases of budding yeast: Six score and more. *Trends Biochem. Sci.* **22**, 18–22.

Keller, A., Nesvizhskii, A. I., Kolker, E., and Aebersold, R. (2002). Empirical statistical model to estimate the accuracy of peptide identifications made by MS/MS and database search. *Anal. Chem.* **74**, 5383–5392.

Keller, A., Eng, J., Zhang, N., Li, X. J., and Aebersold, R. (2005). A uniform proteomics MS/MS analysis platform utilizing open XML file formats. *Mol. Syst. Biol.* **1**(2005), 0017.

Larsen, M. R., Thingholm, T. E., Jensen, O. N., Roepstorff, P., and Jorgensen, T. J. (2005). Highly selective enrichment of phosphorylated peptides from peptide mixtures using titanium dioxide microcolumns. *Mol. Cell. Proteomics* **4**, 873–886.

Macek, B., Mann, M., and Olsen, J. V. (2009). Global and site-specific quantitative phosphoproteomics: Principles and applications. *Annu. Rev. Pharmacol. Toxicol.* **49**, 199–221.

Molina, H., Horn, D. M., Tang, N., Mathivanan, S., and Pandey, A. (2007). Global proteomic profiling of phosphopeptides using electron transfer dissociation tandem mass spectrometry. *Proc. Natl. Acad. Sci. USA* **104**, 2199–2204.

Mueller, L. N., Rinner, O., Schmidt, A., Letarte, S., Bodenmiller, B., Brusniak, M. Y., Vitek, O., Aebersold, R., and Muller, M. (2007). SuperHirn—A novel tool for high resolution LC–MS-based peptide/protein profiling. *Proteomics* **7**, 3470–3480.

Mueller, L. N., Brusniak, M. Y., Mani, D. R., and Aebersold, R. (2008). An assessment of software solutions for the analysis of mass spectrometry based quantitative proteomics data. *J. Proteome Res.* **7**, 51–61.

Olsen, J. V., Blagoev, B., Gnad, F., Macek, B., Kumar, C., Mortensen, P., and Mann, M. (2006). Global, in vivo, and site-specific phosphorylation dynamics in signaling networks. *Cell* **127**, 635–648.

Perkins, D. N., Pappin, D. J., Creasy, D. M., and Cottrell, J. S. (1999). Probability-based protein identification by searching sequence databases using mass spectrometry data. *Electrophoresis* **20**, 3551–3567.

Pinkse, M. W. H., Uitto, P. M., Hilhorst, M. J., Ooms, B., and Heck, A. J. R. (2004). Selective isolation at the femtomole level of phosphopeptides from proteolytic digests using 2D-nanoLC–ESI-MS/MS and titanium oxide precolumns. *Anal. Chem.* **76**, 3935–3943.

Ptacek, J., Devgan, G., Michaud, G., Zhu, H., Zhu, X., Fasolo, J., Guo, H., Jona, G., Breitkreutz, A., Sopko, R., McCartney, R. R., Schmidt, M. C., *et al.* (2005). Global analysis of protein phosphorylation in yeast. *Nature* **438**, 679–684.

Rinner, O., Mueller, L. N., Hubalek, M., Muller, M., Gstaiger, M., and Aebersold, R. (2007). An integrated mass spectrometric and computational framework for the analysis of protein interaction networks. *Nat. Biotechnol.* **25**, 345–352.

Salih, E. (2004). Phosphoproteomics by mass spectrometry and classical protein chemistry approaches. *Mass Spectrom. Rev.* 2005, Nov-Dec, **24**(6), 828–846. Review.

Schroeder, M. J., Shabanowitz, J., Schwartz, J. C., Hunt, D. F., and Coon, J. J. (2004). A neutral loss activation method for improved phosphopeptide sequence analysis by quadrupole ion trap mass spectrometry. *Anal. Chem.* **76**, 3590–3598.

Syka, J. E. P., Coon, J. J., Schroeder, M. J., Shabanowitz, J., and Hunt, D. F. (2004). Peptide and protein sequence analysis by electron transfer dissociation mass spectrometry. *Proc. Natl. Acad. Sci. USA* **101**, 9528–9533.

Tan, C. S., Bodenmiller, B., Pasculescu, A., Jovanovic, M., Hengartner, M. O., Jorgensen, C., Bader, G. D., Aebersold, R., Pawson, T., and Linding, R. (2009).

Comparative analysis reveals conserved protein phosphorylation networks implicated in multiple diseases. *Sci. Signal.* **2,** ra39.

Tao, W. A., Wollscheid, B., O'Brien, R., Eng, J. K., Li, X. J., Bodenmiller, B., Watts, J. D., Hood, L., and Aebersold, R. (2005). Quantitative phosphoproteome analysis using a dendrimer conjugation chemistry and tandem mass spectrometry. *Nat. Methods* **2,** 591–598.

Thingholm, T. E., Jensen, O. N., Robinson, P. J., and Larsen, M. R. (2008). SIMAC (sequential elution from IMAC), a phosphoproteomics strategy for the rapid separation of monophosphorylated from multiply phosphorylated peptides. *Mol. Cell. Proteomics* **7,** 661–671.

Thingholm, T. E., Jensen, O. N., and Larsen, M. R. (2009). Enrichment and separation of mono- and multiply phosphorylated peptides using sequential elution from IMAC prior to mass spectrometric analysis. *Methods Mol. Biol.* **527,** 67–78, xi.

Urban, J., Soulard, A., Huber, A., Lippman, S., Mukhopadhyay, D., Deloche, O., Wanke, V., Anrather, D., Ammerer, G., Riezman, H., Broach, J. R., De Virgilio, C., *et al.* (2007). Sch9 is a major target of TORC1 in *Saccharomyces cerevisiae*. *Mol. Cell* **26,** 663–674.

Zhou, H. L., Watts, J. D., and Aebersold, R. (2001). A systematic approach to the analysis of protein phosphorylation. *Nat. Biotechnol.* **19,** 375–378.

CHAPTER FOURTEEN

A Toolkit of Protein-Fragment Complementation Assays for Studying and Dissecting Large-Scale and Dynamic Protein–Protein Interactions in Living Cells

Stephen W. Michnick,* Po Hien Ear,* Christian Landry,[†] Mohan K. Malleshaiah,* and Vincent Messier*

Contents

* Département de Biochimie, Université de Montréal, C.P. 6128, Succursale Centre-ville, Montréal, Québec, Canada
† Département de Biologie, Institut de Biologie Intégrative et des Systèmes, PROTEO-Québec Research Network on Protein Function, Structure and Engineering, Pavillon Charles-Eugène-Marchand, Université Laval, Québec, Canada

Methods in Enzymology, Volume 470
ISSN 0076-6879, DOI: 10.1016/S0076-6879(10)70014-8

Abstract

Protein-fragment complementation assays (PCAs) are a family of assays for detecting protein–protein interactions (PPIs) that have been developed to provide simple and direct ways to study PPIs in any living cell, multicellular organism or *in vitro*. PCAs can be used to detect PPI between proteins of any molecular weight and expressed at their endogenous levels. Proteins are expressed in their appropriate cellular compartments and can undergo any posttranslational modification or degradation that, barring effects of the PCA fragment fusion, they would normally undergo. Applications of PCAs in yeast have been limited until recently, simply because appropriate expression plasmids or cassettes had not been developed. However, we have now developed and reported on several PCAs in *Saccharomyces cerevisiae* that cover the gamut of applications one could envision for studying any aspect of PPIs. Here, we present detailed protocols for large-scale analysis of PPIs with the survival-selection dihydrofolate reductase (DHFR) reporter PCA and a new PCA based on a yeast cytosine deaminase reporter that allows for both survival and death selection. This PCA should prove a powerful way to dissect PPIs. We then present a method to study spatial localization and dynamics of PPIs based on fluorescent protein reporter PCAs and finally, two luciferase reporter PCAs that have proved useful for studies of dynamics of PPIs.

1. INTRODUCTION

In the protein-fragment complementation assay (PCA) strategy, protein–protein interactions (PPIs) are measured by fusing each of the proteins of interest to complementary N- or C-terminal peptides of a reporter protein that has been *rationally* dissected using protein engineering strategies (Michnick *et al.*, 2000; Pelletier and Michnick, 1997; Pelletier *et al.*, 1998). The reporter protein fragments are brought into proximity by interaction of the two interacting proteins, allowing them to fold together into the three-dimensional structure of the reporter protein, thus reconstituting the activity of the reporter (Fig. 14.1). PCAs have been created with many different reporter proteins and thus provide for different types of readouts, depending on the desired application. This generality means that PCA is not a single reporter assay, but rather a toolkit. PCAs have also been developed to study spatial and temporal changes in PPIs under different conditions and also survival-selection assays that provide a simple readout for large-scale systematic analyses of protein interaction networks or directed evolution experiments (reviewed in Michnick *et al.*, 2007). Finally, there are two unique features of PCAs we must note: first, by nature of the fact that interactions between two proteins must occur in such a way that the reporter protein can fold, PCAs can provide structural and topological

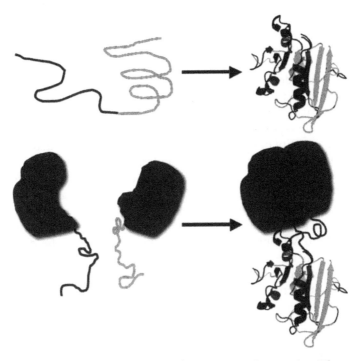

Figure 14.1 Conceptual basis of protein-fragment complementation. The spontaneous unimolecular folding of a protein from its nascent polypeptide (upper) can be made a protein–protein interaction-dependent bimolecular process by fusing two interacting proteins to one or the other complementary N- or C-terminal peptides into which a protein has been dissected (lower). PPI-mediated folding of a reporter protein from its complementary fragments results in reconstitution of reporter protein activity.

details of how a PPI is formed or if such complexes undergo conformation changes under specific conditions (Remy *et al.*, 1999; Tarassov *et al.*, 2008). Second, contrary to intuition, most PCAs are fully reversible, allowing for direct studies of the dynamics of both formation and disruption of PPIs.

 2. GENERAL CONSIDERATIONS IN USING PCA

Measuring PPI in living cells by any method entails that one reconsider any suppositions that we may have about the nature of a PPI, particularly if it has only been studied with *in vitro* methods and most importantly by indirect methods such as affinity or immunopurification. PCAs detect direct binary or indirect proximal interactions between proteins and thus, if it is assumed that there is such an interaction based on experiments that only suggest association of proteins in a complex, it is possible that no interaction will be detected. Our advice is: "life is short, experiment." However, we can make some general statements about what to consider when setting up any PCA experiment in order to maximize the probability of a successful outcome.

First, we consider the sensitivity of PCAs. Like any analytical technique, the sensitivity of the assay depends on the sensitivity of the detection method and background signal that may arise from cells. Regardless of the properties of the reporters, the range of signal detectable will depend in all cases on the quantity of complexes formed, which in turn is determined by the abundances of the proteins studied and their affinity for each other. We have only explored these parameters in great detail for the dihydrofolate reductase (DHFR) PCA (Section 3). We have demonstrated that for this simple survival-selection assay, the number of complexes needed to support survival under the selection conditions was as low as approximately 25 per cell (Remy and Michnick, 1999) for a complex for which the dissociation constant was in the range of 1 nM. We recently showed that we could generalize this result across a proteome, demonstrating that the distribution of detected interactions covered the range of protein abundances down to less than 100 molecules per cell (Tarassov *et al.*, 2008). We have also shown that an upper limit of the dissociation constant for detection of PPI is likely in the range of 10 μM–100 μM for the DHFR (Campbell-Valois *et al.*, 2005) and OyCD (Section 4) PCAs (Ear and Michnick, 2009). These observations suggest that PPI can be detected by PCA within ranges of protein abundances and complex affinities that are commonly observed. However, PPI may or may not bedetected depending on the PCA reporter used. For instance, a PPI studied with a fluorescent protein-PCA reporters (Section 5) might not be detected if the abundance of complexes is lower than necessary to reconstitute enough fluorescent proteins. In this case, signal will not be high enough to overcome background fluorescence of cells in the range of wavelengths over which the fluorophore emits. On the other hand, there are no background issues for luciferase-based PCAs (Section 6) and thus detection is limited only by the sensitivity of the detector used. Finally, an issue of particular importance to studies in yeast where the complementary PCA fragments are fused to gene open reading frames (ORFs) by homologous recombination is whether the genes are hetero- or homozygous for the fusions in diploid cells. In this case, the untagged proteins (A and B) will compete for binding with those that are tagged (A$'$ and B$'$), resulting in a reduced number of reconstituted PCA reporter proteins and thus, reporter signal. Only the A$'$B$'$ complex (out of the four possible AB, AB$'$, A$'$B, and A$'$B$'$) results in a reconstituted PCA reporter protein, leading to a fourfold reduction in signal. The number of reconstituted complexes necessary for signal detection in assays performed in diploid cells (Tarassov *et al.*, 2008) is, therefore, much lower than what is expected for the abundances of the interacting partners alone.

A second set of considerations in using PCA is how the fusion of complementary PCA reporter fragments could affect the proteins of interest and the ability to detect PPI. First, as with any fusion construct, it is critical to test the fusions in established functional assays in order to assure that the tags themselves do not impair the function of the protein or lead to gain of function. One should also not assume that a functional fusion protein with a particular tag ensures that other PCA tags will lead to functional fusions. Different tags may have different effects. Second, we can ask if the orientation of fusion (N- or C-terminus) or identity of the fragment may affect the outcome of a PCA experiment. This can only be determined empirically. We have tested all possible combinations and permutations of tagging individual test proteins that are known to interact (8 total per protein pair) and found that in some cases it made no difference how the proteins were tagged while for others, only an individual arrangement worked (unpublished results).

As we described above, PCAs are sensitive to whether the complementary N- and C-terminal fragments can find each other in space and this depends on the distances between the termini of the interacting proteins to which the fragments are fused. To assure that PCA can occur, we typically insert a 10–15 amino acid flexible polypeptide linker consisting of the sequences (Gly.Gly.Gly.Gly.Ser)$_n$ between the proteins of interest and the PCA reporter protein fragments. We chose the (Gly.Gly.Gly.Gly.Ser)$_n$ linker because it is the most flexible possible and we have empirically observed that linkers of these lengths are sufficiently long to allow for fragments to find each other and fold, regardless of the sizes of the interacting proteins to which the fragments are fused (Remy and Michnick, 2001).

3. DHFR PCA SURVIVAL-SELECTION FOR LARGE-SCALE ANALYSIS OF PPIs

The DHFR PCA was previously developed for *Escherichia coli*, plant protoplasts and mammalian cell lines (Pelletier *et al.*, 1998, 1999; Remy and Michnick, 1999; Subramaniam *et al.*, 2001) and has recently been adapted for large-scale screening of PPIs in yeast (Tarassov *et al.*, 2008). The principle of the DHFR PCA survival-selection assay is that cells lacking endogenous DHFR activity, here achieved by inhibiting the *Saccharomyces cerevisiae* scDHFR with methotrexate, are enabled to proliferate by simultaneously expressing PCA fragments of a methotrexate-resistant DHFR mutant that are fused to interacting proteins or peptides. If the proteins interact and thus allow refolding of the DHFR reporter, cells that are grown in the presence of methotrexate can proliferate (Fig. 14.2) (Remy and Michnick, 1999). To adapt the DHFR PCA for high-throughput screening

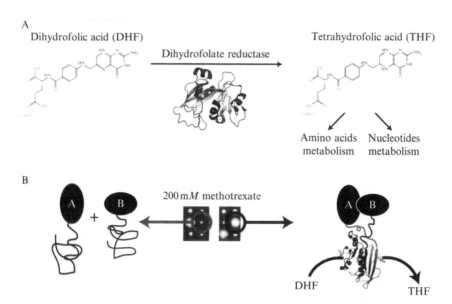

Figure 14.2 (A) DHFR catalyzes the reduction of dihydrofolate to tetrahydrofolate, which is required for nucleotide and in some organisms, amino acid synthesis. This reaction can be inhibited by an antifolate, methotrexate. (B) In the DHFR PCA strategy, the two proteins of interest are fused to complementary fragments of a mutant DHFR that is insensitive to methotrexate. The PCA fragments are inactive in the absence of an interaction. If the proteins interact, the DHFR fragments are brought together in space and fold into the native structure, thus reconstituting the activity of the mutant DHFR and allowing cells to proliferate in the presence of methotrexate.

in *S. cerevisiae,* we created a double mutant (L22F and F31S) that is 10,000 times less sensitive to methotrexate than wild-type scDHFR, while retaining full catalytic activity (Ercikan-Abali *et al.*, 1996). The assay can be used with strains harboring yeast expression vectors of the target protein ORF fused to PCA fragment coding sequence. It is also sensitive enough to be used with genomic recombinant strains, expressing proteins fused PCA fragment under the control of their endogenous promoters. We created two universal oligonucleotide cassettes encoding each complementary DHFR PCA fragment and two unique antibiotic resistance enzymes to allow for selection of haploid strains that have been successfully transformed and recombined with one or the other homologous recombination cassettes (Tarassov *et al.*, 2008). The resulting universal templates were used to create homologous recombination cassettes for most budding yeast genes by PCR using 5′ and 3′ oligonucleotides consisting of 40-nucleotide sequences homologous to the 3′ end of each ORF (prior to the stop codon) and a region approximately 20 nucleotides from the stop codon. Below are protocols to perform DHFR PCA at a large scale with recombinant strains or with yeast transformed with expression plasmids.

3.1. Materials

3.1.1. Reagents

- Glycerol stocks of *MATa* recombinant strains in which ORFs are fused to the complementary DHFR PCA F[1,2] fragment (Open Biosystems).
- Glycerol stocks of *MATα* recombinant strains in which ORFs are fused to the complementary DHFR F[3] PCA fragment (Open Biosystems).
- 3% agar solidified YPD medium in Nunc omniplates.
- 3% agar solidified YPD medium with 100 µg/ml nourseothricin for *MATa* recombinant strains (WERNER BioAgents, Jena, Germany) or 250 µg/ml hygromycin B for *MATα* recombinant strains (Wisent Corporation, Quebec, Canada) in omniplate.
- 3% agar solidified YPD medium with both 100 µg/ml nourseothricin (WERNER BioAgents) and 250 µg/ml hygromycin B (Wisent Corporation) in omniplate.
- 4% noble agar (purified Agar, Bioshop) solidified synthetic complete (SC) medium with 200 µg/ml methotrexate (prepared from a 10-mg/ml methotrexate in DMSO stock solution) in omniplate.

3.1.2. Facultative

Antibodies against DHFR fragments: anti-DHFR polyclonal antibody that specifically recognizes an epitope in the N-terminal F[1,2] fragment (Sigma D1067, 1:6000; Sigma-Aldrich, St. Louis, MO) and an anti-DHFR polyclonal antibody that specifically recognizes an epitope in the C-terminal F[3] fragment (Sigma D0942, 1:5000; Sigma-Aldrich).

3.1.3. Equipment

- *Pintool*: Robotically manipulated (96 pintool) (0.910 mm flat round-shaped pins, AFIX96FP4, V&P Scientific Inc., San Diego, CA), 384 pintool (0.356 mm flat round-shaped pins, custom AFIX384FP8 BMP Multimek FP8N, V&P Scientific Inc.) and a 1536-pintool (0.229 mm flat round-shaped pins, custom AFIX1536FP9 BMP Multimek FP9N, V&P Scientific Inc.) or manually manipulated (96 pintool) (1.58 mm, 1 µl slot pins, 45 mm, VP 408Sa, V&P Scientific Inc.).
- *Plate imaging*: At least a 4.0-mega pixel camera (Powershot A520, Canon), a stationary arm (70 cm mini repro, Industria Fototecnica Firenze, Italy) and a plate-shooting platform.

3.2. Procedure

The general strategy for performing a screen is to generate an array of "prey" strains as indexed colonies grown in a regular grid on agar and then mate them with individual "bait" strains of the opposite mating type to

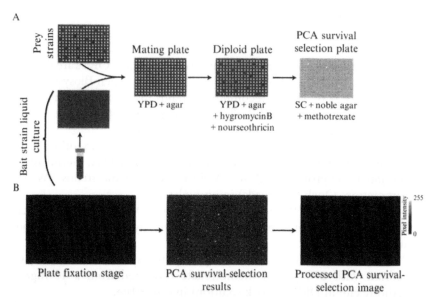

Figure 14.3 (A) The DHFR PCA screen is performed as show in this schematic. The bait reporter strain is incubated in liquid culture. The prey reporter strains are printed on solid medium and incubated to be used on multiple assay plates. The mating plate is produced by sequentially printing the bait strain and the prey strains on sold agar containing rich medium, allowing strains to mate and grow. Resulting haploids and diploid mixture strains are transferred to solid agar plates containing diploid selective medium. The resulting diploid strains can be transferred onto plates containing PCA survival-selection medium (containing methotrexate). (B) The resulting PCA survival selection plate, here a 6144-density plate grown for 2 weeks, can be imaged using a black velvet covered plate fixation platform and a basic digital camera. The image can be processed to remove plate sides, allowing image analysis to be performed only on the region containing colonies and images can be corrected for nonuniform illumination as described in (http://www.mathworks.com/products/image/demos.html?file=/products/demos/shipping/images/ipexrice.html) and small objects, correspond to bubble, gel background and other anomalies can removed using the imopen function. Finally, the integrated pixel density is computed using pixel intensity, represented here as a color- or gray-coded scale, integrated on the area of each colony.

select for diploids and then transfer these to a methotrexate-containing plate for survival selection (Fig. 14.3). The choice of whether to use the *MAT***a** or *MAT*α strains as bait or prey is arbitrary. Here we describe a procedure in which the *MAT***a** strains are bait and *MAT*α are the prey strains. Baits can also be expressed as fusions to DHFR PCA fragments from expression plasmids available in our lab and transformed into appropriate strains.

3.2.1. Experimental procedure

(1) Incubate individual bait strains picked from glycerol stocks into 45 ml liquid culture of strain selective media (YPD with 100 μg/ml nourseothricin for *MAT***a** recombinant strains or 250 μg/ml hygromycin

B for *MATα* recombinant strains) and allow culture to reach saturation at 30 °C.

(2) Print prey strains picked from glycerol stocks onto a 35-ml agar solidified omniplate of strain selective media (3% agar + YPD with 100 μg/ml nourseothricin for *MATa* recombinant strains or 250 μg/ml hygromycin B for *MATα* recombinant strains) using four 96 manual or robotic pintool prints for a total of 384 prints per plates and incubate 16 h at 30 °C.

Note: For the prey strain, step 2 can be repeated from the 384 prints to be transferred to a maximum of four other 1536 pintool prints per omniplate to achieve a density of up to 6144 colonies.

Critical steps:

• Centrifuge a saturated culture of bait strain at 500 × *g* for 5 min and resuspend in 15 ml of YPD.

• Bait culture must be saturated to print enough cells for efficient mating on solid phase.

• Pintool must be cleaned between each cell transfer. We soak the pins twice in a solution of 10% bleach containing glass beads followed by a 10% bleach wash and two sterile water bath washes.

(3) Transfer bait strain suspension in to an empty omniplate.

(4) Print the bait strain suspension from the empty omniplate to a 35-ml agar solidified rich medium omniplate (YPD + 3% agar) at the same density as the prey strains using a pintool appropriate for the desired colony array density.

(5) Transfer prey strains onto the bait strains on an omniplate containing 35 ml-solid agar containing rich medium (YPD + 3% agar) using the appropriate pintool. Allow mating to occur and incubate for 16 h at 30 °C.

(6) Transfer the mixed haploid and diploid colonies from Step 5 onto an omniplate containing solid agar containing diploid selective medium (3% agar + YPD with both 100 μg/ml nourseothricin and 250 μg/ml hygromycin B) using the appropriate pintool. Incubate for 16 h at 30 °C.

(7) Transfer diploid selected strains onto a solid noble agar solidified synthetic minimal media omniplate with methotrexate (4% noble agar + SC + 2% glucose + 200 μg/ml methotrexate) using an appropriate pintool. Incubate at 30 °C and acquire pictures of the colony array every 96 h for approximately 2 weeks.

3.2.1.2. Timeline

Endogenous recombinant strain screen setup (steps 1–2): 8 h to 3 days (depending on the screen density achieved).

Transferring haploids for mating (steps 3–6): 6 min or more per bait strain (depending on screen density and robotic routine efficiency) + 16 h incubation.

Diploid cell selection (step 7): 5 min or more per bait (depending on screen density and robotic routine efficiency) + 16 h incubation.

DHFR PCA survival selection (step 8): 5 min or more per bait (depending on screen density and robotic routine efficiency) + 2 weeks incubation (maximum).

3.2.1.3. Anticipated results and controls To evaluate a DHFR PCA screen, both positive controls (known PPIs) and negative controls (fragments alone or non-interacting protein partners) should be tested on every plate. These non-interacting protein partner strain colonies exhibit background growth that should stop after a few days of incubation on methotrexate-containing plates. Colonies containing interacting baits and preys will continue to grow. The PCA fragment fusions expressed alone should not result in cell proliferation because the individual PCA fragments have no activity, thus if individual strain colonies do grow for unknown reasons, they should not be considered for further analysis. The most critical controls to do are those for spontaneous PCA; cases where a protein-PCA fragment fusion interacts with the complementary fragment alone. We found in our own screen that about 5% of bait or prey protein-expressing strains would grow in the presence of methotrexate when mated to a strain harboring an expression vector encoding the complementary fragment alone (Tarassov *et al.*, 2008). These complementary DHFR PCA fragment expression vectors are available upon request form our lab. Other controls can be included to test how the PCA screen performs. For instance, we have used the engineered heteromeric SspB$_{\text{YGMF}}$: SspB$_{\text{LSLA}}$ interaction as a positive control to validate DHFR PCA activity as suggested in the troubleshooting section (Table 14.1). Another elegant control to examine the range of dissociation constants for which the DHFR PCA is sensitive is to use a complex for which single-point mutations are known by other methods disrupt the interaction to different degrees. To this end, we have used in our own work, mutants of the Ras binding domain of Raf (Campbell-Valois *et al.*, 2005). A potential source of false positives in a PCA screen could be through trapping of nonspecific complexes due to irreversible folding of the DHFR fragments. However, we have used the adenosine 3′,5′-monophosphate dependent dissociation of the yeast protein kinase A complex as a control (Stefan *et al.*, 2007) to show that the DHFR PCA is fully reversible, and thus the trapping of complexes is unlikely (Tarassov *et al.*, 2008). Another control one could use is a condition-dependent PPI. We have used in our own work, the FK506-binding protein that binds to rapamycin and this complex then binds the target of rapamycin (TOR) (Pelletier *et al.*, 1998). All of these reagents are available upon request.

Table 14.1 Trouble shooting large-scale DHFR PCA screen

Step	Problem	Possible reason	Solution
1–2	Strains are not growing or incomplete prey array growth	Erroneous haploid selection	Verify protocol for appropriate culture conditions
		Low glycerol viability	Strains can be streaked on solid agar-selective medium Petri dishes prior to inoculation to increase viability
		Technical problem	Verify that all pins of the pintool touch glycerol stocks and the recipient omniplate
7	Low number or no colonies on diploid selective plates	Erroneous haploid strains type	Verify mating type of haploid strains
		Technical problem	Pintool alignment might have changed. No modifications to the pintool positioning should be done between transfers
8	No colony growth on DHFR PCA survival-selective medium	Erroneous selective conditions	Use heteromeric complex SspB$_{YGMF}$:SspB$_{LSLA}$ as a positive control to validate DHFR PCA activity
		Erroneous DHFR PCA	Verify by a strain diagnostic PCR the complementarity of PCA fragments
			Verify DHFR PCA fragment recombinant insertion by genomic sequencing
		DHFR PCA fragment expression	Verify DHFR PCA fragment expression by western blot
	All colonies grow at the same rate on DHFR PCA survival-selective medium	Erroneous selective conditions	Use DHFR PCA fragment controls alone as negative control
		Methotrexate solubility	Verify methotrexate solubility under conditions used. Stock solution should not exceed 10 mg/ml in DMSO and final concentration in solid agar plates should not exceed 200 mg/ml

3.2.2. Analysis of large-scale DHFR PCA screens

The goal of this section is to turn the size of the colonies on the selection plate into binary data that will represent PPIs. First, the digital images have to be transformed into tables containing colony intensities. Second, these colony intensities have to be turned into PPI confidence scores.

3.2.2.1. Image analysis Several bioinformatics tools are available to perform colony size measurements from digital images of high colony density plates (Carpenter *et al.*, 2006; Collins *et al.*, 2006; Memarian *et al.*, 2007). Alternatively, tools developed for analysis of spotted DNA microarrays can be modified to estimate the sizes of the colonies spaced on regular grids (Dudley *et al.*, 2005). Globally, the analysis consists of measuring the number of pixels per colony position. In cases where high density plates are used (above 1536 position grid), more involved analyses methods have to be utilized to separate adjacent colonies that may touch each other (Tarassov *et al.*, 2008). However, because PPIs are rare, most colonies will have a very slow growth rate and this problem is mostly negligible at lower densities. Thus, when lower densities are used for the screens, simple macros can be implemented in publicly accessible image analysis software such as ImageJ (http://rsb.info.nih.gov/ij/). In this case, digital images of plates are first converted to 8-bit grayscale format and colonies are measured by positioning the measurement tool on a colony center and estimating the integrated pixel intensity in an area that corresponds to the maximal colony size allowed. The process is iterated over all the grid positions and then all the plates, and the grid positions and intensity values are exported to text files for further processing in your favorite spreadsheet or statistical analysis software (Example ImageJ scripts that we use are available at our web site: http://michnick.bcm.umontreal.ca). It is important to note that colonies should always have the same positions on the images. If this is not the case, some of the tools cited above include a step that positions the analysis grid onto the colony positions prior to colony size measurements.

3.2.2.2. Statistical analysis of raw colony data: From continuous to binary data A PCA screen based on survival assay will only be useful if there is a confidence score attached to each of the putative interactions. Raw colony intensity data are continuously distributed, that is, they cover a wide range of values and cannot be directly turned into "yes" or "no" binary scores. Further, not all the colonies that can grow due to protein-fragment complementation will do so at exactly the same rate. As described above, every PCA experiment should include a set of positive controls consisting of pairs of baits and preys that interact with each other, and negative controls, consisting of pairs or baits and preys that do not interact with each other. These will be used for quality control in order to detect mis-positioning of the grid and batch effects (variation in media, incubation, drug concentration) that affect global growth rate of the different plates. Finally, the positive controls can provide a first, visual analysis of the data, whereby the growth rate of the positive controls indicate roughly the intensity threshold above which we expect strains with interacting bait–prey pairs to grow. Beyond these "qualitative" controls, a statistical analysis should be used to separate the interacting pairs from the noninteracting pairs.

The statistical analysis globally includes two steps. First, it has to be determined whether there is a significant difference in growth rates among the plates before applying a global analysis to the data. If there is significant variation, the data should be normalized such that all the plates have the same average colony size. Alternatively, data could be transformed into relative scores, such as Z-scores, whereby each data point is transformed to become the number of standard deviations that data point is from the average of the plate. We found that combining the Z-score and the raw intensity worked best for our large-scale screen (Tarassov et al., 2008). Then continuous values must be turned into binary values by setting a threshold of intensity above which proteins are inferred to interact, and establishing a confidence score for this particular threshold. One way to assign confidence values to PCA interactions is to benchmark the intensity values against a set of data containing interactions that should be detected in the screen (a set of real positives) and others that should not (a set of real negatives). The real positives set can be derived from a set of known and well-supported interactions. The real negative set has however to be approximated because it is impossible to show that two proteins never interact. Sets of proteins that are most likely not interacting can be used for this purpose, for instance proteins that are not localized in the same cell compartments and that have negatively correlated expression profiles (Collins et al., 2007). One can then predict, for a given intensity threshold, what should be the proportion of true positive interactions and false positive interactions. In order to decide on the threshold, the ratio of true positive interactions divided by to the total number of inferred positives (true positives + predicted false positives)—known as the positive predictive value (PPV)—is calculated as a function of threshold of intensities. For instance, at a PPV of 95% percent, one expects 5% of positives to be false. Lower and higher thresholds can be used depending on how stringent one wants the analysis to be. It is important to note that the estimated PPV is only accurate if the relative occurrence of positives and negatives in the reference sets is similar to that of the real positives and negatives (Jansen and Gerstein, 2004). In the case of a genome-wide, comprehensive screen, this fraction corresponds to a very low prior probability of finding interactions among all pairwise possibilities. On the other hand, a small-scale screen of a specific biological process will contain a greater proportion of real positives than a random screen. The reference set therefore needs to be tailored for the actual screen being performed, that is, the space of the interactome that is covered. For a formal treatment of these issues, refer to Jansen and Gerstein (2004). Beyond these statistical considerations, analysis such as Gene Ontology enrichment and visualization of interaction clusters should be used to further assess the confidence in the data set being produced. For instance, the matrix of binary interactions can be clustered to identify groups or complexes of interacting proteins. Finally, sets of true positives and negatives are not a panacea and the functional and evolutionary characterization of PPIs is the only way to provide a definitive answer as to whether an interaction is functionally relevant or not for the cell (Levy *et al.*, 2009).

4. A LIFE AND DEATH SELECTION PCA BASED ON THE PRODRUG-CONVERTING CYTOSINE DEAMINASE FOR DISSECTION OF PPIs

In this section, we present a PCA based on an optimized mutant form of the reporter enzyme yeast cytosine deaminase (OyCD). The choice of yCD as a reporter was based on its role in a pyrimidine salvage pathway and the availability of a prodrug 5-fluorocytosine (5-FC), which is converted to 5-fluorouracil (5-FU) by yCD. Bacteria and yeast can convert cytosine to uracil and use it for the synthesis of UTP and TTP, which are required for cell survival (Kurtz *et al.*, 1999). In *S. cerevisiae*, yCD is encoded by the *FCY1* gene and is the enzyme that catalyzes this reaction. In addition to deaminating cytosine, yCD can also deaminates 5-FC to 5-FU. 5-FU will be further processed by enzymes of the pyrimidine salvage pathway to 5-FUTP, a toxic compound that causes cell death. These particular properties of yCD make it an ideal reporter for a life and death selection PCA (Fig. 14.4A) (Ear and Michnick, 2009).

The OyCD PCA allows death and survival assay to be performed without changing the reporter system. In a two-step selection process, we can engineer mutant forms of a protein in order to dissect its different functions; disrupting interactions with one partner, while retaining interaction with others (Fig. 14.4B). For example, protein A interacts with both protein B and protein C. First, we can screen for mutant forms of protein A that disrupt interaction with protein B. Second, we select for protein A mutants that still interact with protein C. Using OyCD PCA, neither of these selection steps requires replica plating. In addition, no expensive reagent or equipment is required. Specific mutants can be obtained in about 4 weeks.

Both the survival and death selection assays are performed in *fcy1* deletion strains. For the survival selection assay, uracil must be removed from the selection medium. Only cells that have OyCD PCA activity will be able to synthesize uracil and survive. For the death selection assay, cells

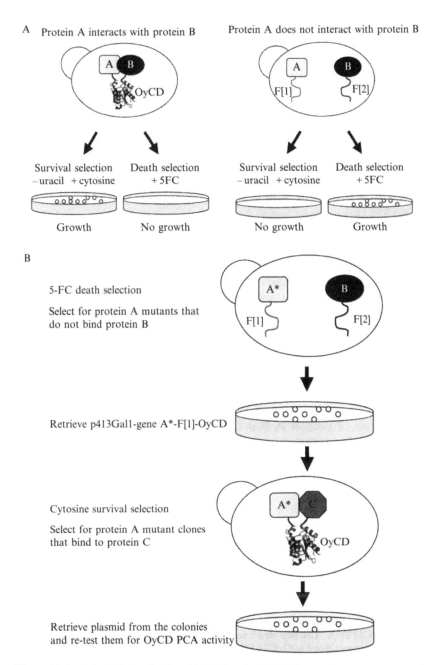

Figure 14.4 (A) A dual selection PCA. The OyCD PCA can serve as a reporter for formation of a protein–protein interaction provided that the reconstituted reporter enzyme supports growth under one condition (survival assay) or no growth under another condition (death assay). In the case where the two test proteins do not interact,

are grown in a selection medium in the presence of 5-FC. In this death assay, cells that have OyCD PCA activity will be sensitive to 5-FC.

4.1. Preparation for a two-step OyCD PCA screen

The proteins of interest are fused to the N-terminal of OyCD fragment 1 or fragment 2 (protein A-OyCD-F[1], protein B-OyCD-F[2] and protein C-OyCD-F[2]). For some proteins, PPIs can only be detected when they are fused at the C-terminal of OyCD fragment 1 (e.g., OyCD-F[1]-protein A). OyCD-F[1] corresponds to amino acid residues 1–77 of yCD with an A23L point mutation. OyCD-F[2] corresponds to amino acid residues 57–158 of yCD with the following point mutations: V108I, I140L, T95S, and K117E (Ear and Michnick, 2009). The proteins of interest and OyCD fragments are separated by a 15-amino acid flexible polypeptide linker (Gly. Gly.Gly.Gly.Ser)$_3$. The nucleotide sequences encoding these fusion proteins are cloned into yeast expression vectors. We used the p413Gal1 and p415Gal1 expression plasmids (Mumberg *et al.*, 1995). Before proceeding with a screen, verify that interactions of your proteins of interest are detected by OyCD PCA. Titrate the amount of cytosine and 5-FC required for detecting your interactions. We normally try a range between 50 and 1000 μg/ml of cytosine or 5-FC. For many proteins, 100 μg/ml of either substrate is sufficient for detecting OyCD PCA activity. Use well known interacting and non-interacting proteins as controls. For the two-step OyCD PCA screen, a library of your gene of interest can be generated by methods such as error-prone PCR. For example, if the goal is to engineer mutant forms of protein A that can specifically disrupt interaction with protein B while preserving interaction with protein C, an ideal library of protein A carrying 1–3 mutations is desired.

4.2. Materials

4.2.1. Reagents

- BY4741, BY4742, or BY4743 strains with a deletion in the *FCY1* gene (*fcy1Δ*) (Giaever *et al.*, 2002) that are resistant to G418.

---

the reverse scenarios are observed. (B) Screen for mutants of protein A that do not bind to protein B but retain binding to protein C using sequential death followed by survival selection OyCD PCA. The first death selection screen consists of screening the library of protein A mutants fused to OyCD-F[1] (A*-F[1]) with protein B fused to OyCD-F[2] (B-F[2]) and screen for clones that show loss of OyCD PCA activity (growth in the presence of 5-FC). The second survival selection step consists of screening A*-F[1] clones harvested from the first death selection screen against protein C fused to OyCD-F[2] (C-F[2]) for clones that show OyCD PCA activity using the life assay (growth in presence of cytosine).

- SC medium with the appropriate amino acid drop out according to the chosen expression plasmids.
- Your genes of interest fused to the OyCD fragments in yeast expression vectors.
- Sorbitol buffer (1 M sorbitol, 1 mM EDTA, 10 mM Tris (pH 8.0), 100 mM lithium acetate)
- PLATE solution (40% PEG 3350, 100 mM lithium acetate, 10 mM Tris (pH 7.5), and 0.4 mM EDTA)
- Dimethylsulfoxide (Fisher)
- Sterile distilled water
- G418 (Wisent)
- Cytosine (Sigma)
- 5-Fluorocytosine (Sigma)
- Agar (Bioshop)
- Noble agar (Bioshop)
- DH5α or MC1061 *E. coli* electrocompetent cells
- LB medium
- DNeasy Tissue Kit (Qiagen)

4.2.2. Facultative

- Antibodies against yCD fragments: anti-yCD polyclonal (Biogenesis).

4.2.3. Equipment

- Genepulser II electroporator system (Bio-Rad) or Electroporator 2510 (Eppendorf)
- Electroporation cuvette with 1 mm wide slot (Sigma)
- Glass spreader
- 100 mm Petri dishes
- Shaking incubators, preset to 30 and 37 °C
- Incubator, preset to 30 and 37 °C

4.2.4. Experiment preparation

10 mg/ml stock solution of cytosine: Dissolve 100 mg of cytosine in 10 ml of distilled water. Vortex the solution and incubate at 37 °C to make it dissolve. Filter the solution and store at room temperature. It is better to make this solution fresh and use it within a week.

10 mg/ml stock solution of 5-FC: Dissolve 100 mg of 5-FC in 10 ml of distilled water. Vortex the solution and incubate at 37 °C to make it dissolve. Filter the solution and use it right away or aliquot in sterile tubes and store at − 20 °C.

Control plates: Make SC plates for selection of clones harboring the expression plasmids. We used the p413Gal1 and p415Gal1 expression vectors, therefore, our control plates contain SC medium without histidine and leucine, with 2% agar, 2% raffinose, and 2% galactose.

Cytosine survival selection plate: Make SC plates without uracil and selection for the expression plasmids. We used the p413Gal1 and p415Gal1 expression vectors, therefore our selection plates contain SC medium without uracil, histidine, and leucine, with 3% noble agar, 2% raffinose, 2% galactose, and cytosine (we use 100 μg/ml of 5-FC for our proteins of interest).

5-FC death selection plates: Make SC plates with 5-FC and selection for the expression plasmids. We used the p413Gal1 and p415Gal1 expression vectors, therefore our selection plates contain SC medium without histidine and leucine, with 2% noble agar, 2% raffinose, 2% galactose, and 5-FC (we use 100 μg/ml of 5-FC for our proteins of interest).

4.3. Procedure

4.3.1. Death selection screen

(1) Transform (Knop *et al.*, 1999) 1 μg of the library encoding mutant forms of protein A (protein A*) in BY4741 *fcy1*Δ strain that already carry a plasmid expressing protein B (Fig. 14.4B).

   Critical step: Make sure that the efficiency of the transformation gives enough colonies to cover six times the size of the library in order to have good coverage of potential mutants. For example, if the size of the library of protein A* is 5000 clones, make sure to obtain more than 30,000 clones.

(2) Plate half of the transformation on the control plates to select for the presence of both expression plasmids (p413Gal1-gene A*-OyCD-F[1] + p415Gal1-gene B-OyCD-F[2]). These plates serve as controls for reporting the efficiency of the transformation. Plate the other half of the transformation on 5-FC death selection plates.

   Critical step: Test the efficiency of your competent yeast cells to determine how many cells to plate per 100 mm Petri dish. Do not plate more than 5000 cells per 100 mm Petri dish.

   Pause point: Make glycerol stock of the pooled yeast colonies obtained on the control plates as a backup source or for future screens if required.

(3) Incubate plates at 30 °C for 2–3 days. Compare the number of colonies obtained on the 5-FC death selection plates to the control plates.

   Critical step: We should expect 10–50% less colonies on the 5-FC death selection plates in comparison to the control plates. This variability depends on the pair of interaction chosen and the number of mutations per clone in the library.

(4) Colonies that grow on 5–FC selection plates are pooled and harvested for DNA extraction (Qiagen DNeasy Tissue Kit or a genomic DNA purification protocol using phenol-choroform) in order to recover the plasmids that express protein A★.

Pause point: Yeast cell pellet can be store at − 20 °C for months.

(5) Digest the extracted DNA with enzyme(s) that cut in the plasmids expressing protein B-OyCD-F[2] but not the plasmids expressing protein A★-OyCD-F[1] library. We use AflII, BspmI, HpaI, MunI, NarI, or XcmI since they cut in p415Gal1 and not in p413Gal1 plasmid or the gene of interest. This step is not required if the two expression plasmids do not have the same antibiotic resistance gene.

(6) Use 2 μl of extracted DNA for electroporation into electrocompetent *E. coli* cells. We use the MC1061 *E. coli* strain since it has higher transformation efficiency than the DH5α strain. Plate the *E. coli* on LB plates with appropriate antibiotic selection. We use LB with ampicillin for the p41XGal1 plasmids.

(7) Pool *E. coli* colonies and extract the plasmid DNA using your miniprep kit of choice.

Pause point: *E. coli* cell pellet can be store at − 20 °C for months.

4.3.2. Survival selection screen

(8) Transform according to Knop *et al.* (1999) the library encoding for mutant forms of protein A (protein A★) retrieved after the death selection screen in BY4741 *fcy1Δ* strain that already carry a plasmid expressing protein C (Fig. 14.4B).

Critical step: Make sure that the efficiency of the transformation gives enough colonies to cover six times the size of the library in order to have a good coverage of potential mutant clones.

(9) Plate half of the transformation on the control plates to select for the presence of both expression plasmids (p413Gal1-gene A★-OyCD-F[1] + p415Gal1-gene C-OyCD-F[2]). These plates serve as control for reporting the efficiency of the transformation. Plate the other half of the transformation on cytosine survival-selection plates.

Critical step: Test the efficiency of your competent yeast cells to have an idea how much cells to plate per 100 mm Petri dish. Do not plate more than 2000 cells per 100 mm Petri dish.

Pause point: Make glycerol stock of the pooled yeast colonies obtained on the control plates as a backup source or for future screens if required.

(10) Incubate plates at 30 °C for 3–7 days.

Critical step: We can expect to obtain from a few to hundreds of colonies at this step. This variability depends mostly on the pair of interaction that was chosen and the complexity of the library.

(11) If the screen resulted in less than 50 colonies, inoculate each yeast colony separately in 5 ml of selection medium and harvest cells for DNA extraction (Qiagen DNeasy Tissue Kit or a genomic DNA purification protocol using phenol–choroform). If over 50 colonies were obtained, pooled all the colonies and extract DNA from the pooled cells.

   Pause point: Make glycerol stock of the single or pooled yeast colonies as a backup source.

(12) Digest the extracted DNA with enzyme(s) that cut in the plasmid expressing protein C-OyCD-F[2] but not the protein A*-OyCD-F[1] library. We use AflII, BspmI, HpaI, MunI, NarI, or XcmI since they cut in p415Gal1 and not in p413Gal1 plasmid or the gene of interest. This step is not required if the two expression plasmids do not have the same antibiotic resistance gene.

(13) Use 2 μl of extracted DNA for electroporation into electrocompetent MC1061 *E. coli* cells. Plate the *E. coli* on LB plates with the appropriate antibiotic selection.

(14) For samples obtained from a single yeast colony in step 11, inoculate one or two *E. coli* colonies for plasmid DNA extraction. For samples obtained from pooled yeast colonies in step 11, inoculate over 90 *E. coli* colonies for plasmid DNA extraction.

   Pause point: Make glycerol stock of the single or pooled bacterial colonies as a backup source.

(15) Digest the isolated plasmids with appropriate restriction enzymes or perform diagnostic PCR to confirm the presence of gene A*-OyCD-F[1].

(16) Retransform individually the purified plasmids expressing protein A mutants in BY4741 *fcy1Δ* strain carrying a plasmid expressing protein B and C, respectively, and test for OyCD PCA activity.

(17) Send the purified plasmids expressing protein A mutants for sequencing in order to identify the mutation(s).

4.3.3. Timeline

5-FC death selection (steps 1–3): 2–3 days.

Cytosine survival selection (steps 8–10): 3–7 days.

Isolation of DNA from yeast (steps 4 and 11): almost 1 day.

Further characterization of the individual clones (steps 14–17): several days depending on the number of clones obtained.

Troubleshooting advice can be found in Table 14.2.

4.3.4. Additional information

We have also generated destination vectors carrying the OyCD fragments that are compatible with the Gateway cloning system. With these plasmids, we can take advantage of the existing Gateway expression clones

Table 14.2   Troubleshooting an OyCD PCA screen

Step	Problem	Possible reason	Solution
3	Less than 10% of colonies died on the 5–FC selection plates	Too many yeast plated on the selection plate	Plate less than 1000 cells per 100 mm Petri dish
			Increase 5–FC concentration
6 and 13	No *E. coli* colonies or very few colonies	Electrocompetent *E. coli* cells not very competent	Use freshly prepare electrocompetent MC1061 *E. coli* cells
10	Several hundreds of colonies grew on the cytosine selection plates	Too many yeast plated on the selection plate	Plate less than 1000 cells per 100 mm Petri dish
		Uracil can diffuse out of cells that have OyCD PCA activity and allow for cells that do not have OyCD PCA activity to grow	Decrease cytosine concentration
	Small colonies form around the initial large colony after 4 days of incubation		Pick only the large colony at the center

(distributed by Open Biosystems) to facilitate the process of generating the fusion between the genes of interest and OyCD fragments.

5. Visualizing the Localization of PPIs with GFP Family Fluorescent Protein PCAs

The first fluorescent protein PCA was described by Lynne Regan's group for GFP (Ghosh *et al.*, 2000; Magliery *et al.*, 2005; Wilson *et al.*, 2004) and we and others have described different color and behavioral variants (Cabantous *et al.*, 2005; Hu *et al.*, 2002; Macdonald *et al.*, 2006; Nyfeler *et al.*, 2005; Remy and Michnick, 2004; Remy *et al.*, 2004). PCAs based on fluorescent proteins have both unique features, but also the most caveats to their application. Notably, and unlike other PCAs, those based on these fluorescent proteins are irreversible, which can be both useful (trapping and visualizing rare and transient complexes) but also require care in interpretation of turnover or localization of interacting proteins (Hu *et al.*, 2002; Magliery *et al.*, 2005). It is important that the kinetics of relocalization of protein interactions observed with fluorescence PCAs be confirmed by immunofluorescence or by monitoring the localization of the same proteins fused to full-length fluorescent proteins. Fluorescent protein PCAs are also limited to the temporal range of dynamics that can be studied. Because different variants of these proteins take minutes to hours to fold and mature, they are obviously not appropriate for studying most dynamic processes in a quantitative way, though many important slower processes can be studied. PCAs based on luciferase enzyme reporters are, like the DHFR PCA, fully reversible and can be used to capture kinetics on the second time scale (see Section 6) (Remy and Michnick, 2006; Stefan *et al.*, 2007). As we previously demonstrated, PPIs that occur within a specific biochemical pathway can be modulated in predicted ways by conditions or molecules that activate or inhibit the pathway. We and others have shown that at least changes in the formation of complexes can be detected with the GFP and YFP PCAs (Remy and Michnick, 2004). Further, the subcellular location of stable complexes and changes in their locations following perturbation can also be detected in intact living cells with the YFP PCA (Macdonald *et al.*, 2006; Remy and Michnick, 2004; Remy *et al.*, 2004). It is this ability to detect the location and intracellular movements of protein complexes that make fluorescent protein-based PCAs unique. Because GFP/YFP-based PCAs do not require additional substrates or cofactors for emission of fluorescence, they are particularly simple to implement. We have shown that PPIs can be monitored by fluorescence microscopy, flow cytometry, and spectroscopy using GFP- and YFP-based PCAs (Macdonald *et al.*, 2006; Remy and Michnick, 2004; Remy *et al.*, 2004). We have applied these assays to the detection and quantification of protein interactions, localization of complexes in living cells, and cDNA library screening in mammalian cells (Benton *et al.*, 2006; Ding *et al.*, 2006; Macdonald *et al.*, 2006; Nyfeler *et al.*, 2008; Remy and Michnick, 2004; Remy *et al.*, 2004). In addition, we have used the YFP-based PCA to detect protein interactions in specific subcellular compartments of *S. cerevisiae*, such as cytoplasm, nucleus, plasma membrane and the bud neck (Fig. 14.5) (Manderson *et al.*, 2008). In the following protocol we describe methods for studying PPI with the "Venus" mutant of YFP (Nagai *et al.*, 2002).

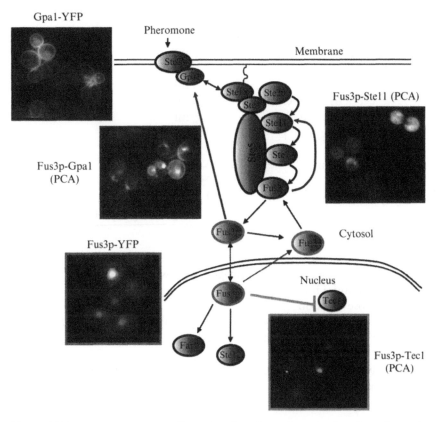

Figure 14.5 Venus YFP PCA allows for detection of precise location of protein complexes within living cells. Illustration for visualization of protein complexes in different regions within cells using yeast pheromone response mitogen activated protein kinase pathway as an example. Images show the location of interactions of Fus3p with Gpa1 (Metodiev *et al.*, 2002) to the membrane, with Ste11 (Choi *et al.*, 1994) to the cytoplasm and with Tec1 (Chou *et al.*, 2004) to the nucleus. As controls for different localizations, Gpa1 fused to full-length Venus YFP protein is shown to be at the membrane while Fus3-Venus YFP is found in both cytoplasm and the nucleus. Cells containing Fus3-venus YFP were treated with 1 μM alpha-factor pheromone for 2–3 h to induce its translocation to membrane and nucleus.

5.1. Materials

5.1.1. Reagents

- Competent *MATa* or diploid yeast prepared according to Knop *et al.* (1999).
- SD medium (6.7 g/l yeast nitrogen base, without amino acids)
- SD agar (SD medium with 2% agar)
- 10× amino acid mix –his, –leu, –lys:

Adenine sulfate	0.4 g/ml
Uracil	0.2 g/ml
L-Tryptophan	0.4 g/ml
L-Arginine HCl	0.2 g/ml
L-Tyrosine	0.3 g/ml
L-Phenylalanine	0.5 g/ml
L-Glutamic acid	1.0 g/ml
L-Asparagine	1.0 g/ml
L-Valine	1.5 g/ml
L-Threonine	2.0 g/ml
L-Serine	3.75 g/ml
Methionine	0.2 g/ml (do not include when growing diploid yeast)

- 10× low fluorescence yeast nitrogen base without riboflavin and folic acid (Sheff and Thorn, 2004)
- 20% glucose solution
- PLATE solution (40% polyethylene glycol 3350, 100 mM LiOAc, 10 mM Tris (pH 7.5), 0.4 mM EDTA)
- DMSO
- poly-L-lysine mol. wt. 30,000–70,000 (Sigma P2636) or concanavalin A (Sigma L7647)

5.1.2. Equipment

- Fluorescence microscope (Nikon Eclipse TE2000U inverted microscope with a CoolSnap HQ Monochrome CCD camera (Photometrics))
- 96-well black, glass-bottom plate (Molecular Machines)
- 6-well culture plate or Petri dish
- Appropriate sterile tubes to grow yeast
- Spectrophotometer spectra MAX GEMINI XS (Molecular Devices)

5.1.3. Plasmids

The proteins to test for interaction are fused to the N- and C-terminal fragments of an enhanced YFP (Venus YFP; Nagai et al., 2002), in 5′ or 3′ of the fragments (protein A-vYFP-F[1], vYFP-F[1]-protein A, protein B-vYFP-F[1], vYFP-F[1]-protein B). vYFP-F[1] (N-terminal) corresponds to amino acids 1–158, and vYFP-F[2] (C-terminal) corresponds to amino acids 159–239 of Venus YFP. The fusions are subcloned into yeast expression vectors p413ADH for the vYFP-F[1] fusion and p415ADH for the vYFP-F[2] fusion (Mumberg et al., 1995). We typically insert a 10-amino acid flexible polypeptide linker consisting of (Gly.Gly.Gly.Gly.Ser)$_2$ between the protein of interest and the vYFP fragments (Table 14.3). In yeast fragments can also be fused to the genes of interest at their chromosomal loci using a homologous recombination method (Ghaemmaghami et al., 2003). For this purpose the PCA fragments are cloned into nonexpression vectors that provide a selection marker (e.g., antibiotic resistance).

5.2. Procedure

5.2.1. Cotransformation of competent yeast

(1) Thaw competent yeast cells on ice.
(2) Mix 10 μl of cells with 1 μl ($\sim$250 ng) of each yeast expression plasmid (e.g., p413ADH and p415ADH, Mumberg et al., 1995) encoding the Venus YFP PCA fusion partners (protein A fused to vYFP-F[1] and protein B fused to vYFP-F[2]), 60 μl of PLATE solution and 8 μl DMSO.
(3) Heat shock yeast at 42 °C for 20 min.

Table 14.3 Troubleshooting vYFP PCA experiments

Step	Problem	Possible reason	Solution
5.2.1. (6)	No colonies after transformation	DNA or cells used is less	Increase quantity of cells and DNA. Increase the volume of cells plated on the Petri dish or six-well plate
	Too many colonies after transformation	Plated lot of cells	Dilute cells before plating on the Petri dish or six-well plate
5.2.2. (3)	Fusion protein is not functioning correctly	Fragment fusion interferes with protein expression/function/stability	Fuse the PCA fragment to the other end of the protein

Critical step: Shorter or longer incubation times at higher or lower temperatures can result in decreased efficiency of transformation.

(4) Centrifuge at 2000 rpm for 3 min. Remove supernatant and resuspend cells in 500 μl SD medium without amino acids or glucose.

(5) Plate 20 μl of cell suspension per well on SC agar (SD agar + 2% glucose + 1× amino acids (-his, -leu, -lys for *MATa*; -his, -leu, -lys, -met for diploids)) in a 6-well plate.

(6) Incubate at 30 °C for 48–72h.

5.2.2. Preparation of cells for fluorescence microscopy

(1) Inoculate a fresh colony for each sample into 3 ml of SC medium (SD medium + 2% glucose + 1× amino acids (-his, -leu, -lys for *MATa*; -his, -leu, -lys, -met for diploids)) and grow overnight at 30 °C with shaking.

(2) The following day, measure the OD_{600} of the overnight culture and inoculate a fresh culture of LFM (1× low fluorescence yeast nitrogen base + 2% glucose + 1× amino acids (-his, -leu, -lys for Mat A; -his, -leu, -lys, -met for diploids)) with enough cells to obtain an OD_{600} of approximately 0.1–0.3 at the time of analysis.

Critical step: It is particularly important for the cells to be in the log phase of growth in order to avoid including dead and unhealthy cells. These cells are highly autofluorescent and thus would confound quantitative analysis. Cells in the lag phase can be used if they are appropriate to study a particular interaction(s) as long as the condition of the cells is verified by bright field microscopy.

(3) Coat the wells of a glass bottom 96-well plate (Molecular Machines) with a solution of 1 mg/ml poly-L-lysine, or 50 μg/ml concanavalin A for 10 min, rinse with distilled water and allow to dry. Transfer 70 μl of cell suspension to each well. Wait 10 min to allow the cells to settle in the wells. Acquire images with a fluorescence microscope equipped with a CCD camera, using a YFP filter cube and ~750 ms of exposure time.

Critical step: It is best to use a 60× or 100× objective to discriminate subcellular structures. Bright field or phase contrast images can be acquired for each field of view to compare the morphology of the yeast with fluorescent PCA signal. Specific functional assays to further characterize a PPI might be performed here.

5.2.3. Timeline

Cotransformation of competent yeast (steps 1–5): 30–45 min (depending on the number of samples) plus 48–72 h for cell growth (step 6).

Preparation of cells for fluorometric analysis (steps 1 and 2): 24 h.

Fluorescence microscopy (step 3): 30 min to hours, depending on the number of samples.

Microplate reader analysis (step 3): a few minutes or more, depending on the number of samples.

5.2.4. Anticipated results

The fluorescence intensity of the reassembled Venus YFP PCA varies with the expression levels and the interaction dissociation constants for the protein pairs attached to the PCA fragments. In the case of our simplest positive control (GCN4 leucine zipper pair fused to the PCA fragments: Zip-vYFP-F[1] + Zip-vYFP-F[2]), the reconstituted PCAs represent approximately 10–20% of the activity of the full-length Venus YFP. The PCA fusions expressed alone should not result in detectable fluorescence (compared to nontransformed cells) because the individual PCA fragments have no activity. For each study, positive (known interaction) and particularly negative (noninteracting proteins) controls should always be performed in parallel. A PCA response should not be observed if non-interacting proteins are used as PCA partners.

6. STUDYING DYNAMICS OF PPIs WITH LUCIFERASE REPORTER PCAs

It has been a major challenge to measure and quantify the dynamics of protein complexes in their native state within living cells. Here, we describe protocols for implementing two luciferase enzyme based PCAs; *Renilla* luciferase (Rluc) and *Gaussia* luciferase (Gluc) that are designed specifically to investigate the dynamics of assembly and disassembly of protein complexes. We have applied these assays to the detection and quantification of protein interactions in mammalian cells as well as yeast. These assays are sensitive enough to detect interactions among proteins expressed at endogenous levels *in vivo* and to study dynamic changes in both the formation and disruption of PPIs over seconds without altering the kinetics of binding (Remy and Michnick, 2006; Stefan *et al.*, 2007). Both of these luciferases catalyze the oxidation of substrate coelenterate luciferins (coelenterazines) in a reaction that emits blue light (at a peak of 480 nm) and requires no cofactors (Tannous *et al.*, 2005). The substrates readily diffuse through cell membranes and into all cellular compartments, enabling quantitative analysis in live cells. Rluc and Gluc are monomeric proteins of 312 (36 kDa) and 185 amino acids (19.9 kDa). Gluc PCA has some advantage in that the reporter protein is smaller and has 10 times higher activity to native coelanterizine than Rluc. However, at present, Rluc has the advantage that stable substrates (e.g., benzyl-coelenterizine) can be used with this reporter allowing for easier handling and integration of signal over longer times. In contrast to fluorescent protein-based PCAs, both Rluc and Gluc are fully reversible; a prerequisite to study signaling events by the dynamics of protein complex assembly and disassembly (Remy and Michnick, 2006; Stefan *et al.*, 2007). Both Rluc and Gluc PCAs provide for extremely high signal-to-background ratio due to lack of any cellular luminescence and can easily be measured spectroscopically on whole cell populations or by imaging single cells. Finally, the luciferase PCAs allow for accurate measurements of time- (for time constants greater than 10 s) and dose–dependence of pharmacologically induced alterations of protein complexes.

6.1. Materials

6.1.1. Reagents

- cDNAs encoding the Rluc and Gluc PCA fusion partners in suitable expression vectors
- Coelenterazine and benzyl-coelenterazine (Nanolight)
- Competent *MATa* or diploid yeast prepared according to Knop *et al.* (1999).
- SD medium (6.7 g/l yeast nitrogen base, without amino acids)
- SD agar (SD medium with 2% agar)
- 10× amino acid mix –his, –leu, –lys:

Adenine sulfate	0.4 g/ml
Uracil	0.2 g/ml
L–Tryptophan	0.4 g/ml
L–Arginine HCl	0.2 g/ml
L–Tyrosine	0.3 g/ml
L–Phenylalanine	0.5 g/ml
L–Glutamic acid	1.0 g/ml
L–Asparagine	1.0 g/ml
L–Valine	1.5 g/ml
L–Threonine	2.0 g/ml
L–Serine	3.75 g/ml
Methionine	0.2 g/ml (do not include when growing diploid yeast)

- 10× low fluorescence yeast nitrogen base without riboflavin and folic acid (Sheff and Thorn, 2004)
- 20% glucose solution
- PLATE solution (40% polyethylene glycol 3350, 100 mM LiOAc, 10 mM Tris (pH 7.5), 0.4 mM EDTA)

- DMSO
- poly-L-lysine mol. wt. 30,000–70,000 (Sigma P2636) or concanavalin A (Sigma L7647)

6.1.2. Equipment

- Luminescence microplate reader (LMax II[384] Luminometer, Molecular Devices)
- Luminescence microscope (Nikon Eclipse TE2000U inverted microscope with a CoolSnap HQ Monochrome CCD camera (Photometrics))
- 96-well white plates (Molecular Machines)
- 6-well culture plate or Petri dish
- Appropriate sterile tubes to grow yeast
- Spectrophotometer

6.1.3. Plasmids

The proteins to test for interaction are fused to the coding sequences for N- and C-terminal fragments of Rluc or Gluc, in 5′ or 3′ of the fragments (e.g., protein A-Rluc-F[1], Rluc-F[1]-protein A, protein B-Rluc-F[2], Rluc-F[2]-protein B). Rluc-F[1] (N-terminal) corresponds to amino acids 1–110, and Rluc-F[2] (C-terminal) corresponds to amino acids 111–312 of Rluc (Stefan et al., 2007). Similarly Gluc-F[1] corresponds to amino acids 1–63 and Gluc-F[1] to amino acids 64–185 of Gluc (Remy and Michnick, 2006). The fusions are subcloned into yeast expression vectors, for example, p413ADH for the Rluc-F[1] or Gluc-F[1] fusion and p415ADH for the Rluc-F[2] or Gluc-F[2] fusion (Mumberg et al., 1995). In yeast fragments can also be fused to the genes of interest at their chromosomal loci using a homologous recombination method (Ghaemmaghami et al., 2003). For this purpose the PCA fragments are cloned into nonexpression vectors that provide a selection marker (e.g., antibiotic resistance). For example, pAG25-Rluc-F[1] and pAG32-Rluc-F[2] plasmids are used for the Rluc PCA fragment fusions.

6.2. Procedure

6.2.1. Cotransformation of competent yeast

(1) Thaw competent yeast cells on ice.
(2) Mix 10 μl of cells with 1 μl ($\sim$250 ng) of each yeast expression plasmid (e.g., p413ADH and p415ADH, Mumberg et al., 1995) encoding the Rluc or Gluc PCA fusion partners (protein A fused to Rluc-F[1] or Gluc-F[1] and protein B fused to Rluc-F[2] or Gluc-F[2], 60 μl of PLATE solution and 8 μl DMSO).
(3) Heat shock yeast at 42 °C for 20 min.

Critical step: Shorter or longer incubation times at higher or lower temperatures can result in decreased efficiency of transformation.

(4) Centrifuge at 2000 rpm for 3 min. Remove supernatant and resuspend cells in 500 μl SD medium without amino acids or glucose.

(5) Plate 20 μl of cell suspension per well on SC agar (SD agar + 2% glucose + 1× amino acids (-his, -leu, -lys for *MATa*; -his, -leu, -lys, -met for diploids)) in a 6-well plate.

(6) Incubate at 30 °C for 48–72 h.

6.2.2. Fusion of PCA fragments at the chromosomal loci

(1) PCR amplify the Rluc or Gluc PCA fragment cassettes (containing the PCA fragment followed by a terminator that is followed by an antibiotic selection marker; available upon request)

(2) Transform the PCR product in to suitable competent cells: mix 10 μl of thawed competent cells with 10 μl ($\sim$1–2 μg) of each PCR amplified cassette DNA encoding the Rluc or Gluc PCA fragments along with a resistance marker, add 85 μl of PLATE solution; incubate for 30 min at room temperature; add 9.5 μl DMSO followed by heat shock at 42 °C for 20 min; centrifuge at 2000 rpm for 3 min, remove supernatant and resuspend cells in 500 μl YPD medium and incubate at 30 °C with shaking for 4 h; centrifuge the cells, remove supernatant and resuspend cells in 200 μl of YPD; plate 60 μl per well in 6-well plate or entire 200 μl on Petri dish that contain the suitable antibiotic; incubate the plates at 30 °C for 48–72h; the colonies can be further verified by colony PCR methods.

6.2.3. Preparation of cells for bioluminescence assay

(1) Inoculate a fresh colony for each sample with plasmids into 3 ml of SC medium (SD medium + 2% glucose + 1× amino acids (-his, -leu, -lys for *MATa*; -his, -leu, -lys, -met for diploids)) and grow overnight at 30 °C with shaking. For cells with fragments fused at chromosomes, grow them in SC medium with suitable antibiotic.

(2) The following day, measure the OD_{600} of the overnight culture and inoculate a fresh culture of LFM (1× low fluorescence yeast nitrogen base + 2% glucose + 1× amino acids (-his, -leu, -lys for Mat A; -his, -leu, -lys, -met for diploids)) or LFM complete with suitable antibiotics with enough cells to obtain an OD_{600} of approximately 0.1–0.3 at the time of analysis.
Critical step: It is particularly important for the cells to be in the log phase of growth in order to avoid including dead and unhealthy cells.

(3) Transfer 160–180 μl of cell suspension (cells equivalent to 0.1– 0.3 OD_{600}) to each well. Manually add or inject 20–40 μl of suitable substrate using the Luminometer injector and initiate the

Table 14.4 Troubleshooting Rluc or Gluc luciferase PCAs

Step	Problem	Possible reason	Solution
6.2.1. (6)	No colonies after transformation	Not enough DNA or cells	Increase quantity of cells and DNA. Increase the number of cells plated on the Petri dish or six-well plate
	Too many colonies after transformation	Too many cells plated	Dilute cells before plating on the Petri dish or six-well plate
6.2.2. (2)	Fusion protein is not functioning correctly	Fragment fusion interferes with protein expression/function/stability	Fuse the PCA fragment to the other end of the protein
6.2.3. (3)	Poor Luminescence signal of Luminescence assay	Signal integration time is too short	Optimize the signal Integration times
		Not enough substrate	Increase the substrate concentration
		Not enough cells used	Increase the number of cells used per assay
	No or low signal modulation after Stimulus or Inhibitor treatment	Number of cells and signal integration times are not optimal	Optimize the number of cells and signal integration times
		Stimulus or Inhibitor concentration are too low or duration of treatment is not optimal	Try different stimulus or inhibitor treatment times and or concentrations
		Signal detection time is not optimal	Peak signal occurs immediately after addition of colelantrezines. Try optimizing the beginning of signal integration after substrate addition
		Signal-to-background ratio is low	If the signal is very low, find an optimal way to extract the meaningful signal from background. Test appropriate positive and negative controls for the interaction you are studying

bioluminescence analysis. Optimize the signal integration times depending on the bioluminescence signal strength. For real time kinetics experiments, add or inject the substrate, immediately initiate the bioluminescence readings with the optimized signal integration time continuously for the desired period. Then, background correct the bioluminescence signals to obtain meaningful signal. Afterwards, normalize the data to total protein concentration in cell lysates if desired (Bio-Rad protein assay).

Critical step: Specific functional assays to further characterize a PPI might be performed here. For example, incubation of cells with agents, such as specific enzyme or transport inhibitors, can be performed for various amount of time, prior to the Luminometric analysis. Troubleshooting advice can be found in Table 14.4.

6.2.4. Timeline

Cotransformation of competent yeast (steps 1–5): 30–45 min (depending on the number of samples) plus 48–72 h for cell growth (step 6).

Fusion of PCA fragments at the chromosomal loci: 5 h (depending on the number of samples) plus 48–72 h for cell growth.

Preparation of cells for Luminometric analysis (steps 1–3): 24 h.

Microplate reader analysis: a few minutes to hours, depending on the signal integration time and the number of samples.

6.2.5. Anticipated results

The luminescence intensity of the reassembled Rluc and Gluc PCAs vary with the strength of interaction between the protein pairs attached to the PCA fragments. In the case of our simplest positive control (GCN4 leucine zipper pair fused to the PCA fragments: e.g., Zip-Rluc-F[1] + Zip-Rluc-F[2]), the reconstituted PCAs represent approximately 10–30% of the activity of the full-length Rluc or Gluc enzymes. The PCA fusions expressed alone should not result in detectable luminescence (compared to nontransfected cells) because the individual PCA fragments have no activity. For each study, positive (known interaction) and particularly negative (non-interacting proteins) controls should always be performed in parallel. A PCA response should not be observed if noninteracting proteins are used as PCA partners.

REFERENCES

Benton, R., *et al.* (2006). Atypical membrane topology and heteromeric function of Drosophila odorant receptors in vivo. *PLoS Biol.* **4**, e20.

Cabantous, S., Terwilliger, T. C., *et al.* (2005). Protein tagging and detection with engineered self-assembling fragments of green fluorescent protein. *Nat. Biotechnol.* **23**(1), 102–107.

Campbell-Valois, F. X., Tarassov, K., et al. (2005). Massive sequence perturbation of a small protein. Proc. Natl. Acad. Sci. USA 102(42), 14988–14993.

Carpenter, A. E., Jones, T. R., et al. (2006). CellProfiler: Image analysis software for identifying and quantifying cell phenotypes. Genome Biol. 7(10), R100.

Choi, K. Y., Satterberg, B., et al. (1994). Ste5 tethers multiple protein kinases in the MAP kinase cascade required for mating in S. cerevisiae. Cell 78(3), 499–512.

Chou, S., Huang, L., et al. (2004). Fus3-regulated Tec1 degradation through SCFCdc4 determines MAPK signaling specificity during mating in yeast. Cell 119(7), 981–990.

Collins, S. R., Schuldiner, M., et al. (2006). A strategy for extracting and analyzing large-scale quantitative epistatic interaction data. Genome Biol. 7(7), R63.

Collins, S. R., Kemmeren, P., et al. (2007). Toward a comprehensive atlas of the physical interactome of Saccharomyces cerevisiae. Mol. Cell Proteomics 6(3), 439–450.

Ding, Z., et al. (2006). A retrovirus-based protein complementation assay screen reveals functional AKT1-binding partners. Proc. Natl. Acad. Sci. USA 103, 15014–15019.

Dudley, A. M., Janse, D. M., et al. (2005). A global view of pleiotropy and phenotypically derived gene function in yeast. Mol. Syst. Biol. 1(2005), 0001.

Ear, P. H., and Michnick, S. W. (2009). A general life-death selection strategy for dissecting protein functions. Nat. Methods 6(11), 813–816.

Ercikan-Abali, E. A., Waltham, M. C., et al. (1996). Variants of human dihydrofolate reductase with substitutions at leucine-22: Effect on catalytic and inhibitor binding properties. Mol. Pharmacol. 49(3), 430–437.

Ghaemmaghami, S., Huh, W. K., et al. (2003). Global analysis of protein expression in yeast. Nature 425(6959), 737–741.

Ghosh, I., Hamilton, A. D., et al. (2000). Antiparallel leucine zipper-directed protein reassembly: Application to the green fluorescent protein. J. Am. Chem. Soc. 122(23), 5658–5659.

Giaever, G., Chu, A. M., et al. (2002). Functional profiling of the Saccharomyces cerevisiae genome. Nature 418(6896), 387–391.

Hu, C. D., Chinenov, Y., et al. (2002). Visualization of interactions among bZIP and Rel family proteins in living cells using bimolecular fluorescence complementation. Mol. Cell 9(4), 789–798.

Jansen, R., and Gerstein, M. (2004). Analyzing protein function on a genomic scale: The importance of gold-standard positives and negatives for network prediction. Curr. Opin. Microbiol. 7(5), 535–545.

Knop, M., Siegers, K., et al. (1999). Epitope tagging of yeast genes using a PCR-based strategy: More tags and improved practical routines. Yeast 15(10B), 963–972.

Kurtz, J. E., Exinger, F., et al. (1999). New insights into the pyrimidine salvage pathway of Saccharomyces cerevisiae: Requirement of six genes for cytidine metabolism. Curr. Genet. 36(3), 130–136.

Levy, E. D., Landry, C. R., et al. (2009). How perfect can protein interactomes be? Sci. Signal. 2(60), pe11.

Macdonald, M. L., Lamerdin, J., et al. (2006). Identifying off-target effects and hidden phenotypes of drugs in human cells. Nat. Chem. Biol. 2(6), 329–337.

Magliery, T. J., Wilson, C. G., et al. (2005). Detecting protein–protein interactions with a green fluorescent protein fragment reassembly trap: Scope and mechanism. J. Am. Chem. Soc. 127(1), 146–157.

Manderson, E. N., Malleshaiah, M., et al. (2008). A novel genetic screen implicates Elm1 in the inactivation of the yeast transcription factor SBF. PLoS ONE 3(1), e1500.

Memarian, N., Jessulat, M., et al. (2007). Colony size measurement of the yeast gene deletion strains for functional genomics. BMC Bioinformatics 8, 117.

Metodiev, M. V., Matheos, D., et al. (2002). Regulation of MAPK function by direct interaction with the mating-specific Galpha in yeast. Science 296(5572), 1483–1486.

Michnick, S. W., Remy, I., *et al.* (2000). Detection of protein–protein interactions by protein fragment complementation strategies. *Methods Enzymol.* **328,** 208–230.

Michnick, S. W., Ear, P. H., *et al.* (2007). Universal strategies in research and drug discovery based on protein-fragment complementation assays. *Nat. Rev. Drug Discov.* **6**(7), 569–582.

Mumberg, D., Muller, R., *et al.* (1995). Yeast vectors for the controlled expression of heterologous proteins in different genetic backgrounds. *Gene* **156**(1), 119–122.

Nagai, T., Ibata, K., *et al.* (2002). A variant of yellow fluorescent protein with fast and efficient maturation for cell-biological applications. *Nat. Biotechnol.* **20**(1), 87–90.

Nyfeler, B., Michnick, S. W., *et al.* (2005). Capturing protein interactions in the secretory pathway of living cells. *Proc. Natl. Acad. Sci. USA* **102**(18), 6350–6355.

Nyfeler, B., *et al.* (2008). Identification of ERGIC-53 as an intracellular transport receptor of {alpha}1-antitrypsin. *J. Cell. Biol.* **180,** 705–712.

Pelletier, J. N., and Michnick, S. W. (1997). A Protein Complementation Assay for Detection of protein–protein interactions *in vivo. Protein Eng.* **10**(Suppl.), 89.

Pelletier, J. N., Campbell-Valois, F. X., *et al.* (1998). Oligomerization domain-directed reassembly of active dihydrofolate reductase from rationally designed fragments. *Proc. Natl. Acad. Sci. USA* **95**(21), 12141–12146.

Pelletier, J. N., Arndt, K. M., *et al.* (1999). An in vivo library-versus-library selection of optimized protein–protein interactions. *Nat. Biotechnol.* **17**(7), 683–690.

Remy, I., and Michnick, S. W. (1999). Clonal selection and in vivo quantitation of protein interactions with protein-fragment complementation assays. *Proc. Natl. Acad. Sci. USA* **96**(10), 5394–5399.

Remy, I., and Michnick, S. W. (2001). Visualization of biochemical networks in living cells. *Proc. Natl. Acad. Sci. USA* **98**(14), 7678–7683.

Remy, I., and Michnick, S. W. (2004). A cDNA library functional screening strategy based on fluorescent protein complementation assays to identify novel components of signaling pathways. *Methods* **32**(14), 381–388.

Remy, I., and Michnick, S. W. (2006). A highly sensitive protein–protein interaction assay based on Gaussia luciferase. *Nat. Methods* **3**(12), 977–979.

Remy, I., Wilson, I. A., *et al.* (1999). Erythropoietin receptor activation by a ligand-induced conformation change. *Science* **283**(5404), 990–993.

Remy, I., Montmarquette, A., *et al.* (2004). PKB/Akt modulates TGF-beta signalling through a direct interaction with Smad3. *Nat. Cell Biol.* **6**(4), 358–365.

Sheff, M. A., and Thorn, K. S. (2004). Optimized cassettes for fluorescent protein tagging in *Saccharomyces cerevisiae. Yeast* **21**(8), 661–670.

Stefan, E., Aquin, S., *et al.* (2007). Quantification of dynamic protein complexes using Renilla luciferase fragment complementation applied to protein kinase A activities in vivo. *Proc. Natl. Acad. Sci. USA* **104**(43), 16916–16921.

Subramaniam, R., Desveaux, D., *et al.* (2001). Direct visualization of protein interactions in plant cells. *Nat. Biotechnol.* **19**(8), 769–772.

Tannous, B. A., Kim, D. E., *et al.* (2005). Codon-optimized Gaussia luciferase cDNA for mammalian gene expression in culture and in vivo. *Mol. Ther.* **11**(3), 435–443.

Tarassov, K., Messier, V., *et al.* (2008). An in vivo map of the yeast protein interactome. *Science* **320**(5882), 1465–1470.

Wilson, C. G., Maglery, T. J., *et al.* (2004). Detecting protein–protein interactions with GFP-fragment reassembly. *Nat. Methods* **1**(3), 255–262.

CHAPTER FIFTEEN

YEAST LIPID ANALYSIS AND QUANTIFICATION BY MASS SPECTROMETRY

Xue Li Guan,[*] Isabelle Riezman,[‡] Markus R. Wenk,[*,†] and Howard Riezman[‡]

Contents

Abstract

The systematic and quantitative analysis of the different lipid species within a cell or an organism has recently become possible and the general approach has been termed "lipidomics." Traditional methods of identification and quantification of lipid species were laborious processes and it was necessary to use a wide variety of techniques to analyse the different lipid species, especially concerning the assigning of particular acyl chain lengths, hydroxylations, and desaturations to the diverse lipid species. While it is still not possible to quantitatively analyze all lipid species in one fell swoop, great progress has been made with the intensive use of quantitative mass spectrometry approaches. It is now relatively simple to quantify most of the lipid species, including all of the major ones, in a yeast cell. Different degrees of sophistication of mass spectrometric analysis exist and the available techniques and instrumentation are evolving rapidly. Therefore, we have decided to present robust, simple methods to quantify the major yeast lipids by mass spectrometry that should be accessible to anyone who has access to a standard mass

* Department of Biochemistry, Yong Loo Lin School of Medicine, National University of Singapore, Singapore
† Department of Biological Sciences, National University of Singapore, Singapore
‡ Department of Biochemistry, University of Geneva, Switzerland

Methods in Enzymology, Volume 470
ISSN 0076-6879, DOI: 10.1016/S0076-6879(10)70015-X

spectrometry equipment. The methods to identify and quantify yeast glycero-phospholipids and sphingolipids involve electrospray ionization mass spectrometry using fragmentation to characterize the lipid species. A simplified gas chromatographic method is used to quantify the major sterols that occur in wild-type yeast cells and ergosterol biosynthesis mutants.

1. Introduction

Studies in yeast, notably *Saccharomyces cerevisiae*, have played a particularly important role in the advancement of our knowledge in biology. An indispensable aspect of yeast research, in addition to genetics and molecular biology, is lipid biochemistry and analysis. Lipids make up a bulk of cellular membrane and have gained immense interest due to their emerging new cellular functions other than their structural roles.

Three major classes of membrane lipids in all eukaryotic organisms include glycerophospholipids, sphingolipids, and sterols and each class is structurally diverse, arising from both nonpolar chains and head group substitutions (Fig. 15.1). The glycerophospholipids are diversified by their head group and fatty acyl compositions and the major head groups are conserved among all eukaryotes. Lipid composition varies between organelles (Schneiter *et al.*, 1999). On the other hand, the sphingolipids of *S. cerevisiae* are distinct and relatively simple compared to mammals. There

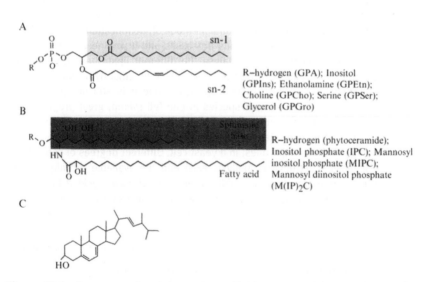

Figure 15.1 Structures of major membrane lipids in *S. cerevisiae*: (A) glycerophospholipids, (B) sphingolipids, (C) ergosterol. R refers to possible head group modifications.

are three classes of inositol-containing sphingolipids; inositol-phosphorylceramide (IPC), mannose-inositol-phosphorylceramide (MIPC), and mannose-(inositol-P)$_2$-ceramide (MIP$_2$C), localized mainly to the plasma membrane where they constitute 20–30% of all the lipids (Patton and Lester, 1991). Their biosynthesis begins in the endoplasmic reticulum (ER) and is completed in the Golgi apparatus (Dickson et al., 2006). Another distinct characteristic of the lipid composition of a normal yeast cell is that instead of cholesterol, it contains approximately 75% ergosterol (ergosta-5,7,22-trienol) and a variety of other sterols, most of which are intermediates in the sterol biosynthesis pathway (Munn et al., 1999). Most of the ergosterol is found in the plasma membrane and many of the other sterol molecules are enriched in the ER where sterol synthesis occurs (Zinser et al., 1993). Sterols can be esterified and stored in lipid bodies from where they can be rapidly mobilized (Wagner et al., 2009). Although the major lipids and the biosynthetic machineries in S. cerevisiae have been extensively characterized (Daum et al., 1998; Ejsing et al., 2009; Guan and Wenk, 2006), novel chemical entities may have escaped discovery in previous analyses and warrants development of high resolution and sensitive methodologies for detection and characterization. Furthermore, the molecular signature of individual lipids is increasingly recognized to encode for highly specific, though not necessarily unique, functions. This has been further supported by recent revelations that the intricate interactions between specific lipids and proteins are important for cellular functions (Guan et al., 2009; Valachovic et al., 2004). High resolution analysis of lipids therefore is critical for enhancing our understanding of the biological roles of lipids.

Traditional methods of lipid analysis include metabolic labeling, thin-layer chromatography (TLC) and gas chromatography (GC) (Wenk, 2005). Metabolic labeling using lipid precursors (such as radiolabeled inositol, ethanolamine, or fatty acyls) have been widely used to (selectively) label certain classes of lipids, which will typically be separated by TLC and visualized by autoradiography. However, these techniques only deliver mass levels of lipids under conditions of steady-state incorporation of the label. Moreover, generally, TLC separation is of low resolution and does not provide molecular information of the diversity of fatty acyl compositions, which would require subsequent analysis, often by GC. Nonetheless, although these methods suffer from limited sensitivity, selectivity, and resolution, they are still commonly used today due to the relative ease and low investment cost in instrumentation.

In the last 15–20 years, the field of lipid research has been spurred by the development of mass spectrometry (MS), and in particular soft ionization such as electrospray ionization (ESI) and matrix-assisted laser desorption ionization (MALDI). These methods have allowed rapid and sensitive profiling, characterization as well as quantification of lipids in complex lipid mixtures (Han et al., 2008; Shui et al., 2007). However, despite continuous advancement in the technology, such as achieving enhanced sensitivity by microfluidics-based ionization (Han *et al.*, 2008) and high resolution by MS such as Fourier transform ion cyclotron resonance (FT-ICR) or Orbitrap MS (Schwudke *et al.*, 2007), there is no one method to probe the entire lipidome of a cell. *S. cerevisiae* has a relatively simple lipidome compared to other eukaryotic cells, yet it is only recently that a fairly comprehensive analysis of its lipidome has been described, and such a detailed catalogue is still currently unavailable for the relatively more complex lipidomes of cell types in other organisms. Ejsing and coworkers described the absolute quantification of 250 lipid molecular species in *S. cerevisiae* and estimated that the coverage of the lipids analyzed encompasses 95% of the yeast lipidome (Ejsing *et al.*, 2009). Failure in the complete annotation and measurement of an entire cellular lipidome is attributed to the complex chemistry between different classes of lipids as well as the dynamic range of their concentration, which limits their extractability and ionization using a single method. In other words, depending on the lipid(s) of interest, employing multiple extraction protocols, as well as different analytical tools are still required to examine the lipidome for any given biological sample.

In this chapter, we describe the extraction and mass spectrometric analysis of various classes of lipids, including glycerophospholipids, sphingolipids, and sterols from yeast, specifically *S. cerevisiae*. Important considerations for isolation of lipids from the cellular milieu include recovery and potential activation of lipases during the extraction procedure. While protocols based on Folch and Bligh and Dyer methods (Bligh and Dyer, 1959; Folch *et al.*, 1957) are seemingly efficient for isolating lipids from mammalian cells and tissues, these methods are effective for isolating glycerophospholipids and sphingolipids from broken yeast cells but not from intact cells (Hanson and Lester, 1980). Furthermore, a technical challenge is recovery of the entire cellular lipidome due to the diverse chemistry of lipids in cells. Polarity of lipids differ depending on the lipid backbone as well as the head group modification and therefore the selection of organic solvents is critical. For instance, phosphorylation of sphingolipids renders them highly polar, and thus they may escape into the aqueous milieu during phase separation using Folch or Bligh and Dyer's methods. Here, we describe a modified protocol for isolation of lipids from intact yeast cells based on Angus and Lester's method (Angus and Lester, 1972). This involves the use of a slightly alkaline mixture of ethanol–diethyl ether–water–pyridine–ammonium hydroxide at elevated temperatures, which leads to effective extraction of inositol-containing lipids and glycerophosphatidylcholine from intact *S. cerevisiae* and *Neurospora crassa* (Hanson and Lester, 1980). However, for analysis of subcellular organelles, the modified Bligh and Dyer method may be applicable. An additional isolation method will be added to analyze total sterols, including an alkaline hydrolysis and extraction with petroleum ether.

Various MS-based methods will be described for the analysis of the different classes of lipids, including (1) the direct analysis of major yeast glycerophospholipids and sphingolipids in complex lipid mixtures using ESI-MS and tandem mass spectrometry (MS/MS) and (2) analysis of sterols and metabolites by GC–MS (Fig. 15.2). Lipid analysis of yeast using ESI-MS is not a novel technique and dates back to the 1990s (Hechtberger *et al.*, 1994; Schneiter *et al.*, 1999) but comprehensive characterization and quantification of multiple classes of lipids in an "omic-centric" fashion has only been reported in recent years (Ejsing *et al.*, 2006, 2009; Guan and Wenk, 2006). For successful MS analysis, often, prior knowledge of the chemistry and the fragmentation patterns of lipids of interest is required. Numerous works have been reported, particularly for the analysis of glycerophospholipids and sphingolipids in mammalian cells and tissues (Brugger *et al.*, 1997; Han and Gross, 2005; Sullards, 2000; Taguchi *et al.*, 2005) and inference can be drawn for lipid analysis in yeast. Moreover, often instruments have accompanying software that allow automated acquisition (termed data-dependent acquisition) (Schwudke *et al.*, 2006). Despite the wealth of information available and the ease of automation, we nonetheless aim to provide a detailed description of the operation to allow the reader to reproduce the techniques in probing yeast lipidomes as it should be noted that the lipid inventory depends on culture conditions (Tuller *et al.*, 1999) and further differs between yeast species (Jungnickel *et al.*, 2005), for example, *Schizosaccharomyces pombe* versus *S. cerevisiae* (Shui *et al.*, under revision).

Previous methods have been published to analyze yeast sterols by GC with or without MS (van den Hazel *et al.*, 1999; Xu and Nes, 1988; Zinser *et al.*, 1991). These methods usually involve silylation of the isolated sterol sample before GC, which improves separation, but which also has some disadvantages. In this chapter, we will present a simple protocol for the semiquantitative analysis of free and total yeast sterols by GC–MS. The term semiquantitative is used because appropriate standards are not commercially available for the large variety of sterols present in wild-type (WT) and mutant yeast cells and therefore, their exact quantities cannot be measured by GC–MS.

2. METHODS

2.1. Sample preparation

Cells (10 OD_{600} units) are harvested by addition of trichloroacetic acid (TCA) to the media to a final concentration of 5% (w/v) and the cells are put onto ice. The cooled cells are pelleted by centrifugation at $3000 \times g$ for 5 min and the pellets are washed once with ice-cold 5% TCA, followed by

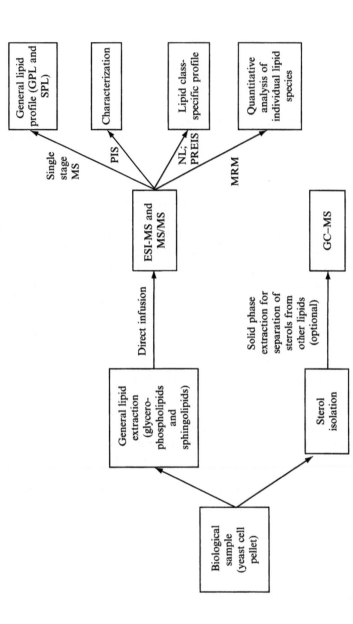

Figure 15.2 General strategy for the extraction and analysis of major membrane lipids from yeast. Glycerophospholipids, sphingolipids, and sterols are extracted according to respective procedures. The former are analyzed by ESI-MS using various scan modes including single stage MS, product ion scan (PIS), precursor ion scan (PREIS), neutral loss (NL) scan, and multiple-reaction monitoring (MRM) for profiling, species identification, and quantification. Sterols are analyzed by GC–MS.

another wash with ice-cold deionized water. The cells are transferred to screw cap tubes and they can be snap-frozen in liquid nitrogen and stored at $-80\ ^{\circ}\text{C}$ or immediately used for extraction. Except for Teflon, all plastic should be avoided in subsequent procedures involving organic solvents because they leach contaminants into the solution. Furthermore, handling of organic solvents should be carried out in a fume hood.

2.2. Normalization for starting amount of material

Two parameters can be used for normalizing the amount of lipids:

(1) *Dry weight*: In this case, cells are freeze-dried and weighed before extraction.
(2) *Protein amount*: An equivalent OD of cells used for lipid extraction is used for protein determination using standard assays.

2.3. Lipid extraction (for glycerophospholipids and sphingolipids)

Cells (10 OD_{600} units) are resuspended in 1 ml of extraction solvent containing 95% ethanol, water, diethyl ether, pyridine, and 4.2 N ammonium hydroxide in the ratio of 15:15:5:1:0.18 by volume. A cocktail of lipid standards (Table 15.1) and $100\mu L$ of glass beads are added. It should be noted that lipid materials removed from storage should be allowed to reach room temperature before opening for use. In addition, the standards are stored as concentrated stock (mM to M) for stability, and are diluted in ethanol prior to use (Moore *et al.*, 2007). We routinely sonicate the lipids to ensure complete solubilization.

The sample is vortexed vigorously for several minutes and incubated at 60 °C for 20 min. Cell debris are pelleted by centrifugation at $10,000 \times g$ for 10 min and the supernatant is collected. One milliliter of extraction solvent

Table 15.1 List of synthetic standards for quantitative analysis of major yeast lipids

Standard	Source	Ionization mode	MRM transition (Q1/Q3)
Didodecanoyl GPEtn	Avanti	Negative	578.6/196.1
Didodecanoyl GPSer	Avanti	Negative	622.6/535.5
Dioctanoyl GPIns	Echelon	Negative	585.5/241.1
d18:1/19:0 Ceramide	Matreya	Positive	580.7/264.2
Dodecanoyl, tridecanoyl GPA	Avanti	Negative	549.5/153.1
Didecanoyl GPGro	Avanti	Negative	553.6/153.1
Dinonadecanoyl GPCho	Avanti	Positive	818.7/184.1
Cholesterol	Avanti	Negative	386.3/368.3

is added to the pellet, which is vortexed again, and the extraction is repeated. The supernatant is pooled together and dried under a stream of nitrogen or in a Centrivap (Labconco Corporation, Kansas City, MO). In the event of using a Centrivap, the temperature should be raised gradually to 60 °C to avoid boiling.

For complex lipids including glycerophospholipids, sphingolipids, and neutral lipids, the extract is desalted using butanol and water. The lipid film is resuspended in 300 μl of water-saturated butanol and 150 μl of water is added. The mixture is vortexed vigorously, followed by centrifugation at 10,000×g for 2 min. The top butanol phase is recovered and the aqueous phase is reextracted again with 300 μl of water-saturated butanol. The butanolic phase is pooled together and dried under a stream of nitrogen or with a Centrivap (50–60 °C). The lipid film is resuspended in 400 μl of chloroform–methanol (1:1, v/v) for MS analysis.

2.3.1. Alkaline methanolysis (optional)

For analysis of ceramides and complex sphingolipids, an optional step of mild alkaline hydrolysis is recommended to reduce ion suppression by other lipids such as glycerophospholipids during MS analysis. This treatment hydrolyzes acyl bonds (and hence the majority of glycerophospholipids) but leaves amide linkages largely intact (Brockerhoff, 1963). Two different methods are available. In the first method, the lipid film from the pyridine extraction is reconstituted in 400 μl of a mixture of chloroform, methanol, and water in the ratio of 16:16:5 by volume. Four hundred microliters of 0.2 N of methanolic NaOH is added and the mixture is incubated for 1 h at 37 °C on a thermoshaker with mild shaking. Eighty microliters of 1 N acetic acid is then added to neutralize the mixture, followed by the addition of 400 μl of 0.5 M EDTA and 400 μl of chloroform. The sample is vortexed vigorously for 1 min, followed by centrifugation at 10,000×g for 2 min. The lower organic phase is recovered and the aqueous phase is reextracted with 600 μl of chloroform. The organic phase is pooled together and dried under a stream of nitrogen or in the Centrivap (Guan and Wenk, 2006). In the second method, the lipid film is reconstituted in 400 μl of monomethylamine reagent (methanol:H$_2$O:n-butanol:methylamine = 4:3:1:5 by volume) and incubated for 1 h at 53 °C (Cheng *et al.*, 2001). The mixture is then dried under a stream of nitrogen or in a Centrivap (50–60 °C). The lipid extract is then desalted using butanol and water as described above.

2.4. Sterol isolation

In order to quantify free, nonesterified sterols, the procedure used for the isolation of glycerophospholipids and sphingolipids cannot be used because it results in partial hydrolysis of sterol esters. Therefore, cells are extracted with a chloroform/methanol procedure. Cells (10 OD$_{600}$ units) are harvested as above. Cholesterol (2 nmol) and 100 μl glass beads are added to the cell pellet, which is resuspended in 50 μl water and 50 μl methanol is added. After vortexing, 100 μl of chloroform is added and the sample is vortexed for 6 min followed by centrifugation at $100 \times g$ for 5 min. The supernatant is transferred to a new tube and the pellet is vortexed again with 100 μl of chloroform/methanol (2:1). After centrifugation the supernatants are combined. Thirty-four microliters of 0.034% $MgCl_2$ is added, vortexed, and centrifuged at $800 \times g$ for 5 min. The aqueous phase is removed and 34 μl of 2 M KCl/methanol (4:1) is added to the lower phase. After vortexing and centrifugation the aqueous phase is removed. This procedure is repeated two times with 34 μl of artificial upper phase (chloroform: methanol:water, 3:48:47). The organic phase is then removed without taking the protein interface and dried. Sterol esters are present in this preparation, but are difficult to quantify by GC–MS. To determine the amounts of total cellular sterols (esterified and nonesterified) the following protocol is used. Two nanomoles of cholesterol are added to the cell pellet as internal standard. Cells are resuspended with 300 μl 60% KOH to which is added 300 μl methanol containing 0.5% pyrogallol and 450 μl methanol in a screw cap glass tube. The tube is placed at 85 °C and once hot, the cap is tightly closed. After 2-h of incubation with occasional mixing, the sample is returned to room temperature and the sterols are extracted three times with 1 mL of petroleum ether. The combined petroleum ether phases are dried and analyzed by GC–MS.

2.4.1. Solid phase extraction for separation of sterols from other lipids (optional)

A column of Chromabond® SiOH (Macherey-Nagel, Germany) (0.1 g) is washed with two times 1 ml chloroform. Total lipid extract from 50 OD_{600} units of cells is resuspended in 0.25 ml chloroform by vortexing and sonication. The extract is then applied to the column and eluted with two times 0.65 ml chloroform. The flow through and chloroform elutions are combined and dried (sterol extract). The column is then eluted with three times 0.5 ml methanol. The combined methanol elutions are dried and can be used as a total lipid extract as above.

2.5. Lipid analysis

2.5.1. Direct analysis of glycerophospholipids and sphingolipids in complex lipid mixtures by ESI-MS and ESI-MS/MS

Glycerophospholipids, ceramides, and inositol sphingolipids are analyzed by direct infusion using ESI-MS and MS/MS (Fig. 15.3). The samples from the butanolic extraction or after alkaline hydrolysis are reconstituted in 400 μl of chloroform–methanol (1:1, v/v). Samples are centrifuged at maximum speed for 10 min to remove any residues. A high-performance

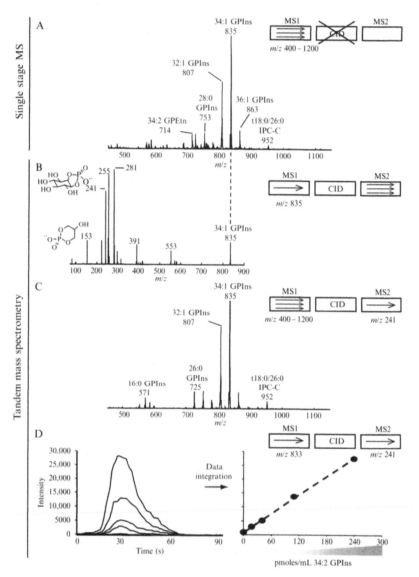

Figure 15.3 Analysis of yeast lipids by electrospray ionization–mass spectrometry (ESI-MS) and tandem mass spectrometry (MS/MS). (A) Single stage ESI-MS in the negative ion mode. The majority of glycerophospholipids and sphingolipids are detected in the mass range of 400–1200. The ions can be tentatively assigned by their mass-to-charge (m/z) ratio. Characterization of ions can be achieved by collision-induced dissociation (CID) and MS/MS. (B) MS/MS spectra of ions with m/z 835. (C) Precursor ion scans for lipids containing inositol phosphate head group (m/z 241). Samples can be spiked with internal standards, which are typically not found naturally in the samples under investigation, to allow for semiquantitative profiling. (D) Overlay of chromatogram (left panel) and standard curve (right panel) obtained from

liquid chromatography system (Agilent Technologies, Santa Clara, CA) is coupled to a 4000 Qtrap triple quadrupole mass spectrometer (Applied Biosystems, Foster City, CA) with an ESI source. The mass spectrometer is operated in both the positive and negative ion modes, depending on the propensity of the class of lipids of interest to ionize (Table 15.2). The spray voltage in the positive ion mode is 5.5 and 4.5 kV in the negative mode. The MS is operated with a curtain gas of 20 psi, source temperature of 250 °C and both ion source gas 1 and 2 set to 30 psi. For all MS/MS experiments, each individual ion dissociation pathway has to be optimized with regard to collision energy and declustering potential to minimize variations in relative ion abundance due to differences in rates of dissociation (Table 15.3 and Fig. 15.5). Samples are introduced into the mass spectrometer by loop injections with chloroform–methanol (1:1 v/v) as a mobile phase at a flow rate of 200 μl/min. Typically, 25 μl of samples is injected for analysis.

2.5.1.1. Single stage (nontargeted) profiling
The mass spectrometer is operated in enhanced MS mode to obtain the profiles of the total lipid extract. Measurement of lipids by ESI-MS is based on the ability of each class of lipids to acquire positive or negative charges when in solution during ionization and the structures of these lipids entail their inherent ionization property. Table 15.2 summarizes the charge states of major glycerophospholipids and sphingolipids. An example of a single stage MS profile, in the negative ion mode, of an extract obtained from S. cerevisiae grown in rich media (yeast–peptone–dextrose) to logarithmic phase is shown in Fig. 15.3A. Typically, the scan range is between m/z 400 and 1200, which includes the detection of various classes of lipids including glycerophosphatidylethanolamine (GPEtn), glycerophosphatidylserine (GPSer), glycerophosphatidic acid (GPA), glycerophosphatidylglycerol (GPGro), glycerophosphatidylinositol (GPIns), ceramide, and inositol sphingolipids.

The profiling of complex lipid mixtures using single stage MS may serve as an initial screen when different conditions or strains are to be compared. The comparison of profiles can be achieved by multivariate analysis or simply generating a differential profile which displays differences in ion response between the two conditions of interest and softwares for analysis, alignment, and comparison of full scan MS, are increasingly available (De Vos et al., 2007; Song et al., 2009; Wong et al., 2005). Such a nontargeted survey may serve as a starting point to reveal perturbations in the lipid

---

quantification of varying concentrations of a commercially available 34:2 GPIns by multiple-reaction monitoring (MRM). Selective quantification can be attained with a reasonably good linearity. Note that 34:2 GPIns is a minor ion in the complex lipid mixture and MRM offers a selective and sensitive method for quantification.

Table 15.2 Precursor ions, MS/MS scan modes and associated parameters for analysis of major glycerophospholipids and sphingolipids in *S. cerevisiae*

Lipid	Precursor ion	MS/MS modes	Fragment	Mass range (m/z)	Declustering potential (V)	Collision energy (V)
Glycerophospholipid						
GPA	$[M - H]^-$	PREIS 153	Glycerophosphate derivative	370–800	−75	−40 to −60
GPGro	$[M - H]^-$	PREIS 153	Glycerophosphate derivative	400–800	−75	−40 to −60
	$[M + NH4]^+$	NL 189	Phosphoglycerol			
GPCho	$[M + H]^+$	PREIS 184	Phosphocholine	400–900	70	45–65
GPEtn	$[M - H]^-$	PREIS 196	Glycerophosphoethanolamine derivative	400–800	−75	−40 to −65
	$[M + H]^+$	NL 141	Phosphoethanolamine			
GPSer	$[M - H]^-$	NL 87	Serine	400–800	−75	−25
	$[M - H]^-$	PREIS 153	Glycerophosphate derivative	400–800	−75	−40 to −60
	$[M + H]^+$	NL 185	Phosphoserine			
GPIns	$[M - H]^-$	PREIS 241	Cyclic inositol phosphate	450–900	−90	−45 to −60
	$[M - H]^-$	PREIS 153	Glycerophosphate derivative	450–900	−75	−45 to −65
	$[M + NH4]^+$	NL 277	Phosphoinositol			
Sphingolipid						
Phytoceramide	$[M + H]^+$	PREIS 266	Double dehydration product of d18:0 sphingoid base	600–800	80	40–50
	$[M + H]^+$	PREIS 282	Double dehydration product of t18:0 sphingoid base		80	40–50
	$[M + H]^+$	PREIS 294	Double dehydration product of d20:0 sphingoid base		100	40–50
	$[M + H]^+$	PREIS 310	Double dehydration product of t20:0 sphingoid base		100	40–50

IPC	[M – H]⁻	PREIS 241	Cyclic inositol phosphate	800–1000	–120	–60 to –70
	[M + H]⁺	PREIS 266	Double dehydration product of d18:0 sphingoid base		100	60–70
	[M + H]⁺	PREIS 282	Double dehydration product of t18:0 sphingoid base		100	60–70
	[M + H]⁺	PREIS 294	Double dehydration product of d20:0 sphingoid base		100	60–70
	[M + H]⁺	PREIS 310	Double dehydration product of f20:0 sphingoid base		100	60–70
MIPC	[M – H]⁻	PREIS 421	Cyclic inositol phosphate	950–1200	–160	–60 to –70
	[M + H]⁺	PREIS 266	Double dehydration product of d18:0 sphingoid base		100	75–80
	[M + H]⁺	PREIS 282	Double dehydration product of t18:0 sphingoid base		100	75–80
	[M + H]⁺	PREIS 294	Double dehydration product of d20:0 sphingoid base		100	75–80
	[M + H]⁺	PREIS 310	Double dehydration product of f20:0 sphingoid base		100	75–80
M(IP)₂C	[M – 2H]2⁻	PREI 241	Cyclic inositol phosphate	600–750	–180	–65 to –75

Table 15.3 Precursor/product ion m/z's for multiple-reaction monitoring of major species of *S. cerevisiae* glycerophospholipids

Molecular species	GPA		GPGro		GPEtn		GPSer		GPIns		GPCho	
	Q1 [M − H]⁻	Q3	Q1 [M − H]⁻	Q3	Q1 [M + H]⁺	Q3	Q1 [M − H]⁻	Q3	Q1 [M − H]⁻	Q3	Q1 [M + H]⁺	Q3
14:0 (Lyso)	379.4	153.1	453.4	153.1	426.4	285.5	466.4	379.4	543.4	241.1	468.4	184.1
16:1 (Lyso)	407.4	153.1	481.4	153.1	452.4	311.6	494.4	407.4	569.4	241.1	494.4	184.1
16:0 (Lyso)	409.4	153.1	483.4	153.1	454.4	313.5	496.4	409.4	571.4	241.1	496.4	184.1
18:1 (Lyso)	435.4	153.1	509.4	153.1	480.4	339.6	522.4	435.4	597.4	241.1	522.4	184.1
18:0 (Lyso)	437.4	153.1	511.4	153.1	482.4	341.5	524.4	437.4	599.4	241.1	524.4	184.1
26:1	561.6	153.1	635.6	153.1	606.7	465.7	648.7	561.7	723.7	241.1	648.6	184.1
26:0	563.6	153.1	637.6	153.1	608.7	467.7	650.7	563.7	725.7	241.1	650.6	184.1
28:1	589.6	153.1	663.7	153.1	634.7	493.7	676.7	589.7	751.7	241.1	676.7	184.1
28:0	591.6	153.1	665.7	153.1	636.7	495.7	678.7	591.7	753.7	241.1	678.7	184.1
30:1	617.6	153.1	691.7	153.1	662.7	521.7	704.7	617.7	779.7	241.1	704.7	184.1
30:0	619.6	153.1	693.7	153.1	664.7	523.7	706.7	619.7	781.7	241.1	706.7	184.1
32:2	643.7	153.1	717.7	153.1	688.7	547.7	730.7	643.7	805.7	241.1	730.7	184.1
32:1	645.7	153.1	719.7	153.1	690.7	549.7	732.7	645.7	807.7	241.1	732.7	184.1
32:0	647.7	153.1	721.7	153.1	692.7	551.7	734.7	647.7	809.7	241.1	734.7	184.1
34:2	671.7	153.1	745.7	153.1	716.7	575.7	758.7	671.7	833.7	241.1	758.7	184.1
34:1	673.7	153.1	747.7	153.1	718.7	577.7	760.7	673.7	835.7	241.1	760.7	184.1
34:0	675.7	153.1	749.7	153.1	720.7	579.7	762.7	675.7	837.7	241.1	762.7	184.1
36:2	699.7	153.1	773.7	153.1	744.7	603.7	786.7	699.7	861.7	241.1	786.7	184.1
36:1	701.7	153.1	775.7	153.1	746.7	605.7	788.7	701.7	863.7	241.1	788.7	184.1
36:0	703.7	153.1	777.7	153.1	748.7	607.7	790.7	703.7	865.7	241.1	790.7	184.1

profiles which cannot be easily anticipated (Guan *et al.*, 2006). For instance, Fig. 15.4 represents profiles obtained from a total lipid and sphingolipid extract obtained from WT cells and cells deficient in the acyltransferase, Slc1p. Note in panel A that the signal of a prominent inositol-phosphorylceramide (IPC, *m/z* 952, asterisk) is substantially increased after alkaline hydrolysis.

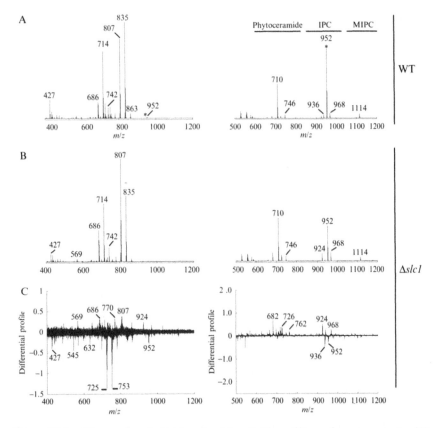

Figure 15.4 Glycerophospholipid and sphingolipid profiling of yeast mutants. (A) Typical phospholipid (left panels) and sphingolipid profiles (right panels) of wild-type (WT) yeast and (B) deletion mutants of the *slc1* gene. Mass spectra from at least four independent samples were averaged for each condition (*n* = 4). (C) Differential lipid profile, which are ratios of single stage mass spectrometry scans plotted as log 10 ratios, are used to compare differences in glycerophospholipid and sphingolipid composition between Δ*slc1* and WT. Note that this approach does not require knowledge of the underlying lipid species for a given ion of interest. Instead, it serves as an "unbiased" screening tool for discovery of lipids which are present in different amounts between two conditions. Reproduced and modified from Guan and Wenk (2006).

The spectra are aligned and the differences between ion intensity between the two conditions (Δ*slc1* vs. wt) are computed to generate the differential profiles as shown in Fig. 15.4C. The major differences lie in ions at m/z 725 and 753, which, based on the mass, can tentatively be assigned to GPIns with a total fatty acyl carbon number of 26 and 0 double bonds (26:0-GPIns) as well as 28:0-GPIns. The precise identity of ions can be confirmed by MS/MS.

A nontargeted profiling approach offers the advantage of discovering a previously uncharacterized lipid moiety or an unexpected lipid mediator under a given condition. However, the likelihood of missing a relevant lipid cannot be ruled out because of the incomplete detection of all lipid species due to the dynamic range and complex chemistries of lipids in a mixture, which is beyond the capacity of existing instrumentations. For instance, sterol is not detected under this condition and requires an alternative method (see Section 2.5.2).

2.5.1.2. Characterization by tandem mass spectrometry

To characterize an ion of interest, the mass spectrometer is set to product ion scan (PIS) mode, in which the first quadrupole is set to transmit a selected precursor ion, for instance, m/z 835 in negative ion mode. These ions enter the second quadrupole, often referred as the collision cell, where they are fragmented by collision-induced dissociation (CID) and the fragments are resolved by the third quadrupole which scans over a mass range, typically starting from m/z 70 to the m/z of the precursor ion selected. In this instance, CID of m/z 835 with collision energy of 55 V and declustering potential of $-$ 100 V yields fragments with m/z 153, 241, 255, and 281, which are dehydrated glycerophosphate ions, inositol phosphate ions, and decarboxylate ions of palmitic acid (16:0), and oleic acid (18:1), respectively (Fig. 15.3B). This identifies the ion to be a 34:1 GPIns. Public databases to facilitate interpretation of MS/MS data for a range of lipids are available, including LipidSearch (Taguchi *et al.*, 2007), LipidMaps MS tool (Fahy *et al.*, 2007), and LipidQA (Song *et al.*, 2007).

2.5.1.3. Targeted profiling and semiquantitative analysis of lipids by class using precursor ion and neutral ion scans

For selective detection and semiquantification of specific class of lipids, the mass spectrometer is operated in precursor ion (PREIS) or neutral loss (NL) scan. Table 15.2 summarizes the PREIS and NL scan modes for major glycerophospholipids and sphingolipids in *S. cerevisiae*. In the PREIS scan mode, the first quadrupole scans over a selected mass range and the ions sequentially enter the collision cell where collision energy is applied to induce fragmentation. The third quadrupole is set to transmit a selected single product ion, which is typically the diagnostic fragment of a specific class of lipid. For instance, a PREIS of m/z 241 in the negative mode scans ions with a phosphoinositol group which results in a mass spectrum that contains all precursor ions that decompose to produce the fragment with m/z 241 and in yeast, this generates a profile of GPIns, IPC, as well as doubly charged ions of M(IP)$_2$C (Fig. 15.3C). It should however be noted that a 15:0 fatty acyl shares the same m/z as the dehydrated phosphoinositol fragment and such data should be handled with care. Odd chain fatty acids are uncommon in *S. cerevisiae* cultured in rich media, and therefore the fragment of m/z 241 is considered specific for phosphoinositol-containing lipids.

An NL scan is used to profile several classes of glycerophospholipids when the charge of the head group does not localize to the lipid head group after fragmentation. In this mode, the first and third quadrupoles are linked and scanned at the same speed over the same mass range with a constant mass difference, for instance 87 amu in negative mode for the loss of serine, between the two analyzers. Because of the mass offset at any time, the third quadrupole will transmit product ions with a fixed lower m/z value than the mass selected precursor ions transmitted by the first quadrupole. The resultant mass spectrum contains all the precursor ions that lose a neutral species of selected mass, in the case of an NL of 87 amu, a glycerophosphatidylserine profile is generated.

2.5.1.4. *Quantitative analysis by multiple-reaction monitoring* Multiple-reaction monitoring (MRM) is a highly selective and sensitive method for quantification of specific lipids of interest. Quantitative analysis requires the use of internal standards, which are typically stable-isotope incorporated lipids or unnatural lipids which are usually chemically synthesized (Table 15.1). In MRM experiments, the first quadrupole is set to pass the precursor ion of interest to the collision cell where it undergoes CID. The third quadruple, Q3, is set to pass the structure-specific product ion characteristic of the lipid of interest. Again, these transitions require user-defined input of the parent ion (Q1) and product ion (Q3). Figure 15.5 and Table 15.3 summarize the MRM transitions for several major yeast sphingolipids and glycerophospholipids for quantification of these lipids in our previous work (Guan *et al.*, 2009; Guan and Wenk, 2006). A chromatogram is generated (Fig. 15.3D) and the area under each curve is integrated for each lipids measured. Lipid concentrations are calculated relative to the relevant internal standards which have been spiked in based on the amount of starting material and is expressed as moles (M)/mg dry weight or M/mg protein. It should be noted that the absolute quantification of inositol-containing lipids was limited by the availability of pertinent standards but such is not the case with the production of nonnatural lipids by genetic manipulation of yeast strains (Ejsing *et al.*, 2009).

2.5.2. Sterol analysis by GC–MS
Extracts are analyzed by GC–MS as follows. Samples are injected into a VARIAN CP-3800 gas chromatograph equipped with a Factor Four Capillary Column VF-5ms 15 m × 0.32 mm i.d. DF = 0.10 and analyzed by a

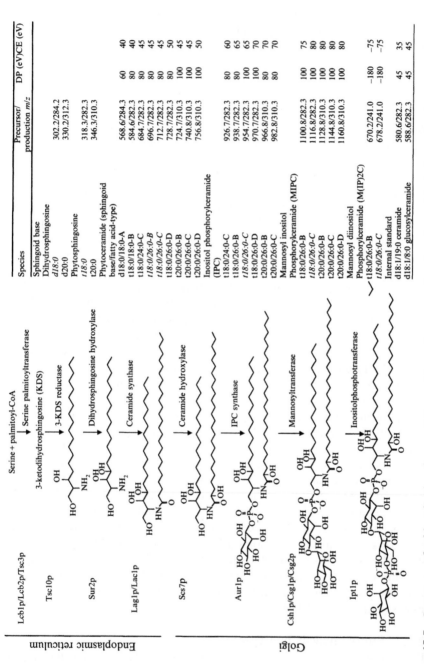

Figure 15.5 Summary of sphingolipid pathway of *S. cerevisiae* and precursor/product ion *m/z*'s and associated parameters for MRM detection of individual molecular species of sphingoid bases and complex ceramides. Reproduced and modified from Guan and Wenk (2006). Abbreviations: DP, declustering potential; CE, collision energy.

Varian 320 MS triple quadrupole with electron energy set to −70 eV at 250 °C. Samples are applied with the column oven at 45 °C, held for 4 min, then raised to 195 °C (20 °C/min). Sterols are eluted with a linear gradient from 195 to 230 °C (4 °C/min), followed by raising to 320 °C (10 °C/min). Finally, the column temperature is raised to 350 °C (6 °C/min) to elute sterol esters. This protocol allows sufficient separation of sterols as judged by elution of standards and extracts from ergosterol biosynthesis mutants of known sterol composition. An example of a chromatogram of a sample of WT yeast is shown (Fig. 15.6) and the profiles of a crude extract and one purified over SPE are compared. Standard curves can be constructed by

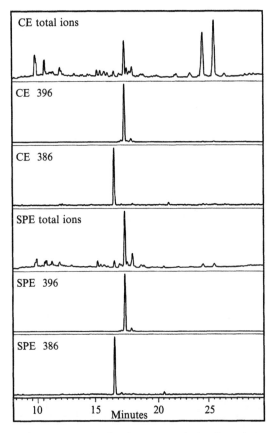

Figure 15.6 Use of SPE separation for GC–MS analysis of sterols. Total lipid extracts (top three panels) or the chloroform eluate from an SPE column of the same extract were analyzed by GC–MS as described. The total ion intensity or intensities of the 396 (ergosterol) or 386 (cholesterol standard) are shown. Separation by SPE eliminates the peaks appearing at around 25 min, which are products of glycerophospholipids and other substances and slightly improves the resolution in the region where sterols are found (15–20 min).

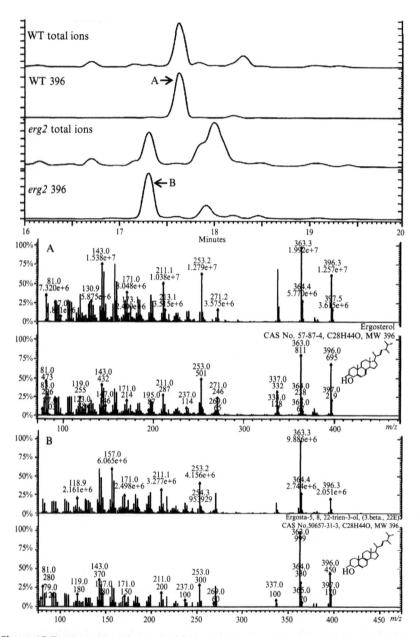

Figure 15.7 GC–MS analysis of wild-type (WT) and *erg2* mutant strain. The portion of the GC–MS profile of the WT and *erg2* mutant strains where sterols elute is shown. The major ion at 396 is ergosta-5,7,22-trienol in WT (A), as seen by its fragmentation pattern (panel A) compared to the pattern found in the NIST library. Exiting the GC column 0.5 min earlier is the major ion at 396 in the *erg2* mutant, ergosta-5,8,22-trienol (B), whose fragmentation pattern is very similar, but not identical to ergosterol (panels B), also similar to the profile from the NIST library. Fragmentation occurs in the source.

extracting data for the relevant ions for known amounts of cholesterol (386) and ergosterol (396). Compounds are identified by their retention times (compared to standards) and fragmentation patterns, which are compared to the NIST library or previously characterized sterols from WT and mutant yeast cells. Most stereoisomers can be separated by this method. In general, stereoisomers with a double bond at position 8 elute before the corresponding isomer with a double bond at position 7 (Fig. 15.7). It is important to pay attention to the ion used for quantification because the major ions may not be of similar intensities for different sterols. For example, the 396 ion for ergosterol is quite a bit more intense than the same ion for ergosta-5,8,22-trienol. To compare these sterols it is best to use a unique, but more representative ion (see Fig. 15.7, e.g., 363) or total ion counts, if sterols are sufficiently well separated. We have recently published data on the major sterols found in five isogenic ergosterol biosynthesis mutants (Guan et al., 2009) and these strains can be used as sources to determine the retention times of yeast sterols that are not commercially available.

ACKNOWLEDGMENTS

Research in the laboratory of Markus Wenk is supported in part by the Singapore National Research Foundation under CRP Award No. 2007-04, the Biomedical Research Council of Singapore (R-183-000-211-305), the National Medical Research Council (R-183-000-224-213), and the Swiss SystemsX.ch initiative, evaluated by the Swiss National Science Foundation. Howard Riezman is funded by the Swiss National Science Foundation, the Human Frontiers Science Program Organization, the Swiss SystemsX.ch initiative, evaluated by the Swiss National Science Foundation, and the University of Geneva. We thank Karl Abele at Varian AG, Basel, for help in setting up sterol analysis.

REFERENCES

Angus, W. W., and Lester, R. L. (1972). Turnover of inositol and phosphorus containing lipids in Saccharomyces cerevisiae; extracellular accumulation of glycerophosphorylinositol derived from phosphatidylinositol. Arch. Biochem. Biophys. 151, 483–495.

Bligh, E. G., and Dyer, W. J. (1959). A rapid method of total lipid extraction and purification. Can. J. Biochem. Physiol. 37, 911–917.

Brockerhoff, H. (1963). Breakdown of phospholipids in mild alkaline hydrolysis. J. Lipid Res. 35, 96–99.

Brugger, B., Erben, G., Sandhoff, R., Wieland, F. T., and Lehmann, W. D. (1997). Quantitative analysis of biological membrane lipids at the low picomole level by nano-electrospray ionization tandem mass spectrometry. Proc. Natl. Acad. Sci. USA 94, 2339–2344.

Cheng, J., Park, T. S., Fischl, A. S., and Ye, X. S. (2001). Cell cycle progression and cell polarity require sphingolipid biosynthesis in Aspergillus nidulans. Mol. Cell. Biol. 21, 6198–6209.

Daum, G., Lees, N. D., Bard, M., and Dickson, R. (1998). Biochemistry, cell biology and molecular biology of lipids of Saccharomyces cerevisiae. Yeast 14, 1471–1510.

De Vos, R. C., Moco, S., Lommen, A., Keurentjes, J. J., Bino, R. J., and Hall, R. D. (2007). Untargeted large-scale plant metabolomics using liquid chromatography coupled to mass spectrometry. *Nat. Protoc.* **2,** 778–791.

Dickson, R. C., Sumanasekera, C., and Lester, R. L. (2006). Functions and metabolism of sphingolipids in *Saccharomyces cerevisiae*. *Prog. Lipid Res.* **45,** 447–465.

Ejsing, C. S., Moehring, T., Bahr, U., Duchoslav, E., Karas, M., Simons, K., and Shevchenko, A. (2006). Collision-induced dissociation pathways of yeast sphingolipids and their molecular profiling in total lipid extracts: A study by quadrupole TOF and linear ion trap-orbitrap mass spectrometry. *J. Mass Spectrom.* **41,** 372–389.

Ejsing, C. S., Sampaio, J. L., Surendranath, V., Duchoslav, E., Ekroos, K., Klemm, R. W., Simons, K., and Shevchenko, A. (2009). Global analysis of the yeast lipidome by quantitative shotgun mass spectrometry. *Proc. Natl. Acad. Sci. USA* **106,** 2136–2141.

Fahy, E., Cotter, D., Byrnes, R., Sud, M., Maer, A., Li, J., Nadeau, D., Zhau, Y., and Subramaniam, S. (2007). Bioinformatics for lipidomics. *Methods Enzymol.* **432,** 247–273.

Folch, J., Lees, M., and Stanley, G. H. S. (1957). A simple method for the isolation and purification of total lipides from animal tissues. *J. Biol. Chem.* **226,** 497–509.

Guan, X. L., and Wenk, M. R. (2006). Mass spectrometry-based profiling of phospholipids and sphingolipids in extracts from *Saccharomyces cerevisiae*. *Yeast* **23,** 465–477.

Guan, X. L., He, X., Ong, W. Y., Yeo, W. K., Shui, G., and Wenk, M. R. (2006). Non-targeted profiling of lipids during kainate-induced neuronal injury. *FASEB J.* **20,** 1152–1161.

Guan, X. L., Souza, C. M., Pichler, H., Dewhurst, G., Schaad, O., Kajiwara, K., Wakabayashi, H., Ivanova, T., Castillon, G. A., Piccolis, M., Abe, F., Loewith, R., *et al.* (2009). Functional interactions between sphingolipids and sterols in biological membranes regulating cell physiology. *Mol. Biol. Cell* **20,** 2083–2095.

Han, X., and Gross, R. W. (2005). Shotgun lipidomics: Electrospray ionization mass spectrometric analysis and quantitation of cellular lipidomes directly from crude extracts of biological samples. *Mass Spectrom. Rev.* **24,** 367–412.

Han, X., Yang, K., and Gross, R. W. (2008). Microfluidics-based electrospray ionization enhances the intrasource separation of lipid classes and extends identification of individual molecular species through multi-dimensional mass spectrometry: Development of an automated high-throughput platform for shotgun lipidomics. *Rapid Commun. Mass Spectrom.* **22,** 2115–2124.

Hanson, B. A., and Lester, R. L. (1980). The extraction of inositol-containing phospholipids and phosphatidylcholine from *Saccharomyces cerevisiae* and *Neurospora crassa*. *J. Lipid Res.* **21,** 309–315.

Hechtberger, P., Zinser, E., Saf, R., Hummel, K., Paltauf, F., and Daum, G. (1994). Characterization, quantification and subcellular localization of inositol-containing sphingolipids of the yeast, *Saccharomyces cerevisiae*. *Eur. J. Biochem.* **225,** 641–649.

Jungnickel, H., Jones, E. A., Lockyer, N. P., Oliver, S. G., Stephens, G. M., and Vickerman, J. C. (2005). Application of TOF-SIMS with chemometrics to discriminate between four different yeast strains from the species *Candida glabrata* and *Saccharomyces cerevisiae*. *Anal. Chem.* **77,** 1740–1745.

Moore, J. D., Caufield, W. V., and Shaw, W. A. (2007). Quantitation and standardization of lipid internal standards for mass spectroscopy. *In* "Methods in Enzymology Lipidomics and Bioactive Lipids: Mass[Hyphen (True Graphic)]Spectrometry-Based Lipid Analysis," (H. A. Brown, ed.), pp. 351–367. Academic Press, San Diego.

Munn, A. L., Heese-Peck, A., Stevenson, B. J., Pichler, H., and Riezman, H. (1999). Specific sterols required for the internalization step of endocytosis in yeast. *Mol. Biol. Cell* **10,** 3943–3957.

Patton, J. L., and Lester, R. L. (1991). The phosphoinositol sphingolipids of *Saccharomyces cerevisiae* are highly localized in the plasma membrane. *J. Bacteriol.* **173,** 3101–3108.

Schneiter, R., Brugger, B., Sandhoff, R., Zellnig, G., Leber, A., Lampl, M., Athenstaedt, K., Hrastnik, C., Eder, S., Daum, G., Paltauf, F., Wieland, F. T., *et al.* (1999). Electrospray ionization tandem mass spectrometry (ESI-MS/MS) analysis of the lipid molecular species composition of yeast subcellular membranes reveals acyl chain-based sorting/remodeling of distinct molecular species en route to the plasma membrane. *J. Cell Biol.* **146,** 741–754.

Schwudke, D., Oegema, J., Burton, L., Entchev, E., Hannich, J. T., Ejsing, C. S., Kurzchalia, T., and Shevchenko, A. (2006). Lipid profiling by multiple precursor and neutral loss scanning driven by the data-dependent acquisition. *Anal. Chem.* **78,** 585–595.

Schwudke, D., Hannich, J. T., Surendranath, V., Grimard, V., Moehring, T., Burton, L., Kurzchalia, T., and Shevchenko, A. (2007). Top-down lipidomic screens by multivariate analysis of high-resolution survey mass spectra. *Anal. Chem.* **79,** 4083–4093.

Shui, G., Bendt, A. K., Pethe, K., Dick, T., and Wenk, M. R. (2007). Sensitive profiling of chemically diverse bioactive lipids. *J. Lipid Res.* **48,** 1976–1984.

Song, H., Hsu, F. F., Ladenson, J., and Turk, J. (2007). Algorithm for processing raw mass spectrometric data to identify and quantitate complex lipid molecular species in mixtures by data-dependent scanning and fragment ion database searching. *J. Am. Soc. Mass Spectrom.* **18,** 1848–1858.

Song, H., Ladenson, J., and Turk, J. (2009). Algorithms for automatic processing of data from mass spectrometric analyses of lipids. *J. Chromatogr. B* **877,** 2847–2854.

Sullards, M. C. (2000). Analysis of sphingomyelin, glucosylceramide, ceramide, sphingosine, and sphingosine 1-phosphate by tandem mass spectrometry. *Methods Enzymol.* **312,** 32–45.

Taguchi, R., Houjou, T., Nakanishi, H., Yamazaki, T., Ishida, M., Imagawa, M., and Shimizu, T. (2005). Focused lipidomics by tandem mass spectrometry. *J. Chromatogr. B Analyt. Technol. Biomed. Life Sci.* **823,** 26–36.

Taguchi, R., Nishijima, M., and Shimizu, T. (2007). Basic analytical systems for lipidomics by mass spectrometry in Japan. *Methods Enzymol.* **432,** 185–211.

Tuller, G., Nemec, T., Hrastnik, C., and Daum, G. (1999). Lipid composition of subcellular membranes of an FY1679-derived haploid yeast wild-type strain grown on different carbon sources. *Yeast* **15,** 1555–1564.

Valachovic, M., Wilcox, L. I., Sturley, S. L., and Bard, M. (2004). A mutation in sphingolipid synthesis suppresses defects in yeast ergosterol metabolism. *Lipids* **39,** 747–752.

van den Hazel, H. B., Pichler, H., Matta, M. A. D. V., Leitner, E., Goffeau, A., and Daum, G. (1999). PDR16 and PDR17, two homologous genes of *Saccharomyces cerevisiae*, affect lipid biosynthesis and resistance to multiple drugs. *J. Biol. Chem.* **274,** 1934–1941.

Wagner, A., Grillitsch, K., Leitner, E., and Daum, G. (2009). Mobilization of steryl esters from lipid particles of the yeast *Saccharomyces cerevisiae*. *Biochim. Biophys. Acta* **1791,** 118–124.

Wenk, M. R. (2005). The emerging field of lipidomics. *Nat. Rev. Drug Discov.* **4,** 594–610.

Wong, J. W. H., Cagney, G., and Cartwright, H. M. (2005). SpecAlign—Processing and alignment of mass spectra datasets. *Bioinformatics* **21,** 2088–2090.

Xu, S. H., and Nes, W. D. (1988). Biosynthesis of cholesterol in the yeast mutant erg6. *Biochem. Biophys. Res. Commun.* **155,** 509–517.

Zinser, E., Sperka-Gottlieb, C. D., Fasch, E. V., Kohlwein, S. D., Paltauf, F., and Daum, G. (1991). Phospholipid synthesis and lipid composition of subcellular membranes in the unicellular eukaryote *Saccharomyces cerevisiae*. *J. Bacteriol.* **173,** 2026–2034.

Zinser, E., Paltauf, F., and Daum, G. (1993). Sterol composition of yeast organelle membranes and subcellular distribution of enzymes involved in sterol metabolism. *J. Bacteriol.* **175,** 2853–2858.

CHAPTER SIXTEEN

MASS SPECTROMETRY-BASED METABOLOMICS OF YEAST

Christopher A. Crutchfield,[*,†] Wenyun Lu,[*,†] Eugene Melamud,[*,†] and Joshua D. Rabinowitz[*,†]

Contents

[*] Lewis-Sigler Institute for Integrative Genomics, Princeton University, Princeton, New Jersey, USA
[†] Department of Chemistry, Princeton University, Princeton, New Jersey, USA

Methods in Enzymology, Volume 470
ISSN 0076-6879, DOI: 10.1016/S0076-6879(10)70016-1

Abstract

Driven by the advent of metabolomics, recent years have seen renewed interest in the investigation of yeast metabolism. Here we provide a practical guide to metabolomic analysis of yeast using liquid chromatography–mass spectrometry (LC–MS). We begin with background on LC–MS and its utility in studying yeast metabolism. We then describe key issues involved at each step of a typical yeast metabolomics experiment: in experimental design, cell culture, metabolite extraction, LC–MS, and data processing and analysis. Throughout, we highlight interdependencies between the steps that are relevant to developing an integrated workflow which effectively leverages LC–MS to reveal yeast biology.

1. INTRODUCTION

Liquid chromatography–mass spectrometry (LC–MS) has emerged as a preferred tool for measuring the small molecule components of cellular metabolism. LC–MS enables simultaneous analysis of dozens to hundreds of chemical species. It has already facilitated significant discoveries in yeast, including identification of metabolites that oscillate in phase with a metabolic cycle (Tu *et al.*, 2007), or that respond to shifting glucose or ammonia availability (Brauer *et al.*, 2006; Wu *et al.*, 2006a). Future applications hold promise for decoding regulation of yeast metabolism, as well as central mysteries in yeast physiology, such as the molecular control of growth rate and cell size.

To achieve these benefits, it is important for metabolomic capabilities to be widely available to the yeast community. Here we aim to provide a practical resource for initiating LC–MS-based metabolomic studies of yeast. The focus is on quantitating the concentrations of metabolites, although many of the concepts also apply to probing metabolic flow ("flux"). We begin with a brief description of LC–MS fundamentals. The order of the subsequent sections then follows roughly the steps of a basic metabolomics experiment: experimental design, cell culture, metabolite extraction, LC–MS, and data processing and analysis.

2. LC–MS BASICS

LC–MS involves three fundamental steps: LC separation, ionization, and separation and quantitation of the ions by MS. The main benefits of LC–MS as an analytical tool are its specificity and sensitivity. Specificity arises from two orthogonal dimensions of separation: chromatographic retention time (RT) and mass-to-charge ratio (m/z). In modern instruments, specificity is enhanced by high mass resolution and/or multiple rounds of MS ("tandem mass spectrometry" or "MS/MS").

The main downside of LC–MS is imperfect quantitative capabilities. On the positive side, for a given analyte and LC–MS method, signal intensity and concentration are often linear over a broad dynamic range (> 100-fold). On the downside, measurements are intrinsically noisy, with typical relative standard deviations (i.e., the standard deviation of signal intensity across repeated runs of the same sample, divided by the average signal intensity) in the range of 10–20%. Moreover, the relationship between MS signal and concentration depends strongly on the analyte structure. Differences in response factor (i.e., the ratio of signal intensity to analyte concentration) of > 100-fold are common across analytes, meaning that equimolar solutions of two different compounds can vary in measured signal by orders of magnitude. Response factors also vary across instruments, and, for a given instrument and analyte, tend to drift over time. Relative ion intensities across samples can therefore be used only to approximate relative concentrations, and do so reliably only when the samples are analyzed in close succession.

3. EXPERIMENTAL DESIGN

The most straightforward application of LC–MS is measurement of relative concentration changes in known metabolites: a set of samples is generated, analyzed by LC–MS, and signals compared for different known metabolites. LC–MS sensitivity fluctuates even within a day. As such, biological comparators, not replicates, should be run in direct succession (e.g., the order A-B-A-B, not A-A-B-B). Such running order also mitigates errors due to metabolite instability, which can be particularly problematic for unstable species such as nucleotide triphosphates, CoA compounds, folates, and NAD(P)H. Even for species that are stable on their own in solution, reactions can occur with other metabolome components (e.g., nucleophiles like amines can react with electrophiles like carbonyls). To minimize systematic errors due to metabolic instability, it is standard practice in our lab to analyze samples within 24–36 h of their generation.

The most rigorous way of ruling out analytical artifacts is to use isotopic internal standards. However, pure isotope-labeled standards for most metabolites are not available. An alternative is to generate an isotope-labeled standard mixture from cells fed uniformly ^{13}C-glucose (Mashego *et al.*, 2004). Unfortunately, such cell-derived standards introduce a myriad of new species into already complex metabolome samples. Accordingly, we typically spike our samples with a limited set of pure isotope-labeled internal standards, which we use to check analytical performance. Assuming adequate results for the standards (i.e., no systematic variation across biological conditions), we draw biological conclusions based on ratios of signal intensities across samples without correction for internal standard signals. The most biologically significant results are then validated through alternative experimental approaches.

Beyond relative quantitation of known metabolites, LC–MS can enable discovery of novel metabolites. Measurement of unanticipated metabolites ("untargeted analysis") can be done in parallel with quantitation of known ("targeted analysis"), but introduces some additional considerations:

(1) Certain LC–MS techniques, especially those relying on MS/MS to achieve specificity (e.g., triple quadrupole MS), are better suited to targeted analysis than metabolite discovery. For untargeted analysis, instruments with good mass resolution and full scan sensitivity (i.e., sensitivity when scanning over a range of molecular weights using MS, without relying on MS/MS) are preferred (e.g., time-of-flight (TOF) or Orbitrap). See Section 11 for details.

(2) Untargeted analysis generates a large amount of raw data. Appropriate computational tools are needed to find compounds of interest.

(3) Each analyte present in the samples may generate multiple adduct ions. These vary depending on the ionization mode, that is, whether the ion source is set to generate, and the mass spectrometer set to detect, positive or negative ions. In positive ion mode LC–MS, quantitation is typically based on $[M + H]^+$; however, one may also see $[M + Na]^+$, $[M + K]^+$, etc. In negative mode LC–MS, quantitation is typically based on $[M - H]^-$; however, one may see $[M + Na - 2H]^-$, $[M + K - 2H]^-$, etc. For a more comprehensive list of ionization adducts, see Table 16.2, under "Untargeted Analysis."

(4) Although accurate mass is a powerful tool for identifying molecular formulas (and often for pulling up candidate metabolite structures through publically available compound databases), compound purification and nuclear magnetic resonance (NMR) analysis are typically required to assign structures to genuinely novel compounds.

LC–MS can also be used for absolute metabolite quantitation. Absolute quantitation enables calculation of mass action ratios, and thereby the thermodynamically favored direction of net metabolite flux (Henry *et al.*, 2007;

Kümmel *et al.*, 2006). Absolute metabolite concentrations also dictate the extent of saturation of enzyme sites, and thus the sensitivity of reaction rates to changes in substrate, product, and competitive inhibitor concentrations (Bennett *et al.*, 2009).

The optimal approach to absolute metabolite quantitation involves spiking isotope-labeled internal standard compounds into the yeast extraction solvent. The ratio of unlabeled-to-labeled compound signal, multiplied by standard concentration, then gives the endogenous cellular compound concentration. When isotope-labeled standard is not available, one can instead label the cells by feeding isotope-labeled nutrient. More readily available unlabeled compounds can then be used as the internal standards (Mashego *et al.*, 2004; Wu *et al.*, 2006b). In such cases, it is advisable to correct for incomplete labeling of cellular metabolites (Bennett *et al.*, 2008). With either approach, it is important to add the internal standard directly into the extraction solution, to correct for the substantial compound losses (e.g., to adsorption or degradation) that often occur during extraction.

An alternative use of isotope labels involves probing metabolic flux (Moxley *et al.*, 2009; Sauer *et al.*, 2007). While full flux quantitation involves experimental and data analysis challenges that are beyond the scope of this chapter, useful information can be achieved through relatively simple LC–MS experiments. These break down into two main categories: studies of labeling kinetics and studies of steady-state isotope patterns.

To investigate labeling kinetics, cells are transferred from normal media to media containing an isotope-labeled nutrient (e.g., ^{15}N-ammonia or ^{13}C-glucose). Samples are taken at various time points thereafter, and the rate of labeling of cellular metabolites quantitated by LC–MS. Fast labeling generally implies high flux from the labeled nutrient to the measured metabolite. If the structure of the metabolic network is known, then quantitative flux determination is often possible (Munger *et al.*, 2008; Yuan *et al.*, 2008). If not, qualitative interpretation is still useful. For example, one can categorize potential novel metabolites into high flux species (potentially central players in metabolism) versus low flux ones (potentially degradation products of limited biological importance). One can also qualitatively assess major flux changes between conditions, for example, whether synthesis of a particular metabolite shuts off during starvation (Brauer *et al.*, 2006).

Studies of steady-state isotope patterns involve feeding a mixture of labeled and unlabeled carbon source, and then assessing the isotopomer patterns of different cellular species. This approach is particularly useful for determining which of two alternative pathways is the main route to a particular metabolic product. For example, in yeast, oxaloacetate can be produced by either malate dehydrogenase (turning of TCA cycle) or pyruvate carboxylase (anapleurosis). Feeding a mixture of unlabeled and labeled glucose results in different isotopomer patterns for malate versus pyruvate; the extent to which oxaloacetate, or its transamination product, aspartate, mirrors one versus the other provides the relative pathway fluxes. Similar logic applies at other points of metabolic conversion, and can be used to deduce system-wide fluxes (Antoniewicz *et al.*, 2007; Sauer, 2006; Wiechert and Nöh, 2005).

4. STRAINS

As is typical in cell biology, selection of strain and culture conditions is critical. Although most yeast strains are suitable for metabolomic analysis, metabolic differences between strains are common and should be considered. For example, S288c and its derivations are HAP1 deficient (Gaisne *et al.*, 1999). Accordingly, although numerous stain collections exist in an S288c background (Brachmann *et al.*, 1998; Ho *et al.*, 2009; Sopko *et al.*, 2006), CEN.PK may be preferred for some metabolomic applications (van Dijken *et al.*, 2000).

Many strain collections contain auxotrophies as selectable markers. While auxotrophies have only minor impact in many experiments, they typically have a large impact on the metabolome, changing the concentrations of compounds both up- and downstream of the block. For example, knockout of the URA3 gene (a common selectable marker) dramatically alters the cellular concentrations of pyrimidines and their precursors. Moreover, it can also impact the concentrations of more distantly related metabolites like glutamate (Boer *et al.*, 2010). Accordingly, antibiotic resistance markers are generally preferable to auxotrophic markers.

5. CULTURE CONDITIONS

Typical means of culturing yeast for metabolomic studies include batch liquid culture, continuous culture in a chemostat, and batch culture on a filter support. The major advantages of batch liquid culture are its common use and ease; the major advantages of chemostats are control of culture conditions and reproducibility; and the major advantages of filter culture are facile manipulation of the cellular nutrient environment and ease of sampling. Irrespective of the culture approach selected, use of a defined media (such as yeast nitrogen base) is recommended, as rich medium contains many small molecule components that can interfere with analysis of cellular metabolites.

5.1. Batch liquid culture and chemostats

Batch liquid culture involves allowing cells to reproduce in a fixed volume of liquid. Due to consumption of nutrients and excretion of waste products, the culture conditions change continually as the cells grow. When more consistent culture conditions are desired, chemostats provide a valuable tool. Chemostats are stirred vessels into which fresh medium (typically having a low concentration of one essential nutrient) is continually being flowed. Addition of the medium provides cells in the vessel with the limiting nutrient, while washing out a portion of the cells. In such a culture system, cells will grow until their replication rate exactly matches the rate of their washout from the vessel. Thus, the experimenter can set the steady-state growth rate of the cells based on the rate of medium addition to the chemostat (which, when normalized by the chemostat volume, is referred to as the dilution rate). The Dunham lab maintains a useful manual detailing chemostat operation (Dunham, 2010).

Beyond generating highly reproducible steady-state cultures, the chemostat approach is valuable for enabling studies of cellular composition as a function of growth rate and nature of the limiting nutrient (Brauer et al., 2008). It allows mutants that differ in growth rate in batch culture to be compared at a constant growth rate. Furthermore, nutrient-limited cultures can be pulsed with the limiting species, and dynamic response to relief of nutrient limitation measured (Wu et al., 2006a,b). In such studies, the spiked nutrient can be isotope labeled, enabling direct tracing of its assimilation by MS (Aboka et al., 2009).

Sampling metabolites from either liquid batch or chemostat culture presents similar challenges: capturing a discrete volume of culture fluid; separating cells from the surrounding medium without altering their metabolome; and effectively extracting metabolites from the isolated cells. The standard literature approach involves mixing culture media directly with cold (≤ -40 °C) methanol to quench metabolism, centrifuging in a prechilled rotor (≤ -20 °C) to isolate the cells, and subsequently extracting the cell pellet (Canelas et al., 2008). In this method, the initial mixing of cells with cold methanol aims to quench metabolism, by cold-induced slowing of reaction rates and/or organic-induced enzyme denaturation. It aims to avoid, however, leakage of the cells due to membrane disruption (e.g., due to cold-induced formation of ice crystals that puncture membranes, or organic-induced membrane dissolution). After the quenched cells are isolated by centrifugation, further exposure to organic solvent results in membrane disruption and extraction of metabolites from the cells. The main risk in this method is metabolite losses (e.g., due to cell leakage) during the centrifugation step. Another downside is the somewhat laborious nature of the process. An alternative approach involves separation of the cells from medium prior to quenching metabolism, using fast filtration. The cell-loaded filter is then simultaneously quenched and extracted by placing it into cold extraction solvent. The main deficiency of this approach is the potential for metabolome changes during filtration.

As the deficiencies of the two approaches do not overlap, if similar results are obtained using both approaches, one can have high confidence in their reliability. We have recently found that both approaches lead to identical biological conclusions for steady-state chemostat cultures limited for a diversity of nutrients

(Boer *et al.*, 2010), suggesting that either approach is acceptable for steady-state yeast cultures. The major difference was higher absolute signal for many compounds, especially nucleotide triphosphates, using the filtration-based approach (see Fig. 16.2). Thus, for steady-state or steadily growing yeast cultures, filtration provides a convenient option. (*Note*: For *Escherichia coli*, in contrast, we have observed metabolome changes during filtration and do not recommend its use without further validation; in addition, *E. coli* leak quickly upon exposure to organic, so the approach of quenching and then centrifuging is also problematic for them.) Even for steady-state yeast, key biological conclusions should still be confirmed using an approach involving faster quenching. For cultures undergoing fast metabolome changes (e.g., due to an acute nutrient perturbation), filtration is unacceptably slow and a faster quenching approach is needed. Relevant harvesting protocols are provided below.

5.2. Protocol for harvesting yeast by vacuum filtration

1. Construct a filtration apparatus as shown in Fig. 16.1.
2. Check for leaks.
3. Thoroughly rinse the apparatus with purified water.
4. Place a 25-mm 0.45-μm pore size nylon filter on the filter base and pre-wet with purified water.
5. Connect a 15-ml centrifuge tube to rubber stopper at bottom of filtration apparatus, forming a tight seal (see Fig. 16.1). This tube will be used to collect the extracellular media.
6. Measure 10 ml of culture, using either a 15-ml centrifuge tube or a volumetric pipette. Recommended culture density at time of extraction is $\sim 4 \times 10^7$ cells/ml (Klett ~ 130 or $OD_{650} \sim 0.5$).
7. Pour culture immediately into the glass cylinder at top of filtration apparatus. Filtration should occur rapidly.
8. When filtration appears complete, wait ~ 1 s and then remove the clamp to free the filter.
9. Immediately place the filter, cells side down, into a 35-mm Petri dish containing 700 μl prechilled (-20 °C) extraction solvent (for details, see Section 6.1). Time from initiation of sampling to quenching should not exceed 30 s. For subsequent steps, see Section 6.1.

5.3. Protocol for harvesting yeast by centrifugation after methanol quenching

1. Prechill a centrifuge rotor capable of handling 50-ml centrifuge tubes to -80 °C.
2. Directly quench 10 ml culture broth into 20 ml -80 °C methanol in a 50-ml centrifuge tube.

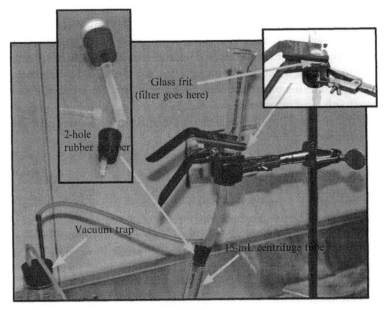

Figure 16.1 Example of a filtration apparatus. The two-hole rubber stopper seals the top of a 15-ml centrifuge tube that will be used to collect the culture medium. One hole connects to vacuum, the other to the filter support. The filter sits on top of the glass (or metal) frit, with the open-bottom graduated cylinder attached by a clamp. The filter must cover the entire frit, or otherwise cells will be lost during filtration and quantitation will be unreliable. To initiate filtration, cells are poured into the graduated cylinder at the top of the apparatus. Once filtration is complete, the clamp is removed and the filter quickly transferred to the quenching solvent.

3. Spin down at $\sim 2000 \times g$ in centrifuge cooled to $-10\,^{\circ}\text{C}$ for 5 min.
4. Discard supernatant.
5. For subsequent steps, see Section 6.2.

5.4. Filter culture

Yeast readily form colonies on agar plates. Such colonies involve a heterogeneous mixture of cells experiencing different nutrient environments. Moreover, quantitative sampling of them is difficult. Accordingly, this culture approach is not well suited to metabolomic studies. A variant, however, can be useful: growing yeast on the surface of a filter atop an agarose-medium support. In the filter culture technique, a modest number of cells is spread evenly across the filter surface, resulting in almost all cells having direct contact with both the filter (and thus the underlying nutrients) and the atmosphere. The number of cells is selected to cover $\sim 10\text{--}20\%$ of the filter surface. This avoids inhomogeneities in nutrient access, and enables cells to replicate on the filter for several doublings at a rate similar to their growth in comparable liquid medium.

The principal virtue of this technique, compared to liquid batch culture, is the ability to quickly manipulate the culture's environment. For example, filter culture enables rapid induction of nutrient starvation (by switching to a plate lacking one medium component). It also allows rapid replacement of unlabeled with labeled nutrient, as required for kinetic flux profiling (Yuan *et al.*, 2008). (*Note*: The filter culture approach also works well for bacteria including *E. coli*.)

5.5. Protocol for growth of yeast on filters atop agarose support

1. Prepare 2×concentrate of minimal media of interest.
2. Autoclave 30 g/l of triply washed ultrapure agarose in purified water.
3. Mix in 1:1 ratio to yield desired minimal media containing 15 g/l of agarose. Equilibrate temperature to ~55 °C in a water bath.
4. Pour 10 ml of mixture per 60 mm Petri dish. Best results are obtained if plates are used within 48 h after pouring.
5. Grow yeast to ~1 × 10^7 cells/ml in liquid batch culture.
6. Place a nylon filter (47 mm diameter, 0.45 μm pore size) on top of a filtration apparatus, under weak vacuum.
7. Using a 2-ml pipette spread 1.6 ml of cell culture evenly across the filter surface. Steadying the pipette with a second hand will make distributing the cells evenly easier. The vacuum is on the correct intensity if the culture pools very briefly before the medium is pulled through the filter.
8. Place filters on individual agarose plates with cells facing away from the agarose surface. Use of broad head forceps (Millipore/#XX6200006) will reduce risk of puncturing the filter.
9. Allow cells to grow on the filter surface for ~2.5 h. After this time the cultures are generally ready for perturbation or extraction.

For further details on filter culture, see Bennett *et al.* (2008).

6. Metabolite Extraction

Once metabolism is quenched, the next step is harvesting metabolites from the yeast. The goal is to obtain the most complete extraction possible, while avoiding degradation of metabolites, including conversion of one metabolite into another. An early approach involved freeze-drying the quenched cells, grinding the frozen cell pellet with a mortar and pestle, and extracting using strong acid or base in water (Sáez and Lagunas, 1976). Later, the grinding steps were eliminated and replaced with a single-step extraction into boiling ethanol (Castrillo *et al.*, 2003; Gonzalez *et al.*, 1997). While this method performs well for most metabolites, some carboxylic acids such as pyruvate and fumarate are lost (Loret *et al.*, 2007). Based on data from *E. coli* showing the superiority of cold methanol to hot ethanol (Prasad Maharjan and Ferenci, 2003), cold methanol was tested in yeast, with generally favorable results (Villas-Boas *et al.*, 2005). Based on our finding that acetonitrile:methanol:water mixtures extract nucleotide triphosphates from *E. coli* much better than methanol:water alone (Rabinowitz and Kimball, 2007), we have tested acetonitrile:methanol:water (40:40:20, v/v/v) also for yeast (Fig. 16.2). For most compounds, results are similar for both mixtures. For nucleotide triphosphates, however, we find a large improvement with acetonitrile, but only for cells harvested by filtration, not methanol quenching. Our interpretation is that initial exposure of the cells to methanol (in the absence of acetonitrile) results in irreversible triphosphates losses. Note that acetonitrile can degrade nitrocellulose; accordingly, if extracting using acetonitrile, use nylon filters.

To keep quenching solvent and extraction mixtures cold, we have found it convenient to use a cold metal surface. An easy way to maintain such a surface is to fill a 13 in. × 9 in. × 2 in. baking tray with gel–packs. With the gel-pack sitting in the bottom of the tray, overflow the remainder with paper towels. Using packing tape, tightly compress paper towels and ice packs into the baking tray. The tray can then be frozen. After freezing, invert the tray. The tray bottom (now facing upward) provides a useful cold working surface. We find it helpful to build a border (e.g., using paper towels and tape) around the bottom of the tray (the cold working surface) to prevent Petri dishes from slipping off.

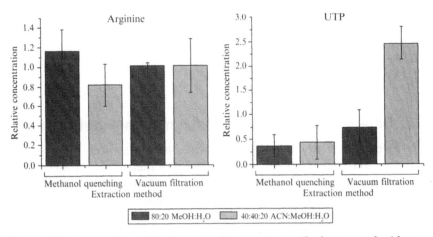

Figure 16.2 Metabolite yields for two cell harvesting methods extracted with two different solvent mixtures. Arginine yield is insensitive to the harvesting and extraction method, whereas UTP yield is increased when cells are harvested by vacuum filtration and extracted with acetonitrile:methanol:water mixture. Error bars reflect standard error of $N = 3$ experimental replicates.

6.1. Extraction protocol for cells on filters (from vacuum filtration of liquid culture, or from filter cultures)

1. Prepare extraction solution: by volume, 40% acetonitrile, 40% methanol, and 20% water. All solvents should be highest purity available (minimum HPLC grade).
2. Cool extraction solution to $-20\,°C$. Shake to mix prior to use.
3. When ready to extract, pipette cold extraction solvent ($-20\,°C$) into 60 mm Petri dish (1 ml extraction solvent/dish). Store at $-20\,°C$ until use, avoiding prolonged storage (e.g., >1 h) due to risk of water condensation in the solvent.
4. Place filter, cell-side down, into the cold extraction solvent.
5. Allow extraction to proceed (stirring not required) at $-20\,°C$ for 15 min.
6. Collect as much solvent, cells, and debris as possible into a 1.5-ml Eppendorf tube. To remove cells and debris from the filter, flip it cell-side up and wash 10 times with the pooled solvent at the bottom of the Petri dish. To release adherent solvent from the filter, dab it on a dry part of the Petri dish approximately five times. From this point forward the sample can be kept on ice.
7. Spin down the mixture in a microcentrifuge (highest speed, $\sim 16{,}000 \times g$) for 5 min to pellet cell debris.
8. Transfer the supernatant to a separate 1.5-ml Eppendorf tube and resuspend the remaining pellet in 100 μl fresh extraction solvent. The pellet is typically difficult to resuspend. Prefilling the pipette tip with the 100 μl of extraction solvent and gently perturbing the pellet with the pipette tip before depressing to release the extraction solvent can aid in resuspension. Take care not to clog the pipette tip with sticky cell debris.
9. Keep the resuspended mixture on ice for additional 15 min.
10. Spin down sample and pool the supernatant with the prior fraction.
11. Vortex pooled mixture and analyze.

6.2. Extraction protocol for cells after methanol quenching

1. Prepare extraction solution: by volume, 40% acetonitrile, 40% methanol, and 20% water, or alternatively 80% methanol and 20% water (both give roughly equivalent results for methanol-quenched cells). All solvents should be highest purity available (minimum HPLC grade).
2. Cool extraction solution to $-20\,°C$. Shake to mix prior to use.
3. When ready to extract (i.e., after pouring off supernatant following the initial quenching step), pipette 400 μl cold extraction solvent ($-20\,°C$) directly onto pellet.
4. After pipetting up and down (do not vortex), transfer mixture to a 1.5-ml Eppendorf tube and let sit on ice for 15 min.

5. Spin down the mixture in a microcentrifuge (highest speed, $\sim 16,000 \times g$) for 5 min to pellet cell debris.
6. Transfer the supernatant to a separate 1.5-ml Eppendorf tube, recording the volume recovered.
7. Resuspend the pellet in a volume of extraction solvent equal to the difference of the recovered supernatant and 800 μl. The pellet is typically difficult to resuspend. Prefilling the pipette tip with the 100 μl of extraction solvent and gently perturbing the pellet with the pipette tip before depressing to release the extraction solvent can aid in resuspension. Take care not to clog the pipette tip with sticky cell debris.
8. Keep the resuspended mixture on ice for additional 15 min.
9. Spin down sample and pool the supernatant with the prior fraction.
10. Vortex pooled mixture and analyze.

7. CHEMICAL DERIVATIZATION OF METABOLITES

While LC–ESI–MS enables measurement of most metabolites directly, separation and/or ionization of some compounds can be enhanced by chemical derivatization. Many useful derivatization procedures are available in the literature (Carlson and Cravatt, 2007a,b; Dettmer *et al.*, 2007; Lamos *et al.*, 2007; Shortreed *et al.*, 2006; van der Werf *et al.*, 2007). Here we highlight two examples: amino acids and thiol-containing compounds. For amino acids, the goal of derivatization is to enhance retention on reversed-phase chromatography. We do this by reaction of the amine to generate a carboxybenzyl, or Cbz, derivative (Scheme 16.1). Such derivatized amino acids ionize preferentially in negative mode due to their free carboxylic acid moieties. For thiol-containing compounds, the goal of derivatization is to convert both free thiols and disulfides to a common form that is readily measured by LC–ESI–MS. To this end, we use a reagent that reacts with both free thiols and disulfides, methyl methanethiosulfonate (Scheme 16.2). This reaction sacrifices information regarding cellular thiol oxidation state, with the benefit of avoiding analytical complexities due to disulfide formation during extraction or sample storage. Relevant protocols are provided in the following sections.

Reaction Scheme 16.1   Generation of carboxybenzyl (Cbz)–derivatized amino acids.

Reaction Scheme 16.2 Generation of methyl disulfide-derivatized thiols.

7.1. Amino acid derivatization

1. Add 4 μl triethylamine to 95 μl of metabolome extract.
2. Vortex to mix.
3. Add 1 μl benzylchloroformate.
4. Vortex to mix.
5. Analyze.

7.2. Thiol and disulfide derivatization

Derivatization stock preparation:

1. Add 795 μl of ultrapure water to a 1.5-ml Eppendorf tube.
2. Add 200 μl of 1 *M* ammonium acetate.
3. Vortex to mix.
4. Add 4.7 μl methyl methanethiosulfonate.

Derivatization:

5. Add 10 μl of derivatization stock to a 90-μl of metabolome extract.
6. Vortex and let sit for at least an hour at 4 °C *in the dark*.
7. Analyze.

8. LC–MS FOR MIXTURE ANALYSIS

Metabolite extracts are chemically complex mixtures. The leading techniques for complex mixture analysis are gas chromatography–MS (GC–MS) (Fiehn, 2008), LC–MS (Dunn WB 2008), and NMR (Lenz *et al.*, 2005; Lewis *et al.*, 2007; Mashego *et al.*, 2007). One key advantage of the MS-based techniques is more sensitive detection of low abundance species. While GC–MS provides unsurpassed measurement of volatile analytes, it cannot readily detect unstable metabolites like ATP and NADH, as they decompose, rather than vaporize, upon heating (Büscher *et al.*, 2009). These chemical species sit at the center of yeast metabolism. As such, LC–MS has become a first-line approach for yeast metabolomics.

One of the key issues in LC–MS analysis is the interface between LC (where the analytes are in solution) and MS (where the analytes must be in the gas phase). Analytes need to be ionized in order to be detected in the mass spectrometer. The two leading methods of converting LC output into gas-phase ions are atmospheric pressure chemical ionization (APCI) and electrospray ionization (ESI). For small molecule analytes containing functional groups that are charged in solution (i.e., most metabolites), ESI generally provides the more efficient ionization (Kostiainen and Kauppila, 2009). ESI is, however, a competitive process, wherein an abundant, strongly ionizing species in solution can suppress ionization of a less abundant one, a problem known as "ion suppression." Minimizing ion suppression is critical for effective detection of low abundance analytes and for avoiding quantitative artifacts, where the signal for a species rises or falls, due not to a change in its own concentration, but to a change in the concentration of a coeluting, ion-suppressing species. Such quantitative artifacts due to ion suppression are among the most likely causes of major errors in LC–ESI-MS analysis. To minimize ion suppression, high-quality LC separation is critical. Performance of an LC–ESI-MS method therefore depends strongly on the integrated functioning of the three steps: chromatographic separation, ESI, and mass spectrometry detection (Fig. 16.3). The following sections will examine each step in turn.

9. LIQUID CHROMATOGRAPHY

Metabolite extracts contain a diversity of small molecules that differ in their physical chemical properties of size, polarity/hydrophobicity, and charge. As such, no LC method is ideal for all classes of metabolites.

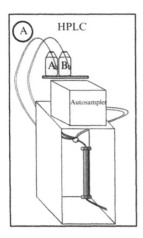  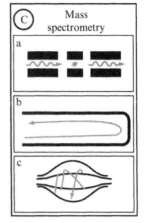

Figure 16.3 Core components of LC–MS. Mass spectrometers schematically depicted in (C) are (a) triple quadrupole, (b) time-of-flight, and (c) Orbitrap.

While various LC methods exist in the literature, most of them deal with only a limited set of analytes. Nevertheless, progress has been made in recent years towards comprehensive metabolite separation (van der Werf *et al.*, 2007). Use of two or more complementary LC methods increases the fraction of the metabolome that can be reliably detected and quantified (Sabatine *et al.*, 2005).

In general, the most robust LC approach to small molecule separation involves reversed-phase chromatography using a nonpolar stationary phase, for example, C18 RP-HPLC. Gradients begin with high water content and gradually add methanol or acetonitrile to elute hydrophobic compounds. Polar molecules elute earlier and nonpolar molecules later. Although C18 RP-HPLC is a good starting point for metabolome analysis (Trauger *et al.*, 2008), many polar metabolites do not retain adequately, eluting near the void volume during the beginning of the chromatographic run. In addition, nucleotide triphosphate compounds like ATP often do not elute as well-defined peaks. These problems tend to be especially acute if the metabolome extract contains organic solvent.

To mitigate this problem, one can strip all organic solvent from the metabolome extract by evaporation, for example, drying under nitrogen flow, vacuum centrifugation, or freeze drying. The dried sample is then redissolved in pure water, resulting in improved chromatographic performance. The downsides are risks of sample alteration: labile metabolites may degrade during sample concentration (a particular concern as reactions between metabolites accelerate as sample concentration increases); more hydrophobic metabolites may fail to redissolve in pure water; and refolding and thereby reactivation of some enzymes may occur when the sample is redissolved in water (these enzymes may then trigger metabolic reactions). A recent exploration of yeast extraction methods included examination of metabolite loss due to lyophilization (Villas-Boas *et al.*, 2005). Loss of certain metabolite classes, including lipids, nucleotides, and basic amino acids, was observed.

Another approach is to enhance retention of polar analytes using an ion-pairing agent: a volatile charged compound that pairs with oppositely charged analytes in solution, resulting in an ion–ion complex. The ion-pairing agent contains hydrophobic moieties that enhance binding of the ion–ion complex to the C18 column. The ion-pairing agent is typically added only to the aqueous mobile phase. As the organic fraction increases during the LC gradient, the concentration of ion-pairing agent drops, favoring effective column elution. Due to its volatility, the pairing agent dissociates into the gas phase during ESI.

To date, metabolomic studies have used amine-containing ion-pairing agents coupled with negative mode ESI (Coulier *et al.*, 2006; Lu *et al.*, 2008; Luo *et al.*, 2007). In our lab, we use reversed-phase ion-pairing chromatography with tributylamine as the pairing agent as our first-line metabolome analysis approach. This LC approach provides reliable separation of a broad range of negatively charged metabolites, including constituents of glycolysis, the pentose phosphate pathway, and the tricarboxylic acid, as well as nucleotides and Cbz-derivatized amino acids.

Due to their cationic nature, amine-based ion-pairing agents will cause ion suppression in positive ion mode. Therefore, they are not suitable for the analysis of positively charged metabolites such as underivatized amino acids. Direct analysis of amino acids and other highly polar metabolites can be achieved using an alternative chromatography approach: hydrophilic interaction chromatography (HILIC), which involves a polar stationary phase (Tolstikov and Fiehn, 2002). Gradients begin with high acetonitrile content and gradually add water to elute polar compounds. Nonpolar molecules elute earlier and polar molecules later. A general challenge with HILIC is retention time variability, as compound retention is often impaired by salts or other components of the biological matrix. Accordingly, for HILIC methods, it is especially important to judge results based on repeated running of actual samples.

The HILIC method that we commonly use involves an aminopropyl stationary phase, which retains metabolites through both hydrogen bonding and ionic interactions (Bajad et al., 2006). At the running pH of 9.45, approximately half of the amino groups on the column are protonated. The method separates a broad range of metabolites including amino acids, nucleosides, nucleotides, coenzyme A derivatives, carboxylic acids, and sugar phosphates. Among HILIC methods, it is relatively robust to the sample matrix. An advantage of this approach is its compatibility with both positive and negative ESI. Accordingly, it is an attractive analytical tool when only a single LC–MS system is available.

In our labs, we routinely apply both HILIC and C18 RPIP-HPLC methods for yeast metabolome analysis. Cellular extracts are aliquoted into two different HPLC autosampler vials and run separately on two LC–MS systems. The first uses HILIC chromatography with positive mode ionization, the second involves reversed-phase chromatography with tributylamine as an ion-pairing agent with negative mode ionization. As ion-pairing agents can be difficult to wash out of LC–MS systems completely, it is not convenient to run both of these methods in alternating fashion on a single LC–MS system; a dedicated system for the ion-pairing chromatography is preferred. For chromatographic details, see Table 16.1.

Looking forward, we anticipate that standard HPLC methods, such as those described above, will be replaced by related methods involving smaller particle-size columns (e.g., ≤ 2 μm instead of ~ 5 μm) and higher pressures (e.g., > 400 bar, instead of ~ 200 bar). In such ultra-performance liquid chromatography (UPLC), the smaller particle size results in greater particle surface area, and thereby faster equilibration of analytes with the column. This in turn enables faster column elution without peak tailing.

Table 16.1   Two complementary LC methods

Approach	HILIC	C18 RPIP-HPLC
ESI	Positive	Negative
Column	Luna NH$_2$ column (5 μm particle size, 250 mm $\times$ 2 mm, from Phenomenex, Torrance, CA)	Synergi Hydro column (4 μm particle size, 150 mm $\times$ 2 mm, from Phenomenex, Torrance, CA)
Solvent A	20 mM ammonium acetate + 20 mM ammonium hydroxide in 95:5 water: acetonitrile, pH 9.45	10 mM tributylamine + 15 mM acetic acid in 97:3 water: methanol
Solvent B	Acetonitrile	Methanol
Flow rate	150 μl/min	200 μl/min
Running time	40 min	50 min
Gradient time	0, 15, 28, 30, 40	0, 5, 10, 20, 35, 38, 42, 43, 50
% B	85, 0, 0, 85, 85	0, 0, 20, 20, 65, 95, 95, 0, 0

The result is increased chromatographic resolution: chromatographic peaks are narrower with higher signal-to-noise (S/N) ratios (Dunn *et al.*, 2008; Nguyen *et al.*, 2006; Wilson *et al.*, 2005). Sample running time is generally shorter, increasing throughput. The fast elution of analytes requires faster MS scanning, an area in which MS systems continue to improve.

10. Electrospray Ionization

The past 20 years has seen a dramatic rise in use of LC–MS as an analytical technique due in large part to the advent of ESI: applying a strong voltage to the liquid stream exiting the tip of a needle. This seemingly simple trick enables efficient conversion of charged molecules from the liquid phase into gas-phase ions. These ions can subsequently be analyzed by MS. The physical mechanisms of ESI remain only partially understood. Charged droplets are initially produced by electrostatic dispersion when liquid emerges from the tip of the metal needle (Nguyen and Fenn, 2007). Solvent then evaporates from the charged droplets. As the droplets become smaller and smaller, the like-charged ions within them repel due to coulombic forces, eventually resulting in release of gas-phase ions.

A major pitfall of ESI is the competitive nature of ion formation. If too many ions are present during their expulsion from the charged capillary, ion production will not increase linearly with concentration. This results in concentration underestimation. No undisputed method eliminating "ion suppression" exists. Instead, one needs to determine the extent of ion suppression and correct for it. This is best done by using isotopic internal standard, which will experience identical ion suppression to the analyte of interest. When isotopic standards are not available (as is typical in metabolome analysis), a simple alternative involves serial dilution of the sample. A linear response suggests the absence of ion suppression, while a strongly nonlinear one points to a problem.

Frequently in metabolome analysis, ion suppression is a major problem, but only during a particular LC retention time window when many species coelute, or when salt from the sample comes off the column. In such cases, it is especially useful to include isotopic standards for the analytes eluting during the problematic chromatographic window. If such standards are not available, it may be necessary to analyze a diluted sample, to enable rough quantitation of compounds during the problematic LC interval.

11. MASS SPECTROMETRY

The ability to resolve and quantitate ions by MS provides a powerful tool for investigating yeast metabolism. Particularly, useful MS approaches for metabolite analysis are triple quadrupole MS in multiple reaction monitoring (MRM) mode, high resolution MS in full scan mode, and hybrid MS in data-dependent MS/MS mode. The essential features of these approaches are described in the following sections. The best approach depends on the particular experimental objectives.

11.1. Triple quadrupole mass spectrometers

Quadrupoles function as mass filters. At any instant, they allow a particular m/z to pass through, sending all other ions to waste. Mass resolution is "low," meaning that, although ions differing by 1 amu are essentially completely separated, those differing by <0.1 amu are not. Due to this low mass resolution, for complex mixtures, a single mass filtration step (single quadrupole MS) is often insufficient to isolate the analyte of interest. Triple quadrupole MS increases specificity by using two mass filtration steps in series, separated by a collision cell. The first quadrupole selects for the parent m/z. The second quadrupole serves as the collision cell, where ions selected in the first quadrupole are collided with a noble gas (e.g., argon) to produce fragment ions. The third quadrupole selects for the fragment m/z. These steps remove most environmental interferences. With careful selection of fragment ions, they are often sufficient to distinguish closely related metabolites.

Effective utilization of a triple quadrupole mass spectrometer requires optimization of two parameters for each targeted metabolite: the fragment ion to select in the third quadrupole, and the collision energy to produce that fragment in the collision cell. These parameters are best determined using compound standards. Once the parameters are determined, each compound is measured via a "selected reaction monitoring" (SRM) scan event, where the "selected reaction" refers to the parent ion forming a specific fragment ion at a defined collision energy in the MS instrument's collision cell. In metabolomics, SRM scan events for different compounds are performed in series, an approach called MRM. MRM can also refer to monitoring production of multiple different product ions for a single parent ion. The main advantages of MRM are sensitivity and linear dynamic range (which result from the efficiency of quadrupoles and ion detectors) and specificity (due to the two MS steps). There are two major disadvantages: data is limited to targeted analytes and sensitivity and quantitative precision fall when analyzing large numbers of compounds, as the scan time becomes divided over many SRM events. To mitigate the latter problem, MRM scans for a given metabolite are typically performed only during the retention time interval in which the compound elutes.

Performance of an MRM-based method depends strongly on selecting the optimal product ion. This requires striking a balance between signal intensity (which is optimized by choosing the most abundant product ion) and specificity (which involves picking as unique a product ion as possible). Generally, it is desirable to avoid product ions that are formed by loss of H_2O or NH_3, as such losses occur for a wide variety of parent ion structures. MS^2 spectra of glucose-1-phosphate and glucose-6-phosphate can be seen in Fig. 16.4. For glucose-6-phosphate, a fragment ion of 199 m/z is specific, so one may use the SRM $259 \rightarrow 199$ to differentiate it from glucose-1-phosphate. This nomenclature describes fragmenting the parent ion of 259 m/z and selecting a product fragment ion of 199 m/z for quantification. On the other hand, no specific product ion was observed for glucose-1-phosphate (fragment ions 79, 97, 139, and 241 m/z were also seen from glucose-6-phosphate). In this case, separation in a second dimension (such as HPLC) would be needed to differentiate signal resulting from glucose-1-phosphate from that resulting from glucose-6-phosphate. An alternative is to subtract the signal at fragment ion 79 m/z arising from glucose-6-phosphate (whose concentration is estimated based on the specific product ion of 199 m/z) from the total signal, and to assign the residual signal to glucose-1-phosphate or other hexose phosphate isomers.

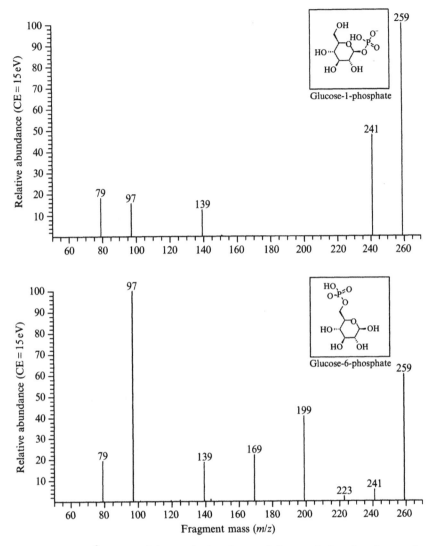

Figure 16.4 MS^2 spectra of glucose-1-phosphate and glucose-6-phosphate in negative ion mode.

11.2. High-resolution mass analyzers

Resolution refers to the ability to separate compounds based on a small mass differences. Resolution can be quantified based on Δm, the smallest mass difference that can be resolved. As Δm typically increases with analyte mass (m), resolution more often reported as $m/\Delta m$.

A common and cost-effective means of gathering high mass resolution data is TOF MS. Each TOF scan involves measuring the time that ions, which have been accelerated through a fixed voltage, take to traverse a flight path. Ions of low m/z fly faster, and thus reach the detector earlier. Modern TOF instruments provide $\sim 10,000$ $m/\Delta m$, which is adequate to differentiate, for example, protonated lysine ($C_6H_{15}N_2O_2^+$, 147.1128 m/z) and protonated glutamine ($C_5H_{11}N_2O_3^+$, 147.0765 m/z), with $m/\Delta m \sim 4000$. Resolution of commercial TOF instruments is expected to increase further in the near future, perhaps to $\sim 50,000$ $m/\Delta m$.

A Fourier transform ion cyclotron resonance (FTICR) mass spectrometer provides yet higher mass resolution ($\sim 100,000$–$1,000,000$ $m/\Delta m$) (Breitling *et al.*, 2006). It involves the circular motion of ions in a magnetic field, where the frequency of rotation relates to the ion's m/z. Its key disadvantage is its very high cost. Orbitrap mass analyzers involve some similar principles to ion cyclotron resonance, but without the need for a magnet. Ions circulate around a central, roughly cylindrical electrode (depicted schematically in Fig. 16.3), with the attractive electric field from the electrode providing the required centripetal force. This is roughly analogous to the attractive gravitational field from the sun proving the required centripetal force to hold the planets in orbit. However, the central electrode of the Orbitrap is not actually cylindrical. Instead, it is shaped so as to result in the ions oscillating along the electrode's long axis. The motion parallel to the electrode's axis depends on the ion's m/z, and the measured oscillation frequency can be Fourier deconvoluted to reveal ion mass with up to $\sim 350,000$ resolution (Makarov *et al.*, 2009), with $\sim 100,000$ resolution reliably obtained using commercially available instruments. The resolving power of $\sim 100,000$ has several uses; for example, one can differentiate singly [13]C- and singly ^{15}N-labeled glutamine ($^{12}C_4{}^{13}CH_{11}{}^{14}N_2O_3^+$, 148.0803 m/z, vs. $^{12}C_5H_{11}{}^{14}N^{15}NO_3^+$, 148.0740 m/z, $m/\Delta m \sim 24,000$), which is helpful in following isotope tracers. Note that the different masses of the [13]C- and ^{15}N-labeled forms reflects the mass difference for a neutron in different nuclei; that is, different nuclear energies result in measurable mass differences.

Among these mass analyzers, we consider both TOF and Orbitrap reasonable choices for typical metabolomics users. In our hands, both offered similar sensitivity, linear dynamic range, and quantitative reproducibility, with the Orbitrap providing greater specificity due to its higher resolving power. Note that we use "Orbitrap" to refer to the Orbitrap mass analyzer, not the hybrid ion trap-Orbitrap MS/MS instrument that is widely used in proteomics and discussed in Section 11.3. In addition to being a component of this hybrid instrument, the Orbitrap is currently sold as a stand-alone mass analyzer by Thermo under the brand name "Exactive" at prices roughly comparable to high-quality TOF instruments. Irrespective of the instrument employed, the main data output of high-resolution MS are full scan MS spectra, collected typically at a rate of ~ 1 spectrum/s. With appropriate software (discussed below), the data can be used both to quantitate known compounds and to identify other compounds that vary across biological conditions.

11.3. Hybrid instruments

Hybrid instruments involve the combination of two or more mass analyzers within a single instrument. Typically, this involves a low-resolution analyzer that filters ions on the front end (e.g., quadrupole or ion trap) and a high-resolution mass analyzer on the back end (e.g., TOF or Orbitrap). At present, the most commercially important hybrid instruments are quadrupole-TOF (Q-TOF) and linear ion trap-Orbitrap. These instruments provide full scan MS capabilities similar to stand-alone high-resolution mass analyzers, although typically with some decrement in sensitivity due to ion losses in the front end. They also enable MS/MS analysis: ions can be selected and fragmented in the front end, and the fragment ions passed to the back end for high resolution analysis.

An important capability of hybrid instruments is data-dependent MS/MS analysis: collecting a full scan MS spectrum, and then running MS/MS on the most prevalent ion(s) present in the full scan. (Such data-dependent analysis can also be run on a standard ion trap instrument, but without the benefits of high resolution). Typically, the full scan data is used for quantitation, with the MS/MS data used for compound identification. This approach is well-proven in proteomics, where MS/MS spectra are searched against genome sequences to identify peptides. In metabolomics, the need for MS/MS is arguably less, as the combination of accurate mass and retention time is often adequate to identify known analytes. For unknown analytes, while MS/MS data is a valuable first step in identification, it is not typically sufficient. Thus, while it is now routine practice to use data-dependent MS/MS for proteomics applications (Yates *et al.*, 2009), we do not routinely use this approach for metabolomics, which involves additional data management challenges and higher MS instrument cost. Others, however, have been successful through routine use of MS/MS (Lawton *et al.*, 2009).

12. TARGETED DATA ANALYSIS

Raw LC–MS data involves the three dimensions of retention time, m/z, and signal intensity (Fig. 16.5). In triple quadrupole data, the m/z domain is discrete: MRM scan events. In high-resolution mass spectrometry data, the m/z domain is continuous if collected in "profile" mode and

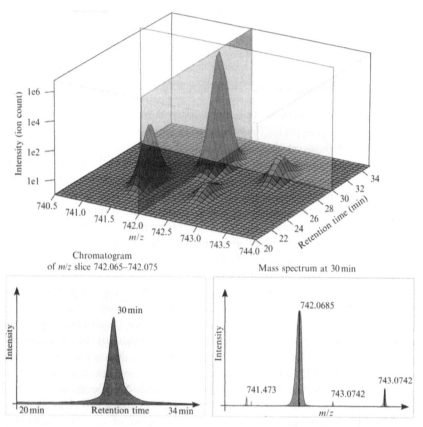

Figure 16.5 Chromatograms and mass spectra are two-dimensional slices of 3D LC–MS data. The upper 3D plot was generated using a small subset of data collected from LC–MS analysis of a yeast extract on an Orbitrap mass analyzer in negative ionization mode. The plot is centered around the anion formed by deprotonation of nicotinamide adenine dinucleotide phosphate, oxidized form: $[NADP-H]^-$ ($m/z = 742.069$, RT = 30 min). Four distinct peaks can be observed. The peak at $m/z = 743.077$, RT = 30 min is the ^{13}C M + 1 isotopic variant of $[NADP-H]^-$. The identities of for the other two peaks are not known. Chromatograms are generally used for quantitation, while mass spectra are used for the identification of coeluting compounds, fragments, adducts, and isotopic variants.

semicontinuous if collected in "centroid" mode, where the instrument records the central m/z for each mass spectral peak, instead of its full distribution.

Targeted metabolomics data are typically visualized as ion-specific chromatograms: signal plotted versus retention time for a specified m/z, that is, for a specific MRM scan in triple quadrupole data or for an m/z range centered about the m/z of the targeted metabolite in high-resolution full scan MS data (Fig. 16.5). The main goal of targeted metabolomics is to quantitate known compounds. This involves reducing ion-specific chromatograms to a scalar value proportional to the compound concentration. There are two main requirements: finding the correct peak and quantitating the peak intensity.

If several peaks are present in an ion-specific chromatogram, selecting the correct peak for a targeted analyte requires prior knowledge of the compound's retention time. A challenge is that retention time can vary from sample-to-sample (e.g., due to differences in sample matrix, environmental factors, or column aging). Accordingly, it is desirable to align chromatograms prior to looking for the peak corresponding to the analyte of interest. Alignment relies on the fact that retention time shifts tend to be consistent across compounds. A number of computational algorithms for alignment that have been developed (Lange et al., 2008; Smith et al., 2006). Once samples are aligned, peaks eluting at the same time are grouped. Peaks are then quantitated, and the peak size for each group in each sample reported. This yields a matrix with peaks as rows, samples as columns, and peak intensities as the entries.

Peak intensity can be measured a number of ways, with the two most common being peak height or peak area. For a fixed-width Gaussian peak, peak height and area are linearly related; thus both approaches are equivalent. If peak width varies (i.e., is broader in one sample than another), then area is the more reliable metric. However, calculations of area depend on where one draws the peak boundaries and baseline. Accordingly, if S/N is poor, peak height can be a more robust metric. A compromise that we sometimes employ involves summing the highest few points in a peak.

We have developed an open-source software package that automates the analysis of targeted metabolomics data (Melamud, 2009). The package, "mzROCK," is written in the statistical software package R and freely available at http://code.google.com/p/mzrock/. It is designed for MRM methods, but the concepts (and most of the code) apply to any targeted LC–MS data. The package includes features that help to cope with the imperfections of real LC–MS data. For example, it reports not only peak sizes, but also a statistical estimate of the peak quality. The estimate of peak quality takes into account factors like retention time consistency and S/N, integrating them via a random forest algorithm. The peak quality estimate is a valuable guide as to peaks to ignore (those with very low quality scores) and those most in need of careful manual review (those with intermediate scores). To facilitate manual data review, the mzROCK software package produces a PDF file that presents all raw ion-specific chromatograms in any easy-to-read, searchable manner. Manual review of raw ion-specific chromatograms is standard practice in LC–MS labs. We encourage yeast biologists working at the interface with metabolomics to devote some time to examining ion-specific chromatograms.

13. UNTARGETED DATA ANALYSIS

A major advantage of high-resolution mass spectrometers is the ability to quantitate known analytes while simultaneously collecting untargeted data on all ions present, including ones arising from novel compounds. Although involving greater data management challenges due to larger file sizes, untargeted data analysis involves the same basic steps as targeted data analysis, except with the added complication of selecting which m/z are interesting. The motivation to focus on specific m/z reflects the otherwise excessive number of ion-specific chromatograms generated by current high-resolution instruments. For example, for $m/z < 1000$, an Orbitrap mass analyzer can reliably distinguish analytes differing by <0.01 m/z, resulting in $>100,000$ differentiable mass slices. Most of these mass slices are empty, however, as no chemically feasible combination of common elements generates analytes of that m/z. In addition, some mass slices contain only noise. Eliminating such slices tends to yield a manageable number of ion-specific chromatograms, from which peaks can be identified, aligned, and quantitated. As for targeted analysis, the net outcome is a matrix with peaks as rows, samples as columns, and peak intensities as the entries. The difference is that peaks do not necessarily correspond to known metabolites, but may instead be identified solely based on m/z and retention time. The Siuzdak lab has developed software called "XCMS" that effectively conducts the above steps (Smith *et al.*, 2006). XCMS accepts data in the common open-exchange format "mzXML" (Pedrioli *et al.*, 2004); other open formats are "mzData" (Orchard *et al.*, 2004) and "mzML" (Deutsch, 2008). Moving to a common open-exchange format will facilitate sharing of metabolomics data going forward.

MS/MS data can contribute to untargeted data analysis in two ways. Peaks with identical MS/MS spectra provide useful benchmarks for chromatogram alignment (Benton *et al.*, 2008). Also, MS/MS spectra provide additional data for peak identification.

A complication in untargeted LC–MS is that the number of observed peaks may markedly exceed the number of actual metabolites present: signal can arise from different adduct ions, isotopic variants, and in-source fragmentation.

13.1. Adducts

Common adduct ions are listed in Table 16.2. Adduct formation occurs during ionization. Accordingly, different adducts of the same metabolite coelute chromatographically, which provides a useful means for their identification. In practice, we typically quantitate based on $M + H^+$ and $M - H^-$ peaks, as formation of other adduct ions may vary strongly based on the

Table 16.2 Some common adduct species formed by positive and negative mode electrospray ionization

Adduct species	Mol ratio	Charge	ΔM	Polarity
$[M + H]^+$	1	1	1.0072	+
$[M + NH_4]^+$	1	1	18.0338	+
$[M + H_3O]^+$	1	1	19.0184	+
$[M + Na]^+$	1	1	22.9899	+
$[M + CH_3OH + H]^+$	1	1	33.0335	+
$[M + K]^+$	1	1	38.9632	+
$[M + CH_3CN + H]^+$	1	1	42.0338	+
$[M + 2Na - H]^+$	1	1	44.9726	+
$[M + CH_3CN + Na]^+$	1	1	64.0165	+
$[M + HCOOH + Na]^+$	1	1	68.9954	+
$[M + 2K - H]^+$	1	1	76.9191	+
$[M + CH_3COOH + Na]^+$	1	1	83.0110	+
$[M + 2CH_3CN + H]^+$	1	1	83.0603	+
$[M + tributylamine + H]^+$	1	1	186.2222	+
$[M + 2H]^{2+}$	1	2	2.0145	+
$[M + H + NH_4]^{2+}$	1	2	19.0410	+
$[M + H + Na]^{2+}$	1	2	23.9971	+
$[M + H + K]^{2+}$	1	2	39.9704	+
$[M + CH_3CN + 2H]^{2+}$	1	2	43.0410	+
$[M + 2Na]^{2+}$	1	2	45.9798	+
$[M + Na + K]^{2+}$	1	2	61.9531	+
$[M + 2K]^{2+}$	1	2	77.9263	+
$[M + 2CH_3CN + 2H]^{2+}$	1	2	84.0675	+
$[M + 3CH_3CN + 2H]^{2+}$	1	2	125.0941	+
$[M + tributylamine + 2H]^{2+}$	1	2	187.2288	+
$[M + 3H]^{3+}$	1	3	3.0217	+
$[M + 2H + Na]^{3+}$	1	3	25.0044	+
$[M + H + 2Na]^{3+}$	1	3	46.9871	+
$[M + Fe]^{3+}$	1	3	55.9344	+
$[M + 3Na]^{3+}$	1	3	68.9698	+
$[M + 3K]^{3+}$	1	3	116.8895	+
$[2M + H]^+$	2	1	1.0072	+
$[2M + NH_4]^+$	2	1	18.0338	+
$[2M + Na]^+$	2	1	22.9899	+
$[2M + K]^+$	2	1	38.9632	+
$[2M + CH_3CN + H]^+$	2	1	42.0338	+
$[2M + 3H_2O + 2H]^+$	2	1	56.0461	+
$[2M + CH_3CN + Na]^+$	2	1	64.0165	+

(continued)

Table 16.2 (*continued*)

Adduct species	Mol ratio	Charge	ΔM	Polarity
$[M - H]^-$	1	1	-1.0072	$-$
$[M - OH]^-$	1	1	17.0067	$-$
$[M - H_2O - H]^-$	1	1	-20.0250	$-$
$[M + Na - 2H]^-$	1	1	20.9755	$-$
$[M + Na - 2H]^{2-}$	1	1	20.9755	$-$
$[M + CH_3OH - H]^-$	1	1	31.0190	$-$
$[M + Cl]^-$	1	1	34.9694	$-$
$[M + K + 2H]^-$	1	1	36.9487	$-$
$[M + CH_3CN + H]^-$	1	1	40.0193	$-$
$[M + HCOO]^-$	1	1	44.9982	$-$
$[M + CH_3COO]^-$	1	1	59.0133	$-$
$[M + NaO_2CH - H]^-$	1	1	66.9803	$-$
$[M + {}^{79}Br]^-$	1	1	78.9183	$-$
$[M + {}^{81}Br]^-$	1	1	80.9168	$-$
$[M + NaO_2CCH_3 - H]^-$	1	1	80.9960	$-$
$[M + CF_3COO]^-$	1	1	112.9856	$-$
$[M + tributylamine - H]^-$	1	1	184.2071	$-$
$[M - 2H]^{2-}$	1	2	-2.0145	$-$
$[M - 3H]^{3-}$	1	3	-3.0217	$-$
$[2M - H]^-$	2	1	-1.0072	$-$
$[2M + HCOO]^-$	2	1	44.9982	$-$
$[3M - H]^-$	3	1	-1.0072	$-$

The relative ratios of the adduct species depend on the analyte concentration, solvent composition, pH, and concentrations of other ions like sodium. To calculate adduct m/z given parent m/z use the following formula: $\text{adduct}[m/z] = ((\text{parent}[m/z]) \times (\text{mol ratio}) + \Delta M)/(\text{charge})$.

biological matrix (e.g., sodium concentration in the sample). An alternative approach is to identify all adduct ions of a given parent and sum their intensities; however, this requires additional computational tools.

13.2. Isotopic variants

In the absence of isotopic labeling, the most common isotopic variant is ^{13}C, with the abundance of the ^{13}C M + 1 peak ($S_{M + 1}$) given by

$$\frac{S_{M+1}}{S_M} = \frac{N \times 0.011(0.989)^{N-1}}{0.989^N} \approx 0.011N$$

where N is the number of carbon atoms in the metabolite and S_M is the intensity of the M + 0 peak. Other isotopic variants that can often be detected include ^{15}N, ^2H, ^{34}S, etc.

13.3. In-source fragmentation

With modern ESI sources, in-source fragmentation is less of an issue than adduct formation; however, it is still worth being aware of the possibility for in-source losses of water, ammonia, carbon dioxide, phosphate, etc.

13.4. Unknown identification

Once the myriad of peaks found in untargeted LC–MS analysis are reduced to the more finite set of probable underlying metabolites, the question becomes determining their identities. A first-pass approach involves searching of known metabolite databases, such as KEGG, Metacyc, Human Metabolome Database, and METLIN for known compounds of the correct exact mass (Smith *et al.*, 2005; Wishart *et al.*, 2009). When a match is found, the question then becomes whether it is a true positive or false positive. Three criteria are of immediate assistance: isotopic variants (e.g., if the candidate structure contains sulfur, then there should be an $\sim$5% M + 2 peak); retention time (e.g., if the compound is a monophosphate, it should elute close to other related monophosphates); and MS/MS spectrum if available. More definitive structure assignment typically requires obtaining pure standard (which should have the same MS/MS spectrum as the endogenous compound, and coelute by at least two orthogonal chromatography methods) and/or purification and NMR analysis.

   When no database match is found, structure elucidation becomes yet more challenging. Exact mass information provides a useful starting point for determining the compound's molecular formula, and a variety of molecular formula calculators, including ones that take into account isotopic variants present, are available (Jarussophon *et al.*, 2009; Koch *et al.*, 2007; Sana *et al.*, 2008). Additional information of molecular formula can be obtained by feeding ^{13}C-, ^{15}N-, or other isotope-labeled nutrients, and observing labeling of the unknown compounds of interest. If one obtains from such analyses a guess as to the metabolic origin of the novel species, feeding more downstream metabolic precursors in labeled form can clarify the precise metabolic origin. Throughout the above steps, MS/MS information is quite useful (Ashline *et al.*, 2007; Kurimoto *et al.*, 2006). For example, one could demonstrate that a novel species contains adenine by (1) finding a characteristic adenine fragment ion and (2) feeding labeled adenine and observing label incorporation.

14. FUTURE OUTLOOK

   Yeast metabolomics is still a young field with many advances in instrumentation and computation on the horizon. This makes it prime for both technical innovation and biological discovery. Much can be learned from the examples of genomics and transcriptomics, in terms of building common resources (e.g., SGD) and developing data visualization tools (e.g., clustered heat maps). A unique feature of metabolomics is the well-defined connections between metabolites: the metabolic map that was in many respects the first great achievement in systems biochemistry. The ability to conveniently relate data, both metabolomic and fluxomic, to this map is an immediate need. Another immediate need is experimental and computational tools for flux profiling that are simple and robust enough to be widely used throughout the yeast community. Looking further into the future, the challenges of metabolomics will likely coalesce with those of systems biology: integration of diverse types of data to gain insight into the biochemical mechanisms underlying the complex emergent properties that distinguish living systems.

REFERENCES

Aboka, F. O., Heijnen, J. J., and Winden, W. A. V. (2009). Dynamic 13C-tracer study of storage carbohydrate pools in aerobic glucose-limited *Saccharomyces cerevisiae* confirms a rapid steady-state turnover and fast mobilization during a modest stepup in the glucose uptake rate. *FEMS Yeast Res.* **9,** 191–201.

Antoniewicz, M. R., Kraynie, D. F., Laffend, L. A., González-Lergier, J., Kelleher, J. K., and Stephanopoulos, G. (2007). Metabolic flux analysis in a nonstationary system: Fed-batch fermentation of a high yielding strain of *E. coli* producing 1,3-propanediol. *Metab. Eng.* **9,** 277–292.

Ashline, D. J., Lapadula, A. J., Liu, Y., Lin, M., Grace, M., Pramanik, B., and Reinhold, V. N. (2007). Carbohydrate structural isomers analyzed by sequential mass spectrometry. *Anal. Chem.* **79,** 3830–3842.

Bajad, S. U., Lu, W., Kimball, E. H., Yuan, J., Peterson, C., and Rabinowitz, J. D. (2006). Separation and quantitation of water soluble cellular metabolites by hydrophilic interaction chromatography–tandem mass spectrometry. *J. Chromatogr. A* **1125,** 76–88.

Bennett, B. D., Yuan, J., Kimball, E. H., and Rabinowitz, J. D. (2008). Absolute quantitation of intracellular metabolite concentrations by an isotope ratio-based approach. *Nat. Protoc.* **3,** 1299–1311.

Bennett, B. D., Kimball, E. H., Gao, M., Osterhout, R., Van Dien, S., and Rabinowitz, J. D. (2009). Absolute metabolite concentrations and implied enzyme active site occupancy in *Escherichia coli*. *Nat. Chem. Biol.* **5,** 593–599.

Benton, H. P., Wong, D. M., Trauger, S. A., and Siuzdak, G. (2008). XCMS2: Processing tandem mass spectrometry data for metabolite identification and structural characterization. *Anal. Chem.* **80,** 6382–6389.

Boer, V. M., Crutchfield, C. A., Bradley, P. H., Botstein, D., and Rabinowitz, J. D. (2010). Growth-limiting intracellular metabolites in yeast under diverse nutrient limitations. *Mol. Biol. Cell* **21,** 198–211.

Brachmann, C. B., Davies, A., Cost, G. J., Caputo, E., Li, J., Hieter, P., and Boeke, J. D. (1998). Designer deletion strains derived from *Saccharomyces cerevisiae* S288C: A useful set of strains and plasmids for PCR-mediated gene disruption and other applications. *Yeast* **14,** 115–132.

Brauer, M. J., Yuan, J., Bennett, B. D., Lu, W., Kimball, E., Botstein, D., and Rabinowitz, J. D. (2006). Conservation of the metabolomic response to starvation across two divergent microbes. *Proc. Natl. Acad. Sci. USA* **103**, 19302–19307.

Brauer, M. J., Huttenhower, C., Airoldi, E. M., Rosenstein, R., Matese, J. C., Gresham, D., Boer, V. M., Troyanskaya, O. G., and Botstein, D. (2008). Coordination of growth rate, cell cycle, stress response, and metabolic activity in yeast. *Mol. Biol. Cell* **19**, 352–367.

Breitling, R., Pitt, A. R., and Barrett, M. P. (2006). Precision mapping of the metabolome. *Trends Biotechnol.* **24**, 543–548.

Büscher, J. M., Czernik, D., Ewald, J. C., Sauer, U., and Zamboni, N. (2009). Cross-platform comparison of methods for quantitative metabolomics of primary metabolism. *Anal. Chem.* **81**, 2135–2143.

Canelas, A., Ras, C., ten Pierick, A., van Dam, J., Heijnen, J., and van Gulik, W. (2008). Leakage-free rapid quenching technique for yeast metabolomics. *Metabolomics* **4**, 226–239.

Carlson, E. E., and Cravatt, B. F. (2007a). Enrichment tags for enhanced-resolution profiling of the polar metabolome. *J. Am. Chem. Soc.* **129**, 15780–15782.

Carlson, E. E., and Cravatt, B. F. (2007b). Chemoselective probes for metabolite enrichment and profiling. *Nat. Methods* **4**, 429–435.

Castrillo, J. I., Hayes, A., Mohammed, S., Gaskell, S. J., and Oliver, S. G. (2003). An optimized protocol for metabolome analysis in yeast using direct infusion electrospray mass spectrometry. *Phytochemistry* **62**, 929–937.

Coulier, L., Bas, R., Jespersen, S., Verheij, E., van der Werf, M. J., and Hankemeier, T. (2006). Simultaneous quantitative analysis of metabolites using ion-pair liquid chromatography–electrospray ionization mass spectrometry. *Anal. Chem.* **78**, 6573–6582.

Dettmer, K., Aronov, P. A., and Hammock, B. D. (2007). Mass spectrometry-based metabolomics. *Mass Spectrom. Rev.* **26**, 51–78.

Deutsch, E. (2008). mzML: A single, unifying data format for mass spectrometer output. *Proteomics* **8**, 2776–2777.

Dunham, M. (2010). Experimental evolution in yeast. A practical guide. *In* "Methods in Enzymology," (J. Weissman, C. Guthrie, and G. R. Fink, eds.), pp. 393–426. Academic Press, New York. (this volume).

Dunn, W. B., Broadhurst, D., Brown, M., Baker, P. N., Redman, C. W., Kenny, L. C., and Kell, D. B. (2008). Metabolic profiling of serum using ultra performance liquid chromatography and the LTQ-Orbitrap mass spectrometry system. *J. Chromatogr. B* **871**, 288–298.

Fiehn, O. (2008). Extending the breadth of metabolite profiling by gas chromatography coupled to mass spectrometry. *TrAC Trends Anal. Chem.* **27**, 261–269.

Gaisne, M., Bécam, A., Verdière, J., and Herbert, C. J. (1999). A 'natural' mutation in *Saccharomyces cerevisiae* strains derived from S288c affects the complex regulatory gene HAP1 (CYP1). *Curr. Genet.* **36**, 195–200.

Gonzalez, B., Francois, J., and Renaud, M. (1997). A rapid and reliable method for metabolite extraction in yeast using boiling buffered ethanol. *Yeast* **13**, 1347–1355.

Henry, C. S., Broadbelt, L. J., and Hatzimanikatis, V. (2007). Thermodynamics-based metabolic flux analysis. *Biophys. J.* **92**, 1792–1805.

Ho, C. H., Magtanong, L., Barker, S. L., Gresham, D., Nishimura, S., Natarajan, P., Koh, J. L. Y., Porter, J., Gray, C. A., Andersen, R. J., Giaever, G., Nislow, C., *et al.* (2009). A molecular barcoded yeast ORF library enables mode-of-action analysis of bioactive compounds. *Nat. Biotech.* **27**, 369–377.

Jarussophon, S., Acoca, S., Gao, J., Deprez, C., Kiyota, T., Draghici, C., Purisima, E., and Konishi, Y. (2009). Automated molecular formula determination by tandem mass spectrometry (MS/MS). *Analyst* **134**, 690–700.

John, Y., Cristian, I. R., and Aleksey, N. (2009). *Proteomics by Mass Spectrometry: Approaches, Advances, and Applications.* Available at:http://arjournals.annualreviews.org/doi/abs/10.1146/annurev-bioeng-061008-124934?url_ver=Z39.88-2003&rfr_id=ori:rid:crossref.org&rfr_dat=cr_pub%3dncbi.nlm.nih.gov, (accessed May 4, 2009).

Koch, B. P., Dittmar, T., Witt, M., and Kattner, G. (2007). Fundamentals of molecular formula assignment to ultrahigh resolution mass data of natural organic matter. *Anal. Chem.* **79,** 1758–1763.

Kostiainen, R., and Kauppila, T. J. (2009). Effect of eluent on the ionization process in liquid chromatography–mass spectrometry. *J. Chromatogr. A* **1216,** 685–699.

Kümmel, A., Panke, S., and Heinemann, M. (2006). Putative regulatory sites unraveled by network-embedded thermodynamic analysis of metabolome data. *Mol. Syst. Biol.* **2,** 2006–2034.

Kurimoto, A., Daikoku, S., Mutsuga, S., and Kanie, O. (2006). Analysis of energy-resolved mass spectra at MSn in a pursuit to characterize structural isomers of oligosaccharides. *Anal. Chem.* **78,** 3461–3466.

Lamos, S. M., Shortreed, M. R., Frey, B. L., Belshaw, P. J., and Smith, L. M. (2007). Relative quantification of carboxylic acid metabolites by liquid chromatography–mass spectrometry using isotopic variants of cholamine. *Anal. Chem.* **79,** 5143–5149.

Lange, E., Tautenhahn, R., Neumann, S., and Gröpl, C. (2008). Critical assessment of alignment procedures for LC–MS proteomics and metabolomics measurements. *BMC Bioinformatics* **9,** 375.

Lawton, K. A., Beebe, K., Berger, A., Guo, L., Rose, D., Roulston, N., Tsutsui, N., Ryals, J. A., and Milburn, M. V. (2009). Technology and applications of metabolomics: Comprehensive biochemical profiling to solve complex biological problems. *In* Computational and Systems Biology. Methods and Applications. Available at http://www.ressign.com/UserArticleDetails.aspx?arid=7705, (accessed May 8, 2009).

Lenz, E. M., Weeks, J. M., Lindon, J. C., Osborn, D., and Nicholson, J. K. (2005). Qualitative high field 1H-NMR spectroscopy for the characterization of endogenous metabolites in earthworms with biochemical biomarker potential. *Metabolomics* **1,** 123–136.

Lewis, I. A., Schommer, S. C., Hodis, B., Robb, K. A., Tonelli, M., Westler, W. M., Sussman, M. R., and Markley, J. L. (2007). Method for determining molar concentrations of metabolites in complex solutions from two-dimensional ^{1}H–^{13}C NMR spectra. *Anal. Chem.* **79,** 9385–9390.

Loret, M. O., Pedersen, L., and Francois, J. (2007). Revised procedures for yeast metabolites extraction: Application to a glucose pulse to carbon-limited yeast cultures, which reveals a transient activation of the purine salvage pathway. *Yeast* **24,** 47–60.

Lu, W., Bennett, B. D., and Rabinowitz, J. D. (2008). Analytical strategies for LC–MS-based targeted metabolomics. *J. Chromatogr. B Anal. Technol. Biomed. Life Sci.* **871,** 236–242.

Luo, B., Groenke, K., Takors, R. W. C., and Oldiges, M. (2007). Simultaneous determination of multiple intracellular metabolites in glycolysis, pentose phosphate pathway and tricarboxylic acid cycle by liquid chromatography–mass spectrometry. *J. Chromatogr. A* **1147,** 153–164.

Makarov, A., Denisov, E., and Lange, O. (2009). Performance evaluation of a high-field Orbitrap mass analyzer. *J. Am. Soc. Mass Spectrom.* **20,** 1391–1396.

Mashego, M. R., Wu, L., Dam, J. C. V., Ras, C., Vinke, J. L., Winden, W. A. V., Gulik, W. M. V., and Heijnen, J. J. (2004). MIRACLE: Mass isotopomer ratio analysis of U-13C-labeled extracts. A new method for accurate quantification of changes in concentrations of intracellular metabolites. *Biotechnol. Bioeng.* **85,** 620–628.

Mashego, M. R., Rumbold, K., De Mey, M., Vandamme, E., Soetaert, W., and Heijnen, J. J. (2007). Microbial metabolomics: Past, present and future methodologies. *Biotechnol. Lett.* **29,** 1–16.

Melamud, E. (2009). *mzRock—Google Code.* Available at:http://code.google.com/p/mzrock/, (accessed April 20, 2009).

Moxley, J. F., Jewett, M. C., Antoniewicz, M. R., Villas-Boas, S. G., Alper, H., Wheeler, R. T., Tong, L., Hinnebusch, A. G., Ideker, T., Nielsen, J., and Stephanopoulos, G. (2009). Linking high-resolution metabolic flux phenotypes and transcriptional regulation in yeast modulated by the global regulator Gcn4p. *Proc. Natl. Acad. Sci. USA* **106**, 6477–6482.

Munger, J., Bennett, B. D., Parikh, A., Feng, X., McArdle, J., Rabitz, H. A., Shenk, T., and Rabinowitz, J. D. (2008). Systems-level metabolic flux profiling identifies fatty acid synthesis as a target for antiviral therapy. *Nat. Biotechnol.* **26**, 1179–1186.

Nguyen, S., and Fenn, J. B. (2007). Gas-phase ions of solute species from charged droplets of solutions. *Proc. Natl. Acad. Sci. USA* **104**, 1111–1117.

Nguyen, D. T., Guillarme, D., Rudaz, S., and Veuthey, J. (2006). Chromatographic behaviour and comparison of column packed with sub-2 μm stationary phases in liquid chromatography. *J. Chromatogr. A* **1128**, 105–113.

Orchard, S., Taylor, C. F., Hermjakob, H., Weimin-Zhu, R. K. J. Jr., and Apweiler, R. (2004). Advances in the development of common interchange standards for proteomic data. *Proteomics* **4**, 2363–2365.

Pedrioli, P. G. A. , et al. (2004). A common open representation of mass spectrometry data and its application to proteomics research. *Nat. Biotech.* **22**, 1459–1466.

Prasad Maharjan, R., and Ferenci, T. (2003). Global metabolite analysis: The influence of extraction methodology on metabolome profiles of *Escherichia coli. Anal. Biochem.* **313**, 145–154.

Rabinowitz, J. D., and Kimball, E. (2007). Acidic acetonitrile for cellular metabolome extraction from *Escherichia coli. Anal. Chem.* **79**, 6167–6173.

Sabatine, M. S., Liu, E., Morrow, D. A., Heller, E., McCarroll, R., Wiegand, R., Berriz, G. F., Roth, F. P., and Gerszten, R. E. (2005). Metabolomic identification of novel biomarkers of myocardial ischemia. *Circulation* **112**, 3868–3875.

Sáez, M. J., and Lagunas, R. (1976). Determination of intermediary metabolites in yeast. Critical examination of the effect of sampling conditions and recommendations for obtaining true levels. *Mol. Cell. Biochem.* **13**, 73–78.

Sana, T. R., Roark, J. C., Li, X., Waddell, K., and Fischer, S. M. (2008). Molecular formula and METLIN personal metabolite database matching applied to the identification of compounds generated by LC/TOF-MS. *J. Biomol. Technol.* **19**, 258–266.

Sauer, U. (2006). Metabolic networks in motion: ^{13}C-based flux analysis. *Mol. Syst. Biol.* **2**, 62.

Sauer, U., Heinemann, M., and Zamboni, N. (2007). GENETICS: Getting closer to the whole picture. *Science* **316**, 550–551.

Shortreed, M. R., Lamos, S. M., Frey, B. L., Phillips, M. F., Patel, M., Belshaw, P. J., and Smith, L. M. (2006). Ionizable isotopic labeling reagent for relative quantification of amine metabolites by mass spectrometry. *Anal. Chem.* **78**, 6398–6403.

Smith, C. A., O'Maille, G., Want, E. J., Qin, C., Trauger, S. A., Brandon, T. R., Custodio, D. E., Abagyan, R., and Siuzdak, G. (2005). METLIN: A metabolite mass spectral database. *Ther. Drug Monit.* **27**, 747–751.

Smith, C. A., Want, E. J., O'Maille, G., Abagyan, R., and Siuzdak, G. (2006). XCMS: Processing mass spectrometry data for metabolite profiling using nonlinear peak alignment, matching, and identification. *Anal. Chem.* **78**, 779–787.

Sopko, R., Huang, D., Preston, N., Chua, G., Papp, B., Kafadar, K., Snyder, M., Oliver, S. G., Cyert, M., Hughes, T. R., Boone, C., and Andrews, B. (2006). Mapping pathways and phenotypes by systematic gene overexpression. *Mol. Cell* **21**, 319–330.

Tolstikov, V. V., and Fiehn, O. (2002). Analysis of highly polar compounds of plant origin: Combination of hydrophilic interaction chromatography and electrospray ion trap mass spectrometry. *Anal. Biochem.* **301**, 298–307.

Trauger, S. A., Kalisak, E., Kalisiak, J., Morita, H., Weinberg, M. V., Menon, A. L., Poole, F. L. II, Adams, M. W. W., and Siuzdak, G. (2008). Correlating the transcriptome, proteome, and metabolome in the environmental adaptation of a hyperthermophile. *J. Proteome Res.* **7,** 1027–1035.

Tu, B. P., Mohler, R. E., Liu, J. C., Dombek, K. M., Young, E. T., Synovec, R. E., and McKnight, S. L. (2007). Cyclic changes in metabolic state during the life of a yeast cell. *Proc. Natl. Acad. Sci. USA* **104,** 16886–16891.

van der Werf, M. J., Overkamp, K. M., Muilwijk, B., Coulier, L., and Hankemeier, T. (2007). Microbial metabolomics: Toward a platform with full metabolome coverage. *Anal. Biochem.* **370,** 17–25.

van Dijken, J. P., *et al.* (2000). An interlaboratory comparison of physiological and genetic properties of four *Saccharomyces cerevisiae* strains. *Enzyme Microb. Technol.* **26,** 706–714.

Villas-Boas, S. G., Hojer-Pedersen, J., Akesson, M., Smedsgaard, J., and Nielsen, J. (2005). Global metabolite analysis of yeast: Evaluation of sample preparation methods. *Yeast* **22,** 1155–1169.

Wiechert, W., and Nöh, K. (2005). From stationary to instationary metabolic flux analysis. *Adv. Biochem. Eng. Biotechnol.* **92,** 145–172.

Wilson, I. D., Nicholson, J. K., Castro-Perez, J., Granger, J. H., Johnson, K. A., Smith, B. W., and Plumb, R. S. (2005). High resolution "Ultra Performance" liquid chromatography coupled to oa-TOF mass spectrometry as a tool for differential metabolic pathway profiling in functional genomic studies. *J. Proteome Res.* **4,** 591–598.

Wishart, D. S., *et al.* (2009). HMDB: A knowledgebase for the human metabolome. *Nucleic Acids Res.* **37,** D603–D610.

Wu, L., van Dam, J., Schipper, D., Kresnowati, M. T. A. P., Proell, A. M., Ras, C., van Winden, W. A., van Gulik, W. M., and Heijnen, J. J. (2006a). Short-term metabolome dynamics and carbon, electron, and atp balances in chemostat-grown *Saccharomyces cerevisiae* CEN.PK 113-7D following a glucose pulse. *Appl. Environ. Microbiol.* **72,** 3566–3577.

Wu, L., Mashego, M. R., Proell, A. M., Vinke, J. L., Ras, C., van Dam, J., van Winden, W. A., van Gulik, W. M., and Heijnen, J. J. (2006b). In vivo kinetics of primary metabolism in *Saccharomyces cerevisiae* studied through prolonged chemostat cultivation. *Metab. Eng.* **8,** 160–171.

Yuan, J., Bennett, B. D., and Rabinowitz, J. D. (2008). Kinetic flux profiling for quantitation of cellular metabolic fluxes. *Nat. Protoc.* **3,** 1328–1340.

SYSTEMS ANALYSIS

CHAPTER SEVENTEEN

IMAGING SINGLE MRNA MOLECULES IN YEAST

Hyun Youk,* Arjun Raj,‡ *and* Alexander van Oudenaarden*,†

Contents

Abstract

Yeast cells in an isogenic population do not all display the same phenotypes. To study such variation within a population of cells, we need to perform measurements on each individual cell instead of measurements that average out the behavior of a cell over the entire population. Here, we provide the basic concepts and a step-by-step protocol for a recently developed technique enabling one such measurement: fluorescence *in situ* hybridization that renders single mRNA molecule visible in individual fixed cells.

1. INTRODUCTION

Within an isogenic population of yeast cells, the behavior of any individual cell can differ markedly from the average behavior of the population (Raj and van Oudenaarden, 2008). For example, it has been shown

* Department of Physics, Massachusetts Institute of Technology, Cambridge, Massachusetts, USA
† Department of Biology, Massachusetts Institute of Technology, Cambridge, Massachusetts, USA
‡ Department of Bioengineering, University of Pennsylvania, Philadelphia, Pennsylvania, USA

Methods in Enzymology, Volume 470
ISSN 0076-6879, DOI: 10.1016/S0076-6879(10)70017-3

that random partitioning of proteins during cell division leads to variability in the number of proteins in individual cells (Rosenfeld *et al.*, 2005), while random bursts of transcription results in variability in number of mRNAs (Chubb *et al.*, 2006; Golding *et al.*, 2005; Raj *et al.*, 2006). These are just a few examples that highlight the importance of studying the behavior of a single cell rather than that of the whole population. One primary tool for studying the behavior of a single cell is the fluorescent protein such as GFP (green fluorescent protein). The most straightforward application of a fluorescent protein is to have it either driven by the promoter of interest or fused to the protein of interest to study variability in gene expression. Yet while the use of fluorescent proteins has certainly been pivotal in monitoring gene expression, fluorescent proteins suffer from a number of limitations. One such limitation is their low sensitivity: fluorescence from GFP and its variants is typically undetectable at the small number of molecules involved in studying gene expression. In yeast, fluorescence from GFP is typically detectable only when many hundreds of GFPs are present in a cell; the abundance of many transcription factors, for example, falls below this limit. Since the effects of expression variability are magnified when the number of molecules is low, the sensitivity limitation may preclude effective study of these processes. Another issue is that it is difficult to quantify the exact number of fluorescent proteins in individual cells because it is difficult to measure the amount of fluorescence emitted by a single GFP molecule. In addition, the slow decay time of fluorescent proteins (due to their relatively high stability) means that fluorescence is only diluted by cell division but not through other degradation mechanisms. This prevents observation of rapidly varying changes in gene activation, effectively averaging temporal fluctuations.

While having a fluorescent protein expressed by the promoter of interest or fused to a protein of interest suffers from a number of setbacks, other applications of the fluorescent protein led to powerful techniques enabling the detection of a single mRNA molecule in a single cell. mRNA of a given gene in a single cell has been difficult to detect in the past because each cell has very small copy numbers of it at any one time. One such technique is the MS2 mRNA detection scheme (Beach *et al.*, 1999; Bertrand *et al.*, 1998). One way to implement this technique is to engineer a gene so that its mRNA contains 96 copies of a particular RNA hairpin in its untranslated region. These hairpins then tightly bind to a coat protein of the bacteriophage MS2. Therefore, by also having a gene expressing the MS2 coat protein fused to GFP in the cell, a single mRNA with the 96 copies of RNA hairpin will now emit high enough fluorescence to be resolved as a single diffraction-limited spot under a fluorescence microscope. This method can help measure the transcription of a gene in real-time in a single-cell, as was done in *Escherichia coli* (Golding *et al.*, 2005). Despite the vast improvement in resolution the MS2 method provides over conventional methods using

GFP and its variants, it has a disadvantage in that mRNAs tend to aggregate together and that the regulation of the endogenous mRNA may change (thus one monitors this altered regulation rather than the endogenous one) because it has now been engineered to have the long artificial sequence for hairpin formation.

In this chapter, we describe fluorescence *in situ* hybridization (FISH) method (Gall, 1968; Levsky and Singer, 2003) for detecting single endogenous mRNA molecules in individual yeast cells (Raj *et al.*, 2008). Since the target gene sequence does not have to be modified to use this method, it bypasses the aforementioned problems associated with engineering the mRNA to have hairpin forming sequences in the MS2 mRNA detection scheme. It is also highly sensitive and allows for the counting of mRNA molecules in single cells, thus obviating many of the issues associated with using GFP as either a fusion to a protein of interest or driven by a promoter of interest mentioned before. In this method, we utilize a large collection (at least 30) of oligonucleotides, each labeled with a single fluorophore, that binds along the length of the target mRNA (Fig. 17.1A). The binding of so many fluorophores to a single mRNA results in a signal bright enough to be detectable with a microscope as a diffraction-limited spot. The method we describe is a modification of the RNA FISH method described by Singer and coworkers (Femino *et al.*, 1998), in which the authors use a smaller number (~5) of longer oligonucleotides (~50 bp), each of which contains up to five fluorophores (Fig. 17.1B). While that method has been used successfully to count mRNAs in single cells (Long *et al.*, 1997; Maamar *et al.*, 2007; Sindelar and Jaklevic, 1995; Zenklusen *et al.*, 2007), it has not been widely adopted. This may be due to the difficulties and costs associated

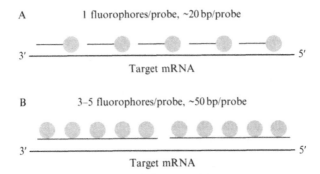

Figure 17.1 Comparison between two *in situ* hybridization methods for imaging a single mRNA molecule. (A) Method of Raj *et al.* (2008) involves about 30 or more singly labeled probes, each about 20 bases long, that bind along the stretch of a target mRNA molecule. (B) Method of Femino *et al.* (1998) involves multiple fluorophores (between 3 and 5) coupled to a single oligonucleotide probe of about 50 bases long that bind along the stretch of a target mRNA molecule.

with synthesizing and purifying several oligonucleotides with the internal modifications required to label those oligonucleotides with multiple fluorophores. Another potential issue is self-quenching between tightly spaced fluorophores. We anticipate that the simplicity of the method described herein will allow many researchers to utilize single-molecule RNA FISH in their own studies.

2. RNA FISH PROTOCOL

A brief overview of our method is as follows. A set of short (between 17 and 22 bases long) oligonucleotide probes that bind to a desired target mRNA are designed and are coupled to a fluorophore (such that one oligonucleotide probe is bound to a single fluorophore) with desired spectral properties. After fixing the yeast cells, these probes are hybridized to the target mRNA molecule. This results in multiple (typically about 48) singly labeled probes bound to a single mRNA molecule. In turn, the mRNA molecule can give off enough fluorescence to be detected as a diffraction–limited spot using a standard fluorescent microscope. Below we describe a step-by-step procedure for implementing RNA FISH in *Saccharomyces cerevisiae*.

2.1. Designing oligonucleotides

The first step is the design of a collection of oligonucleotide probes that together are complimentary to a large part of the open read frame of the target mRNA (one can also utilize the untranslated regions of the mRNA, if necessary). Each probe is between 17 and 22 bases long and we have generally found that 30 or more such probes are sufficient to give a detectable signal. We have also found that our signals are sometimes clearer when the GC content of each probe is close to 45%. We also leave a minimum of two bases as a spacer between two adjacent probes that cover the mRNA, although it is possible that one can relax this requirement without any adverse effects. A program that facilitates the designing of probes meeting the constraints mentioned above is available freely at http://www.singlemoleculefish.com. Sometimes it is not possible to design probes that meet all the constraints mentioned above, and these criteria should not be viewed as absolutes, but more as guidelines we try to adhere to when possible. After designing the probes, we order them from companies with parallel synthesis capabilities (we use BioSearch Technologies based in Novato, CA, USA) with 3′-amine modifications. Since the synthesis typically results in a much larger number of oligonucleotides than are necessary, one should have them synthesized on the smallest possible scale (we typically have them synthesized on the 10 nmol (delivered) scale). The 3′-amine then serves as a reactive group for the succinomidyl-ester coupling of the fluorophore described in Section 2.2.

2.2. Coupling fluorophores to oligonucleotides

The next step is the attachment of a fluorophore with desired spectral properties to the commercially synthesized oligonucleotides (we will describe which fluorophores we use in Section 2.2.1.) We do this by pooling the oligonucleotides and coupling them *en masse*, thus reducing the labor involved. In all the steps we describe below, we use RNase free water (Ambion) to prepare our solutions and use filtered pipette tips to prevent aerosol contaminations.

Procedure:

1. From the commercially synthesized set of oligonucleotides, each at a concentration of 100 μM in RNase free water (we find this is a practical starting concentration to work with), pipette around 1 nmol/10 μl of each oligonucleotide probe into a single microcentrifuge tube (i.e., if there are 48 probes, then 1 nmol of each of the 48 probe solutions should be combined into a single tube with a final volume of 480 μl).

2. Add 0.11 volumes (v/v) of 1 M sodium bicarbonate (prepared with RNase free water) to this probe mixture, resulting in a final sodium bicarbonate concentration of 0.1 M. If the total volume of the mixture at this stage is less than 0.3 ml, add enough 0.1 M sodium bicarbonate to bring the final volume of the mixture to 0.3 ml.

3. Dissolve roughly 0.2 mg of the desired fluorophore (functionalized with a succinimidyl ester group) separately into a tube containing 50 μl of 0.1 M sodium bicarbonate. If using tetramethylrhodamine (TMR), first dissolve it in about 5 μl of dimethyl sulfoxide (DMSO) and then add 50 μl of 0.1 M sodium bicarbonate to it. This is because TMR does not readily dissolve in aqueous solutions.

4. Add the dissolved fluorophore to the 0.3 ml of probe mixture, vortex, and cover this tube in aluminum foil to prevent photobleaching from unwanted exposure to ambient light. Leave the tube in the dark overnight.

5. Next day, precipitate the probes out of solution by adding 12% (v/v) of sodium acetate at pH 5.2 followed by 2.5 volumes of ethanol (95% or 100%).

6. Place the tube at −70 °C for at least 1 h, then spin the sample down at 16,000 rpm for at least 15 min at 4 °C.

7. A small colored pellet should have collected at the bottom of the tube at this stage. This pellet contains both the coupled and uncoupled oligo-nucleotides. The vast majority of the uncoupled fluorophore, however,

remains in the supernatant, and so aspirate as much of this supernatant away as possible without disturbing the pellet (one should take care to aspirate soon after removal from the centrifuge, since oligonucleotides have a tendency to redissolve rapidly at room temperature.

Note: Many precipitation protocols now call for another washing step in 70% ethanol. We have found this step unnecessary.

8. The pellet is stable and can be stored in $-20\ ^{\circ}$C for up to 1 year. This concludes the coupling step.

2.2.1. Choice of fluorophore and appropriate filter sets

In order to perform imaging of multiple different RNA species at the same time, one needs to select fluorophores with excitation and emission properties that can be distinguished by appropriately chosen bandpass filters; otherwise, the signal from one channel may potentially bleed into another channel. We describe here the fluorophore and filter set combination that we use for our microscopy. Other combinations are no doubt feasible as well.

The fluorophores we utilize are TMR, Alexa 594, and Cy5. TMR has proven to be exceptionally photostable in our hands, and its excitation maximum of 550 nm aligns nicely with the excitation maxima of mercury and metal-halide light sources. Alexa 594 is also quite photostable, and while its spectral properties are similar to those of TMR (absorption at 594 nm), we are able to distinguish its presence using appropriate filters. The third fluorophore we use is Cy5, which is rather bright and is spectrally separated from the other two fluorophores (Cy5 absorbs at 650 nm). Cy5 does, however, suffer from photobleaching effects, thus requiring the use of a glucose oxidase oxygen scavenging system to make imaging feasible. We have not tried any dyes that are further redshifted than Cy5. However, we have experimented with Alexa 488, which absorbs at a lower wavelength than TMR. While we were sometimes able to detect signals, the higher cellular background at these lower wavelengths lead to weaker signals, so we generally avoid the use of fluorophores bluer than TMR.

The filter combinations we use are typical bandpass filter and dichroic sets mounted in cubes that the microscope can place in the fluorescence light path. For TMR, we use a standard XF204 filter from Omega Optical. For Alexa 594, we use a custom filter from Omega Optical with a 590DF10 excitation filter, a 610DRLP dichroic, and a 630DF30 emission filter. For Cy5, we use the 41023 filter from chroma, which is designed for Cy5.5. It is likely that a filter more appropriate for Cy5 would work even better. These filters do a good job of preventing any signals from one fluorophore from being detected in another channel (Raj *et al.*, 2008). Sometimes a very bright Alexa 594 signal can bleed somewhat into the TMR channel (we estimate the bleedthrough to be about 10%) but practically this bleed-through is impossible to detect owing to the low signal intensities of the mRNA spots.

2.3. Purification of probes using HPLC

We now describe a purification procedure for separating the coupled oligonucleotides from the uncoupled oligonucleotides. We purify the coupled oligonucleotides using HPLC (high-performance liquid chromatography): the addition of the fluorophore makes the normally hydrophilic oligonucleotide significantly more hydrophobic, allowing for separation by chromatography. The HPLC should be equipped with a dual wavelength detector for a simultaneous measurement of absorption by DNA (at 260 nm) and fluorophore (depends on the fluorophore: e.g., 555 nm for TMR and 594 nm for Alexa 594). In our lab, we have used an Agilent 1090 equipped with Chemstation software and a C18 column suitable for oligonucleotide purification (218TP104). The two buffers used for HPLC are: 0.1 M triethylammonium acetate ("Buffer A") and acetonitrile ("Buffer B").

Procedure:

1. Before running the purification program on the HPLC, equilibrate the column by flowing 93% Buffer A/7% Buffer B through for about 10 min; if the column is not equilibrated, then the oligonucleotides will simply flow straight through without any separation.
2. Resuspend the oligonucleotide pellet in an appropriate volume of water (we use 115 μl) and then inject this into the HPLC inlet.
3. Run an HPLC program in which the percentage of Buffer A varies from 7% to 30% over the course of about 45 min with a flow rate of 1 ml/min. During the execution of the program, carefully monitor the two absorption curves, one for DNA (at 260 nm) and the other for the coupled fluorophore (e.g., 555 nm for TMR and 594 nm for Alexa 594). Generally speaking, one will observe two broad peaks over time. The first peak, containing the more hydrophilic material, consists of the uncoupled oligonucleotides and will only exhibit absorption in the 260 nm channel (Fig. 17.2A). This peak may appear relatively ragged due to the presence of multiple oligonucleotides, each of which has a slightly different retention time in the HPLC. The second peak, often narrower than the first, will appear some time after the first peak and contains the coupled oligonucleotides; thus, it will show absorption in both the 260 nm and the fluorescent (e.g., 555 nm) channels (Fig. 17.2B). The duration of time between the first and second peaks varies depending on the hydrophobicity of the fluorophore; we have found that oligonucleotides coupled to Cy5 have a long

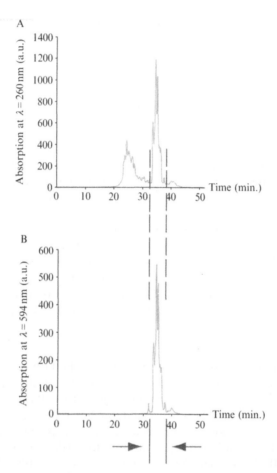

Figure 17.2 Chromatographs obtained during the HPLC purification of oligonucleo-tides coupled to the fluorophore (Alexa 594) from uncoupled oligonucleotides. (A) Absorption (at 260 nm, for DNA) curve as a function of time monitored during purification of probes coupled to Alexa 594 using HPLC. The first peak that appears between 20 and 30 min in this channel correspond to oligonucleotide probes that do not have Alexa 594 coupled to them. Eluate is not collected for the duration of this peak. (B) Absorption (at 594 nm, for Alexa 594) curve as a function of time. Both absorption curves (A) and (B) are obtained simultaneously for the duration of the HPLC run. Only one distinct peak appears in this channel, representing absorption by probes with Alexa 594 successfully coupled to them. This peak coincides with the second peak in the 260 nm channel shown in (A). Eluate is collected for the entire duration of this peak in the 594 nm channel.

retention time of almost 20 min after the first peak, whereas TMR and Alexa 594 result in shorter retention shifts (Fig. 17.2B).

4. Collect the contents of this peak (in the flurophore absorption channel) manually into clean, RNase free tubes. It is important to collect all the

solution that is coming out of the outlet, starting from the beginning of the left shoulder of this second peak and stopping the collection just at the tail-end of the right shoulder of this second peak (Fig. 17.2B), because the different coupled oligonucleotides will have slightly different retention times; do not just "collect the peak." This collection typically lasts around 3–7 min in our experience. With the volumes we mentioned for our HPLC setup above, we typically collect between 5 and 14 ml in this step with 0.5 ml/tube. The program we use then typically flows 70% Buffer B through the column for about 10 min. This step will "strip" the column of any impurities that may have stuck to the column and is especially important if you plan to purify additional probes. Be sure, however, to allow sufficient time for the column to reequilibrate to 7% B/93% A before injecting another sample.

5. After collecting the solution of coupled probes, dry the collection in a SpeedVac rated for acetonitrile until the liquid is fully evaporated (about 3–5 h). It is important to keep light out of the SpeedVac to avoid photobleaching of dyes, especially for highly photolabile cyanine dyes such as Cy3 and especially Cy5.

6. Resuspend the contents in a total of 50–100 μl of TE (10 mM Tris with HCl to adjust pH, 1 mM EDTA, Ambion) at pH 8.0. This final suspension solution is now the "probe stock."

7. From the "probe stock," create dilutions of 1:10, 1:20, 1:50, and 1:100 in TE to make "working stocks." This dilution series is used to determine which concentration of probes yields the best signals for RNA FISH.

8. Store these probes in dark at -20 °C until sample is ready to be prepared. We found that the probes can be stored for years in this way.

2.4. Fixing *S. cerevisiae*

Having isolated the coupled probes, it is now time to fix the yeast cells so that these probes can be hybridized to their target mRNAs in these cells. In the following procedure, we have adopted the procedure for fixing *S. cerevisiae* from Long *et al.* (1995).

Procedure:

1. Grow the yeast cells to an OD of around 0.1–0.2 (corresponding to about 1–2 $\times$ 10^6 cells/ml) in a 45-ml volume of minimal media with appropriate supplements (depending on the auxotroph) in a batch shaker at 30 °C (we use 225 rpm).

2. Add 5 ml of 37% formaldehyde (i.e., 100% formalin) directly to the growth media containing the cells and let it sit for 45 min at room temperature to fix the cells. One should take safety precautions when using the carcinogen, formaldehyde (i.e., use chemical fume hood, gloves, and long-sleeved protective clothing).

3. Concentrate the cells in this 50 ml into a single microcentrifuge tube. We found that one way to concentrate the cells was to run the above 50 ml mixture through a vacuum filter (with a filter paper having 0.2 μm pores: VWR vacuum filtration system "PES 0.2 μm") once, then shake the filter paper into an RNase free water. Alternately, one may simply centrifuge the content at 2300 rpm for about 5 min and then resuspend in 1 ml Buffer B to transfer to a microcentrifuge tube.

4. Wash these concentrated cells in the microcentrifuge tube twice with 1 ml ice-cold Buffer B (Long *et al.*, 1995).

5. Add 1 ml of spheroplasting buffer (from a stock made by adding 100 μl of 200 mM vanadyl-ribonucleoside complex to 10 ml Buffer B), and transfer the mixture to a new RNase free microcentrifuge tube.

6. Add 1 μl of zymolyase and incubate at 30 °C for about 15 min; this spheroplasting step removes the cell wall and is important for probe penetration.

7. Wash the solution twice with 1 ml ice-cold Buffer B, with centrifuging the content at 2000 rpm for 2 min in between.

8. Add 1 ml of 70% ethanol (diluted in RNase free water) to the cells and leave them for an hour or even overnight at 4 °C.

The yeast cells have now been fixed and are ready for hybridization. These cells can be stored in ethanol for up to a week after fixation and perhaps even longer.

2.5. Hybridizing probes to target mRNA

The hybridization step contains three key parameters that may be varied to optimize the FISH signal. These are the temperature at which hybridization takes place, the concentration of formamide used in the hybridization and wash, and the concentration of the probe. The first two parameters essentially set the stringency of the hybridization; that is, the higher the temperature or the concentration of formamide, the lower the likelihood of nonspecific binding of the probes. We usually elect to adjust the formamide concentration rather than temperature and thus perform all FISHs at 30 °C. Typically, we have found that hybridization and wash buffers containing 10% formamide work quite nicely for most probes, yielding a fairly low background while also producing clear particulate signals. However, when the GC content of the probes is relatively high ($>$ 55%), we have found that we sometimes have to employ formamide concentrations up to 20% or sometimes higher. However, care must be taken in these instances, since the use of higher formamide concentrations can sometimes lead to a greatly diminished signal. Generally, we try to obtain signals at a standard concentration of formamide, because this greatly facilitates the simultaneous detection of multiple mRNAs: if the hybridization conditions are the same, multiplex detection is simply a matter of mix and match.

The concentration of probe used is also very important in obtaining clear, low background signals. Typically, the optimal probe concentration must be found empirically, but we have found that concentrations can vary over roughly an order of magnitude and still produce satisfactory results. We typically start by using a 1:1000, 1:2500, and 1:5000 dilution of the original stock into hybridization buffer. One of these concentrations will usually yield good signals, but sometimes one must use drastically lower concentrations (100-fold lower) in order to obtain signals.

2.5.1. Preparation of hybridization and wash buffers

The following procedure describes preparation of 10 ml of hybridization buffer with the desired formamide concentration. Be sure to adjust the volumes appropriately if you are preparing a different total volume of hybridization buffer.

Procedure:

1. Dissolve 1 g of high molecular weight dextran sulfate ($>50,000$) in approximately 5 ml of nuclear free water. Depending on the particular preparation of dextran sulfate used, the powder may dissolve quite rapidly with a bit of vortexing or may require rocking for several hours at room temperature. In the end, the solution should be clear and fairly viscous, although some preparations are far less viscous but still appear to work.

2. Add 10 mg of *E. coli* tRNA (Sigma, 83854), vortexing to dissolve.

3. Add 1 ml of $20\times$ SSC (RNase free, Ambion), 40 μl (to get 0.02% in 10 ml) of RNase free BSA (stock is 50 mg/ml = 5% solution from Ambion, AM261), 100 μl of 200 mM vanadyl-ribonucleoside complex (NEB S1402S), formamide to the desired concentration (10–30%), and then water to a final volume of 10 ml. When using formamide, one must first warm the solution to room temperature before opening to avoid oxidation; also, care must be taken when using formamide (i.e., use in the hood, wear protection, etc.) because it is a suspected carcinogen and teratogen and is readily absorbed through the skin.

4. Once the solution is thoroughly mixed, filter the buffers into small aliquots; this removes any potential clumps that can yield a spotty background. We simply filter the solution in 500 μl aliquots using cartridge filters from Ambion.

5. Store the solution at $-20\,^{\circ}$C for later use; solution is typically good for several months to a year.

6. Prepare the wash buffer by combining 5 ml of $20\times$ SSC (Ambion), 5 ml of formamide (to final concentration of 10% (v/v); this is adjusted if the hybridization buffer has a different formamide concentration), and 40 ml of RNase free water (Ambion) into one solution.

2.5.2. Hybridizing probes to yeast cells in solution

Procedure:

1. Warm the hybridization solution to room temperature before opening its cap to prevent oxidation of the formamide.
2. Add 1–3 μl of desired concentration of probes to 100 μl of the hybridization buffer. To determine what the desired concentration of probes is, we initially perform hybridizations with four dilutions of probes: 1:10, 1:20, 1:50, and 1:100 (mentioned in Section 2.3), and see which dilution gives the clearest signal.
3. Centrifuge the fixed sample and aspirate away the ethanol, then resuspend the fixed cells in a 1-ml wash buffer containing the same formamide concentration as the hybridization buffer.
4. Let the resuspension stand for about 2–5 min at room temperature.
5. Centrifuge the sample and aspirate the wash buffer. Then add the hybridization solution.
6. Incubate the sample overnight in the dark at 30 °C.
7. Next morning, add 1 ml of wash buffer to this sample, vortex, centrifuge, then aspirate away the supernatant.
8. Resuspend in 1 ml of wash buffer, then incubate in 30 °C for 30 min.
9. Repeat the wash in another 1 ml of wash buffer for another 30 min at 30 °C, this time adding 1 μl of 5 mg/ml DAPI for a nuclear stain.
10A. *If using photostable fluorophores such as TMR or Alexa 594*: then there is no need to add the GLOX solution. Just resuspend the sample in an appropriate volume (larger than 0.1 ml) of 2× SSC and proceed to imaging.
10B. *If using a highly photolabile fluorophore such as Cy5*: resuspend the fixed cells in the GLOX buffer (used as an oxygen-scavenger that removes oxygen from the medium to prevent light-initiated fluorophore destroying-reactions; see Section 2.5.3) without the enzymes and incubate it for about 2 min for equilibration (see Section 2.5.3 for details). Then centrifuge, aspirate away the buffer and resuspend the cells in a 100-μl of GLOX buffer with the enzymes (glucose oxidase and catalase). These cells are now ready to be imaged.

We found that our samples (either with or without the antibleach solution) can be kept at 4 °C for a day's worth of imaging. Keeping the samples at 4 °C prevents the probe-target hybrids from dissociating and thus degrading the signals.

2.5.3. Preparation of antibleach solution and enzymes ("GLOX solution")

During imaging, we typically take several vertical stacks ("z-stacks") of images through a cell in a field of view, causing a hybridized fluorophore in a fixed cell to be excited by intense light several times. More importantly, when more than one type of fluorophore is used for imaging two or three species of mRNA, such z-stacks must be repeated to excite each of the different fluorophores, leading to even more exposure of the fluorophores. In our experience, only TMR and Alexa 594 could withstand such repeated excitations, whereas Cy5 signal would rapidly degrade due to its especially high rate of photobleaching. To decrease the photolability of Cy5, we used an oxygen-scavenging system consisting of catalase, glucose oxidase, and glucose (GLOX solution) that is slightly modified from that used by Yildiz *et al.* (2003). This GLOX solution acts as an oxygen-scavenger that removes oxygen from the medium. Since the light-initiated reactions that destroy fluorophores require oxygen, the GLOX buffer thus prohibits these reactions from taking place. Indeed, we found that Cy5 was able to withstand nearly 10 times more exposure with the GLOX solution than without it. The following is a procedure for preparing the GLOX solution.

Procedure:

1. Mix together 0.85 ml of RNase free water with 100 μl of 20× SSC, 40 μl of 10% glucose, and 5 μl of 2 M Tris–Cl (pH 8.0). This is the GLOX buffer (without glucose oxidase and catalase).
2. Vortex the mixture, and then aliquot 100 μl of it into another tube.
3. To this 100 μl aliquot of GLOX buffer (glucose–oxygen-scavenging solution without enzymes), add 1 μl of glucose oxidase (from 3.7 mg/ml stock, dissolved in 50 mM sodium acetate, pH 5.2, Sigma) and 1 μl of catalase (Sigma). Before pipetting the catalase, vortex it a bit, since the catalase is kept in suspension (also, care should be taken when handling the catalase, since it has a tendency to get contaminated). This 100 μl will be referred to as "GLOX solution with enzymes." The GLOX solution without the enzyme will later be used as an equilibration buffer.

2.5.4. Imaging samples using fluorescent microscope

The fixed cells with probes properly hybridized are now ready for imaging. Our microscopy system is relatively standard: we use a Nikon TE2000 inverted widefield epifluorescence microscope. It is important to use a fairly bright light source. For instance, a standard mercury lamp will suffice, although the newer metal-halide light sources (e.g., Lumen 200 from Prior) tend to produce a more intense and uniform illumination. Another important factor is the camera. It is important to use a cooled CCD camera that is optimized for low-light imaging rather than acquisition speed; we use a Pixis camera from Roper. Also, the camera should have a pixel size of 13 μm or less. We should point out that the signals from the newer EMCCD cameras are no better than these more standard (and cheaper) cooled CCD cameras. We typically use a 100× DIC objective. If one is interested in imaging with Cy5, one must be sure that the objective has sufficient light transmission at those longer wavelengths; this can sometimes require an IR coating. When mounting the cells, it is important to make sure that one uses #1 coverglass (18 mm × 18 mm, 1 ounce) and that the yeast are directly on the coverglass: do not adhere the yeast to the slide and then cover with coverglass. One can enhance the adherence of the yeast to the coverglass by coating the coverglass with poly-L-lysine (put fresh 1 mg/ml poly-L-lysine solution on the coverglass for 20 min, then suction off) or concanavalin A. It is also important to use #1 coverglass: we have found that even though most objectives are corrected for #1.5 coverglass, the mRNA spots are usually fuzzier and less distinct when imaged through #1.5 coverglass.

There are two somewhat standard procedures often employed during fluorescence microscopy that we have found interferes with our single mRNA signals. One of these is the use of commercial antifade mounting solutions, which tend to introduce a large background while also decreasing the fluorescent signals from target mRNAs. We recommend instead using the custom made GLOX solution or 2× SSC for imaging, being careful not to let the sample dry out. We also discourage using the standard practice of using a nail polish to seal the sample, as it introduces a background autofluorescence in the red channels that interferes with fluorescence from mRNA.

2.6. Image processing: Detecting diffraction-limited mRNA spot

We have devised an algorithm that automates some fraction of the work involved in analyzing images obtained from the samples (Raj et al., 2008). The first step in our algorithm is applying a three-dimensional linear filter that is approximately a Gaussian convolved with a Laplacian to remove the nonuniform background while enhancing the signals from individual mRNA particles, thus enhancing the signal-to-noise ratio (SNR) (Fig. 17.3B). The full width at half maximum of this Gaussian corresponds to the optimal bandwidth of our filter, and depends on the size of the observed particle. This width is a fit parameter that we empirically adjust to maximize the SNR. However, even after filtering the images, they will contain some noise that requires thresholding to remove. In order to make a principled choice of threshold, we sweep over a range of possible values of the threshold, and plot the number of mRNAs detected at each value (Fig. 17.3C). Here, a single mRNA is defined as a collection of localized pixels (in the series of z-stacks) that form a connected component (Fig. 17.3D). We then typically find a plateau in this plot of the number of mRNAs counted as a function of the value of the threshold (i.e., increasing the threshold does not change the number of mRNAs counted) as seen in Fig. 17.3C. This implies that the signals from mRNAs are well separated from the background noise rather than a smooth

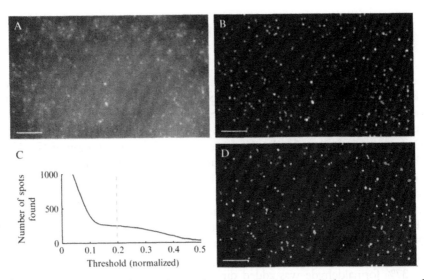

Figure 17.3 Example of mRNA spot detection algorithm applied to raw images of *FKBP5* mRNA particles in A549 cells induced with dexamethasone. (A) Raw image data showing *FKBP5* mRNA particles. (B) Upon applying a three-dimensional linear filter that is approximately a Gaussian convolved with a Laplacian to remove the nonuniform background while enhancing the signals from individual mRNA particles on the raw image shown in (A) the SNR is increased. (C) The number of spots counted as a function of the threshold value of the background after the application of the linear filter shows an existence of a plateau. This indicates a clear distinction between background fluorescence and actual mRNA spots. (D) Using the value of threshold shown as the gray line in (C), the raw image (A) has been transformed to an image in which each distinct computationally identified spot has been assigned a random color to facilitate visualization. Reprinted with permission from Raj *et al.* (2008).

"blending" in of the mRNA signals with the background noise. Indeed, the value of threshold chosen in this plateau range yielded mRNA counts nearly equal to the mRNA counts we obtained through an independent method in which we count by eye without the aid of automation. The software used for this purpose is available for download on *Nature method*'s supplementary information site for Raj *et al.* (2008). One can also make measurements based on mRNA spot intensity, although we feel that great care must be taken in these situations. One issue is that the intensity depends on how precisely focused the spot is, although this can be ameliorated by taking a large number of closely spaced fluorescent stacks. Another problem with computing total or mean intensity is that the boundary of the mRNA is hard to define, and the ultimate intensity measurement will depend heavily on this somewhat arbitrary choice. One way to skirt the issue is to use the maximum intensity within a given spot, since this is independent of the size of the spot.

3. EXAMPLE: *STL1* mRNA DETECTION IN RESPONSE TO NaCl SHOCK

As an application of the FISH technique we just outlined, and we now show an example of this technique applied to *S. cerevisiae*. One mRNA of interest in yeast is that of the *STL1* gene, whose expression level dramatically increases when the cell is subjected to an osmotic shock (Rep *et al.*, 2000). One way to induce such a shock is by increasing the concentration of NaCl in the cell's growth medium. For this purpose, a strain based on the common laboratory strain BY4741 (Mat a, his3Δ1 leu2Δ0 met15Δ0 ura3Δ0, YER118^c:: kanMX^R) was grown to an OD of 0.56 ($\sim$0.7 $\times$ 10^7 cells/ml) in a 50-ml volume of complete supplemental media without histidine and uracil. We then shocked them osmotically by transferring the cells to a medium with 0.4 M NaCl and leaving them there for 10 min. We fixed these shocked cells along with their unshocked counterparts using the method we outlined before (5 ml of 37% formaldehyde was added directly to the medium for 45 min). We adopted the fixation and spheroplasting procedures were from Long *et al.* (1995), but with the exception that after spheroplasting, the cells were incubated in concanavalain A (0.1 mg/ml, Sigma) for about 2 h before letting them settle onto a coverglass with a chamber that was coated with concanavalin A overnight. We used concanavalin A because it helped the yeast cells stick to a cover glass, although as mentioned earlier, it is possible also to simply use poly-L-lysine coated coverglass without incubating the cells in concanavalin A. The resulting images of RNA FISH performed on unshocked and shocked cells can be seen in Fig. 17.4A and B, respectively. As seen in these figures, the RNA

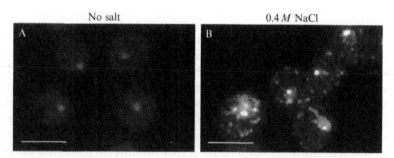

No salt   0.4 M NaCl

Figure 17.4 Single mRNA molecules imaged in *S. cerevisiae* using the fluorescence *in situ* hybridization method of Raj *et al.* (2008). Scale bars (white lines) indicate 5 μm. (A) *STL1* mRNA particles in yeast cells before being subjected to osmotic shock (0 M NaCl in the growth medium). (B) *STL1* mRNA particles in yeast cells 10 min after they have been growing in the presence of a high level of salt (0.4 M NaCl), thus inducing osmotic shock. DAPI was used to stain the nucleus of the cells shown in purple. The *STL1* gene expression increases dramatically after the osmotic shock. Reprinted with permission from Raj *et al.* (2008).

FISH technique of these workers (Raj *et al.*, 2008) allows one to not only resolve individual *STL1* mRNAs but also to extract spatial information on their whereabouts (helped by DAPI staining of the cell's nucleus). In addition, taking snapshots of *STL1* mRNAs at two different time points as shown in Fig. 17.4A and B illustrates how one can construct dynamics of the mRNA distribution in a population of cells by performing FISH on the cells at different time points.

4. CONCLUSIONS

Although we have limited our description of RNA FISH to just *S. cerevisiae*, this method has so far been applied to *E. coli*, *Caenorhabditis elegans*, *Drosiphila melanogaster*, and rat hippocampus neuronal cell cultures (Raj *et al.*, 2008). In fact, the protocol we described requires just a few adjustments in order to be applicable to these organisms. The method is likely to be applicable to other organisms as well. Studying how individual yeast cells behave through single cell measurements and using the distributions constructed through those measurements to look at how populations of cells behave remains a vital field of research today. We believe that the FISH method for visualizing a single mRNA molecule in yeast will play an important role in such endeavors.

ACKNOWLEDGMENTS

We thank G. Neuert for sharing with us his unpublished *STL1* RNA FISH data. A. v. O. was supported by NSF grant PHY-0548484 and NIH grant R01-GM077183. A. R. was supported by an NSF fellowship DMS-0603392 and a Burroughs Wellcome Fund Career Award at the Scientific Interface. H. Y. was partly supported by the Natural Sciences and Engineering Research Council of Canada's Graduate Fellowship PGS-D2.

REFERENCES

Beach, D., Salmon, E., and Bloom, K. (1999). Localization and anchoring of mRNA in budding yeast. *Curr. Biol.* **9**, 569–578.

Bertrand, E., Chartrand, P., Schaefer, M., Shenoy, S., Singer, R., and Long, R. (1998). Localization of ASH1 mRNA particles in living yeast. *Mol. Cell* **2**, 437–445.

Chubb, J., Trcek, T., Shenoy, S., and Singer, R. (2006). Transcriptional pulsing of a developmental gene. *Curr. Biol.* **16**, 1018–1025.

Femino, A., Fay, F., Fogarty, K., and Singer, R. (1998). Visualization of single RNA transcripts in situ. *Science* **280**, 585–590.

Gall, J. (1968). Differential synthesis of the genes for ribosomal RNA during amphibian oogenesis. *Proc. Natl. Acad. Sci. USA* **60**, 553–560.

Golding, I., Paulsson, J., Zawilski, S., and Cox, E. (2005). Real-time kinetics of gene activity in individual bacteria. *Cell* **123,** 1025–1036.

Levsky, J., and Singer, R. (2003). Fluorescence in situ hybridization: Past, present and future. *J. Cell Sci.* **116,** 2833–2838.

Long, R., Elliott, D., Stutz, F., Rosbash, M., and Singer, R. (1995). Spatial consequences of defective processing of specific yeast mRNAs revealed by fluorescent in situ hybrdization. *RNA* **1,** 1071–1078.

Long, R., Singer, R., Meng, X., Gonzalez, I., Nasmyth, K., and Jansen, R. (1997). Mating type switching in yeast controlled by asymmetric localization of ASH1 mRNA. *Science* **277,** 383–387.

Maamar, H., Raj, A., and Dubnau, D. (2007). Noise in gene expression determines cell fate in *bacillus subtilis. Science* **317,** 526–529.

Raj, A., and van Oudenaarden, A. (2008). Nature, nurture, or chance: Stochastic gene expression and its consequences. *Cell* **135,** 216–226.

Raj, A., Peskin, C., Tranchina, D., Vargas, D., and Tyagi, S. (2006). Stochastic mRNA synthesis in mammalian cells. *PLoS Biol.* **4,** e309.

Raj, A., van den Bogaard, P., Rifkin, S., van Oudenaarden, A., and Tyagi, S. (2008). Imaging individual mRNA molecules using multiple singly labeled probes. *Nat. Methods* **5,** 877–879.

Rep, M., Krantz, M., Thevelein, J., and Hohmann, S. (2000). The transcriptional response of *Saccharomyces cerevisiae* to osmotic shock. Hot1p and Msn2p/Msn4p are required for the induction of subsets of high osmolarity glycerol pathway-dependent genes. *J. Biol. Chem.* **275,** 8290–8300.

Rosenfeld, N., Young, J., Alon, U., Swain, P., and Elowitz, M. (2005). Gene regulation at the single-cell level. *Science* **307,** 1962–1965.

Sindelar, L., and Jaklevic, J. (1995). High-throughput DNA synthesis in a multichannel format. *Nucleic Acids Res.* **23,** 982–987.

Yildiz, A., Forkey, J., McKinney, S., Ha, T., Goldman, Y., and Selvin, P. (2003). Myosin V walks hand-over-hand: Single fluorophore imaging with 1.5 nm localization. *Science* **300,** 2061–2065.

Zenklusen, D., Wells, A., Condeelis, J., and Singer, R. (2007). Imaging real-time gene expression in living yeast. *Cold Spring Harb. Protoc.* DOI: 10.1101/pdb.prot4870.

CHAPTER EIGHTEEN

RECONSTRUCTING GENE HISTORIES IN *ASCOMYCOTA* FUNGI

Ilan Wapinski*,‡ *and* Aviv Regev*,†

Contents

Abstract

Whole-genome sequencing allows researchers to study evolution through the lens of comparative genomics. Several landmark studies in yeast have showcased the utility of this approach for identifying functional elements, tracing the evolution of gene regulatory sites, and the revelation of an ancestral

* Howard Hughes Medical Institute, Broad Institute of MIT and Harvard, Cambridge, Massachusetts, USA
† MIT Department of Biology, Cambridge, Massachusetts, USA
‡ Department of Systems Biology, Harvard Medical School, Boston, Massachusetts, USA

Methods in Enzymology, Volume 470

ISSN 0076-6879, DOI: 10.1016/S0076-6879(10)70018-5

whole-genome duplication event. Such studies first require an accurate and comprehensive mapping of all orthologous loci across all species. In this chapter, we present a computational framework for systematic reconstruction of all gene orthology relations across multiple yeast species. We then discuss how to use the resulting genome- and species-wide catalogue of gene phylogenies to study the histories of gene duplications and losses from a functional genomics perspective. We show how these methods allowed us to uncover the functional constraints underlying gene duplications and losses within *Ascomycota* fungi, and to highlight the importance and limitations of these evolutionary processes. The analytical framework we present here is generalizable and scalable, and can be applied to an array of comparative genomics needs.

1. INTRODUCTION

Comparative genomics is a powerful tool for evolutionary and functional studies of biological systems (Cliften *et al.*, 2003; Kellis *et al.*, 2003). Such studies require a reconstruction of the evolutionary history of individual genes, and their relation to one another through speciation (orthologs) or duplication (paralogs) events (Fitch, 1970). Applying these concepts at a genome-wide scale (*phylogenomics*; Eisen, 1998; Eisen and Fraser, 2003; Eisen and Wu, 2002) allows us to functionally characterize and classify genes (Engelhardt *et al.*, 2005; Tatusov *et al.*, 1997), and to understand the evolutionary impact of genomic events (Blomme *et al.*, 2006; Dietrich *et al.*, 2004; Kellis *et al.*, 2004a; Scannell *et al.*, 2006).

There are two broad categories of methods for the identification of orthologous and paralogous genes. The first class of methods infer homology relations based on *hit-clustering*, using the results ("hits") from sequence similarity searches such as BLAST or FASTA (Altschul *et al.*, 1990; Pearson and Lipman, 1988) between all the proteins in different species to output an orthology assignment between the genes. The most widely used variant of this approach is "reciprocal (bidirectional) best hits" (RBH) where two genes in two different species are identified as orthologs if each is the others' best "hit" in that species (Fitch, 1970; Wall *et al.*, 2003). Related approaches include more inclusive clustering methods (*e.g.*, COGs; Tatusov *et al.*, 1997), and algorithms designed to distinguish between recent and ancient gene duplications (*e.g.*, InParanoid, Remm *et al.*, 2001; OrthoMCL, Li *et al.*, 2003). One recent extension of a hit-clustering algorithm incorporated information on orthologous chromosomal regions (synteny) to guide orthology assignments (*e.g.*, BUS; Kellis *et al.*, 2004b). Synteny-based methods are particularly helpful in handling orthology assignments that are ambiguous based on hit-clustering alone, but they cannot be applied between distantly related species, where gene order is not sufficiently conserved. Hit-clustering methods are easy to implement and fast, but they do not explicitly reconstruct the evolutionary history of orthologous genes, as they either ignore paralogs altogether by assuming that orthology is a one-to-one relationship (*e.g.*, RBH) or do not resolve exact orthology and paralogy relations when identifying genes with shared ancestry (*e.g.*, COGs).

A complementary set of approaches identifies homology relations in light of the *phylogenetic gene tree* of a related group of genes. These allow us to infer lineage-specific duplications and losses by comparison to the corresponding species tree (*reconciliation*; Goodman *et al.*, 1979; Zmasek and Eddy, 2001; Fig. 18.1). The main limitation of these methods is that

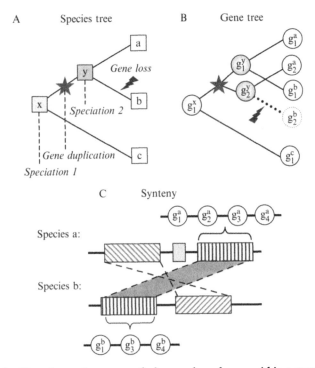

Figure 18.1 Homology subtypes—orthology and paralogy—within a group of orthologous genes. (A) Species tree. Each node (square) in the tree is a species—either extant (leaf node) or ancestral (internal node). In this toy example, speciation events 1 and 2 have resulted in extant species a, b, and c. (B) A gene tree describing the evolutionary events for the genes g_1^a, g_2^a, g_1^b, and g_1^c (*informally* denoted on the species tree for illustrative purposes). Each node in the tree is a gene (circle) or a duplication event (star). The tree shows the evolutionary descent of the ancestral gene g_1^x to paralogs and orthologs following gene duplication before species y, and the subsequent speciation yielding species a and b. Gene g_2^b was lost (strike and dashed lines) after the duplication event, but its paralog, g_1^b, was retained. (C) Synteny between chromosomal regions in species a and b. Each chromosome has several similar (syntenic) blocks (hatched boxes) comprised of multiple genes. Some regions in one genome (gray box) do not have a syntenic counterpart in the other. The synteny similarity score for a pair of genes is the fraction of their neighbors that are orthologous to each other. For example, the score for g_3^a and g_3^b is 2/3.

single-locus phylogenies are rarely accurate, especially when considering a large number of taxa, where there are many possible tree topologies (as discussed in Felsenstein, 2004). The resulting gene trees may, therefore, require a high number of duplication and loss events to be reconciled, even when there is only a single copy of a particular locus in each taxa; an unlikely scenario assuming such events are relatively rare. Recent methods attempt to balance the number of inferred duplications and losses with evidence derived from sequence alignments (Arvestad et al., 2003; Durand et al., 2006). While such approaches often result in high-quality reconstructions of gene histories, they are computationally intensive and have, therefore, been typically restricted to predefined families of genes rather than applied on a genomic scale.

Recent efforts to apply phylogenetic methods toward large-scale resolution of orthologies (Goodstadt and Ponting, 2006; Hahn et al., 2007; Jothi et al., 2006), handle the task in a sequential way: first, they use hit-clustering methods to identify coarse gene families and then construct gene trees to refine these assignments. With few exceptions (Hahn et al., 2007), the latter phylogenetic step does not employ the more sophisticated but computationally intensive phylogenetic algorithms. Thus, it does not account for gene tree distortions that induce large numbers of unlikely duplication and loss events (Blomme et al., 2006; Dufayard et al., 2005; Jothi et al., 2006). Such distortions are common as genes within families often evolve at variable rates, especially following gene duplication events (Kellis et al., 2004a; Lynch and Katju, 2004). Consequently, laborious manual curation by experts may be required to achieve reasonable results (Li et al., 2006), or more complicated families must be ignored a priori (Blomme et al., 2006). Further, these approaches have not incorporated synteny, an important source of evidence for determining gene ancestry.

Here, we describe a framework for the genome-wide reconstruction of homology relations across multiple eukaryotic genomes and present a fully automatic and scalable implementation of this framework in the Synergy algorithm (Wapinski et al., 2007a,b). Given a set of genomes and the known species phylogeny, Synergy resolves the orthology and paralogy relations for all the protein-coding genes in those genomes, while *simultaneously* reconstructing the phylogenetic gene trees for each group of orthologs. Our approach combines the scalability and automation of hit clustering approaches with the detailed phylogenetic reconstruction of tree-based methods, resulting in a robust resolution of homology relations. By simultaneously reconstructing gene trees while identifying orthologous groups, our approach avoids many of the pitfalls of sequential methods. This method is flexible and can incorporate additional types of data whenever available (e.g., synteny). To automatically assess the quality of our assignments, we also develop a jackknife-based method for measuring their robustness to perturbations in the included genes and species.

We demonstrate the quality and utility of such reconstructions in an analysis of gene histories in *Ascomycota* fungi. In particular, we show how to evaluate the quality of gene orthologies and how to apply an array of functional genomics data and techniques to study the functional constraints on gene duplications and losses and the impact of gene duplications on cellular networks. The Synergy algorithm and these analytical approaches are broadly applicable to the study of gene and genome evolution in other microbial and eukaryotic genomes.

2. SYNERGY

2.1. Overview

Given a set of species, their protein-coding genes, and their phylogenetic tree, Synergy partitions the genes into disjoint subsets, where each subset contains all and only those genes that descended from a single gene in the species' last common ancestor. Synergy simultaneously reconstructs the phylogenetic gene tree for each such subset of genes. Briefly, Synergy performs this task in a step-wise bottom-up fashion, solving it sequentially for each ancestral node in a species tree from the leaves of the tree to the root. At each stage (*i.e.*, node in the species tree), Synergy first clusters together the genes or groups of orthologs from previous stages that share significant sequence similarity. It then reconstructs a phylogenetic gene tree for each of these intermediate groups of orthologs, and uses this tree to partition the clusters such that each contains only genes that are descended from a single hypothetical gene in the ancestral species corresponding to the current stage. Thus, after each stage, Synergy has made a complete orthology assignment and gene tree reconstruction for the complement of genes below the corresponding node in the species tree. These are then passed up to the next stage. Once Synergy reaches the root of the species tree, it has made a full partitioning of groups of orthologs that are descended from a single ancestral gene along with a corresponding gene tree for each such group. Below we discuss each step in this procedure.

2.2. Defining orthogroups

There are two major classes of homology relations between genes (Fitch, 1970). Orthologs are genes that share a common ancestor at a speciation event, while paralogs are related through duplication events (Fig. 18.1A and B). These are not necessarily simple one-to-one relationships. For example, two paralogous genes that resulted from a duplication after a speciation event, are both orthologous to the same gene in another species (Fig. 18.1). Conversely,

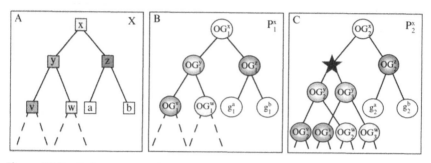

Figure 18.2 Orthogroups and their phylogenetic gene trees. (A) A species tree X rooted at the ancestral species **x**. Only a fraction of the tree is shown. (B, C) A gene tree (*e.g.*, P_1^x) represents the evolutionary history of all the genes that descended from the gene g_1^x in species **x**. Each internal node in the gene tree defines a corresponding orthogroup (*e.g.*, OG_1^x, OG_1^y, ...), whose members are the genes below that node in the tree. The gene tree can track duplication events (star, panel C, OG_2^y and OG_3^y).

when genes are lost in a particular species or lineage, orthology may be a one- or many-to-none relationship.

Such relations are captured by phylogenetic trees (Fig. 18.2A–C, Section 1). We denote a species tree **T** where internal nodes (**x, y, ...**) represent ancestral species, and leaf nodes (**a, b, ...**) represent extant species (Fig. 18.2A). We denote as g^a a gene g in species **a**. The exact orthology and paralogy relations between genes are represented in a gene tree P (Fig. 18.1B).[1] The leaves in P are the genes which descended from a single common ancestor gene at the root of P. Its internal nodes represent the speciation and duplication events that occurred within the course of the genes' evolution (Fig. 18.1B).

We define an *orthogroup* as the set of genes that descended from a single common ancestral gene. An orthogroup OG_i^x is defined with respect to an ancestral species **x** in **T** and includes only and all of those genes from the extant species under **x** that descended from a single common ancestral gene, g_i^x, in **x**. We therefore define:

Definition 1. (Orthogroup soundness)
An orthogroup OG_i^x under the ancestral species $\mathbf{x} \in \mathbf{T}$ is *sound* if there existed a gene g_i^x in **x** such that for every gene $g_j^a \in OG_i^x$, g_j^a is a descendant of g_i^x.

Definition 2. (Orthogroup completeness)
OG_i^x is *complete* if every gene g_j^a that descended from the ancestral gene g_i^x is in OG_i^x.

---

[1] We assume that gene fusion and horizontal transfer events are rare and that therefore genes are descended from single genes, allowing us to represent gene phylogenies as trees.

Each orthogroup OG_i^x has a corresponding gene tree P_i^x. The leaves in P_i^x are the genes $g_j^a \in OG_i^x$ (for every extant species **a** at the leaves of **T** under **x**), and its internal nodes denote ancestral genes and the duplication events that occurred along OG_i^x's evolution (Fig. 18.1B). The root of P_i^x represents the ancestral gene g_i^x, the last common ancestor of all $g_j^a \in OG_i^x$.

Since Synergy aggregates orthogroups as it recursively traverses the nodes of the species tree (Section 2.5), it will at times be at nodes whose child nodes are extant species (*e.g.*, node **x** in Fig. 18.2A) and at times be at nodes whose children are also ancestral nodes (*e.g.*, node **y** in Fig. 18.2A). In the former situation, the orthogroups in consideration are equivalent to individual genes in an extant species. In the latter, they correspond to hypothetical ancestral genes in the corresponding ancestral species. Since an orthogroup OG_i^x represents the gene g_i^x whether it be at an ancestral or extant node, we will subsequently refer to orthogroups and genes interchangeably.

2.3. Scoring gene similarity

The common ancestry of homologous proteins suggests that they should retain some sequence similarity. The estimate of the evolutionary distance between pairs of proteins is the basis for our reconstruction method. Although our method can be applied with any method for computing these pairwise distances, much of its success depends on these choices. Here, we use a measure of distance that examines the evolution of both the amino acid sequence of the proteins and the chromosomal organization of genomes.

When comparing amino acid sequences, we use standard models of amino acid evolution. Specifically, our *peptide sequence similarity score* (d^P) between a pair of proteins is based on the JTT amino acid substitution rates (Jones *et al.*, 1992). To compute d^P we first globally align two proteins, then search for the distance that maximizes the likelihood of substitutions in each alignment position.

To capture the information that genome organization conveys about the homology between proteins, our *synteny similarity score* (d^s) quantifies the similarity between the chromosomal neighborhoods of two genes (Fig. 18.1C). A (preliminary) orthology assignment anchors chromosomal regions in two species to one another. Genes that are highly syntenic to each other will share many such anchors between their chromosomal neighborhoods. Since there is currently no agreed-upon evolutionary model of genome organization, we compute the synteny similarity score between two genes as the fraction of their neighbors that are orthologous to one another (Fig. 18.1C). The source of the preliminary orthology assignment will be discussed below.

Both d^P and d^s are scaled and treated as distances for assessing protein and chromosomal evolution between pairs of genes. The protein similarity score scales with evolutionary distance. We scale the synteny score to the same range but we do not make any assumptions about its direct evolutionary interpretation. Two genes with high similarity have scores close to zero, while genes sharing no similarity have scores of 2.0. We combine these two measures to identify potentially orthologous genes.

2.4. Gene similarity graph

Synergy relies on the precomputed distances between genes to make orthology assignments, represented by a *gene similarity graph*. This is a weighted directed graph $\mathcal{G} = (\mathcal{V}, \mathcal{E})$, where $\mathcal{V}$ are all the individual genes in the input genomes, and the edges $\mathcal{E}$ represent potential homology relations. To generate $\mathcal{E}$, we first execute all-versus-all FASTA alignments between all genes in our input (Pearson and Lipman, 1988).

Since we expect most of the gene pairs to share no common ancestor or sequence similarity, we wish to maintain edges only between genes with relevant distances, thus helping to guide the algorithm by identifying the potential homologies. Once we designate gene pairs that are significantly similar, we place an edge between g_i^a in species **a** and g_i^b in species **b** if the FASTA E-value of their alignment is below 0.1 and either g_i^b is the best FASTA hit in species **b** to g_i^a or the percent identity between g_i^a and g_j^b is above 50% of that between g_i^a and its best hit in **b**. These parameter choices are relevant to the implementation that we employed but not to the framework for resolving orthologs that we propose. Our results from varying these choices show that these parameters were relaxed enough to capture a high proportion of putative homology relations, while at the same being restrictive enough to ensure that Synergy runs efficiently. These parameters are given as inputs to our implementation.

Once we place edges between the nodes in the similarity graph, we weigh each them by the d^p score defined above.[2] While this distance is symmetric, the edges are directed, and are placed from the query to the target gene based on the direction of the similarity search. Thus, not all edges are reciprocal; the significant hits from a given gene g_i^a may not all include g_i^a among their significant hits.

2.5. Identifying orthogroups

Identifying orthogroups across multiple species amounts to sequentially reconstructing the shared ancestral relationships between genes at each internal position of a phylogenetic tree. To this end, Synergy (Fig. 18.3) recursively traverses the nodes of the given species tree **T** from its leaves to

---

[2] We rely more heavily on the protein similarity described in Section 2.3 than on bitscores or E-values because the best "hit" is often not the nearest phylogenetic neighbor (Koski and Golding, 2001; Wall *et al.*, 2003).

```
Synergy Algorithm
Input: A species tree node x
Output: A set of orthogroups {OG^x}

if x is an extant species
    {OG^x} ← {g^x}
else
    // Call Synergy recursively
    {OG^r} ← Synergy(x.right)
    {OG^l} ← Synergy(x.left)
    // step 1: match orthogroups; make putative orthogroups {OG^x}
    {OG^x} ← MatchOrthogroups( x, {OG^r}, {OG^l} )
    // step 2: make the phylogenetic gene tree {P^x} for the orthogroups OG^x
    repeat
        Choose an unprocessed OG_i^x ∈ {OG^x}
        // construct the unrooted phylogenetic tree topology
        P_i^x ← MakeTree(OG_i^x)
        // now use equation 1 to select the root
        RootTree(P_i^x)
        // break an orthogroup if it does not resolve to a single ancestral gene
        if P_i^x.root ∉ {g^x}
            (OG_k^x, OG_l^x) ← BreakOrthogroup(OG_i^x, P_i^x)
            // update the set of putative orthogroups
            {OG^x} ← ({OG^x} \ OG_i^x) ∪ (OG_k^x, OG_l^x)
        else
            Mark OG_i^x as processed
    until all orthogroups are processed
    UpdateSimilarityGraph(x, {OG^x})
return {OG^x}
```

Figure 18.3 Overview of the Synergy algorithm. The algorithm is initially called with the root of the species tree **T**.

its root, identifying orthogroups with respect to each node. At each recursive Stage, Synergy assumes that sound and complete orthogroups and their corresponding gene trees are resolved for the lower nodes in the tree. For each internal node $\mathbf{x} \in \mathbf{T}$, Synergy uses the distances between genes (or, equivalently, between orthogroups resolved in previous stages) to determine the orthogroups $\{OG^{\mathbf{x}}\}$ and reconstruct the phylogenetic gene trees $\{P^{\mathbf{x}}\}$ between the member genes of each orthogroup. Once this is completed, the set of newly identified orthogroups and their corresponding gene trees are recorded. At this point the procedure updates the gene similarity graph by replacing the genes in the species below $\mathbf{x}$ by orthogroups in $\{OG^{\mathbf{x}}\}$, and the next stage of the algorithm treats these orthogroups as genes. When the bottom-up recursion reaches the root of $\mathbf{T}$, every gene g_i^a in each species has been assigned uniquely into an orthogroup and located as a leaf in the corresponding gene tree.

We now expand on the details of each step of the procedure.

2.5.1. Matching candidate orthogroups

At each node $\mathbf{x}$ of the species tree $\mathbf{T}$, Synergy considers orthology assignments for orthogroups pertaining to the species directly below $\mathbf{x}$ in the species $\mathbf{T}$ (denoted $\mathbf{y}$ and $\mathbf{z}$). Our goal is to capture as many true orthology relations without substantially compromising specificity. As noted above, the orthogroups from both $\mathbf{y}$ and $\mathbf{z}$ are now vertices in the gene similarity graph. Synergy begins by matching orthogroups in both $\mathbf{x}$ and $\mathbf{y}$ into *candidate* orthogroups. We assign orthogroups into the same candidate orthogroup if they have reciprocal edges between them and apply transitive closure on these reciprocal relations. More precisely, for a pair of orthogroups OG_i, $OG_j \in \{OG^\mathbf{y}\} \cup \{OG^\mathbf{z}\}$, we have that $OG_i \sim_\mathbf{x} OG_j$ (*i.e.*, OG_i and OG_j are *reciprocally connected*) if either both $OG_i \rightarrow OG_j$ and $OG_j \rightarrow OG_i$ are in $\mathcal{E}$ or if there is a third orthogroup $OG_k \in \{OG^\mathbf{y}\} \cup \{OG^\mathbf{z}\}$ such that $OG_i \sim_\mathbf{x} OG_k$ and $OG_k \sim_\mathbf{x} OG_j$. This leads to a partitioning of the orthogroups from species $\mathbf{y}$ and $\mathbf{z}$ into equivalence classes under $\sim_\mathbf{x}$. Each such equivalence class is taken to be a single *candidate* orthogroup for $\mathbf{x}$. By permitting indirectly connected orthogroups, we increase our sensitivity in identifying putative orthologs. However, we stop short of defining these candidate orthogroups as ordinary connected components in the graph, since this generates candidate orthogroups too coarsely. We nonetheless find this partitioning in a linear time (in the number of edges) using a standard connected component algorithm.

This step is similar to many hit-based methods (*e.g.*, COGs; Tatusov *et al.*, 1997). Due to our lenient inclusion policy and the promiscuity of edges in the gene similarity graph, candidate orthogroups may contain genes (orthogroups) that are related through duplication events that predate $\mathbf{x}$, and in fact descend from multiple genes in the ancestral species $\mathbf{x}$. Such violations of the orthogroup soundness condition (Definition 1) are handled after each candidate orthogroup is arranged into a phylogenetic tree.

2.5.2. Phylogenetic tree reconstruction

Given a candidate orthogroup $OG_i^\mathbf{x}$, we reconstruct a phylogenetic tree $P_i^\mathbf{x}$ whose leaves are the orthogroups from $\mathbf{y}$ and $\mathbf{z}$ that comprise $OG_i^\mathbf{x}$ (Fig. 18.4A). Recall that since the trees $\{P^\mathbf{y}\}$ and $\{P^\mathbf{z}\}$ were already resolved in previous stages, we can treat the root of each of these trees as extant genes in the phylogenetic reconstruction.

When only a pair of orthogroups $OG_j^\mathbf{y}$ and $OG_k^\mathbf{z}$ are matched into the candidate orthogroup $OG_i^\mathbf{x}$, there is a clear one-to-one orthology relation, making this task trivial: the gene tree would appear exactly as the species tree appears at the point $\mathbf{x}$ (Fig. 18.4A). When an orthogroup $OG_i^\mathbf{x}$ contains one-to-many or many-to-many relationships (due to possible duplications and/or losses), we reconstruct the tree using the neighbor-joining method[3]

---

[3] We could replace neighbor-joining by other phylogenetic reconstruction procedures; our choice was based on the relatively efficiency and effectiveness of the neighbor-joining procedure, but this component of our procedure can be considered modular to the rest.

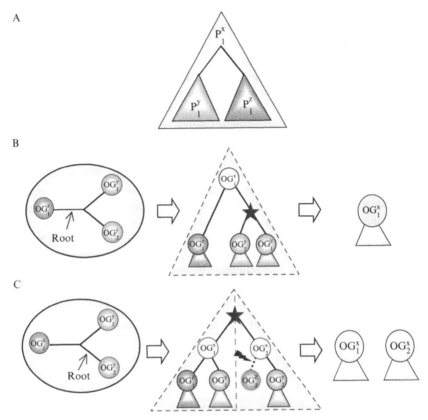

Figure 18.4 Construction of phylogenetic gene trees. (A) A gene tree P_1^x for a candidate orthogroup OG_1^x is constructed by joining the trees P_1^y and P_1^z resolved in Stages y and z. (B, C) When OG_1^x consists of more than two members, there are several alternative rootings. In (B), a root is selected that invokes one duplication between species y and the root of the tree, such that OG_1^y and OG_2^y are paralogs. In (C), the rooting suggests a duplication at the root of the gene tree, such that OG_1^x and OG_2^x are paralagous with respect to a duplication predating x and OG_2^z is lost after the speciation event. If (C) is selected, the orthogroup must be broken.

(Saitou and Nei, 1987) applied to the distance matrix between the orthogroups that comprise OG_i^x. If the leaves of this orthogroup tree are singleton genes, then the distances input to this matrix are drawn from the pairwise distances described above. The case for distances between predefined orthogroups from the lower branches in **T** is discussed below (2.5.5).

The result of this reconstruction is an unrooted phylogenetic tree, whose leaves are the orthogroups that have been matched together.

2.5.3. Tree rooting
The resulting unrooted tree contains all of the orthogroup components that were matched into the candidate orthogroup. To obtain the exact phylogenetic relationships between these components, the tree must first be

rooted. Correct rooting is important since the selected root position may also determine whether all of an orthogroup's members descended from a single gene in species $\mathbf{x}$ or from multiple genes (Fig. 18.4B and C).

Assuming equal rates of evolution among all the leaves in a tree, a tree's root should be approximately equidistant to all the leaves. Given an unrooted tree, we compute the leaf-to-root variances for every possible rooting r at an internal branch in it, and assign a score to each rooting that is proportional to the variance in both amino acid and synteny scores, termed π_r and σ_r, respectively.

Following a gene duplication, one or both of the paralogs are often under relaxed selection, and can evolve at an accelerated rate (Lynch and Katju, 2004; Ohno, 1970). This conflicts with the above assumption that all branches of the tree evolve at an equal rate, and complicates tree-rooting. We therefore introduce a preference for root locations that are more likely in terms of the number of duplications and losses invoked. For each root position r, we compute the number of duplications and losses it implies for each branch $\mathbf{s}$ below $\mathbf{x}$ in the species tree (i.e., either $\mathbf{y}$ or $\mathbf{z}$). We denote these as $\#\text{dups}_r^s$, and $\#\text{loss}_r^s$, respectively. To estimate the probabilities of such events, we assume that they are governed by a Poisson distribution.[4] We define

$$\omega_r = \prod_{s \in \{\mathbf{y}, \mathbf{z}\}} P(\#\text{dups}_r^s = d^s, \#\text{loss}_r^s = l^s)$$

$$= \prod_{s \in \{\mathbf{y}, \mathbf{z}\}} \left(\frac{e^{-\delta_s} \delta_s^{d^s}}{d^s!} \right) \left(\frac{e^{-\lambda_s} \lambda_s^{l^s}}{l^s!} \right)$$

where δ_s and λ_s are the rates of duplication and loss at the branch $\mathbf{s}$, respectively. These rates may either be learned by the algorithm through repeated iterations or be based on prior knowledge of the studied lineages.

We select the root for each candidate orthogroup OG_i by combining the three scores into a single rooting score ρ_r, reflecting the relative importance of each score. We select the rooting that maximizes:

$$\rho_r(OG_i) = -\alpha \pi_r - \beta \sigma_r + \gamma \omega_r \tag{18.1}$$

where α, β, and γ are constants specifying the relative contribution to the rooting score of peptide similarity, synteny similarity, and the likelihood of the invoked duplications and losses.

2.5.4. Breaking candidate orthogroups

Once a rooting r for an orthogroup tree $P_i^{\mathbf{x}}$ is chosen, we may find that the root of $P_i^{\mathbf{x}}$ no longer represents a single gene as the last common ancestor of all the genes present, but rather an earlier duplication event from which two

---

[4] The Poisson model assumes that these events occur as a memoryless process. This is likely true for most duplications and losses, a notable exception being loci with tandemly duplicated genes, where subsequent duplications and losses may occur at higher rates.

ancestral genes were derived (Fig. 18.4C). This violates Definition 1, and we must therefore split the orthogroup's components at the root of its current tree P_i^x. This situation frequently occurs when orthogroups are paralogous with respect to a duplication event that predates **x**.

This step allows us to be very permissive with the edges we include between genes in the gene similarity graph $\mathcal{G}^T$ and in how we match candidate orthogroups. By admitting more edges, we include many spurious ones, but we also include edges that capture the many-to-many relations that may arise from duplications, ensuring that our orthogroups satisfy Definition 2 of orthogroup completeness. If the spurious edges cause nonorthologous orthogroups to be matched, an accurate rooting will subsequently lead the procedure to partition the candidate orthogroup into separate orthogroups. Synergy iterates this until each orthogroup represents a single ancestral gene and no orthogroups need to be partitioned.

2.5.5. Updating the gene similarity graph

Once we constructed orthogroups at the ancestral node **x**, we no longer need to consider the orthogroups in the species below **x** individually. We avoid doing so by removing vertices in the gene similarity graph that correspond to orthogroups in $\{OG^y\}$ and $\{OG^z\}$ and introducing new vertices that correspond to the newly created orthogroups in $\{OG^x\}$. The edges incident to the new vertices are acquired by taking the union of the edges that were incident to its constituent orthogroups (or genes). To weight these new edges, we recall that each new orthogroup represents the root of a tree. Thus, we can use the standard neighbor-joining distance updating rule in the order specified by the topology of orthogroups' corresponding gene trees (Fig. 18.5). When one of the distances in question is not defined in the original similarity graph, we use the maximal distance value.

The edges in this updated similarity graph can always be traced to one (or more) edges between extant genes in the original similarity graph. However, reciprocal edges between two orthogroups (that might lead them to be merged into the same orthogroup in subsequent iterations) may originate from two different pairs of extant genes that are assigned to the two orthogroups. Thus, our matching criteria is able to capture nontrivial paths in relating the extant genes.

3. EVALUATING ORTHOGROUP QUALITY

Many comparative genomics methods were first applied in studies of *Ascomycota* fungal genomes. These organisms are phenotypically variable in substantial ways, and their genomes have undergone major changes, including an ancestral whole-genome duplication (WGD) which resulted in the

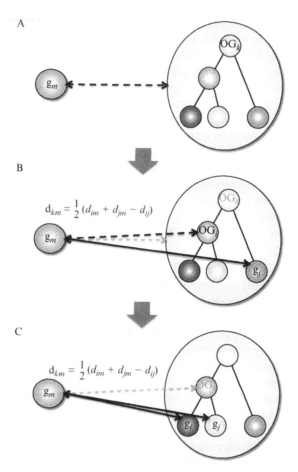

Figure 18.5 Schema of how edge weights are updated in the gene similarity graph. Edge weights between two nodes in the updated graph are calculated by applying the neighbor-joining distance update rule to the constituent nodes within the orthogroups. Dashed lines denote edges induced from updated graphs while solid lines denote edges present in initial graph $\mathcal{G}$. (A) Without loss of generality, the distance between a singleton node g_m and a node representing the orthogroup OG_k is computed according to the tree structure of P_k. (B) To obtain d_{km}, the pairwise distances between g_m, OG_i, and g_j are used. (C) During each step down the tree, the recursive distance updating procedure expands the current orthogroup, indexed as OG_k. This process is repeated until the leaf nodes of P_k are reached. In practice, the distances are cached each time they are calculated to avoid repeated computation.

retention hundreds of paralogous genes in subsequent lineages. They thus offer an excellent opportunity to study evolutionary relations between genes.

We applied our approach to identify orthologs and reconstruct gene trees in over a dozen *Ascomycota* genomes (Fig. 18.6). To assess the quality of

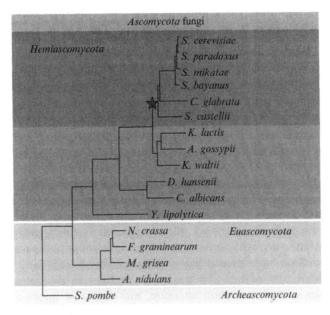

Figure 18.6 Species tree of *Ascomycota* fungi included in this study. Additional classifications for *Hemiascomycota* (top), *Euascomycota* (middle), and *Archeascomycota* (bottom) clades, the WGD event (star), and post-WGD species (darker shade). The branch lengths are proportional to the estimated evolutionary distances. Species in bold were included in our evaluation of orthogroup accuracy by comparison to YGOB (Section 3.2).

our results we relied on three measures. First, we estimated the robustness of the orthology assignments based on a new jackknife procedure. Second, we compared our predictions to a manually curated gold standard and to those of another more traditional hit-based approach. Third, we measured our performance on a simulated dataset of orthogroup evolution. We used a smaller subset of six genomes (Fig. 18.6) to assess the quality of our predictions, and then expand to additional genomes for further biological analysis.

3.1. Fungal orthogroup robustness

To empirically measure Synergy's robustness and to evaluate our confidence in each orthogroup, we developed a jackknife-based approach. We systematically and repeatedly excluded ("held out") different portions of the data, and measured the robustness of orthogroup assignment to (1) the choice of species included and (2) the accuracy of gene predictions within each species. We tested the soundness and completeness of the identified orthogroups under each type of perturbation.

A sound orthogroup (Definition 1) contains *only* the genes that descended from a single common ancestor and thus its genes should not "migrate out" of it in the holdout experiments. To test this, we count the number of orthologous gene pairs (g_j, g_k) in an orthogroup OG_i that remained orthologous in our holdout experiments H.[5] We compute the soundness bootstrap score η_i^s for each orthogroup OG_i by counting the fraction of orthology assignments that remained constant across each holdout experiment $h \in H$:

$$\eta_i^s = \frac{|\{(g_j, g_k) \in OG_i | h(g_j, g_k) = OG_i(g_j, g_k)\}|}{N} \qquad (18.2)$$

where $h(g_j, g_k)$ and $OG_i(g_j, g_k)$ specify the last species in the tree in which g_j and g_k share a common ancestor in the holdout experiment h and the original orthogroup, respectively (this is equal to -1 if g_j and g_k are not members of the same orthogroup), and N is the number comparisons made across all holdout experiments; $|H| \cdot |((OG_i | ((OG_i | - 1))/2)$.

A complete orthogroup (Definition 2) contains *all* the genes that descended from a single common ancestor, and thus new genes should not "migrate into" the orthogroup in the holdout experiments. We use a similar formula to obtain the completeness confidence score η_i^c, except we count the number of pairs of nonorthologous genes $(g_j, g_k), g_j \in OG_i, g_k \notin OG_i$ that became orthologous in each holdout condition h:

$$\eta_i^c = 1 - \frac{|\{(g_j, g_k) \notin OG_i | h(g_j, g_k) \neq -1\}|}{|H| \cdot |g_j \in OG_i, g_k \notin OG_i|} \qquad (18.3)$$

Since pairs of genes that share no protein sequence similarity are highly unlikely to be considered orthologous in h, we restrict our tests to gene pairs that can be loosely regarded as similar ($E < 0.1$), rendering this task computationally feasible.

We compute the confidence measures, η_i^s and η_i^c, for both species- and gene-holdout experiments, generating four measures of robustness for each orthogroup. For the gene-holdout experiments, we set the probability of hiding out each gene at 0.1, and performed 50 holdout experiments. We performed the branch-holdout experiments by removing each branch in the tree separately once (Fig. 18.6), resulting in 31 separate holdout experiments.

Of the nonsingleton orthogroups identified, 79% had all four confidence values above 0.9% and 99% obtained a confidence value above 0.9 in at least one class of experiments (Fig. 18.7). Perturbations to gene content were

---

[5] We must account for the fact that some assignments are expected to change when genes within an orthogroup are among those hidden.

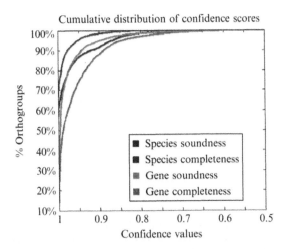

Figure 18.7 Cumulative distributions of confidence scores for orthogroup soundness and completeness under species and gene holdout experiments. Most orthogroups are robustly sound and complete to both types of perturbations.

more disruptive than to species, and soundness was more robust than completeness (*i.e.*, perturbations introduced more new "incorrect" orthologies than loss of "correct" ones).

As expected, orthogroups exhibiting higher frequencies of duplication and loss events tended to be most sensitive to such perturbations, although Synergy's performance was surprisingly robust for even those orthogroups. Lack of such duplication and loss events significantly correlated with higher confidence values ($P < 10^{-4}$ in all four measures). Overall, Synergy was remarkably robust to perturbations in the species phylogeny or noisy gene predictions.

3.2. Comparison to curated resource

We next assessed how Synergy's predictions align with the assignments of a manually curated gold standard, the Yeast Gene Order Browser (YGOB; Byrne and Wolfe, 2005). YGOB contains orthologies for six of the species we investigate based on sequence similarity, chromosomal alignment, and intensive manual curation. This resource is limited by its assumption that the WGD is the only duplication event among this lineage, and relies predominantly on synteny to assign orthology relations. Nonetheless, it provides a "gold standard" of orthology and paralogy relations which we use to evaluate our automated methods.

We found that Synergy's automatic predictions conform very well to those of the YGOB "gold standard" for the relevant species (Fig. 18.8).

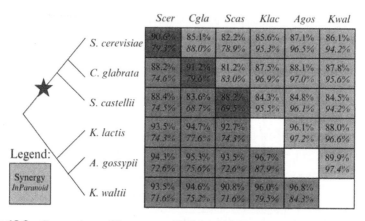

Figure 18.8 Comparison of Synergy and InParanoid (Remm *et al.*, 2001) predictions to the gold standard of YGOB (Byrne and Wolfe, 2005). The matrix displays the sensitivity (lower left cells) and specificity (upper right cells) of orthology assignments in YGOB that were automatically identified by Synergy (top number) and InParanoid (bottom, italicized) for each pair of species. Synergy achieved higher sensitivity at identifying orthology relations than InParanoid, albeit at apparently lower specificity rates. However, because YGOB was designed specifically to study the WGD in these yeast species using syntenic relations, Synergy may include many orthology assignments that were not detected by YGOB due to lack of chromosomal evidence, making this an underestimate of Synergy's specificity performance. The diagonal shows the percent of paralogues reported by YGOB that were detected by Synergy and InParanoid.

For example, Synergy was able to automatically identify over 80% of the orthology assignments between all pairs of species. More significantly, Synergy was able to resolve at a similarly high level of accuracy the precise paralogy relations within orthogroups where both paralogous copies were maintained in at least one species following the whole-genome duplication. This task is challenging since determining which pairs of genes that were retained in duplicate are orthologous requires disambiguating between genes sharing high degrees of sequence similarity.

We also compared the quality of Synergy's paralogy assignments to that of InParanoid (Remm *et al.*, 2001), a hit-clustering method designed to identify paralogous relations between genes within genomes. Synergy identified more known paralogs dating to the WGD than InParanoid did (Fig. 18.8). Unlike InParanoid, Synergy also resolved orthologies and gene trees for *multiple* species simultaneously.

Synergy also showed greater sensitivity than InParanoid when identifying orthology relations, albeit potentially at the cost of lower specificity. Some of the reduced specificity may be the result of a limitation of our gold standard: While YGOB's annotations are highly accurate, their methodology is limited by two assumptions: (1) all duplication events originated in the WGD and thus orthology is at most a 2-to-1 relationship and (2) gene order is nearly always conserved and thus can be used as the primary source of evidence for shared ancestry. These assumptions relegate a greater portion of genes as singletons without orthologs, and a far smaller proportion of YGOB's orthologous loci are ancestral to all of their species than those that Synergy identified (62% vs. 82%). We therefore believe that many of the orthology assignments reported by Synergy but not by YGOB (or InParanoid) (Fig. 18.8, green cells) are likely to be correct assignments.

To estimate the contribution of including synteny in our approach, we reran Synergy on these data while ignoring the genes' locations. We found that synteny plays a relatively minor role in predicting a genes' correct orthologs, but contributed substantially to reconstructing the correct gene trees. For example, over 200 duplication events were detected at the root of the *sensu stricto* species when ignoring synteny, most of which should have been traced to the WGD. The contribution of synteny information to orthology prediction may be most noticeable in cases where genes are undergoing exceptionally slow or fast rates of evolution, as is often the case between paralogs undergoing gene conversion or neofunctionalization (Kellis *et al.*, 2004a; data not shown). It is here that synteny can help the most when deciding how to root the gene tree in Stage 2 of the algorithm (Section 2.5.2).

3.3. Simulated orthogroups

To obtain an objective measure of Synergy's accuracy, we simulated orthogroup evolution including multiple rounds of speciation events and with prespecified rates of gene duplication and loss. At each stage, we used the SEQ-GEN program (Rambaut and Grassly, 1997) to simulate the evolution of protein sequences using the JTT model of amino acid substitution (Jones *et al.*, 1992). In order to make these simulations as true to fungal protein sequences as possible, we initiated the simulations with varying numbers of randomly drawn sequences from *Saccharomyces cerevisiae*. For the purposes of this benchmark, we ignored the chromosomal ordering of the simulated sequences, since to the best of our knowledge there is no general agreed-upon model for chromosomal evolution. As a result, this test evaluates Synergy's performance when no synteny information is considered.

We parameterized our simulations as follows: Using a balanced species tree topology containing 16 species, we gave each orthogroup a probability of 0.1 of incurring a duplication or loss along every branch of the species tree. These duplication and loss rates are relative high, but we were interested in examining how well Synergy performs under such volatile conditions. We specified the rates of amino acid substitutions between orthologs by the branch lengths in the simulated gene trees. These lengths were drawn from an exponential distributed with a mean of 0.36 (approximately the mean branch length in the fungal species tree we used).

Synergy accurately detected over 85% of the orthologous relations in our simulated datasets of various sizes (Fig. 18.2). Further, its specificity was remarkably high—nearly 99%—despite the presence of many of paralogs in the genome from which the simulated sequences were originally drawn. This sensitivity may have been further improved had we included chromosomal order into these simulations, allowing Synergy to predict paralogs more robustly. Importantly, we found no significant trend suggesting that the number of sampled sequences affected Synergy's overall performance in these simulations.

While we recognize that the implications of such benchmarks should be carefully interpreted, we believe that these simulations accurately reflect Synergy's strong performance on data that is based on a reasonable model of fungal sequence evolution.

4. Biological Analysis of Gene Histories

A high-quality set of orthogroups and gene trees (from Synergy or future improved tools) opens the way for comprehensive analysis of the relation between gene evolution and function. These require a host of genomics resources and tools. The rich genomics datasets collected for *S. cerevisiae* (and increasingly for other yeasts) allow for such successful analysis. Here, we highlight several key approaches with which to tackle this challenge by available genomic resources and datasets and the evolutionary categorization of orthogroups.

4.1. Defining orthogroup categories

Each orthogroup's gene tree contains its own set of characteristics that describe its history. We can use these characteristics to categorize them. For instance, the gene tree associated with the orthogroup in Fig. 18.9A has the same topology as the species tree and exhibits no duplications or losses throughout its history. In contrast, the orthogroup in Fig. 18.9B contains a duplication that led to two paralogous loci (*IFH1* and *CRF1*), one of which was subsequently lost in *Candida glabrata* (*CRF1*), as well as a loss in the lineage leading to the *Euascomycota* clade. These features can be used to categorize genes into sets. For example, we say that the *SEN34* gene is "uniform" because it contains a *uniform* history with no duplication and loss. It can also be described as "persistent", because it contains at least gene in all of the species. *IFH1* and *CRF1* are not included in these categories, since they do not have a uniform history within this phylogeny, nor are they represented in at least one copy in all of the genomes. As we will describe, such categorization allows for further analysis with respect to how gene functions and histories might be related.

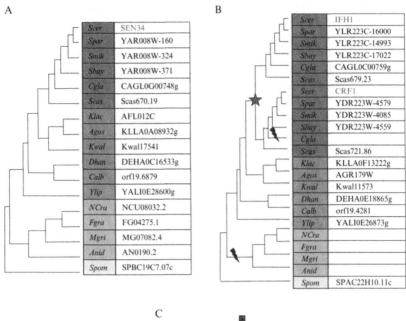

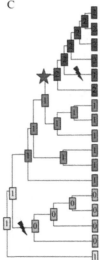

Figure 18.9   Gene trees and copy-number variations. (A) A uniform orthogroup that contains the *S. cerevisiae* gene *SEN34*. The topology of the gene tree (left panel) is identical to that of the species tree. Genes in the orthogroups tree are named on right, next to the four letter abbreviated species names. (B) Orthogroup containing the *S. cerevisiae* genes *CRF1* and *IFH1*. The gene trees topology (left panel) differs from that of the species tree and shows a single duplication event (star) and two loss events (strikes). (C) The extended phylogenetic profile of the orthogroup in (B) summarizes the number of genes in the orthogroup at each extant and ancestral species in the tree (numbered boxes).

Another major type of history-based categorization reflects the particular "age" of a gene. Many genes only have orthologs tracing back to a specific point in the phylogeny, where they seem to "appear" (*e.g.*, Fig. 18.10A). These may indicate points of genomic innovations or elevated mutation rates among these loci. Genes that appear at the same branch of the phylogeny can be categorized into the same set. Similarly, those that are "ancestral" to the phylogeny are an additional gene set. Note that ancestral genes are not always persistent (*e.g.*, Fig. 18.9B). To ensure that appearing genes are not resulting from orthology mispredictions, we can conduct more sensitive remote homology searches to determine if a gene has homology that are more distant than those denoted by its reconstructed tree (Durbin *et al.*, 1998) (Table 18.1).

4.2. Singletons and ORF predictions

We define genes that appear solely in an individual species as "singleton" or orphan genes. These genes have no orthologs and may indicate lineage-specific genes that perform a function unique to that species or, more often, they are mispredicted ORFs. In this way, orthogroups assignments can be used to refine genome annotations. We discarded singleton ORFs for most of our analyses, but recent works by others (*e.g.*, Fedorova *et al.*, 2008; Kasuga *et al.*, 2009; Khaldi and Wolfe, 2008; Li *et al.*, 2008) have explored the origins and roles of these genes in different species.

4.3. Gene sets and orthogroup projections

To study the relations between gene history and gene function, we must first gather functional categories describing the genes' roles within an organism. These gene sets may be derived from experimental data, domain composition, or homology to known genes. Some of the main resources of *S. cerevisiae* gene sets are (the number of gene sets is in brackets): the Gene Ontology (Ashburner *et al.*, 2000) (GO) hierarchy (1794), the Kyoto Encyclopedia of Genes and Genomes (Ogata *et al.*, 1999) (KEGG) (87), the BioCyc database (Karp *et al.*, 2005) (107), the MIPS database of manually curated protein complexes (Mewes *et al.*, 2002) (1022), targets genes bound by various transcription factors (Harbison *et al.*, 2004) (310), targets genes harboring a given *cis*-regulatory element in their promoters (Harbison *et al.*, 2004) (70), and targets of RNA binding proteins (Gerber *et al.*, 2004) (5).[6] Additional resources which are important for the assessment of the relation between gene history and functions include datasets

---

[6] These 3395 gene sets were used in most of the analyses below, and in particular to construct transcriptional modules.

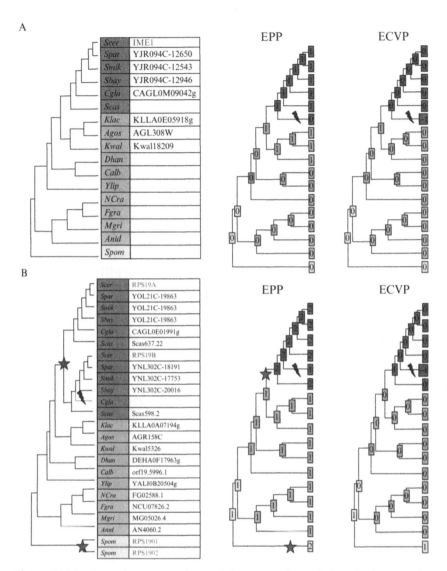

Figure 18.10 Appearing genes and extended copy-number variation. (A) An appearing orthogroup that contains the *S. cerevisiae* gene *IME1*. The gene tree topology (left, black lines) differs from that of the species tree (right) as it is only supported by genes from the clade spanning *K. waltii* and *S. cerevisiae*. The extended phylogenetic profile (EPP, center, numbered boxes) shows the gene copy number for all the species. The copy-number variation profile (ECVP, right, numbered boxes) indicates the changes in gene copy number. An increase (+1) is placed at the first ancestral species where this orthogroup is traced to (appears in). The *S. castellii* ortholog in this orthogroup was lost (−1, blue strike). (B) A persistent orthogroup that contains the *S. cerevisiae* genes RPS19A and RPS19B, which encode proteins of the small ribosomal subunit. The gene tree topology (left) differs from that of the species tree (center, right) as it includes both duplication (star) and loss (strike) events. EPP and ECVP for the orthogroup are show the center and right panels, respectively. The ECVP indicates an increase in copy number (+1) due to gene duplication at the WGD and along the branch leading to *S. pombe* (stars). One of the WGD paralogs was subsequently lost in *C. glabrata* (−1, strike). Despite the loss event, the orthogroup contains at least one member gene in each extant species.

Table 18.1 Summary of terms used to characterize orthogroups

Uniform	No retained duplications or losses across all lineages
Persistent	Retain *at least* one ortholog in all lineages
Volatile	Much greater number of duplication and loss events than expected by chance
Ancestral	Present at the last common ancestor of all the species studied
Appearing	Could only be traced up to a certain point in the phylogeny; the opposite of ancestral
Singleton	Individual genes that appear at a leaf node in the phylogeny, having no orthologs identified in any other lineage

measuring genes controlled by the SAGA and/or TFIID transcription complexes (Huisinga and Pugh, 2004), genes with and without TATA box control (Tirosh *et al.*, 2006), genes with large levels of expression variation between yeast species (Tirosh *et al.*, 2006), genes with high and low levels of noise in protein abundance (Newman *et al.*, 2006), haploinsufficient genes (Deutschbauer *et al.*, 2005), genes whose overexpression reduces fitness (Sopko *et al.*, 2006), and genes belonging to complex cores, attachments and modules based on high-throughput assays (Gavin *et al.*, 2006). To streamline our analysis and build on the extensive expression data available for *S. cerevisiae*, we used these gene sets to determine a hierarchy of transcriptional modules following the procedure presented by Segal *et al.* (2004). Each module contains functionally related genes with coherent expression patterns. We applied this procedure to the 3395 gene sets described above and a compendium of 1216 arrays and followed it by manually selecting which modules to use in the hierarchy (for full details see Wapinski *et al.*, 2007a).

These gene sets allow us to analyze orthogroups in the context of their cellular functions, regulatory relations, and condition-dependent expression programs. We can readily create orthogroup sets from these gene sets by assigning orthogroups to sets according their genes' annotations. We then use Fisher's exact test to measure the statistical significance of the overlap between orthogroup sets (*e.g.*, uniform orthogroups) and functional categories (*e.g.*, meiosis orthogroups). Significant enrichments allow us to evaluate the global constraints that influence the evolutionary trajectories of genes.

For instance, when applying this approach to our *Ascomycota* orthogroups we found that the ancestral orthogroups are strongly enriched in *S. cerevisiae* genes that are essential; 1008 of 1047 genes essential for growth in rich YPD medium are ancestral ($P = 1.3 \times 10^{-31}$). On the other hand, the clade spanning *S. cerevisiae* and *K. waltii* is marked by appearing orthogroups that contain *S. cerevisiae* genes whose annotations are significantly enriched for meiosis and sporulation genes (51/166 sporulation genes, $P = 2.49 \times 10^{-7}$), including the master meiosis regulator IME1.

Notably, we can extend this functional analysis from gene sets to interaction networks. Previous studies have assembled physical and genetic interaction networks from several existing manually curated and high-throughput data sources (Reguly *et al.*, 2006). The networks can be represented as graphs where the genes or proteins are nodes and an edge is placed between interacting genes or proteins. As discussed below, we can annotate such networks with the evolutionary categories (*e.g.*, *uniform*, *persistent*, *appearing*) and use them to test the relation between evolutionary history and gene function.

4.4. Copy-number variation profiles

The orthogroup categories described above provide a concise but simple measure of a gene's history, but obscure some of the finer details. For example, two orthogroups can both be persistent, but include very different patterns of gene duplication and loss, which are represented in the associated gene trees. To capture this, we expand on the well-known *phylogenetic profiling* approach that considers the profile of species in which each gene is present or absent (Marcotte *et al.*, 1999).

We define an *extended* phylogenetic profile that includes the number of gene copies present at each extant and ancestral species in the phylogeny (Fig. 18.10). From these profiles, we can readily derive an *extended copy-number variation profile* (ECVP) that measures that gene copy number changes through duplications and losses. We compute ECVPs by inspecting the extended phylogenetic profile of an orthogroup and counting the number of duplications and losses that occur along each index of the species tree. We subtract the number of losses from the number of duplications at each index to generate the ECVP, thus summarizing these events in a numerical vector that is amenable for further analysis (Fig. 18.10, right). We increment this copy-number variation profile at the last common ancestor identified for the orthogroup, indicating its age.

4.4.1. ECVPs within orthogroup classes

Using ECVPs we can ask whether genes that share a function also share similar (coherent) histories. To assess the coherence in gene copy-number variation across a class of orthogroups, we first calculated the class centroid ECVP by averaging the ECVPs from all the orthogroups belonging to a class. This centroid is then applied to estimate the degree of deviation between the orthogroups belonging to a class by summing the L1 distance from each of the class orthogroups to it. We compare this deviation to that of 10,000 randomly assigned orthogroup classes, each containing the same number of ECVPs. The fraction of times the deviation is equal to or less than that of the orthogroup class is the measure (P-value) we use to evaluate the coherence of that class. Since copy-number variation occurs at each individual branch of the species tree, we similarly define a coherence profile for an orthogroup class by evaluating the copy-number variation coherence for each position along the species tree.

In our catalog, functional constraints on copy-number variation manifest in remarkably similar patterns of specific duplications and losses in functionally related orthogroups. To show this, we compared the extended copy-number variation profiles (ECVPs, which track these events) among sets of orthogroups harboring functionally related *S. cerevisiae* genes. We found that functionally related orthogroups, in particular those related to growth, often show coherent profiles of duplication and loss events (Fig. 18.11), indicating that such events occur in concert for functional counterparts. Despite this general pattern, some classes of orthogroups, especially those related to stress, do not show such coherence: these are often related to known sources of phenotypic variation between different yeasts (*e.g.*, the cell wall organization and biosynthesis class; Fig. 18.11E) or to very general functions that would not be expected to have shared evolutionary constraints (*e.g.*, oxidoreductases). Classes of orthogroups defined by genes with a common regulatory mechanism (*i.e.*, shared *cis*-regulatory motif (Harbison *et al.*, 2004), transcription factor (Harbison *et al.*, 2004), or RNA binding protein (Gerber *et al.*, 2004)) are typically not coherent, suggesting that regulatory mechanisms do not impose strong evolutionary constraints on copy-number variation. Coherence is not merely a reflection of the prevalence of uniform orthogroups, since these observations are maintained even when all uniform orthogroups are omitted from the analysis.

4.4.2. ECVPs within interaction networks

We can also use ECVPs to assess the evolutionary constraints imposed by physical and genetic interactions. To test the relation between proximity in interaction networks and similarity in copy-number variation, we first computed the difference (using the L1 distance) between the ECVPs for each pair of proteins in the network, ignoring pairs that belong to the same orthogroup (hence sharing the same profile). Next we averaged these differences among all proteins within a given radius in the network. To determine whether these averages were significant, we repeated this computation by shuffling profile assignments to proteins in the network 1000 times, obtaining the expected range of average differences between pairs of proteins in the network for each radius.

Indeed, when we examined the ECVPs of pairs of orthogroups containing (nonparalogous) *S. cerevisiae* genes, we found that those that are nearby in either a biochemical network or a genetic network tend to have

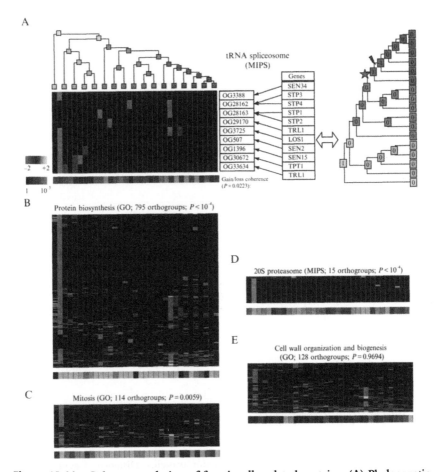

Figure 18.11 Coherent evolution of functionally related proteins. (A) Phylogenetic coherence of the tRNA spliceosome orthogroup class. The set of *S. cerevisiae* tRNA spliceosome genes (MIPS) was mapped to the set of orthogroups that contain these genes (black arrows, middle panel). Some orthogroups (e.g., #28162) contain multiple paralogs from the gene set. The ECVPs of all the orthogroups in the set are compiled into a matrix (left panel). Each row denotes one profile, and each column the copy number changes in one species (red, increase; blue, decrease; black, no change). The species (extant and ancestral) are ordered according to the order of the nodes in the species tree (top). The bottom row shows the coherence score for each column (purple, coherent), as evaluated by comparing the number of events to the distribution of events in the specific node within a random set of orthogroups of the same size. The overall significance of the coherence is reported in a *P*-value. (B–E) Phylogenetic coherence of the protein biosynthesis, Mitosis, 20S proteasome, and cell wall organization and biogenesis orthogroup sets. Copy-number variation coherence is presented as in (A). The copy-number changes observed in the protein biosynthesis (B), Mitosis (C), and 20S proteasome (D) orthogroup classes are coherent. Those in the cell wall organization and biogenesis orthogroup set are not coherent (E). Orthogroup classes are projected from the GO gene classes. (See Color Insert.)

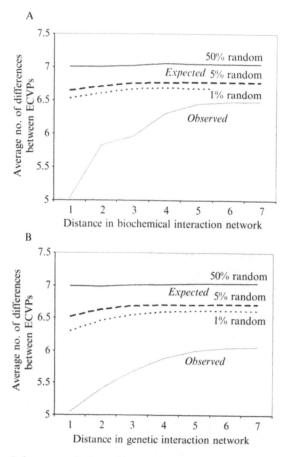

Figure 18.12 Coherent evolution of interacting proteins. Distance of genes in biochemical (A) and genetic (B) interaction networks (x-axis) is plotted versus the average difference between the ECVPs of pairs of genes of that distance or less in each network (y-axis). Pairs of paralogous genes are excluded from the computation of the averages. Black lines show the 1%, 5%, and 50% of the distribution of average distances from repetitions of this computation in networks with the same topology obtained by random reshuffling of the gene to profile associations. In both cases, similarity in ECVPs inversely scales with distance in the interaction network. Each network combines literature-curated results and high-throughput measurements. Similar results are obtained when only one source of data is used (data not shown).

significantly similar patterns of gene gain and loss (Fig. 18.12). The similarity decreases as the distance in interaction networks increases.

These findings extend the well-established observation that functionally related or interacting genes tend to show correlated co-occurrence (and coabsence) across species. However, unlike conventional phylogenetic profiles the ECVP approach examines the full phylogenetic history of each gene. It therefore allows us to recognize that the persistent orthogroups related to tRNA splicing have coherent patterns duplication and loss while the persistent ones related to cell wall organization and biogenesis do not (both have an identical conventional phylogenetic profile).

4.4.3. Copy-number volatility

Orthogroups show wide variation with respect to the number of duplication and loss events: many are uniform with no such events (*e.g.*, Fig. 18.9A), while many others are highly dynamic with many events (*e.g.*, Fig. 18.9B). To benchmark the *volatility* in gene copy number for each orthogroup, we used the estimated rates of duplications and losses along each branch of the species tree to calculate the log-probability of the observed number of such events in each orthogroup, assuming that they occur independently according to a standard Poisson distribution. This statistic is used as a measure of volatility for each orthogroup. We compare this volatility metric to those of 10,000 hypothetical orthogroups with randomly generated duplications and losses (based on the empirical rates). We label orthogroups whose volatility deviates more than three standard deviations from the mean of the random distribution as being significantly "volatile." We find that the distribution of these events is inconsistent with a uniform rate of duplication and loss across orthogroups, as can be seen by comparing log-likelihood ratios for the observed occurrences with those expected by chance (Fig. 18.13).

We analyzed the 1018 uniform and 313 volatile orthogroups at the opposite extremes of this copy-number volatility scale, and found that they show diametrically opposed patterns with respect to a wide range of biological properties defined in *S. cerevisiae* (Fig. 18.13). For example, volatile orthogroups are enriched ($P < 10^{-5}$) for functional (GO) (Ashburner *et al.*, 2000) categories reflecting peripheral transporters, receptors, and cell wall proteins and genes that participate in stress responses. By contrast, uniform orthogroups are enriched ($P < 10^{-5}$) in genes involved in essential growth processes; genes residing in the nucleus, nucleolus, mitochondrion, endoplasmic reticulum, and Golgi apparatus. This evolutionary dichotomy is also aligned with the transcriptional program of *S. cerevisiae*, as reflected by regulatory modules (Section 4.3). Growth and cell cycle modules are overwhelmingly enriched for uniform and persistent orthogroups (*e.g.*, modules for ribosome biogenesis, ER protein modification and targeting, morphogenesis). By contrast, development, stress, and carbohydrate metabolism modules are strongly enriched for volatile orthogroups (*e.g.*, redox–detox, mating and filamentous growth, reserve carbohydrates, trehalose). There are a few notable exceptions to this general trend, including the proteasome and mitochondrial ribosome modules, that have stress-like expression pattern but are enriched for uniform orthogroups. The uniform orthogroups are enriched in *S. cerevisiae* genes

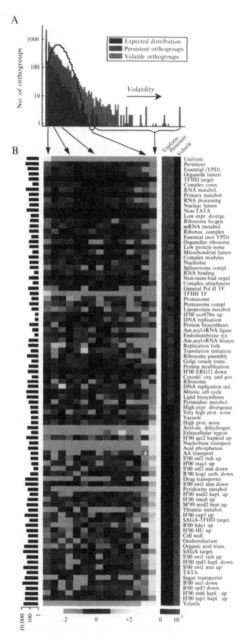

Figure 18.13 A functional dichotomy of uniform, persistent, and volatile orthogroups. (A) Orthogroup volatility. The histogram shows the number of orthogroups in each bin of volatility scores. The expected distribution when sampling random orthogroups from the evolutionary model is shown as a black line. Uniform orthogroups (leftmost blue column) are the lowest scoring orthogroups. Persistent

that are essential for viability in YPD (405/1047 essential genes are in uniform orthogroups, $P < 2.59 \times 10^{-70}$), suggesting that not only are such genes rarely lost in other species (as would be expected), but that they are rarely duplicated. This may indicate a limit on the ability to maintain duplicate copies of essential genes or a latent functional redundancy between duplicate genes, making yeast genes with paralogs less essential overall. The former hypothesis is consistent with the observation that volatile orthogroups are enriched in genes whose expression changes significantly in response to many single-gene knockouts (Hughes *et al.*, 2000), genes with noisy levels of protein abundance within *S. cerevisiae* (Newman *et al.*, 2006), and genes with variable RNA expression across species (Tirosh *et al.*, 2006). On the other hand, the uniform orthogroups show complimentary attributes and are enriched in genes whose expression is largely unchanged in the same systems.

These results highlight a general bipolar principle in which certain types of genes readily undergo duplication and loss, while others rarely tolerate such events. Copy-number variation in stress responsive genes may not only be tolerate but beneficial, allowing fungal cells to adapt to changing ecological niches. On the other hand, genes essential for cell growth, including but not limited to those necessary for the operation of intricate multiprotein complexes (Papp *et al.*, 2003), cannot readily tolerate such noise and tend not to evolve by gradual duplication and loss. The correspondence we observe extends to orthogroups with moderately low- and high-volatility scores (and not only to extreme ones); these groups show similar patterns to the uniform and volatile orthogroups, suggesting a general principle imposed on the vast majority of genes in the genome. Overall, the evolutionary dichotomy in the rate of duplication and loss aligns closely with a clear bipolarity in gene function, transcriptional program and noise level in expression levels across cells, strains, and species (Newman *et al.*, 2006; Tirosh *et al.*, 2006). These features likely reflect similar functional constraints on the amount of gene products available in the cell under different conditions.

orthogroups usually receive low scores as well (blue columns). All orthogroups with a score above three standard deviations from the expected mean are volatile (red columns). (B) Gene class annotations enriched in uniform, persistent and volatile orthogroups. Annotations that were significant (at most $P < 10^{-5}$ after FDR correction, purple significance) among either uniform, persistent or volatile orthogroups are shown (the size of each annotation is shown on the left). The functional and mechanistic dichotomy between volatile and nonvolatile orthogroups largely reproduces along the full range of volatility scores (columns—bins of orthogroups with similar volatility scores; Rows—significant annotations. Yellow/blue higher/lower relative enrichment compared to the expected enrichment in the class). (See Color Insert.)

5. ANALYSIS OF PARALOGOUS GENES

There are two basic processes by which gene duplications can give rise to functional innovation: (Conant and Wolfe, 2006), *neofunctionalization*, when one copy of the gene acquires a new functional role, and *subfunctionalization*, wherein a gene's functional roles are divided between paralogs. A catalog of evolutionary histories and functional annotations provides an outstanding resource with which to study the degrees to which each of these processes have occurred. We consider two measures of functional divergence between paralogs: (1) the degree of similarity in functional annotation and (2) the degree of similarity in (genetic or physical) interactions.

5.1. Estimating functional divergence between paralogous genes

To estimate the functional divergence between a pair of paralogous genes, we consider the gene set annotations assigned to each paralog. Ordinarily, we assume that paralogs have conserved their functions if they are both contained within the same functional annotation. Conversely, we estimate that paralogs have diverged with respect to a given function if one of them is not annotated in the same gene set as the other. For each gene class, we calculated the fraction of paralogous pairs that are retained within the class. To avoid confounding factors, we studied only cases in which both paralogs had been annotated and in which the annotation had not been inferred solely from sequence similarity (*i.e.*, computationally). In cases where the gene sets are organized in an annotation hierarchy (*e.g.*, GO), we regard a pair of genes as functionally diverged only if both genes are assigned to at least one annotation class and they are not both assigned to the most specific annotations of either of the two genes.

When applying this approach to *S. cerevisiae* gene sets we find that paralogous pairs rarely migrate between functional GO categories. The vast majority of paralogous pairs have not migrated at all on the three GO hierarchies The retention rate is highest for GO molecular function categories (92%) and somewhat lower for biological process and cellular component categories (85% and 81%, respectively).

By contrast, paralogous pairs frequently migrate with respect to gene classes defined by shared regulatory mechanisms (such as genes that are targets of a transcription factor, or contain the same *cis*-regulatory motif (Harbison *et al.*, 2004) or RNA-binding motif (Gerber *et al.*, 2004)). Indeed, in the majority of cases (70%) regulatory gene classes contain no retained paralogy relations within them, reflecting either novel regulation or regulatory specialization. The transcriptional modules exhibit an intermediate behavior, with 26% of the paralogous gene pairs having migrated between modules, with a quarter of the paralog migrations occurring across the two main growth and stress groups.

Our analysis shows that paralogs diversify most frequently at the level of regulation, less frequently through changes in cellular component and biological process and very rarely at the level of biochemical function. This highlights inherent limitations of gene duplication in accomplishing molecular innovation. It also emphasizes the overwhelming importance of regulatory divergence in reconfiguring molecular systems following duplication (Carroll, 2008; Gu *et al.*, 2004; Makova and Li, 2003). Changes in transcriptional regulation are clearly the foremost force that drives functional divergence after duplication.

5.2. Estimating divergence based on degree of conserved interactions

While functional categories provide extensive information, they are also crude, and hence cannot capture subtle divergence of two paralogs within the same general process. Functional characterization based on interaction networks addresses this limitation. In particular, annotation based on genetic interaction reflects the degree of shared biological process, whereas annotation based on biochemical (physical) interactions reflects the degree of shared molecular function.

We employed two statistics to compute the degree of conserved interactions between pairs of paralogous proteins. The first was simply the fraction of shared interactions between both proteins. For this we counted the number of interactions each protein takes part in ($a1$ and $a2$ for proteins 1 and 2, respectively) and the number of interactions they both share (s). The fraction of shared interactions is thus

$$f = \frac{s}{\min(a1, a2)} \qquad (18.4)$$

We also used the subfunctionalization index (I_{sf}) as previously described (He and Zhang, 2005) to characterize how diverged a pair of paralogs interactions are. This is calculated as

$$I_{sf} = \frac{1 - (s + |a1 - a2|)}{t} \qquad (18.5)$$

where s is as above and t is equal to ($a1 + a2s$). This statistic gives a reasonable estimate of the degree of subfunctionalization in the absence of neofunctionalization, since subfunctionalization would reduce the number of shared interactions. This measure considers the proportion of ancestral interactions that are no longer shared between the paralogs and the extent of subfunctionalization for these interactions. Other methods for quantifying the degree of conservation between paralogous proteins have also been proposed by others and we refer to these publications for further reading (*e.g.*, Musso *et al.*, 2007).

We find that the paralogous pairs can be partitioned into two distinct categories of roughly equal size. The pairs in the first category share a high and significant proportion of their interacting partners (136/318 pairs in the genetic network and 225/543 in the biochemical network). This is much higher than the proportion observed in comparable random networks, as estimated by comparing the degree of conserved interactions between the two paralogs in the real network to that in a degree-preserving randomized network (obtained by swapping edges between random pairs of nodes 10^6 times). We repeated this procedure 10,000 times, and assigned and empirical *P*-value to the shared protein interaction neighborhood of a pair of proteins according to the number of times the fraction of shared interacting partners between paralogs is equal to or greater than the fraction in the real network. These paralogs that exhibit such a high overlap between their interacting partners thus show little migration from their initial configuration immediately following their duplication.

By contrast, the pairs in the second category share no interacting partners whatsoever, even though they themselves may frequently interact. Intermediate examples are rare (< 9% of the total). Notably, the second category includes many pairs of paralogous genes that belong to the same functional classes. In the biochemical network, these disjoint paralog pairs tend to be highly dispersed in the network (often at distance of four or greater) implying that they act in distinct physical environments. In the genetic network, however, these paralogs are often neighbors. This suggests that they have some overlapping function or can compensate for each other, despite the apparent divergence in roles. This pattern of diversification suggests a partial division of labor involving specialization (rather than adoption of a new function) of two paralogous proteins that become physically or temporally separated but can still have adverse consequences when both are compromised (Kafri *et al.*, 2005).

6. Discussion and General Applicability

In this chapter, we presented a framework for identifying groups of orthologous genes in a step-wise manner, while simultaneously reconstructing a corresponding phylogenetic gene tree for each group. We describe a novel algorithm—Synergy—that uses this framework to reconstruct a genome-wide catalog of gene trees across species by incorporating multiple sources of information, including sequence similarity and conserved gene order when relevant. Synergy's gene trees reflect the evolutionary history of each group of genes, allowing us to accurately identify orthology and paralogy relations between genes, and the duplication, loss and divergence events that underly these relations.

This approach has several important benefits. Its accurate automatic genome-wide resolution is unprecendented—it is typically absent from automatic "hit-clustering" methods applied on a genomic scale, which either ignore paralogs altogether (RBH) or do not make detailed distinctions between orthologs and paralogs (Tatusov *et al.*, 1997). Since Synergy's gene tree reconstruction is constrained *a priori* by the topology of the species tree, we do not have to apply extra reconciliation steps (*e.g.*, Durand *et al.*, 2006). For example, in orthogroups that have no duplication and loss events, our algorithm is guaranteed to yield the correct gene tree.

Synergy strikes an important balance between orthogroup completeness (sensitivity) and soundness (specificity). We ensure completeness by allowing many edges (candidate homology relations) into the input gene similarity graph and by applying a lenient criterion to derive candidate orthogroups. Then, we achieve soundness by refining these coarse relations as we progress through the species tree, breaking orthogroups using phylogenetic principles at each Stage. The bottom-up design of our algorithm also renders it scalable, allowing us to handle a large number of species and genes.

While our bottom-up approach provides high-quality results, it is nevertheless a greedy algorithm and can occasionally mis-assign genes. This greediness could be relaxed by adding top-down re-assignments after the bottom-down reconstruction is completed. Formulating the orthology resolution problem within the framework of bottom-up orthogroup identification should provide an important paradigm for additional algorithmic solutions. In addition, reconstructing the orthogroups' ancestral sequences at each Stage of the algorithm may allow it to deduce valuable information that can also be included in identifying orthologs more accurately. For instance, estimating the gene-specific mutation rates might also be considered when determining orthology assignments. Not doing so might currently be hindering Synergy in cases where genes are particularly fast- or slow-evolving. These top-down and ancestral sequence reconstruction steps are both strategies that may yield improved results as well as intriguing insights about gene family evolution.

Synergy opens the way to a host of comparative genomics studies. As comparative genomics gains widespread popularity, scores of groups of species have been extensively sequenced, all of which can be tackled with our scalable algorithm. We describe several methods for incorporating functional genomics data, including gene set annotations and interaction networks, to study the effects of and constraints on gene duplication and loss on molecular networks. This approach can be broadly applied to study the evolutionary patterns within any group of organisms whose genomes have been sequences, allowing us to study how gene histories and functions can be closely interrelated.

REFERENCES

Altschul, S. F., Gish, W., Miller, W., Myers, E. W., and Lipman, D. J. (1990). Basic local alignment search tool. *J. Mol. Biol.* **215,** 403–410.

Arvestad, L., Berglund, A. C., Lagergren, J., and Sennblad, B. (2003). Bayesian gene/species tree reconciliation and orthology analysis using MCMC. *Bioinformatics* **19**(Suppl. 1), 7–15.

Ashburner, M., Ball, C. A., Blake, J. A., Botstein, D., Butler, H., Cherry, J. M., Davis, A. P., Dolinski, K., Dwight, S. S., Eppig, J. T., Harris, M. A., Hill, D. P., *et al.* (2000). Gene ontology: Tool for the unification of biology. The gene ontology consortium. *Nat. Genet.* **25,** 25–29.

Blomme, T., Vandepoele, K., De Bodt, S., Simillion, C., Maere, S., and Van de Peer, Y. (2006). The gain and loss of genes during 600 million years of vertebrate evolution. *Genome Biol.* **7,** R43.

Byrne, K. P., and Wolfe, K. H. (2005). The Yeast Gene Order Browser: Combining curated homology and syntenic context reveals gene fate in polyploid species. *Genome Res.* **15,** 1456–1461.

Carroll, S. B. (2008). Evo-devo and an expanding evolutionary synthesis: A genetic theory of morphological evolution. *Cell* **134,** 25–36.

Cliften, P., Sudarsanam, P., Desikan, A., Fulton, L., Fulton, B., Majors, J., Waterston, R., Cohen, B. A., and Johnston, M. (2003). Finding functional features in *Saccharomyces* genomes by phylogenetic footprinting. *Science* **301,** 71–76.

Conant, G. C., and Wolfe, K. H. (2006). Functional partitioning of yeast co-expression networks after genome duplication. *PLoS Biol.* **4,** e109.

Deutschbauer, A. M., Jaramillo, D. F., Proctor, M., Kumm, J., Hillenmeyer, M. E., Davis, R. W., Nislow, C., and Giaever, G. (2005). Mechanisms of haploinsufficiency revealed by genome-wide profiling in yeast. *Genetics* **169,** 1915–1925.

Dietrich, F. S., Voegeli, S., Brachat, S., Lerch, A., Gates, K., Steiner, S., Mohr, C., Phlmann, R., Luedi, P., Choi, S., Wing, R. A., Flavier, A., *et al.* (2004). The *Ashbya gossypii* genome as a tool for mapping the ancient *Saccharomyces cerevisiae* genome. *Science* **304,** 304–307.

Dufayard, J. F., Duret, L., Penel, S., Gouy, M., Rechenmann, F., and Perrire, G. (2005). Tree pattern matching in phylogenetic trees: Automatic search for orthologs or paralogs in homologous gene sequence databases. *Bioinformatics* **21,** 2596–2603.

Durand, D., Halldrsson, B. V., and Vernot, B. (2006). A hybrid micro-macroevolutionary approach to gene tree reconstruction. *J. Comput. Biol.* **13,** 320–335.

Durbin, R., Eddy, S., Krogh, A., and Mitchison, G. (1998). Biological sequence analysis. Cambridge University Press, Cambridge.

Eisen, J. A. (1998). Phylogenomics: Improving functional predictions for uncharacterized genes by evolutionary analysis. *Genome Res.* **8,** 163–167.

Eisen, J. A., and Fraser, C. M. (2003). Phylogenomics: Intersection of evolution and genomics. *Science* **300,** 1706–1707.

Eisen, J. A., and Wu, M. (2002). Phylogenetic analysis and gene functional predictions: Phylogenomics in action. *Theor. Popul. Biol.* **61,** 481–487.

Engelhardt, B. E., Jordan, M. I., Muratore, K. E., and Brenner, S. E. (2005). Protein molecular function prediction by Bayesian phylogenomics. *PLoS Comput. Biol.* **1,** e45.

Fedorova, N. D., Khaldi, N., Joardar, V. S., Maiti, R., Amedeo, P., Anderson, M. J., Crabtree, J., Silva, J. C., Badger, J. H., Albarraq, A., Angiuoli, S., Bussey, H., et al. (2008). Genomic islands in the pathogenic filamentous fungus *Aspergillus fumigatus*. *PLoS Genet.* **4**, e1000046.

Felsenstein, J. (2004). *Inferring Phylogenies*. Sinauer Associates, Inc., Sunderland, MA.

Fitch, W. M. (1970). Distinguishing homologous from analogous proteins. *Syst. Zool.* **19**(2), 99–113.

Gavin, A. C., Aloy, P., Grandi, P., Krause, R., Boesche, M., Marzioch, M., Rau, C., Jensen, L. J., Bastuck, S., Dmpelfeld, B., Edelmann, A., Heurtier, M. A., et al. (2006). Proteome survey reveals modularity of the yeast cell machinery. *Nature* **440**, 631–636.

Gerber, A. P., Herschlag, D., and Brown, P. O. (2004). Extensive association of functionally and cytotopically related mRNAs with Puf family RNA-binding proteins in yeast. *PLoS Biol.* **2**, E79.

Goodman, M., Czelusniak, J., William Moore, G., Romero-Herrera, A. E., and Matsuda, G. (1979). Fitting the gene lineage into its species lineage, a parsimony strategy illustrated by Cladorams constructed from globin sequences. *Syst. Zool.* **28**(2), 132–163.

Goodstadt, L., and Ponting, C. (2006). Phylogenetic reconstruction of orthology, paralogy, and conserved synteny for dog and human. *PLoS Comput. Biol.* **2**(9), e133.

Gu, Z., Rifkin, S. A., White, K. P., and Li, W. H. (2004). Duplicate genes increase gene expression diversity within and between species. *Nat. Genet.* **36**, 577–579.

Hahn, M. W., Han, M. V., and Han, S. G. (2007). Gene family evolution across 12 Drosophila genomes. *PLoS Genet.* **3**, e197.

Harbison, C. T., Gordon, D. B., Lee, T. I., Rinaldi, N. J., Macisaac, K. D., Danford, T. W., Hannett, N. M., Tagne, J. B., Reynolds, D. B., Yoo, J., Jennings, E. G., Zeitlinger, J., et al. (2004). Transcriptional regulatory code of a eukaryotic genome. *Nature* **431**, 99–104.

He, X., and Zhang, J. (2005). Rapid subfunctionalization accompanied by prolonged and substantial neo-functionalization in duplicate gene evolution. *Genetics* **169**, 1157–1164.

Hughes, T. R., Marton, M. J., Jones, A. R., Roberts, C. J., Stoughton, R., Armour, C. D., Bennett, H. A., Coffey, E., Dai, H., He, Y. D., Kidd, M. J., King, A. M., et al. (2000). Functional discovery via a compendium of expression profiles. *Cell* **102**, 109–126.

Huisinga, K. L., and Pugh, B. F. (2004). A genome-wide housekeeping role for TFIID and a highly regulated stress-related role for SAGA in *Saccharomyces cerevisiae*. *Mol. Cell* **13**, 573–585.

Jones, D. T., Taylor, W. R., and Thornton, J. M. (1992). The rapid generation of mutation data matrices from protein sequences. *Comput. Appl. Biosci.* **8**, 275–282.

Jothi, R., Zotenko, E., Tasneem, A., and Przytycka, T. M. (2006). COCO-CL: Hierarchical clustering of homology relations based on evolutionary correlations. *Bioinformatics* **22**, 779–788.

Kafri, R., Bar-Even, A., and Pilpel, Y. (2005). Transcription control reprogramming in genetic backup circuits. *Nat. Genet.* **37**, 295–299.

Karp, P. D., Ouzounis, C. A., Moore-Kochlacs, C., Goldovsky, L., Kaipa, P., Ahrn, D., Tsoka, S., Darzentas, N., Kunin, V., and Lpez-Bigas, N. (2005). Expansion of the BioCyc collection of pathway/genome databases to 160 genomes. *Nucleic Acids Res.* **33**, 6083–6089.

Kasuga, T., Mannhaupt, G., and Glass, N. L. (2009). Relationship between phylogenetic distribution and genomic features in Neurospora crassa. *PLoS One* **4**, e5286.

Kellis, M., Patterson, N., Endrizzi, M., Birren, B., and Lander, E. S. (2003). Sequencing and comparison of yeast species to identify genes and regulatory elements. *Nature* **423**, 241–254.

Kellis, M., Birren, B. W., and Lander, E. S. (2004a). Proof and evolutionary analysis of ancient genome duplication in the yeast *Saccharomyces cerevisiae*. *Nature* **428**, 617–624.

Kellis, M., Patterson, N., Birren, B., Berger, B., and Lander, E. S. (2004b). Methods in comparative genomics: Genome correspondence, gene identification and regulatory motif discovery. *J. Comput. Biol.* **11,** 319–355.

Khaldi, N., and Wolfe, K. H. (2008). Elusive origins of the extra genes in Aspergillus oryzae. *PLoS One* **3,** e3036.

Koski, L. B., and Golding, G. B. (2001). The closest BLAST hit is often not the nearest neighbor. *J. Mol. Evol.* **52,** 540–542.

Li, L., Stoeckert, C. J., and Roos, D. S. (2003). OrthoMCL: Identification of ortholog groups for eukaryotic genomes. *Genome Res.* **13,** 2178–2189.

Li, H., Coghlan, A., Ruan, J., Coin, L. J., Hrich, J. K., Osmotherly, L., Li, R., Liu, T., Zhang, Z., Bolund, L., Wong, G. K., Zheng, W., *et al.* (2006). TreeFam: A curated database of phylogenetic trees of animal gene families. *Nucleic Acids Res.* **34,** D572–D580.

Li, Q. R., Carvunis, A. R., Yu, H., Han, J. D., Zhong, Q., Simonis, N., Tam, S., Hao, T., Klitgord, N. J., Dupuy, D., Mou, D., Wapinski, I., *et al.* (2008). Revisiting the *Saccharomyces cerevisiae* predicted ORFeome. *Genome Res.* **18,** 1294–1303.

Lynch, M., and Katju, V. (2004). The altered evolutionary trajectories of gene duplicates. *Trends Genet.* **20,** 544–549.

Makova, K. D., and Li, W. H. (2003). Divergence in the spatial pattern of gene expression between human duplicate genes. *Genome Res.* **13,** 1638–1645.

Marcotte, E. M., Pellegrini, M., Ng, H. L., Rice, D. W., Yeates, T. O., and Eisenberg, D. (1999). Detecting protein function and protein-protein interactions from genome sequences. *Science* **285,** 751–753.

Mewes, H. W., Frishman, D., Gldener, U., Mannhaupt, G., Mayer, K., Mokrejs, M., Morgenstern, B., Mnsterktter, M., Rudd, S., and Weil, B. (2002). MIPS: A database for genomes and protein sequences. *Nucleic Acids Res.* **30,** 31–34.

Musso, G., Zhang, Z., and Emili, A. (2007). Retention of protein complex membership by ancient duplicated gene products in budding yeast. *Trends Genet.* **23,** 266–269.

Newman, J. R., Ghaemmaghami, S., Ihmels, J., Breslow, D. K., Noble, M., DeRisi, J. L., and Weissman, J. S. (2006). Single-cell proteomic analysis of *S. cerevisiae* reveals the architecture of biological noise. *Nature* **441,** 840–846.

Ogata, H., Goto, S., Sato, K., Fujibuchi, W., Bono, H., and Kanehisa, M. (1999). KEGG: Kyoto encyclopedia of genes and genomes. *Nucleic Acids Res.* **27,** 29–34.

Ohno, S. (1970). *Evolution by Gene Duplication.* George Allen and Unwin, London.

Papp, B., Pl, C., and Hurst, L. D. (2003). Dosage sensitivity and the evolution of gene families in yeast. *Nature* **424,** 194–197.

Pearson, W. R., and Lipman, D. J. (1988). Improved tools for biological sequence comparison. *Proc. Natl. Acad. Sci. USA* **85**(8), 2444–2448.

Rambaut, A., and Grassly, N. C. (1997). Seq-Gen: An application for the Monte Carlo simulation of DNA sequence evolution along phylogenetic trees. *Comput. Appl. Biosci.* **13**(3), 235–238.

Reguly, T., Breitkreutz, A., Boucher, L., Breitkreutz, B. J., Hon, G. C., Myers, C. L., Parsons, A., Friesen, H., Oughtred, R., Tong, A., Stark, C., Ho, Y., *et al.* (2006). Comprehensive curation and analysis of global interaction networks in *Saccharomyces cerevisiae. J. Biol.* **5,** 11.

Remm, M., Storm, C. E., and Sonnhammer, E. L. (2001). Automatic clustering of orthologs and in-paralogs from pairwise species comparisons. *J. Mol. Biol.* **314,** 1041–1052.

Saitou, N., and Nei, M. (1987). The neighbor-joining method: A new method for reconstructing phylogenetic trees. *Mol. Biol. Evol.* **4,** 406–425.

Scannell, D. R., Byrne, K. P., Gordon, J. L., Wong, S., and Wolfe, K. H. (2006). Multiple rounds of speciation associated with reciprocal gene loss in polyploid yeasts. *Nature* **440,** 341–345.

Segal, E., Friedman, N., Koller, D., and Regev, A. (2004). A module map showing conditional activity of expression modules in cancer. *Nat. Genet.* **36**, 1090–1098.

Sopko, R., Huang, D., Preston, N., Chua, G., Papp, B., Kafadar, K., Snyder, M., Oliver, S. G., Cyert, M., Hughes, T. R., Boone, C., and Andrews, B. (2006). Mapping pathways and phenotypes by systematic gene overexpression. *Mol. Cell* **21**, 319–330.

Tatusov, R. L., Koonin, E. V., and Lipman, D. J. (1997). A genomic perspective on protein families. *Science* **278**, 631–637.

Tirosh, I., Weinberger, A., Carmi, M., and Barkai, N. (2006). A genetic signature of interspecies variations in gene expression. *Nat. Genet.* **38**, 830–834.

Wall, D. P., Fraser, H. B., and Hirsh, A. E. (2003). Detecting putative orthologs. *Bioinformatics* **19**, 1710–1711.

Wapinski, I., Pfeffer, A., Friedman, N., and Regev, A. (2007a). Natural history and evolutionary principles of gene duplication in fungi. *Nature* **449**, 54–61.

Wapinski, I., Pfeffer, A., Friedman, N., and Regev, A. (2007b). Automatic genome-wide reconstruction of phylogenetic gene trees. *Bioinformatics* **23**, i549–i558.

Zmasek, C. M., and Eddy, S. R. (2001). A simple algorithm to infer gene duplication and speciation events on a gene tree. *Bioinformatics* **17**, 821–828.

CHAPTER NINETEEN

EXPERIMENTAL EVOLUTION IN YEAST: A PRACTICAL GUIDE

Maitreya J. Dunham

Contents

Department of Genome Sciences, University of Washington, Seattle, Washington, USA

Methods in Enzymology, Volume 470
ISSN 0076-6879, DOI: 10.1016/S0076-6879(10)70019-7

Abstract

Experimental evolution refers to a broad range of studies in which selection pressures are applied to populations. In some applications, particular traits are desired, while in others the subject of study is the mechanisms of evolution or the different modes of behavior between systems. This chapter will explore the range of studies falling under the experimental evolution umbrella, and their relative merits for different types of applications. Practical aspects of experimental evolution will also be discussed, including commercial suppliers, analysis methods, and best laboratory practices.

1. INTRODUCTION

Experimental evolution is a generic term for laboratory selection experiments beyond those requiring simple one-step mutagenesis but perhaps more restricted in scale than the longer term pressures associated with domestication or geological timescale evolution. As our ability to analyze whole genome sequences improves via microarray and sequencing-based methods, we can expect more problems to become accessible through experimental selection approaches.

In this chapter, I will cover the different types of selections typically performed under the guise of experimental evolution, citing a limited selection of example cases, and then move to the practical considerations involved in undertaking a subset of these. The many scientific contributions of experimental evolution in viral, microbial, and animal systems will not be covered in this chapter. However, the reader will find many excellent reviews that cover these systems (Adams, 2004; Buckling *et al.*, 2009; Burke and Rose, 2009; Elena and Lenski, 2003; Garland and Kelly, 2006; Philippe *et al.*, 2007; Zeyl, 2006).

2. EXPERIMENT RATIONALE

Rationales for experimental evolution approaches in yeast are as numerous as the practitioners. Many research groups see the promise in explicitly testing many of the tenets of modern evolutionary theory. Experiments with sex and ploidy are exemplars of this approach, and have been reviewed thoroughly elsewhere (Zeyl, 2004). Other evolutionary questions subjected to experimental testing include the role of mutation rates (e.g., Thompson *et al.*, 2006), mechanisms of assortative mating (Leu and Murray, 2006), cooperation (Shou *et al.*, 2007), and clonal interference (Kao and Sherlock, 2008).

In its simplest guise, such as selection experiments on drug resistance, laboratory populations can mimic the types of long-term adaptation that occur in chronic infections or cancer progression. Because the selection pressures can be carefully controlled under laboratory conditions, however, mutations can be more carefully assigned to a variable than by examining clinical samples. Acquisition of fluconazole resistance is a fine example from the Candida literature (e.g., Cowen *et al.*, 2000). These experiments serve as both a model for the development of drug resistance, and for the unraveling of the molecular mechanisms underlying resistance to particular drugs.

Other types of experiments essentially extend this concept of the mutant or suppressor screen. With longer term, less severe selection pressures than with viability-based selection schemes, more subtle mutations, including combinations of such mutations, may be recovered. Although the yeast genome deletion collection provides an interesting set of mutants for the assay of phenotypes, the strains represent only null alleles. To make progress in further dissecting genetic pathways, the field may benefit from a return to the rare but interesting alleles generated by spontaneous mutation, particularly for essential genes.

This style of experimentation can also shed light on larger questions of systems biology, such as evolution of gene expression, the relative merits of regulatory versus structural mutation, whether mutations affecting control points in a network are more wide-acting, and the mutability of different gene targets. Metabolic selection pressures have been particularly useful for studies along these lines (e.g., Ferea *et al.*, 1999; Francis and Hansche, 1972; Gresham *et al.*, 2008; Hansche *et al.*, 1978).

Interesting questions about genome structure and organization can also be answered using the results of experimental evolutions. For example, point mutations, transposon insertions, and copy number variants have all been recovered from selection experiments (e.g., Blanc and Adams, 2003; Brown *et al.*, 1998; Dunham *et al.*, 2002; Gresham *et al.*, 2008). The types of effects generated by these different classes of genetic alterations, and the relative accessibility of different gene targets to each type of lesion, are still unexplored but accessible by these techniques. For example, copy number changes may be the most effective route by which to change gene expression level, but the ability of a gene to change copy number may in turn be determined by the proximity to repetitive DNA segments that facilitate copy number change, and further complicated by the pleiotropic effects of additional neighboring genes on an amplicon. Point mutation, on the other hand, represents a more surgical approach to perturbing gene function, but single point mutations may rarely provide large expression changes.

Finally, experimental evolution can provide a facile technique to optimize or even create desired traits, for example in bioproducts, food, and beverage production (reviewed in Verstrepen *et al.*, 2006). Industrial yeast geneticists have long used this strategy successfully, often with strains of unknown genotype. Although recombinant DNA techniques in food-related industries such as wine and beer production are becoming more widespread (reviewed in Schuller and Casal, 2005), there is consumer reluctance to use such products. In such cases selection is frequently the only acceptable tool for improving extant yeast strains.

For more industrial processes, introduction of recombinant DNA into a strain is less of a concern, but there is to date no clear recipe for the rational design of metabolic networks. Selection without prior knowledge of the mechanism can instead improve the performance of yeast strains involved in processes such as production of fuel ethanol and biotechnology products. Transfer of exogenous synthesis pathways into yeast, for example, could be followed by selection for more efficient integration into the yeast network, or higher product production.

Applications to synthetic biology (one such approach is reviewed in Saito and Inoue, 2007) and synthetic ecology (e.g., Shou *et al.*, 2007) will also put experimental evolution techniques in the forefront. Allowing the processes of mutation and selection to tune synthetic constructs may be the most efficient way both to create such circuits and to better understand what is required to achieve optimal performance. Clever selection schemes will no doubt be necessary to push these systems in the right direction.

3. EXPERIMENTAL EVOLUTION APPROACHES

Laboratory evolution experiments fall into two broad categories: serial batch transfer and continuous culture. The most suitable approach depends on technical, practical, and scientific considerations, covered in the following sections.

3.1. Serial dilution

Serial dilution generally refers to selection preformed in the standard growth regimes typically used in the lab: flasks, test tubes, solid media, or 96-well plates. Cultures are usually allowed to grow through a normal growth curve, with daily transfer of a small volume of the expanded culture into fresh medium. Serial dilution has many advantages: the materials necessary are typically already present in the lab and require no special engineering.

Conditions can be adjusted as the experiment progresses (e.g., drug concentrations increased as drug resistance improves). Selection pressures of a number of types can be accommodated. The easiest selections to understand are improvements to growth when maximal performance is attenuated either by exogenous or genetic means. In these cases, full growth curves may not be desired, as improved performance with respect to nutrient exhaustion or stationary phase may be separate outcomes unrelated to the main selection applied by the experimenter.

Nutrient exhaustion is a popular selection scheme for batch transfer experiments, brought to prominence by experiments in bacteria by Lenski and colleagues (reviewed in Elena and Lenski, 2003) and adapted for yeast by Zeyl (2005). Here, one nutrient is lowered to the point that it uniquely runs out first and limits the saturated biomass of the culture. The relative amount of time cells spend in each phase of growth may change over the course of one of these experiments, particularly as lag phase shortens and maximal growth rate improves.

Plate-based selection allows even more control over the transfer step, with visual identification of colonies. Either obviously larger or otherwise morphologically desired (e.g., Kuthan et al., 2003) candidates can be serially inoculated to fresh plates, or, on the other end of the spectrum, as little selection as possible can be imposed by selecting random colonies. The latter approach has been used to generate mutation accumulation lines (Zeyl and DeVisser, 2001).

3.2. Chemostats

Chemostats have long been another favored platform for experimental evolution (reviewed in Dykhuizen and Hartl, 1983), and were, in fact, invented for this application (Monod, 1950; Novick and Szilard, 1950a,b). A chemostat is a growth vessel into which fresh medium is delivered at a constant rate and cells and spent medium overflow at that same rate. Thus, the culture is forced to divide to keep up with the dilution, and the system exists in a steady state where inputs match outputs. The chemostat is attractive due to the enormous amount of control that is possible: growth rate, cell density, and selection pressure are all independently set. Because of these advantages, chemostats are also being used as tools for studying aspects of cell biology such as ammonium toxicity (Hess et al., 2006), growth rate control (e.g., Brauer et al., 2008), and comparative gene expression for mutants that would otherwise be difficult to compare due to profound growth rate differences (e.g., Hayes et al., 2002; Torres et al., 2007).

Unlike serial transfer, chemostats require more specialized equipment, which can range from rather inexpensive (< USD$10,000) systems assembled from available parts to elaborate custom fermenters costing upward of USD$100,000. A table of suppliers and plans is provided (Table 19.1).

Table 19.1 Fermenter suppliers

Maker	Web site or citation
Commercial	
ATR	http://www.atrbiotech.com/
Infors HT	http://www.infors-ht.com/
New Brunswick	http://www.nbsc.com/
Dasgip	http://www.dasgip.com/
Applikon	http://www.applikon-bio.com/
Sartorius	http://www.sartorius-stedim.com
Homemade or custom-made	
Tom Gibson and Ted Cox	Plans available in Gibson 1970 thesis, "The Fitness of an *E. coli* Mutator Gene," available through interlibrary loan from Princeton University
Reeves Glass (custom-made glassware)	http://www.reevesglass.com/
Gavin Sherlock	Plans available upon inquiry
G. Finkenbeiner (custom-made glassware)	http://finkenbeiner.com/

In choosing equipment, several experiment-specific questions are important to consider, including volume/population size, number of parallel experiments required, space constraints, and measurement and control needs. The simplest chemostat experiment requires a media-feed, volume-metering device, and growth vessel with overflow. pH control, dissolved oxygen monitoring, real-time data feeds, and other features, may be required for more complex experiments. Large volume (> 1 l) fermenters, such as those available from New Brunswick, Applikon, and ATR, offer in-line probes for such measurements. Small vessels can be made from modified laboratory glassware or with the assistance of a glass blower. Glass-blown designs may include a glass frit for aeration and a water jacket for temperature control in addition to sampling and media flow ports.

3.3. Turbidostats

Another continuous culture system, the turbidostat, first introduced by Bryson and Szybalski (1952), combines some properties of serial dilution and chemostats. Instead of adding new medium at a constant rate, in a turbidostat, cell density is held constant. This is achieved by a feedback loop allowing adjustment of the nutrient addition rate in response to changes in density, usually measured via light transmittance. Few commercial options appear to exist currently, but turbidostats can be built using modern microprocessor controlled peristaltic pumps. Designs using simple light-measurement devices can be found in textbooks (e.g., Norris and Ribbons, 1970), and variations using LED and photodiode components would be straightforward extensions.

The turbidostat provides selection on maximal growth rate while simultaneously maintaining other conditions constant. The media composition defines the selection pressure as in other systems. Very little has been published on yeast grown in turbidostats, although that is likely to change given the benefits that this system provides.

3.4. More specialized systems

Other continuous culture systems have also been invented to control cell growth in various ways, via feedback at the level of pH or dissolved oxygen (generally known as auxostats, or, depending on implementation, accelerostats, see Kasemets *et al.*, 2003), dielectric permittivity (the permittistat, Mark *et al.*, 1991), or carbon dioxide (e.g., Lane *et al.*, 1999). Undoubtedly, many other variations are possible.

3.5. Miniaturization

Both serial dilution and chemostat culture can be greatly miniaturized. For example, microfluidic chemostats have been reported by Groisman *et al.* (2005). This can be a huge advantage when large numbers of replicates or single-cell resolution are required. Volume reduction can greatly affect the population size, though, which can in turn change how evolution proceeds (see the following section for further discussion). For now, microscale chemostats might best be used as a phenotyping tool to better characterize clones isolated from larger chemostat experiments.

4. Experimental Design

There are a number of design considerations in planning any experimental evolution project. Several of the most important ones are covered in the following sections.

4.1. Growth conditions

The growth rate and selection pressures at which experimental evolutions are performed should be given careful thought. Selection upon maximal growth rate is preferred for many suppressor-type experiments, and may best be accomplished in serial dilution or turbidostat approaches because chemostat cultures are difficult to operate near this value. To maintain a constant selection pressure, dilution should occur before the onset of growth limitation. In the event that entire growth curves are allowed each day (e.g., Zeyl, 2005), subpopulations with adaptations relevant to the different parts of the growth curve can be isolated. For example, strains with quicker resumption in lag, with faster maximal growth, or with additional ability to divide in stationary phase may all coexist in the culture. In such cases, the nutrient that runs out first is frequently termed limiting, but may be only one of several selection pressures.

In chemostat cultures, the selection pressure is mainly defined by the limiting nutrient. Limiting nutrient should always be explicitly tested in the chemostat conditions but can be prototyped in batch cultures by measuring the saturation density of cultures grown with varying amounts of the limiting nutrient. Confirmatory experiments should always be done in the exact conditions under which the real experiments will be performed. Complex selection pressures may not be perfectly modeled by batch experiments; for example, ammonium toxicity is only apparent under limiting potassium in the chemostat (Hess *et al.*, 2006). Micronutrients are another common culprit as hidden limitations (e.g., de Kock *et al.*, 2000). A comprehensive test of limiting nutrient would include demonstrating that density varies linearly with the nutrient of interest and not at all with other additives. The exact profile of limiting concentrations for various nutrients is strain-dependent and should be tested explicitly when working in different backgrounds.

The concentration of limiting nutrient is another key parameter since it determines cell density. Population size can greatly affect evolutionary parameters such as mutation supply, importance of drift, and time required for advantageous mutations to rise to detectable frequency (see further discussion in the following section). Also, density may affect gene expression to some degree. Given recent findings on quorum sensing in yeast cultures, more work remains to be done to understand density-dependent effects. There are also practical considerations such as sample volume required for accurate measurements: dry weight yield, for example, may require more material than an expression microarray using an amplification procedure. For continuous cultures in particular, sampling too much volume at once from the fermentor vessel can perturb the system, disrupting the steady state. When possible, passive sampling from the outflow is preferred, though this is not always possible, especially for time-sensitive applications such as expression measurements.

4.2. Population size

Beyond practical constraints, population size is a critical parameter for experimental evolution. In large populations, a modest adaptive mutation will take a long time to reach reasonable frequency in the population, and clonal interference could be generating competitor clones simultaneously. In a small population, adaptation may be limited by the supply of beneficial mutations, and thus dominated by the highest frequency class. (See Desai *et al.*, 2007 for one treatment of these issues.) Per base mutation rate is on the order of 10^{-9} or 10^{-10} per site per generation, while per gene mutation rate to a null allele is closer to 10^{-6}. Given these estimates, populations of different sizes will sample vastly different subsets of the mutation landscape.

In serial dilution, two population size parameters must be determined: the saturation density and the bottleneck size. Severe bottlenecks may eliminate the vast majority of small-to-intermediate fitness variants that have not had time to reach appreciable frequency in a single day growth curve. In practice, the bottleneck population size is typically on the order of 10^6–10^7 cells.

Chemostat populations may be much larger, 10^{10} cells or more. Even in chemostat cultures, the initial phases may be dominated by variation generated as the culture grows from a single cell to the final population size. In this regime, the stochastic factor of when a mutation occurs can affect the allele frequency. For example, mutations of $\sim 10\%$ fitness advantage that become detectable starting around 100 generations of growth were hypothesized to fit this model (Gresham *et al.*, 2008).

4.3. Experiment duration

The number of generations or length of time to allow a culture to evolve is both a practical and theoretical matter. In some cases, a particular desired outcome may be reached. In others, the end of an experiment may be governed by unfortunate circumstances such as contamination, user error, or infrastructure breakdown. Steps can be taken to prevent some of these events, such as careful attention to sterile practices. Backup power and aeration systems can also be implemented if necessary. Clumping of the culture is another less catastrophic, though perhaps less avoidable, endpoint (see below). Practical considerations concerning the number of manipulations and the necessity for lab worker attention past work hours may also limit experiment length.

Barring errors, however, experiments can run for days to decades. Whether experiments really require such long timescales is purely a scientific question. Early events in the chemostat may determine to a large extent what direction the population takes. Subsequent events may in fact be modifying mutations that optimize early events as opposed to "primary" events that may be more interesting. For example, where genome rearrangement is operative, amplicon size may shrink to contain relatively more causative genes and fewer copy number sensitive genes, or second-site suppressors of these sensitivities may arise.

Restarting interesting evolutions is also always an option, though there are likely added complications from loss of population complexity, plus added selection constraints on freeze tolerance and outgrowth from the stock. Evolved clones could alternatively be used as new founders.

5. PRACTICAL CONSIDERATIONS

5.1. Strains and markers

The strain used for experimental evolutions should be considered very carefully. Auxotrophies should be avoided for any metabolic selection, since selection would be strong for harnessing supplements as nutrient sources. Also, the network biology may be rather different in metabolically blocked strains. Marker genes can also cause fitness differences. Baganz et al. (1997) explicitly tested the fitness consequences of drug and nutritional markers in chemostat competition and found variation by selection pressure and marker used. In general, drug resistance markers were neutral while nutritional markers frequently caused fitness costs. These results are likely to depend strongly on the particulars of the growth regime and should be explicitly tested in novel environments.

Strain genotype beyond engineered genetic markers should also be considered. The classic S288C strain background, though a workhorse for decades of yeast genetics and biochemistry, has a number of probably lab-selected traits, including Ty insertions in *HAP1* and *CTR3*, increased petite frequency, and abnormal nitrogen source preferences. Almost all the lab strain alternatives (e.g., W303, sigma 1278b, and CEN.PK) share a large proportion of their genomes with this strain, though the exact alleles carried vary. These other backgrounds may also carry additional mutations, such as an adenylate cyclase (*CYR1*) mutation in CEN.PK. All lab strains are likely to contain some signatures of their domestication.

In addition, new mutations are generated spontaneously during lab cultivation and strain construction, and may result in hidden problems. One cautionary example can be found in a series of glucose-limited chemostat evolutions (Ferea et al., 1999). During creation of the prototrophic ancestor strain, a loss of function mutation occurred in the gene *AEP3*, which stabilizes the RNA coding for subunits of ATP synthetase in the mitochondria. Not surprisingly, this mutation was detrimental in glucose-limited cultures, and reversion of this mutation was later found in evolved strains from two independent experiments (Brauer et al., 2006; Dunham et al., 2002; Gresham et al., 2006).

Although such problems cannot be completely avoided, they can be mitigated by pairing compatibility of a particular strain with a particular selection pressure. For example, as long as a particular mutation is not limiting, it may not be the target of beneficial mutation. van Dijken *et al.* (2000) undertook a comparison between strain backgrounds and found variation for all parameters tested, leaving no single strain with the "best" array of desired characteristics.

Flocculant growth is another strain feature that can present problems. In many experimental evolution regimes biofilm formation and clumping provide an advantage unrelated to the selection pressure of interest. For example, groups of cells may sink to the bottom of a fermentor, or biofilms may form on any surfaces, allowing subpopulations to avoid being diluted out of the culture. Besides complicating measurements of cell density, these subpopulations can contribute a constant supply of minority genotypes and interfere with population genetic measurements. Also, since many evolution experiments are designed around evaluating particular selection pressures, generic responses to the growth apparatus can be a confounding result.

For serial dilution-type experiments, some of this effect can be eliminated by transferring to a new vessel at each dilution rather than pouring off excess culture and adding new medium to the original vessel. In chemostats, transferring vessels is a riskier process, but can still be accomplished with care. Using strains with genotypes that limit their flocculation potential is another approach. Most lab strains carry at least one such mutation, and engineered FLO gene deletions (e.g., *flo8*) will not revert. Operation at high cell densities may further aggravate this phenotype, though much of this data is anecdotal. In practice, severe flocculation typically ends an experiment. Experiments can be prolonged by briefly sonicating culture samples before analysis.

Strains resulting from experimental evolution may also have a number of characters limiting their further use. Selection upon purely mitotic growth may relax selection on the rest of the yeast life cycle. Aneuploids, for example, may have trouble sporulating or segregate lethality resulting from the heterozygous deletion of essential genes. The mating pathway may also be abrogated as a means of conserving cellular resources (Lang *et al.*, 2009). Since most samples are archived through cryopreservation, freeze tolerance may be another hidden variable affecting later analysis. In addition, some cultures evolved in poor nutrient conditions may actually show growth deficiencies on rich media, though the reverse may also be true for the purposes of recovery from frozen stocks.

5.2. Media

Media requirements will depend on the desired selection pressure, but must be consistent no matter what the application. For this reason, rich media made from coarse or technical grade ingredients is not recommended due to batch variation. High-quality chemicals and water are required, especially for nutrient-limited cultures, since trace contamination may provide a nontrivial amount of the nominal limiting nutrient. When possible, direct measurement of the limiting nutrient in media samples is recommended. Very sensitive spectrophotometric, enzyme-based assays are available commercially for many carbon sources. Phosphate can also be reliably measured by colorimetric assay. Other chemical analysis techniques such as inductively coupled plasma and mass spectrometry can more generally measure the elemental or metabolite profile of a sample, though sensitivity should generally be tested explicitly.

Preparation of media for experimental evolution requires more care than most microbiology experiments. Even small measurement errors can have profound effects on culture density. All materials used for media preparation should be rinsed thoroughly to prevent cross contamination. Variation in volume levels due to evaporation may be introduced by unanticipated differences in autoclave pressure, temperature, or timing and these may dramatically affect the outcome of the experiment. Filter sterilization of media into autoclaved carboys is one way of mitigating this effect. Some media components may be light or temperature labile. Also, in large volumes, viscous additives such as glucose may settle to create a gradient in the media vessel. Extra effort may be required to ensure that all such additives are thoroughly dissolved.

The growth apparatus itself may also leech chemicals into the medium. Metal fittings are one example supported by anecdotal evidence. Low-reactivity plastics and glass are typically a better choice for experiments that would be sensitive to such fluctuations. Plastics may pose their own problems if paired with incompatible solvents. Drug solutions requiring such reagents require particular care.

5.3. Growth rate

Growth rate is explicitly set by the experimenter for chemostat cultures, but the allowable range is dependent on media, temperature, and strain background, and growth rate differences introduce differences in many parameters. One important example is the ratio of respiration to fermentation in glucose-limited cultures grown at different rates. At growth rates below a strain-specific critical growth rate parameter, respiration dominates, but above this threshold, fermentation predominates. Evolutions performed very close to this boundary may shift thresholds as a mechanism of increasing efficiency. In chemostat and batch conditions, growth rate correlates strongly with a large gene expression pattern that overlaps that of the environmental stress response (Brauer et al., 2008). Incorrect dilution settings can thus easily lead to spurious gene expression differences. A working rule of thumb is to keep settings within 10% of the target growth rate.

5.4. Good sterile practices

With some experiments running for years, contamination is a threat to experiment integrity. Contamination can be introduced during any break in continuity, most commonly during changing media supplies, or sampling from inside the vessel. Sampling should be done passively from the overflow if possible, though contaminants may also grow in exposed tubing. Periodic changes of this tubing may be required, particularly if drug-resistant contaminants interfere with detection of low-frequency variants from the main culture. This design also helps to prevent retrograde colonization of the main culture with contaminants from the effluent. Positive pressure provided by vigorous aeration is also recommended to limit contamination opportunities.

When changing media carboys, leakage should be avoided as much as possible. Droplets of media left around connectors provide a rich growth opportunity for microbes, and a risk of transfer to the inside of the tube during the next carboy transfer. Self-closing connectors are one preventative, though not entirely fail safe. Ethanol or bleach can remove most material, though again, such treatment is not fail safe.

Visual inspection of culture under the microscope or of colonies can detect high-frequency contaminants with morphological differences, such as bacteria or filamentous fungi. Experimental evolution frequently leads to morphological changes, so this is only appropriate for obviously different species. Also, some contaminants may not grow on solid medium. Checking that strain markers are constant over a time course can provide experimental assurance that other strains or species have not invaded. Strains with drug markers are more amenable to this test, but PCR-based markers can also be developed to differentiate between common strains. Obviously, if the contaminant is of the same strain background, problems will be more difficult to detect.

Another type of contamination is from the growth chamber into the media supply. Aerosolized yeast droplets can be pushed up the tubing if air pressure is forced through. Also, variants with improved flocculation capacity can lodge in crevasses at junction points. Clear tubing is recommended so that such colonization can be detected and problem spots eliminated. In extreme cases, the media input port may need to be heated to kill any back-contaminants. Wrappable heated tape is one common approach.

5.5. Good strain hygiene

Because mutations that arise during experimental evolution are generally not cloned using a functional assay, and because multiple mutations are typically present after sufficiently long timescales, strains should undergo limited passaging before introduction into the growth vessel to limit mutation accumulation. Even minor handling of strains can introduce lesions (see *AEP3* example above). Preservation of a time zero sample of the population for comparison is important to eliminate these possibilities when performing *post hoc* analysis. Exact records of strain stock of origin for each population are also recommended, since ambiguities introduce uncertainly later in analysis.

5.6. Record-keeping

Because experimental evolutions may run for long periods of time, and be reanalyzed by many people within and between labs, good record-keeping practices are essential. An example system uses index cards for recording daily data, plus a digital copy of these records for analysis and archiving. A master database in Filemaker or some other software package can assist in keeping track of many experiments and their related data files, which should be backed up at regular intervals. Parameters such as strain background, media formulations, growth conditions, and other important details should be recorded. To identify potential problems with media composition, addition of new media supplies should be tracked.

Freezer stocks must be maintained in a very ordered way, particularly when lab personnel turn over. Systematic naming conventions are essential for long-term continuity. Freezer maintenance is also crucial to the long-term viability of evolved cultures. If possible, duplicates of glycerol stocks may be stored off site for backup purposes. Complex population samples are impossible to perfectly duplicate, so preplanning is required for this approach.

Sharing populations with other labs is another problem when unique samples are involved. Dense lawns or patches, or large numbers of isolated colonies, can be scraped from plates and frozen in glycerol culture to attempt to maintain population frequencies.

6. ANALYSIS TECHNIQUES

6.1. Sampling regimen

Obviously the details of what to measure day-to-day will depend heavily on what experiment is being performed. Parameters that may be recorded include cell density as surveyed by Klett colorimeter, spectrophotometer, or cell count; viable cell count on rich- or low-nutrient plates; notes on cell morphology, colony morphology, and flocculation status; and even changes in aroma. Frozen stocks may be collected daily. Small samples for processing into RNA and DNA may also be collected at intervals. If these samples are collected on a filter, a filtrate sample is generated simultaneously that can be assayed for residual nutrient or metabolite levels.

6.2. Population genetics

Population-scale expression and CGH measurements generated from these samples can be useful for interrogating the frequencies of copy number changes, though with the caveat that frequency changes and extra copy number changes in a subpopulation may look identical. Allele frequency measurements can also be made via quantitative PCR (Kao and Sherlock, 2008) or quantitative sequencing (Gresham et al., 2008). Mutations of a variety of types can be measured via microarray (reviewed in Gresham et al., 2008), though next generation sequencing is sure to contribute in the near future. With strong enough phenotypes, mutations can also be linkage-mapped using classical approaches or by bulk segregant analysis (Brauer et al., 2006; Segre et al., 2006).

Phenotype characterization methods can also vary. By definition, only phenotypes present in the selective conditions are relevant to the experiment at hand. However, particularly in the chemostat, it may be impossible to survey large numbers of samples for fitness or growth parameters. Bulk competition experiments provide one solution (e.g., Gresham et al., 2008). If individual genotypes can be marked, or detected directly by sequencing, their frequency over time can be used to calculate their fitness. Assays on plates or nonselective media can also be used as a screen to narrow down the search space that needs to be covered in a more tedious growth state. However, plate phenotypes do not always behave as expected in the milieu of the population.

6.3. Fitness

Fitness is generally the most relevant phenotype in an experimental evolution, and can conveniently be assayed by direct competition experiments. Fitness measurements in particular should be optimally performed not just in the conditions under which the strain evolved, but even in the exact population context since fitness may be highly dependent on the competitors (Paquin and Adams, 1983a,b). This condition is almost impossible to recreate. In practice, fitness is generally assayed by direct competition with the ancestral strain or with other evolved clones. One or both strains may be marked to facilitate frequency measurements, or the relative frequency can be sampled by following mutations via quantitative sequencing or some other means.

Strain tagging with drug resistance markers is the most common way of performing mixed fitness assays (e.g., Gresham et al., 2008; Paquin and Adams, 1983a,b). Fluorescence markers have also been used successfully (e.g., Kao and Sherlock, 2008; Thompson et al., 2006), and are attractive because of both their ease of use and improved accuracy. While only hundreds to thousands of colonies can be easily assayed for drug resistance, orders of magnitude more cells can be assayed by FACS. For both methods, accuracy improves as sampling density increases. One disadvantage of fluorescent markers is that expression of these proteins may impose a selective cost, which must be assayed in controls and subtracted out from all further measurements. Whether this cost is constant across conditions and strain backgrounds must be tested in each situation.

7. EXAMPLE PROTOCOL

Working from these general recommendations, this section describes an example glucose-limited chemostat evolution experiment. Related detailed protocols with photographs and recipes are available at http://dunham.gs. washington.edu/.

7.1. Medium formulation

Chemostat glucose-limited synthetic minimal media contains (per liter) 0.1 g calcium chloride, 0.1 g sodium chloride, 0.5 g magnesium sulfate, 1 g potassium phosphate monobasic, 5 g ammonium sulfate, 500 μg boric acid, 40 μg copper sulfate, 100 μg potassium iodide, 200 μg ferric chloride, 400 μg manganese sulfate, 200 μg sodium molybdate, 400 μg zinc sulfate, 1 μg biotin, 200 μg calcium pantothenate, 1 μg folic acid, 1 mg inositol, 200 μg niacin, 100 μg p-aminobenzoic acid, 200 μg pyridoxine, 100 μg riboflavin, 200 μg thiamine, and 0.08% glucose.

Medium is prepared in 10 l quantities, mixed thoroughly, and filter sterilized into an autoclaved glass carboy. Carboy has an outlet port at bottom, leading to a small piece of tubing with a luer lock connector at the end. All entry and exit ports are covered with foil before autoclaving. Outflow tubing is sealed with a metal clamp before filling. Carboy is placed on a shelf above chemostat area.

7.2. Chemostat preparation

A glass-blown chemostat apparatus (Reeves Glass) is outfitted with input tubing including an in-line segment of peristaltic pump tubing and an appropriate luer fitting for connection to the media supply. The overflow port is connected to tubing leading through a bored cork to an effluent collection bottle which drains into a larger reservoir. All free tubing ends are foil-wrapped, and the entire assembly is placed in a tray and autoclaved.

7.3. Chemostat assembly

Autoclaved elements are assembled using sterile technique. Airflow is provided by an aquarium pump, via a water diffuser for humidification, and sterilized by two in-line autoclaved filters. Temperature control at 30 °C is provided by a circulating waterbath attachment to the water jacket of the chemostat vessel.

Once assembled, carboy outflow is connected to chemostat inflow and the tubing is unclamped to allow chemostat to fill by gravity flow. When chemostat begins to overflow (working volume ∼200 ml), flow is clamped off via loading of the pump tubing into the peristaltic pump head. The pump has been precalibrated to supply a dilution rate of 0.17 chemostat volumes per hour.

7.4. Inoculation

The strain FY4, a prototroph haploid of the S288C background, is streaked for single colonies from a glycerol stock to a YPD plate and grown at 30 °C for 2 days. A single colony is inoculated into 2.5 ml of glucose-limited chemostat medium and grown overnight at 30 °C. One milliliter of the culture is used to inoculate the chemostat and 1 ml is frozen in glycerol stock as the time 0 sample. Chemostat is grown to saturation overnight and then the pump is started.

7.5. Daily sampling

Daily, the effluent volume is measured and any necessary modification is made to the pump settings. The cork is removed from the effluent bottle and placed in a small tube to passively collect 10 ml culture. One milliliter is frozen in glycerol stock. The sample is measured for A_{600} and Klett density, then briefly sonicated for cell counting in a hemacytometer. Diluted samples are plated to YPD and minimal media agar plates. Notes are also recorded about cell morphology, colony morphology, and chemostat vessel observations (e.g., wall growth, aroma). Carboy supply is monitored and new carboys of sterile media are supplied as necessary. A 10 ml sample from each retired bottle is collected and frozen for analysis of media composition.

In the first 2–3 days after inoculation, the culture has not yet reached steady state. Steady state is usually defined operationally as occurring once all measurements have been equal for 2 days in a row. This should occur at approximately generation 10–15.

7.6. Weekly sampling

Once or twice a week, 25 ml is passively collected from the effluent port, pelleted, and resuspended in glycerol stock for later DNA preparation. Ten milliliters of culture is removed from the main vessel via a port in the chemostat lid using a sterile pipette. This sample is collected on a filter and snap-frozen for later RNA preparation. Filtrate from the RNA collection is frozen for later metabolite and residual nutrient analysis.

Sampling continues until an error occurs, or the culture develops a clumping phenotype, as defined by clumps that cannot be broken up by light sonication and are observed in most microscope fields.

7.7. Analysis

Data from the experiment are recorded in the database and the raw data index cards are filed in the master system.

Collected samples are processed for DNA and RNA and assayed via microarrays and/or Solexa sequencing for genotype and phenotype differences. Collected media and filtrate samples are analyzed for limiting nutrient concentrations to ensure constant nutrient source and to detect increased consumption.

Representative clones are isolated from population glycerol stocks and assayed for growth phenotypes and mutations. Clones are regrown in new chemostats just until reaching steady state, and then harvested for expression analysis versus the ancestral strain grown in the same conditions.

Once mutations are discovered, their gross frequency can be retrospectively assayed by performing PCR directly on small samples of cells obtained from the population glycerol stocks. The mixed PCR product is sequenced to determine the relative amount of each allele. Time 0 samples are included to ensure mutation was not already present in the inoculum.

Clones may also be subjected to competition versus a marked wild-type ancestor strain. In this case, both strains are grown to steady state in individual chemostats and then mixed. In the null exception, 50% of each strain should be present, but this often gives insufficient survey time for evolved strains with 5–50% fitness increases. When the strain is known or suspected to carry such an advantage, more useful data can be collected using a starting frequency of 5–10%.

8. Conclusions

Experimental testing of evolutionary questions is almost as old as evolutionary theory itself. The use of these techniques in yeast is yielding exciting results in evolutionary genomics, systems biology, and theory, complementing the excellent comparative genomic and ecological tools also maturing in yeast concurrently. The use of evolution as a tool will also help tune synthetic systems and generate new and useful strains and constructs. This guide is only to be taken as touching on the highlights of this exciting field, and, hopefully, lowering the bar to entry for new researchers.

ACKNOWLEDGMENTS

The author thanks Frank Rosenzweig for being the source of much of her chemostat training and techniques. She also thanks Matt Brauer, David Botstein, and other members of the Dunham and Botstein labs for helping develop these methods. Finally, thanks to the other experimental evolution laboratories who contributed equipment recommendations found in Table 19.1. Please forgive omissions in citations.

REFERENCES

Adams, J. (2004). Microbial evolution in laboratory environments. *Res. Microbiol.* **155,** 311–318.

Baganz, F., Hayes, A., Marren, D., Gardner, D. C., and Oliver, S. G. (1997). Suitability of replacement markers for functional analysis studies in *Saccharomyces cerevisiae*. *Yeast* **13,** 1563–1573.

Blanc, V. M., and Adams, J. (2003). Evolution in *Saccharomyces cerevisiae*: Identification of mutations increasing fitness in laboratory populations. *Genetics* **165,** 975–983.

Brauer, M. J., Christianson, C. M., Pai, D. A., and Dunham, M. J. (2006). Mapping novel traits by array-assisted bulk segregant analysis in *Saccharomyces cerevisiae*. *Genetics* **173,** 1813–1816.

Brauer, M. J., Huttenhower, C., Airoldi, E. M., Rosenstein, R., Matese, J. C., Gresham, D., Boer, V. M., Troyanskaya, O. G., and Botstein, D. (2008). Coordination of growth rate, cell cycle, stress response, and metabolic activity in yeast. *Mol. Biol. Cell* **19,** 352–367.

Brown, C. J., Todd, K. M., and Rosenzweig, R. F. (1998). Multiple duplications of yeast hexose transport genes in response to selection in a glucose-limited environment. *Mol. Biol. Evol.* **15,** 931–942.

Bryson, V., and Szybalski, W. (1952). Microbial selection. *Science* **116,** 45–51.

Buckling, A., Craig Maclean, R., Brockhurst, M. A., and Colegrave, N. (2009). The Beagle in a bottle. *Nature* **457,** 824–829.

Burke, M. K., and Rose, M. R. (2009). Experimental evolution with *Drosophila*. *Am. J. Physiol. Regul. Integr. Comp. Physiol.* **296,** R1847–R1854.

Cowen, L. E., Sanglard, D., Calabrese, D., Sirjusingh, C., Anderson, J. B., and Kohn, L. M. (2000). Evolution of drug resistance in experimental populations of *Candida albicans*. *J. Bacteriol.* **182,** 1515–1522.

de Kock, S. H., du Preez, J. C., and Kilian, S. G. (2000). The effect of vitamins and amino acids on glucose uptake in aerobic chemostat cultures of three *Saccharomyces cerevisiae* strains. *Syst. Appl. Microbiol.* **23,** 41–46.

Desai, M. M., Fisher, D. S., and Murray, A. W. (2007). The speed of evolution and maintenance of variation in asexual populations. *Curr. Biol.* **17,** 385–394.

Dunham, M. J., Badrane, H., Ferea, T., Adams, J., Brown, P. O., Rosenzweig, F., and Botstein, D. (2002). Characteristic genome rearrangements in experimental evolution of *Saccharomyces cerevisiae*. *Proc. Natl. Acad. Sci. USA* **99,** 16144–16149.

Dykhuizen, D. E., and Hartl, D. L. (1983). Selection in chemostats. *Microbiol. Rev.* **47,** 150–168.

Elena, S. F., and Lenski, R. E. (2003). Evolution experiments with microorganisms: The dynamics and genetic bases of adaptation. *Nat. Rev. Genet.* **4,** 457–469.

Ferea, T. L., Botstein, D., Brown, P. O., and Rosenzweig, R. F. (1999). Systematic changes in gene expression patterns following adaptive evolution in yeast. *Proc. Natl. Acad. Sci. USA* **96,** 9721–9726.

Francis, J. C., and Hansche, P. E. (1972). Directed evolution of metabolic pathways in microbial populations. I. Modification of the acid phosphatase pH optimum in S. cerevisiae. Genetics **70**, 59–73.

Garland, T., and Kelly, S. A. (2006). Phenotypic plasticity and experimental evolution. J. Exp. Biol. **209**, 2344–2361.

Gresham, D., Ruderfer, D. M., Pratt, S. C., Schacherer, J., Dunham, M. J., Botstein, D., and Kruglyak, L. (2006). Genome-wide detection of polymorphisms at nucleotide resolution with a single DNA microarray. Science **311**, 1932–1936.

Gresham, D., Dunham, M. J., and Botstein, D. (2008). Comparing whole genomes using DNA microarrays. Nat. Rev. Genet. **9**, 291–302.

Gresham, D., Desai, M. M., Tucker, C. M., Jenq, H. T., Pai, D. A., Ward, A., DeSevo, C. G., Botstein, D., and Dunham, M. J. (2008). The repertoire and dynamics of evolutionary adaptations to controlled nutrient-limited environments in yeast. PLoS Genet. **4**, e1000303.

Groisman, A., Lobo, C., Cho, H., Campbell, J. K., Dufour, Y. S., Stevens, A. M., and Levchenko, A. (2005). A microfluidic chemostat for experiments with bacterial and yeast cells. Nat. Methods **2**, 685–689.

Hansche, P. E., Beres, V., and Lange, P. (1978). Gene duplication in Saccharomyces cerevisiae. Genetics **88**, 673–687.

Hayes, A., Zhang, N., Wu, J. Y., Butler, P. R., Hauser, N. C., Hoheisel, J. D., Lim, F. L., Sharrocks, A. D., and Oliver, S. G. (2002). Hybridization array technology coupled with chemostat culture: Tools to interrogate gene expression in Saccharomyces cerevisiae. Methods **26**, 281–290.

Hess, D. C., Lu, W., Rabinowitz, J. D., and Botstein, D. (2006). Ammonium toxicity and potassium limitation in yeast. PLoS Biol. **4**, e351.

Kao, K. C., and Sherlock, G. (2008). Molecular characterization of clonal interference during adaptive evolution in asexual populations of Saccharomyces cerevisiae. Nat. Genet. **40**, 1499–1504.

Kasemets, K., Drews, M., Nisamedtinov, I., Adamberg, K., and Paalme, T. (2003). Modification of A-stat for the characterization of microorganisms. J. Microbiol. Methods **55**, 187–200.

Kuthan, M., Devaux, F., Janderova, B., Slaninova, I., Jacq, C., and Palkova, Z. (2003). Domestication of wild Saccharomyces cerevisiae is accompanied by changes in gene expression and colony morphology. Mol. Microbiol. **47**, 745–754.

Lane, P. G., Oliver, S. G., and Butler, P. R. (1999). Analysis of a continuous-culture technique for the selection of mutants tolerant to extreme. Biotechnol. Bioeng. **65**, 397–406.

Lang, G. I., Murray, A. W., and Botstein, D. (2009). The cost of gene expression underlies a fitness trade-off in yeast. Proc. Natl. Acad. Sci. USA **106**, 5755–5760.

Leu, J. Y., and Murray, A. W. (2006). Experimental evolution of mating discrimination in budding yeast. Curr. Biol. **16**, 280–286.

Mark, G. H., Davey, C. L., and Kell, D. B. (1991). The permittistat: A novel type of turbidostat. J. Gen. Microbiol. **137**, 735.

Monod, J. (1950). La Technique de culture continue. Theorie et applications. Ann. Inst. Pasteur **79**, 390–410.

Norris, J. R., and Ribbons, D. W. (1970). Turbidostats. Methods Microbiol. **2**, 349–376.

Novick, A., and Szilard, L. (1950a). Description of the chemostat. Science 715–716.

Novick, A., and Szilard, L. (1950b). Experiments with the chemostat on spontaneous mutations of bacteria. Proc. Natl. Acad. Sci. USA **36**, 708–719.

Paquin, C., and Adams, J. (1983a). Relative fitness can decrease in evolving asexual populations of S. cerevisiae. Nature **306**, 368–371.

Paquin, C., and Adams, J. (1983b). Frequency of fixation of adaptive mutations is higher in evolving diploid than haploid yeast populations. *Nature* **302,** 495–500.

Philippe, N., Crozat, E., Lenski, R. E., and Schneider, D. (2007). Evolution of global regulatory networks during a long-term experiment with *Escherichia coli*. *Bioessays* **29,** 846–860.

Saito, H., and Inoue, T. (2007). RNA and RNP as new molecular parts in synthetic biology. *J. Biotechnol.* **132,** 1–7.

Schuller, D., and Casal, M. (2005). The use of genetically modified *Saccharomyces cerevisiae* strains in the wine industry. *Appl. Microbiol. Biotechnol.* **68,** 292–304.

Segre, A. V., Murray, A. W., and Leu, J. Y. (2006). High-resolution mutation mapping reveals parallel experimental evolution in yeast. *PLoS Biol.* **4,** e256.

Shou, W., Ram, S., and Vilar, J. M. G. (2007). Synthetic cooperation in engineered yeast populations. *Proc. Natl. Acad. Sci. USA* **104,** 1877–1882.

Thompson, D. A., Desai, M. M., and Murray, A. W. (2006). Ploidy controls the success of mutators and nature of mutations during budding yeast evolution. *Curr. Biol.* **16,** 1581–1590.

Torres, E. M., Sokolsky, T., Tucker, C. M., Chan, L. Y., Boselli, M., Dunham, M. J., and Amon, A. (2007). Effects of aneuploidy on cellular physiology and cell division in haploid yeast. *Science* **317,** 916–924.

van Dijken, J. P., Bauer, J., Brambilla, L., Duboc, P., Francois, J. M., Gancedo, C., Giuseppin, M. L., Heijnen, J. J., Hoare, M., Lange, H. C., Madden, E. A., Niederberger, P., *et al.* (2000). An interlaboratory comparison of physiological and genetic properties of four *Saccharomyces cerevisiae* strains. *Enzyme Microb. Technol.* **26,** 706–714.

Verstrepen, K. J., Chambers, P. J., and Pretorius, I. S. (2006). The development of superior yeast strains for the food and beverage industries: Challenges, opportunities and potential benefits. *In* "Yeasts in Food and Beverages," (A. Querol and G. H. Fleet, eds.), Springer-Verlag, Berlin Heidelberg.

Zeyl, C. (2004). Experimental studies on ploidy evolution in yeast. *FEMS Microbiol. Lett.* **233,** 187–192.

Zeyl, C. (2005). The number of mutations selected during adaptation in a laboratory population of *Saccharomyces cerevisiae*. *Genetics* **169,** 1825–1831.

Zeyl, C. (2006). Experimental evolution with yeast. *FEMS Yeast Res.* **6,** 685–691.

Zeyl, C., and DeVisser, J. A. (2001). Estimates of the rate and distribution of fitness effects of spontaneous mutation in *Saccharomyces cerevisiae*. *Genetics* **157,** 53–61.

CHAPTER TWENTY

ENHANCING STRESS RESISTANCE AND PRODUCTION PHENOTYPES THROUGH TRANSCRIPTOME ENGINEERING

Felix H. Lam,*,† Franz S. Hartner,* Gerald R. Fink,† *and* Gregory Stephanopoulos*

Contents

Abstract

As *Saccharomyces cerevisiae* is engineered further as a microbial factory for industrially relevant but potentially cytotoxic molecules such as ethanol, issues of cell viability arise that threaten to place a biological limit on output capacity and/or the use of less refined production conditions. Evidence suggests that one naturally evolved mode of survival in deleterious environments involves the

* Department of Chemical Engineering, Massachusetts Institute of Technology, Cambridge, Massachusetts, USA
† Whitehead Institute for Biomedical Research, Cambridge Center, Cambridge, Massachusetts, USA

Methods in Enzymology, Volume 470
ISSN 0076-6879, DOI: 10.1016/S0076-6879(10)70020-3

complex, multigenic interplay between disparate stress response and homeostasis mechanisms. Rational engineering of such resistance would require a systems-level understanding of cellular behavior that is, in general, not yet available. To circumvent this limitation, we have developed a phenotype discovery approach termed global transcription machinery engineering (gTME) that allows for the generation and selection of nonphysiological traits. We alter gene expression on a genome-wide scale by selecting for dominant mutations in a randomly mutagenized general transcription factor. The gene encoding the mutated transcription factor resides on a plasmid in a strain carrying the unaltered chromosomal allele. Thus, although the dominant mutations may destroy the essential function of the plasmid-borne variant, alteration of the transcriptome with minimal perturbation to normal cellular processes is possible via the presence of the native genomic allele. Achieving a phenotype of interest involves the construction and diversity evaluation of yeast libraries harboring random sequence variants of a chosen transcription factor and the subsequent selection and validation of mutant strains. We describe the rationale and procedures associated with each step in the context of generating strains possessing enhanced ethanol tolerance.

1. INTRODUCTION

In addition to its central position in the histories of baking and brewing, *Saccharomyces cerevisiae* has played a prominent role in biotechnology as a production platform for numerous therapeutics and industrial small molecules. Typically, strain engineering has involved manipulating biochemical fluxes in a rational manner for the maximal accumulation of either commercially relevant native metabolites or products synthesized by heterologously expressed genes (Nevoigt, 2008). However, as the budding yeast is modified further to expand its output capacity or product spectrum (Hadiji-Abbes *et al.*, 2009; Rao *et al.*, 2008; Waks and Silver, 2009), the likelihood of encountering conditions that prove stressful to cytotoxic increases greatly. The ability to withstand such deleterious circumstances is likely to be a complex phenotype impinging on multiple stress response mechanisms simultaneously. For example, physiological tolerance to ethanol—a natural product of glycolytic fermentation—is thought to be imparted through the coordinated action of proteins involved in functions as diverse as vacuolar maintenance, phosphatidyl inositol metabolism, and histone acetylation (van Voorst *et al.*, 2006). A rational approach to engineering tolerance would thus likely necessitate the development of a comprehensive genome-scale model of homeostasis, and the directed, concurrent manipulation of multiple pathways. Unfortunately, a quantitative wiring diagram of such breadth does not currently exist.

In the absence of such a detailed, systems-level understanding of cellular control, we have developed a mutagenesis and selection approach termed global transcription machinery engineering (gTME) that generates non-physiological traits through the alteration of transcription on a genome-wide scale. In contrast to traditional protein engineering where individual enzyme or binding partner variants are screened for modified activity *in vitro*, gTME involves the introduction of a randomly mutagenized, plasmid-borne copy of a general (i.e., gene nonspecific) transcription factor into yeast and the selection for cellular phenotypes created *in vivo*. Novel phenotypes are thus produced through the differential regulation of transcriptional programs elicited by the presence of the mutated allele. Indeed, a reconstruction of evolutionary divergence across a large group of yeast species has shown that much natural phenotypic innovation emerges not from changes to protein function, but rather from gene duplication and subsequent changes to transcriptional regulation (Wapinski *et al.*, 2007). Likewise, gTME involves two copies of a target transcriptional regulator: a mutagenized variant on a plasmid and the intact, native allele on the chromosome. Thus, the isolation of desired phenotypes arising from altered function of even essential genes is possible as the presence of the chromosomal allele continues to provide that function.

Although published later, perturbation of the transcriptome via modification of a global transcription factor was originally conceived in *Escherichia coli* as an engineering strategy to unlock multigenic phenotypes (Alper and Stephanopoulos, 2007). Variants of the *rpoD* gene were shown to confer enhanced lycopene production and combined ethanol and sodium dodecyl sulfate (SDS) tolerance in bacteria. Subsequently, the technique was adapted for use in *S. cerevisiae* and demonstrated to be successful in imparting elevated ethanol resistance and rates of fermentation (Alper *et al.*, 2006). This chapter thus details the various steps associated with implementing gTME in the budding yeast using the goal of enhanced ethanol tolerance as a template. In brief, the overall approach can be organized into five major steps: (1) identification of a relevant transcription factor, (2) selection, design, and construction of expression constructs for mutant plasmid libraries, (3) creation of yeast libraries and evaluation of phenotypic diversity, (4) selection for phenotypes of interest, and (5) validation of mutated transcription factor alleles.

2. TRANSCRIPTION FACTOR SELECTION

The transcription machinery component(s) chosen for mutagenesis will perhaps have the largest influence on the likelihood of obtaining the desired phenotype. Depending on the biology thought to underlie that phenotype, the choice may be obvious or speculation. Since it is a decision that may ultimately call for some informed guesswork, we furnish several examples here in an attempt to assist in the selection process.

Knowledge of the processes that could potentially contribute to the desired phenotype should provide the best hint of the transcriptional complexity and relevant players involved. On the simpler end of the spectrum, for example, a few cellular functions can be reduced to gene expression programs dominated by a handful of regulators. If the desired phenotype is enhanced galactose metabolism, the *GAL3*, *GAL4* activators, and/or the *GAL80* repressor would be clear targets (Lohr *et al.*, 1995); for enhanced phosphate utilization, obvious choices would be the *PHO2* and/or *PHO4* activators (Ogawa *et al.*, 2000). In dissecting more complex cellular traits, even what appears to be sophisticated behavior can often derive from the combinatorial interplay of a small number of factors. For example, if the goal is greater tolerance to osmotic stress—a response that integrates multiple signals and modulates the expression of hundreds of genes—the *SKO1*, *HOT1*, and/or *MSN2/4* transcription factors would serve as appropriate candidates (Capaldi *et al.*, 2008). In our efforts to improve ethanol tolerance, proteins with widely disparate functions were known to be involved yet no transcriptional pathways presented themselves as obvious targets. Thus, to maximize the number of transcripts perturbed, *SPT15* (the *S. cerevisiae* homolog of the TATA-box binding protein) was chosen for its putative ability to control as much as 90% of the yeast transcriptome by its association with the general transcription complex TFIID (Huisinga and Pugh, 2004; Kim and Iyer, 2004).

In general, the more nonspecific and globally acting the transcription component, the higher the number of genes that will be affected and, in principle, the wider the sector of adaptation space that will be explored. However, readers are encouraged to evaluate transcription factors more specific to their pathways of interest to increase the probability of sampling advantageous phenotypes. In addition to surveying the literature and the numerous internet-based resources available (e.g., Saccharomyces Genome Database, MIPS Comprehensive Yeast Genome Database), candidates may also be identified through consideration of mechanisms underlying similar traits in other organisms. Nevertheless, if no clear targets exist, one of the many subunits of TFIID may serve as a suitable starting point.

3. Plasmid Library Construction

3.1. Promoter selection

The manner in which the plasmid-borne, mutated allele is expressed is also an important variable having a large influence on the likelihood of success. A natural choice is to drive the mutagenized transcription factor by its endogenous promoter. This strategy may be appropriate for applications where native regulation of the gene is desired; indeed, numerous enhancements were generated in *E. coli* from variants of *rpoD* coupled to its upstream intergenic sequence (Alper and Stephanopoulos, 2007). However, preserving the physiological control of a transcription factor creates the possibility of its repression in the potentially nonphysiological conditions of the selection process. Thus, the use of a constitutive promoter is recommended, at least until further constraints or data in a specific application call for an alternative.

Although constitutive expression may avert possible downregulation of the plasmid-borne allele, promoter strength is a variable that must be optimized. Low transcription levels could result in unaltered phenotypes, while transcription rates that are too high risk out-competition of the native transcription factor and possible cytotoxicity. The ratio of mutated to wild-type proteins that would result in the maximal number of beneficial traits is believed to be influenced by many factors including the expression level of the chromosomal allele, redundantly acting paralogs, and the ploidy and genetic background of the host strain. Consequently, an optimal expression level must be determined empirically. A reasonable approach is to titrate promoter strength on a plasmid copy of the wild-type transcription factor and use growth as a quantifiable proxy for the phenotype of interest (a proxy particularly applicable to tolerance phenotypes). This is supported by preliminary data showing that survivability trends (in stress and nonstress conditions alike) are somewhat correlated between strains carrying a gTME allele and strains carrying an additional native allele (Alper *et al.*, unpublished).

In principle, any series of yeast promoters with varying constitutive strengths may be used. For example, the p413–p426 family of shuttle expression vectors offers a $\sim$1000-fold range of transcription rates through low (*CEN/ARS*) and high (2 μ) copy number plasmids containing either a *CYC1*, *TEF1* (referred to originally as *TEF2*, from before the availability of the complete *S. cerevisiae* genome sequence), *ADH1*, or *TDH3* (often referred to as *GPD*) promoter (Funk *et al.*, 2002; Mumberg *et al.*, 1995). It is worth noting that the "constitutive" *ADH1* promoter is, in fact, repressible on nonglucose medium (Denis *et al.*, 1983), and that select versions of these plasmids carrying the kanamycin dominant drug resistant gene have been developed for use with prototrophic yeast strains (Dualsystems Biotech AG, Switzerland). Alternatively, one could opt for a promoter series derived from a common sequence background offering finer grain increments in transcription rate. For example, a collection of 11 *TEF1* promoter mutants based on p416TEF was developed that features expression levels between $\sim$8% and 120% of the wild type (Alper *et al.*, 2005; Nevoigt *et al.*, 2006). Furthermore, the TEF1 promoters offer constitutive behavior is preserved on both glucose and nonfermentable carbon sources.

After selecting a promoter family, insert the coding sequence of the wild-type transcription factor in front of the yeast promoter in each member of the series using established DNA cloning and manipulation techniques (Ausubel, 2001). Typically, these yeast expression vectors have multiple cloning sites flanked by the promoter and a common transcriptional terminator (e.g., *CYC1*). Thus, one generally does not need to be concerned with any endogenous untranslated regulatory sequences and can directly clone in the open reading frame of interest (e.g., using polymerase chain reaction (PCR) and primers incorporating the appropriate restriction sites). To minimize the possibility of DNA sequence variation across various *S. cerevisiae* strain backgrounds, genomic DNA prepared from the host strain should be used as the source of the coding sequence. Additionally, as a functional control for expression, it is recommended that an identical plasmid series containing a transcriptional reporter as the insert (e.g., yeast-enhanced green fluorescent protein) be obtained or constructed at the same time so that promoter strengths can be verified independently by fluorescence or another method as appropriate (Cormack *et al.*, 1997).

After preparation and sequence validation of the plasmid family, assess the *in vivo* impact of the additional transcription factor copy as follows. Using standard DNA uptake and strain selection techniques (Ausubel, 2001), transform the plasmids into yeast to create a family of strains where each member contains a unique construct from the expression series. From individual colonies, create starter cultures by inoculating each transformant into laboratory standard liquid selection medium and growing cells to saturation. Prepare liquid medium containing the condition of interest; if such a condition is toxic, determine a dilution that is amenable to growth beforehand. Since it is the relative performance between the differentially promoted transcription factors that is important, any dilution allowing growth will suffice. Using established cell density measurement methods, inoculate the liquid medium of interest to a common density from the saturated cultures and follow the increase in cell number until stationary phase (Ausubel, 2001). Determine the member of the expression series that confers the highest performance by using either the mid-logarithmic phase growth rate or final stationary phase cell density as a metric. It is also important to evaluate the expression series driving the transcriptional reporter under the condition of interest, especially if it is very different from common laboratory conditions where these promoters are usually assayed.

For goals of enhanced production rather than tolerance, studies in *E. coli* have shown that biomass formation is often correlated with product yields, particularly for metabolites or recombinant molecules synthesized through fermentative pathways (Babaeipour *et al.*, 2008). Thus, growth may also serve as an appropriate proxy for production. However, conflicting data also exist suggesting that output may *not* necessarily be coupled to growth under certain cultivation conditions (Lutke-Eversloh and Stephanopoulos, 2008). Given these uncertainties, it is best to directly read out product formation whenever possible. For example, visual or absorbance-based assays may be easily developed that allow for fast and specific detection of the product of interest (Santos and Stephanopoulos, 2008; Yu *et al.*, 2008).

3.2. Random mutagenesis by PCR

Packaged solutions for generating pools of sequence variants by error-prone PCR are commercially available and simplify much of the handling and reagent optimization previously needed to achieve specific mutation frequencies. Enzymes mixtures have been further optimized for less mutational bias, and one need only supply primers and a plasmid template to obtain a large library of sequence variants ($\sim 10^5$–10^6) within several days' time. In the past, our laboratory has extensively used the GeneMorph II (Stratagene, La Jolla, CA) random mutagenesis kit where PCR fragments are manually gel-purified, restriction-digested, and ligated into the backbone of the plasmid template. Recently, we have begun using the newer GeneMorph II EZClone (Stratagene) domain mutagenesis kit that, in brief, bypasses the restriction and ligation steps by using the mutagenized PCR products themselves as "megaprimers" to anneal to and completely replicate the original plasmid in a second iteration of PCR. As a tradeoff, the EZClone kit (unlike the original) is formally restricted to the mutagenesis of sequences up to 3.5 kb; fortunately, the majority of *S. cerevisiae* coding sequences are well under this limit. The procedure outlined here thus closely follows and refers to the official GeneMorph II EZClone protocol; however, we provide commentary and a few changes customized to our application. Furthermore, because equivalent steps can, in fact, be performed using material from the original GeneMorph II kit plus reagents supplemented separately, these substitutions will be documented, as well.

The plasmid selected from the expression series containing the optimally promoted wild-type transcription factor serves as the template for error-prone PCR. The template DNA must be methylated to allow for digestion by *Dpn*I in a subsequent step; therefore, the use of dam$^+$ *E. coli* strains for preparation of the plasmid template is required. Design PCR primers to flank the entire coding sequence of the transcription factor. However, if there is evidence suggesting that particularly advantageous phenotypes may arise from mutations enriched in particular regions (e.g., DNA-binding domain or protein-binding interface), design primers to amplify those specific segments. Furthermore, because errors are generated with each primer extension, mutations accumulate with each additional cycle of PCR and the final average mutation frequency is directly proportional to the total number of fragment duplications. Thus, one can establish the library mutation frequency simply by changing the total number of PCR cycles or, better yet, by the input amount of template DNA (for more consistent product yields).

Finally, combined experimental and modeling studies have shown that an optimal, protein-specific mutation frequency exists that maximizes sequence diversity while retaining protein function (Drummond *et al.*, 2005). Therefore, it is recommended that one follow the suggested practice of constructing multiple libraries with different average mutation frequencies to maximize the probability of isolating enhanced mutants.

To perform mutagenesis by error-prone PCR, follow the Stratagene instructions and prepare 50 μl reactions corresponding to low (0–4.5 mutations/kb; use $\sim$1 μg initial DNA), medium (4.5–9 mutations/kb; use $\sim$300 ng initial DNA), and high (9–16 mutations/kb; use $\sim$50 ng initial DNA) mutation frequencies. The values listed in parentheses indicate the initial *target* amount of DNA—not the total input amount of plasmid. For example, to produce full-length variants of a 2 kb transcription factor cloned into a plasmid with a total length of 8 kb at the medium mutation frequency, use 1.2 μg of plasmid in the sample reaction. Since it is the nontemplating components of the PCR mixture that ultimately becoming limiting, final product yields will typically all be similar regardless of the different initial target amounts of DNA used to induce the different proportions of substitutions.

After proceeding with the Stratagene-suggested 30 cycle program, it is important both to quantify the yield and inspect homogeneity of the PCR product (also referred to from here on as megaprimers) by gel electrophoresis. Separate the mutagenized fragments from the plasmid template on 1% agarose by running out 5 μl of each reaction along with 50 ng of the included 1.1 kb standard or other appropriate standards for DNA quantification. To verify that the expected mutation frequencies were achieved, it is generally sufficient to visually confirm that the quantity in the product band lies between 50 ng and 1 μg. However, for a more accurate estimate, perform the quantification by densitometry of the product band and a standard curve of known DNA concentrations (of a fragment length similar to the amplicon), and interpolate off the manufacturer's published linear regression of mutation frequency versus duplication (usually located in the instruction appendix).

Once average mutation frequencies have been confirmed, purify the remaining 45 μl of megaprimers by standard gel extraction methods. It is recommended that each sample be divided across at least two lanes to avoid loss through overloading of the gel or saturation of the DNA binding columns. The Stratagene protocol indicates that direct column purification, as an alternative, is permissible for PCR samples that use miniscule amounts of template (i.e., high mutation frequency reactions using <50 ng initial target DNA). Technically, this step is justified for our application since the second (megapriming) PCR will, in fact, be using the same plasmid template. However, it is better practice to remove all contaminating traces of the original plasmid in order to have stricter control over its amount in the second PCR reaction mixture. This issue is particularly relevant for the eventual elimination of methylated DNA by DpnI digestion: because the plasmid template is methylated (i.e., prepared from dam$^+$ E. coli strains), it is important to minimize its amount and the resulting proportion of hemimethylated—but correctly megaprimed—plasmids generated by the second PCR that will also be subject to digestion.

Next, quantify the concentration of the gel-extracted megaprimers by agarose gel electrophoresis. As before, visualize a 1–5 μl aliquot (depending on the initial concentration) alongside 50 ng of the included gel or other appropriate standard, or perform densitometry accompanied by a standard curve of known DNA quantities. As an alternative to gel-based quantification, megaprimer concentrations—now free from plasmid contaminants—may also be assessed by an ultralow volume spectrophotometer (e.g., Nanodrop—Thermo Fisher Scientific, Wilmington, DE).

For the second iteration of PCR ("EZClone Reaction"), follow the Stratagene instructions and prepare as many reactions as necessary (typically, at least five samples per mutation frequency) to exhaust the entire volume of purified megaprimers. Use the original plasmid as template, and proceed with the recommended 25 cycle plasmid amplification program. If the supply of EZClone enzyme and solution mix have been depleted, or the megaprimers were produced using a means other than the GeneMorph II EZClone kit, one can obtain similar results using a substitute high-fidelity and high-processivity enzyme such as the Phusion^{TM} High-Fidelity DNA Polymerase (New England Biolabs, Ipswich, MA). Be sure to make adjustments to the temperature cycling program (e.g., extension times) as necessary.

After completing the plasmid amplification, the product pool will be a collection of nicked plasmids, the majority of which will be unmethylated, doubly/stagger-nicked, and contain the mutagenized fragments. In the minority will be the methylated template and singly nicked, hemimethylated plasmids containing one extended megaprimer. Prepare the reactions for DpnI digestion of the methylated and hemimethylated species by pooling all PCR samples corresponding to the same mutation frequency and cooling to below 37 °C. Add 10–20 U of DpnI restriction enzyme per 50 μl PCR reaction directly to the pooled samples (i.e., no need for a DpnI-specific buffer), mix thoroughly, and incubate at 37 °C for 2 h. Heat inactivate each sample at 80 °C for 20 min, and concentrate and purify the mutated plasmids by standard DNA column purification.

As repair and replication of nicked DNA are carried out in E. coli, introduce the purified, mutated plasmids into the included XL10-Gold ultracompetent cells by following the Stratagene instructions. To maximize the diversity of the resulting library, prepare as many transformations as necessary to consume each stock of plasmid while attempting to use similar DNA concentrations and volumes for all reactions. As an alternative to

XL10-Gold, any ultracompetent (chemically or electro-) cells will serve the same purpose provided that the highest yield recipients are used. Electroporation generally gives the highest efficiencies, but be reminded that high volumes of DNA will alter the overall salt concentration and electrical resistance of the sample, and thus dramatically decrease transformation efficiencies (Ausubel, 2001).

After DNA uptake, add outgrowth medium to each reaction, and again pool all samples corresponding to the same mutation frequency. Estimate the total outgrowth culture volume V_o as best as possible (for determination of library size), and incubate at 37 °C for 1 h with agitation to allow for expression of the resistance gene but minimal cell replication. Plate 50 μl of a 1:10–1:100 dilution (or an appropriate dilution for reliable colony counts) on each of three LB agar plates containing the relevant antibiotic, and allow colonies to form by incubating at 37 °C for at least 16 h.

For the remaining outgrowth culture ($\approx V_o$), inoculate into 500 ml of liquid LB medium containing the appropriate antibiotic and immediately measure the diluted cell density by absorbance ($OD_{600,o}$). Grow at 37 °C with shaking until saturation and measure the final cell density $OD_{600,f}$ to estimate the fold-increase in cell number. This culture expansion will homogeneously amplify all mutants in the library under the assumption that variants of a yeast transcription factor will not elicit significant differential rates of growth in *E. coli*.

3.3. Quantification of total sequence diversity and library maintenance

Count the number of individual bacterial colonies on each of the three agar plates to determine the mean number of successful transformants per 50 μl of diluted outgrowth culture. As these platings have been performed after outgrowth but before significant cell replication, each colony should represent a unique transcription factor variant. Scale this average up to the total outgrowth volume V_o to arrive at the total library size. Based on previous efforts, sequence diversity in the range of 10^6 or above is satisfactory. Furthermore, it is informative to pick at least 10 random colonies from these agar plates for small-scale preparations of plasmid DNA and to sequence the transcription factor inserts to confirm that the desired mutation frequencies were attained.

To prepare frozen bacterial stocks that contain adequate representation of each library, a statistical model based on a Poisson distribution of variants predicts that a minimum threefold excess of clones is needed to ensure 95% coverage (Patrick *et al.*, 2003; Reetz *et al.*, 2008). The amount of saturated culture V_s needed to sample the library once over is simply the total culture volume divided by the fold-increase in cell density:

$$V_s = V_o \frac{V_f/V_o}{OD_{600,f}/OD_{600,o}} = 500\,ml \left(\frac{OD_{600,o}}{OD_{600,f}}\right) \qquad (20.1)$$

For example, an outgrowth culture diluted to $OD_{600,o} = 0.03$ and grown to $OD_{600,f} = 3$ would require $V_s = 5$ ml of the final culture to encapsulate a postamplification equivalent of clones. Incidentally, because V_s is the proportion of V_o that was diluted and subsequently amplified, the changes in cell density are directly counterbalanced and thus V_o cancels in the final calculation.

This calculation also assumes that 100% of the cells in the outgrowth culture take up plasmid and contribute to expansion of the culture; in reality, this is unlikely to be the case and only a smaller fraction is viable. Fortunately, this means that the fold-increase is likely larger than $OD_{600,f}/OD_{600,o}$, and V_s is, in fact, a conservative estimate of the volume needed for $1\times$ library sampling. To achieve the target minimum $3\times$ over-sampling, concentrate the saturated culture by centrifugation for 10 min at $4000 \times g$, 4 °C, and resuspend the cells in a volume such that 1 ml will contain a threefold equivalent of the library (i.e., concentrate the culture by a factor of $3 \times V_s$). From the example above where $V_s = 5$ ml, pellet the 500 ml culture, and resuspend in 33.3 ml of LB containing antibiotic. To several 2 ml cryogenic vials, aliquot 1 ml of the concentrated culture, add sterile glycerol to produce a final concentration of 15%, and freeze immediately at -80 °C. Each frozen stock will thus provide at least $3\times$ coverage of the library, and may be thawed and used as an inoculum for further propagation of the plasmid library in the future.

Centrifuge the remaining volume of culture and perform a large-scale preparation of plasmid DNA. Based on a typical plasmid size of 5–15 kb, a final DNA concentration on the order of 1 $\mu g/\mu l$ is equivalent to $\sim 10^{12}$ plasmids/μl. Thus, for a library of size 10^6, 1 μl of plasmid should provide approximately 10^6 copies of each transcription factor variant. However, the transformation efficiency specific to the protocol and strain of yeast will eventually be the limiting factor, and must therefore be determined (outlined in the next section) before proceeding to the selection phase.

4. ASSESSMENT OF PHENOTYPIC DIVERSITY

Although one can estimate the number of unique clones contained within a plasmid library, its sequence diversity is unlikely to translate into the same number of unique cellular phenotypes in yeast. For example, in addition to synonymous substitutions, there may be combinations of muta-tions that elicit such minor structural changes that the protein is effectively unchanged in function. Conversely, there may be mutations so deleterious that the cell is rendered inviable. Depending on the ultimate phenotype of interest, the selection process may also require a large investment of time and/or reagents. Thus, some strategy of accessing and preevaluating the range of phenotypic consequences contained within the various libraries would be extremely valuable by allowing comparison and vetting of the best candidates to proceed with.

A method and corresponding metric developed specifically to quantify a library's inherent adaptation potential is the phenotypic diversity (Klein-Marcuschamer and Stephanopoulos, 2008). In brief, a cellular property is chosen based on two criteria: (1) it is the culmination of the complex interplay between numerous cellular processes, and (2) it is quantifiable in a relatively high-throughput manner. For example, quantities such as growth rate or cytosolic pH can be easy readouts that integrate a multitude of intracellular signals pertaining to cellular health (Karagiannis and Young, 2001; Klein-Marcuschamer *et al.*, 2009). The property is then measured separately in populations of untransformed host cells and cells transformed with a library, and the phenotypic diversity is calculated by comparing the means and dispersions of the two distributions. The phenotypic diversity is formally a statistical measure of the additional behavioral heterogeneity that has been manifested over the host population. Performing this procedure with libraries of different mutation frequencies can provide an indicator for which collections have been under- or overmutagenized and a method for ranking which libraries demonstrate the greatest phenotypic potential. Indeed, it has been shown in bacteria that this value is highly correlated with the probability of isolating improved strains (Klein-Marcuschamer and Stephanopoulos, 2008). Furthermore, these rankings can be revised or strengthened by redetermining phenotypic diversity values under different conditions, perhaps even those approximating the eventual selection process. Finally, because phenotypic diversity is a normalized score—it is always calculated relative to the untransformed population under a specific condition—it is, in theory, valid to extend the use of this metric to the comparison and rating of libraries constructed from different transcription factors.

4.1. Determination of yeast transformation efficiency

To obtain statistically meaningful numbers of yeast clones for the evaluation of phenotypic diversity (and ultimately, selection), the transformation efficiency specific to the particular host strain of *S. cerevisiae* intended for the selection process must be estimated. Many researchers typically use laboratory standard strains; however, one can also use industrial or feral strains if selection for the plasmid can be arranged. Unfortunately, nondomesticated strains often transform poorly, and even laboratory strains can vary widely in transformation efficiency, particularly if premodified extensively for specific applications.

As a starting point, use a high-efficiency yeast transformation protocol (Ausubel, 2001; Gietz and Schiestl, 2007) and 1 μg of the plasmid containing the wild-type transcription factor. Since transformation efficiencies do not necessarily scale linearly with plasmid DNA concentrations, it is important that this value is determined using DNA quantities similar in range to what will eventually be used for transforming the libraries. Assuming that approximately 2–5 OD_{600} units ($\sim 10^7$–10^8 cells) of mid-logarithmic phase cells are used (per typical protocols), plate 100 μl of 1:100 and 1:500 dilutions in triplicate on agar selection plates. Incubate all plates at 30 °C for 24–48 h until colonies are visible. Count the number of individual clones and determine the mean number of transformants resulting from the fraction of the total reaction plated (e.g., 100 μl from a 1:500 dilution of a 1-ml cell resuspension is 0.02%). Scale this mean to the total size of the transformation reaction to arrive at the number of transformants/μg of plasmid DNA.

This value, specific to the host strain and protocol, is particularly important at the selection stage for ensuring that all transcription factor variants in the library are represented. For example, given an efficiency of 10^5 yeast transformants/μg of plasmid and a library size of 10^6 mutants, it would be wise to use at least 30 μg of plasmid DNA (e.g., distributed across 15 reactions using 2 μg each) to cover 95% of the library with confidence (Reetz et al., 2008).

4.2. Evaluation of mutant libraries

The following section outlines the procedure of assessing phenotypic diversity using growth rate as the per-clone variable, and specifically, the quantification of colony areas from agar plates as a proxy. From the transformation efficiency determined previously, individually transform each library into the host strain of S. cerevisiae using an amount of plasmid DNA sufficient to generate a minimum of $\sim 10^3$–10^4 clones. For the purpose of assessing phenotypic diversity, it is acceptable to dramatically undersample a library because one is interested only in population parameters that can be estimated from smaller, random samples. Include two controls: a sample with untransformed host cells and one transformed with the plasmid containing the wild-type transcription factor. In triplicate, plate fractions of each reaction on solid selection medium appropriate to yield approximately 100–200 colonies per plate. For the untransformed host cell control, plate a dilution to yield roughly the same number of colonies on an equivalent agar plate without selection. Incubate all plates at 30 °C for 24–48 h until individual colonies are visible. It is imperative, however, that colonies are not allowed to overgrow. Otherwise, the variation in colony areas will be minimized and the utility of this assessment will be diminished. If, for unexpected reasons, the number of colonies is too high such that the density precludes accurate area quantification, or too low such that the analysis will lack statistical power, repeat the transformations using a corrected amount of plasmid DNA for improved data.

Photograph all agar plates immediately using a digital imaging system (e.g., Alpha Innotech, San Leandro, CA), or place at 4 °C to prevent significant further growth until ready. Perform computational segmentation on each image to distinguish individual colonies. Numerous image processing options are available including such packages as MetaMorph (Molecular Devices, Sunnyvale, CA) or the Image Processing Toolbox for MATLAB (The Mathworks, Natick, MA).

4.3. Calculation of phenotypic diversity

From each segmented image, extract the area in pixels of each identified object/colony. Numbers deriving from the same transformation or control sample (e.g., those belonging to the same triplicate plating) may be merged into a single data set. Within each set, take the positive differences of the natural logarithms over all i, j pairs of areas:

$$D = \left\{ \left| \ln\left(\frac{A_i}{A_j}\right) \right| \right\} \quad \forall i, j \qquad (20.2)$$

where A_i is the area of colony i and D is a set consisting of computed unitless values termed "phenotypic distances."

If D is far from approximating a normal distribution or its shape is sufficiently jagged due to insufficient data points, bootstrap resampling methods may be employed to generate a better-behaved distribution of population parameters which encapsulates the variability contained in the parent population. For example, subsets of D may be randomly selected with replacement (bootstrap samples) and their means computed. The resulting set d of "average phenotypic distances" will then be normally distributed (by the central limit theorem). Furthermore, the population dispersion contained in D will be manifested in the distribution width of d given a sufficient number of subsets that randomly sample the outer quantiles (Klein-Marcuschamer and Stephanopoulos, 2008). The set d may then used in place of D in all subsequent calculations. However, if D appears sufficiently Gaussian or is symmetrically distributed (which may be the case if another clonal property is measured instead of colony area), it is not necessary to perform bootstrapping as the resulting phenotypic diversity rankings will be similar.

A variety of indices can be used to quantify the dissimilarity/divergence between the variability exhibited by cells transformed with a library over that exhibited by a control population (the control can either be untransformed cells or those transformed with the wild-type transcription factor). The

Bhattacharyya distance (BD) was chosen as the phenotypic diversity metric, which is computed between the two distributions as:

$$BD = \frac{1}{8}(\mu_{lib} - \mu_{cont})^2 \left(\frac{2}{\sigma_{lib}^2 + \sigma_{cont}^2}\right) + \frac{1}{2}\ln\left[\frac{(\sigma_{lib}^2 + \sigma_{cont}^2)/2}{\sqrt{\sigma_{lib}^2 \sigma_{cont}^2}}\right] \quad (20.3)$$

Here, μ_{lib} and μ_{cont} are the means of the phenotypic distances of the library and control populations, respectively, and σ_{lib}^2 and σ_{cont}^2 are the corresponding variances. As an illustration, the analysis above can be implemented easily with just a few lines of code in MATLAB:

```
>> Dlib = pdist(log(lib)); % Calculate all pairwise log
   distances from library population
>> Dlib = bootstrp(length(Dlib), @mean, Dlib); % If
   bootstrapping
>> mulib = mean(Dlib); varlib = var(Dlib); % Compute
   library mean and variance
>> Dcont = pdist(log(cont)); % Calculate all pairwise log
   distances from control population
>> Dcont = bootstrp(length(Dcont), @mean, Dcont); % If
   bootstrapping
>> mucont = mean(Dcont); varcont = var(Dcont); % Compute
   control mean and variances
>> BD = 0.125*(mulib--mucont)^2*2/(varlib+varcont) +
   0.5*log((varlib+varcont)/2/sqrt(varlib*varcont)) %
   Compute Bhattacharyya distance
```

Here, lib and cont are column vectors that contain colony areas derived directly from image segmentation of the library and control platings, respectively. Calculating BDs between the same control population and each of the libraries with different mutation frequency thus provides the quantitative scores needed to rank the libraries.

Equation (20.3) is a simplified, univariate form of the BD. A more general, matrix form also exists that can be used to calculate BD when data is collected for each library and the control under multiple conditions (e.g., synthetic media, synthetic media + 5% ethanol):

$$BD = \frac{1}{8}(\mu_{lib} - \mu_{cont})^T \left(\frac{\Sigma_{lib} + \Sigma_{cont}}{2}\right)^{-1}(\mu_{lib} - \mu_{cont})$$
$$+ \frac{1}{2}\ln\left[\frac{|(\Sigma_{lib} + \Sigma_{cont})/2|}{\sqrt{|\Sigma_{lib}||\Sigma_{cont}|}}\right] \quad (20.4)$$

Here, μ_{lib} and μ_{cont} are vectors that contain the means of the phenotypic distances in condition 1, condition 2, etc. for the library and control populations, respectively. The corresponding covariance matrices Σ_{lib} and Σ_{cont} are derived from matrices of phenotypic distances where column 1 refers to condition 1, column 2 to condition 2, etc. A MATLAB implementation of a multivariate scenario can also be fairly straightforward (illustrated here with two conditions and bootstrapping omitted):

```
>> Dlib = [pdist(log(lib1))'pdist(log(lib2))'];
>> mulib = mean(Dlib); covlib = cov(Dlib);
>> Dcont = [pdist(log(cont1))'pdist(log(cont2))'];
>> mucont = mean(Dcont); covcont = cov(Dcont);
>> BD = 0.125*(mulib--mucont)*inv((covlib+covcont)/2)
   *(mulib--mucont)'+0.5*log(det((covlib+covcont)/2)/
   sqrt(det(covlib)*det(covcont)))
```

Here, lib1 and lib2 are column vectors with the same number of rows containing data derived from platings of a library under two different conditions, and cont1 and cont2 are column vectors with its own matching number of rows containing data from platings of a control under the same two conditions. Of course, *in lieu* of Eq. (20.4), it is also appropriate to compute Eq. (20.3) independently for each condition; it is likely, then, that the scores between conditions will covary. However, if a condition is identified that breaks the correlation in rankings, it may be indicative of a fundamentally different transcriptional program mediated by the mutated transcription factor and warrant further investigation once specific mutants are isolated.

5. SELECTING FOR PHENOTYPES OF INTEREST

Since one generally gets what one selects for in a mutant search, it is important to ponder the selection strategy to consider potential outcomes. In our experience, enhanced tolerance is typically more straightforward to achieve as the selection pressure (i.e., survival) is directly coupled to the resulting phenotype. In contrast, enhanced production, particularly of toxic products, requires the abilities to both withstand and continue output in increasingly harsh conditions and, therefore, may be more difficult to attain. For example, yeast isolated after prolonged exposure to a combination stress such as elevated ethanol and glucose are, by definition, more resistant than the parental strain. However, they may survive by upregulating a variety of membrane remodeling and osmoadaptive mechanisms that allow them to continue metabolizing and growing, or they may enter a quiescent phase that allows them to fortify and lie dormant until more favorable conditions are detected (Gray *et al.*, 2004). The former scenario would likely allow the continued output of fermentative products, such as ethanol, while the latter would not. As such, the following steps describe a selection primarily for gaining improved tolerance (using elevated ethanol and glucose as an example), and discussion is offered on how improved production phenotypes may also be achieved.

5.1. Creation and maintenance of yeast library

Choose the one or two libraries with the highest phenotypic diversity scores, and transform the host yeast strain with an amount of plasmid DNA corresponding to at least $3\times$ library coverage. To recover the maximal number of variants, plate the entire reaction on a series of large-format dishes ($\sim 50 \times 150$ mm (D) $\times 10$ mm (H), or $\sim 15 \times 245$ mm (L) $\times 245$ mm (W) $\times 18$ mm (H)) containing solid selection medium. Unlike typical transformations requiring comfortable margins between colonies, higher densities are acceptable here as the entire population of plasmid-containing cells will ultimately be pooled. Allow growth at 30 °C for a duration sufficient to *minimize* the difference in colony size between slower and faster growing clones (~ 96 h). Incidentally, the rationale for performing plasmid selection on agar plates instead of in (more convenient) liquid medium arises from the phenotypic diversity analysis and the potentially large range of growth rates known to result from the expression of transcription factor variants. Clonal expansion in liquid medium would generate widely different proportions of each mutant and, thus, result in disproportionate representation in the selection process.

Mechanically harvest all colonies off the entire series of plates and pool into liquid selection medium. If necessary, concentrate the cells by centrifugation, and resuspend to a cell density encapsulating the total number of variants by at least threefold. For example, again assuming a library size of $\sim 10^6$ variants, resuspend cells thoroughly to an $OD_{600} \approx 1$ ($\sim 10^7$ cells/ml). Prepare 15% glycerol stocks containing at least 1 OD_{600} unit of cells per cryogenic vial, and freeze immediately at -80 °C (Ausubel, 2001). Cells may be revived by thawing, inoculating into liquid plasmid-selection medium, and growing briefly at 30 °C (4–6 h maximum) to allow for cell recovery but minimal replication. Cultures are then ready for treatment appropriate to the phenotype of interest. These glycerol stocks should *not* be used as inocula for subsequent propagation of the yeast library. Regenerating the full population of yeast variants should be done by retransforming library plasmid DNA into fresh host cells and expanding the clones on solid selection medium.

5.2. Basic selection on liquid versus solid media

To perform a basic selection in liquid medium, prepare a 25–30 ml culture containing the plasmid-selection condition and the condition of interest (e.g., YSC -URA +100 g/l glucose +5% ethanol). Inoculate the yeast library to an

$OD_{600} \approx 0.05$ and incubate for several days to a week at 30 °C. Follow growth by OD_{600} and subculture into fresh selection medium as necessary. To recover surviving mutants, streak for discrete colonies on solid medium containing the plasmid-selection condition but without the condition of interest (e.g., YSC–URA).

As an alternative to liquid culture, selection may also be performed on solid medium if the condition of interest is amenable to a 2% agar mixture. In this scenario, spread the entire yeast library across several plates and depending on the density of colonies that form, restreak or velvet-copy onto fresh selection plates as necessary. Allow growth at 30 °C for a total of several days to a week. If the condition of interest contains a volatile ingredient such as ethanol, placing a liquid drop ($\sim$100 μl) of the pure component faceup in the plate can compensate for evaporation over the course of selection. Although liquid cultures ensure greater homogeneity of conditions, agar plates have the benefit of allowing for observation of individual clonal behavior such as growth rates.

From yeast isolates emerging from the selection process, randomly select 10–20 individual clones for further analysis. An important first step is to quantify the actual improvement in behavior imparted by the mutated alleles over the wild-type transcription factor. For a phenotype like enhanced tolerance, this involves characterizing individual growth rates and/or maximum cell densities in the condition of interest. First, generate two reference strains by individually transforming an empty plasmid and the plasmid containing the wild-type transcription factor into the host yeast strain and selecting for individual transformants. Next, create liquid starter cultures of all strains by inoculating from single colonies and growing cells until saturation. Prepare fresh cultures with medium containing the condition of interest, inoculate to a common density from the starter cultures, and follow the increase in biomass by OD_{600} until stationary phase. Parameters extracted from the growth curves will thus allow ranking of the mutants showing the most improved performance over the wild type. In addition, the data should reveal if the extra copy of the wild-type transcription factor is sufficient to influence the phenotype of the host yeast strain under the condition of interest.

5.3. Alternative selection strategies and postselection screening

Readers are encouraged to consider variations on the basic selection to increase the probability of isolating higher performing mutants. For example, instead of conditions that remain static, one scheme is to gradually increase the strength of the perturbation over the course of selection. Certain response mechanisms may operate by remodeling cellular structures or processes and require longer adaptation periods; thus, cells may be able to withstand smaller, incremental stresses instead of a single large shock. For example, glucose- and ethanol-tolerant yeast strains were isolated through a selection starting at 5% ethanol and 100 g/l glucose and ending at 6% ethanol and 120 g/l glucose through multiple subcultures (Alper *et al.*, 2006).

Particularly for phenotypes involving combined tolerance to multiple, distinct conditions, alternative adaptation sequences—instead of a single selection that presents all stresses simultaneously—may also yield higher performing mutants. In an effort to generate combined tolerance to ethanol and SDS in *E. coli* using variants of *rpoD*, various search trajectories were directly investigated to identify the selection path giving rise to the most doubly resistant strains (Alper and Stephanopoulos, 2007). The following strategies were compared: (a) a doubly tolerant strain was directly selected in ethanol + SDS (reference scenario); (b) an ethanol-tolerant strain was first isolated, its specific allele used as a template for a new round of mutagenesis, and a second selection was done in ethanol + SDS; (c) an SDS-tolerant strain was first isolated, its specific allele used as a template for a new round of mutagenesis, and a second selection was done in ethanol + SDS; and (d) ethanol- and SDS-tolerant strains were isolated individually in parallel, and their respective alleles subsequently isolated and coexpressed. Surprisingly, it was observed that scenario (d) resulted in the greatest improvement, and that coexpression conferred the two full phenotypes independently of one another. To our knowledge, such a parallel adaptation and combination strategy has yet to be attempted in *S. cerevisiae*, and it is formally possible that the pure additivity of traits was a consequence of functional decoupling specific to the two *rpoD* alleles isolated (one full-length, one truncated). Nonetheless, these results from *E. coli* serve as a successful demonstration that variations on a selection strategy can have considerable influence on the resulting behaviors.

While enhanced tolerance to a condition of interest is a direct outcome of the environmental pressure applied during selection, improved production phenotypes typically are not. For example, it is generally believed that yeast cells able to endure prolonged exposure to elevated ethanol do not necessarily experience a concurrent demand to increase ethanol output (as it would contribute little to survivability). There may even be a small pressure to inhibit production if continued output begins to increase ethanol concentrations). Likewise, any hypothetical selection scheme that is able to couple higher production to survival would likely culminate in respiration-enhanced strains that can easily transition to ethanol metabolism—scenarios that would all minimize the accumulation of ethanol.

In general, it is thought that there is neither strong selection for nor against ethanol production during prolonged exposure to elevated concentrations. Assuming that ethanol remains a neutral by-product of glucose metabolism, yields may be loosely associated with growth rates.

Strains endowed with improved resistance may thus harbor a potential for producing more ethanol as higher levels are, presumably, less disruptive to their normal metabolic processes. Therefore, it is suggested that enhanced production phenotypes be screened for among the tolerance-improved isolates. For example, enzymatic assay kits are available for ethanol (R-Biopharm, Darmstadt, Germany) that enable feasible comparison of dozens of candidates. Mutants with improved tolerance can be individually fermented in small cultures containing the condition of interest, and specific ethanol titers from the media easily quantified *in vitro*. If commercial solutions do not exist for the product of interest, it may be possible to develop visual or dye-based approaches that offer both specific detection and reasonable throughput (Santos and Stephanopoulos, 2008; Yu *et al.*, 2008).

6. Validation

Most of the strains emerging from selection will have achieved their unique properties through an altered transcriptome mediated by a mutated transcription factor. However, it is possible that some phenotypes were partially underwritten by mutations in the genome incurred during the course of adaptation, a possibility that increases if the selection was particularly harsh. Additionally, transformation itself can be mutagenic. Dissecting the source of the enhanced behavior requires isolation of the plasmids carried by these strains, and testing for transferability of the phenotype to fresh hosts of the parent genetic background.

To rescue the plasmids containing the mutated transcription factor from the improved strains, perform a basic genomic DNA preparation (e.g., "smash and grab" protocol) and directly transform competent *E. coli* cells (Ausubel, 2001). Alternatively, timesaving commercial kits are available that yield relatively pure plasmid DNA straight from yeast cultures (e.g., Zymoprep™ II—Zymo Research, Orange, CA). Retransform these constructs into fresh cultures of the host yeast strain, and determine from growth rate or product yield assays if the phenotype of interest can be recapitulated. For constructs showing partial or full transferability of phenotype, repurify the plasmids and sequence the inserts to identify the responsible sets of mutations.

In the event that the observed phenotype is different from that of the selected strain, it is likely that chromosomal mutations contributed to the improved behavior either in conjunction with, or independently of, the plasmid-borne transcription factor variant. Fortunately, identifying these potentially obscure changes in the vastness of the host genome is becoming increasingly tractable with current and next-generation DNA profiling technologies. For example, microarray-based comparative genomic hybridization can detect rearrangements (e.g., deletions and amplifications) at the single-gene level (Dunham *et al.*, 2002; Watanabe *et al.*, 2004) while whole genome deep sequencing can resolve alterations at the single-nucleotide level (Bentley, 2006; Liti *et al.*, 2009). By comparing the evolved strain with the parental strain, these approaches promise insight into how complex, nonphysiological cellular behaviors can be synthesized with potentially minimal genetic changes.

7. CONCLUDING REMARKS

Although we have not tested *in vivo* cloning methods in our laboratory extensively, we wish to apprise readers of the possibility of creating gTME yeast libraries directly from mutagenized linear DNA fragments. Given the ease of homologous recombination in *S. cerevisiae*, PCR products containing 50 bp of sequence complementary to the 5' and 3' ends of a cotransformed, linearized vector can effectively be "ligated" *in vivo* by the endogenous gap repair machinery. The time savings can be substantial as the steps involving whole plasmid amplification, library preparation in *E. coli*, and plasmid transformation into yeast are all bypassed. Furthermore, efficiencies of approximately 10^4–10^5 transformants/μg of insert DNA and yeast library sizes of $\sim 10^7$ have been reported (Swers *et al.*, 2004). Although one would likely need to return to PCR mutagenesis to regenerate a yeast library containing transcription factor variants of the same mutation frequency (e.g., in another host background), the benefits of this approach are many and should warrant consideration.

With one or a set of phenotype-enhancing transcription factor variants isolated and characterized, numerous avenues are available postselection for gaining potentially further improvements (Neylon, 2004; Wong *et al.*, 2007). The sampling of options mentioned here are all strategies typical in the practice of directed molecular evolution—strategies designed to extend the adaptation trajectory to superior properties by finer grain sampling of the sequence space in the region of the previous enhancement.

Three example subsequent steps are provided here that can further enhance the desired phenotype. First, gTME can be applied in a directed evolution manner where beneficial transcription factor variants are used as templates in multiple iterations of mutant library creation and strain selection. Indeed, work in *E. coli* has shown that additional mutagenesis and selection cycles result in fine-tuning of the altered transcriptome: the enhanced phenotype is either maintained or improved while the number of genes perturbed actually decreases (Alper and Stephanopoulos, 2007). Second, a collection of phenotype-enhancing variants can indicate sites within the protein enriched for mutations and thus favorable for further mutagenesis. Targeted saturation mutagenesis of key codons can quickly reveal combinations (including nonconservative substitutions) offering significant phenotypic increases that would otherwise have been inaccessible by random point mutagenesis (Miyazaki and Arnold, 1999; Reetz *et al.*, 2008). Third, a set of positive transcription factor variants can be subjected to *in vitro* DNA recombination techniques that allow for the coupling of beneficial mutations and the simultaneous elimination of deleterious or neutral substitutions. In a method such as the staggered extension process (StEP), the coding sequences of both the variants and wild-type serve as a mixed pool of templates in a thermal cycling program modified with very abbreviated primer extension phases. By randomly annealing incompletely extended fragments to different templates in each cycle, a collection of chimeric full-length sequences is generated that, when subjected to selection, can reveal optimized sets of mutations that display significantly improved phenotypes (Zhao, 2004; Zhao *et al.*, 1998).

This overview has described gTME as a mutagenesis and selection technique for generating industrially relevant phenotypes in *S. cerevisiae*. The capacity to elicit enhanced behaviors is increasingly valuable as the budding yeast is engineered further as a production platform for nonnative or ever more toxic molecules.

REFERENCES

Alper, H., and Stephanopoulos, G. (2007). Global transcription machinery engineering: A new approach for improving cellular phenotype. *Metab. Eng.* **9**, 258–267.

Alper, H., Fischer, C., Nevoigt, E., and Stephanopoulos, G. (2005). Tuning genetic control through promoter engineering. *Proc. Natl. Acad. Sci. USA* **102**, 12678–12683.

Alper, H., Moxley, J., Nevoigt, E., Fink, G. R., and Stephanopoulos, G. (2006). Engineering yeast transcription machinery for improved ethanol tolerance and production. *Science* **314**, 1565–1568.

Ausubel, F. M. (2001). *Current Protocols in Molecular Biology*. Wiley, New York.

Babaeipour, V., Shojaosadati, S. A., Khalilzadeh, R., Maghsoudi, N., and Tabandeh, F. (2008). A proposed feeding strategy for the overproduction of recombinant proteins in *Escherichia coli. Biotechnol. Appl. Biochem.* **49**, 141–147.

Bentley, D. R. (2006). Whole-genome re-sequencing. *Curr. Opin. Genet. Dev.* **16**, 545–552.

Capaldi, A. P., Kaplan, T., Liu, Y., Habib, N., Regev, A., Friedman, N., and O'Shea, E. K. (2008). Structure and function of a transcriptional network activated by the MAPK Hog1. *Nat. Genet.* **40**, 1300–1306.

Cormack, B. P., Bertram, G., Egerton, M., Gow, N. A., Falkow, S., and Brown, A. J. (1997). Yeast-enhanced green fluorescent protein (yEGFP) a reporter of gene expression in *Candida albicans. Microbiology* **143**(Pt 2), 303–311.

Denis, C. L., Ferguson, J., and Young, E. T. (1983). mRNA levels for the fermentative alcohol dehydrogenase of *Saccharomyces cerevisiae* decrease upon growth on a nonfermentable carbon source. *J. Biol. Chem.* **258**, 1165–1171.

Drummond, D. A., Iverson, B. L., Georgiou, G., and Arnold, F. H. (2005). Why high-error-rate random mutagenesis libraries are enriched in functional and improved proteins. *J. Mol. Biol.* **350**, 806–816.

Dunham, M. J., Badrane, H., Ferea, T., Adams, J., Brown, P. O., Rosenzweig, F., and Botstein, D. (2002). Characteristic genome rearrangements in experimental evolution of *Saccharomyces cerevisiae*. *Proc. Natl. Acad. Sci. USA* **99**, 16144–16149.

Funk, M., Niedenthal, R., Mumberg, D., Brinkmann, K., Ronicke, V., and Henkel, T. (2002). Vector systems for heterologous expression of proteins in *Saccharomyces cerevisiae*. *Methods Enzymol.* **350**, 248–257.

Gietz, R. D., and Schiestl, R. H. (2007). High-efficiency yeast transformation using the LiAc/SS carrier DNA/PEG method. *Nat. Protoc.* **2**, 31–34.

Gray, J. V., Petsko, G. A., Johnston, G. C., Ringe, D., Singer, R. A., and Werner-Washburne, M. (2004). "Sleeping beauty": Quiescence in *Saccharomyces cerevisiae*. *Microbiol. Mol. Biol. Rev.* **68**, 187–206.

Hadiji-Abbes, N., Borchani-Chabchoub, I., Triki, H., Ellouz, R., Gargouri, A., and Mokdad-Gargouri, R. (2009). Expression of HBsAg and preS2-S protein in different yeast based system: A comparative analysis. *Protein Expr. Purif.* **66**, 131–137.

Huisinga, K. L., and Pugh, B. F. (2004). A genome-wide housekeeping role for TFIID and a highly regulated stress-related role for SAGA in *Saccharomyces cerevisiae*. *Mol. Cell* **13**, 573–585.

Karagiannis, J., and Young, P. G. (2001). Intracellular pH homeostasis during cell-cycle progression and growth state transition in *Schizosaccharomyces pombe*. *J. Cell Sci.* **114**, 2929–2941.

Kim, J., and Iyer, V. R. (2004). Global role of TATA box-binding protein recruitment to promoters in mediating gene expression profiles. *Mol. Cell. Biol.* **24**, 8104–8112.

Klein-Marcuschamer, D., and Stephanopoulos, G. (2008). Assessing the potential of mutational strategies to elicit new phenotypes in industrial strains. *Proc. Natl. Acad. Sci. USA* **105**, 2319–2324.

Klein-Marcuschamer, D., Santos, C. N., Yu, H., and Stephanopoulos, G. (2009). Mutagenesis of the bacterial RNA polymerase alpha subunit for improvement of complex phenotypes. *Appl. Environ. Microbiol.* **75**, 2705–2711.

Liti, G., Carter, D. M., Moses, A. M., Warringer, J., Parts, L., James, S. A., Davey, R. P., Roberts, I. N., Burt, A., Koufopanou, V., Tsai, I. J., Bergman, C. M., et al. (2009). Population genomics of domestic and wild yeasts. *Nature* **458**, 337–341.

Lohr, D., Venkov, P., and Zlatanova, J. (1995). Transcriptional regulation in the yeast GAL gene family: A complex genetic network. *FASEB J.* **9**, 777–787.

Lutke-Eversloh, T., and Stephanopoulos, G. (2008). Combinatorial pathway analysis for improved L-tyrosine production in *Escherichia coli*: Identification of enzymatic bottlenecks by systematic gene overexpression. *Metab. Eng.* **10**, 69–77.

Miyazaki, K., and Arnold, F. H. (1999). Exploring nonnatural evolutionary pathways by saturation mutagenesis: Rapid improvement of protein function. *J. Mol. Evol.* **49**, 716–720.

Mumberg, D., Muller, R., and Funk, M. (1995). Yeast vectors for the controlled expression of heterologous proteins in different genetic backgrounds. *Gene* **156**, 119–122.

Nevoigt, E. (2008). Progress in metabolic engineering of *Saccharomyces cerevisiae*. *Microbiol. Mol. Biol. Rev.* **72**, 379–412.

Nevoigt, E., Kohnke, J., Fischer, C. R., Alper, H., Stahl, U., and Stephanopoulos, G. (2006). Engineering of promoter replacement cassettes for fine-tuning of gene expression in *Saccharomyces cerevisiae*. *Appl. Environ. Microbiol.* **72**, 5266–5273.

Neylon, C. (2004). Chemical and biochemical strategies for the randomization of protein encoding DNA sequences: Library construction methods for directed evolution. *Nucleic Acids Res.* **32**, 1448–1459.

Ogawa, N., DeRisi, J., and Brown, P. O. (2000). New components of a system for phosphate accumulation and polyphosphate metabolism in *Saccharomyces cerevisiae* revealed by genomic expression analysis. *Mol. Biol. Cell* **11**, 4309–4321.

Patrick, W. M., Firth, A. E., and Blackburn, J. M. (2003). User-friendly algorithms for estimating completeness and diversity in randomized protein-encoding libraries. *Protein Eng.* **16,** 451–457.

Rao, Z., Ma, Z., Shen, W., Fang, H., Zhuge, J., and Wang, X. (2008). Engineered *Saccharomyces cerevisiae* that produces 1,3-propanediol from D-glucose. *J. Appl. Microbiol.* **105,** 1768–1776.

Reetz, M. T., Kahakeaw, D., and Lohmer, R. (2008). Addressing the numbers problem in directed evolution. *Chembiochem* **9,** 1797–1804.

Santos, C. N., and Stephanopoulos, G. (2008). Melanin-based high-throughput screen for L-tyrosine production in *Escherichia coli*. *Appl. Environ. Microbiol.* **74,** 1190–1197.

Swers, J. S., Kellogg, B. A., and Wittrup, K. D. (2004). Shuffled antibody libraries created by *in vivo* homologous recombination and yeast surface display. *Nucleic Acids Res.* **32,** e36.

van Voorst, F., Houghton-Larsen, J., Jonson, L., Kielland-Brandt, M. C., and Brandt, A. (2006). Genome-wide identification of genes required for growth of *Saccharomyces cerevisiae* under ethanol stress. *Yeast* **23,** 351–359.

Waks, Z., and Silver, P. A. (2009). Engineering a synthetic dual-organism system for hydrogen production. *Appl. Environ. Microbiol.* **75,** 1867–1875.

Wapinski, I., Pfeffer, A., Friedman, N., and Regev, A. (2007). Natural history and evolutionary principles of gene duplication in fungi. *Nature* **449,** 54–61.

Watanabe, T., Murata, Y., Oka, S., and Iwahashi, H. (2004). A new approach to species determination for yeast strains: DNA microarray-based comparative genomic hybridization using a yeast DNA microarray with 6000 genes. *Yeast* **21,** 351–365.

Wong, T. S., Roccatano, D., and Schwaneberg, U. (2007). Steering directed protein evolution: Strategies to manage combinatorial complexity of mutant libraries. *Environ. Microbiol.* **9,** 2645–2659.

Yu, H., Tyo, K., Alper, H., Klein-Marcuschamer, D., and Stephanopoulos, G. (2008). A high-throughput screen for hyaluronic acid accumulation in recombinant *Escherichia coli* transformed by libraries of engineered sigma factors. *Biotechnol. Bioeng.* **101,** 788–796.

Zhao, H. (2004). Staggered extension process *in vitro* DNA recombination. *Methods Enzymol.* **388,** 42–49.

Zhao, H., Giver, L., Shao, Z., Affholter, J. A., and Arnold, F. H. (1998). Molecular evolution by staggered extension process (StEP) *in vitro* recombination. *Nat. Biotechnol.* **16,** 258–261.

ADVANCES IN CYTOLOGY/ BIOCHEMISTRY

CHAPTER TWENTY-ONE

VISUALIZING YEAST CHROMOSOMES AND NUCLEAR ARCHITECTURE

Peter Meister, Lutz R. Gehlen, Elisa Varela,[1] Véronique Kalck, *and* Susan M. Gasser

Contents

Abstract

We describe here optimized protocols for tagging genomic DNA sequences with bacterial operator sites to enable visualization of specific loci in living budding yeast cells. Quantitative methods for the analysis of locus position relative to the nuclear center or nuclear pores, the analysis of chromatin dynamics and the relative position of tagged loci to other nuclear landmarks are described.

Friedrich Miescher Institute for Biomedical Research, Basel, Switzerland
[1] Present address: Fundacion Centro Nacional de Investigaciones Oncologicas Carlos IIII, Melchor Fernandez Almagro 3, Madrid, Spain

Methods in Enzymology, Volume 470
ISSN 0076-6879, DOI: 10.1016/S0076-6879(10)70021-5

Methods for accurate immunolocalization of nuclear proteins without loss of three-dimensional structure, in combination with fluorescence *in situ* hybridization, are also presented. These methods allow a robust analysis of subnuclear organization of both proteins and DNA in intact yeast cells.

1. INTRODUCTION

Quantitative imaging techniques have improved dramatically in the last 15 years, reflecting both the rapid adaptation of naturally fluorescent proteins to cellular applications and improvements in fluorescence microscopy itself. Methods are also being continually optimized for the analysis and localization of endogenous proteins and chromosomal loci in living yeast cells. This involves novel microscope systems as well as improved computational tools for image analysis. Crucial to this process are tools for the rapid processing of the high-resolution digital-image stacks, since megabytes of data are produced in a single 3D time-lapse experiment on either a deconvolution widefield microscope or spinning disk (SD) confocal instrument (Hom *et al.*, 2007).

While techniques of live microscopy are powerful, it is not trivial to perform them correctly. Specifically, accurate visualization of more than two fluorophores at the same time can be difficult, and care must be taken to avoid damage by the light that is used for imaging. This can be particularly problematic when dealing with mutants that enhance sensitivity to damage or stress. Maintenance of unperturbed growth conditions and minimization of exposure time and light intensity are essential for meaningful results. Because high-resolution time-lapse microscopy often captures only one or a few cells per 3D stack, the imaging step can itself take considerable time, rendering it difficult to obtain sufficient numbers of cells or to carry out large time-course experiments. If several strains are to be analyzed in parallel, it is recommended that cells be fixed by formaldehyde at the desired time points, so that the localization of proteins or DNA can be achieved later by immunofluorescence (IF) and/or fluorescent probe *in situ* hybridization (FISH).

This chapter contains two sets of optimized protocols for the visualization of specific proteins and/or DNA sequences in budding yeast. The first set describes the targeting and analysis of proteins fused to the fluorescent protein GFP or its derivatives. The second section describes more classical methods for IF and/or FISH, which are sometimes the methods of choice for visualizing different types of macromolecules at once. Basic methods for quantitative analysis of subnuclear position and chromatin dynamics are described. These methods have been optimized for the localization of one or several targets in the nucleus relative to DNA or the nuclear envelope (NE). We note that improvements are continually being made in these procedures and that future users should seek updates on the methodology in the literature.

2. STRAIN CONSTRUCTIONS AND IMAGE ACQUISITION FOR NUCLEAR ARCHITECTURE ANALYSIS IN LIVING CELLS

2.1. Tagging chromatin *in vivo* with *lac* and *tet* operator arrays

The study of chromatin organization in live budding yeast cells often exploits the recognition of integrated arrays by fluorescently labeled bacterial DNA binding factors, usually the LacI or TetR repressor (reviewed in Belmont, 2001; Hediger *et al.*, 2004; Neumann *et al.*, 2006). The target arrays consist of anywhere from 100 to 256 copies of the recognition consenses (*lacO* or *tetO*). As few as 24 binding sites are usually sufficient to allow the formation of a visible spot, although the signal-to-noise ratio depends on the expression level of the fluorescently tagged binding protein.

Tagging chromatin *in vivo* is a two-step process. The first step involves the expression of a fusion between a DNA-binding protein, a fluorescent protein, and a nuclear localization signal. Both integrative and episomal plasmids have been used to express these proteins (Michaelis *et al.*, 1997; Straight *et al.*, 1996). Integrative plasmids give more reproducible levels of the fluorescently tagged proteins. The DNA binding Lac repressor is expressed as a fusion with green fluorescent protein (GFP), the cyan and/or yellow variants (CFP, YFP), and the Tet repressor exists as fusion proteins with GFP, CFP, YFP, and the monomeric variant of the red fluorescent protein, mRFP (Lisby *et al.*, 2003). To increase the fluorescence signal, a fluorescent protein can be introduced as a tandem array (3× CFP, Bressan *et al.*, 2004). Expression levels of these proteins have to be kept low, as overexpression elevates the background fluorescence, enhances non-specific binding, and can cause slow growth.

The binding site arrays recognized by the fluorescently labeled repressors are repetitive and unstable by nature in both bacteria and yeast. To avoid recombination and loss of copy number, the bacteria (either DH5α or recombination-deficient strains like SURE (Stratagene)) should be grown at 25 or 30 °C. When thawing bacterial strains, several colonies have to be tested for the size of the array by digestion of plasmid preparations with enzymes encompassing the array. The binding sites are inserted as an array in strains expressing the fluorescent DNA-binding proteins. For unknown reasons, expression of the DNA-binding protein in yeast stabilizes the array; therefore, it is recommended to transform yeast with the fusion protein construct prior to introducing the *lacO* or *tetO* sites.

To date, three techniques have been used to insert arrays at specific loci in the yeast genome. The first technique is based on the cloning of a small PCR-generated fragment of genomic DNA (about 400–800 bp) into the array-containing plasmid (Fig. 21.1A, Heun *et al.*, 2001a,b). This fragment

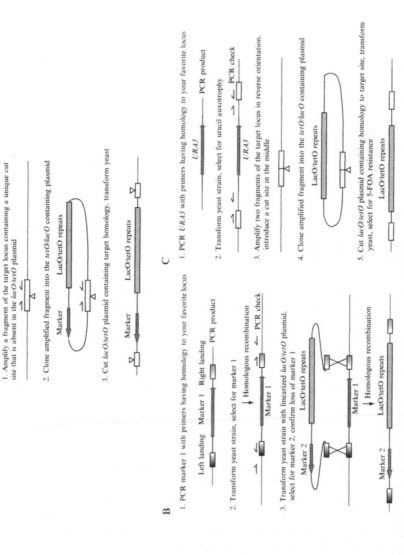

Figure 21.1 Outline of the methods for site-specific integration of *lacO/tetO* repeats in the genome cloning–free chromatin tagging. For details see text.

is chosen so that it contains a unique restriction site that is not present in the *lacO/tetO* plasmid. Once cloned into the array, digestion with this single-cutter enzyme will linearize the plasmid, which can be used for homologous recombination. The homology created by the small genome segment targets the plasmid to the desired genomic locus. It also creates direct repeats flanking the array, which might be detrimental to the stability of the array, as these allow popping-out of the whole plasmid by recombination between the two direct repeats. Positive transformants are selected by resistance to a selective marker present on the plasmid then correct insertion is tested by PCR and/or southern blotting. During the transformation process, some binding site repeats may be lost, therefore transformed yeast colonies have to be screened microscopically for the presence of a bright spot. One should not store the resulting yeast strains at room temperature for more than a week and positive clones should be frozen immediately. Spot presence has to be reconfirmed after thawing.

The second technique was developed to avoid tedious cloning steps with large plasmids containing *lacO/tetO* repeats (Rohner *et al.*, 2008). It is a two-step process involving first PCR-based integration of a marker flanked by 100-bp tags at the locus of interest. Once the tags are integrated into the genome at the locus of interest, they can be used for homologous recombination to integrate *lacO/lexA* repeats and a second selectable marker (Fig. 21.1B). To this end, the tags are cloned into the *lacO/tetO* repeat plasmid in reverse orientation with a rare cutting site in between them. When cut with this enzyme, the two adaptamers encompass the *lacO/tetO* repeats and can therefore be aligned with the tags flanking the marker in the genome. This technique is more flexible in terms of markers and allows one to tag the same locus with different binding sites without the need to reclone a PCR fragment into an array-containing plasmid.

A third technique combines the previous two and has been developed to avoid integrating a marker gene next to the repeats (Fig. 21.1C; Kitamura *et al.*, 2006). In a first step, a *URA3* gene is inserted at the locus of interest using long primer PCR-mediated recombination. To achieve replacement of the *URA3* gene, a fragment of about 700 bp corresponding to the *URA3* insertion site is cloned in the *lacO/tetO* repeat plasmid. As for the first technique described above, the recipient plasmid contains a single cut site in the middle of the cloned fragment. Transformation of the cut plasmid leads to replacement of *URA3* by the repeats. In this case, positive colonies are selected by their ability to grow on 5-fluoro-orotic acid (FOA), which is toxic in the presence of *URA3*. The main drawback of this technique is that it does not allow selection of the colonies which still contain the array, for example, after freezing. Direct replacement of *URA3* using an adaptamer-based technique with a marker-free plasmid is impossible, as FOA-resistant colonies arise more frequently than recombination events.

It is often useful to insert a low number of binding sites for another DNA-binding protein next to the *lacO* or *tetO* sites integrated at specific loci. This allows one to target another protein to the site of interest, which can be used to manipulate the locus. For example, integrated lexA sites allow binding of lexA fusion proteins, such as a lexA–Yif1 fusion, that anchors the tagged chromatin locus to the NE (Taddei *et al.*, 2004). Plasmids for the tagging methods described above are available with lexA binding sites located next to the *lacO/tetO* repeats (Rohner *et al.*, 2008; Taddei *et al.*, 2004). Other locus-tagging systems in development include a lambda repressor/operator system (K. Bystricky, A. Taddei, personal communication).

2.2. Determining the position of the nucleus

For precise localization studies, as well as for studying chromatin dynamics, the nuclear volume has to be defined. This can be achieved either by expression of a nucleoporin fused to a fluorescent protein (commonly Nup49-GFP) or by using the nuclear background fluorescence created by the unbound TetR protein. LacI-GFP tends to give very little background even in the absence of a *lacO* array, probably due to its low expression level.

2.3. Immobilizing cells for microscopy

To obtain images which allow the reliable measurement of chromatin position and dynamics there are two central concerns. First, one must immobilize the yeast cells and second one must prevent distortion of cell shape by pressure from the coverslip or objective. Both are achieved by the following methods.

For "snapshot" exposures where yeasts will be imaged only once, living cells are mounted on pad of agarose in synthetic medium. Immobilizing cells between agarose and a coverslip does not flatten or distort cells, while coverslip pressure on a glass slide does. Optimal agarose patches are created on depression slides, which have a concave depression in which the agarose and cells are placed. The agarose (1.4%) is dissolved in an appropriate medium (YPD gives more background than SD), and if imaging or cell maintenance lasts more than 20–30 min, it is recommended to use higher than usual levels of glucose (4% instead of 2%). Glucose can be locally depleted by cells in the agarose pad, while they are being imaged, and this reduces chromatin mobility within nuclei (Heun *et al.*, 2001b). Agarose prepared with yeast medium can be distributed in aliquots and kept for months at room temperature.

1. Prior to use, agarose is dissolved in growth media at 95 °C for several minutes. The agarose should be liquid, but prolonged maintenance at high temperature increases background fluorescence.

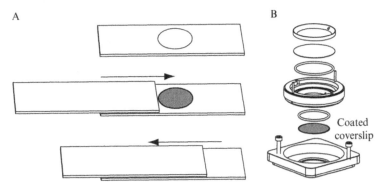

Figure 21.2 Means to immobilize yeast cells for imaging. (A) Formation of a flat-topped pad of agarose dissolved in media on a depression slide. (B) Cell observation chamber (Ludin chamber, Life Imaging Services) with cells immobilized on the lectin-coated bottom glass coverslide is shown.

2. Melted agarose is then poured into the depression of the slide.
3. A normal slide is immediately placed across the top to remove excess agarose and create a flat surface on the pad (Fig. 21.2A). While the agarose solidifies, 1 ml of an exponentially growing culture (at concentrations $<0.5 \times 10^7$ cells/ml) is spun in a microcentrifuge and resuspended in 20 μl of appropriate medium. Cells can be grown in synthetic medium or YPD, but YPD cultures show more autofluorescence. Note that high cell density or glucose depletion alter chromatin dynamics (Heun et al., 2001b).
4. After removal of the upper slide by sliding along the depression slide surface, 5 μl of the concentrated cells are placed on the agarose, and the pad is covered by a fresh coverslip. Capillary forces are generally strong enough to hold the coverslip in place. One should avoid fixing the coverslip with nail polish as some brands of nail polish contain solvents that inhibit yeast growth.

For live imaging over longer periods of time, cells can be noncovalently immobilized on a coverslip coated with lectin and visualized in media in an observation Chamber (Ludin Chamber, Life Imaging Services, Fig. 21.2B) as described below.

1. For budding yeast, Concanavalin A (Sigma) is used at 1 mg/ml, while for fission yeast a lectin from *Neisseria gonorrhoeae* (Sigma, 1 mg/ml) is optimal. Coverslips (18 mm ∅) are covered with 100 μl lectin solution which is immediately removed (the solution can be reused and kept at $-20\ ^\circ$C).
2. Coverslips are left to dry at room temperature (>20 min) and can be kept for months protected from dust.

3. These coverslips are used in an observation chamber (Ludin chamber) that allows cells to be immersed in media that can be exchanged by continuous flow or at defined intervals (Life Imaging Services, Fig. 21.2B).
4. Cells are sedimented on the coverslip before removal of excess media. One milliliter of fresh preheated medium is then added to the cells before sealing of the chamber. If needed, a flow of medium can be used, although very slow rates (flow < 1 ml/min) should be used as pressure changes induced by liquid pumping can cause movement of the coverslips or cells in the chamber.

2.4. Controlling temperature

Stable conditions for microscopy are best achieved by temperature-controlled rooms (± 1 °C). The microscope stage itself can then be heated to the desired temperature (30 °C for wild-type strains) using a Plexiglas box that encloses the entire microscope stage (many providers now offer this option adapted to the specific instrument). Another method only heats the stage, but temperature control is less precise as an unheated objective can act as a heat sink and cool the sample during observation. Heated objectives are also available.

2.5. Image acquisition set-ups

The appropriate choice of microscope depends on the aim of the experiment. Whatever system is used, it is essential to check first that the cells survive the high-intensity light used for fluorescence illumination without damage or cell cycle arrest. The more subtle the monitored phenomenon is, the more extensive the controls must be for light-induced changes in cell physiology. The simplest assay is to compare the kinetics of cell cycle progression in cells subjected to the experimental pattern of illumination with nonimaged cells. Various time intervals, intensities of light, wavelengths and/or gray filters should be tested; unbudded cells should rebud within 120 min at room temperature after imaging on YPD.

Every microscopic system is a compromise between speed of acquisition (the higher the speed, the lower the amount of light that can be recorded), the field of acquisition (in general, the bigger the field, the slower the acquisition), and resolution (higher resolution decreases speed and signal, since each pixel on the image corresponds to a smaller part of the sample and more pixels take more time to acquire). Since the haploid yeast nucleus is only 1 μm in radius, it is recommended that the objective magnification is at least 63×, or ideally 100×, with a numerical aperture (NA) as high as possible (between 1.3 and 1.45). This allows a high-resolution camera to obtain maximal detail from the sample (resolution power is inversely proportional to NA).

The first image acquisition setup described here is based on an improved widefield microscope, with a monochromator that regulates the light source, combined with rapid, high-precision Z motor, and a rapid and highly sensitive CCD camera for image capture. Since there is no pinhole, light from out-of-focus planes will be recorded, which can be later used by deconvolution algorithms that recalculate position of the emitted light based on an ideal or measured light spread function. The main drawback of this system is the phototoxicity due to whole cell illumination.

A second, widely available system is the laser–scanning microscope (Zeiss LSM510/710, Leica SP5). These systems have been proven very useful for acquiring very fast time-lapse recordings. Their limitation is the scanning speed, which is only fast enough for live imaging of chromatin dynamics if the field of scanning (region of interest or ROI) is reduced to a minimum (e.g., one cell). These confocals allow manual minimization of the beam intensity and pinhole. Again, there is a compromise between laser power (which increases phototoxicity, but allows more rapid image capture) and scanning speed (essential for the identification of rapid movements observed for chromatin *in vivo*).

A third system that we strongly recommend is based on a rapid, wide-field high precision microscope, although the light source is a laser whose beam is focused on a rotating disk with thousands of pinholes. This disk spins at high speed dispersing the laser beam such that the whole laser power is never focused on a single point in the sample. This reduces phototoxicity and bleaching of the fluorochrome; moreover, the speed of capture is faster than that of a scanning laser system. Out-of-focus light is filtered through the pinholes, and entire fields of cells can be captured at once. In the following sections, we discuss the critical points of each of these setups.

2.5.1. Rapid high-precision widefield microscopy

For the imaging of a large number of cells at a single time point, best results are obtained with a high-precision widefield microscope. These microscopes are equipped with a piezoelectric focus either with the objective mounted on it directly (e.g., PiFoc, Physik Instrumente) or a piezoelectric table (e.g., ASI MS2000, Prior), which allows one to capture stacks of focal planes. Z distances between planes is carefully controlled and highly reproducible, and movement from one plane to the next is nearly instantaneous. The light source is very important, as the classical mercury bulbs show phototoxicity. The light source of choice for maximum versatility is a monochromator (Xenon light source coupled with Polychrome, TillVision), which allows excitation wavelength choice in nanometer steps (320–680 nm continuous spectrum, 20 nm window). Switching wavelengths is rapid (<1 ms). A cheaper though less flexible illumination alternative is a LED-based illumination (CoolLED, precisExcite), where up to four wavelength (fixed) can be chosen at the time of order. LEDs are very long lived (3 years guaranteed by the supplier), which makes it a cost-effective solution. Switching time is even faster than with the monochromator (around 300 μs). From a performance point of view, we found no significant differences between a monochromator and a LED-based illumination system.

Acquisition is achieved with a high-resolution CCD camera. To detect subnuclear or subcellular details, one needs a final pixel size between 60 and 80 nm with a 100× objective. The readout of the camera by the computer is often the rate limiting factor of the system. Typically, high-speed CCD cameras (Roper Scientific Coolsnap HQ, Andor IKon, Hamamatsu ORCA) achieve about 30 frames/s, which makes exposure times shorter than 30 ms impossible. These systems are relatively inexpensive and are easier to setup than confocal microscopes. Several proprietary software can drive the entire system (microscope, camera, shutters, monochromator) such as MetaMorph (Universal Imaging).

This modified widefield microscopy is well-suited for scoring the position of a locus relative to another locus, or relative to a fixed structure (spindle pole body, nuclear periphery and nucleolus) in a large number of cells on an agarose pad. It is less well-suited for rapid, high-resolution time-lapse imaging, due to the high sampling and deconvolution that is needed for highest resolution data. If the position of two loci is to be monitored, either two different excitation colors have to be used (which increases the resolution power) or the spots have to be of significantly different sizes. 3D stacks of images are needed to evaluate the spatial positioning of the locus relative to another spot or to the nuclear periphery (see below). Optimal parameters for GFP imaging are excitation 475 nm, z-spacing 200 nm with 20 plane stacks, 100–200 ms exposure time per slice. Due to the optical resolution of the microscope, it is not useful to sample more in the z-axis, which would also increase the acquisition time, and impair accuracy if the imaged locus is moving.

For dual color imaging using CFP and YFP chromophores, optimal wavelengths are 432 and 514 nm, respectively, with exposure times of about 200 ms. The two wavelengths should be acquired successively at each focal plane. Note that the wavelengths and exposure times depend greatly on the filters present on the microscope, and should be optimized for each system (monochromators allow nm-scale changes in wavelength). A phase image is useful for determining cell cycle stage, and can be taken before or after acquisition of the fluorescence stack of images.

In widefield microscopy, an entire field of cells is illuminated during exposure. The camera records both the in-focus and out-of-focus photons. While this creates a higher background than confocal microscopy, it allows more photons to be recorded by the camera, and these signals are used for image restoration algorithms. Deconvolution is particularly powerful when applied to widefield imaging to reassign signal to the right plane. Several software packages propose deconvolution solutions, including Metamorph, DeltaVision, and Huygens. Similarly, denoising of the images (which removes optical and electronic noise from the digitalized images) can be applied to increase the signal-to-noise ratio. Although no commercial package is available to date, development of such denoising solutions is a very active field in image processing and could lead to significant reduction of both light intensity and illumination time in the near future.

Live cell time-lapse imaging is used to record the dynamics of tagged chromatin or other subnuclear structures. Since repetitive illumination of the sample is involved, it is important to keep in mind that excitation light can stress the organism, and control experiments must be carried out to ensure that the level of illumination does not have deleterious cellular consequences. Parameters to optimize include image resolution (pixel size), the number of frames along the z-axis, excitation light intensity and exposure time.

Widefield high-precision microscopy is useful for low-frequency time-lapse imaging over fairly long periods of time (hours). The excitation light from the monochromator or the LED should be filtered using gray filters to reduce phototoxicity. Limits are set by the intensity of light used, the number of planes acquired for each time point and the time between each acquisition. In our experience, up to 300 stacks of 5 sections (1500 frames, 50 ms exposure per frame, 1 min interval between stacks) can be acquired without affecting cell cycle and with only moderate bleaching. Increasing sampling frequency will increase bleaching and damage the cells. Confocal or SD-systems are better choices for rapid time-lapse imaging, as acquisition speed is faster and photo-induced damage can be reduced by limiting the excitation time.

2.5.2. Laser-scanning microscopy

Laser-scanning systems are based on the rapid scanning of the sample by an excitation laser and recording of the emitted signal by photomultipliers (PMTs). The out-of-focus light is blocked by a pinhole which should be closed as far as possible. While these systems are well-suited to discriminate wavelengths and capture several at once, the scanning speed is often the limiting factor for image acquisition. Nonetheless, to track chromatin in individual cells at intervals of 1.5 s over timescales of 5–10 min, commercially available systems such as the Zeiss LSM510 system are well suited. This system, although slower than the newer SDs (see below), is fast enough to track significant changes in chromatin movements (jumps >0.5 μm in 10 s; Heun et al., 2001b). Useful settings are described below (see also Neumann et al., 2006). Note that pixel size is set by the user, and to track chromatin in vivo pixel size should be ≤ 100 nm. A high-resolution piezo table is essential to achieve speed and reproducibility in z position:

Laser	Argon/2 458, 488, or 514 tube current 4.7 A. Output 25%
GFP acquisition	Channel 1 LP 505 nm
YFP/CFP	Single track Channel 1 LP 530 nm
	Channel 3 BP 470–500 nm
Channel settings	Pinhole 1–1.2 Airy unit (optical slice 700–900 nm); detector gain 930–999; amplifier gain 1–1.5; amplifier offset 0.2–0.1 V; laser transmission AOTF 0.1–1% for GFP excitation, 1–15% for YFP, and 10–50% for CFP single track acquisition
Scan settings	Speed 10 (0.88 μs/pixel), 8 bits one scan direction; 4 average line scans; zoom 1.8 (pixel size 100 × 100 nm)
Imaging intervals	1.5 s

2.5.3. Spinning-disk confocal microscopy

As mentioned above, an attractive alternative to widefield and laser-scanning microscopes is the SD confocal. SD microscopes look similar to widefield systems yet the excitation light is provided by lasers, the beams of which are focussed on pinholes located on a disk rotating at high speed. Every point of the focal plane is illuminated several thousand times per second, but only for a fraction of a microsecond. The emitted light is filtered by passing through the pinholes to remove out-of-focus photons. Acquisition is achieved on a CCD camera, as for widefield systems. The overall quality of the picture is improved due to the confocality of the system: there is no haze as observed in widefield images. For example, nuclei which appear elongated along the z-axis in widefield stacks will appear more round using an SD confocal (Fig. 21.3A and B). Moreover, due to the "intermittent" excitation of fluorophores by the SD, these systems show less bleaching and phototoxicity. This allows higher frequency sampling, at a rate that is generally limited only by the acquisition rate of the camera. Where a laser-scanning confocal can record only a single nucleus with five planes and a 0.45-μm z-spacing, with one stack every 1.5 s, an SD system is able to record 20 planes at 0.2 μm spacing every 1.5 s, on a whole field of view.

Many systems are now available (Roper, Perkin Elmer, Andor, Zeiss provide full setups), all of which are based on Yokogawa scan heads. This head is the part which contains the SD itself, as well as filters for excitation/emission and the dichroic filter. As most of the light is stopped by the SD, powerful lasers ($>$15 mW for 488 nm excitation line) have to be used, which increases the cost of such setups. The camera can be either a classical CCD system (see above) or a more sensitive (but more noisy) EM-CCD.

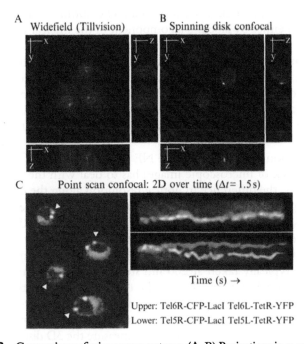

Figure 21.3 Comparison of microscope systems. (A, B) Projections in *x–y*, *x–z*, and *y–z* for yeast cells tagged with Nup49-GFP and a locus bearing a *lacO* array bound by a GFP-LacI fusion: (A) the image stack was taken with a high-resolution widefield microscope equipped with a monochromator and piezo (Tillvision®), while in (B) the image was taken with a spinning disk confocal. The images were not further treated or deconvolved. (C) Zeiss LSM510 confocal image of yeast cells growing in an agarose pad, bearing the following markers and fusion proteins: Nuclear envelope (Nup49-GFP, white ellipse) and the spindle pole body (Spc42-CFP, lighter spot on the nuclear envelope, indicated by a white arrow), nucleolus (Nop1-CFP, gray internal crescent), and a tagged telomere, Tel5R::*lacO* bound by GFP-LacI (gray arrow). Alongside are examples of time-lapse 2D confocal imaging on a Zeiss LSM 510 confocal microscope of two differently tagged telomeres relative to each other. They are displayed orthogonally and rotated such that the time axis (*z*) is horizontal. Top panel:Tel6L-TetR-YFP (lighter gray), Tel6R-CFP-LacI (darker gray); bottom panel (these two telomeres have been shown to colocalize (Schober *et al.*, 2009)): Tel5L-TetR-YFP (lighter gray), Tel5R-CFP-LacI (darker gray). The green background staining of the nucleus is due to the TetR-YFP diffuse in the nuclear volume (reproduced with permission from Bystricky *et al.*, 2005, see this paper for color images).

3. DATA ANALYSIS AND QUANTITATIVE MEASUREMENTS

3.1. Accurate determination of the 3D position of a tagged locus

To determine the position of a tagged locus inside the nucleus, the position of the center of the nucleus and of the locus have to be reconstructed from the microscopic images. As described before, the locus is usually labeled by

LacI or TetR fused to a fluorescent protein. The outline of the nucleus can be determined either by labeling a component of the nuclear pore complex or by using the background fluorescence given by unbound repressor proteins filling the nuclear volume. The latter method allows reliable identification of the center of the nucleus, yet it is difficult to measure its exact size since background fluorescence fades at the boundary. Whenever the size of the nucleus or the exact location of the NE is required, nuclear pore staining is recommended, as the boundaries of the nucleus are sharper.

The extraction of the shape of the NE and the position of a fluorescent spot from a stack of microscopic images has to deal with the anisotropy of the data, that is, the difference in optical resolution along the optical axis of the microscope (z-axis) and perpendicular to it (x/y-axes). One image (x/y-direction) has a typical optical resolution of 200 nm (with a $100\times$ objective) and is sampled with a pixel size of 50–100 nm. In contrast, the resolution in z is not better than 300 nm even for a confocal microscope, and the images of a stack are typically taken at 200 nm steps. In addition, the fluorescent signal from the nuclear pores close to the top and bottom of the nucleus is diffuse and poorly resolved, impairing reconstruction of the NE.

We discuss here two methods to measure the position of a spot relative to the NE. Ideally, one would want to directly measure the 3D distance between the nuclear rim and the tagged locus. A budding yeast nucleus can be accurately represented by an ellipsoid or even a sphere. One possibility is therefore to fit an ellipsoid to the nuclear pore staining and use it as a model for the NE. Analogously, a 3D Gaussian distribution can be fitted to the staining of the locus to determine its position with high accuracy. The distance between the locus and the NE (or the center of the nucleus) can then be calculated using the ellipsoid and the position of the spot. However, due to the limited microscopic resolution in the z-direction ($\sim$0.6 μm for green light in widefield and $\sim$0.45 μm for a confocal), and the small size of the yeast nucleus, precise definition of the NE is particularly difficult within 0.4 μm of the top or bottom of the nuclear sphere. Attempts to solve this problem require custom-tailored multistep processing of highly sampled image stacks (Berger *et al.*, 2008), and to date no standard software has been established.

Once the position of the locus and of a second nuclear structure, such as the nucleolus or the spindle pole body, have been determined accurately, a more detailed analysis of nuclear organization can be performed based on determination of an axis within the nucleus. If only the distance of a locus to the nuclear center is measured, the nucleus is treated as spherically symmetric, which is, of course, not the case. Since the nucleolus and spindle pole body are located at opposite ends of the nucleus, they define an axis that can be exploited as a landmark for locus position. This allows one to score deviations of locus distribution from spherical symmetry (Berger *et al.*, 2008).

To deal with the poor z resolution of microscopic stacks an alternative method exploits the fact that resolution is better in x–y and a spot can be assigned to a specific plane of an image stack. Instead of calculating the 3D distance between the spot and the spherical NE directly, one measures position in the plane where the spot is brightest. In this plane, the nucleus is a circle, which can be partitioned into three concentric zones of equal area (Fig. 21.4B). The spot position is then sorted into the outermost (zone 1), the intermediate (zone 2), or the innermost zone (zone 3). To obtain equal areas for the three zones, the boundaries between zones 1 and 2 and between zones 2 and 3 are at radii of $\sqrt{2/3}R$ and $\sqrt{1/3}R$, respectively, where R is the radius of the nucleus in the chosen plane. Then it follows from the principle of Cavalieri that each zone represents one third of the nuclear volume, justifying the use of this approach.

For practical applications, we use the following procedure:

1. Measure the distance between the spot and the periphery along a nuclear diameter as well as the diameter itself. Several programs can be used to extract the coordinates of points of interest from an image. For this task, the freely available pointpicker plug-in for ImageJ is particularly useful (http://bigwww.epfl.ch/thevenaz/pointpicker/).
2. Normalize the spot pore distance to the radius (not diameter!) of the circle.
3. Sort the spot into zone 1 (if the normalized distance is $<1 - \sqrt{2/3}$), zone 2 (if it is between $1 - \sqrt{2/3}$ and $1 - \sqrt{1/3}$), or zone 3 ($>1 - \sqrt{1/3}$).
4. Compare the measured distribution to another one (different strain, condition, etc.) or to a uniform distribution using, for example, a χ^2 test. If only percentages of one zone (e.g., the outermost zone) are compared, a proportional test should be used.

A locus whose position is uniformly distributed will be found with an equal probability of 1/3 in each of the three zones. It should be noted, however, that the three zones do not coincide exactly with three concentric shells of equal volume, which is the desired partitioning of the nucleus, if one wishes to assess whether a locus is enriched at the nuclear periphery (Fig. 21.4D). We have calculated the error incurred by this method, and plotted it against the true distribution of spots in Fig. 21.4E. Whereas the zone measurement is no longer precise when there is strong enrichment in any of the three zones, it accurately monitors a uniform distribution of spots. Moreover, the zone method consistently underestimates enrichment or depletion, which means that any measured enrichment in one zone did not arise from an artifact of the measurement method (Gehlen, 2009).

As mentioned above, measuring spot position with respect to the NE is particularly difficult close to the poles of the nucleus. This is aggravated if the NE and spot are both tagged with GFP. To avoid severe errors that arise

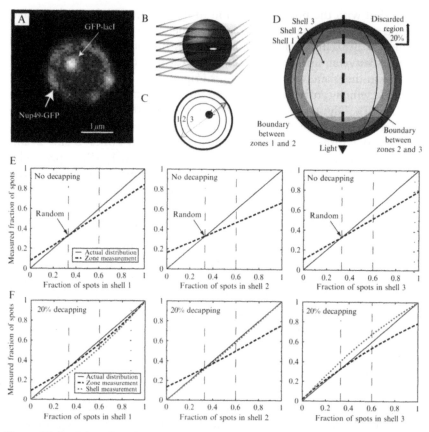

Figure 21.4 Subnuclear localization relative to the nuclear envelope: the zoning method. (A) Fluorescence microscopy image of a yeast nucleus (one plane of a 3D stack of images) bearing GFP-Nup49, a component of the nuclear pore complex, and a *lacO* array integrated into the genome and bound by a LacI-GFP fusion (fluorescent spot). (B, C) For quantification, the ring representing the nuclear envelope in the plane where the spot is brightest is partitioned into three zones of equal area. The nuclear diameter in this plane (gray arrow) and the distance of the spot to the periphery (black arrow) are measured and the ratio, which defines the localization of the spot, is scored as falling into zone 1, 2, or 3. (D) Vertical cut through the nucleus. Three shells of equal volume are shown in shades of gray. The division of the nucleus into three zones based on equal area in each plane also results in three equal volumes (the boundaries are shown as black lines), but these do not coincide exactly with shells of equal volume. Because of lack of resolution in the top and bottom slices of an image stack (see text), we remove samples in which the locus falls into the upper or lower 20% of the nuclear sphere. This so-called "decapping" is indicated in darker gray. Removal does not affect the zones and shells equally. (E) The deviation from the actual distribution in each zone when foci are scored using the zoning method with no decapping. Without decapping, the shell measurement is exact and coincides with the solid line. (F) The deviation from actual distribution is shown for foci monitored by either the zoning method or the shell method, after removal of 0.4 μm from each pole. Both types of measurements deviate from the true enrichment, although the zoning method is most accurate for zone 1. A fraction of one-third corresponds to a uniform distribution, 0.6 is a typical fraction, for example, for an anchored yeast telomere.

from such poorly resolved signals, we do not score cells in which the tagged locus is positioned within 0.4 μm of the top or bottom of the nucleus. This so-called decapping can include 3–4 planes (up to 20% of the focal planes) from each pole. While it removes questionable signals, it also affects the distribution determined by both shell (ideal 3D distance measurement) and the zoning method, because peripheral spots are more likely to be discarded than interior ones (Fig. 21.4D). In Fig. 21.4F we plot the error incurred by zoning and shell measurements as a function of spot enrichment, under decapping conditions. Intriguingly, decapping by 20% actually improves the accuracy of the zone measurements, while the shell measurements suffer from removal of these planes. Our analysis shows that the shell measurement method performs best in cases of extreme enrichment or depletion while the zoning is more accurate for moderate enrichments (35–60%), particularly in the outermost zone (zone 1). In principle, it is possible to compensate for these errors but one needs to know the exact size of the caps removed. On a practical level, it is important to remember that the zoning method accurately scores both uniformly distributed loci and distributions close to a uniform, independently of the amount of decapping performed.

3.2. Colocalization of a DNA locus with a subnuclear structure

To further investigate the function of DNA position, it is interesting to know if a fluorescently tagged locus colocalizes with other structural components of the nucleus. This can be investigated by tagging the locus in one fluorophore and the structure of interest with another, and monitoring their colocalization. Correction for chromatic shift must be made for each instrument and imaging session, by alignment of signals from small beads that emit fluorescence at multiple wavelengths.

Unless a locus is actively excluded from a subnuclear structure, a certain level of random colocalization will be detected. The amount of this background overlap will depend on the size and form of the structures monitored. To assess whether experimentally obtained colocalization values are significant or not, one must determine the expected degree of non-specific colocalization for a uniformly distributed locus. This can be calculated as the ratio between the volume of the region in which the spot is considered as colocalizing with the structure, and the total volume available to the spot.

As an example we take the binding of a chromatin locus (gene or telomere) to nuclear pores (Schober et al., 2009). The diffraction limited resolution of a light microscope is not sufficient to distinguish the binding of a locus to nuclear pores from its binding to other components at the NE. A genetic trick to circumvent this problem is to examine a yeast strain with an N-terminal deletion of the nuclear pore component *NUP133* (nup133ΔN; Schober et al., 2009). In this mutant the pores are not distributed all over the NE, but are clustered on one side of the nucleus

(Fig. 21.5A). A high degree of colocalization of a locus with the pore cluster may indicate specific affinity for a nuclear pore component.

To determine the colocalization arising from a uniform distribution of the locus, we first model the pore cluster as a conical disk at the nuclear periphery, whose dimensions are set based on empirical measurements. The spot is considered to colocalize with the cluster if it at least touches it (Fig. 21.5B). For the center of the chromatin spot, this defines a region that is larger than the pore cluster, which represents spot and pore colocalization. The predicted degree of background coincidence is the ratio between this colocalization volume and the total volume that is available to the spot. In the calculation of nonspecific colocalization, one can include other parameters, such as an exclusion of the spot from a subnuclear volume like the nucleolus, or a nonpore-associated enrichment at the NE. The significance of any experimental enrichment in colocalization is then determined by a proportional analysis test with a Bonferroni multiple test component.

3.3. Quantification of locus mobility

A stretch of chromatin (or any other object) inside the nucleus is exposed to numerous hits of water or other small molecules, proteins, and other macromolecules, as well as other chromatin fibers. Due to these interactions, it inevitably performs a seemingly random movement called Brownian motion. This motion is limited by the NE, but in many cases locus diffusion is even more constrained, either confined to a certain area or

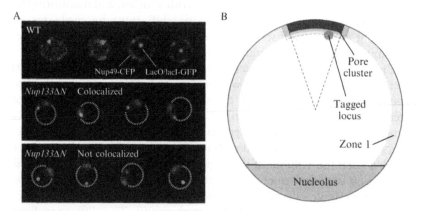

Figure 21.5 Determining the significance of colocalization. (A) Nuclear pores tagged with Nup49-GFP (red) and a LacI-tagged locus (green). The two left images in the upper panel are not deconvolved, all other images are. In the *nup133ΔN* mutant, the nuclear pores form a cluster (Schober *et al.*, 2009). (B) The expected colocalization for a randomly positioned spot and the pore cluster can be calculated as a ratio of volumes (see text). The figure shows a cut through the nucleus. The pore cluster is modeled as a conical layer shown in red. The spot is considered as colocalizing if it at least touches the pore cluster, which results in the colocalization zone (green). (See Color Insert.)

obstructed by obstacles. The random movement can also be temporarily or continuously superimposed by active displacement which possibly expresses itself as increased speed and/or directionality of movement.

The first step of the quantitative analysis of chromatin movement is the determination of the position of the locus and the nuclear center for each time point of the time-lapse series. Indeed, since the nucleus itself is moving inside the cytoplasm, one must compensate for its displacement to measure the movement of a locus relative to the nucleus. Several general purpose software packages like Imaris (http://www.bitplane.com) offer object tracking functionality but usually require uniformly high-contrast images. The algorithms are mostly based on threshold principles, and it is difficult to correct insufficient results by hand. In collaboration with D. Sage and M. Unser, a dynamic programming algorithm was developed which is dedicated to the tracking of single spots in noisy images and can be applied to 2D or 3D time-lapse movies (Sage *et al.*, 2005). The algorithm is implemented as a publicly available plug-in for the free software ImageJ (http://bigwww.epfl.ch/sage/soft/spottracker/).

This tracking works in two steps: first, the images are aligned with respect to the center of the nucleus to compensate for the movement of the entire nucleus throughout the time-lapse series. A Mexican hat filter can be applied to enhance spot-like structures in the images. Next, the spot tracking is performed using three different properties of the spot:

1. *Spot intensity*: the spot fluorescence is more intense than that of the background.
2. Within one time step the spot can only travel a limited distance.
3. In contrast to nuclear pores, the spot can be located in the nuclear interior.

To reflect these properties the tracking algorithm uses four different criteria to determine the spot position at a given time point:

1. Pixel intensity
2. Displacement from the location at the previous time point
3. Displacement from the last user-defined position (see below)
4. Distance from the nuclear center

The user can give different weights to these criteria to optimize the performance of the algorithm for different situations or image qualities. Most importantly, the plug-in offers the possibility to correct the trajectory manually by forcing it to pass through a given pixel at a certain time point. The output of the plug-in is the position of spot and nuclear center for each time point.

Because of individual differences between cells it is inevitable to analyze at least 8–10 movies with a total time of more than 40 min for each strain or condition. We discuss three parameters that can be extracted from the trajectories to compare different samples.

3.3.1. Track length

A simple and robust parameter of chromatin dynamics is the track length over a time-lapse series of fixed duration. This parameter monitors average mobility of a locus and can be used for comparison of movies with the same time step and duration. It is, however, a very artificial parameter because the true trajectory of the spot is inaccessible due to the lack of temporal and spatial resolution and is much longer than the measured track length (see Fig. 21.6A for illustration).

3.3.2. Step size and large steps

The average step size of the chromatin locus within its "walk" is another useful characteristic. Like the track length, this parameter depends on the time step used for image acquisition, but can be used to compare differences in mobility in identically imaged samples. Directed movement does not necessarily reveal itself in large single steps but rather in several successive correlated steps. Therefore, it is also useful to look for exceptionally high displacements ("large steps") within a certain time window. Empirically we find that a useful parameter for distinguishing patterns of mobility is the frequency of steps larger than 500 nm during 10.5 s (7 × 1.5 s steps; Heun *et al.*, 2001b).

3.3.3. Mean-squared displacement analysis

A robust method to analyze the global properties of an object's movement is the mean-squared displacement (MSD) analysis. An object in solution changes its direction when it bumps into solvent molecules and moves linearly in between, generating a random walk. If a number of objects would be initially confined in a small volume and then released, they would spread over time. It can be derived mathematically that for free diffusion the mean of the squared distance from one point on the trajectory to another is proportional to the time difference Δt: $\langle (\mathbf{r}(t + \Delta t) - \mathbf{r}(t))^2 \rangle \sim \Delta t$, where $\mathbf{r}(t)$ is the position of the object at time t (Berg, 1993). The proportionality constant is usually written as $2dD$ where d is the number of dimensions and D is called the diffusion coefficient of the object. Thus, for three-dimensional free diffusion we get $\langle (\mathbf{r}(t + \Delta t) - \mathbf{r}(t))^2 \rangle = 6Dt$ (Fig. 21.6).

However, in a cellular environment there is no free diffusion. The free movement of an object can be impaired by confinement, obstacles, and the binding to immobile or actively moving structures. The most inevitable restriction is the confinement of the object's movement to a nuclear or cellular compartment. This implies that the distance of any two points of the trajectory cannot exceed the maximal extension of the confining volume. Therefore, the MSD curve has to reach a plateau for large time windows (Fig. 21.6B). In the case of a spherical confinement, the value of the plateau

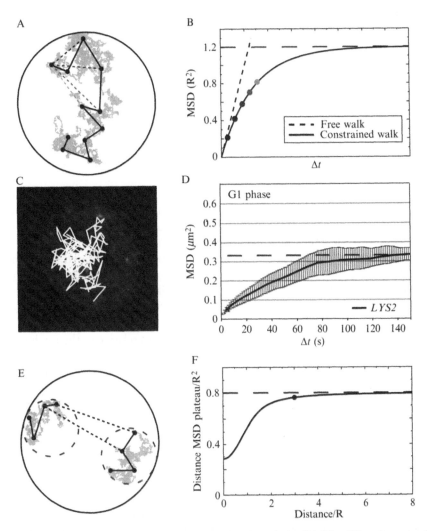

Figure 21.6 Mean-squared displacement (MSD) analysis. (A) The full trajectory of microscopic movement (light gray) cannot be detected by fluorescence microscopy due to limited resolution in time and space. A coarser trajectory (black) is recorded instead. (B) The mean of all squared spatial distances between each two points at a given time difference results in one point on the MSD graph. The mean-squared distance between a point and its successor on the trajectory is the first point on the MSD graph (black). The mean-squared distance between a point and its second successor yields the second point (dark gray) and so on. Compare the gray tones of the example distances in (A) with those of the points on the MSD graph in (B). (C) Analysis of DNA locus dynamics. The projected trace of 200 images of a movie of the *LYS2* locus is in white. The average track length in 5 min is 37.4 μm. Bar: 1 μm. (D) MSD analysis on an average of eight movies of the *LYS2* locus. All cells were observed in G1 phase. (E) The mean-squared change of spot–spot distance. In contrast to a classical MSD analysis, the mean-squared

can be calculated as $6/5R^2$, where R is the radius of the sphere (Neumann *et al.*, submitted). Thus, the so-called radius of constraint or the plateau value can be directly used as a measure for the size of the region explored by the object.

Due to the difficulties in accurately reconstructing the 3D position of fluorescent spots (see above), the movement is often observed in a 2D projection of the microscopic stacks. It can be calculated that the MSD of projected free 3D diffusion is equal to the MSD of free 2D diffusion: $\langle(\mathbf{r}(t + \Delta t) - \mathbf{r}(t))^2\rangle = 4Dt$. In the case of a spherical confinement, the MSD plateau behaves in the same way and has a value of $4/5R^2$ (see Fig. 21.6C and D; Neumann *et al.*, submitted).

For free diffusion the slope of the MSD line is a measure for the diffusion coefficient of the object, as discussed above. In the case of confined diffusion, the slope of the MSD curve is not constant. The curve is steepest at $\Delta t = 0$, and then the slope decreases monotonously (Fig. 21.6B). This is also true for diffusion with obstacles where—in the unconfined case—the MSD is not proportional to t but to t^α with $\alpha \neq 1$ (reviewed in Bouchaud and Georges, 1990). Nonetheless, one can still use the initial slope of the curve to compare the intrinsic mobility of different objects or one object under different conditions.

It should be noted that the movement of a locus relative to the nucleus is superimposed by the movement of the nucleus itself. Translational movement of the nucleus can be subtracted from locus movement by aligning the nuclear center throughout the time course (see Section 2.8). If two spots are observed, there is the alternative possibility to align one of the spots throughout the movie and analyze the movement of the other spot relative to the first one. This procedure also eliminates the global movement of the nucleus. However, neither the alignment of the nuclear center nor the alignment of one spot eliminates the rotational movement of the nucleus. A possibility to obtain a quantification of locus mobility that is independent of nuclear rotation is to observe the distance between the two loci and calculate the mean-squared change of this distance (see Fig. 21.6E). Since the distance between the two spots is unaffected by both translation and rotation of the nucleus, this "distance MSD" is only influenced by the individual movement of the two spots. The distance MSD curve shows

---

change of the distance between two spots instead of the mean-squared change of the position of one spot is analyzed. (F) The plateau of the distance MSD curve does not only depend on the radius of confinement R of the loci, but also on the distance d between the confining regions. However, this dependency becomes very weak for $d > 3R$. Therefore, the radius of confinement can be reconstructed from the distance MSD plateau only if the confining regions are either equal ($d = 0$) or sufficiently far from each other ($d > 3R$).

similar behavior to a classical MSD curve and has been used to derive diffusion coefficients and radii of constraints (Marshall *et al.*, 1997). However, it is important to note that the authors assumed that both loci are confined to the same region. If this is not the case, the height of the plateau, as well as the initial slope of the curve, does not only depend on the mobility of the loci but also on the distance separating the regions of constraint for the two spots. The distance MSD analysis is a valid technique to determine radius of constraint and diffusion coefficient for two diffusing spots if one of the two following conditions is fulfilled. Either (a) one can assume that the confining regions are identical (e.g., the whole nucleus) or (b) they are sufficiently far from each other. Three times the radius of constraint was found to be a reasonable threshold (Gehlen, 2009).

4. IF AND FISH ON FIXED SAMPLES

Despite the power of live imaging of GFP-tagged foci, the more classical techniques of IF and FISH are recommended in several cases. First, for a scientist working alone, the analysis of multiple samples at one time point is cumbersome by live imaging. Second, if more than two DNA loci need to be imaged at once, or multiple foci in one background, FISH is more efficient. Finally, these techniques allow colocalization of protein, specific genes, and either genomic DNA or cellular substructures such as the spindle. The combination of three or four fluorochromes in a single labeling experiment is routine. Nonetheless, there are pitfalls in applying this to yeast. First, antibody background and nonspecific fluorescence signal is more often observed with yeast cells than with mammalian cells. Second, one must preserve native 3D structures of both nuclear and cytoplasmic compartments, while eliminating the cell wall to facilitate macromolecular access. To check that this was done, the integrity of an NE and the size of the nucleus can be monitored either by DNA stains or by immunolabeling with an antibody recognizing the nuclear pore (e.g., Mab414 (Abcam) which recognizes yeast Nsp1 and yields a perinuclear ring). The spherical ring structure is lost when spheroplasting conditions are too harsh or if detergent use is too high.

The diameter of an intact haploid yeast nucleus should measure between 1.8 and 2 μm, and this measurement should be monitored regularly to ensure that the nuclei observed are intact. Inappropriate methods produce flattened nuclei with a chromatin mass spanning $\sim$6–8 μm (Heun *et al.*, 2001a; Weiner and Kleckner, 1994). Due to the nature of *in situ* hybridization, accessibility of the DNA probe to the nuclear chromatin is critical and FISH protocols seek the best possible compromise between accessibility of the probe and complete integrity of nuclear and chromatin structure.

To this end, we eliminate treatments that involve protease, nuclease, and/or combinations of ionic and nonionic detergent from our protocol. We find that yeast nuclei collapse when enzymatically digested or if exposed to detergent mixtures (Gotta *et al.*, 1996; Hediger *et al.*, 2002; Heun *et al.*, 2001a). Generally, formaldehyde fixation should be performed prior to the enzymatic removal of the cell wall (spheroplasting). However, if maximal diffusion of fixative or probes is critical, spheroplasting in osmotically buffered medium can be performed prior to fixation. Double *in situ*/immunofluorescence staining often requires this type of fixation. Finally, even though cells and spheroplasts are fixed, we recommend imaging in agarose pads, since pressure on coverslips can distort 3D structure. Confocal microscopy confirms that 3D organization can be maintained by the following procedure (Heun *et al.*, 2001a).

4.1. Yeast strains and media

Diploids yeast strains may facilitate the microscopic localization of chromosomal loci, since the nuclei are nearly twice the size of haploid nuclei. There is a significant variation in the efficiency with which different strains are converted to spheroplasts, probably reflecting differences in the cell wall composition. Diploid strains usually spheroplast faster. Whenever mutants and wild type are compared we recommend using isogenic strains or strains with similar genetic background to avoid differences in the digestion time. Moreover, the efficiency of spheroplasting can be affected by growth conditions, that is, carbon source, rate of growth and stage of growth at the time of harvest. Best results are obtained with cells grown on rich medium (YPD) (Rose *et al.*, 1990) and harvested in early to mid-logarithmic phase ($0.5-1 \times 10^7$ cells/ml). When a strain background is used for the first time, it is useful to do a titration of the spheroplasting enzymes.

4.2. Antibody purification and specificity

Polyclonal antibodies can be an advantage for IF because they can recognize multiple epitopes. However, rabbit sera very often have strong background reactivity with a variety of yeast proteins, besides the desired antigen. This can be avoided in two ways: affinity purification of the specific antibodies or depletion of nonspecific antibodies by incubation with yeast deleted for the gene encoding the antigen. Affinity purification against recombinant antigen is performed as follows:

1. Transfer by Western blotting at least 50 μg of recombinant antigen to a nitrocellulose filter.
2. After staining with Ponceau red (0.05% in 3%TCA), cut out the strip containing the protein band. Wash the nitrocellulose strip 3×10 min in

1× TEN (20 mM Tris–Cl, pH 7.5, 1 mM EDTA, 140 mM NaCl), 0.05% Tween 20. Block excess protein binding sites by incubating in 1× TEN + 0.05% Tween 20 + 1% dry milk powder, at room temperature for 20 min.

3. Incubate the strip with 10–50 μl of serum (depending on antibody titer and amount of antigen loaded) in 1 ml of 1× TEN, 0.05% Tween 20, 1% dry milk powder, overnight at 4 °C with constant agitation (rocker or wheel).

4. Remove the supernatant, wash the strip 3× 10 min in 1× TEN, 0.05% Tween 20 at room temperature. Elute the bound immunoglobulin with 300 μl of cold 0.1 M glycine, pH 3.0, for 2 min.

5. Immediately raise the pH to 7.0 by adding 1 M Tris base (the volume required should be determined empirically before starting), and place on ice.

6. Repeat the elution once or twice and pool the elutions that contain antibody. Note that it may be necessary to drop the pH of the glycine to pH 1.9 for efficient elution.

7. The antibodies can be stored as aliquots at −80 °C. Stabilization is enhanced by addition of 1–2% ovalbumin and 20% glycerol. The antibody is used at a dilution of 1:2 or more for IF. The specificity of the purified antibodies should be demonstrated by Western blot and IF on strains lacking the protein in question.

If recombinant antigen is not available, rabbit sera can be preadsorbed against fixed yeast spheroplasts from a strain lacking the desired antigen. Incubation of antiserum and cells can be performed for several hours, and the nonbound antibodies are used on the test sample after sedimentation of the fixed spheroplasts.

Monoclonal antibodies usually recognize a single epitope which reduces background in yeast, yet some commonly used monoclonals (e.g., anti-HA) do recognize an endogenous yeast protein epitope. This can be tested on Western blots, although SDS denatured antigens are not always equivalent to formaldehyde fixed ones. The obvious disadvantage of staining for a unique epitope is that the risk is greater that it is masked or denatured by the fixation conditions.

The fluorophore-coupled secondary antibodies should always be tested on permeabilized material lacking the primary antibody to assess the background fluorescence created by unspecific binding of the secondary antibodies. To improve signal specificity it is advisable to preabsorb the secondary antibody on fixed yeast cells, and to dilute it maximally to avoid unnecessary background.

4.3. Choice of fluorophores

For efficient visualization of several targets, fluorophores should be chosen that are excited and visualized independently. This depends on the excitation lines and filter sets available in your microscope. If there is overlap between the emission spectra, we recommend attenuating some signals by controlling the intensity of the excitation line (e.g., on a confocal microscope) to avoid "bleed through." Some of the more commonly used fluorophores are Alexa Fluor conjugated antibodies at several excitation/emission wavelengths (Molecular Probes, Invitrogen), Cy3 ($A = 554$ nm, $E = 566$ nm) and Cy5 ($A = 649$ nm, $E = 666$ nm). The Alexa fluorophores offer the advantage of increased photostability, as compared to the older Cy dyes.

4.4. Protocol

We present here one protocol for combined IF/FISH, but the same procedure can be used to perform only IF by omitting Sections E and F, or only FISH by omitting Section D. Follow all Sections A–G for combined IF/FISH.

(A) Fixation

Cells are fixed either before or after conversion to spheroplasts by the addition of freshly dissolved paraformaldehyde (not glutaraldehyde). If preservation of cell shape and cytosolic structures is required, then cells should be fixed before spheroplasting. For detection of low abundance nuclear antigens, postspheroplasting fixation can be used. A fresh stock solution of 20% paraformaldehyde should be prepared before the experiment begins by mixing 5 g of paraformaldehyde, 15 ml H_2O and 25 μl 10 N NaOH. Dissolve at 70 °C in a closed bottle in a fume hood for about 30 min with occasional shaking. Adjust final volume to 25 ml and cool on ice. Note that paraformaldehyde fumes are toxic and care should be taken with this reagent. The commercially available 37% formaldehyde solution, while less toxic, has long formaldehyde polymers that hinder entry into cells. Glutaraldehyde should be avoided since it often masks or destroys antigenic epitopes.

1. Grow yeast cells overnight to about 1×10^7 cells/ml in 50 ml YPD or selective media (Rose *et al.*, 1990).
2. Adjust to 4% paraformaldehyde (final concentration) and incubate 15 min at room temperature. For a 20-ml culture one would add 5 ml of 20% paraformaldehyde. If fixation is performed in synthetic medium, the fixative should be quenched by adjusting to 0.25 M glycine or 0.1 M Tris–Cl, pH 8.0, after 15 min.
3. Centrifuge 5 min at 800×g.
4. Carefully resuspend the pellet in 40 ml of YPD and centrifuge 3 min at 800×g.
5. Repeat step 4.

Resuspend pellet in YPD (1/10 of initial culture volume) and keep it at 4 °C (up to overnight) or proceed to spheroplasting using the protocol below. If epitopes are of low abundance, it may be preferred to spheroplast prior to fixation. In this case start with Section B.

(B) Spheroplasting

6. Harvest cells at $1200 \times g$ for 5 min at room temperature in preweighed 50 ml polypropylene tubes.

7. Decant the supernatant and weigh the cell pellet.

8. Resuspend the cells in 1 ml/0.1 g of cells 0.1 M EDTA–KOH (pH 8.0), 10 mM DTT. DTT has to be added freshly. Use roughly 1/20 culture volume of EDTA–DTT solution.

9. Incubate at 30 °C for 10 min with gentle agitation.

10. Collect the cells by centrifugation at $800 \times g$ for 5 min at room temperature.

11. Carefully resuspend the cell pellet in 1 ml/0.1 g cells YPD + 1.2 M sorbitol (mix 22 g sorbitol with 100 ml YPD). To resuspend evenly, suspend the cell pellet first in 500 μl.

12. Add lyticase (β-glucanase; Verdier et al., 1990) to 250–500 U/ml and predissolved Zymolyase (20T, Seikagaku) to final 10–100 μg/ml. This step is critical; appropriate amounts of enzyme should be determined in a trial experiment with the same cells.

For a 20-ml culture we use 2 ml of solution with 12 μl lyticase (40,000 U/ml) and 4 μl of Zymolyase (20T) freshly dissolved in YPD at 5 mg/ml. Because diploid strains spheroplast faster than haploid strains, we often pretreat with only 1 mM DTT and use half of the final concentration of lyticase and Zymolyase for diploid cells.

13. Incubate at 30 °C in the original Erlenmeyer flask with gentle agitation (150 rpm) and monitor spheroplast formation in the microscope at 5, 10, 15, and 20 min.

The appropriate stage of spheroplasting is determined by microscopic observation with polarized light. Initially cells will have a bright interior and a bright halo. Well spheroplasted cells become dark with a bright halo around the cell shape. When cells are dark inside and do not have the bright halo outside anymore, spheroplasting has been carried out for too long. This leads to a loss of antigen by diffusion and an altered 3D structure of the cell. In a given culture, speed of spheroplasting varies among cell stages, thus it is therefore advisable to stop digestion when 50% of the cells are properly spheroplasted.

14. Dilute with YPD + 1.2 M sorbitol to 40 ml. Centrifuge 5 min at $800 \times g$.

15. Wash twice in 40 ml YPD + 1.2 M sorbitol, resuspending gently using a rubber bulb on the end of pipette (do not vortex or use glass rods). Centrifuge 5 min at $800 \times g$.

If cells were not fixed prior to spheroplasting, resuspend the spheroplasts gently in 0.5× culture volume of YPD + 1.2 M sorbitol, and fix by incubating at room temperature in 4% paraformaldehyde (final

concentration) for 15 min. All washes should be done with YPD +
1.2 *M* sorbitol to avoid cell lysis.

(C) Cell permeabilization

16. Resuspend fixed spheroplasts thoroughly in YPD + 1.2 *M* sorbitol
    (0.5 g in 0.8 ml). Sorbitol can be omitted for cells that were fixed
    first prior to spheroplasting. The concentration of cells in this
    suspension should be such that only one layer of nonconfluent
    cells will adhere to the slide. Leave a drop on each spot of Teflon-
    coated slides (Super-Teflon slides, Milian) for 1–2 min to allow
    adherence, and remove as much liquid as possible using a pipet.
    Superficially air dry 2 min. All the following washes are performed
    by immersing the slide in a Coplin jar containing the indicated
    solution.

17. Place the slides in prechilled methanol at − 20 °C for 6 min.

18. Transfer the slides to prechilled acetone at − 20 °C for 30 s.

19. Air dry 3 min.

(D) Antibody treatment (IF)

20. Incubate slides in 1× PBS (Sambrook *et al.*, 1989) + 1% ovalbu-
    min + 0.1% Triton X-100 for 20 min or more. Shake gently two
    or three times at room temperature. After this step the cells appear
    transparent and nuclei can be seen as a dark spot. This is an
    indication of good spheroplasting. If this is not the case, it may
    help to leave the slides for up to an hour in PBS + 1% ovalbumin
    + 0.1% Triton X-100.

21. Dry the Teflon surfaces and bottom of the slides with a paper
    tissue.

22. Cover each spot on the slide with 25 μl of the appropriate primary
    antibody diluted as required in PBS containing 0.1% Triton
    X-100. For affinity purified antibodies start with a 1/20 dilution
    in 0.5× PBS + 0.1% Triton X-100 to avoid high salt concentra-
    tions. For overnight incubation Triton should be avoided.

23. Incubate for 1 h at 37 °C in a humid chamber or overnight at
    4 °C. In the latter case the slides should be covered with a
    coverslip (but not sealed) to avoid drying of the antibody solution.

24. Preabsorb the secondary antibody on yeast cells. For this purpose,
    use the remaining fixed spheroplasts by washing them 3× in PBS
    containing 0.1% Triton X-100 and resuspending them in 1 ml of
    PBS. Dilute the secondary antibody (stock is usually 1 mg/ml)
    1:250 in this spheroplast suspension and incubate for 30 min on a
    rotating wheel at 4 °C in the dark. Centrifuge at top speed. Store
    on ice until needed.

25. After the primary antibody incubation, wash the slides 3× 5 min
    by immersion in PBS + 0.1% Triton X-100 in a Coplin jar at
    room temperature.

26. Dry the Teflon surfaces and bottom of the slides. Cover each slide with 25 μl/spot of the fluorescent secondary antibody after pre-adsorption and incubate for 1 h either at room temperature or 37 °C in a dark, humid chamber.

27. After the secondary antibody, wash the slides 3× 5 min in PBS + 0.1% Triton at room temperature.

(E) *In situ* hybridization probes

To label probes for FISH, plasmids containing the target sequence can be used as well as PCR fragments amplified using appropriate primers. For optimal FISH signals a fragment of 6–10 kb from a genomic locus should be used as a template for nick-translation to prepare probes. Fragments as small as 2 kb can be used, although labeling efficiency will be lower. Final probe length should be between 200 and 300 nucleotides after nick-translation. This can be checked by running the final probe on a 2% agarose gel. Probes for FISH are labeled by a nick-translation protocol for which kits are commercially available (e.g., Nick Translation Mix, Roche). The fluorescent labeling can be carried out either during the nick-translation reaction or indirectly using an antibody against modified nucleotides. Detailed protocols for probe preparation have been published previously (Gotta *et al.*, 1999; Heun *et al.*, 2001a).

Direct labeling of the probe is achieved by using a fluorescently labeled dUTP (Alexa fluor dUTP, Invitrogen) in place of dTTP in the nick-translation reaction. Efficiency of the Alexa-dUTP incorporation into the probe can be quantified using a Fluorimeter (NanoDrop), or the fluorescence in the dried probe pellet can be directly visualized under a fluorescent microscope. Alternatively, commercially available kits offer a two-step labeling using amine-modified dUTP, which will then be cross-linked to fluorochromes (FISH-Tag, Invitrogen). Since the amine modification is small compared to Alexa molecules, the nick-translation reaction is more efficient and the resulting probe is brighter. Finally, probe labeling can also be achieved using digoxigenin-derivatized dUTP (dig-dUTP, Roche). Note that the detection of the digoxigenin-derivatized dUTP will require an antidigoxigenin fluorescent primary antibody or an antidigoxigenin primary antibody and a fluorescently labeled secondary antibody. This approach can be used to amplify weak signals.

(F) FISH

If only FISH is to be performed go directly from step 19 (cell permeabilization) to step 30. For a combined IF/FISH protocol continue here with step 28, which prevents primary or secondary antibody dissociation under the harsh conditions used for FISH

28. Postfix the cells in 4× SSC, 4% paraformaldehyde 20 min at room temperature after the last wash. Rinse 3× 3 min in 4× SSC.
29. Immerse cells in 4× SSC, 0.1% Tween 20, 20 μg/ml preboiled RNaseA (optional). Incubate overnight at room temperature (in the dark if IF was performed).
30. Wash in H_2O.
31. Dehydrate in ethanol: 70%, 80%, 90%, and 100% consecutively at -20 °C, 1 min each bath.
32. Air dry.
33. Add 10 μl/spot of 2× SSC, 70% formamide. Cover with a coverslip. Leave 5 min at 72 °C (place the slide on top of an aluminum block which is partially submerged in a 72 °C waterbath. On the narrow edges of the slide, place few drops of water, which will spread between the aluminum block and the slide, improving the heat conductance).
34. Dehydrate in ethanol: 70%, 80%, 90%, and 100% consecutively at -20 °C, 1 min/bath.
35. Air dry.
36. Apply hybridization solution, 3 μl for each spot. The optimal concentration of probe depends on the sequence and must be determined empirically. Place a coverslip on top avoiding air bubbles, seal with nail polish.
37. Incubate 10 min at 72 °C.
38. Incubate 24–60 h at 37 °C.
39. Remove the coverslip and wash twice in 0.05× SSC, 5 min at 40 °C.
40. Incubate in BT buffer (0.15 M $NaHCO_3$, 0.1% Tween 20, pH 7.5) 0.05% BSA, 2× 30 min at 37 °C in the dark.
41. If FISH probe was labeled using digoxigenin-derivatized dUTP continue with step 42.
    If probe was done using a fluorescent dUTP go directly to visualization or stain DNA by following step 45.
42. Add mouse antidigoxigenin diluted 1:50 in BT buffer without BSA + the secondary goat–anti-mouse or rabbit antibody 1:50 (for refreshing the IF signal, if necessary; Boehringer Mannheim). Stock solutions are usually 1 mg/ml. At this point you can either use derivatized sheep anti-Dig (rhodamine or FITC derivatized) or detect the protein two steps, first with a nonderivatized primary mouse–anti-DIG, and then with a secondary fluorescent antibody to amplify the anti-DIG signal. For two-step labeling, repeat steps 42–44 twice.
43. Incubate 1 h at 37 °C in a humid chamber.
44. Wash 5× 3 min in BT buffer.

(G) DNA visualization

To visualize DNA, you must avoid the wavelengths of excitation and emission relevant for the fluorophores used. The most frequent staining agents used are ethidium bromide (diluted to 1 μg/ml in antifade reagent excitation 518 nm/emission 605 nm), DAPI (1 μg/ml, excitation 358 nm/emission 461 nm), or cyanine nucleic acid dyes (TOTO/POPO/YOYO/BOBO family of dyes, Molecular Probes).

45. Add 25 μl/spot of the DNA stain agent diluted in 1× PBS + 0.1% Triton X-100 for 10min at room temperature.
46. Wash in 1× PBS + 0.1% Triton X-100.
47. Dry the black Teflon surface and bottom of the slides and add one drop of antifade solution (Prolong antifade, Invitrogen). An alternative antifade is 1× PBS, 50% glycerol, 24 μg diazabicyclo-2,2,2-octane or DABCO, pH 7.5.
48. Cover with a coverslip avoiding air bubbles. Slides can be examined immediately or kept at 4 °C in the dark overnight. For longer storage, seal the coverslip with nail polish and keep at 4 °C in the dark or at -80 °C.

4.5. Special notes

To monitor different targets at a time, primary antibodies from different species must be used (e.g., mouse, rabbit, sheep) and species-specific secondary antibodies. To reduce incubation times, we recommend mixing primary or secondary antibodies. However, it is essential to pretest the secondary antibody with each of the primary antibodies separately to ensure that they do not cross react.

An alternative way to localize proteins both in living and in fixed cells is to generate a GFP fusion (Shaw et al., 1997), although proteins fused to GFP must be tested for proper functionality. When the GFP-fluorescence signal is very strong (abundant or overexpressed proteins), it can sometimes be visualized after the IF protocol. For weaker signals, or CFP fusions, samples should be fixed with 1% paraformaldehyde for 3 min and washed at least three times with 1× PBS. These samples need to be visualized by microscopy as quickly as possible. Epifluorescence (particularly for CFP) will not last long than a week at 4 °C. For visualization of a strong GFP signal cells can also be fixed with ethanol 80% for 5 min and washed with 1× PBS containing DAPI. Alternatively GFP fusions can be detected using the IF protocol and anti-GFP antibodies.

It is not always necessary to preserve 3D nuclear structure, for example, for scoring mitotic or meiotic chromosome pairing (Guacci et al., 1994; Weiner and Kleckner, 1994). However, one must be careful not to draw conclusions about nuclear architecture from results obtained with flattened or spread preparations.

For time-course experiments, or when a large number of samples need to be handled (20 or more), we recommend to fixing overnight at 4 °C, and performing spheroplasting the next day. For unexplained reasons, one uses half the amount of lyticase and Zymolyase under these circumstances. Moreover, spheroplasts can be spotted on glass, permeabilized and kept at 4 °C in blocking solution without Triton X-100 for extended periods of time. Prolonged exposure to Triton X-100 should be avoided. Some protocols recommend coating slides with poly-lysine (Sigma, P8920) to promote spheroplast or cell attachment, but we avoid it because it increases background fluorescence. Plastic multiwell slides (μ-Slide, Ibidi) can be used to spot multiple samples on one slide, reducing the number of slides needed.

This protocol is not only useful for *S. cerevisiae*, but has also been successfully used for *Neurospora crassa*. We used Novozyme 234 (Novo Biolabs) instead of Zymolyase to digest the *Neurospora* cell wall, and incubation times with the antibodies tested were increased to 48 h.

REFERENCES

Belmont, A. S. (2001). Visualizing chromosome dynamics with GFP. *Trends Cell Biol.* **11**, 250–257.

Berg, H. C. (1993). *Random Walks in Biology*. Princeton University Press, Princeton, NJ.

Berger, A. B., Cabal, G. G., Fabre, E., Duong, T., Buc, H., Nehrbass, U., Olivo-Marin, J. C., Gadal, O., and Zimmer, C. (2008). High-resolution statistical mapping reveals gene territories in live yeast. *Nat. Methods* **5**, 1031–1037.

Bouchaud, J.-P., and Georges, A. (1990). Anomalous diffusion in disordered media: Statistical mechanisms, models and physical applications. *Phys. Rep.* **195**, 127–293.

Bressan, D. A., Vazquez, J., and Haber, J. E. (2004). Mating type-dependent constraints on the mobility of the left arm of yeast chromosome III. *J. Cell Biol.* **164**, 361–371.

Bystricky, K., Laroche, T., van Houwe, G., Blaszczyk, M., and Gasser, S. M. (2005). Chromosome looping in yeast: Telomere pairing and coordinated movement reflect anchoring efficiency and territorial organization. *J. Cell Biol.* **168**, 375–387.

Gehlen, L. R. (2009). *Biophysical analysis of diffusion controlled processes in the budding yeast nucleus*. PhD thesis, University of Basel.

Gotta, M., Laroche, T., Formenton, A., Maillet, L., Scherthan, H., and Gasser, S. M. (1996). The clustering of telomeres and colocalization with Rap1, Sir3, and Sir4 proteins in wild-type Saccharomyces cerevisiae. *J. Cell Biol.* **134**, 1349–1363.

Gotta, M., Laroche, T., and Gasser, S. M. (1999). Analysis of nuclear organization in Saccharomyces cerevisiae. *Methods Enzymol.* **304**, 663–672.

Guacci, V., Hogan, E., and Koshland, D. (1994). Chromosome condensation and sister chromatid pairing in budding yeast. *J. Cell Biol.* **125**, 517–530.

Hediger, F., Dubrana, K., and Gasser, S. M. (2002). Myosin-like proteins 1 and 2 are not required for silencing or telomere anchoring, but act in the Tel1 pathway of telomere length control. *J. Struct. Biol.* **140**, 79–91.

Hediger, F., Taddei, A., Neumann, F. R., and Gasser, S. M. (2004). Methods for visualizing chromatin dynamics in living yeast. *Methods Enzymol.* **375**, 345.

Heun, P., Laroche, T., Raghuraman, M. K., and Gasser, S. M. (2001a). The positioning and dynamics of origins of replication in the budding yeast nucleus. *J. Cell Biol.* **152,** 385–400.

Heun, P., Laroche, T., Shimada, K., Furrer, P., and Gasser, S. M. (2001b). Chromosome dynamics in the yeast interphase nucleus. *Science* **294,** 2181–2186.

Hom, E. F., Marchis, F., Lee, T. K., Haase, S., Agard, D. A., and Sedat, J. W. (2007). AIDA: An adaptive image deconvolution algorithm with application to multi-frame and three-dimensional data. *J. Opt. Soc. Am. A Opt. Image Sci. Vis.* **24,** 1580–1600.

Kitamura, E., Blow, J. J., and Tanaka, T. U. (2006). Live-cell imaging reveals replication of individual replicons in eukaryotic replication factories. *Cell* **125,** 1297–1308.

Lisby, M., Mortensen, U. H., and Rothstein, R. (2003). Colocalization of multiple DNA double-strand breaks at a single Rad52 repair centre. *Nat. Cell Biol.* **5,** 572–577.

Marshall, W. F., Straight, A., Marko, J. F., Swedlow, J., Dernburg, A., Belmont, A., Murray, A. W., Agard, D. A., and Sedat, J. W. (1997). Interphase chromosomes undergo constrained diffusional motion in living cells. *Curr. Biol.* **7,** 930–939.

Michaelis, C., Ciosk, R., and Nasmyth, K. (1997). Cohesins: Chromosomal proteins that prevent premature separation of sister chromatids. *Cell* **91,** 35–45.

Neumann, F. R., Hediger, F., Taddei, A., and Gasser, S. M. (2006). Tracking individual chromosomes with integrated arrays of lac^{op} sites and GFP-lacI repressor: Analysing position and dynamics of chromosomal loci in *Saccharomyces cerevisiae*. *In* "Cell Biology A Laboratory Handbook." (J. E. Celis, ed.), Vol. 2, pp. 359–367. Elsevier Science.

Neumann, F. R., Dion, V., Gehlen, L. R., Tsai-Pflugfelder, M., Taddei, A., and Gasser, S. M. INO80 promotes chromatin movement and functionally impacts homologous recombination. submitted.

Rohner, S., Gasser, S. M., and Meister, P. (2008). Modules for cloning-free chromatin tagging in Saccharomyces cerevisiae. *Yeast* **25,** 235–239.

Rose, M. D., Winston, F., and Hieter, P. (1990). Methods in Yeast Genetics. Cold Spring Harbor Laboratory Press, New York.

Sage, D., Neumann, F. R., Hediger, F., Gasser, S. M., and Unser, M. (2005). Automatic tracking of individual fluorescence particles: Application to the study of chromosome dynamics. *IEEE Trans. Image Process.* **14,** 1372–1383.

Sambrook, J., Fritsch, E. F., and Maniatis, T. (1989). *Molecular Cloning: A Laboratory Manual* 2nd ed. Cold Spring Harbor Laboratory Press, Cold Spring Harbor, NY.

Schober, H., Ferreira, H., Kalck, V., Gehlen, L. R., and Gasser, S. M. (2009). Yeast telomerase and the SUN domain protein Mps3 anchor telomeres and repress subtelomeric recombination. *Genes Dev.* **23,** 928–938.

Shaw, S. L., Yeh, E., Bloom, K., and Salmon, E. D. (1997). Imaging green fluorescent protein fusion proteins in Saccharomyces cerevisiae. *Curr. Biol.* **7,** 701–704.

Straight, A. F., Belmont, A. S., Robinett, C. C., and Murray, A. W. (1996). GFP tagging of budding yeast chromosomes reveals that protein-protein interactions can mediate sister chromatid cohesion. *Curr. Biol.* **6,** 1599–1608.

Taddei, A., Hediger, F., Neumann, F. R., Bauer, C., and Gasser, S. M. (2004). Separation of silencing from perinuclear anchoring functions in yeast Ku80, Sir4 and Esc1 proteins. *EMBO J.* **23,** 1301–1312.

Verdier, J. M., Stalder, R., Roberge, M., Amati, B., Sentenac, A., and Gasser, S. M. (1990). Preparation and characterization of yeast nuclear extracts for efficient RNA polymerase B (II)-dependent transcription in vitro. *Nucleic Acids Res.* **18,** 7033–7039.

Weiner, B. M., and Kleckner, N. (1994). Chromosome pairing via multiple interstitial interactions before and during meiosis in yeast. *Cell* **77,** 977–991.

CHAPTER TWENTY-TWO

QUANTITATIVE LOCALIZATION OF CHROMOSOMAL LOCI BY IMMUNOFLUORESCENCE

Donna Garvey Brickner, William Light, *and* Jason H. Brickner

Contents

Abstract

DNA within the yeast nucleus is spatially organized. Yeast telomeres cluster together at the nuclear periphery, centromeres cluster together near the spindle pole body, and both the rDNA repeats and tRNA genes cluster within the nucleolus. Furthermore, the localization of individual genes to subnuclear compartments can change with changes in transcriptional status. As such, yeast researchers interested in understanding nuclear events may need to determine the subnuclear localization of parts of the genome. This chapter describes a straightforward quantitative approach using immunofluorescence and confocal microscopy to localize chromosomal loci with respect to well characterized nuclear landmarks.

Chromosomes within the yeast nucleus are spatially organized. Parts of chromosomes are associated with different subnuclear compartments such as the nucleolus, the nuclear envelope or the spindle pole body (Berger *et al.*,

Department of Biochemistry, Molecular Biology and Cell Biology, Northwestern University, Evanston, Illinois, USA

Methods in Enzymology, Volume 470
ISSN 0076-6879, DOI: 10.1016/S0076-6879(10)70022-7

2008; Dorn *et al.*, 2007; Hueun *et al.*, 2001; Jin *et al.*, 2000). Localization to these subnuclear organelles is not static. Chromosomal DNA is in constant motion and exhibits varying degrees of constrained diffusion (Cabal *et al.*, 2006; Casolari *et al.*, 2004; Chuang *et al.*, 2006; Schmid *et al.*, 2006; Shav-Tal *et al.*, 2004). Sophisticated techniques have been developed to understand the dynamics of chromosomal elements in living cells (Chapter 21). However, to determine where a gene localizes within the nucleus and how its localization is affected by inputs of interest (such as transcriptional activation), simpler methods can be used. Here, we describe a quantitative method for localizing chromosomal loci with respect to subnuclear landmarks using established immunofluorescence methods.

1. YEAST STRAIN CONSTRUCTION

This protocol is based on binding of the lac repressor from *Escherichia coli* to an array of lac repressor-binding sites integrated near the chromosomal locus of interest (Fig. 22.1A; Robinett *et al.*, 1996; Straight *et al.*, 1996). Similar experiments can be carried out using the Tet repressor array (Abruzzi *et al.*, 2006; Cabal *et al.*, 2006; Chekanova *et al.*, 2008; Dundr *et al.*, 2007; Fischer *et al.*, 2004; Köhler *et al.*, 2008; Kumaran ad Spector, 2008). Depending on the sensitivity of the microscope being used, this method requires ≥ 100 binding sites. In our experiments, we readily visualize an array of ∼ 128 lac repressor-binding sites (lac operators/LacO array). To introduce this array at a locus of interest, the LacO array is first cloned into a plasmid with an appropriate selective marker (Fig. 22.1A). The LacO array was originally cloned in plasmid pAFS52, an integrating *TRP1*-marked plasmid (Brickner and Walter, 2004; Straight *et al.*, 1996). We have also moved the LacO array (as a *Hin*DIII–*Xho*I fragment) into pRS306 (*URA3*-marked integrating plasmid; Sikorski and Heiter, 1989) to create p6LacO128 (Brickner and Walter, 2004). Whenever cloning the LacO array, it is important to confirm that the array is the expected size (> 5 kb) by restriction digestion, as the array is sometimes lost or reduced in size after propagation in *E. coli*.

The next step is to introduce a targeting sequence into this plasmid so that the LacO array can be integrated at a locus of interest by homologous recombination. Because the sequences in the plasmid will be duplicated upon recombination (Fig. 22.1A), it is important to consider carefully which targeting sequences to clone. When localizing genes, we usually use sequences at the 3′ end of the gene to introduce the LacO array downstream of the 3′UTR and to avoid duplicating the promoter. Also, to allow for homologous recombination, the targeting sequence must include a

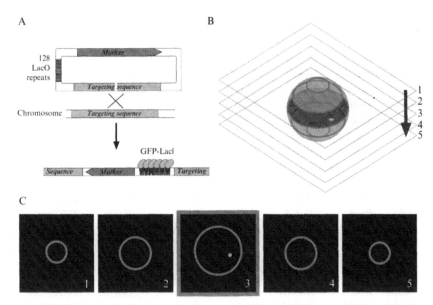

Figure 22.1 Schematic of the chromatin localization assay. (A) Strategy for marking a locus of interest with GFP. A plasmid containing the lac repressor array (LacO) and both a marker and a targeting sequence is digested at a unique restriction site within the targeting sequence. Transformation and homologous recombination introduce the marker and the lac repressor array into the yeast genome, flanked by the targeting sequence. The strain into which the plasmid is transformed also expresses the Lac repressor fused to GFP (GFP-LacI). (B) Confocal slices through a nucleus. The nuclear envelope is stained and a series of slices (numbered 1–5) are collected along the z-axis. Not shown in this representation is the staining of the cortical endoplasmic reticulum, which is visible when the Sec63-13myc marker is used. (C) Selection of the optimal slice. Slices 1–5 from panel (B) are shown. Slice #3 has the brightest, most focused GFP-LacI spot and is selected for scoring. The GFP-LacI spot in this cell is scored as nucleoplasmic.

restriction site that will be unique in the context of the LacO array plasmid. Table 22.1 shows a list of sites that are absent from the LacO array.

Once a targeting sequence has been introduced into the LacO array plasmid, it is digested at a unique restriction site within the targeting sequence and transformed into yeast. We typically start with a yeast strain that has previously been transformed with the GFP-Lac repressor (GFP-LacI) and, where necessary, additional tagged proteins localizing to other subnuclear domains (e.g., Sec63-13myc for the nuclear envelope/endoplasmic reticulum). The LacO array should be introduced last because not all transformants will possess a full-length array. To identify transformants that possess the full-length array, we screen through four or five clones to identify those that exhibit a clear green dot of GFP fluorescence. Once we have confirmed that the array is intact, we create a frozen stock of this strain.

Table 22.1 Sites absent from the LacO array

AarI	BcefI	BsmAI	ClaI	Hpy188I	PshAI	SspD5I
AatII	BcgI	BsmBI	Csp6I	HpyCH4III	PsiI	Sth132I
Acc65I	BciVI	BsmFI	CstMI	HpyCH4IV	Psp03I	StsI
AceIII	BclI	Bsp24I	DdeI	KpnI	PspGI	StuI
AclI	BfrBI	Bsp1286I	DpnI	MaeIII	PspOMI	StyI
AcuI	BglI	BspCNI	DraI	MboI	PsrI	StyD4I
AfeI	BglII	BspEI	DraIII	MboII	PssI	SwaI
AflII	BlpI	BspHI	DrdI	MfeI	PvuI	TaiI
AgeI	Bme1580I	BsrI	EagI	MluI	PvuII	TaqII
AhdI	BmgBI	BsrDI	EarI	MmeI	RleAI	TatI
AleI	BmrI	BsrFI	EciI	MseI	RsaI	TauI
AlfI	BmtI	BsrGI	EcoHI	MslI	RsrII	TfiI
AloI	BplI	BssHII	EcoICRI	MspA1I	SacI	TseI
AlwI	BpmI	BssSI	Eco57MI	MwoI	SacII	Tsp45I
AlwNI	Bpu10I	BstAPI	EcoNI	NaeI	SanDI	TspDTI
ApaI	BpuEI	BstBI	EcoO109I	NciI	SapI	TspGWI
ApaBI	BsaI	BstEII	FalI	NcoI	Sau96I	TspRI
ApaLI	BsaAI	BstF5I	FauI	NdeI	ScaI	Tth111I
AscI	BsaBI	BstKTI	FmuI	NgoMIV	ScrFI	Tth111II
AseI	BsaJI	BstNI	Fnu4HI	NheI	SelI	UnbI
AsiSI	BsaWI	BstUI	FokI	NotI	SexAI	UthSI
AvaII	BsaXI	BstXI	FseI	NruI	SfaNI	VpaK11AI
AvrII	BscAI	BstYI	FspI	NsiI	SfiI	XcmI
BaeI	BseMII	BstZ17I	FspAI	PacI	SgrAI	XmaI
BamHI	BseRI	Bsu36I	GdiII	PasI	SimI	ZraI
BanII	BseYI	BtgI	HaeIV	PflMI	SmaI	
Bbr7I	BsgI	BtgZI	HgaI	PfoI	SnaBI	
BbsI	BsiEI	BthCI	Hin4I	PmeI	SpeI	
BbvI	BsiHKAI	BtsI	HpaI	PmlI	SrfI	
BbvCI	BsiWI	CdiI	HpaII	PpiI	Sse232I	
BccI	BslI	ChaI	HphI	Ppu10I	Sse8647I	
BceAI	BsmI	CjePI	Hpy99I	PpuMI	SspI	

2. IMMUNOFLUORESCENCE

Having tagged a locus of interest, it can be colocalized with respect to subnuclear structures either in live cells or in fixed cells. Chapter 21 describes methods for imaging chromosomal loci in live cells and defining their dynamic behavior. Here, we describe how to use immunofluorescence methods with populations of fixed cells to determine the localization of genes. The resulting localization represents a dynamic distribution and is expressed as the fraction of the population in which the chromosomal locus and the protein marker. We focus on the localization of chromosomal loci with respect to the nuclear periphery or the nucleolus. We have successfully used several markers for the nuclear periphery. We have used the 9E10 anti-myc monoclonal antibody (Santa Cruz Biotechnology) to localize a 13myc-tagged Sec63 as a marker for the nuclear envelope (Brickner and Walter, 2004). This protein localizes throughout the endoplasmic reticulum (Gilmore, 1991). We have also used the 32D6 anti-Nsp1 monoclonal antibody from EnCor Biotechnology (Gainsville, FL) to stain nuclear pore complexes. This has the advantage that it does not require expression of a tagged protein. To stain the nucleolus, we have used the 37C12 monoclonal anti-Nop5/6 antibody from EnCor Biotechnology and we have used epitope-tagged versions of Spc42 as a marker for the spindle pole body.

3. FIXING CELLS

Most immunofluorescence protocols use formaldehyde fixation. This works well for certain proteins and certain organelles. However, we have found that the shape of the nucleus can be poorly preserved by formaldehyde fixation (Fig. 22.2). Therefore, we use methanol fixation. This fixation

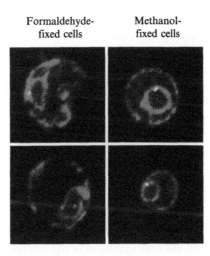

Formaldehyde-fixed cells   Methanol-fixed cells

Figure 22.2 Methanol fixation versus formaldehyde fixation. Shown are representative examples of cells fixed using either methanol (as described in the protocol) or 4% formaldehyde (twice for 30 min). The ER/nuclear envelope was stained with Sec63-myc (red) and the chromosomal locus (in this case, the *INO1* gene) was stained with the GFP-Lac repressor (green). Note the nonspherical shape of the nucleus in the formaldehyde-fixed cells. (See Color Insert.)

method causes cells to shrink somewhat but it preserves the spherical shape of the nucleus better than in formaldehyde-fixed cells.

1. Grow cells to $OD_{600} \sim 0.3$–0.7.

Note: This protocol has been used to examine cells from many different media, including both rich and synthetic media. Overgrown cells are more difficult to spheroplast and stain.

2. Harvest 10^7–10^8 cells by centrifugation and discard the supernatant. We typically harvest a culture of 10–15 ml at an $OD_{600} \sim 0.1$–0.7.
3. Suspend the cells in 1 ml of chilled methanol (store at $-20\,^\circ C$).
4. Incubate at $-20\,^\circ C$ for ≥ 20 min.
5. Harvest cells by centrifugation and resuspend in 1 ml SHA (1 *M* sorbitol, 50 m*M* HEPES, pH 6.8, 1 m*M* NaN$_3$). The fixed cells can be stored at $4\,^\circ C$ (good for 4 or 5 days).

4. SPHEROPLASTING

Before permeablizing cells, the cell wall must be removed. Once the cells are converted to fixed spheroplasts they should be processed for immunofluorescence immediately.

1. Harvest 5×10^7 cells by centrifugation, 30 s.
2. Resuspend in 1 ml SHA + 0.2% β-mercaptoethanol.
3. Add 2.5 μl (50 units) lyticase (Sigma catalog #L4025).
4. Incubate at $30\,^\circ C$ for 30 min. Invert tubes occasionally.
5. Harvest by centrifugation, 30 s.
6. Resuspend in 0.5 ml SHA + 0.1% Triton X-100. Incubate 10 min at room temperature.
7. Harvest by centrifugation, 30 s.
8. Resuspend in 250 μl SHA.

5. PREPARING SLIDES

Ten well slides from Carlson Scientific (Peotone, IL; catalog # 101007) are used. Depending on the configuration of the microscope, we usually do not use the column of wells closest to the edge of the slides because they are more difficult to image. Between strains or treatments, we leave a column of wells empty to avoid cross-contamination. For washes, we add buffers with a Pasteur pipette and carefully remove the buffers by aspiration (using low vacuum and holding the tip at an angle beside the well).

1. Coat each well with 20 μl 0.1% polylysine (Sigma-Aldrich, catalog #P 8920).
2. Incubate $\geq$ 15 min at room temperature.
3. Remove by aspiration and let dry completely.
4. Add 20 μl fixed spheroplasts to each well and let settle for 15 min. Aspirate to remove excess liquid.
5. Wash cells twice with Buffer WT (1% nonfat dry milk 0.5 mg/ml BSA 200 mM NaCl 50 mM HEPES–KOH (pH 7.5) 1 mM NaN$_3$ 0.1% Tween-20)

6. ANTIBODY INCUBATIONS

1. Dilute the primary antibodies into Buffer WT.

Dilutions for antibodies we have used: anti–myc (1:200), anti–Nsp1 (1:200), anti–GFP (1:1000), anti–Nop5/6 (1:200), anti–HA (1:200).

2. Add 15 μl/well of diluted 1° Ab in Buffer WT.
3. Incubate 60–90 min at RT.

This incubation can be carried out overnight at 4 °C.

4. Wash five times with 20 μl Buffer WT.
5. Dilute secondary antibodies in Buffer WT.

We use Alexa Fluor® 594 goat antimouse IgG (Invitrogen catalog #A-11032) and Alexa Fluor® 488 goat antirabbit IgG (Invitrogen catalog #A-11008), diluted 1:200.

6. Add 15 μl of diluted secondary antibody in Buffer WT.
7. Incubate 60–90 min at room temperature in the dark.

This incubation can be carried out overnight at 4 °C.

8. Wash five times with 20 μl Buffer WT.

To stain for DNA, include 0.3 μg/ml DAPI in the third wash, incubate for 1 h at 4 °C (in the dark), and wash twice more.

7. MOUNTING AND STORAGE OF SLIDES

1. After aspirating most of the liquid from each well on a slide (without letting them dry completely), quickly add 1–2 μl of Vectashield mounting solution (Vector Laboratories, catalog #H-1000).

Aspirate and add mounting medium to each well before moving on to the next well.

2. Carefully cover wells with a clean 50 mm coverslip.

The mounting medium should fill the wells without overflowing into neighboring wells.

3. Seal the slide by painting the seams with clear fingernail polish. Let dry in the dark.
4. The slides may be stored in the refrigerator, but they look best when they are fresh.

8. MICROSCOPY AND ANALYSIS

We use a Zeiss LSM510 confocal microscope with a 100× 1.4NA objective to image the cells. We have found that confocal imaging is the best system for accurate localization of genes within the nucleus. We routinely visualize fixed and stained cells using both a 30-mW 458/488/514 nm Argon laser and a 1-mW 543 nm Helium Neon laser. The pinhole is set to 146 nm, with the detector gain set to 750–900 and the amplifier offset is ~ -0.3. It is not necessary to collect three-dimensional stacks or to reconstruct whole cells. We use the motorized stage control to step through the nuclear volume to find a confocal slice (~ 0.7 μm in z dimension) that displays the most intense and most focused spot for the lac repressor-GFP (schematized in Fig. 22.1C; slice #3). Then we collect data from this slice in both channels. For experiments in which we are assessing peripheral localization, we only include cells in which the selected slice displays a clear ring staining for the nucleus (i.e., not at the top or the bottom of the nucleus, nor cells in which the nuclear envelope is incompletely stained).

For each cell, we compare the localization of the GFP-Lac repressor (GFP-LacI) spot and the marker of interest (e.g., nuclear envelope or nucleolus). In the case of the nuclear envelope, we bin cells into one of two classes (Brickner and Walter, 2004). If the center of the GFP spot is overlapping with the nuclear envelope, we classify the spot's localization as peripheral. If the center of the GFP spot is not overlapping with the nuclear envelope, we classify the spot's localization as nucleoplasmic. For each biological replicate, we collect data from 30 to 50 cells. For a given population, we determine the percentage of cells in which the spot localizes to the nuclear periphery.

The light microscope has a resolution of 100–200 nm in the X–Y dimensions (Born and Wolf, 1998; Pawley, 2006; Schermelleh *et al.*, 2008). The radius of the haploid yeast nucleus is approximately 1 μm. Therefore, GFP spots within the shell corresponding to the outermost $\sim 10\%$ of the radius will be scored as peripheral in our assay. The fraction of cells in which the GFP spot would be expected to localize at the periphery by chance can be calculated from the fraction of the total volume represented by this outermost 10% of the radius. The volume of sphere is biased to the outside; comparing spheres having radii of 0.9 μm versus 1 μm, the small sphere has a volume (3.1 μm^3), that is 74% of the volume of the larger sphere (4.2 μm^3). Therefore, this outermost shell corresponds to 26% of the volume of the nucleus and an unbiased distribution within the yeast nucleus should result in ~26% colocalization with the nuclear membrane. This level of peripheral localization is the baseline for this assay. Peripherally localized chromosomal loci score as peripheral in 60–85% of cells using the scoring criterion above. For example, genes such as INO1 and GAL1 that are recruited to the nuclear periphery upon activation, localize at the periphery in ~30% of cells when they are repressed and in ~65% of cells when they are activated (Brickner and Walter, 2004; Brickner et al., 2007; Casolari et al., 2004, 2005; Schmid et al., 2006). More stably associated peripheral loci such as telomeres localize at the nuclear periphery in ~85% of cells using this assay (our unpublished data).

To determine the variance in the peripheral localization for a particular locus, we use three or more biological replicates (i.e., cells harvested from independent cultures). From these measurements we determine the mean value and the standard error of the mean. As a negative control for peripheral localization, we use the URA3 gene, a nucleoplasmic locus that localizes at the nuclear periphery in ~30% of cells (Brickner and Walter, 2004). To determine if peripheral localization is statistically significant, we use an unpaired t-test to compare the percentage of cells with peripheral localization of the locus of interest with the percentage of cells with peripheral localization of the URA3 gene. In a typical experiment, the URA3 gene localizes at the nuclear periphery in 30% $\pm$ 5% of cells and a peripheral gene like INO1 localizes at the nuclear periphery in 65% $\pm$ 5% of cells ($P = 0.0078$).

We have used a similar strategy to localize chromosomal loci with respect to the nucleolus. The nucleolus in yeast is a single, crescent-shaped organelle that usually aligns opposite the spindle pole body (Stone et al., 2000; Yang et al., 1989). The rDNA repeats and tRNA genes localize within the nucleolus (Thompson et al., 2003). Recent work using high-resolution probabilistic analysis has shown that, although most RNA polymerase II-transcribed genes are excluded from the nucleolus, some localize within the nucleolus (Berger et al., 2008). In particular, the GAL2 gene localizes within the nucleolus and becomes localized at the periphery of the nucleolus upon activation. We have also observed the nucleolar localization of GAL2 using immunofluorescence (Fig. 22.3). Immunofluorescence using commercially available antibodies against the Nop5 and Nop6 proteins defines the nucleolus, which typically occupies approximately one-third of the nuclear volume and appears as a crescent or spherical shape. Using the same confocal microscopy analysis, we have quantified the colocalization of GAL2 or INO1 with the nucleolus. The GAL2 gene colocalizes

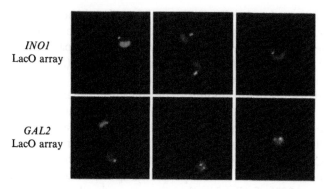

Figure 22.3   Nucleolar colocalization. Cells were fixed and probed with anti-GFP and anti-Nop5/6 as described in the protocol. In the cells shown in the top panels, the LacO array was integrated beside the *INO1* gene. In the cells shown in the bottom panels, the LacO array was integrated beside the *GAL2* gene. Note the clear separation of the *INO1* gene from the nucleolus and the localization of *GAL2* within the nucleolus.

with the nucleolus in most (~90%) of the cells in the population (Fig. 22.3; Gard *et al.*, 2009). In contrast, the *INO1* gene localizes within the nucleolus in ~10% of cells (Fig. 22.3; our unpublished data). Therefore, although the nucleolus occupies a significant fraction of the nuclear volume, the baseline colocalization of chromosomal loci with this structure is lower than expected. This suggests that cells in which a nonnucleolar chromosomal locus appears nucleolar because it is above or below the nucleolus is not a major source of background in these measurements. Therefore, this method readily distinguishes between two classes of RNA polymerase II-transcribed genes that differ in their localization with respect to the nucleolus.

The methods described in this chapter allow colocalization of chromosomal loci with respect to two major subnuclear compartments. It remains to be seen what fraction of the yeast genome localizes to particular subnuclear domains. It is conceivable that there are few parts of the genome that are randomly localized. As the spatial organization of the nucleus becomes better understood, we anticipate that these methods can be adapted and improved to allow the localization of loci with respect to additional subnuclear compartments.

ACKNOWLEDGMENTS

The authors thank John Sedat for sharing the GFP-Lac repressor and the LacO array plasmids, Francoise Stutz for the *GAL2* LacO plasmid, Kevin Redding for developing the immunofluorescence protocol on which these methods are based and members of the Brickner laboratory for helpful comments on this chapter. J. H. B. is supported by NIH grant GM080484, the W. M. Keck Foundation and the Baldwin Fund for Biomedical Research.

REFERENCES

Abruzzi, K. C., Belostotsky, D. A., Chekanova, J. A., Dower, K., and Rosbash, M. (2006). 3′-End formation signals modulate the association of genes with the nuclear periphery as well as mRNP dot formation. *EMBO J.* **25**, 4253–4262.

Berger, A. B., Cabal, G. G., Fabre, E., Duong, T., Buc, H., Nehrbass, U., Olivo-Marin, J., Gadal, O., and Zimmer, C. (2008). High-resolution statistical mapping reveals gene territories in live yeast. *Nat. Methods* **5**, 1031–1037.

Born, M., and Wolf, E. (eds.) (1998). *In* "Principles of Optics", Cambridge University Press, Cambridge.

Brickner, J. H., and Walter, P. (2004). Gene recruitment of the activated *INO1* locus to the nuclear membrane. *PLoS Biol.* **2**, 1843–1853.

Brickner, D. G., Cajigas, I., Fondufe-Mittendorf, Y., Ahmed, S., Lee, P. C., Widom, J., and Brickner, J. H. (2007). H2A.Z-mediated localization of genes at the nuclear periphery confers epigentic memory of previous transcriptional state. *PLoS Biol.* **5**, 0704–0716.

Cabal, G. G., Genovesio, A., Rodriguez-Navarro, S., Zimmer, C., Gadal, O., Lesne, A., Buc, H., Feuerbach-Fournier, F., Olivo-Marin, J., Hurt, E. C., and Nehrbass, U. (2006). SAGA interacting factors confine sub-diffusion of transcribed genes to the nuclear envelope. *Nature* **441**, 770–773.

Casolari, J. M., Brown, C. R., Komili, S., West, J., Hieronymus, H., and Silver, P. A. (2004). Genome-wide localization of the nuclear transport machinery couples transcriptional status and nuclear organization. *Cell* **117**, 427–439.

Casolari, J. M., Brown, C. R., Drubin, D. A., Rando, O. J., and Silver, P. A. (2005). Developmentally induced changes in transcriptional program alter spatial organization across chromosomes. *Genes Dev.* **19**, 1188–1198.

Chekanova, J. A., Abruzzi, K. C., Rosbash, M., and Belostotsky, D. A. (2008). Sus1, Sac3, and Thp1 mediate post-transcriptional tethering of active genes to the nuclear rim as well as to non-nascent mRNP. *RNA* **14**, 66–77.

Chuang, C. H., Carpenter, A. E., Fuchsova, B., Johnson, T., de Lanerolle, P., and Belmont, A. S. (2006). Long-range directional movement of an interphase chromosome site. *Curr. Biol.* **16**, 825–831.

Dorn, J. K., Rines, D. R., Jelson, G. S., Sorger, P. K., and Danuser, G. (2007). Yeast kinetochore microtubule dynamics analyzed by high-resolution three-dimensional microscopy. *Biophys. J.* **89**, 2835–2854.

Dundr, M., Ospina, J. K., Sung, M. H., John, S., Upender, M., Ried, T., Hager, G. L., and Matera, G. A. (2007). Actin-dependant intranuclear repositioning of an active gene locus in vivo. *J. Cell Biol.* **179**, 1095–1103.

Fischer, T., Rodrigues-Navarro, S., Pereira, G., Racz, A., Schiebel, E., and Hurt, E. (2004). Yeast centrin Cdc31 is linked to the nuclear mRNA export machinery. *Nat. Cell Biol.* **6**, 840–848.

Gard, S., Light, W., Xiong, B., Bose, T., McNairn, A. J., Harris, B., Ruan, C., Zueckert-Gaudenz, K., Gogol, M., Fleharty, B., Seidel, C., Brickner, J., and Gerton, J. L. (2009). Cohesinopathy mutations disrupt the subnuclear organization of chromatin. *J. Cell Biol.* **187**, 455–462.

Gilmore, R. (1991). The protein translocation apparatus of the rough endoplasmic reticulum, its associated proteins, and the mechanism of translocation. *Curr. Opin. Cell Biol.* **3**, 580–584.

Hueun, P., Laroche, T., Shimada, K., Furrer, P., and Gasser, S. M. (2001). Chromosome dynamics in the yeast interphase nucleus. *Science* **294**, 2181–2186.

Jin, Q. W., Fuchs, J., and Loidl, J. (2000). Centromere clustering is a major determinant of yeast interphase nuclear organization. *J. Cell Sci.* **113**, 1903–1912.

Köhler, A., Schneider, M., Cagal, G. G., Nehrbass, U., and Hurt, E. (2008). Yeast Ataxin-7 links histone deubiquitination with gene gating and mRNA export. *Nat. Cell Biol.* **10,** 707–715.

Kumaran, R. I., and Spector, D. L. (2008). A genetic locus targeted to the nuclear periphery in living cells maintains its transcriptional competence. *J. Cell Biol.* **180,** 51–65.

Pawley, J. B. E. (2006). *Handbook of Biological Confocal Microscopy* 3rd edn. Springer, New York.

Robinett, C. C., Straight, A., Li, G., Willhelm, C., Sudlow, G., Murray, A., and Belmont, A. S. (1996). In vivo localization of DNA sequences and visualization of large-scale chromatin organization using lac operator/repressor recognition. *J. Cell Biol.* **135,** 1685–1700.

Schermelleh, L., Carlton, P. M., Haase, S., Shao, L., Winoto, L., Kner, P., Burke, B., Cardoso, M. C., Agard, D. A., Gustafsson, M. G. L., Leonhardt, H., and Sedat, J. W. (2008). Subdiffraction multicolor imaging of the nuclear periphery with 3D structured illumination microscopy. *Science* **320,** 1332–1336.

Schmid, M., Arib, G., Laemmli, C., Nishikawa, J., Durussel, T., and Laemmli, U. (2006). Nup-PI: The nucleopore-promoter interaction of genes in yeast. *Mol. Cell* **21,** 379–391.

Shav-Tal, Y., Darzacq, X., Shenoy, S., Fusco, D., Janicki, S. M., Spector, D. L., and Singer, R. H. (2004). Dynamics of single mRNPs in nuclei of living cells. *Science* **304,** 1797–1800.

Sikorski, R. S., and Heiter, P. (1989). A system of shuttle vectors and yeast host strains designed for efficient manipulation of DNA in *Saccharomyces cerevisiae. Genetics* **122,** 19–27.

Stone, E. M., Huen, P., Laroche, T., Pilus, L., and Gasser, S. M. (2000). MAP kinase signaling induces nuclear reorganization in budding yeast. *Curr. Biol.* **10,** 373–382.

Straight, A. F., Belmont, A. S., Robinett, C. C., and Murray, A. W. (1996). GFP tagging of budding yeast chromosomes reveals that protien-protein interactions can mediate sister chromatid cohesion. *Curr. Biol.* **6,** 1599–1608.

Thompson, M., Haeusler, R. A., Good, P. D., and Engelke, D. R. (2003). Nucleolar clustering of tRNA genes. *Science* **302,** 1399–1401.

Yang, C. H., Lambie, E. J., Hardion, J., Craft, J., and Snyder, M. (1989). Higher order structure is present in the yeast nucleus: Autoantibody probes demonstrate that the nucleolus lies opposite to the spindle pole body. *Chromosoma* **98,** 123–128.

CHAPTER TWENTY-THREE

SPINNING-DISK CONFOCAL MICROSCOPY OF YEAST

Kurt Thorn

Contents

Abstract

Spinning-disk confocal microscopy is an imaging technique that combines the out-of-focus light rejection of confocal microscopy with the high sensitivity of wide-field microscopy. Because of its unique features, it is well suited to high-resolution imaging of yeast and other small cells. Elimination of out-of-focus light significantly improves the image contrast and signal-to-noise ratio, making it easier to resolve and quantitate small, dim structures in the cell. These features make spinning-disk confocal microscopy an excellent technique for studying protein localization and dynamics in yeast. In this review, I describe the rationale behind using spinning-disk confocal imaging for yeast, hardware considerations when assembling a spinning-disk confocal scope, and methods

Department of Biochemistry and Biophysics, University of California at San Francisco, San Francisco, California, USA

Methods in Enzymology, Volume 470
ISSN 0076-6879, DOI: 10.1016/S0076-6879(10)70023-9

for strain preparation and imaging. In particular, I discuss choices of objective lens and camera, choice of fluorescent proteins for tagging yeast genes, and methods for sample preparation.

1. INTRODUCTION

In recent years, fluorescent protein tagging combined with microscopy has become a powerful tool for imaging subcellular localization in live yeast cells (Davis, 2004; Kohlwein, 2000; Rines et al., 2002). A particularly popular imaging method is spinning-disk confocal microscopy, which provides a way around one of the fundamental limitations of fluorescence microscopy: the objective lens captures light not only from the region of the sample that is in focus but also from regions in the sample above or below the focus plane. This out-of-focus light reduces contrast in the image and obscures in-focus information.

Spinning-disk confocal microscopy has become such a powerful tool for imaging yeast because it provides much better optical sectioning than a conventional fluorescence microscope, but much better sensitivity that a laser-scanning confocal. In general, laser-scanning confocal microscopes lack the sensitivity required to image small dim objects, and are unlikely to produce good images of any yeast sample that is not very bright. Conversely, conventional fluorescence microscopes can be very sensitive but do not reject out-of-focus light and so cannot acquire optical sections or produce high-resolution 3D reconstructions of the cell. Spinning-disk confocals can acquire high-resolution optical sections with good sensitivity and thus fill a niche between conventional fluorescence microscopes and laser-scanning confocals. They are particularly good when imaging small dynamic processes in vivo, such as cytoskeletal dynamics or vesicle and organelle movement. For imaging requiring less resolution, such as looking at translocation between the cytoplasm and nucleus or the abundance of a transcriptional reporter, they have relatively little advantage over a conventional fluorescence microscope.

There are two general ways to deal with out-of-focus light; it can either be blocked from reaching the detector, as in confocal microscopy, or it can be computationally removed after the fact, as in deconvolution microscopy (Shaw, 2006). In confocal microscopy, a pinhole in the excitation light path excites a single spot in the sample. A confocal pinhole in the detection path then blocks all light that does not originate from the excited spot. Traditional laser-scanning confocal microscopes suffer from two drawbacks for live cell work. First, because they require scanning a single illumination spot over the sample, they are typically slow, acquiring approximately 1 frame per second (fps). Second, and more importantly, due to the detectors used and to the very short integration times at each pixel in the image, they are relatively insensitive.

Spinning-disk confocal microscopes (reviewed in Toomre and Pawley, 2006) avoid both of these problems by replacing the single pinhole in a laser-scanning confocal with multiple pinholes that simultaneously illuminate many points in the sample. These pinholes are arranged in a spiral pattern on a disk so that as the disk rotates, the pinholes sweep over every point in the sample, illuminating it uniformly. This greatly improves the speed of the spinning-disk confocal (imaging at 30 fps is routine) and, as the confocal now illuminates the entire field of view in a short period of time (typically 33 ms), it forms an image that can be recorded on a highly sensitive CCD camera (as opposed to an insensitive photomultiplier tube, as in a laser-scanning confocal). The net result is that spinning-disk confocal microscopy systems are both faster and more sensitive than laser-scanning confocals. The increase in sensitivity is striking; a spinning-disk confocal system can collect as many as 50 times more photons for a given exposure than a conventional laser-scanning confocal (Murray *et al.*, 2007). Other confocal techniques, such as slit or line scanners, share some of the same advantages as spinning-disk confocal but have not been as extensively validated and so will not be discussed further.

One common misconception about confocal microscopes, including spinning-disk confocals, is that a confocal system has higher resolution than a nonconfocal system. In theory, this is not true, in that a point-like object will be imaged to a blurry disk of the same width in both systems. That is, the width of the point spread function of both a confocal and a nonconfocal system will be the same. In practice, however, the achievable resolution of a microscope is often lower than theoretically expected due to a low signal-to-noise ratio of the image. Low signal-to-noise ratios can arise from weakly fluorescent samples, but out-of-focus light can also reduce the signal-to-noise ratio even on bright samples. Out-of-focus light contributes background to the image that reduces contrast and adds additional noise, which reduces the overall signal-to-noise ratio of the image. Thus, while the spinning-disk confocal does not improve the diffraction-limited resolution of the microscope, it may improve the achievable resolution in real samples if this is limited by out-of-focus light contributing background fluorescence and noise to the images.

2. BUILDING A SPINNING-DISK CONFOCAL MICROSCOPE

A photograph of the spinning-disk confocal in the Nikon Imaging Center at UCSF, with components labeled, is shown in Fig. 23.1A. A schematic of the major components in the optical path is shown in Fig. 23.1B.

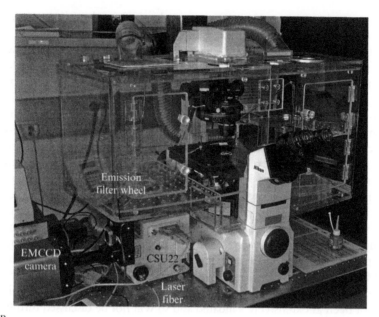

B

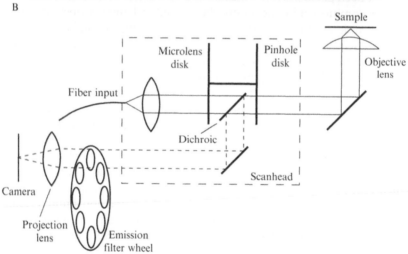

Figure 23.1 Overview of a spinning-disk confocal system. (A) Photograph of the spinning-disk confocal system in the Nikon Imaging Center at UCSF. It consists of a Nikon TE2000 microscope with a temperature controlled chamber from *In Vivo* Biosciences. The Yokogawa CSU-22 spinning-disk scanhead is labeled, along with the optical fiber bringing excitation light to the scanhead from the laser launch. A Sutter filter wheel holding emission filters is coupled to the exit of the scanhead; a projection lens is mounted after the filter wheel. The Cascade II EMCCD camera (Photometrics) is labeled as well. (B) Schematic diagram of the spinning-disk optical path. Solid lines show the excitation path, and the short dashed lines show the emission path. Elements within the dashed box are contained within the spinning-disk confocal scanhead.

2.1. Microscope base

Spinning-disk confocal systems can be built on microscopes from any of the major manufacturers, and on either an upright or inverted system, although mounting components is somewhat simpler on an inverted scope. Microscopes from all of the major manufacturers (Leica, Nikon, Olympus, and Zeiss) have excellent optical performance so the major considerations in choosing one are software support (not all software packages support all microscopes) and the availability of other features on the microscope. Most of the major microscope manufacturers have recently introduced hardware autofocus systems on their microscopes, such as the Nikon Perfect Focus System, the Zeiss Definite Focus, or the Olympus Zero Drift. These are devices that measure the position of the sample coverslip and adjust the focus to compensate for any movement of the sample, thereby maintaining focus at all times. These autofocus systems greatly improve focal stability and are considerably more precise and faster than image-based autofocus techniques. They also greatly simplify the acquisition of time-lapse movies, and if you will be acquiring time-lapse data for more than approximately 30 min, are well worth the added expense.

2.2. Scanhead

While a number of disk-scanning confocal systems exist, spinning-disk confocals from the Yokogawa Electric Corp. have come to dominate the market because they include microlenses that focus the laser light through the pinhole disk, dramatically increasing the excitation efficiency of the system (Toomre and Pawley, 2006). Other spinning-disk confocal systems do not include these microlenses, making them much less bright at equivalent laser powers. There are two different versions of the Yokogawa spinning-disk confocal scanhead: the older CSU10 and the newer CSU-X1. The CSU-X1 has several improvements over the CSU10, including a redesigned excitation path that increases the amount of excitation light reaching the sample and a higher disk rotation rate that allows for frame rates faster than 30 fps. The CSU-X1 also has an optional bypass path that allows bypassing the spinning disk for conventional wide-field imaging, and an optional motorized filter changer to allow automated filter switching in the scanhead. This is helpful if you intend to build a system with a large number of laser lines.

2.3. Lasers and filters

Typically, the vendor that provides the spinning-disk scanhead will also provide the lasers and laser launch for the system. The laser launch contains the optics to combine the beams from multiple lasers together and to launch them into the single mode optical fiber attached to the spinning–disk scanhead. It will also typically have an acousto–optic tunable filter (AOTF) for rapidly switching laser lines. For spinning–disk confocal microscopy, it is probably best to avoid gas lasers (with the possible exception of Ar–ion lasers). The latest solid state lasers typically have sufficient power for confocal microscopy and are smaller, produce less heat, and have longer lifetimes than gas lasers. Typically, a 50–mW laser will supply sufficient power for spinning–disk applications, although some applications (particularly rapid imaging) may benefit from increased laser power. While the exact laser wavelengths used will depend on which fluorescent tags you wish to image, commonly used wavelengths include 405 nm (DAPI), 440 nm (CFP), 473 nm, 491 nm (GFP), 532 nm, 561 nm (RFP), and 640 nm (Cy5).

You will need a dichroic beamsplitter(s) in the scanhead that matches the laser lines you are using. The dichroic beamsplitter separates the emission light from the excitation light, allowing the emitted light to be imaged onto the camera. Additionally, for multicolor imaging, you will want a filter wheel at the exit of the scanhead with emission filters that define the wavelength band you wish to detect for each channel (Fig. 23.1B). This can either be a third-party filter wheel or a CSUX optional add-on filter wheel. Alternatively, if you are using a small number of wavelengths this can be replaced with a single multipass emission filter but this may result in cross talk between different channels. Multipass emission filters are, however, useful for rapid multicolor imaging by changing the excitation wavelength as they eliminate the need for moving any mechanical parts to change wavelength. Filters with very high transmission in the passband ($> 90\%$) are now available from the major vendors (Chroma Technology, Omega Optical, and Semrock) and are highly desirable to maximize the amount of light detected by the camera.

2.4. Choice of objective

Typically, for spinning-disk confocal microscopy of yeast cells one wishes to maximize the resolution of the images acquired. Ultimately, the achievable resolution is limited by the numerical aperture (NA) of the objective lens used to image the sample; the larger the numerical aperture, the higher the achievable resolution. The standard measure of resolution is given by the Rayleigh criterion, which measures the minimum distance two point sources must be separated by to be distinguished, and is given by $r_{min} = 0.61$ λ/NA_{obj}, where λ is the wavelength of light emitted by the sample (Inoue, 2006). Maximizing resolution, therefore, suggests the use of oil-immersion objectives with a numerical aperture of 1.4 or higher. Plan Apo objectives (corrected for both field flatness and chromatic aberration at multiple wavelengths; Keller, 2006) are available from all major manufacturers with a numerical aperture of 1.4. An objective with a numerical aperture of 1.4 will have a resolution limit of $\sim$220 nm for GFP.

Such objectives are also well matched to the pinholes in the spinning-disk confocal. Because an objective has a finite resolution, the image of a point object smaller than the diffraction limit of the objective is blurred into a disk (the Airy disk) of radius $r_{min} = 0.61\lambda/NA_{obj}$. For optimum confocality, the diameter of the confocal pinhole should match the diameter of this Airy disk times the magnification of the objective. When these two diameters are equal, the confocal pinhole will pass all of the in-focus light while rejecting the maximum amount of out-of-focus light. If the pinhole diameter is smaller than the Airy disk diameter, in-focus light will be blocked by the pinhole, reducing the overall efficiency of the system. If the pinhole is larger than the Airy disk, additional out-of-focus light will pass through the pinhole, reducing the confocality and contrast of the system. The Yokogawa CSU spinning-disk confocal has pinholes of diameter 50 μm, which are well matched to the Airy disk diameter of $\sim$44 μm produced by a 100$\times$/1.4 NA objective imaging GFP.

One drawback of using oil-immersion objectives for imaging yeast is that using an oil-immersion objective to image into a yeast cell in an aqueous environment introduces spherical aberration due to the refractive index difference between the immersion oil and the aqueous medium. An oil-immersion objective is designed to produce aberration-free images immediately adjacent to the coverslip. As one images deeper into the sample, the thickness of the immersion oil layer will shrink, and the thickness of the water layer will increase, resulting in a change in the optical properties of the system which gives rise to spherical aberration (Fig. 23.2A). When viewed through a wide-field system (e.g., the eyepieces) spherical aberration manifests as an axial asymmetry in the point spread function, and can be recognized as haloing around point-like objects on one side of focus, with no haloing on the other side of focus (Fig. 23.2B). Spherical aberration results in the detected light from the sample no longer coming to a tight focus which reduces contrast and intensity as the broader focal spot of the spherically aberrated light from deeper in the sample will be partially cut off by the spinning-disk pinholes. This effect can be quite noticeable even for objects as thin as a yeast cell. Indeed, measurable intensity falloff can be seen after imaging as little as 2 μm into the sample (Fig. 23.2C).

This spherical aberration can be eliminated in one of two ways. First, using a water-immersion objective instead of an oil-immersion objective will eliminate the change in spherical aberration with imaging depth as now the oil layer is replaced with a water layer having the same refractive index as the specimen, so that as the optical properties of the system no longer change as you image deeper into the specimen. A water-immersion objective will have a lower numerical aperture than an oil objective due to the

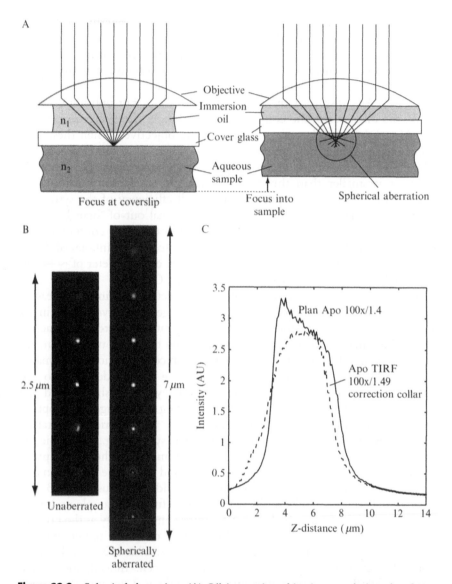

Figure 23.2 Spherical aberration. (A) Oil-immersion objectives are designed to focus at the coverslip (redrawn from a figure by Mats Gustafsson). Under these conditions, shown in the left-hand portion of the figure, all light rays entering the sample focus to a single point. When imaging into an aqueous sample, as shown on the right, refraction occurs at the boundary between the coverslip and the solution. This additional refraction differs for rays entering at different angles, and as a result the light is no longer tightly focused to a single spot. This is spherical aberration. (B) Unaberrated and aberrated images of beads, acquired on an epifluorescence microscope. Spherical aberration can be recognized by haloing around point objects on one side of focus. On the left is shown a series of Z slices of a 100 nm bead, taken at 0.5 μm intervals.

lower refractive index of the immersion medium which will result in lower resolution for the objective. However, as the spherical aberration introduced by using an oil objective also reduces resolution, the unaberrated water-immersion objective may have higher resolution than the aberrated oil objective. As the sample gets thicker, the benefit from using the water-immersion objective instead of the oil-immersion objective increases, as the induced spherical aberration increases with the distance from the coverslip.

An alternate approach to eliminate spherical aberration is to add additional optics that correct for the spherical aberration. In principle, this can be done with adaptive optics approaches, but this remains highly challenging (Booth, 2007). A more practical approach is to introduce additional lenses that move as the objective moves in Z to compensate for the induced spherical aberration. One such device is the MID/SAC spherical aberration corrector available from Intelligent Imaging Innovations, which has been used to great effect for imaging samples up to 80 μm thick with a spinning-disk confocal microscope (McAllister et al., 2008). For imaging thin yeast samples, however, such an approach is probably overkill. An alternative approach is to use an objective with a correction collar, which allows the optics inside the objective to be adjusted to eliminate spherical aberration at any single focal plane. While such a correction collar (unless continuously adjusted) cannot correct for spherical aberration at all focal planes in the sample, setting it to minimize spherical aberration in the middle of the yeast cell minimizes spherical aberration over the thickness of the cell and has been found to give improved quantitation of fluorescent intensities over the full thickness of the cell (Fig. 23.2C). Oil-immersion objectives with correction collars are typically designed for total internal reflection fluorescence microscopy and so additionally benefit from having a slightly higher numerical aperture (up to 1.49). For applications where accurate three-dimensional

---

These were taken under conditions of minimal spherical aberration, and the image of the bead (the point spread function) is relatively symmetric above and below focus. On the right is an image of the same bead with spherical aberration induced; the Z slices are now taken at 1 μm intervals. The pronounced asymmetry between focusing above and below the bead is now apparent. Observing small structures in a sample and adjusting the correction collar to minimize the asymmetry in the point spread function above and below focus is a good way to minimize spherical aberration in your images. (C) The effect of spherical aberration on quantitative intensity measurements with a spinning-disk confocal (data provided by Susanne Rafelski). A 2-μm bead was imaged with both a Plan Apo 100×/1.4 NA objective (without a correction collar) and an Apo TIRF 100×/1.49 NA objective, with the correction collar set to minimize spherical aberration. The average intensity from a small region in the center of the bead is plotted as a function of Z. The intensity falls off rapidly when imaging into the sample with the Plan Apo objective, but is much more symmetric when using the TIRF objective with correction collar. The image of the bead acquired with the correction collar adjusted to minimize spherical aberration better matches the actual brightness of the bead, which should be uniform throughout its thickness.

intensity measurements are important, an oil- or water-immersion objective with a correction collar is highly recommended. Setting the correction collar is done by adjusting it to achieve symmetry in the point spread function above and below focus around the middle of the cell. This is done by observing the point spread function of a small object in the cell through the eyepieces while adjusting the correction collar. One of the drawbacks to the use of objectives with correction collars is that setting the collar accurately takes practice, and if the collar is set inaccurately it will introduce additional spherical aberration. For this reason, we have both a $100\times/1.4$ NA oil lens (without a correction collar) and a $100\times/1.49$ NA oil lens (with a correction collar) on our spinning-disk confocal. The 1.4 NA lens is used for routine applications and the 1.49 NA lens is used for more demanding 3D reconstructions. For thicker specimens, such as yeast bio-films, a water-immersion lens may give more satisfactory performance. While the specific objectives mentioned here are Nikon objectives, other microscope manufacturers have generally similar objectives.

For applications where high resolution and high contrast are critical, it is probably best to compare both oil- and water-immersion objectives. The achievable resolution in a spinning-disk confocal experiment depends on the numerical aperture, magnification, and aberrations of your microscope in a complex way, and the solution with the best performance is not always predictable *a priori*. In particular, while water-immersion objectives have lower magnification and numerical aperture, and would therefore be expected to have lower resolution, they also have lower spherical aberration, which reduces contrast and degrades resolution. In a real experiment, the achievable resolution depends not only on the theoretical resolution of the objective but also on the intensity of the signal and the background, and on any aberrations, and so can only be determined empirically.

2.5. Cameras

For any spinning-disk confocal system, you will need a camera to acquire images for analysis and publication. The camera should be highly sensitive, so as to capture as many photons as possible that arrive at it, and low noise, so as to contribute as little extra noise to the image as possible. Camera sensitivity is measured by the quantum efficiency of the camera, which measures the fraction of photons incident on the CCD that are recorded. For spinning-disk confocal microscopy, particularly on dim specimens, cameras with very high quantum efficiencies are desirable, so that nearly every photon collected by the microscope is recorded on the camera. The highest possible quantum efficiencies are achieved by back-thinning of the CCD chip, whereby the substrate the chip is grown on is physically ground away and the CCD is illuminated from the back. Illuminating the CCD chip from the back eliminates absorption and scattering by the electronics fabricated on the front face of the CCD and allows quantum efficiencies in the visible wavelength range of $\sim 90\%$. By contrast, nonback-thinned CCDs (front-illuminated CCDs) have typical quantum efficiencies of $\sim 60\%$ (Art, 2006; Janesick, 2001).

Noise in the image comes from a number of sources, of which two are dominant: photon shot noise and camera read noise. Photon shot noise occurs because photons are collected in integer numbers (so that while a given pixel may record 10 photons on average, sometimes it will record more or less) and is unavoidable. The standard deviation due to the photon shot noise is given by the square root of the number of photons collected, and so the signal-to-noise ratio is given by one over the square root of the number of photons collected. Thus, the only way to reduce the photon shot noise is by collecting more photons, for example, by exposing longer.

Camera read noise is noise introduced by the digitization process in the camera. This introduced noise is proportional to the readout speed of the camera—the faster the camera is read out, the higher the read noise. A typical camera that can be readout at ~ 10 fps, such as the Coolsnap HQ2 (Photometrics) or Orca-R2 (Hamamatsu) has ~ 6–8 e^- of read noise. Very low read noise can be achieved with a very slow readout speed. For instance, the Orca-II-ER (Hamamatsu) has a read noise of 4 e^-, but takes 1.2 s to read out a single image. For many applications of spinning-disk confocal microscopy, such a frame rate is prohibitively slow.

To achieve a fast frame rate while maintaining low noise, the solution has been to amplify the signal before reading it out. Since the signal-to-noise ratio depends on both the signal strength and the amount of noise, the signal-to-noise ratio can be increased either by amplifying the signal or by reducing the noise. Either amplifying the signal or reducing the noise by the same amount results in the same increase in the signal-to-noise ratio. Therefore, amplification can be thought of as reducing the read noise by the same factor, which is why amplified cameras often quote an effective "read noise" of < 1 e^-. This was first done by using an image intensifier attached to the CCD, a so-called intensified CCD or ICCD. The image intensifier greatly amplifies the number of photons prior to their arrival at the CCD (by up to a million-fold) rendering even the high read noise typical of a fast camera negligible. However, ICCDs suffer from a number of drawbacks including low quantum efficiency and potential damage due to brief exposure to bright light, so they have been largely supplanted for routine imaging by a newer technology, the electron multiplying CCD (EMCCD).

An EMCCD is essentially a normal CCD with a gain register located prior to the readout electronics. The photoelectrons from the CCD are shifted through the wells of this register at higher than normal voltages, resulting in a gain of $\sim 1\%$ for each transfer step. That is, after a single shift event through this register, 100 electrons will on average be amplified to

101 electrons. The gain register consists of several hundred to a thousand wells, so shifting through the entire register results in very high gains, up to 1000-fold (Art, 2006; Janesick, 2001). This method of amplifying the signal has several advantages over that of an ICCD. First, because the gain is located after the light-sensitive CCD, rather than in front of it like an image intensifier, the high quantum efficiency (> 90%) of a back-thinned CCD can be realized. Second, the amplification process introduces remarkably little extra noise. The gain register does introduce additional shot noise, however, as the amplification process is random. As this gain noise has the same statistical properties as the photon shot noise, the resulting shot noise in the signal is increased, as if fewer photons had been collected. The net effect of this is to make it appear as if half as many photons had been gathered, so that the camera's quantum efficiency appears to be halved.

Imaging quickly necessitates short exposure times (hence few photons) and cameras with fast readout (hence high read noise). Under these conditions, noise in the image will be dominated by the camera read noise, and amplification, which reduces the effective read noise, will substantially improve the final image quality. This is often the regime that spinning-disk confocal microscopy experiments fall in, as a typical experiment involves acquiring rapid Z-stacks in time-lapse, often necessitating exposures of 100 ms or less. For this reason, most spinning-disk confocal systems are now paired with EMCCD cameras. However, if you are imaging bright samples and do not need to image quickly, it may also be worth considering conventional CCD cameras as well.

EMCCDs are made by three major manufacturers: Roper Scientific, Andor Technologies, and Hamamatsu. The cameras from all three companies tend to be similar, as they all utilize the same EMCCD chips. The most popular EMCCDs (e.g., Photometrics Evolve, Andor Ixon, and Hamamatsu ImageEM) use a 512×512 pixel back-thinned EMCCD chip from e2v, which can be read out at video rate (30 fps). There is also a $1k \times 1k$ pixel back-thinned EMCCD available, but it can be only read out at 8 fps and it is substantially more expensive than the 512×512 pixel EMCCDs. There is also a front illuminated EMCCD available, but this should probably be avoided due to its substantially lower quantum efficiency compared to the back-thinned EMCCDs.

To achieve optimal resolution of a microscopy system, the magnification of the image on the camera must be chosen to match the resolution limit of the objective. Specifically, for the camera to be able to accurately resolve an object of a given size, that object must span at least two pixels on the camera; this is known as Nyquist–Shannon sampling (Pawley, 2006). Therefore, for your camera to achieve the full resolution that your objective is capable of, you need to magnify the image on the camera such that one camera pixel covers a distance of less than half the resolution limit of the objective when referred to the object plane. For example, a $100\times/1.4$ NA objective has a resolution limit of $\sim$220 nm for GFP, so one pixel should span < 110 nm, referred to the object plane, to achieve maximum resolution. For a camera with 16 μm pixels, one pixel will correspond to a size of 160 nm in the object plane with no magnification beyond that of the 100× objective. Therefore, to achieve maximal resolution, additional magnification must be introduced; this is most easily done by replacing the projection lens of the confocal scanhead with a lens of longer focal length. It is often useful to include additional magnification beyond that required by Nyquist–Shannon sampling; often 2.5–3 pixels per resolution unit are helpful, and for some digital image analysis procedures, additional magnification beyond this ("empty magnification") may be helpful.

2.6. Other hardware considerations

For doing time-lapse measurements of yeast cells, a few additional components will be helpful. Most likely, you will want some kind of temperature control system so that your cells can be grown at a constant temperature. Temperature control can either be achieved by enclosing the entire microscope in a thermostatted plexiglass box, or by using a stage top heater combined with an objective heater. While enclosing the microscope in a temperature controlled box makes access to the microscope and the sample somewhat cumbersome, it does maintain a very stable environment around the entire microscope (typically stable to within ±0.1 °C). This temperature stability helps minimize focus drift due to thermal fluctuations and makes switching samples and objectives simple. Stage top incubators are much smaller and cheaper and do not encumber the microscope at all. However, for use with oil-immersion objectives, they must be paired with an objective heater as otherwise the objective acts as a heat sink and will chill the sample. As use of an objective heater makes changing objectives difficult, and a stage top incubator will typically only hold a single size of sample dish, this approach adds its own set of difficulties.

Another addition which is desirable for time-lapse data acquisition is a motorized X–Y stage. This greatly increases the number of cells that can be recorded in time-lapse by recording multiple fields of view in succession. As typically it will only take 10–15 s to move to a new field of view, autofocus with a hardware autofocus device, and acquire images, a motorized stage will allow tens of fields to be collected during the typical 5 min interval of a time-lapse movie.

For cases where fast Z-stack acquisition is required, a piezoelectric Z-stage is a necessary addition. Addition of a piezoelectric Z-stage to either a motorized or a manual stage allows Z-stack acquisition at up to 30 fps if your camera is fast enough and your exposure time is short enough.

To minimize vibration and sample drift, you will want to place the entire system on a vibration isolation table.

2.7. Software

As the spinning-disk scanhead is a relatively passive device (you can set the rotation speed and internal filters), it can be controlled by a large number of software packages. Therefore, the software package you choose to control your microscope will depend more on the other components in your system, particularly the microscope body. Most microscope manufacturers now have their own software packages for controlling their microscopes, which should also be capable of controlling the other components in their systems. In addition, there are a number of third-party software packages that can control most instrumentation. Metamorph, from Molecular Devices, has been a standard third-party software package for many years, and controls a large variety of hardware. More recently, a free, open-source, microscope control package called Micro-Manager has been released. This is an appealing choice for labs looking to save money or who wish to customize their software. Microscope control software will usually include image analysis tools as well and so is generally a good starting point for data analysis.

2.8. System integration

In most cases, it does not make sense to purchase all the items described here individually. You will want to work with a system integrator who will sell you the scanhead, lasers and optics, and additional microscope components you need and will assemble the system and get it working. You may need to purchase the microscope stand separately and then add the spinning-disk confocal from a separate vendor. Two system integrators we have had good luck with are Solamere Technologies and Andor Technology.

3. SAMPLE PREPARATION

3.1. Fluorescent tagging and choice of fluorescent protein

One of the major strengths of budding yeast as a cell biological model organism is the ease with which genes can be tagged with fluorescent proteins. Homologous recombination is extremely efficient in budding yeast allowing a fluorescent protein sequence to be integrated nearly any-where in the genome using a 40-bp sequence to provide targeting. This enables tagging of a specific gene by amplification of a template with PCR primers containing targeting sequences for the gene to be tagged. Providing methods for performing gene tagging is beyond the scope of this review, but protocols are widely available (Amberg *et al.*, 2005; Gauss *et al.*, 2005; Knop *et al.*, 1999; Longtine *et al.*, 1998; Petracek and Longtine, 2002). Because of the ease of fluorescent protein tagging in yeast, these are probably the most common fluorescent probes used, although other probes can be used as well (Giepmans et al., 2006).

A wide variety of fluorescent proteins (FPs) have been discovered and engineered recently. While a comprehensive discussion of available fluorescent proteins and their properties is not possible given space constraints, several excellent reviews are available (Shaner et al., 2005, 2007; Straight, 2007; Zacharias and Tsien, 2006). Here, I focus on FPs readily available in Saccharomyces cerevisiae tagging vectors and newer proteins with desirable properties (Table 23.1). Tagging vectors are readily available for CFP, GFP, and YFP, and vectors are also available for mCherry, mTFP1, and Sapphire; unpublished vectors also exist for other fluorescent proteins.

When choosing a fluorescent protein for spinning-disk confocal imaging of yeast, there are several factors to take into account. Most important is the detectability of the protein—how bright will a tagged protein be relative to the background fluorescence of the cell? Additional factors are the availability of a laser line well matched to the excitation spectrum of the fluorescent protein, the photostability of the fluorescent protein, and whether or not the fluorescent protein perturbs the function of the protein to which it is fused.

Detectability of a fluorescent protein is a function of both the brightness of the tagged protein and the autofluorescence of the cell. The intrinsic brightness of the FP plays a major role as, if all other factors are held constant, the brighter the protein, the more detectable it will be. This is easily calculated as the product of the extinction coefficient, which measures how efficiently the fluorophore absorbs light, and quantum yield, which measures how many absorbed photons are reemitted as fluorescence, of the FP; values are given in Table 23.1. Additionally, the maturation rate of the FP will directly affect the brightness of the tagged protein, as a protein with a slow maturation rate will be degraded and diluted by cell growth faster than it matures, leading to a majority of the protein being nonfluorescent. While maturation rates are not known for all FPs, and are frequently measured in vitro or in E. coli, conditions that may not be relevant to yeast expression, the proteins listed in Table 23.1 generally have reasonably fast maturation rates (30 min or less). Similarly, translation efficiency will influence the brightness of a fusion protein; I have found that codon optimization of the FP can give a twofold increase in brightness (Sheff and Thorn, 2004), although mRNA secondary structure may play a larger role (Kudla et al., 2009).

The other component of detectability is cellular autofluorescence; the brighter the cellular autofluorescence, the brighter a tagged protein will have to be detected above this background. Careful choice of yeast strain and growth conditions can reduce autofluorescence (see below), but it cannot be completely eliminated. Much of the autofluorescence in yeast is due to flavins, which absorb broadly in the violet–blue range (400–500 nm) and emit in the green ($\sim$530 nm). Yeast autofluorescence is therefore

Table 23.1 Fluorescent proteins

Protein	References	λ_{ex}	λ_{em}	$\varepsilon\,(M^{-1}\,cm^{-1})$	QY	Brightness[a]	Tagging vectors
TagBFP	Subach et al. (2008)	402	457	52,000	0.63	32.8	
Sapphire	Cubitt et al. (1999)	399	511	29,000	0.64	18.6	Sheff and Thorn (2004)
T-Sapphire	Zapata-Hommer and Griesbeck (2003)	399	511	44,000	0.6	26.4	
ECFP	Tsien (1998)	433	475	32,500	0.4	13.0	Sheff and Thorn (2004), Janke et al. (2004), Hailey et al. (2002)
mTFP1	Ai et al. (2006)	462	492	64,000	0.85	54.0	Deng et al. (2009)
EGFP	Tsien (1998)	488	507	56,000	0.6	33.6	Sheff and Thorn (2004), Janke et al. (2004), Deng et al. (2009), Longtine et al. (1998), Gauss et al. (2005), Wach et al. (1997)
Citrine	Griesbeck et al. (2001)	516	529	83,400	0.76	58.5	Sheff and Thorn (2004), Deng et al. (2009)
Venus	Nagai et al. (2002)	515	529	92,200	0.57	52.5	Sheff and Thorn (2004)
mKOκ	Tsutsui et al. (2008)	551	563	105,000	0.61	64.0	
mCherry	Shaner et al. (2004)	587	610	72,000	0.22	15.8	Deng et al. (2009)
TagRFP	Merzlyak et al. (2007)	555	584	100,000	0.48	49.0	
mPlum	Wang et al. (2004)	590	649	41,000	0.1	4.1	
mKate2	Shcherbo et al. (2009)	588	633	62,500	0.4	25.0	

An expanded table is available online at http://thornlab.org/gfps.htm.
[a] Product of ε and QY, divided by 1000.

strongly wavelength dependent, being much less intense for longer wavelength imaging than shorter wavelength imaging. Because of this, orange and red fluorescent proteins are more detectable than would be expected given their intrinsic brightness.

The photostability of the fluorescent protein is important when doing time-lapse imaging; more photostable proteins will bleach more slowly and so will allow more frames to be acquired before the signal drops to an unacceptable level. For acquiring single images, photobleaching rate is relatively unimportant. Photobleaching rates have been measured for many of the proteins suggested here and are available in the original publications and in several reviews (Shaner et al., 2005, 2007; Zacharias and Tsien, 2006).

Finally, of critical importance is whether or not fusion of a fluorescent protein to your protein of interest will perturb its function. This will of course depend on the protein being tagged and on where it is tagged, but it seems that GFP fusions to the C-terminus of proteins in yeast is generally well tolerated, as 87% of essential yeast genes could be successfully tagged with GFP in a systematic tagging effort (Howson et al., 2005; Huh et al., 2003). In my experience, other GFP variants (e.g., CFP and YFP) are equally well tolerated, and mCherry is well tolerated also. Many of the newer fluorescent proteins mentioned have not been extensively studied in the context of fusion proteins, and so little information is available about how they are tolerated.

For general purpose imaging, tagging with GFP is a good place to start. It is readily available in a monomeric form, reasonably bright, and generally well behaved. For imaging a second color, mCherry has minimal cross talk with GFP and is also generally well behaved. For multicolor imaging, CFP/ YFP/mCherry is a good combination, and Sapphire can be added with the introduction of a small amount of cross talk (Sheff and Thorn, 2004). TagBFP is another possibility for a fourth color, and it may be possible to multiplex mKOκ for a fifth color with some small amount of cross talk. For imaging low-abundance proteins, it may be worth exploring less common options to maximize signal-to-noise ratio. In general, moving to longer wavelengths is advantageous as yeast autofluorescence is less intense at longer wavelengths. For proteins that are rapidly turned over, using a fast-folding protein such as Venus may help (Yu et al., 2006). For difficult imaging cases, I would expect that it may be necessary to try several proteins before identifying an optimal tag. Finally, imaging in diploid cells may be preferred to imaging in haploid cells as diploid cells are larger.

3.2. Minimizing autofluorescence

Yeast cells are somewhat autofluorescent, and the commonly used ade2⁻ strains accumulate a highly fluorescent red pigment (Ishiguro, 1989; Stotz and Linder, 1990). If possible, it is best to avoid ade2⁻ strains, although accumulation of this pigment can be avoided by supplementing the growth

medium with 20 μg/ml adenine. This supplementation can also be beneficial for strains that are ADE$^+$. Cells tend to be less fluorescent in early log
phase (OD$_{600}$ $\sim$ 0.2). Yeast media can also be autofluorescent. Rich media
such as YPD are highly autofluorescent and should be avoided when
imaging. Cells grown in rich media can be washed into a less fluorescent
media or buffer prior to imaging; however, these fluorescent media seem to
increase the autofluorescence of cells slightly even after washing. In synthetic (minimal) media, the major sources of autofluorescence are riboflavin
and folic acid, and omitting these components eliminates the media autofluorescence if the medium is made from scratch. Most *S. cerevisiae* strains
can synthesize both riboflavin and folic acid, so eliminating these vitamins
from the medium does not appear to have drawbacks. Commercial yeast
nitrogen base is often fluorescent even when it is lacking these two vitamins,
so I recommend testing its fluorescence before use or making your own
from scratch (Sheff and Thorn, 2004). Commercial CSM supplements do
not seem to be autofluorescent and so can be safely used. Growing cells in
this low-fluorescence medium allows direct imaging of cultures without
washing, and substantially reduces backgrounds for cells grown on agarose
pads or in a perfusion system.

3.3. Mounting

The simplest way to prepare yeast for imaging for short periods of time
(30–60 min) is by immobilizing them on Concanavalin A coated coverslips.
Concanavalin A binds to sugars in the yeast cell wall and will stick the yeast
tightly to the coverslip so that they will not move during imaging.
Concanvalin A coated coverslips can easily be prepared as follows:

1. Prepare Concanavalin A solution by dissolving Concanavalin A (Sigma
   #L 7647) in distilled water to 0.5 mg/ml. Refrigerate.
2. Rack 22 mm #1.5 coverslips (Racks: Electron Microscopy Sciences
   #72241-01).
3. Soak coverslips overnight in 1 M NaOH with gentle shaking. Sterile
   filter NaOH before use to remove dust.
4. Pour off NaOH and save for reuse. Wash coverslips 3× with distilled
   water.
5. Add Concanavalin A solution and soak for 20 min with gentle shaking.
6. Pour off Concanavalin A solution and save for reuse.
7. [Optional] Rinse coverslips once with distilled water.
8. [Optional] Spin coverslips dry on microplate carriers in Sorvall RT7.
   Place racks on a piece of paper towel to catch liquid and spin 1 min at
   700 rpm.
9. Place racks in hood until absolutely dry. Store in coverslip box at RT.
   Coverslips should be good for at least 1 month.

To use the coverslips, simply place a 6–8 μl drop of yeast cell suspension on a slide, and drop a coated coverslip on it. The cells should stick to the coverslip within a few minutes. The coverslip can be sealed with petroleum jelly or VALAP (Vaseline:Lanolin:Paraffin, 1:1:1) for longer term imaging, or left unsealed for short-term studies (~10 min). If the cells are grown in low-fluorescence media, they can be directly placed on coverslips for imaging; if they are grown in more fluorescent media, particularly in YPD or other rich media, you will substantially reduce the background fluorescence by first washing them into PBS or some other nonfluorescent buffer or medium. Pelleting and resuspending cells can also be used to concentrate the cells if the cell density in the initial culture is not high enough to get a sufficient number of cells in the field of view.

For longer term imaging (up to 4–6 h), cells can be grown on agarose pads containing low-fluorescent yeast medium. These can be prepared by dissolving 1.2% agarose in low-fluorescence SC + carbon source. The agarose solution is then cast in a 1-mm slab in a device used for casting polyacrylamide gels. If kept in a sealed container with moist paper towels, such a gel can be kept for roughly a week. Pieces of agarose (~15 × 15 mm) are then cut out with a sterilized razor blade and placed on a slide, a drop of yeast suspension is placed on the agarose block, and a coverslip is placed on top. The gap between the coverslip and slide can then be filled with petroleum jelly to prevent evaporation. This is easily done by filling a syringe with petroleum jelly and injecting it through a large gauge needle into the space between the coverslip and slide. Alternatively, these pads can be sealed with VALAP. Agarose pads can also be made by filling depression slides with agarose, or sandwiching a drop of agarose between two slides, but I find the casting method described here to be the easiest.

For even longer term imaging (overnight or longer), it is probably best to use a microfluidic device to provide constant nutrient flow to the cells and remove waste products. We have had good luck with the CellASIC ONIX™ system (www.cellasic.com). This system consists of a control device paired with a special microfluidic plate which has the same footprint as an ordinary 96-well plate. The plate contains viewing areas where cells can be trapped and immobilized for long-term imaging while being continuously perfused with solution. The cells are trapped and held in place by being loaded under pressure (6–8 psi) into a viewing area which traps the cells under an elastomeric membrane 4.2 μm above the coverslip. The downward pressure applied by the ceiling membrane holds the cells in place indefinitely. Media can be perfused from one of two wells, allowing rapid media switching. Cells have been kept in this device for up to 3 days and continued to divide (Lee *et al.*, 2008).

ACKNOWLEDGMENTS

Thanks to S. Rafelski, C. Mrejen, and G. Peeters for helpful discussions. Thanks to the many users of the Nikon Imaging Center spinning-disk confocal for discussions and insight into how to optimize spinning-disk confocal microscopy for specific applications. The images in Fig. 23.2 were acquired in the Nikon Imaging Center at UCSF/QB3.

REFERENCES

Ai, H. W., *et al.* (2006). Directed evolution of a monomeric, bright and photostable version of Clavularia cyan fluorescent protein: Structural characterization and applications in fluorescence imaging. *Biochem. J.* **400**(3), 531–540.

Amberg, D. C., Burke, D., and Strathern, J. N. Cold Spring Harbor Laboratory (2005). Methods in yeast genetics: A Cold Spring Harbor Laboratory course manual. 2005 edn. Cold Spring Harbor Laboratory Press, Cold Spring Harbor, N.Y.

Art, J. (2006). Photon detectors for confocal microscopy. *In* "Handbook of Biological Confocal Microscopy," (J. Pawley, ed.), 3rd edn. Springer Science+Business Media, New York, NY.

Booth, M. J. (2007). Adaptive optics in microscopy. *Phil. Trans. A Math. Phys. Eng. Sci.* **365**, 2829–2843.

Cubitt, A. B., Woollenweber, L. A., and Heim, R. (1999). Understanding structure-function relationships in the Aequorea victoria green fluorescent protein. *Methods Cell Biol.* **58**, 19–30.

Davis, T. N. (2004). Protein localization in proteomics. *Curr. Opin. Chem. Biol.* **8**, 49–53.

Deng, C., Xiong, X., and Krutchinsky, A. N. (2009). Unifying fluorescence microscopy and mass spectrometry for studying protein complexes in cells. *Mol. Cell Proteomics.* **8**, 1413–1423.

Gauss, R., Trautwein, M., Sommer, T., and Spang, A. (2005). New modules for the repeated internal and N-terminal epitope tagging of genes in *Saccharomyces cerevisiae*. *Yeast* **22**, 1–12.

Giepmans, B. N., Adams, S. R., Ellisman, M. H., and Tsien, R. Y. (2006). The fluorescent toolbox for assessing protein location and function. *Science* **312**, 217–224.

Griesbeck, O., *et al.* (2001). Reducing the environmental sensitivity of yellow fluorescent protein. Mechanism and applications. *J. Biol. Chem.* **276**(31), 29188–29194.

Hailey, D. W., Davis, T. N., and Muller, E. G. (2002). Fluorescence resonance energy transfer using color variants of green fluorescent protein. *Methods Enzymol.* **351**, 34–49.

Howson, R., Huh, W. K., Ghaemmaghami, S., Falvo, J. V., Bower, K., Belle, A., Dephoure, N., Wykoff, D. D., Weissman, J. S., and O'Shea, E. K. (2005). Construction, verification and experimental use of two epitope-tagged collections of budding yeast strains. *Comp. Funct. Genomics* **6**, 2–16.

Huh, W. K., Falvo, J. V., Gerke, L. C., Carroll, A. S., Howson, R. W., Weissman, J. S., and O'Shea, E. K. (2003). Global analysis of protein localization in budding yeast. *Nature* **425**, 686–691.

Inoue, S. (2006). Foundations of confocal scanned imaging in light microscopy. *In* "Handbook of Biological Confocal Microscopy," (J. Pawley, ed.), 3rd edn. Springer Science+Business Media, New York, NY.

Ishiguro, J. (1989). An abnormal cell division cycle in an AIR carboxylase-deficient mutant of the fission yeast Schizosaccharomyces pombe. *Curr. Genet.* **15**, 71–74.

Janesick, J. R. (2001). Scientific Charge-Coupled Devices. SPIE Press, Bellingham, Washington.

Janke, C., et al. (2004). A versatile toolbox for PCR-based tagging of yeast genes: New fluorescent proteins, more markers and promoter substitution cassettes. *Yeast* **21**(11), 947–962.

Keller, H. E. (2006). Objective lenses for confocal microscopy. *In* "Handbook of Biological Confocal Microscopy," (J. Pawley, ed.), 3rd edn. Springer Science+Business Media, New York, NY.

Knop, M., Siegers, K., Pereira, G., Zachariae, W., Winsor, B., Nasmyth, K., and Schiebel, E. (1999). Epitope tagging of yeast genes using a PCR-based strategy: More tags and improved practical routines. *Yeast* **15**, 963–972.

Kohlwein, S. D. (2000). The beauty of the yeast: Live cell microscopy at the limits of optical resolution. *Microsc. Res. Tech.* **51**, 511–529.

Kudla, G., Murray, A. W., Tollervey, D., and Plotkin, J. B. (2009). Coding-sequence determinants of gene expression in Escherichia coli. *Science* **324**, 255–258.

Lee, P. J., Helman, N. C., Lim, W. A., and Hung, P. J. (2008). A microfluidic system for dynamic yeast cell imaging. *Biotechniques* **44**, 91–95.

Longtine, M. S., McKenzie, A. 3rd, Demarini, D. J., Shah, N. G., Wach, A., Brachat, A., Philippsen, P., and Pringle, J. R. (1998). Additional modules for versatile and economical PCR-based gene deletion and modification in *Saccharomyces cerevisiae*. *Yeast* **14**, 953–961.

McAllister, R. G., Sisan, D. R., and Urbach, J. S. (2008). Design and optimization of a high-speed, high-sensitivity, spinning disk confocal microscopy system. *J. Biomed. Opt.* **13**, 054058.

Merzlyak, E. M., et al. (2007). Bright monomeric red fluorescent protein with an extended fluorescence lifetime. *Nat. Methods* **4**(7), 555–557.

Murray, J. M., Appleton, P. L., Swedlow, J. R., and Waters, J. C. (2007). Evaluating performance in three-dimensional fluorescence microscopy. *J. Microsc.* **228**, 390–405.

Nagai, T., et al. (2002). A variant of yellow fluorescent protein with fast and efficient maturation for cell-biological applications. *Nat. Biotechnol.* **20**(1), 87–90.

Pawley, J. (2006). Points, pixels, and gray levels, digitizing image data. *In* "Handbook of Biological Confocal Microscopy," (J. Pawley, ed.), 3rd edn. Springer Science+Business Media, New York, NY.

Petracek, M. E., and Longtine, M. S. (2002). PCR-based engineering of yeast genome. *Methods Enzymol.* **350**, 445–469.

Rines, D. R., He, X., and Sorger, P. K. (2002). Quantitative microscopy of green fluorescent protein-labeled yeast. *Methods Enzymol.* **351**, 16–34.

Shaner, N. C., et al. (2004). Improved monomeric red, orange and yellow fluorescent proteins derived from Discosoma sp. red fluorescent protein. *Nat. Biotechnol.* **22**(12), 1567–1572.

Shaner, N. C., Steinbach, P. A., and Tsien, R. Y. (2005). A guide to choosing fluorescent proteins. *Nat. Methods* **2**, 905–909.

Shaner, N. C., Patterson, G. H., and Davidson, M. W. (2007). Advances in fluorescent protein technology. *J. Cell Sci.* **120**, 4247–4260.

Shaw, P. J. (2006). Comparison of widefield/devonvolution and confocal microscopy for three-dimensional imaging. *In* "Handbook of Biological Confocal Microscopy," (J. Pawley, ed.), 3rd edn. Springer Science+Business Media, New York, NY.

Shcherbo, D., et al. (2009). Far-red fluorescent tags for protein imaging in living tissues. *Biochem. J.* **418**(3), 567–574.

Sheff, M. A., and Thorn, K. S. (2004). Optimized cassettes for fluorescent protein tagging in *Saccharomyces cerevisiae*. *Yeast* **21**, 661–670.

Stotz, A., and Linder, P. (1990). The ADE2 gene from *Saccharomyces cerevisiae*: Sequence and new vectors. *Gene* **95**, 91–98.

Straight, A. F. (2007). Fluorescent protein applications in microscopy. *Methods Cell Biol.* **81**, 93–113.

Subach, O. M., *et al.* (2008). Conversion of red fluorescent protein into a bright blue probe. *Chem. Biol.* **15**(10), 1116–1124.

Toomre, D., and Pawley, J. B. (2006). Disk-scanning confocal microscopy. *In* "Handbook of Biological Confocal Microscopy," (J. Pawley, ed.), 3rd edn. Springer Science+Business Media, New York, NY.

Tsien, R. Y. (1998). The green fluorescent protein. *Annu. Rev. Biochem.* **67**, 509–544.

Tsutsui, H., *et al.* (2008). Improving membrane voltage measurements using FRET with new fluorescent proteins. *Nat. Methods* **5**(8), 683–685.

Wach, A., *et al.* (1997). Heterologous HIS3 marker and GFP reporter modules for PCR-targeting in *Saccharomyces cerevisiae*. *Yeast* **13**(11), 1065–1075.

Wang, L., *et al.* (2004). Evolution of new nonantibody proteins via iterative somatic hypermutation. *Proc. Natl. Acad. Sci. USA* **101**(48), 16745–16749.

Yu, J., Xiao, J., Ren, X., Lao, K., and Xie, X. S. (2006). Probing gene expression in live cells, one protein molecule at a time. *Science* **311**, 1600–1603.

Zacharias, D. A., and Tsien, R. Y. (2006). Molecular biology and mutation of green fluorescent protein. *Methods Biochem. Anal.* **47**, 83–120.

Zapata-Hommer, O., and Griesbeck, O. (2003). Efficiently folding and circularly permuted variants of the Sapphire mutant of GFP. *BMC Biotechnol.* **3**, 5.

CHAPTER TWENTY-FOUR

Correlative GFP-Immunoelectron Microscopy in Yeast

Christopher Buser* *and* Kent McDonald[†]

Contents

Abstract

Correlative light and electron microscopy represents the ultimate goal for the visualization of cell biological processes. In theory, it is possible to combine the strengths of both methods, that is, the live-cell imaging of the movement of GFP-tagged proteins captured by fluorescence microscopy with an image of the fine structural context surrounding the tagged protein imaged and localized by immunoelectron microscopy. In practice, inherent technical limitations of the two individual methods and their combination make the technique very

* Department of Molecular and Cell Biology, University of California, Berkeley, California, USA
† Electron Microscope Laboratory, University of California, Berkeley, California, USA

Methods in Enzymology, Volume 470
ISSN 0076-6879, DOI: 10.1016/S0076-6879(10)70024-0

complex to handle. Here, we present a high-pressure freezing and freeze-substitution protocol which fulfills the key criterion for correlative microscopy, namely, the ability to achieve excellent visibility of fine structures without disrupting the antigens recognized by the immunolabeling protocol. This is achieved by a fixative-free freeze-substitution and low-temperature embedding.

1. INTRODUCTION

With the ease of genetic manipulation and the availability of online databases on the localization patterns of GFP-fusion proteins in yeast (Huh et al., 2003) our knowledge of the localization and function of yeast proteins at the light microscope (LM) level has increased dramatically. Also structural and immunolabeling studies by electron microscopy (EM) are adding data on the cellular architecture and the distribution of proteins on the nanometer scale. Still, the electron microscopic characterization of yeast cells is lagging behind when compared to tissue culture cells. This is mainly due to the difficulties encountered when preparing yeast for transmission electron microscopy (TEM), which are mainly characterized by infiltration problems and the inherent density of the yeast cytoplasm. Furthermore, there is a lack of preparation protocols yielding both good visibility of the organelles of interest together with immunolabeling in the same sample. The goal of the research presented here is to establish a protocol that ultimately should allow the direct correlation of live-cell fluorescent imaging of GFP-fusion proteins with structural- and immuno-EM data from the same cell.

Unfortunately, key publications in the yeast field still extensively rely on chemical fixation protocols in spite of the multitude of known artifacts induced by these methods. This is especially astonishing since already in the late 1980s, Baba and Osumi noted the superior structure of yeast that were plunge frozen and freeze-substituted (Baba and Osumi, 1987; Baba et al., 1989). The images published in those articles still hold up to today's standards 20 years later. Today, high-pressure freezing (HPF) and freeze-substitution (FS) are on their way to becoming routine techniques in many EM facilities, since all the necessary equipment is commercially available and the know-how to successfully prepare biological samples has increased dramatically.

The scope of this chapter is to update the information presented in the last edition (McDonald and Mueller-Reichert, 2002) and to introduce a new protocol to prepare yeast TEM samples with excellent structural visibility which are at the same time suitable for immunolabeling. The following paragraphs will first cover the rationale and pitfalls of the HPF//FS approach and the immunolabeling, with the step-by-step protocols at the end.

2. Recent Advances in High-Pressure Freezing and Freeze-Substitution

The list of commercially available equipment for both HPF and FS has grown considerably in recent years, although it can be expensive to purchase and maintain. If state of the art equipment is not available, keep in mind that the first trials in using cryo-methods can also be done with far cheaper homemade equipment as mentioned in the previous edition (McDonald and Mueller-Reichert, 2002).

2.1. High-pressure freezing

Both the Bal-Tec HPM 010 and the Leica EMPact have been updated. Leica Microsystems now sells the EMPact2, which follows the same design as the EMPact (i.e., a mobile HPF machine with separated pressurizing and cooling lines), but now adds the option of the so-called RTS (rapid transfer system) designed for correlative LM/EM projects. The RTS allows the observation of the sample in the LM and high-pressure freezing within approximately 4 s (Verkade, 2008).

A similar move toward instrument mobility and correlative microscopy has been made by Bal-Tec with the new HPM100. Leica Microsystems has recently acquired Bal-Tec and is now selling this system as well under the name Leica HPM100. The machine formerly known as the Bal-Tech HPM 010 is now being sold by Boeckler Instruments, Inc. (Tucson, AZ) as the HPM 010.

A new high-pressure freezer, the HPF Compact 01, has been introduced recently by Wohlwend Engineering (distributed by Technotrade International in the USA). It operates under the same principle as the Bal-Tec HPM 010, but has an improved liquid nitrogen streaming and pressurizing system. Due to that it has significantly reduced liquid nitrogen consumption and an increased reproducibility in freezing.

Both the Leica EMPact2 and the Wohlwend HPF Compact 01 machines have a range of loading solutions for different types of specimens and applications. It has to be noted that the maximal sample diameter for the Leica EMPact is only 1.5 mm, whereas the Wohlwend HPF Compact 01 and the Bal-Tec HPM 010 use carriers with a cavity of 2 mm in diameter. Ultimately, the question of what HPF to buy is a matter of preference and budget, since all produce well-frozen samples.

2.2. Freeze-substitution

The basic principle underlying FS systems is simple: A metal stage which is constantly cooled from below by liquid nitrogen contains a heating system to regulate the temperature. The temperature is monitored and the heating is triggered if it drops below the desired value. Various inexpensive home-made devices are used in EM facilities. While for most standard FS and embedding procedure these systems are perfectly sufficient, handling some toxic resins at low temperature is tedious and usually exposes the user to fumes. The only commercially available FS machines are the AFS series by Leica Microsystems and the FS 7500 by Boeckler Instruments, Inc. Leica has recently added the automated AFS, in which a pipeting robot can be programmed to exchange the solutions, thus greatly reducing the exposure of the user to fumes.

An important recent shift in thinking has also occurred in the composition of FS media: the addition of water. Previously, water was thought to prevent complete FS, based on observations by Humbel and Müller (1985). In their elegant quantitative assay, the addition of small amounts of water reduced the substitution capacity of various solvents. For long this added water was believed to cause a delay of the substitution process beyond the recrystallization temperature of the cellular water, which would lead to ice crystal damage. In contrast to this hypothesis stands the observation that adding 5% water to the FS media dramatically increases the visibility of the cellular fine structure (Walther and Ziegler, 2002). Intriguingly, the water only needs to be present at a temperature around $-60\,^{\circ}C$ during the FS to exert this effect (Buser and Walther, 2008). In spite of not knowing the exact mechanism, we now routinely and successfully add 5% water to our FS media for a wide range of specimens. The presence of this amount of water also enables us to dissolve any metal stain in acetone, which in future might be useful for selective staining of cellular structures.

3. How to Prepare Yeast by HPF/FS

In order to be imaged in the TEM, yeast cells require extensive processing which takes several days to complete. The only point at which a quality control can be made is the final observation in the TEM. This makes troubleshooting very difficult and time-consuming, and minor errors can ruin an entire batch of samples. In summary, the yeast cells have to be concentrated, loaded into the HPF hat, high-pressure frozen, freeze-substituted, embedded in resin, and sectioned. Most of these aspects have been covered in the previous edition of this book (McDonald and Mueller-Reichert, 2002). Here, we add a more demanding preparation protocol yielding samples with excellent structural preservation (comparable to heavily fixed samples) and concurrently good preservation of epitopes for immunolabeling.

3.1. Growing and concentrating yeast

The medium used for growing the yeast is exceedingly important and determines the success of this preparation. Any components that are poorly soluble in acetone during FS should be avoided or used at low concentrations for reasons stated below. The most important of these components is the carbon source, that is, glucose in most cases. Accordingly, we used standard YPD medium containing only 1% glucose.

A concentration step is necessary to obtain a sample that has a sufficient number of yeast cells per field of view in the TEM. Additionally, less water in the sample means better freezing quality. Ideally, the cells should be manipulated (and thus stressed) as little as possible to obtain a morphology that is as close to the native state as possible. A simple way is to filter approximately 50 ml of yeast culture as described in McDonald and Mueller-Reichert (2002). Here, we propose an alternative filtration method using small syringe filters. This approach only requires 2 ml of yeast culture in log phase which is concentrated to 50 μl with a nylon membrane syringe filter. The yeast are resuspended from the filter and transferred to a 0.5-ml tube (to avoid drying and osmotic stress) and aspirated into dialysis capillaries as described below. Nylon filters are superior to regenerated cellulose because they clog less and the yeast are easily resuspended from the membrane.

Concentration by brief spinning down in a centrifuge—for example, 10 s in a tabletop centrifuge—should be avoided. Centrifugation has been reported to activate stress responses and depolarization of actin patches (Petersen and Hagan, 2005; Soto et al., 2007). We found that after a brief centrifugation of 10 s the morphology of yeast cells is intact, but only very few endocytic sites can be found compared to filtered samples. Since actin patches are sites of endocytosis in yeast, this effect might be linked to the reported actin depolarization.

Irrespective of the method of concentration, it is essential to prevent drying of the cells: filters can be transferred to agar dishes while tubes and syringes with suspensions should be kept closed.

3.2. High-pressure freezing

It is possible to transfer a very concentrated yeast paste from a filter directly into the freezing hats using a toothpick (the "yeast cake" method). This technique requires practice and some amount of drying (i.e., osmotic stress) of the top cell layers will always occur.

Alternatively, less concentrated yeast suspensions can be aspirated into a dialysis capillary and immersed in 1-hexadecene (Studer et al., 1995) (see HPF protocol below). The capillary can then be cut into short pieces that are loaded into the freezing hats using a 200-μm cavity (Hohenberg *et al.*, 1994). The advantage of this approach is that 1-hexadecene has good freezing properties and is not miscible with water. This causes the cell suspension to remain in the capillary during cutting and protects it from drying. Additionally, filling capillaries requires a less concentrated yeast suspension and a smaller culture volume (2–5 ml) than the yeast cake method (30–50 ml).

3.3. Freeze-substitution

Once frozen, it is vital to prevent the samples from premature warming to avert recrystallization. Yeast cells are reportedly easy to vitrify and crystal damage seen in the final sections in the TEM originates frequently from improper handling of the frozen samples or a problem in FS rather than in HPF. Accordingly, the samples should always be manipulated with precooled tweezers. It is also important to know the temperature variations of your FS setup. The temperature measured by the FS device is not necessarily identical with the temperature of the FS medium. Especially, when using a system in which the sample vials are immersed in ethanol, the temperature gradient along the sample tubes can be up to 15 °C from top to bottom (this is less of a problem if you are using a cooled metal block to hold your samples). Also keep in mind that when removing the tubes to transfer the samples the FS medium warms at a rate of 0.5–1 °C/s. These temperature variations are avoided easily by immersing the vial with the FS medium in liquid nitrogen until it freezes completely (i.e., <-95 °C for acetone or <-117 °C for ethanol), dropping the HPF hat containing the sample on the surface and transferring the vial back to the FS device to warm it up to the starting temperature. Make sure the freezing hat really is immersed in the FS solvent once it melts and drops to the bottom of the tube.

A key point is the composition of the FS medium. Acetone is the most suitable solvent to retain the fine structure. Ethanol and methanol can be used as well, but appear to cause more extraction of the sample structure which is partially balanced by the addition of water (Buser and Walther, 2008). A major pitfall in FS is that certain components of yeast growth media and also fillers used for HPF are dissolved poorly by some solvents during FS, especially when infiltration is done at sub-zero temperatures. Two combinations are important when dealing with yeast: First and most importantly, sugars are very poorly soluble in acetone. High concentrations of glucose (and probably glycerol) encase the yeast cells during FS and cause incomplete FS and poor infiltration of the cell periphery (Fig. 24.1). In extreme cases, this results in the cells being ripped out during sectioning. The second problem is that 1-hexadecene is poorly soluble in ethanol. Using hexadecane as filler to load capillaries for HPF combined with substitution in ethanol or methanol leads to incomplete FS and

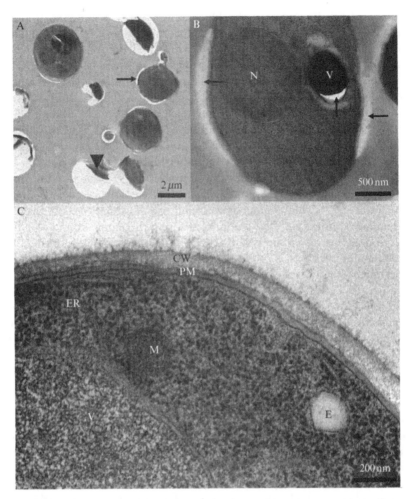

Figure 24.1 Infiltration problems frequently observed in yeast grown in media containing >2% glucose. (A) Overview of a yeast culture grown in YPD and 2% glucose. Many cells are ripped out of section during the sectioning process (arrowhead). Those that remain in section show a poorly infiltrated cell periphery with obvious gaps (arrows). (B) Yeast cell grown in YPD and 2% glucose with characteristics of poor infiltration at the cell periphery and commonly the vacuole (arrows). The gaps observed here tend to widen during immunolabeling causing further distortions of the ultrastructure. (C) Yeast cell grown in YPD and only 1% glucose. Note the perfectly infiltrated periphery and vacuole (V). The cell wall (CW) and plasma membrane (PM) are well retained, and mitochondria (M), endoplasmic reticulum (ER), and endosomes (E) are clearly visible.

recrystallization damage. If substitution in ethanol or methanol is essential the excess 1-hexadecene around the capillaries can either be scratched off carefully after HPF or different filler should be used.

Fixatives can be added to the FS medium to better preserve the fine structure. The most commonly used are osmium tetroxide or glutaraldehyde, but the former is highly oxidative while the latter may cross-link and mask epitopes. Accordingly, both fixatives are known to interfere with immunolabeling. When labeling GFP-fusion proteins as described below, labeling was visible at a glutaraldehyde concentration of 0.1%, but was lost at 0.4% (not shown). Since the aim of the preparation protocol presented here is to prepare samples optimal for immunolabeling, fixatives were omitted entirely in the FS mixture. To increase contrast, uranyl acetate was added to a final concentration of 0.1% (w/v) together with 5% (v/v) water. The FS temperature protocol is less critical in comparison. We suggest and initial incubation step at $-90\ °C$ for a few hours, followed by a slow warming of $3\ °C/h$ to $-60\ °C$ and a quicker warming of $8\ °C/h$ to $-18\ °C$ for low-temperature infiltration. The duration of the initial incubation at $-90\ °C$ does not appear to influence the FS process and can be adjusted for convenience from 2 to >24 h (e.g., for FS over the weekend). Based on experiments with tissue culture cells the critical temperature range for adequate FS appears to be between -90 and $-60\ °C$ (Buser and Walther, 2008) and the warming rate should be kept low at this point. Since warmer temperatures correlate with stronger extraction, the remaining steps of FS and infiltration are performed as quickly as possible (the LR and Lowicryl resins are potent solvents).

3.4. Embedding

Based on previous observations, fixative-free FS with water requires low-temperature embedding to prevent disruption of the fine structure which appears to occur rapidly at temperatures above $0\ °C$ (Buser and Walther, 2008).

In many articles the yeast cell wall is considered a major infiltration barrier and accordingly the suggested infiltration times are fairly long considering the small size of yeast cells. Here, we propose that the actual infiltration barrier is formed by the aggregation of glucose around the cells during FS. In agreement with this hypothesis, a reduction in the glucose content of the medium to 1% and three washes in ethanol (which dissolves sugars better than acetone) before infiltration make it possible to infiltrate yeast cells with LR-Gold in as little as 5 h even at $-18\ °C$.

Several resins suitable for low-temperature embedding and polymerization are available; the Lowicryl and LR resins are commonly used. The different Lowicryl resins offer a wide range of properties, but are very volatile and toxic which makes their use in nonautomated FS systems difficult. The LR resins LR-Gold and LR-White are considered less toxic and are thus a good choice for low-temperature embedding with manual solution exchanges. LR-Gold is specially designed for low-temperature UV polymerization down to − 20 °C, while LR-White is commonly polymer-ized at higher temperatures, usually around room temperature. Still, LR-White will also polymerize under UV at temperatures down to at least − 10 °C if a UV catalyst is added. In our hands, the main difference is the better visibility of cellular structures in LR-Gold, which is an important aspect for the application proposed here. Samples embedded in LR-Gold generally show a well-preserved ultrastructure, including filaments which probably represent f-actin (Fig. 24.2).

3.5. Sectioning

Both LR-resins are more brittle and difficult to section than standard Epon blocks. A common problem is the wetting of the block face during section-ing, which can be reduced by lowering the water level in the knife trough. In addition, the resins often do not bond strongly with the dialysis capil-laries, which can sometimes cause the sections to split on the water surface. Still, the ease in handling of the capillaries during HPF outweighs this disadvantage.

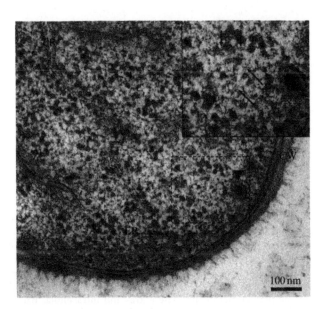

Figure 24.2 High-magnification view of the tip of a small-budded yeast cell showing the high structural quality of the samples obtained with this preparation protocol. Two ER cisterns (ER) are extending toward the bud tip and a secretory vesicle is visible (arrowhead). Further magnification of the boxed region reveals weakly visible filaments (arrows), probably actin.

4. IMMUNOLABELING

The aim of our protocol is to label GFP–fusion proteins by immuno-EM and to correlate the ultrastructural data with dynamic data acquired by live-fluorescence microscopy (Fig. 24.3). The ideal sample for the localization of proteins by immuno-EM possesses both an excellent visibility of cellular structures and a good retention of epitopes to be bound by the antibody. The dilemma is that good structural preservation usually requires strong fixation, while the retention of epitopes asks for as little fixation as possible. The aim of any protocol thus has to be to find the optimal middle way. The use of low-temperature embedding enables us to completely avoid fixatives while still retaining the fine structure. Low-temperature embedding is especially crucial for this approach, since the transition to room temperature causes strong extraction in weakly fixed samples. We chose LR–Gold as resin because it can be polymerized at $-18\,^{\circ}\mathrm{C}$ and is less

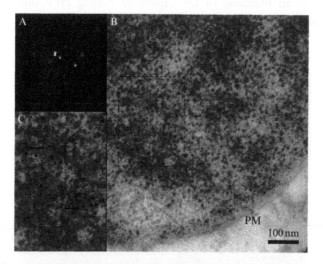

Figure 24.3 Correlation of live-fluorescence images of a yeast strain expressing Abp1-GFP with immuno-EM data after anti-GFP labeling. (A) Live-fluorescence image of Abp1-GFP expressing yeast cells. The actin-binding protein Abp1 is involved in the final steps of endocytosis and remains on the primary endosomes for several seconds after internalization (Kaksonen *et al.*, 2003). (B) Anti-GFP immuno-EM on sections of Abp1-GFP expressing yeast cells. The label is visible in small clusters within approximately 1 μm of the plasma membrane. The extra- and intracellular background was negligible. The ultrastructure is slightly degraded during the labeling process, probably due to a combination of section swelling and extraction of the unfixed cellular material (compare with Figs. 24.1C and 24.2). (C) Magnification of the boxed area reveals that the gold clusters around small spherical structures (arrows), which probably represent tangentially sectioned primary endosomes that are still surrounded by a cloud of Abp1.

toxic than the comparable Lowicryl resins. Furthermore, the visibility of fine structures is better in LR-Gold than in the similar LR-White. The samples were freeze-substituted in acetone containing 0.1% uranyl acetate and 5% water, low-temperature embedded in LR-Gold at −18 °C and labeled by a two-step immuno-EM protocol. While the yeast fine structure was well retained in unlabeled sections, the longer incubation in the blocking and labeling solutions caused some deterioration of the sections especially in poorly infiltrated cells. This was probably due to a combination of swelling and extraction of the unfixed cellular material. Still, the structural quality of the samples is sufficient to detect Abp1-GFP-labeling (Kaksonen *et al.*, 2003) of fine structures like primary endosomes with negligible background (Fig. 24.3B). The difference in the structural detail visible in Figs. 24.2 and 24.3B is a consequence of the above mentioned extraction during labeling of the sections, since extended on-section staining with uranyl acetate and lead citrate does not improve the visibility of the vesicles. This could possibly be improved by either reducing the incubation times with the antibody, a stronger cross-linking of the resin, or the addition of low amounts of fixatives to the FS mixture (e.g., <0.1% glutaraldehyde or osmium tetroxide). Unfortunately, all those changes can decrease the labeling efficiency and need to be optimized individually for every epitope–antibody combination. By generally omitting fixatives the present protocol is a good starting point for this endeavor. Furthermore and in contrast to mammalian tissue culture cells, the high density of the yeast cytoplasm often obscures fine structures and poses a major challenge in the study of endocytic trafficking in yeast.

5. CONCLUSIONS

The preparation scheme presented here aims at preserving yeast cells in a close-to-native state and to allow on-section immunolabeling with good visibility of their fine structure. The immunolabeling of GFP is of particular interest due to the vast number of GFP-tagged proteins being studied by fluorescence microscopy and because it allows standardization of the immunolabeling protocol for different GFP-tagged proteins. Importantly, the fixative-free approach presented here combines excellent fine structural preservation and retention of epitopes for immunolabeling in one sample, allowing a direct correlation of protein distributions with the cellular ultrastructure. The most damage to the ultrastructure is apparently done during the immunolabeling process when the sections are immersed in aqueous buffers for approximately 1 h. It remains to be seen if shorter labeling times can improve the structure without reducing the signal. Alternatively, the damage can also be avoided by serial sectioning and using complementary sections for ultrastructural and labeling purposes, respectively. Nevertheless, it cannot be overemphasized that the key aspects for successfully preparing yeast lie in the growth and the handling of the cells before HPF, that is, low sugar content in the growth medium and prevention of osmotic and centrifugal stress.

The ultimate goal in correlative microscopy of biological systems is to directly correlate the movement of GFP-tagged proteins in living cells, to arrest them in a defined state and process the same cell for immuno-EM to reveal the distribution of the GFP-tagged component within the fine structural context. The protocol presented here fulfills the last of the above mentioned conditions, namely the ability to immunolocalize a protein while still preserving a clearly visible cellular ultrastructure. Still, to realize true correlative microscopy (i.e., observation of the same cell by both LM and EM), several technical aspects of our preparative protocol need to be adapted. Most importantly, a sample holder system has to be developed to meet restrictions imposed by both the fluorescence microscopy (e.g., transparency to light) and the freezing method (e.g., good cooling rates). While the Leica EMPact2-RTS meets these criteria, the time resolution is too low (4–5 s) for some of the endocytic events that interest us.

In summary, we show that correlative immuno-EM is technically achievable without sacrificing ultrastructural resolution or epitopes. The ability to label GFP in such a way also allows us to build a GPF-immuno-EM localization database similar to the database available for fluorescence microscopy (Huh et al., 2003).

6. Protocols

The following protocols describe the process and materials needed for high-pressure freezing with a Wohlwend HPF Compact 01 and freeze-substitution in a Leica AFS.

6.1. Filtration and HPF

1. Special materials needed:
   - syringe filters, 4 mm diameter, 0.45 μm pore size, nylon membrane, National Scientific, Cat# F2504-1
   - 3 ml syringes with Luer-Lok tip
   - dialysis capillaries, Spectra/Por RC, MWCO 13,000, 200 μm i.d., Spectra/Por, Cat#132 290
   - Number 10 scalpel (curved blade)
   - 1-hexadecene, Sigma-Aldrich
   - HPF platelets, recess 0.3 mm (Cat# 242) and recess 0.1/0.2 mm (Cat# 241), Technotrade International

2. Cut the dry dialysis capillary to pieces of approximately 4 cm in length with a scalpel. Fill a 6-cm plastic dish with 1-hexadecene so that the HPF platelets are fully immersed. Turn the platelets with the 0.2 and 0.3 mm cavity upward, respectively.

3. Pull up 2 ml yeast culture (OD_{600} = 0.2–0.5) in the syringe and filter it through the syringe filter until the entire volume is filtered or the filter clogs. Remove the syringe filter and carefully resuspend the yeast stuck on the filter in the remaining 50 μl of medium. Transfer the suspension to a 0.5-ml microcentrifuge tube and drop in the dialysis capillary so that the suspension is pulled in by capillary action. Immerse the filled capillary in 1-hexadecene and cut it into >2 mm short pieces with a scalpel (curved blades work best). Carefully fill a 0.2-mm cavity platelet with 3–5 pieces using a fine tipped forceps, then transfer the platelet into the HPF sample holder. Make sure there are no air bubbles and cover the cavity with the other platelet (flat side down), close the sample holder and freeze in the HPF.

4. Store the samples in cryovials under liquid nitrogen.

6.2. Freeze-substitution

1. Special materials needed:
   - EM-grade acetone
   - EM-grade ethanol
   - 0.5 ml microcentrifuge tubes
   - 2% (w/v) uranyl acetate (UA) in distilled water

2. Dilute the 2% aqueous UA 1:20 in acetone to prepare the final FS medium containing 0.1% UA and 5% water in acetone. Aliquot 0.5 ml per microcentrifuge tube (labeled with pencil, as permanent markers easily wash off in the ethanol bath).

3. Fill the AFS with liquid nitrogen and program as follows (see above for modifications): start at −90 °C for 4 h; warm at a rate of 3 °C/h; hold at −60 °C for 2 h; warm 8 °C/h; hold at −18 °C for 72 h (duration of the actual FS is only 24 h). Start the program and hit "pause" to precool to −90 °C. Fill the AFS cups halfway with ethanol and allow them to cool for approximately 10–15 min.

4. Separate the two HPF platelets enclosing the capillaries under liquid nitrogen. Take one microcentrifuge tube filled with the FS medium, open it and immerse it in the liquid nitrogen until it freezes up completely (this makes sure that your FS medium is −95 °C when it melts so that the sample is not warmed prematurely). Transfer the HPF platelet containing the cells (i.e., the 0.1/0.2 mm platelet) on top of the frozen FS medium. Make sure any drops of liquid nitrogen that may have spilled in the tube have evaporated (otherwise the microcentrifuge

tubes can explode!), close the tube and put it in the AFS cup (it is useful to make a small rack that fits in the cups). Repeat with all samples and start the AFS by hitting "pause" again.

5. Used HPF platelets can be cleaned in acetone and reused multiple times until they show signs of deformation such as bulging.

6.3. Embedding

1. Special materials needed:
   - Wheaton Snap-Cap Specimen Vials, Ted Pella, Inc.
   - EM-grade ethanol
   - LR-Gold resin
   - benzoyl peroxide (thermal initiator)
   - benzoin methyl ether (UV initiator)
2. Fill snap cap vials with ethanol (1.5 ml per FS sample). Mix 50% (v/v) LR-Gold (without initiators) in ethanol (0.5 ml per FS sample) in snap cap vials. When mixing LR-Gold always exclude air by flushing with dry nitrogen gas and mix by gentle pipeting (avoid bubbles). Precool the vials in the AFS for > 20 min.
3. When the samples have reached − 18 °C gently wash them three times with the precooled ethanol and remove the HPF platelets with a forceps. The capillaries should fall out by themselves. If not, gently shake the microcentrifuge tube. Do not let the sample become dry, always leave a layer of liquid when exchanging solutions.
4. Infiltrate with the 50% LR-Gold in ethanol for 2 h at − 18 °C.
5. Prepare and precool 100% LR-Gold (without initiators) and infiltrate at − 18 °C for 2 h.
6. Prepare and precool 100% LR-Gold with 0.1% (w/w) benzoyl peroxide and 0.1% (w/w) benzoin methyl ether (mix by gentle pipeting, avoid air bubbles). Infiltrate at − 18 °C for 1 h, partially cover the top of the tubes with aluminum foil (for indirect UV), then mount the UV lamp and polymerize over night at − 18 °C.
7. The next day check if the samples polymerized completely. Any unpolymerized resin with a reddish color came in contact with oxygen and will not polymerize. Remove the aluminum foil and let the AFS reach room temperature under UV for 3–4 h to make sure the polymerization completed. Remove any unpolymerized resin and leave the samples on a window sill for 1 day to degas.
8. Remove the microcentrifuge tube with a razor blade and proceed to sectioning.

6.4. Anti-GFP immunolabeling

1. Special materials needed:
   - formvar-filmed TEM copper mesh grids
   - blocking buffer (0.2% BSA, 0.2% fish skin gelatin in PBS)
   - 0.5% glutaraldehyde (EM grade) in distilled water
   - anti-GFP goat primary antibody, Fitzgerald Industries, Cat# 70R-GG001
   - rabbit antigoat 10 nm-gold secondary antibody, Ted Pella, Cat# 15796
2. Pick up 70 nm thin sections on formvar-filmed copper TEM grids and allow to dry.
3. Prepare a moist chamber: place a piece of wet paper in your staining dish (either using staining molds or drops on parafilm).
4. Block unspecific binding sites by incubating the sections on a drop of blocking buffer (BB) for 5 min.
5. Incubate with primary antibody (1:100 in BB) for 30 min.
6. Wash on five drops of BB.
7. Incubate with secondary antibody (1:50 in BB) for 30 min.
8. Wash on five drops of PBS.
9. Fix on 0.5% glutaraldehyde for 5 min.
10. Wash on three drops of water.
11. Stain with 2% UA in water and Reynolds lead citrate as needed.

ACKNOWLEDGMENTS

Christopher Buser acknowledges a Fellowship for prospective researchers from the Swiss National Science Foundation. Many thanks go to David Drubin, Georjana Barnes, and the Drubin/Barnes lab for their continuing support and critical discussions.

REFERENCES

Baba, M., and Osumi, M. (1987). Transmission and scanning electron microscopic examination of intracellular organelles in freeze-substituted Kloeckera and *Saccharomyces cerevisiae* yeast cells. *J. Electron. Microsc. Tech.* **5**, 249–261.

Baba, M., Baba, N., Ohsumi, Y., Kanaya, K., and Osumi, M. (1989). Three-dimensional analysis of morphogenesis induced by mating pheromone alpha factor in *Saccharomyces cerevisiae. J. Cell Sci.* **94**(Pt 2), 207–216.

Buser, C., and Walther, P. (2008). Freeze-substitution: The addition of water to polar solvents enhances the retention of structure and acts at temperatures around -60 degrees C. *J. Microsc.* **230**, 268–277.

Hohenberg, H., Mannweiler, K., and Muller, M. (1994). High-pressure freezing of cell suspensions in cellulose capillary tubes. *J. Microsc.* **175**(Pt 1), 34–43.

Huh, W. K., Falvo, J. V., Gerke, L. C., Carroll, A. S., Howson, R. W., Weissman, J. S., and O'Shea, E. K. (2003). Global analysis of protein localization in budding yeast. *Nature* **425,** 686–691.

Humbel, B. M., and Müller, M. (1985). Freeze substitution and low temperature embedding. *In* "Science of Biological Specimen Preparation," (M. Müller, R. P. Becker, A. Boyde, and J. J. Wolosewick, eds.), pp. 175–183. SEM Inc., AMF O'Hare, Chicago.

Kaksonen, M., Sun, Y., and Drubin, D. G. (2003). A pathway for association of receptors, adaptors, and actin during endocytic internalization. *Cell* **115,** 475–487.

McDonald, K. L., and Mueller-Reichert, T. (2002). Cryomethods for thin section electron microscopy. *In* "Methods in Enzymology," (C. Guthrie and G. R. Fink, eds.), Vol. 351, pp. 96–123. Academic Press, San Diego, CA.

Petersen, J., and Hagan, I. M. (2005). Polo kinase links the stress pathway to cell cycle control and tip growth in fission yeast. *Nature* **435,** 507–512.

Soto, T., Nunez, A., Madrid, M., Vicente, J., Gacto, M., and Cansado, J. (2007). Transduction of centrifugation-induced gravity forces through mitogen-activated protein kinase pathways in the fission yeast *Schizosaccharomyces pombe. Microbiology* **153,** 1519–1529.

Studer, D., Michel, M., Wohlwend, M., Hunziker, E. B., and Buschmann, M. D. (1995). Vitrification of articular cartilage by high-pressure freezing. *J. Microsc.* **179**(Pt 3), 321–332.

Verkade, P. (2008). Moving EM: the Rapid Transfer System as a new tool for correlative light and electron microscopy and high throughput for high-pressure freezing. *J. Microsc.* **230,** 317–328.

Walther, P., and Ziegler, A. (2002). Freeze substitution of high-pressure frozen samples: The visibility of biological membranes is improved when the substitution medium contains water. *J. Microsc.* **208,** 3–10.

CHAPTER TWENTY-FIVE

ANALYZING P-BODIES AND STRESS GRANULES IN *SACCHAROMYCES CEREVISIAE*

J. Ross Buchan,* Tracy Nissan,† *and* Roy Parker*

Contents

Abstract

Eukaryotic cells contain at least two types of cytoplasmic RNA–protein (RNP) granules that contain nontranslating mRNAs. One such RNP granule is a P-body, which contains translationally inactive mRNAs and proteins involved in mRNA degradation and translation repression. A second such RNP granule is a stress granule which also contains mRNAs, some RNA binding proteins and several translation initiation factors, suggesting these granules contain mRNAs stalled in translation initiation. In this chapter, we describe methods to analyze P-bodies and stress granules in *Saccharomyces cerevisiae*, including procedures to determine if a protein or mRNA can accumulate in either granule,

* Department of Molecular and Cellular Biology and Howard Hughes Medical Institute, University of Arizona, Tucson, Arizona, USA
† Department of Molecular Biology, Umeå University, Umeå, Sweden

Methods in Enzymology, Volume 470   © 2010 Elsevier Inc.
ISSN 0076-6879, DOI: 10.1016/S0076-6879(10)70025-2   All rights reserved.

if an environmental perturbation or mutation affects granule size and number, and granule quantification methods.

1. INTRODUCTION

An important aspect of the control of gene expression is the range of degradation and translation rates of different mRNAs. Recently, evidence has begun to accumulate that the control of translation and mRNA degradation can involve a pair of conserved cytoplasmic RNA granules (Fig. 25.1). One class of such RNA granules is the cytoplasmic processing body, or P-body. P-bodies are dynamic aggregates of untranslating mRNAs in conjunction with translational repressors and proteins involved in deadenylation, decapping and 5′ to 3′ exonucleolytic decay (Parker and Sheth, 2007). P-bodies and the mRNPs assembled within them are of interest for several reasons. They have been implicated in translational repression (Coller and Parker, 2005; Holmes *et al.*, 2004), normal mRNA decay (Cougot *et al.*, 2004; Sheth and Parker, 2003), nonsense-mediated decay (Sheth and Parker, 2006; Unterholzner and Izaurralde, 2004), miRNA-mediated repression in metazoans (Liu *et al.*, 2005; Pillai *et al.*, 2005), and mRNA storage (Bhattacharyya *et al.*, 2006; Brengues *et al.*, 2005). At a minimum, P-bodies serve as markers that are proportional to the concentration of mRNPs complexed with the mRNA decay/translation repression machinery and may have additional biochemical properties that affect the control of mRNA translation and/or degradation.

Stress granules are a second mRNP granule implicated in translational control, and have been extensively studied in mammalian cells; for reviews, see Anderson and Kedersha (2006, 2008). Stress granules are generally not observable under normal growth conditions in yeast or mammalian cells and greatly increase in response to defects in translation initiation including decreased function of eIF2 or eIF4A (Dang *et al.*, 2006; Kedersha *et al.*, 2002; Mazroui *et al.*, 2006). Because stress responses often involve a transient inhibition of translation initiation, stress granules accumulate during a wide range of stress responses. Stress granules have been argued to function as "triage" centers for mRNAs exiting polysomes during stress, wherein mRNAs are either sorted to P-bodies for decay, maintained in a stored nontranslating state, or returned to translation (Anderson and Kedersha, 2006, 2008).

Recent results have shown that mRNP granules similar to mammalian stress granules can form in budding yeast. This was first suggested by the observation that the translation initiation factors eIF4E, eIF4G, and Pab1p, components of mammalian stress granules, formed foci in yeast during glucose deprivation and high OD conditions, which could either colocalize

A

Pab1-GFP
(stress granules)

Dcp2/Edc3-mCh
(P-bodies)

Merge

B

C

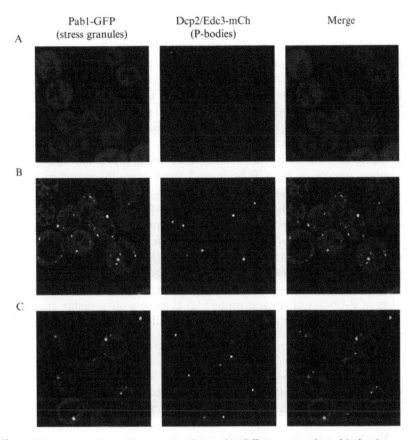

Figure 25.1 P-bodies and stress granules under different growth and induction conditions. (A) Mid-log wild-type cells (yRP840), transformed with pRP1660. Dcp2-mCh serves as the P-body marker. *Note*: Very faint and large foci in P-body column (A and B) are infact vacuolar autofluorescence. (B) Mid-log wild-type cells (yRP840), transformed with pRP1660, and subject to 10 min –Glu deprivation stress. Dcp2-mCh serves as the P-body marker. (C) High OD wild-type cells (BY4741), transformed with pRP1657; 2 days growth in minimal media. Edc3-mCh serves as the P-body marker. *Note*: Pab1 foci in high OD may not be directly equivalent to mid-log stress granules (see main text).

with or be distinct from P-bodies (Brengues and Parker, 2007; Hoyle *et al.*, 2007). These stress granules, also called EGP bodies, also contain mRNAs (Hoyle *et al.*, 2007). Further evidence that these EGP bodies or yeast stress granules are equivalent to mammalian stress granules is that they contain the yeast orthologs of several proteins seen in mammalian stress granules (Table 25.2) and share similar rules of assembly (Buchan *et al.*, 2008). Such assembly rules include a requirement for nontranslating mRNA, stimulation by decreased functional levels of the translation initiation factor eIF2, and a requirement for similar protein assembly factors (Buchan *et al.*, 2008).

P-bodies and stress granules interact and are often docked in mammalian cells, whereas in budding yeast they predominantly overlap (Buchan *et al.*, 2008; Kedersha *et al.*, 2005). Budding yeast provides a good system for analyzing P-body and stress granule interactions as assembly of both granules can be prevented or modified in various mutant strains (Buchan *et al.*, 2008; Coller and Parker, 2005; Decker *et al.*, 2007; Teixeira and Parker, 2007). Indeed, such experiments have suggested that stress granule formation in some cases is enhanced by preexisting P-bodies, suggesting a functional relationship between the two.

Analysis of mRNA turnover and translational repression can involve examining aspects of both P-body and stress granule composition and assembly, given the concentration of mRNAs, decay factors, translational repressors and initiation factors in these granules. In this chapter, we describe methods to analyze P-bodies and stress granules in the budding yeast, *Saccharomyces cerevisiae*. We focus on describing methods to address three common questions: (a) Does a given protein or mRNA accumulate in P-bodies or stress granules? (b) Does a specific perturbation (e.g., mutation, overexpression, or environmental cue) qualitatively change the size or number of P-bodies or stress granules? (c) Is there a quantifiable change in the number and size of P-bodies or stress granules in a given population of cells?

2. DETERMINING IF A SPECIFIC PROTEIN CAN ACCUMULATE IN P-BODIES OR STRESS GRANULES

2.1. Markers of P-bodies and stress granules

A common experimental goal is determining if a given protein accumulates in P-bodies or stress granules. Previous work has identified many proteins enriched in yeast P-bodies (Table 25.1). These include a conserved core of proteins found in P-bodies from yeast to mammals that consists of the mRNA decapping machinery. Core yeast P-body components include the decapping enzyme, Dcp1/Dcp2, the activators of decapping Dhh1, Pat1, Scd6, Edc3, the Lsm1–7 complex, and the 5′–3′ exonuclease, Xrn1. Some proteins observed in yeast P-bodies, such as proteins involved in nonsense-mediated decay (Sheth and Parker, 2006), are only observed in P-bodies under certain mutant, cell-type, stress, or overexpression conditions (Table 25.1). These proteins may normally rapidly transit through P-bodies, but under some conditions accumulate to detectable levels.

Characterization of yeast stress granule composition is at a more nascent state than that of P-bodies; nonetheless a number of factors have been identified including the translation initiation factors eIF4G1, eIF4G2, eIF4E, and Pab1. Additional components include orthologs of factors

Table 25.1 Protein components of P-bodies

Core components	Function
Dcp1	Decapping enzyme subunit[a]
Dcp2	Catalytic subunit of decapping enzyme[a]
Dhh1	DEAD box helicase required for translational repression; decapping activator[a]
Edc3	Decapping activator/P-body assembly factor[b,c]
Lsm1-7	Sm-like proteins involved in decapping[a]
Pat1	Decapping activator and translational repressor[a]
Scd6	Protein containing Sm-like and FDF domain; involved in translation repression[d]
Xrn1	5′ to 3′ exonuclease[a]
Ccr4/Pop2/Not1-5	Major cytoplasmic deadenylase[a,e]
Proteins involved in NMD function	
Upf1	ATP-dependent helicase required for NMD, accumulates in yeast P-bodies in *dcp1Δ*, *xrn1Δ*, *dcp2Δ*, *upf2Δ*, and *upf3Δ* strains[f]
Upf2	Component required for NMD, accumulates in P-bodies in *dcp1Δ*, *dcp2Δ*, and *xrn1Δ* mutants[f]
Upf3	Component required for NMD, accumulates in P-bodies in *dcp1Δ*, *dcp2Δ*, and *xrn1Δ* mutants[f]
Ebs1	Putative ortholog of human Smg7, accumulates in P-bodies during glucose deprivation[g]
Translation and translational repression function	
Cdc33	eIF4E: mRNA m7G cap binding protein[h,i]
Pab1	Binds poly(A) sequences; promotes translation initiation and mRNA stability[h,i]
Ngr1/Rbp1	RNA-binding protein, localizes to P-bodies during stress[j]
Sbp1	Facilitates mRNA decapping[k]
Tif4631	eIF4G1: component of eIF4F initiation factor[h,i]

(continued)

Table 25.1 *(continued)*

Core components	Function
Tif4632	eIF4G2: component of eiF4F, induced at stationary phase[h,i]
Ded1	DEAD box helicase implicated in translational control[l]
Additional components function	
Dsc2	Nutrient stress-dependent regulator of the scavenger enzyme Dcs1[m]
Pby1	Putative tubulin tyrosine ligase[n]
Rpb4	Subunit of RNA polymerase II[o]
Rpm2	Protein component of the mitochondrial RNaseP[p]
Vts1	May recruit Ccr4–Pop2 deadenylation complex to mRNAs; accumulates in P-bodies in *xrn1*Δ strains[q]

[a] Sheth and Parker (2003).
[b] Kshirsagar and Parker (2007).
[c] Decker *et al.* (2007).
[d] Barbee *et al.* (2006).
[e] Muhlrad and Parker (2005).
[f] Sheth and Parker (2006).
[g] Luke *et al.* (2007).
[h] Brengues and Parker (2007).
[i] Hoyle *et al.* (2007).
[j] Jang *et al.* (2006).
[k] Segal *et al.* (2006).
[l] Beckham *et al.* (2008).
[m] Malys and McCarthy (2006).
[n] Sweet *et al.* (2007).
[o] Lotan *et al.* (2005).
[p] Stribinskis and Ramos (2007).
[q] Rendl *et al.* (2008).

implicated in mammalian stress granule assembly, namely Pub1 (TIA-1), Ngr1/Rbp1 (TIA-R), and Pbp1 (Ataxin-2; see Table 25.2 for a complete list), all of which have been implicated in regulation of mRNA stability and translational control of specific yeast mRNAs (Buu *et al.*, 2004; Duttagupta *et al.*, 2005; Ruiz-Echevarría and Peltz, 2000; Tadauchi *et al.*, 2004; Vasudevan *et al.*, 2005).

To date, all proteins identified in yeast stress granules, which we define as foci distinct from P-bodies (as visualized by Edc3- or Dcp2-mCh), are also seen to partially or predominantly overlap with P-bodies during stress (Fig. 25.1; Buchan *et al.*, 2008), and therefore are sometimes classified as being present in P-bodies. This is due to the overlap of P-bodies and stress

Table 25.2   Protein components of stress granules

Core components	Function
Pab1	Binds poly(A) sequences; promotes translation initiation and mRNA stability[a,b]
Tif4631	eIF4G1: component of eIF4F initiation factor[a,b]
Tif4632	eIF4G2: component of eiF4F, induced at stationary phase[a,b]
Cdc33	eIF4E: mRNA m7G cap binding protein[a,b]
Pub1	Poly(A)/(U) binding protein; stabilizes mRNAs; possess QN domain (TIA-1 ortholog)[c]
Ngr1/Rbp1	RNA binding protein; destabilizes mRNAs (TIA-R ortholog)[c]
Pbp1	Polyadenylation regulator; translational regulator (Ataxin-2 ortholog)[c]
Eap1	eIF4E binding protein; role in maintaining genetic stability[c]
Gbp2	Poly(A) binding protein; mRNA export role[c]
Hrp1	Cleavage factor I subunit; mRNA 3' end cleavage and polyadenylation[c]
Nrp1	Putative RNA binding protein[c]
Ygr250c	Putative RNA binding protein[c]

[a] Brengues and Parker (2007).
[b] Hoyle et al. (2007).
[c] Buchan et al. (2008).

granules in budding yeast and highlights the difficulty in unambiguously defining a protein as solely being present in P-bodies or stress granules. More realistically, these types of observations suggests that there is a continuum of mRNP states between P-bodies and stress granules as individual mRNAs in one biochemical state exchange proteins and remodel into the predominant state accumulating in a different granule. Indeed, stress granules which are distinct from bright P-bodies can often exhibit a very faint microscopic signal for either Edc3 or Dcp2-mCh, which often fades with time during stress (Buchan et al., 2008). However, the relative abundance of Dcp2 and Edc3 in "P-body distinct" stress granules is extremely low relative to their concentration in P-bodies, as judged by microscopic intensity measurements.

To determine if a given protein can accumulate in P-bodies or stress granules, one simply needs to examine the subcellular distribution of the protein relative to known P-body or stress granule components. The simplest way to do this is to tag the protein of interest with a fluorescent protein, and then examine its localization relative to another fluorescent protein-tagged P-body or stress granule component. For this type of experiment, many of the core P-body and stress granule components are available as fusions to fluorescent proteins on yeast plasmids (Table 25.3). Alternatively, P-bodies can be visualized in fixed cells by standard immunofluorescent methods using antisera against specific components (Gaillard and Aguilera, 2008). To our knowledge, no one has attempted to detect yeast stress granules via such methods. To examine if a protein accumulates in yeast P-bodies or stress granules using a fluorescent fusion protein one can take the following steps:

(a) Obtain a fluorescent tagged version of the protein of interest. Note that most of the yeast ORFs fused to GFP are available from a genomic collection (Huh *et al.*, 2003) and can be purchased from Invitrogen. Use of the native promoter in fusion strains integrated into the genome should avoid mislocalization due to overexpression of the tagged protein.

(b) Determine if the fusion protein is functional by some criteria as it is problematic to interpret the localization of nonfunctional proteins.

(c) Compare the localization of the tagged protein of interest with a core P-body/stress granule marker tagged with a different fluorescent protein as described below.

In our experience, the most reliable and easily visualized components of yeast P-bodies are Edc3 and Dcp2, whereas Pab1 and Pub1 serve as good stress granule markers (Buchan *et al.*, 2008). Use of more than one marker for P-bodies and stress granules is good practice as specific mutants or stress conditions can specifically affect individual components of P-bodies and stress granules while not necessarily perturbing overall granule assembly (Teixeira and Parker, 2007).

2.2. Preparation of samples

To determine if a protein can accumulate in P-bodies or stress granules, we recommend examining its subcellular distribution in mid-log phase and also in stress conditions such as glucose deprivation, heatshock, hyperosmotic stress, high OD, or when decapping is inhibited (Buchan *et al.*, 2008; Grousl *et al.*, 2009; Sheth and Parker, 2003; Teixeira *et al.*, 2005). While all of these conditions increase P-bodies to varying extents, glucose deprivation and heatshock are the only well-characterized stress conditions identified which strongly induce formation of stress granules, although high OD typically induces 1–2 large Pab1 foci which are distinct from bright P-bodies and may be similar to stress granules (see p16 and Fig. 25.1C). Additionally treatment with 0.5% (v/v) sodium azide for 30 min also induces stress granule-like foci which require nontranslating mRNA, and compositionally resemble glucose deprived stress granules (unpublished data).

Table 25.3 Core P-body and stress granule components available on plasmids as fluorescent protein fusions

Protein	Fluorescent tag	Plasmid marker	Lab plasmid number
Dcp2 full-length	GFP	LEU2	pRP1175[a]
	GFP	TRP1	pRP1316[b]
	GFP	URA3	pRP1315[c]
Dcp2 full-length	RFP	LEU2	pRP1155[d]
	RFP	TRP1	pRP1156[e]
	RFP	URA3	pRP1186[f]
Dcp2 truncated (1–300)	RFP	LEU2	pRP1167[g]
	RFP	TRP1	pRP1165[h]
	RFP	URA3	pRP1152[h]
Dhh1	GFP	LEU2	pRP1151[g]
Edc3	mCherry	URA3	pRP1574[i]
	mCherry	TRP1	pRP1575[i]
Lsm1	GFP	LEU2	pRP1313[a]
	GFP	URA3	pRP1314[j]
Lsm1	mCherry	LEU2	pRP1400[k]
Lsm1	RFP	URA3	pRP1084[l]
	RFP	LEU2	pRP1085[l]
Pat1	GFP	URA3	pRP1501[m]
Pab1	GFP	URA3	pRP1362[n]
	GFP	TRP1	pRP1363[n]
Pub1	mCherry	URA3	pRP1661[i]
	mCherry	TRP1	pRP1662[i]
Pab1 + Edc3	GFP/ mCherry	URA3	pRP1657[i]
	GFP/ mCherry	TRP1	pRP1659[i]
Pab1 + Dcp2	GFP/ mCherry	URA3	pRP1658[i]
	GFP/ mCherry	TRP1	pRP1660[i]

[a] Coller and Parker (2005).
[b] Segal et al. (2006).
[c] Unpublished, Parker Lab.
[d] Teixeira et al. (2005).
[e] Sheth and Parker (2006).
[f] Teixeira et al. (2005).
[g] Sweet et al. (2007).
[h] Muhlrad and Parker (2005).
[i] Buchan et al. (2008).
[j] Tharun et al. (2005).
[k] Beckham et al. (2007).
[l] Sheth and Parker (2003).
[m] Pilkington and Parker (2008).
[n] Brengues and Parker (2007).

It is important to be careful when preparing cells for examining P-bodies or stress granules microscopically as both are dynamic and can change rapidly in response to a variety of stresses (Buchan *et al.*, 2008; Teixeira *et al.*, 2005). If this is a serious issue in an experiment one could in principle use fixed cells although at this time, no one has detected stress granules in yeast by standard immunofluorescence methods. Some tips to consider in this type of experiment are as follows: (i) In general, cells that are growing vigorously prior to the onset of a stress tend to show greater inductions of P-bodies and stress granules, therefore ensuring optimal aeration and temperature of cultures is important. (ii) Care should be taken to reduce centrifugation and handling times as variations can alter P-bodies and stress granules due to their rapid dynamics. (iii) Media conditions may alter P-body and stress granule composition, for example, different results can be obtained from growth in rich versus minimal media for noncore proteins. (iv) Finally, not all lab strains of *S. cerevisiae* should be assumed to behave equally—for example, we have observed that our lab strain yRP840 (cross of S28CC and A364A; Hatfield *et al.*, 1996) induces brighter and slightly larger stress granules than BY4741 during glucose deprivation, while simultaneously exhibiting a lower level of P-bodies than BY4741 during normal mid-log growth conditions (Buchan *et al.*, 2008). Detailed protocols for examining P-bodies and stress granules in mid-log cultures, ±glucose deprivation stress, are described in the following sections.

2.2.1. Examination of P-bodies and stress granule in mid-log glucose-deprived cells

1. In a 50-ml conical flask, placed in a 30 °C shaking water bath, grow a 5 ml yeast culture in YPD★ or minimal media as appropriate to mid-log phase, with an absorbance between 0.3 and 0.5 at 600 nm. *★YPD is not suitable for resuspension of cells prior to microscopic examination due to autofluorescence in the media in the GFP channel.*
2. Decant 1–1.5 ml of culture into a microfuge tube, and centrifuge at approximately 13,000 rpm for 30 s.
3. Remove media without disturbing the cell pellet, and resuspend in 1–1.5 ml minimal media supplemented with the same amino acids that the cells were originally grown in (assuming original culture was in minimal media, otherwise a complete mix for YPD grown cultures is recommended). This media can either contain glucose at the original concentration (typically 2%), which acts as a negative control for P-body and stress granule induction, or lack glucose entirely, which should inhibit translation initiation and hence induce P-bodies and stress granules.
4. Repeat steps 2 and 3 to wash out residual glucose, then decant cells into a fresh 50-ml flask and return to shaking at 30 °C for 10 min.

5. Repeat steps 2 and 3, then concentrate cells in approximately 40–80 μl of minimal media ±glucose as appropriate. Keep in a microfuge tube with constant flicking by hand to maintain aeration.

6. Add a small volume of the suspension to the slide at the microscope, then place a coverslip on the sample, press and wipe down gently with a kimwipe to remove excess volume under the coverslip (prevents cells movement), then examine immediately. When comparing live cell samples, be sure to be consistent about the time period the cells are examined on the microscope. This can be a serious issue as we observe that cells under a coverslip eventually induce a stress response, probably due to lack of aeration, thereby artifactually increasing the presence of P-bodies. In +glucose treated cells, however, this artifactual P-body induction is not as strong as that of glucose deprivation induced P-bodies in wild-type cells, but may nonetheless cloud subtle phenotypic differences between strains. Almost no stress granule protein foci are observed in +glucose cells that sit on slides for elongated periods, even up to 45 min.

If a protein shows substantial overlap in its subcellular distribution with core P-body or stress granule proteins, it can be inferred to be a P-body or stress granule component. Naturally, other methods such as coimmunoprecipitation or yeast 2–hybrid interaction of your protein of interest with a P-body or stress granule marker potentially offer additional support for such conclusions.

2.2.2. Additional tips for optimal yeast fluorescence microscopy of P-bodies and stress granules

1. Capture images rapidly to allow accurate assessment of the P-body and stress granule state of the cells. As at least two channels need to be taken to colocalize the P-body/stress granule markers with the protein of interest, it is optimal to use a microscope that splits the beam to record two/three channels simultaneously. If this is not possible, attempt to reduce the time between channel images as much as possible. Yeast P-bodies and stress granules can move slightly from their original location within a period of a few seconds.

2. If cells are moving on the slide, which can make colocalization difficult, an alternative is to immobilize the cells by coating the coverslips with the lectin concanavalin A, which binds to the yeast cell wall. The protocol we use to coat coverslips is to wash coverslips overnight in sterile filtered 1 M NaOH. After washing well with distilled water, add concanavalin A solution (0.5 g/l, Sigma #L7647, 10 mM phosphate buffer (pH 6), 1 mM CaCl$_2$, 0.02% azide) for 20 min with gentle shaking. After removal of the solution, rinse once in distilled water, pour off the liquid and let dry overnight. The coverslips can be stored at room temperature after

coating. Experiment with different concentrations of yeast suspension to put on these slides as if the cell density is too high, cells have a tendency to clump and not form a single focal plane that is preferable for imaging.

3. If longer exposures or time-lapse data are required, we use an inverted Deltavision microscope. In addition to use of normal microscopic slides, which can dry out over time around the edges of the coverslip, depending on atmospheric conditions, we often use concanavalin A coated glass bottom microwell dishes (MatTek Corporation #P35G-1.5-14-C) with the coverslip immersed in enough minimal media (supplemented as appropriate for the experiment) to fully cover it.

4. Beware of the autofluorescent properties of yeast cells! All strains exhibit modest autofluorescence in the GFP channel (while *ade1/ade2* mutants exhibit strong autofluorescence), which often increases during glucose deprivation and high OD, and which can sometimes form confusing foci-like structures. Additionally, weak vacuolar autofluoresence often shows up in the RFP/mCh channel. Many proteins in yeast are not expressed at levels high enough to easily distinguish between legitimate foci and autofluorescence, therefore, ensure you have a firm idea of the basal threshold level of intensity at which you can trust your signal in your strain background of interest.

5. Ensure you conduct bleedthrough controls for your fluorescently tagged proteins, to ensure that errors in filter set up, or fluorescent protein choice do not lead to experimental artifacts. Simply examine one tagged protein at a time, but image in all channels you eventually wish to examine, ensuring that no signal from one channel is bleeding through into another channel.

6. In order to be confident that the behavior of a given protein changes under different conditions, or in different yeast strains, ensure all microscope settings (e.g., exposure times) and post image-capture manipulations (e.g., image scaling) are consistent across experiments.

3. Monitoring Messenger RNA in P-Bodies

In some cases it is useful or important to determine whether bulk mRNA or a specific transcript is accumulating in P-bodies. There are two general approaches to determine if specific mRNAs are accumulating in P-bodies. First, one can use fluorescence *in situ* hybridization (FISH) techniques to monitor the presence of bulk mRNA using an oligo(dT) probe, or specific mRNAs using sequence specific probes. Such approaches have worked well in mammalian cells (Franks and Lykke-Andersen, 2007; Pillai *et al.*, 2005), and while detection of single mRNA species has been demonstrated in yeast (see Garcia *et al.*, 2007; Zenklusen *et al.*, 2008), it is currently unclear if the majority of yeast P-bodies (or stress granules) remain assembled during FISH protocols (see Brengues and Parker, 2007). Though this can likely be optimized, one can nevertheless no longer observe *in vivo* mRNA dynamics after cell fixation.

In an alternative method, one can use "GFP-tagged" mRNAs to follow the localization of specific transcripts in yeast and determine if they accumulate in P-bodies. To visualize specific mRNAs, multiple binding sites for an RNA binding protein fused to a fluorescent protein are inserted into the 3' UTR of the mRNA of interest allowing its subcellular distribution to be examined by following the location of the RNA binding protein fused to the fluorescent protein. Most commonly, the well-characterized U1A or the MS2 binding sites are inserted into the 3' UTR of the mRNA of interest (for detailed methods review, see Bertrand *et al.*, 1998; Brodsky and Silver, 2000). These mRNA constructs are then coexpressed with either the U1A-GFP (Brodsky and Silver, 2000) or the MS2 coat protein fused to GFP (Bertrand *et al.*, 1998). Both of these have nano- to picomolar affinity for their respective binding sites allowing detection of the mRNA.

Several of these types of engineered mRNAs have been constructed on plasmids and used to demonstrate the accumulation of specific yeast mRNAs in P-bodies. Available plasmids expressing "tagged" versions of the stable *PGK1* and the unstable *MFA2* mRNA are described in Table 25.4. In addition, variants of the tagged *PGK1* mRNAs are available with premature nonsense codons in specific positions, which can be used for examining the accumulation of mRNA in P-bodies due to the action of NMD (Sheth and Parker, 2006). A variety of plasmids expressing the MS2 or U1A proteins fused to GFP are also available (Table 25.4).

4. DETERMINING IF A MUTATION/PERTURBATION AFFECTS P-BODY OR STRESS GRANULE SIZE AND NUMBER

4.1. Conditions to observe increases or decreases in P-bodies and stress granules

A common experimental issue is determining if a mutation, protein overexpression, or an environmental cue affects the size and number of P-bodies or stress granules. To examine if P-bodies or stress granules are altered under a certain condition we make three suggestions. First, one should use multiple markers of P-bodies or stress granules to ensure that any differences seen are not unique to a single protein. Second, since a specific mutant may affect P-bodies or stress granules only under certain conditions, we recommend that P-bodies and stress granules be examined under multiple conditions (e.g., mid-log growth, glucose deprivation, high OD, etc.). Finally,

Table 25.4 Plasmids for localizing mRNA in yeast cells: GFP fusion proteins that bind to specific binding sites in mRNA engineered in their 3' UTR

Protein + tag	Plasmid marker	Promoter	Lab plasmid number
MS2 CP-GFP	HIS3	Met25	pRP1094[a]
U1A-GFP	TRP1	GPD	pRP1187[b]
U1A-GFP	LEU2	GPD	pRP1194[c]

RNA	Binding seq.	Plasmid marker	Promoter	Description[i]	Lab plasmid number
MFA2	pG MS2	URA3	GPD	Two MS2 sites 3' to poly(G) tract in 3' UTR	pRP1081[d]
MFA2	MS2	URA3	GPD	Two MS2 sites in 3' UTR	pRP1083[c]
MFA2	U1A	URA3	GPD	PGK1 3' UTR with 16 U1A binding sites	pRP1193[c]
MFA2	U1A	URA3	Tet-Off	PGK1 3' UTR with 16 U1A binding sites	pRP1291[c]
PGK1	MS2	URA3	PGK1	Two MS2 sites 3' to poly(G) tract in 3' UTR	pRP1086[c]
PGK1	U1A	URA3	PGK1	16 U1A sites in 3' UTR	pPS2037 (pRP1354)[f]
PGK1	U1A	URA3	PGK1	PGK1 U1A with nonsense mutation at position 22	pRP1295[g]
PGK1	U1A	URA3	PGK1	PGK1 U1A with nonsense mutation at position 225	pRP1296[g]
PGK1	U1A	URA3	GAL	16 U1A sites in 3' UTR	pRP1303[h]

[a] Beach and Bloom (2001).
[b] Teixeira et al. (2005).
[c] Brengues et al. (2005).
[d] M. Valencia-Sanchez and R. Parker (unpublished).
[e] Sheth and Parker (2003) have short polyG tract in 3' UTR that does not inhibit exonucleolytic decay.
[f] Brodsky and Silver (2000).
[g] Sheth and Parker (2006).
[h] U. Sheth and R. Parker (unpublished).
[i] All mRNA constructs have their native 5' and 3' UTR except where noted.

we recommend quantification of the number of P-bodies and stress granules by computational methods to allow unbiased calculation of the size and number of P-bodies and stress granules present in a given situation (see Section 5).

In practice, there are different methods for examining if a perturbation reduces or increases granule size or number. To determine if a perturbation reduces P-bodies or stress granules, it is most convincing to examine conditions when both are strongly induced and easily detectable. For example, the intensity and number of P-bodies and stress granules increase during glucose deprivation, particularly in the presence of constitutively active Gcn2c kinase, which ultimately limits translation initiation by reducing functional levels of eIF2 (Buchan et al., 2008). P-bodies are also constitutively induced when decapping or 5′ to 3′ degradation are inhibited by dcp1Δ or xrn1Δ, respectively. This makes dcp1Δ and xrn1Δ possible strains to examine the presence of proteins in P-bodies, although there is the caveat of this being a specific mutant condition that may not reflect the normal situation (Buchan et al., 2008; Teixeira et al., 2005). Stress granules are also increased in dcp1Δ and xrn1Δ strain during glucose deprivation as compared to wild-type cells (Buchan et al., 2008). Thus, the above conditions/strains are possible conditions to examine if P-bodies/stress granules are reduced by a mutation or physiological response (e.g., see Buchan et al., 2008; Decker et al., 2007; Teixeira and Parker, 2007). Conversely, P-bodies are small and infrequent, and stress granules nonexistent when cells are undergoing mid-log growth (Buchan et al., 2008; Teixeira et al., 2005), which makes this condition an ideal situation to see if a given perturbation increases either granule (Teixeira and Parker, 2007). P-bodies are also clearly induced at high OD, and while distinct aggregates of Pab1, eIF4E, and eIF4G are also observed under these conditions (Brengues et al., 2007; Fig. 25.1C), a rigorous analysis of the composition and assembly mechanisms of this granule have not yet been completed, thus its significance is currently unclear.

4.2. Interpreting alterations in P-body/stress granule size and number

Any alteration observed in P-body size and number, due to a specific mutation or alteration in growth, can be due to a variety of mechanisms. This is because various changes in cell physiology will affect the size and number of P-bodies, probably due to an altered flux of mRNP in and out of this structure (Table 25.5). For example, the size and number of P-bodies can be increased by defects in mRNA decapping or 5′ to 3′ degradation, which increase the pool of mRNPs in P-bodies by decreasing the destruction of mRNAs in this compartment (Sheth and Parker, 2003), or by defects in translation initiation, which increase the pool of untranslating mRNPs

Table 25.5 Dissection of effects on P-body size and number

Observation	Possible cause	Follow-up experiments
Increase in P-body size and number	1. Inhibition of decapping or 5′ to 3′ degradation	1. Examine mRNA degradation rate and whether mRNAs are degraded normally
	2. Defects in translation initiation	2. Measure the rate of protein synthesis by polysomes or ^{35}S incorporation
Decrease in P-body size and number	1. Slowed translation elongation rate	1. Polysome analysis to determine if the size of polysomes is increased
	2. Enhanced translation/ decreased repression	2. Measure the rates of protein synthesis by polysomes or ^{35}S incorporation
	3. Reduced expression of marker proteins	3. Western blot to ensure P-body protein levels
	4. Impaired granule assembly mechanism	4. Mutational/protein interaction analyses, genetic analyses, etc.

associated with P-bodies (Brengues *et al.*, 2005). Alternatively, P-bodies can be reduced in size and number by inhibiting translation repression, by preventing disassociation of elongating ribosomes from mRNA, by removing interactions that promote aggregation of the individual mRNPs into larger structures, or by reductions in the level of the P-body marker being examined (Coller and Parker, 2005; Decker *et al.*, 2007). Note that in order to be confident of the underlying mechanism affecting P-body size and number, additional experiments should be performed to identify the true cause of the defect (Table 25.5).

Based on our findings in yeast, which suggest that P-bodies promote yeast stress granule assembly, and existing mammalian cell data, it seems likely that all of the above processes could also affect yeast stress granule numbers. Indeed, preventing ribosomal dissociation (cycloheximide) or limiting translational repression (*dhh1Δ*, *pat1Δ*, *dhh1Δ*, and *pat1Δ* strains) inhibits stress granule assembly, whereas specific blocks in mRNA decay (*dcp1Δ*, *xrn1Δ*), increase stress granules (Buchan *et al.*, 2008). Moreover, because defects in P-body formation can reduce stress granules in budding yeast (Buchan *et al.*, 2008), an observed defect in stress granules could be due to alterations in P-body formation or function.

5. QUANTIFICATION OF P-BODY SIZE AND NUMBER

A common and important issue is determining if quantitative differences exist in the size and number of P-bodies or stress granules under different cellular conditions, or between particular mutant strains. Ideally, one should use unbiased approaches to determine the effects. Thus, having one person prepare the images and another person score them blindly is strongly recommended.

In our experiences to date, P-bodies and stress granules present different quantification challenges, especially when strains bearing plasmid-expressed versions of a stress granule marker are used. This is because most stress granule markers (e.g., Pab1, Pub1) are distributed throughout the cytoplasm, and the intensity of a Pab1 or Pub1 stress granule foci is often not much greater than the intensity of the diffuse cytoplasmic protein (e.g., 1.5–4-fold difference). Given that yeast cells can harbor differing copy numbers of plasmids from cell to cell, in practice, the intensity of a stress granule focus in one cell may not be as bright as the diffuse cytoplasmic signal in another cell, making automated counting methods using thresholds problematic. This issue can be countered by using fluorescent tagged proteins in the chromosome since this reduces, but does not eliminate, cell to cell variability. Such tagged proteins can be made using standard techniques and vectors (Longtine *et al.*, 1998). In contrast, most P-body proteins exhibit a very low diffuse cytoplasmic signal, and a very high foci signal (e.g., 5–30-fold difference under some conditions). Thus, for P-bodies, semiautomated scoring approaches are much more feasible.

In our lab, scoring of both granules is accomplished by using ImageJ, a freely downloadable image analysis package from the NIH (Abramoff *et al.*, 2004), although other software and algorithms have been successfully employed elsewhere for similar purposes (e.g., "MatLab"; Aragón *et al.*, 2008). On the ImageJ web site, there are instructions and downloadable plugins, which allow direct loading of raw images from the microscope. An alternative source of ImageJ, preloaded with many useful image analysis plugins, can be obtained from the Wright Cell Imaging facility web site— http://www.uhnres.utoronto.ca/facilities/wcif/download.php.

The semiautomated quantification of P-bodies is accomplished by setting a threshold mask, which allows regions of the image above a certain intensity to be scored as "on" and the rest of the image "off." The P-bodies are counted by the number of "on" regions and their area, intensity and number can be computed. To further reduce bias, the masking procedure, to determine "real" foci, can be performed automatically using the Otsu Thresholding Filter (Otsu, 1979).

Protocols for quantifying P-bodies and stress granules are detailed in the following sections.

5.1. Semiautomated quantification of P-body size and number

1. Open the image in ImageJ to quantify. If using a Z-stack image, rather than a single plane image, the former of which benefits from capturing the entire thickness of the cell, ensure that the Z-stack is first collapsed (Image menu—Stacks—Z-project—choose Max intensity projection). Collapsed Z-stack images will capture all foci in a cell, and therefore are best for accurate quantitation. In contrast, single-plane images tend to be more "aesthetically pleasing," as the total cellular volume is not averaged into one plane; this also makes colocalization of proteins less problematic.

2. Next, select Process, then Smooth—this subtly averages intensity of each pixel with the intensity of surrounding pixels, helping discriminate real foci from image capture artifacts.

3. Go to Process, select Math and then select Subtract to remove the background. This is a key step in the process, which helps discriminate meaningful foci from background noise caused by the media, the microscope, and the diffuse level of the protein/RNA being examined that is present throughout the cell. This level should be individually tailored according to the use of different microscopes, different exposure conditions, and examination of different proteins. Note that because of threshold subtractions, such quantifications are not necessarily absolute numbers or areas of P-bodies but provide a systematic and unbiased measure of relative P-body number and area within experiments.

4. Go to Plugins, select Filters and then select Otsu Thresholding. This option will only be available if you have downloaded the plugin and placed it in the Plugins directory, within the ImageJ directory, on your computer (see ImageJ web site for plugins).

5. Go to Image, select Lookup Tables and then select Invert LUT. This will reverse the image, allowing the P-bodies to be considered "on."

6. Go to Analyze and select Analyze Particles, and in the size bar, set the pixel size range to be counted. This defines the size at which a focus is considered to be a real P-body, for which the appropriate range will vary depending on image resolution, samples, and strains. Optimization is required so random speckles do not count as P-bodies and large fluorescent regions unrelated to P-bodies are not counted (e.g., autofluorescent debris in the media). After pressing ok, a table is generated that lists the number of foci, average area, and total area of each foci. Modifying the data reported for each focus within the foci list can be achieved by going to Analyze—Set Measurements, then ticking the boxes for each parameter you wish to measure.

7. *Optional:* Prior to hitting ok in step six, select Masks in the pull down menu—in addition to the tabulated data, this will present a graphical

display of the thresholded foci so that you can visually check no erroneous foci have been counted.

8. The number of cells in each image must be calculated manually, however, all steps except cell counting can be automated by writing/recording a macro (Plugins—Macros—Record) to perform them to obtain faster throughput analyses. Copying the data output from step 6 directly into programs such as Microsoft Excel allows easy manipulation of the data.

5.2. Manual quantification of stress granule size and number

Repeat steps 1 and 2 from above

3. As a uniform background level from which foci can be distinguished is harder to obtain, it is easier to divide the image into sections (e.g., group of cells with similar cytoplasmic intensity), and quantify separately. To do this, click any of the four shape buttons (bottom left on ImageJ bar), to either draw rectangles, circles, polygon, or freehand selections on the image.

4. Go to Image—Adjust—Threshold, and choose intensity limits (use slider bars or\set values manually) within which your foci signal, but not diffuse cytoplasmic signal, lie.

5. Repeat steps 6–8 as necessary.

ACKNOWLEDGMENTS

We thank the members of the Parker lab for helpful discussions, particularly Carolyn Decker for proofreading and suggestions. NIH grant (R37 GM45443) and funds from the Howard Hughes Medical Institute supported this work. TN was supported in part by T32 CA09213.

REFERENCES

Abramoff, M. D., Magelhaes, P. J., and Ram, S. J. (2004). Image processing with image. *J. Biophotonics Int.* **11**, 36–42.

Anderson, P., and Kedersha, N. (2006). RNA granules. *J. Cell Biol.* **172**, 803–808.

Anderson, P., and Kedersha, N. (2008). Stress granules: The Tao of RNA triage. *Trends Biochem. Sci.* **33**, 141–150.

Aragón, T., van Anken, E., Pincus, D., Serafimova, I. M., Korennykh, A. V., Rubio, C. A., and Walter, P. (2008). Messenger RNA targeting to endoplasmic reticulum stress signaling sites. *Nature* **457**, 736–740.

Barbee, S., *et al.* (2006). Staufen- and FMRP-containing neuronal RNPs are structurally and functionally related to somatic P bodies. *Neuron* **52**, 997–1009.

Beach, D. L., and Bloom, K. (2001). ASH1 mRNA Localization in three acts. *Mol. Biol. Cell* **12**, 2567–2577.

Beckham, C. J., et al. (2007). Interactions between brome mosaic virus RNAs and cytoplasmic processing bodies. J. Virol. **81**, 9759–9768.

Beckham, C., et al. (2008). The DEAD-box RNA helicase Ded1p affects and accumulates in Saccharomyces cerevisiae P-bodies. Mol. Biol. Cell **19**, 984–993.

Bertrand, E., Chartrand, P., Schaefer, M., Shenoy, S. M., Singer, R. H., and Long, R. M. (1998). Localization of ASH1 mRNA particles in living yeast. Mol. Cell **2**, 437–445.

Bhattacharyya, S. N., Habermacher, R., Martine, U., Closs, E. I., and Filipowicz, W. (2006). Relief of microRNA-mediated translational repression in human cells subjected to stress. Cell **125**, 1111–1124.

Brengues, M., and Parker, R. (2007). Accumulation of polyadenylated mRNA, Pab1p, eIF4E, and eIF4G with P-bodies in Saccharomyces cerevisiae. Mol. Biol. Cell **18**, 2592–2602.

Brengues, M., Teixeira, D., and Parker, R. (2005). Movement of eukaryotic mRNAs between polysomes and cytoplasmic processing bodies. Science **310**, 486–489.

Brodsky, A. S., and Silver, P. A. (2000). Pre-mRNA processing factors are required for nuclear export. RNA **6**, 1737–1749.

Buchan, J. R., Muhlrad, D., and Parker, R. (2008). P bodies promote stress granule assembly in Saccharomyces cerevisiae. J. Cell Biol. **183**, 441–455.

Buu, L. M., Jang, L. T., and Lee, F. J. (2004). The yeast RNA-binding protein Rbp1p modifies the stability of mitochondrial porin mRNA. J. Biol. Chem. **279**, 453–462.

Coller, J., and Parker, R. (2005). General translational repression by activators of mRNA decapping. Cell **122**, 875–886.

Cougot, N., Babajko, S., and Seraphin, B. (2004). Cytoplasmic foci are sites of mRNA decay in human cells. J. Cell Biol. **165**, 31–40.

Dang, Y., Kedersha, N., Low, W. K., Romo, D., Gorospe, M., Kaufman, R., Anderson, P., and Liu, J. O. (2006). Eukaryotic initiation factor 2 alpha independent pathway of stress granule induction by the natural product pateamine A. J. Biol. Chem. **281**, 32870–32878.

Decker, C. J., Teixeira, T., and Parker, R. (2007). Edc3p and a glutamine/asparagine-rich domain of Lsm4p function in processing body assembly in Saccharomyces cerevisiae. J. Cell Biol. **179**, 437–449.

Duttagupta, R., Tian, B., Wilusz, C. J., Khounh, D. T., Soteropoulos, P., Ouyang, M., Dougherty, J. P., and Peltz, S. W. (2005). Global analysis of Pub1p targets reveals a coordinate control of gene expression through modulation of binding and stability. Mol. Cell. Biol. **25**, 5499–5513.

Franks, T. M., and Lykke-Andersen, J. (2007). TTP and BRF proteins nucleate processing body formation to silence mRNAs with AU-rich elements. Genes Dev. **21**, 719–735.

Gaillard, H., and Aguilera, A. (2008). A novel class of mRNA-containing cytoplasmic granules are produced in response to UV-irradiation. Mol. Biol. Cell **19**, 4980–4992.

Garcia, M., Darzacq, X., Delaveau, T., Jourdren, L., Singer, R. H., and Jacq, C. (2007). Mitochondria-associated yeast mRNAs and the biogenesis of molecular complexes. Mol. Biol. Cell **18**, 362–368.

Grousl, T., Ivanov, P., Frýdlová, I., Vasicová, P., Janda, F., Vojtová, J., Malínská, K., Malcová, I., Nováková, L., Janosková, D., Valásek, L., and Hasek, J. (2009). Robust heat shock induces eIF2alpha-phosphorylation-independent assembly of stress granules containing eIF3 and 40 S ribosomal subunits in budding yeast, Saccharomyces cerevisiae. J. Cell Sci. **122**, 2078–2088.

Hatfield, L., Beelman, C. A., Stevens, A., and Parker, R. (1996). Mutations in trans-acting factors affecting mRNA decapping in Saccharomyces cerevisiae. Mol. Cell. Biol. **16**, 5830–5838.

Holmes, L. E., Campbell, S. G., De Long, S. K., Sachs, A. B., and Ashe, M. P. (2004). Loss of translational control in yeast compromised for the major mRNA decay pathway. Mol. Cell. Biol. **24**, 2998–3010.

Hoyle, N. P., Castelli, L. M., Campbell, S. G., Holmes, L. E., and Ashe, M. P. (2007). Stress-dependent relocalization of translationally primed mRNPs to cytoplasmic granules that are kinetically and spatially distinct from P-bodies. *J. Cell Biol.* **179**, 65–74.

Huh, W. K., Falvo, J. V., Gerke, L. C., Carroll, A. S., Howson, R. W., Weissman, J. S., and O'Shea, E. K. (2003). Global analysis of protein localization in budding yeast. *Nature* **425**, 686–691.

Jang, L. T., Buu, L. M., and Lee, F. J. (2006). Determinants of Rbp1p localization in specific cytoplasmic mRNA-processing foci, P-bodies. *J. Biol. Chem.* **281**, 29379–29390.

Kedersha, N., Chen, S., Gilks, N., Li, W., Miller, I. J., Stahl, J., and Anderson, P. (2002). Evidence that ternary complex (eIF2-GTP-tRNA(i)(Met))-deficient preinitiation complexes are core constituents of mammalian stress granules. *Mol. Biol. Cell* **13**, 195–210.

Kedersha, N., Stoecklin, G., Ayodele, M., Yacono, P., Lykke-Andersen, J., Fritzler, M. J., Scheuner, D., Kaufman, R. J., Golan, D. E., and Anderson, P. (2005). Stress granules and processing bodies are dynamically linked sites of mRNP remodeling. *J. Cell Biol.* **169**, 871–884.

Kshirsagar, M., and Parker, R. (2007). Identification of Edc3p as an enhancer of mRNA decapping in *Saccharomyces cerevisiae*. *Genetics* **166**, 729–739.

Liu, J., Rivas, F. V., Wohlschlegel, J., Yates, J. R. 3rd, Parker, R., and Hannon, G. J. (2005). A role for the P-body component GW182 in microRNA function. *Nat. Cell Biol.* **7**, 1261–1266.

Longtine, M. S., McKenzie, A. 3rd, Demarini, D. J., Shah, N. G., Wach, A., Brachat, A., Philippsen, P., and Pringle, J. R. (1998). Additional modules for versatile and economical PCR-based gene deletion and modification in *Saccharomyces cerevisiae*. *Yeast* **14**, 953–961.

Lotan, R., *et al.* (2005). The RNA polymerase II subunit Rpb4p mediates decay of a specific class of mRNAs. *Genes Dev.* **19**, 3004–3016.

Luke, B., *et al.* (2007). *Saccharomyces cerevisiae* Ebs1p is a putative ortholog of human Smg7 and promotes nonsense-mediated mRNA decay. *Nucleic Acids Res.* **35**, 7688–7697.

Malys, N., and McCarthy, J. E. (2006). Dcs2, a novel stress-induced modulator of m7G pppX pyrophosphatase activity that locates to P bodies. *J. Mol. Biol.* **363**, 370–382.

Mazroui, R., Sukarieh, R., Bordeleau, M. E., Kaufman, R. J., Northcote, P., Tanaka, J., Gallouzi, I., and Pelletier, J. (2006). Inhibition of ribosome recruitment induces stress granule formation independently of eukaryotic initiation factor 2 alpha phosphorylation. *Mol. Biol. Cell* **17**, 4212–4219.

Muhlrad, D., and Parker, R. (2005). The yeast EDC1 mRNA undergoes deadenylation-independent decapping stimulated by Not2p, Not4p, and Not5p. *EMBO J.* **24**, 1033–1045.

Otsu, N. (1979). Threshold selection method from gray-level histograms. *IEEE Trans. Syst. Man Cybern.* **9**, 62–66.

Parker, R., and Sheth, U. (2007). P bodies and the control of mRNA translation and degradation. *Mol. Cell.* **25**, 635–646.

Pilkington, G. R., and Parker, R. (2008). Pat1 contains distinct functional domains that promote P-body assembly and activation of decapping. *Mol. Cell. Biol.* **28**, 1298–1312.

Pillai, R. S., Bhattacharyya, S. N., Artus, C. G., Zoller, T., Cougot, N., Basyuk, E., Bertrand, E., and Filipowicz, W. (2005). Inhibition of translational initiation by Let-7 MicroRNA in human cells. *Science* **309**, 1573–1576.

Rendl, L. M., Bieman, M. A., and Smibert, C. A. (2008). *S. cerevisiae* Vts1p induces deadenylation-dependent transcript degradation and interacts with the Ccr4p-Pop2p-Not deadenylase complex. *RNA* **14**, 1328–1336.

Ruiz-Echevarría, M. J., and Peltz, S. W. (2000). The RNA binding protein Pub1 modulates the stability of transcripts containing upstream open reading frames. *Cell* **101**, 741–751.

Segal, S. P., Dunckley, T., and Parker, R. (2006). Sbp1p affects translational repression and decapping in *Saccharomyces cerevisiae*. *Mol. Cell. Biol.* **26,** 5120–5130.

Sheth, U., and Parker, R. (2003). Decapping and decay of messenger RNA occur in cytoplasmic processing bodies. *Science* **300,** 805–808.

Sheth, U., and Parker, R. (2006). Targeting of aberrant mRNAs to cytoplasmic processing bodies. *Cell* **125,** 1095–1109.

Stribinskis, V., and Ramos, K. S. (2007). Rpm2p, a protein subunit of mitochondrial RNase P, physically and genetically interacts with cytoplasmic processing bodies. *Nucleic Acids Res.* **35,** 1301–1311.

Sweet, T. J., *et al.* (2007). Microtubule disruption stimulates P-body formation. *RNA* **13,** 493–502.

Tadauchi, T., Inada, T., Matsumoto, K., and Irie, K. (2004). Posttranscriptional regulation of HO expression by the Mkt1-Pbp1 complex. *Mol. Cell Biol.* **24,** 3670–3681.

Teixeira, D., and Parker, R. (2007). Analysis of P-body assembly in *Saccharomyces cerevisiae*. *Mol. Biol. Cell* **18,** 2274–2287.

Teixeira, D., Sheth, U., Valencia-Sanchez, M. A., Brengues, M., and Parker, R. (2005). Processing bodies require RNA for assembly and contain nontranslating mRNAs. *RNA* **11,** 371–382.

Tharun, S., Muhlrad, D., Chowdhury, A., and Parker, R. (2005). Mutations in the *Saccharomyces cerevisiae* LSM1 gene that affect mRNA decapping and 3′ end protection. *Genetics* **170,** 33–46.

Unterholzner, L., and Izaurralde, E. (2004). SMG7 acts as a molecular link between mRNA surveillance and mRNA decay. *Mol. Cell* **16,** 587–596.

Vasudevan, S., Garneau, N., Tu, Khounh D., and Peltz, S. W. (2005). p38 mitogen-activated protein kinase/Hog1p regulates translation of the AU-rich-element-bearing MFA2 transcript. *Mol. Cell. Biol.* **25,** 9753–9763.

Zenklusen, D., Larson, D. R., and Singer, R. H. (2008). Single-RNA counting reveals alternative modes of gene expression in yeast. *Nat. Struct. Mol. Biol.* **15,** 1263–1271.

CHAPTER TWENTY-SIX

ANALYZING mRNA EXPRESSION USING SINGLE mRNA RESOLUTION FLUORESCENT *IN SITU* HYBRIDIZATION

Daniel Zenklusen *and* Robert H. Singer

Contents

Abstract

As the product of transcription and the blueprint for translation, mRNA is the main intermediate product of the gene expression pathway. The ability to accurately determine mRNA levels is, therefore, a major requirement when studying gene expression. mRNA is also a target of different regulatory steps, occurring in different subcellular compartments. To understand the different

Department of Anatomy and Structural Biology and The Gruss-Lipper Biophotonics Center, Albert Einstein College of Medicine, Bronx, New York, USA

Methods in Enzymology, Volume 470
ISSN 0076-6879, DOI: 10.1016/S0076-6879(10)70026-4

steps of gene expression regulation, it is therefore essential to analyze mRNA in the context of a single cell, maintaining spatial information. Here, we describe a stepwise protocol for fluorescent *in situ* hybridization (FISH) that allows detection of individual mRNAs in single yeast cells. This method allows quantitative analysis of mRNA expression in single cells, permitting "absolute" quantification by simply counting mRNAs. It further allows us to study many aspects of mRNA metabolism, from transcription to processing, localization, and mRNA degradation.

1. INTRODUCTION

The life cycle of an mRNA comprises many different steps. Starting with mRNA synthesis, mRNAs are processed, assembled into mRNPs, exported from the nucleus, sometimes localized, usually translated, and ultimately always degraded. These different steps along the gene expression pathway are tightly regulated and many are subjected to quality control steps that ensure their proper execution (Houseley and Tollervey, 2009; Moore and Proudfoot, 2009). How these different steps are carried out and what proteins are involved in these processes has been the focus of gene expression studies over the last few decades. The ability to detect and quantify mRNA levels was thus the key requirement. Traditionally, mRNA detection is achieved using some kind of hybridization technique. While Northern blots are able to detect only a few mRNAs at the time, array technologies now allow expression studies of an entire organism in a single experiment (Ausubel, 1988; Coppée, 2008; Holstege *et al.*, 1998).

One limitation of arrays or Northern blots, however, is that large numbers of cells are required to isolate sufficient material to perform an experiment. Additionally, cells must be broken up to isolate RNA and RNA get lost or degraded during the isolation procedure. Therefore, spatial information gets lost. The steps along the gene expression pathway occur in different cellular compartment and preserving spatial information is often critical to understand cellular processes. Furthermore, variability among different cells in a population cannot be observed by ensemble measurements. Cells from different cell cycle or developmental stages express unique sets of genes and such alternate expression profiles are obscured when pooling cells. Finally, expression "noise" resulting from stochastic fluctuations in biological processes cannot be observed without single cell analysis (Elowitz *et al.*, 2002; Kaufmann and van Oudenaarden, 2007).

These limitations are circumvented by single cell analysis (Kaufmann and van Oudenaarden, 2007; Zenklusen *et al.*, 2008). Spatial information and cell-to-cell differences become easily observed when molecules are detected in single cells, made possible by the extensive use of fluorescent proteins (Shaner et al., 2007). To analyze mRNA expression, however, single cell techniques are less widely used. Fluorescent in situ hybridization (FISH) is the most robust and straight-forward method for single cell mRNA analysis (Dong et al., 2007; Long et al., 1997; Zenklusen et al., 2008). To detect mRNA in cells, fluorescently labeled probes are hybridized to fixed cells immobilized on glass slides. The technique is noninvasive, as no genetic modifications are necessary. Choosing well-designed probes coupled with bright fluorescent dyes allows the detection of single mRNA molecules in single yeast cells (Dong et al., 2007; Femino et al., 1998; Long et al., 1997; Zenklusen et al., 2008). The applications of this technique in gene expression analysis have a wide range; we have used FISH to study transcription, splicing, and mRNA localization (Dong et al., 2007; Long et al., 1997; Zenklusen et al., 2008).

Yeast is an ideal system to perform single molecule expression analyses. Many genes in yeast are expressed at a very low level of less than 10 copies per cell (Holstege et al., 1998; Zenklusen et al., 2008). Therefore, "absolute" quantification of mRNA expression can be performed; the number of mRNA molecules can be determined simply by counting. The small size of a yeast cell is also advantageous in this case, allowing analysis of expression levels in many single cells simultaneously. Expression and localization studies can, therefore, be performed with unprecedented precision.

In this chapter, we will progress through the different steps of performing a single mRNA resolution FISH experiment. We begin with how probes are designed and labeled before we describe a step-by-step protocol for FISH. Finally, we briefly describe some aspects of data analysis.

2. PROBE DESIGN

A crucial step of a successful FISH experiment is designing FISH probes. To achieve single molecule sensitivity, multiple oligonucleotide probes, each labeled with up to five fluorescent dyes are hybridized to an mRNA (Fig. 26.1). To allow the coupling of multiple dyes onto one probe, a minimal probe length is required. Probes should also be long enough to ensure high specificity and allow stringent hybridization conditions. Probes of around 50 nucleotides (nt) in length with about 50% CG content typically work best, demonstrating high specificity under stringent hybridization conditions. As multiple probes against one gene are used, it is important to design probes with similar melting temperature. Using these standard settings (50 nt/50% CG) during probe design also facilitates the simultaneous use of differentially labeled probes against multiple target mRNAs (Fig. 26.1).

Probes are designed using commercial DNA sequence analysis software such as Oligo (Molecular Biology Insights, Inc.). To find target sites,

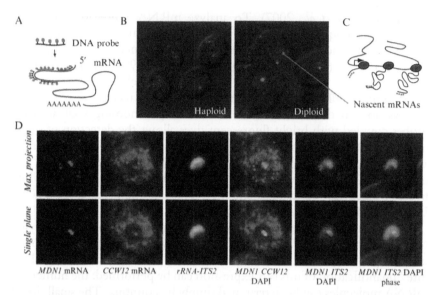

Figure 26.1 Single mRNA sensitivity fluorescent *in situ* hybridization (FISH). (A) Schematic diagram of the FISH protocol. A mix of four 50 nt DNA oligonucleotides, each labeled with five fluorescent dyes, is hybridized to paraformaldehyde fixed yeast cells to obtain single transcript resolution. (B) Single mRNA FISH for *MDN1* mRNA. Single mRNAs are detected in the cytoplasm, higher intensity spot in the nucleus. Haploid and diploid yeast cells are shown. Probes hybridize to the 5' of the mRNA. *MDN1* mRNA (red), DAPI (blue), and DIC. (C) Cartoon illustrating that the number of nascent mRNAs at the site of transcription is used to determine the polymerase loading on a gene using probes to the 5'end of *MDN1*. (D) Nascent transcripts of neighboring genes colocalize at the site of transcription. Diploid cells are hybridized with probes against *MDN1* (red) and *CCW12* (green). Nucleolus is stained with probes against the ITS2 spacer of the rRNA precursor (yellow). Maximum projection of 3D-dataset and single plane containing the transcription sites are shown (Zenklusen *et al.*, 2008). (See Color Insert.)

the gene of interest is scanned for 50 nt complementary sequences with ~50% CG content. If none fitting the criteria can be found, the length of the probe can be adjusted by adding or removing a few bases while keeping a similar melting temperature. It is important not use probes forming stable secondary structures as this may interfere with efficient hybridization. Avoid using probes forming internal stem loops with a $\Delta G > -2.5$ kcal/mol. Probes should also be tested for cross-hybridization to other genes, for example, by using Blast in SGD^{TM} (Saccharomyces Genome database). Strong sequence homology is rare but can challenge probe design, for example, when designing probes for ribosomal protein genes, present in two copies per genome with strong sequence homology.

To incorporate multiple labels into a single DNA oligonucleotide probe, modified bases are inserted during synthesis. Inserting amino-allyl dTs allows efficient coupling with most commercially available dyes after synthesis. To avoid quenching of dyes, modified bases should be spaced by 8–10 nt. Different companies synthesize oligos containing internal labels, but due to relatively high costs it is often preferable to synthesize the probes on site if a DNA synthesis facility or a DNA synthesizer is available. Alternatively, probes containing a single modified base can be used. Such probes are synthesized by most companies and are much cheaper compared to probes bearing multiple labels. However, more probes have to be used to allow single molecule detection (Raj et al., 2008).

3. PROBE LABELING

Single molecule detection requires high-labeling efficiencies. We use cyanine dyes, containing a monofunctional NHS-ester for efficient coupling to amino-allyl Ts. Cy3, Cy3.5, and Cy5 (CyDye^{TM}, GE Healthcare) work well, but other dyes with monofunctional NHS-ester from other companies can be used. Dyes in the green (emission below 500 nm) are less well suited for FISH in yeast as cells show more background fluorescence and single molecule detection becomes difficult.

Labeling is done as described by the manufacturer with minor modifications. We prepurify probes prior to labeling using a QIAquick Nucleotide Removal Column (Quiagen), as this has been shown to increase labeling efficiency. Five micrograms of DNA oligonuleotide is labeled using a single Amersham Cy3, Cy3.5, or Cy5 dye pack. When multiple probes against one gene are used, probes can be pooled in equal molar ratios and the probe mix is labeled together.

Labeling efficiency is determined by measuring absorption in a spectrophotometer. If available, use a NanoDrop (Thermo Fisher), which allows measuring of low volumes (1 μl), therefore, reducing probe loss. Labeling efficiency is calculated using a formula that corrects for absorption of the fluorophore at 260 nm. Labeling of >90% should be obtained. For unknown reasons, labeling efficiency of Cy3.5 is generally lower (75–80%).

3.1. Materials

- DNA oligonucleotide containing amino-allyl modified Ts
- Mono-Reactive CyDye^{TM} Cy3, Cy3.5, and Cy5 (GE Healthcare, #PA23001, PA23501, PA25001)
- QIAquick Nucleotide Removal Kit (Qiagen #28304)
- Spectrophotometer
- Labeling buffer (0.1 M sodium bicarbonate, pH 9.0)

3.2. Protocol

1. Measure concentrations of unlabeled DNA oligonucleotides.
2. When using multiple probes against one gene, combine probes to total of 5 μg of probes per labeling. For example, when using four probes to gene A, use 1.25 μg of each).
3. Add 500 μl of buffer PN from QIAquick Kit, mix.
4. Purify on QIAquick column according to the protocol. To increase binding, load the sample twice onto the same column.
5. Elute probes from columns using 40 μl H$_2$O. Do not use the elution buffer from the kit.
6. Lyophilize probes in a SpeedVac.
7. Resuspend the DNA pellet in 10 μl labeling buffer and add to the dye containing tube.
8. Resuspend the dye by vortexing vigorously and then perform a quick spin to collect the labeling reaction at the bottom of the tube.
9. Incubate in the dark at room temperature overnight.

Purify the probes from the free dye using QIAquick columns:

10. Add 500 μl of buffer PN to the labeling reaction and load onto column.
11. Spin through columns according to the protocol.
12. Load the flow-through a second time onto the same column to increase probe recovery.
13. Spin through columns according to the protocol.
14. Wash column twice with buffer PE to remove all nonincorporated dye.
15. Elute the labeled probes using 100 μl of elution buffer.
16. Measure concentration and labeling efficiency using a spectrophotometer.
17. Store probes at $-20\ °C$ in the dark.

3.3. Measuring labeling efficiency

To calculate the labeling efficiency, the extinction coefficient and the absorbance of the dye and the oligo at 260 nm and the emission peak of the dye have to be considered. The molar extinction coefficient (ε) of the DNA oligonucleotides is calculated as described by Beer–Lambert law (Cavaluzzi and Borer, 2004). A web site from an oligo synthesis company could be used for the calculation (we use http://www.idtdna.com/analyzer/Applications/OligoAnalyzer). To calculate the molecular weight of the amino-modified oligo, add 179.16 g/mol per modified base to the calculated molecular weight of the unmodified oligo.

The exact DNA concentration [DNA] is calculated using Eq. (26.1), the dye concentration [Dye] using formula (26.2). The labeling efficiency is then determined by dividing the [Dye]/[DNA] by the number of modified bases on the probe (26.3). A_{DNA} is the absorption of the sample at 260 nm. A_{dye} is the absorption at absorbance max of the dye. ε_{dye} is extinction coefficient of the dye, ε_{DNA} the extinction coefficient of the DNA:

$$[DNA] = \frac{A_{DNA} - \varepsilon_{dye(260)} \times (A_{dye}/\varepsilon_{dye(max)})}{\varepsilon_{DNA} \times 0.1\,cm} \qquad (26.1)$$

$$[Dye] = \frac{A_{dye(max)}}{\varepsilon_{dye} \times 0.1\,cm} \qquad (26.2)$$

$$\text{Labeling efficiency} = \frac{[Dye]}{[DNA]} \times \frac{1}{5} \qquad (26.3)$$

Extinction coefficients of the dyes at 260 nm (ε_{260}) and their absorption maximum (ε_{max}) are shown in the table as follows:

Dye	ε_{260}	ε_{max}	Absorbance (nm)	Emission (nm)
Cy3	12,000 (8%)	150,000	550	570
Cy3.5	40,800 (24%)	170,000	581	596
Cy5	12,500 (5%)	250,000	649	670

4. CELL FIXATION, PREPARATION, AND STORAGE

To prepare cells for FISH, cells are grown in the appropriate media and fixed by adding paraformaldehyde directly to the media. After extensive washes, the cell wall is removed using lyticase. Cells are digested in an isotonic buffer to prevent cells from bursting after the cell wall has been removed. Cells also become very fragile and strong shearing forces (extensive pipetting and vortexing) will break the cells open, so gentle handling is required. Complete digestion, however, is necessary to obtain optimal FISH results. Progression of the digest is, therefore, observed by visual inspection using phase contrast. Cells will turn dark when the cell wall is digested away, whereas undigested cells look transparent. Avoid digesting cells for too long, as overdigestion can lead to cell lysis.

Following digestion, cells are attached to coverslips. Using round 18 mm cover glass slips allows most subsequent steps to be performed in 12-well tissue culture plates. The cover glass is coated with poly-L-lysine for cells to attach. Alternatively, precoated coverslips can be purchased from different vendors. Cells are spotted on coverslips and allowed to settle by gravity. Unadhered cells are washed off and coverslips are finally stored in 70%

ethanol at $-20\ °C$. Ethanol dissolves membranes allowing better penetration of probes during the hybridization step and serves at the same time as a preservative, permitting cells to be stored for many months.

4.1. Materials

- Paraformaldehyde 32% solution, EM grade (Electron Microscopy Science #15714)
- Lyticase (Sigma # L2524, resuspend in $1\times$ PBS to 25,000 U/ml. Stored at $-20\ °C$)
- Ribonucleoside–vanadyl complex (VRC; NEB #S1402S)
- β-Mercaptoethanol
- Sorbitol
- 1 M KHPO$_4$, pH 7.5
- 70% ethanol
- Noncoated coverslips (Fisherbrand Cover Glasses Circles No. 1: 0.13–0.17 mm thick; size: 18 mm (#12-545-100)) or
- Precoated coverslips (Fisherbrand Coverglass for growth 18 mm (12-545-84))
- Poly-L-lysine (#P8920)
- 12-well cell culture plates

Solutions to be prepared:

• Buffer B	1.2 M sorbitol, 100 mM KHPO$_4$, pH 7.5
• Spheroplast buffer	1.2 M sorbitol
	100 mM KHPO$_4$, pH 7.5
	20 mM ribonucleoside–vanadyl complex (VRC; NEB #S1402S)
	20 mM β-mercaptoethanol
	Lyticase (25 U lyticase per OD of cells)
• Respuspention buffer	1.2 M sorbitol
	100 mM KHPO$_4$, pH 7.5
	20 mM ribonucleoside–vanadyl complex

4.2. Protocol

- *Growth and fixation*
  1. Grow 50 ml BY4741 cells in YPD in a 250-ml flask at 30 °C on an orbital shaker to an optical density at 600 nm (OD 600) of 0.6.
  2. Prepare a 50-ml Falcon tube containing 6.3 ml of 32% (v/v) paraformaldehyde. Paraformaldehyde is toxic, wear gloves and handle in the fume hood!

3. Fix cells by transferring 43.7 ml of culture to a 50-ml tube containing the paraformaldehyde (final concentration of 4%, v/v) and mix.
4. Incubate cells for 45 min at room temperature on a tabletop shaker.
5. Collect cells by centrifugation using a swinging bucket rotor at 3500 rpm at 4 °C.
6. Wash cells three times with 10 ml of cold buffer B.
7. Resuspend cells in 1 ml buffer B and transfer cells to a 1.5-ml Eppendorf tube.
8. Pellet cells using tabletop centrifuge (3 min, 4000 rpm).
- *Digestion*
9. Resuspend cells in 1 ml spheroplast buffer plus 30 μl of lyticase (at 25 U/μl).
10. Incubate cells at 30 °C for 8 min.
11. Check the progression of the digest using a phase contrast microscope. Place 3.5 μl on a microscope slide, cover with a coverglass and inspect digestion using a 20× objective. Undigested cells are transparent while digested cells will turn dark. If >80% of cells are digested proceed to step 12. If fewer cells are digested, continue incubation and check for digestion every 2–3 min.
12. Collect cells by centrifugation for 3 min at 3500 rpm at 4 °C. Do not spin at a higher speed or cells will break.
13. Wash cells with 1 ml of cold buffer B (pipette carefully).
14. Resuspend pellet in 1.5 ml of buffer B, keep on ice.
- *Attaching cells to coverslips*
15. Place poly-L-lysine treated 18 mm round coverslips face up into 12-well tissue culture dishes, one coverslip per well.
16. Drop 150 μl of cells to the center of a coated coverslip.
17. Let cells settle for 30 min at 4 °C.
18. Slowly add 2 ml of buffer B to each well, then remove buffer B using a vacuum aspirator. This will remove cells not attached to the coverslip and leave a monolayer of immobilized cells.
19. Slowly add 2 ml of 70% ethanol of each well.
20. Store cells for at least 3 h at − 20 °C. Cells can be stored at − 20 °C for at least 6 months.
- *Prepare poly-L-lysing coverslips*

Carefully put one box of 18 mm round coverslips into 500 ml 0.1 N HCl and boil for 10 min. Rinse extensively with H_2O, autoclave and store in 70% ethanol.

To coat coverslips with poly-L-lysine, place 100 μl of a 0.01% (w/v) poly-L-lysine solution onto a coverslip, incubate at room temperature for 5 min, remove the solution using a vacuum pump and let the remaining liquid dry. Then wash twice with H_2O and let air dry. The poly-L-lysine coated coverslips can be stored for several months.

5. HYBRIDIZATION

Only very low probe concentrations are needed in the hybridization reaction to allow single mRNA detection. Generally, 0.5 ng per probe per hybridization reaction is sufficient. To block nonspecific binding of the probes, competitor DNA and RNA is added in large excess to the hybridization solution.

The formamide concentration in the hybridization mix and the subsequent wash steps is critical to get optimal hybridization specificity. Generally, we use 40% formamide for standard probes (50 nt/50% CG), but if high background is observed, increasing the formamide concentration from 40% to 50% can reduce background. To detect the entire pool of polyA, mRNAs in the cell can be detected using a 50-nt poly-dT probe, but the formamide concentration has to be reduced to 15%.

For the hybridization step, the coverslip with the immobilized cells are inverted onto a droplet of the hybridization solution. Floating of the coverslip on the hybridization solution leads to even distribution of hybridization solution and the best results. This works much better than using multiwell microscope slides. Hybridization is done in hybridization chamber overnight at 37 °C. The chamber is a simple, self-assembled unit consisting of a glass plate and two Parafilm layers separated by cardboard spacers (Fig. 26.2).

After hybridization, the coverslips are placed back into a 12-well plates and washed extensively to ensure that all unbound probes are removed. After a short wash in a DAPI containing solution, cells are mounted and are ready to be imaged.

5.1. Materials

- Glass plate, about 20 × 20 cm
- Parafilm
- Cardboard spacers
- 12-well cell culture plates
- Glass slide

Solutions to be prepared:

- 40% formamide/2× SSC
- 2× SSC/0.1% Triton X-100
- 1× SSC
- 1× PBS
- Solution F (40% formamide, 2× SSC, 10 mM NaHPO$_4$, pH 7.5)
- Solution H (2× SSC, 2 mg/ml BSA, 10 mM VRC)

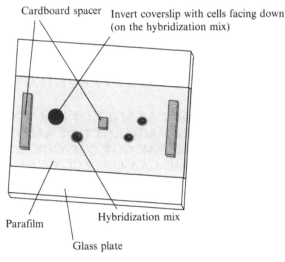

Figure 26.2 Hybridization chamber. The hybridization chamber is assembled using a glass plate, Parafilm and cardboard spacers. The coverslips with cells are inverted onto a drop of the hybridization solution placed onto the first Parafirm layer. To seal the chamber, a second layer of Parafilm is placed on top of the coverslips. To keep the second Parafilm layer from touching the coverslips, cardboard spacers are placed on both sides and in the middle of the first Parafilm layer. The interior volume of the chamber is small and evaporation is not a problem at 37 °C. However, the two layers of Parafilm have to be properly sealed to prevent evaporation.

- *Escherichia coli* tRNA (Roche # 10 109 541 001)
- ssDNA (deoxyribonucleic acid, single stranded from salmon testes, Sigma #D9156)
- DAPI solution (0.5 μg/ml DAPI (Sigma #D9564) in 1× PBS. Store at 4 °C in the dark)
- Mounting solution (ProLong® Gold antifade reagent (Invitrogen # P36934))

5.2. Probes used for the hybridization shown in Figs. 26.1 and 26.3

Bold Ts represent amino modified bases

MDN1 probes (Cy3)	
MDN1-794	TTT GTC GTG GAT AGT GTG GAC CTT AGG GAC GAT AAC GCC ACA GAT TGA CG
MDN1-860	CTC CCG AGT TGA CGA AGA GAG GAA ACC GTT TTA TGA GTA GGG ACA AAG GTT

(continued)

(continued)

MDN1-1104	CTA TAA GTA CCC ATC TCC CTT CTT TGA CCG CGG TAG CGA GAA CAC CAG CTC
MDN1-1210	TTT GCA GCC TTT ACA GTC TCT CCT CTG GAT GGA ATG GTT AGT TCG CGC TT

CCW12 probes (Cy3.5)

CCW12-59	GGT GAC CAA AGT GGT AGA TTC TTG GCT GAC AGT AGC AGT GGT AAC GTT AG
CCW12-140	GTC ATC GAC GGT GAC GGT AGC GGT GGA AAC CAA AGC TGG GGA GAC AGT TT
CCW12-191	CTT TGG GGC TTC AGT GGT CAA TGG GCA CCA GGT GGT GTA TTG AGT GAT AA
CCW12-245	GGT GTT CTT TGG AGC TTC AGT AGA GGT AAC TGG AGC AGC AGT AGA AGT AC

rRNA-ITS2 (Cy5)

ITS2-1	ATA GGC CAG CAA TTT CAA GTT AAC TCC AAA GAG TAT CAC TC

5.3. Protocol

1. Remove the ethanol from the 12-well plate using a vacuum pump and rehydrate samples by adding 2 ml $2\times$ SSC at RT for 5 min. Do this twice.
2. Wash cells once with 40% formamide/$2\times$ SSC at RT for 5 min.

During washes, prepare the hybridization mix:

3. Mix 0.5 ng of each probe per coverslip with 10 μg of *E. coli* tRNA and 10 μg of ssDNA (2 ng of probe mix when using four probes against one gene).
4. Lyophilize using a SpeedVac.
5. Add 12 μl of solution F to probe tube, heat at 95 °C for 3 min.
6. Add 12 μl of solution H to the hybridization mix.
7. Put a drop of 22 μl of hybridization mix onto the Parafilm stretched out on a glass plate. Avoid bubbles in the hybridization mix. (Use the back of a forceps to scratch the edges of the Parafilm so that the Parafilm sticks to the glass plate.)
8. Using forceps, place the coverslip with cells facing down on the hybridization mix. No bubbles should form. Multiple coverslips can be placed next to each other onto a single glass plate, but leave about 1.5 cm space between coverslips.
9. To seal the "hybridization chamber," place two cardboard spacers (2–3 mm thick and 5×0.5 cm in length) on opposite sides of the

glass plate over the Parafilm plus a 0.5 × 0.5 cm place onto the centre of the plate. Cover the glass plate with a second layer of Parafilm, without touching the coverslips. Seal the two layers of Parafilm using the back of the forceps to avoid evaporation. Cover with aluminum foil.

10. Incubate at 37 °C over night in the dark.
11. Preheat 40% formamide/2× SSC at 37 °C, put 2 ml in 12-well tissue culture dish.
12. Place cover slips back in 12-well tissue culture dish containing 40% formamide/2× SSC, cells facing up; incubate 15 min at 37 °C (incubator).
13. Wash once more with 40% formamide/2× SSC at 37 °C (2 ml, 15 min).
14. Wash once with 2× SSC 0.1% Triton X-100 at RT (2 ml, 15 min).
15. Wash once with 1× SSC at RT (2 ml, 15 min).
16. Wash coverslip in 1× PBS plus DAPI (2 ml, 2 min).
17. Wash 1× with 1 × PBS (2 ml, 2 min).
18. Before mounting, dip coverslip in 100% EtOH, let them dry.
19. Invert cells facing down onto a drop of mounting solution placed on a glass slide. Allow the mounting solution to polymerize over night at room temperature in the dark.
20. Seal coverslips with nail polish. Let nail polish dry before imaging, otherwise the objective may be damaged.
21. Go to the microscope and enjoy your images.

Slides can be stored at 4 °C for a few days and at −20 °C for months in the dark.

6. IMAGE ACQUISITION

The need for sensitive imaging equipment was likely one reason why single molecule detection was not approachable in the past. However, since sensitive CCD cameras have become a standard component of most microscopes and dyes are very bright and photostable, signal intensities are not a limiting factor for detection of single mRNAs by FISH. Most epifluorescence microscope setups in imaging facilities are sensitive enough to detect single mRNAs. We use a standard epifluorescent microscope and CCD camera (described below).

When simultaneously imaging mRNAs expressed from multiple genes using probes labeled with different fluorophores, it is crucial to use the correct filter sets to avoid bleedthrough between the different channels. For example, when using Cy3 and Cy3.5, whose absorbance and emission are relatively close to each other (550/570 nm and 581/596 nm) narrow band pass filter sets have to be used. Appropriate filter sets are listed below.

To obtain expression profiles and mRNA distributions, images have to be acquired in 3D. Using a 100× objective, collect z-slices every 200 nm. Using the setup presented below, exposure times of 1 s per z-stack should lead to sufficient signal. If single mRNAs cannot be detected, it is likely that the hybridization did not work or the microscope is not aligned properly.

6.1. Microscope (example)

- Olympus BX61 epifluorescence microscope (Olympus, Center Valley, PA)
- Olympus UPlanApo 100×, 1.35 NA oil-immersion objective
- Olympus U-DICTHC Nomarski prism for DIC
- Chroma Filters 31000 (DAPI), 41001 (FITC), SP-102v1 (Cy3), SP-103v1 (Cy3.5), and CP-104 (Cy5) (Chroma Technology, Rockingham, VT)
- Light source X-Cite 120 PC (EXFO, Mississauga, ON)
- CoolSNAP HQ camera (Photometrics, Tucson, AZ)

7. IMAGE ANALYSIS

Hybridizing four to five FISH probes, each labeled with five fluorescent dyes to an mRNA creates a strong fluorescent signal. Although barely visible by eye, single mRNAs are easily detected using a standard CCD camera. Single mRNAs signals appear as diffraction limited spots within the cell. Sites of transcription often show higher signal intensities and are easily distinguishable as they colocalize with the DAPI signal (Figs. 26.1 and 26.3). *MDN1* transcription sites are visible by eye and being able to see a *MDN1* transcription site by eye is a good first indicator for a successful FISH experiment.

To simplify the data analysis, it is often helpful to reduce the 3D dataset to a 2D image using a maximum projection. The maximum projection displays the maximum value of all images in the z-stack for particular pixel locations and creates a 2D image. As mRNAs for most genes are expressed at low numbers, the probability that two mRNAs are found in the same $x-y$ but a different z position is low, allowing a reduction to 2D to accurately represent the 3D dataset.

To test for specificity of the signal, probes can be hybridized to control cells not expressing the transcript of interest, for example, a deletion strain. Alternatively, a gene can be put under an inducible promoter, like a GAL promoter and transcription turned off long enough that all mRNAs are degraded. Using well-labeled probes and high hybridization efficiency, the difference in signal between cells expressing and not expressing is generally

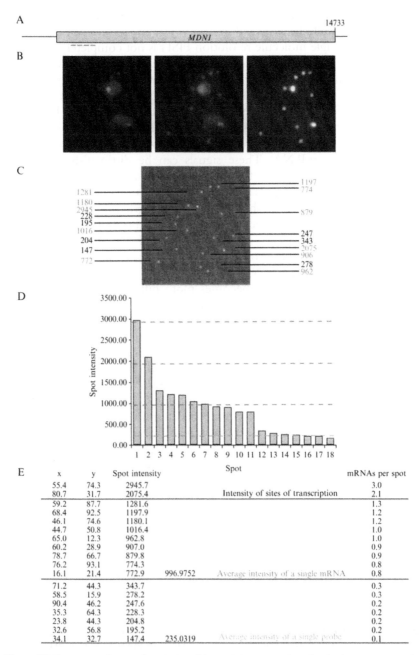

Figure 26.3 Quantifying mRNA signal intensities. Intensity of a single mRNA can be calculated by determining the fluorescence intensity emitted from a single probe. (A) FISH probes hybridizing to the 5′ end of *MDN1* mRNA were used for hybridization shown in (B). A small number of single probes tend to hybridize unspecifically to the cells and can be visualized by changing the contrast levels (B, compare left and

obvious. Spots are observed throughout the cell when a gene is expressed and no signal should be observed in cells where the gene is not expressed.

However, single molecule resolution FISH is not completely devoid of background (Fig. 26.3). When analyzing the images carefully, low-intensity signals are found in the negative control. The weak signals originate from single FISH probes sticking nonspecifically to the cell. Despite sequence specificity and stringent hybridization and washing conditions, a low number of single probes will usually stick to the cell. Their signal intensity is low, and they appear as weaker diffraction limited spots compared to the signal emitted from an mRNA. These signals can be distinguished from an mRNA signal. In most cases, the difference is obvious, mRNA signals are bright and single probe signals are low. However, sometimes this difference is not so evident. Distinguishing between background-sticking and real mRNA signal particularly becomes an issue if hybridization efficiency of the probes is low. In this case, some mRNAs will only have one, two, or three out of four possible probes bound, resulting in signal with variable intensities for different mRNAs within a single cell. When two or less probes are bound, the signal becomes more difficult to separate from a single probe nonspecifically bound to the cell.

There are two ways to determine the signal intensity of a single probe and to distinguish them from mRNA signals. The first uses a rough approximation of the signal intensities of single probes. Similar to nonspecifically sticking to cells, a small number of probes will also stick to the glass surface outside of the cells. When using well-labeled probes, their signal intensity is homogenous and they are easily distinguishable from other "junk" on the glass. Use image acquisition software to determine the brightest pixel of each spot. Signal intensities as low as signals from spots on the glass slide indicate background, while higher intensity signals originate from mRNA signals. However, it is important to notice that using this method, the autofluorescence from the cell, although usually low, is added to the signal emitted from a single probe within a cell but not the one from the glass. Therefore, using the intensity of a single probe from the glass background will underestimate the signals expected from mRNAs inside the cell. This method is simple, although only approximative in distinguishing background spots from real signals.

A better and more quantitative approach is to determine the exact signal intensity emitted from each mRNA. Different spot detection and

middle panel). The intensity is determined using a spot detection program. (C) Signal intensity of each spot corresponding to a single DNA probe is shown in black, signal intensities of single mRNA and sites of transcription are shown in red. Consistent with the four probes used in the hybridization (A), intensity of single mRNA signals in the cytoplasm is four times the intensity of a single probe (D, E). Nascent mRNAs at the site of transcription are two and three times the intensity of a single mRNA in the cytoplasm (Zenklusen et al., 2008). (See Color Insert.)

quantification algorithms exist and one of the most established methods determines the signal intensity emitted from a diffraction limited spot by fitting a 2D Gaussian mask over each spot (Thompson *et al.*, 2002). We have developed custom software to apply this algorithm, which also takes into account a background correction and can be found at http://www.singerlab.org (Zenklusen *et al.*, 2008). Shown in Fig. 26.3 are two cells hybridized with four probes to the $5'$ of the *MDN1* mRNA. The spot detection program identifies 18 spots. Spots containing a single or four probes can easily be distinguished from each other. Single probes intensities are around 230 a.u. and mRNA signals show a mean intensity of 996 a.u., four times the intensity of a single probe. This illustrates how signals of nonspecific probe binding can be distinguished from signals of probes hybridized to mRNA molecules. Determining the intensity of single probes also allows to establish the signal intensity that is expected from an efficient hybridization and thereby allows to determine hybridization efficiency.

Figure 26.3 furthermore illustrates why achieving high hybridization efficiency is crucial. Low hybridization efficiency will lead to datasets that are difficult to analyze, as a clear distinction between signal and background is not possible. When signal intensity of individual mRNAs is highly heterogeneous, it is best to repeat the hybridization to obtain more uniform signals. For some probe sets, efficient hybridization can not be achieved and new probes against different regions in a gene will have to be synthesized.

The ability to determine the intensity of a single mRNA also allows calculation of the number of nascent mRNAs at the site of transcription. Dividing the signal intensity of the two spots colocalizing with the DAPI signal in Fig. 26.3 shows that two respectively three nascent mRNAs are present on the *MDN1* genes. Determining the number of nascent transcripts is a measure of polymerase loading and therefore the most direct assessment for transcriptional activity on a single gene. Importantly, to determine polymerase density on a gene, probes hybridizing to the $5'$ end of the mRNAs have to be used.

Quantification of signals from highly expressed genes is more difficult. As shown in Fig. 26.1, *CCW12* is highly expressed and individual mRNAs overlap each other so that it is not possible to determine the intensity of every single mRNA. Therefore, this technique is better suited to study genes expressed at low levels.

8. SUMMARY AND PERSPECTIVES

Single molecule resolution FISH is a powerful tool to study gene expression. We have applied it to count single mRNAs and determine transcription kinetics, investigate splicing regulation, and study mRNA localization. However, its potential applications are even broader. There are many aspects of gene expression regulation where using single molecule resolution FISH will be a useful tool because it is able to detect and count every individual mRNA molecule in a cell. Even if expressed at only one molecule per cell, mRNAs can be detected and the precise location within the cell can be determined. Studies of transcription networks as well as more classical gene expression processes like mRNA export and degradation can be analyzed with greater detail using single molecule methodologies. The ability to detect single molecules will expand our understanding of these cellular processes.

ACKNOWLEDGMENTS

We thank S. Hocine, S. Gandhi, and M. Oeffinger for suggestions and critical reading of the manuscript. This work was supported by NIH grant GM57071 to R. H. S.

REFERENCES

Ausubel, F. M. (1988). Current Protocols in Molecular Biology. Greene Pub. Associates, Wiley-Interscience, New York, pp. v. (loose-leaf).

Cavaluzzi, M. J., and Borer, P. N. (2004). Revised UV extinction coefficients for nucleoside-5'-monophosphates and unpaired DNA and RNA. *Nucleic Acids Res.* **32,** e13.

Coppée, J.-Y. (2008). Do DNA microarrays have their future behind them? *Microbes Infect.* **10,** 1067–1071.

Dong, S., Li, C., Zenklusen, D., Singer, R. H., Jacobson, A., and He, F. (2007). YRA1 autoregulation requires nuclear export and cytoplasmic Edc3p-mediated degradation of its pre-mRNA. *Mol. Cell* **25,** 559–573.

Elowitz, M. B., Levine, A. J., Siggia, E. D., and Swain, P. S. (2002). Stochastic gene expression in a single cell. *Science* **297,** 1183–1186.

Femino, A. M., Fay, F. S., Fogarty, K., and Singer, R. H. (1998). Visualization of single RNA transcripts in situ. *Science* **280,** 585–590.

Holstege, F. C., Jennings, E. G., Wyrick, J. J., Lee, T. I., Hengartner, C. J., Green, M. R., Golub, T. R., Lander, E. S., and Young, R. A. (1998). Dissecting the regulatory circuitry of a eukaryotic genome. *Cell* **95,** 717–728.

Houseley, J., and Tollervey, D. (2009). The many pathways of RNA degradation. *Cell* **136,** 763–776.

Kaufmann, B. B., and van Oudenaarden, A. (2007). Stochastic gene expression: From single molecules to the proteome. *Curr. Opin. Genet. Dev.* **17**(2), 107–112.

Long, R. M., Singer, R. H., Meng, X., Gonzalez, I., Nasmyth, K., and Jansen, R. P. (1997). Mating type switching in yeast controlled by asymmetric localization of ASH1 mRNA. *Science* **277,** 383–387.

Longo, D., and Hasty, J. (2006). Dynamics of single-cell gene expression. *Mol. Syst. Biol.* **2,** 64.

Moore, M. J., and Proudfoot, N. J. (2009). Pre-mRNA processing reaches back to transcription and ahead to translation. *Cell* **136,** 688–700.

Raj, A., van den Bogaard, P., Rifkin, S. A., van Oudenaarden, A., and Tyagi, S. (2008). Imaging individual mRNA molecules using multiple singly labeled probes. *Nat. Methods* **5,** 877–879.

Shaner, N. C., Patterson, G. H., and Davidson, M. W. (2007). Advances in fluorescent protein technology. *J. Cell Sci.* **120,** 4247–4260.

Thompson, R. E., Larson, D. R., and Webb, W. W. (2002). Precise nanometer localization analysis for individual fluorescent probes. *Biophys. J.* **82,** 2775–2783.

Zenklusen, D., Larson, D. R., and Singer, R. H. (2008). Single-RNA counting reveals alternative modes of gene expression in yeast. *Nat. Struct. Mol. Biol.* **15,** 1263–1271.

CHAPTER TWENTY-SEVEN

THE USE OF *IN VITRO* ASSAYS TO MEASURE ENDOPLASMIC RETICULUM-ASSOCIATED DEGRADATION

Jeffrey L. Brodsky

Contents

Abstract

Approximately one-third of all newly translated polypeptides interact with the endoplasmic reticulum (ER), an event that is essential to target these nascent proteins to distinct compartments within the cell or to the extracellular milieu. Thus, the ER houses molecular chaperones that augment the folding of this diverse group of macromolecules. The ER also houses the enzymes that catalyze a multitude of posttranslational modifications. If, however, proteins misfold or are improperly modified in the ER they are proteolyzed via a process known as ER-associated degradation (ERAD). During ERAD, substrates are selected by molecular chaperones and chaperone-like proteins. They are then delivered to the cytoplasmic proteasome and hydrolyzed. In most cases, delivery and proteasome-targeting require the covalent attachment of ubiquitin. The discovery and underlying mechanisms of the ERAD pathway have been aided by the development of *in vitro* assays that employ components derived from the yeast,

Department of Biological Sciences, University of Pittsburgh, Pittsburgh, Pennsylvania, USA

Methods in Enzymology, Volume 470
ISSN 0076-6879, DOI: 10.1016/S0076-6879(10)70027-6

Saccharomyces cerevisiae. These assays recapitulate the selection of ERAD substrates, the "retrotranslocation" of selected polypeptides from the ER into the cytoplasm, and the proteasome-mediated degradation of the substrate. The ubiquitination of integral membrane ERAD substrates has also been reconstituted.

1. INTRODUCTION

Cells are continuously faced with various forms of stress, including altered temperature, limited nutrient availability, changes in osmotic pressure, and the presence of toxic agents. To surmount such challenges, adaptive pathways are triggered that induce the synthesis of proteins that lessen the effects of cell stress. In model eukaryotes, such as the yeast *Saccharomyces cerevisiae*, many of the stress-induced adaptive pathways have been defined, thanks to the multitude of available genetic, genomic, and biochemical tools.

Stress-responsive pathways can also be triggered from within intracellular compartments. One compartment in which this has been examined in detail is the endoplasmic reticulum (ER), which is the first organelle encountered by newly synthesized secreted proteins. In *S. cerevisiae*, nascent secreted proteins can translocate into the ER either during or soon after translation (Cross *et al.*, 2009; Rapoport *et al.*, 1999). Translocation is facilitated by a multiprotein complex that resides at the ER membrane. The key component of this complex is an aqueous translocation pore, and the pore and its associated partners have been termed the "translocon" (Schnell and Hebert, 2003).

Concomitant with translocation, most secreted proteins are posttranslationally modified and begin to fold, which explains why the ER is stocked with molecular chaperones and enzymes that catalyze protein folding (Vembar and Brodsky, 2008). Because the acquisition of the native or near-native state is a prerequisite for the subsequent delivery of secreted proteins to their ultimate destinations, the ER also contains a rich variety of factors that monitor protein folding. Many of these "quality control" factors are molecular chaperones. In the event that proteins fail to fold, either as a result of the stresses noted above or due to genetic or stochastic errors, an ER stress response, known as the unfolded protein response (UPR), is initiated. One outcome of the UPR is an increase in the machinery that destroys aberrant secreted proteins, thus clearing the ER of potentially toxic protein conformers (Jonikas *et al.*, 2009; Travers *et al.*, 2000).

For many years, the existence of a quality control protease was sought (Vembar and Brodsky, 2008). Early data from mammalian cell systems suggested that the protease resided within the ER, and a candidate protease was eventually purified (Otsu *et al.*, 1995). However, parallel studies suggested the existence of an alternate system to degrade aberrant proteins that had entered the ER. In short, evidence emerged that misfolded ER proteins might employ the services performed by the cytoplasmic ubiquitin–protesome system (UPS). The proteasome is a large (26S), multi-catalytic enzyme that binds and unfolds proteins and then processively degrades substrates to short peptides (Hanna and Finley, 2007; Pickart and Cohen, 2004). Nearly all proteasome-targeted substrates are modified with poly-ubiquitin, which facilitates proteasome-capture. Early evidence indicated that the UPS proteolyzes misfolded, integral membrane proteins in the ER of yeast (Hampton et al., 1996; Sommer and Jentsch, 1993) and mammalian (Jensen et al., 1995; Ward et al., 1995) cells. These substrates were multispanning membrane proteins, which by definition contained cytoplasmic polypeptide loops; therefore, it made sense that the cytoplasmically localized UPS might recognize and destroy misfolded membrane proteins. These data were also consistent with the established function of the UPS in mediating cytoplasmic protein quality control (Sherman and Goldberg, 2001). What remained unknown was how soluble misfolded proteins within the ER lumen were destroyed.

To answer this question, we developed an *in vitro* system that monitored the fates of an ER-localized wild-type secreted protein and a secreted protein that was unable to acquire N-linked oligosaccharides (McCracken and Brodsky, 1996; Werner et al., 1996). Our establishment of this assay built-upon the pioneering *in vitro* yeast systems developed in the Walter, Blobel, Meyer, and Schekman labs that had been co-opted to follow the translocation of nascent secreted proteins into the yeast ER (Deshaies and Schekman, 1989; Hansen et al., 1986; Rothblatt and Meyer, 1986; Waters and Blobel, 1986). In our system, the wild-type substrate was the alpha mating type prepheromone, pre-pro-alpha factor (ppαF). Upon translocating into the ER, ppαF is processed by the signal sequence peptidase, generating pro-alpha factor (pαF). PαF is then triply glycosylated, which generates GpαF (Julius et al., 1984). This substrate is competent for ER-exit through the action of Golgi-targeted COPII vesicles (Belden and Barlowe, 2001). In contrast, the mutated substrate cannot be N-glycosylated so that pαF is the terminal species that forms within the ER (Fig. 27.1). These substrates were chosen for the following reasons. First, wild type and mutant forms of ppαF posttranslationally translocate into the ER *in vitro* (Hansen et al., 1986; Rothblatt and Meyer, 1986; Waters and Blobel, 1986). This feature allowed for the large-scale synthesis and isolation of radiolabeled substrate, which could then be aliquotted and stored prior to use. Second, pαF appeared to be degraded within the yeast secretory pathway (Caplan et al., 1991), indicating the existence of a secretory protein quality control system for soluble proteins. Finally, ppαF-derivatives were also degraded in mammalian cell lines (Su et al., 1993). Thus, a dissection of the pαF biogenic pathway might lead to the elucidation of a conserved quality control machinery.

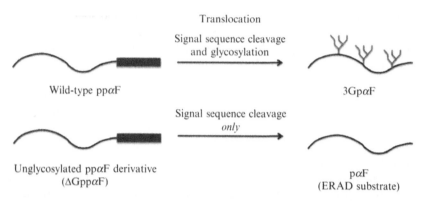

Fig. 27.1 The early biogenic pathways utilized by wild-type ppαF and ΔGppαF, a soluble ERAD substrate. Upon translocation into the ER, the signal sequence in ppαF is liberated and the protein becomes triply glycosylated. The resulting species, 3GpαF, is stable in yeast ER-derived microsomes. In contrast, the three sites required for the addition of N-linked glycans in ΔGppαF have been mutated. Thus, ΔGppαF is converted into pαF, which is an ERAD substrate.

In 1996, we reported that pαF could be selectively exported—or retrotranslocated—from ER-derived vesicles back to the cytoplasm (McCracken and Brodsky, 1996). Once in the cytoplasm, pαF was destroyed by the proteasome (Werner et al., 1996). However, GpαF, which derived from the wild-type precursor (Fig. 27.1), was stable. Based on our results, we named this process ER-associated degradation (ERAD) (McCracken and Brodsky, 1996). In parallel to our efforts, Wolf and colleagues established that another mutated secreted protein, CPY★, was also degraded by the proteasome in yeast (Hiller et al., 1996). Collectively, these data indicated that integral membrane and soluble proteins in the ER were both handled by the UPS (Vembar and Brodsky, 2008). In fact, a subsequent modification of our assay established that the proteasome was necessary and sufficient to retrotranslocate and degrade pαF (Lee et al., 2004). Moreover, the use of ER-derived vesicles, or "microsomes" from mutant strains allowed our laboratory and others to identify the ER lumenal chaperones required for pαF degradation (Brodsky et al., 1999; Gillece et al., 1999; Kabani et al., 2003; Lee et al., 2004; McCracken and Brodsky, 1996; Nishikawa et al., 2001). The in vitro system was also employed by Römisch and Schekman to provide evidence suggesting that the translocon might serve as the conduit for retrotranslocation (Pilon et al., 1997). Each of these in vitro assays is described in detail, in Section 2.

Do integral membrane proteins also retrotranslocate and become solubilized from the ER membrane prior to proteasome-mediated destruction? Data obtained from mammalian cell systems suggested that this might be the case. First, Kopito and colleagues reported that cytoplasmic "aggresomes" accumulated in mammalian cells expressing high levels of a misfolded membrane protein that were simultaneously challenged with proteasome inhibitors (Johnston et al., 1998). The aggresomes contained the misfolded substrate and components of the UPS (Johnston et al., 1998; Wigley et al., 1999). Second, Ploegh and coworkers reported that a human cytomegalovirus gene product catalyzed the "dislocation" of the major histocompatibility class I molecule into cytoplasmic fractions in transfected cells (Wiertz et al., 1996). Because these phenomena were only evident in mammalian cells, further attempts to elucidate the mechanism of membrane protein retrotranslocation have had to rely on pharmacological and RNAi-related technologies.

To better define the pathway by which integral membrane proteins are selected and retrotranslocated for ERAD, we developed an *in vitro* assay in which each step in the degradation pathway could be dissected (Nakatsukasa et al., 2008). The components for this assay—ER-derived microsomes and concentrated cytosol—were again isolated from *S. cerevisiae*. The use of this assay led to the following discoveries. First, we observed that Hsp70 and Hsp40 molecular chaperones help link ERAD substrates to E3 ubiquitin ligases. Second, we found that the E3s required for ERAD exhibit functional redundancy. Third, we were able to evoke the ATP- and cytosol-dependent retrotranslocation of a polytopic integral membrane protein into the cytosolic fraction. Fourth, we determined that membrane protein extraction required the Cdc48p complex, which was previously found to play an important role in ERAD (Jentsch and Rumpf, 2007). Fifth, we established that a cytoplasmic polyubiquitin extension enzyme, or "E4," elongated the polyubiquitin chain on the ERAD substrate and was required for maximal rates of substrate degradation. And sixth, we confirmed that the solubilized substrate was competent for proteasome-dependent degradation. These discoveries were made possible through the use of ER-derived microsomes and cytosol that were prepared from yeast containing specific loss-of-function and thermosensitive mutant alleles. The assays that led to these discoveries are described in Section 3.

2. *In Vitro* ERAD Assays Using a Soluble Substrate, PαF

In this section, first, the isolation of the materials required to monitor the degradation of pαF is described. Next, the assays for pαF retrotranslocation and ERAD are detailed. Throughout the section, comments are added to note how reagents prepared from mutant strains have been used to better define the ERAD pathway.

2.1. Materials

2.1.1. Microsome preparation

The preparation of ER-derived microsomes from *S. cerevisiae* has been well documented (Deshaies and Schekman, 1989; Rothblatt and Meyer, 1986) but is described in outline form with minor revisions, below. When temperature sensitive mutants are employed, the cultures are grown at a permissive temperature and are then shifted to the restrictive temperature (i.e., 37 °C). Although thermosensitive phenotypes can be recapitulated *in vitro*, the growth period at the nonpermissive temperature needed to obtain the mutant defect in the assay must be determined empirically and can range from < 20 min to 5 h (Becker *et al.*, 1996; Brodsky and Schekman, 1993; Brodsky *et al.*, 1993; Latterich *et al.*, 1995; Nakatsukasa *et al.*, 2008). Smaller scale isolations of microsomes have also been used in ERAD studies (Nakatsukasa *et al.*, 2008), but the translocation efficiency is somewhat lower (our unpublished data).

1. Yeast are grown at the desired temperature in rich medium and with vigorous shaking until the culture reaches mid-log to late-log phase (optical density at 600 nm [OD_{600}] of 2.0–3.0). We typically grow 1–2 l of yeast for a microsome preparation.
2. The cell walls are digested with a β-1,3-glucanase hydrolyzing enzyme that either can be purified from recombinant *Escherichia coli* that express the enzyme (Shen *et al.*, 1991) or purchased commercially (e.g., Zymolyase, from MP BioMedicals).
3. The resulting spheroplasts are collected by centrifugation through 20 mM HEPES, pH 7.4, 0.8 M sucrose, 1.5% Ficoll 400 at 6000 rpm, in an HB-6 swinging bucket rotor. It is critical that each of the following steps is performed at 4 °C.
4. The spheroplasts are resuspended to a final OD_{600} of 100/ml in 20 mM HEPES, pH 7.4, 0.1 M sorbitol, 50 mM KOAc, 2 mM EDTA, and a protease inhibitor cocktail, and the solution is transferred to a tight-fitting, Teflon-glass homogenizer that can be driven with a motor.
5. The plasma membrane is then broken by 10 strokes with the motor running at the highest setting.
6. The broken cellular material is layered onto a cushion that contains 20 mM HEPES, pH 7.4, 1.0 M sucrose, 50 mM KOAc, 1 mM DTT.
7. After centrifugation at 6500 rpm in an HB-6 swinging bucket rotor, the crude microsomal fraction is resuspended in an equal volume of B88 (20 mM HEPES, pH 6.8, 250 mM sorbitol, 150 mM KOAc, 5 mM MgOAc).
8. The microsomes are collected by centrifugation at $\sim$15,000$\times g$ for 10 min, resuspended in B88, and recentrifuged.
9. After final resuspension in a small volume of B88, the microsome concentration is adjusted such that the OD_{280} should equal $\sim$40 when a small aliquot of a 1:10 dilution of the resuspended material is assessed in 2% SDS.
10. Microsomes aliquots ($\sim$50 μl) are frozen and stored at -80 °C.

2.1.2. Isolation of ΔGppαF, the precursor of a soluble ERAD substrate, and ppαF, the wild-type control

The substrates required for the *in vitro* ERAD assay are the wild-type ppαF control (which is encoded by the *MFα1* locus; Kurjan and Herskowitz, 1982) and a form of ppαF that contains site-directed mutations in the three sites required for the addition of N-linked oligosaccharides (Caplan *et al.*, 1991). We denote this mutant as ΔGppαF. As described above and in Fig. 27.1, the signal sequence of ppαF is removed in the ER and the resulting species becomes triply glycosylated, ultimately forming 3GpαF in the microsomes. In contrast, ΔGppαF is converted to pαF, which is an ERAD substrate in the yeast microsome-based system (McCracken and Brodsky, 1996; Werner *et al.*, 1996). PαF has also been prepared with a fluorescent tag and is retrotranslocation competent after its entrapment in dog pancreas microsomes (Wahlman *et al.*, 2007).

Using SP6 polymerase, wild-type ppαF is transcribed from plasmid pDJ100 and ΔGppαF is transcribed from plasmid pGEM2alpha36. We next isolate the messages encoding ppαF and ΔGppαF and perform an *in vitro* translation reaction in the presence of ^{35}S-methionine and concentrated, gel-purified yeast lysate to obtain radiolabeled ppαF and ΔGppαF (Lee *et al.*, 2004; McCracken and Brodsky, 1996; Werner *et al.*, 1996). The logic underlying this protocol is that maximal ppαF translocation efficiency requires factors in the yeast lysate—presumably binding stably to ppαF—that are absent in other translation-competent lysates (Chirico *et al.*, 1988; Deshaies *et al.*, 1988).

More recently, we have employed the Promega TnT SP6 Coupled Reticulocyte Lysate System to synthesize radiolabeled ppαF and ΔGppαF and other substrates (Hrizo *et al.*, 2007), and have discovered that the resulting products translocate efficiently into yeast microsomes. In brief, each plasmid template is mixed on ice with the commercial buffer, ribonuclease inhibitor, SP6 polymerase, an amino acid mixture (lacking met), and the supplied rabbit reticulocyte lysate. The reaction is then supplemented with 20 μCi of ^{35}S-labeled amino acid (PerkinElmer EXPRE^{35}S^{35}S Protein Labeling Mix). A 50-μl (total volume) reaction is typically performed according to the manufacturer's instructions at 30 °C for 90 min. Single-use aliquots are then flash frozen and stored at −80 °C.

2.1.3. Yeast cytosol

The preparation of concentrated yeast cytosol using liquid nitrogen was first described by Sorger and Pelham (1987) and modified for the ERAD assay as described (McCracken and Brodsky, 1996). Strains containing temperature sensitive mutants can be used to prepare cytosol but again the conditions required to recapitulate temperature sensitive defects *in vitro* must be determined empirically.

1. Yeast cells are grown with vigorous shaking in rich medium to log phase ($OD_{600} = \sim 2.0$) at 30 °C or at the desired alternate temperature(s). In our experience, at least 6 l of culture are needed for efficient lysis.
2. The cells are collected by centrifugation, resuspended in water, recentrifuged, and resuspended in a minimal amount of B88. Typically, we use <5 ml of B88 for the number of cells obtained from a 6 l yeast culture.
3. The cells are added slowly to 500 ml of liquid nitrogen in a tripour plastic beaker. After the yeast are frozen, the liquid nitrogen is decanted and the cells are stored at −80 °C.
4. Approximately 500 ml of liquid nitrogen is added to a stainless–steel blender, followed by the frozen yeast. The blender is initially turned-on at the lowest setting, but the blade speed is soon increased to the highest setting. The volume must be maintained above the rotating blades by the periodic addition of liquid nitrogen. To prevent spills, the blender must be covered as often as possible with a lid that can withstand low temperatures (e.g., a thick Styrofoam slab).
5. After 10 min, the blender is turned-off, the liquid nitrogen is evaporated, and the resulting powder is transferred to a 50-ml plastic tube, which can be stored at −80 °C.
6. A minimal amount of B88 (e.g., ~ 0.5 ml/40 ml of broken yeast) is added as the lysate begins to thaw at room temperature or on ice, and freshly prepared DTT is added to a final concentration of 1 mM.
7. The lysate is centrifuged at $10,000 \times g$ for 10 min at 4 °C, and the supernatant is collected and recentrifuged.
8. The second supernatant is centrifuged at $300,000 \times g$ for 1 h at 4 °C and is aliquoted, frozen in liquid nitrogen, and stored at −80 °C. A Bio-Rad protein assay is used to determine the protein concentration. In our experience, cytosols at >20 mg/ml work best in ERAD assays as long as they are not thawed and refrozen.

2.1.4. ATP regenerating system

Optimal ERAD efficiency requires an ATP regenerating system. To this end, a 10× stock is made up as follows:

- 10 mM ATP
- 500 mM creatine phosphate
- 2 mg/ml of creatine phosphokinase
- B88 to volume

The solution is then distributed into single-use aliquots, frozen in liquid nitrogen, and stored at −80 °C. Reactions lacking the addition of the ATP regenerating system may support a low level of retrotranslocation/ERAD and ubiquitination. Thus, to fully decipher the ATP-dependence of the following reactions, we have performed incubations with either ATPγS (Lee et al., 2004) or apyrase (Nakatsukasa et al., 2008) in place of the regenerating system.

2.2. The *in vitro* degradation assay for pαF

Prior to examining the retrotranslocation and degradation efficiencies of pαF, the precursor to this substrate and the precursor to the wild-type control, 3GpαF, must be introduced into ER-derived microsomes through an *in vitro* translocation assay. The resulting microsomes are then reisolated to examine the degradation efficiency and retrotranslocation steps during the ERAD of a soluble substrate.

2.2.1. Translocation of ppαF and ΔGppαF into yeast microsomes

1. A translocation reaction is set up in a microcentrifuge tube on ice. Most commonly, the 60 μl reaction contains 45 μl of B88, 6 μl of the 10× ATP regenerating system, 5 μl of yeast microsomes, and 4 μl of ^{35}S-labeled ΔGppαF or ppαF (or the appropriate volume to obtain ∼300,000 cpm per reaction). The reaction is mixed gently and then incubated for 1 h in a 20 °C water bath.
2. Following the incubation, the solution is centrifuged at ∼16,000×g for 3 min at 4 °C.
3. After the tubes are placed on ice, the supernatant is removed with a gel-loading tip. Care must be taken not to disturb the pellet.
4. The pellet is gently resuspended in 60 μl of ice-cold B88 and the solution is recentrifuged, as above.
5. The supernatant is again removed and the pellet is taken up in 5 μl of ice-cold B88.

2.2.2. Reconstitution of cytosol- and ATP-dependent degradation

The following assay has been adapted to assess the contributions of a number of ER lumenal and integral membrane components on the ERAD of a soluble substrate (Brodsky et al., 1999; Gillece et al., 1999; Kabani et al., 2003; Lee et al., 2004; McCracken and Brodsky, 1996; Nishikawa et al., 2001; Pilon et al., 1997). The role of the proteasome during this process can be assessed either through the use of cytosol from a proteasome mutant or through the addition of proteasome inhibitors (Werner et al., 1996). Moreover, the cytosol requirement can be circumvented through the addition of purified 26S proteasomes isolated from yeast or mammals (Lee et al., 2004). Negative controls for this experiment include reactions supplemented with ATPγS or apyrase, and/or reactions lacking cytosol (Lee et al., 2004; McCracken and Brodsky, 1996; Werner et al., 1996).

To a microcentrifuge tube on ice, with the appropriate amount of ice-cold B88 for a final volume of 60 μl, the following reagents are added (in order):

- 5 μl of microsomes containing ^{35}S-labeled pαF (a product of the translocation reaction with ΔGppαF) or 3GpαF (a product of the translocation reaction with ppαF), prepared as described above
- 6 μl of the 10× ATP regenerating system
- An appropriate amount of yeast cytosol to obtain a final concentration of 1–3 mg/ml

1. The reaction is incubated at 30 °C for 20 min or at higher temperatures if the contributions of some temperature-sensitive mutations will be monitored (Brodsky et al., 1999; Gillece et al., 1999; Kabani et al., 2003; Lee et al., 2004; Nishikawa et al., 2001; Pilon et al., 1997). Multiple reactions can also be set up if a time-course will be conducted.
2. The reaction tubes are placed on ice and 12 μl of an ice-cold 100% TCA stock solution is added.
3. The quenched reactions are agitated vigorously on a Vortex mixer for ~3 s and incubated on ice for 15 min.
4. The solutions are centrifuged at 16,000×g for 5 min at 4 °C, and the supernatant is removed with a gel-loading tip.
5. Sufficient ice-cold acetone is added to cover each pellet, and the mixture is again briefly agitated on a Vortex mixer and immediately recentrifuged.
6. The acetone is removed with a gel-loading tip and the pellet is air-dried for 2–3 min.
7. The final pellets are resuspended in SDS–PAGE sample buffer by repetitive pipetting, and the mixture is incubated for 10 min at ~70 °C.
8. The radiolabeled proteins are best resolved through the use of an 18% denaturing polyacrylamide gel that also contains 6 M urea. To further maximize the separation between the signal sequence-containing (i.e., ΔGppαF, ~18 kDa) and signal sequence-cleaved (i.e., pαF, ~16 kDa) species, which is an ERAD substrate (Fig. 27.2), we use 6 cm gels. Once the dye front is near the bottom of the gel, the plates are disassembled, the gels are fixed and dried, and the radioactivity is visualized using a phosphorimager. A 2- to 3-day exposure is usually sufficient.
9. To quantify the amount of degradation, we consider only the translocated pαF (i.e., the pαF species in the "-cytosol" control at $t = 0$ min; Fig. 27.2). Thus, the amount of pαF remaining under conditions that promote degradation should be averaged and compared to the average amount of material in control reactions that lack ATP and/or cytosol. The wild-type substrate (i.e., 3GpαF, ~28 kDa) should be stable regardless of the assembled reaction conditions.

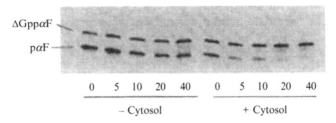

Fig. 27.2 pαF is a soluble ERAD substrate. The cytosol- and time-dependent degradation of pαF is shown. Values represent the time (in min) that microsomes containing pαF were incubated in the presence or absence of cytosol and an ATP-regenerating system at 30 °C. ΔGppαF is membrane associated, untranslocated precursor. Figure taken from (McCracken and Brodsky, 1996). © McCracken and Brodsky (1996). Originally published in *The Journal of Cell Biology*. 132: 291–298.

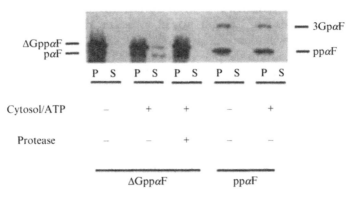

Fig. 27.3 pαF is retrotranslocated from ER-derived yeast microsomes into the cytosolic fraction. Either ΔGppαF or ppαF, as indicated, was translocated into microsomes and after a 25-min incubation in the presence or absence of cytosol and an ATP-regenerating system the reaction was centrifuged to obtain a pellet (P) and supernatant (S) fraction. As noted, in one reaction the supernatant fraction was treated with protease (trypsin at a final concentration of 0.2 mg/ml for 30 min at 4 °C). ΔGppαF and ppαF are membrane associated, untranslocated precursors, as in Fig. 27.2. Figure taken from McCracken and Brodsky (1996). © McCracken and Brodsky (1996). Originally published in *The Journal of Cell Biology*. 132: 291–298.

2.3. The pαF retrotranslocation assay

The degradation assay, described above, can be modified to monitor the retrotranslocation of pαF from yeast ER microsomes (Fig. 27.3). The translocation of wild-type ppαF, which forms 3GpαF, serves as a negative control (i.e., 3GpαF should remain in the microsome fraction). The inclusion of ATPγS instead of the ATP-regenerating system also serves as a negative control. In addition, the assay can be used to assess retrotranslocation efficiency upon the addition of purified cytoplasmic proteins, such as the proteasome, the 19S proteasome "cap," Cdc48p, and purified chaperones (Lee *et al.*, 2004). Of note, the retrotranslocated pαF is protease sensitive, indicating that it is not encapsulated in ER-derived vesicles (Fig. 27.3) (McCracken and Brodsky, 1996).

1. Translocation reactions (60 µl) are assembled as described above.
2. The microsomes are harvested, washed, and used in the degradation assay containing B88, the ATP-regenerating system, and yeast cytosol, as presented in the preceding section.
3. The reaction is incubated for 20 min at 30 °C or higher temperatures in the event that reagents from temperature sensitive mutant strains are being examined. The initial rate of retrotranslocation can also be examined by taking earlier time points (e.g., 5 and 10 min).
4. The microsomes are pelleted in a refrigerated microcentrifuge at 16,000×g for 3 min.
5. The tubes are returned to ice and the supernatant is quickly removed with a gel-loading tip and placed in a prechilled fresh tube. The supernatant will contain the retrotranslocated pαF. Care must be taken not to disturb the pellet.
6. Twelve microliters of ice-cold 100% TCA is immediately added to the supernatant.
7. This solution is agitated on a Vortex mixer for ∼3 s and then incubated at 4 °C for 15–20 min.
8. During this time, the pellet is resuspended in 60 µl B88 and 12 µl of 100% ice-cold TCA is added. The solution, which contains microsome-retained pαF, is also agitated and incubated at 4 °C for 15 min.
9. After 15–20 min, each of the samples (the supernatant/cytosol and pellet/microsomes) is centrifuged at 16,000×g in a refrigerated microcentrifuge, and the supernatants are removed with a gel-loading tip.
10. Sufficient ice-cold acetone is added to cover the pellets, and the solution is briefly agitated on a Vortex mixer.
11. The samples are immediately recentrifuged, as above, and the supernatant is removed with a gel loading tip.
12. The pellets are air-dried for 2–3 min and resuspended in SDS–PAGE sample buffer by repetitive pipetting.
13. The solution is incubated for 10 min at ∼70 °C, and the products are resolved on 6 M urea–18% denaturing polyacrylamide gels, as described above.
14. After the gels are fixed and dried, and the radioactive species are visualized using a phosphorimager, the amount of pαF in the supernatant and pellet are summed to establish the total pαF in the reaction. Then, the amount of pαF in the supernatant is divided by the total pαF to calculate the percentage of pαF retrotranslocated to the supernatant. These values are averaged among triplicate experiments. Most

commonly, the amount of pαF exported in the negative controls is
< 10%. For the data shown in Fig. 27.3, 36% of pαF was retrotranslo-
cated in the presence of cytosol/ATP.

3. *In Vitro* Assays for Integral Membrane Proteins that are ERAD Substrates

To better define the ERAD pathway taken by integral membrane
proteins in the yeast ER, we developed a system in which the selection,
ubiquitination, retrotranslocation, and degradation of Ste6p★, a mutated
form of the Ste6p **a**-mating factor transporter, could be followed. Ste6p★
was chosen because the genetic requirements for the degradation of this
integral membrane protein were relatively well defined (Huyer *et al.*, 2004;
Loayza *et al.*, 1998; Vashist and Ng, 2004). In addition, the mutation in
STE6 (Q1249X) results in a C-terminally truncated form of the protein;
therefore, the quality control "decisions" that trigger the destruction of
Ste6p★ are made posttranslationally. In principle, this simplifies the machinery required for the ERAD of the substrate, and excludes the contribution
of cotranslational (i.e., ribosome-associated) factors. Finally, Ste6p★ is a
member of the large ABC family of transporters, which includes the cystic
fibrosis transmembrane conductance regulator (CFTR): The topologies of
Ste6p★ and CFTR are identical and the proteins share domain-restricted
sequence homology. Previous work had also established many of the
genetic requirements for the ERAD of CFTR after its heterologous expression in yeast (Ahner *et al.*, 2007; Gnann *et al.*, 2004; Youker *et al.*, 2004;
Zhang *et al.*, 2001). Consequently, microsomes prepared from yeast strains
expressing either Ste6p★ or CFTR can be employed for many of the assays
described below (Nakatsukasa *et al.*, 2008).

3.1. *In vitro* ubiquitination assay

3.1.1. Isolation of yeast microsomes containing integral membrane ERAD substrates

Yeast expressing HA-tagged versions of Ste6p★ (Huyer *et al.*, 2004) or
CFTR (Zhang *et al.*, 2001, 2002) have been described previously, and
ER-derived microsomes are prepared from these strains as presented
above. The only difference is that microsomes are isolated from strains
expressing the ERAD substrate; therefore, yeast are grown in selective
media.
    When one wishes to recapitulate a temperature-sensitive mutant defect
in these assays, a more rapid, smaller scale microsome preparation should be
used (Nakatsukasa *et al.*, 2008). The small-scale preparation minimizes the

elapsed time between the *in vivo* temperature shift and the use of the isolated reagents in the following assays. An appropriate negative control for each reaction is the preparation of yeast microsomes from a strain lacking the plasmid for the expression of the ERAD substrate, but that instead contains the expression plasmid without an insert.

3.1.2. Yeast cytosol

The cytosol required for the following experiments is prepared as described in Section 2.1. When the phenotypes associated with temperature-sensitive mutants are to be recapitulated, the *in vivo* shift to the nonpermissive temperature must be determined empirically, and can range substantially (also see above).

3.1.3. Preparation of ^{125}I-labeled ubiquitin

Bovine ubiquitin (Sigma) is dissolved in phosphate-buffered saline at a final concentration of 10 μg/μl. The protein is then labeled with ^{125}I (NEN Research, BioRad) using the ICL method (Helmkamp *et al.*, 1960; McFarlane, 1958). The labeled ubiquitin is enriched and unincorporated ^{125}I is removed with a *D*-salt Excellulose Desalting column (Pierce). The final, isolated product is stored in 20 μl aliquots at -80 °C at a final concentration of 0.2 μg/μl ($\sim1.0\times10^6$ cpm/μl). The reagent must be used within 2 months after its preparation.

3.1.4. The ubiquitination reaction

The following reagents are combined into the appropriate volume of B88 (total volume, 18 μl) on ice:

- 2 μl of yeast microsomes containing the HA-epitope-tagged integral membrane ERAD substrate
- 2 μl of the 10× ATP regenerating system (see Section 2.1)
- Sufficient yeast cytosol to achieve a final concentration of 1–4 mg/ml

The reaction can also be supplemented with apyrase, which serves as a "-ATP" control, or methylated ubiquitin, which inhibits ubiquitin chain extension (Hershko and Heller, 1985) (Fig. 27.4). As noted above, microsomes lacking the ERAD substrate serve as another negative control, as can reactions lacking cytosol.

1. The mixture is preincubated in a 23 °C water bath for 10 min before 2 μl of ^{125}I-labeled ubiquitin are added.
2. The incubation is then continued for up to 1 h at 23 °C.
3. At the desired time point, 80 μl of an SDS stop solution (50 mM Tris–Cl, pH 7.4, 150 mM NaCl, 5 mM EDTA, 1.25% SDS, 1 mM PMSF, 1 μg/ml leupeptin, 0.5 μg/ml pepstatin A, 10 mM NEM) are added.

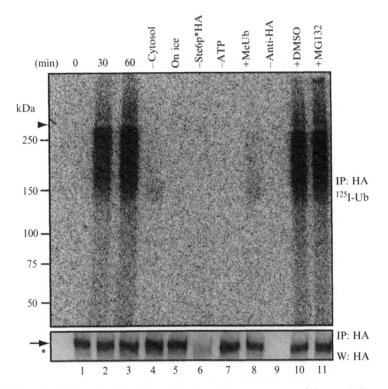

Fig. 27.4 Ste6p★ is polyubiquitinated *in vitro*. Microsomes containing an HA-tagged form of Ste6p★ were incubated with 2 mg/ml cytosol, an ATP-regenerating system, and ^{125}I-labeled ubiquitin at 23 °C or on ice for the indicated times (lanes 1–3) or for 60 min (lanes 4–11). The reactions were then quenched and Ste6p★ was immunoprecipitated with an anti-HA antibody. Samples were processed as described in the text. Where indicated, reactions either lacked cytosol, or were treated with apyrase (final concentration of 0.02 μg/μl), 0.5 mg/ml methylated-ubiquitin, or 100 μM MG132, a proteasome inhibitor. "-anti-HA" denotes a precipitation performed in the absence of antibody and "-Ste6p★" denotes that microsomes were prepared from cells lacking the substrate. Figure taken from Nakatsukasa *et al.* (2008).

4. The solution is briefly agitated on a Vortex mixer and then incubated at 37 °C for 30 min.

5. In preparation for an immunoprecipitation, 400 μl of 50 mM Tris–Cl, pH 7.4, 150 mM NaCl, 5 mM EDTA, 2% Triton X-100 is added and the solution is mixed gently and placed on ice.

6. A 2 μl aliquot ($\sim$10 μg) of anti-HA antibody (Roche) is added and the immunoprecipitation reaction is incubated overnight at 4 °C with mild agitation.

7. A 30 μl, 50% (v/v) suspension of Protein A-Sepharose (GE Healthcare), equilibrated in 50 mM Tris–Cl, pH 7.4, 150 mM NaCl, 5 mM

EDTA, 1 mM azide, is added and the mixture is incubated at 4 °C for another 2–3 h.

8. The beads are harvested by a 10 s, low-speed centrifugation at 4 °C and are then washed four times with 800 μl of ice-cold 50 mM Tris–Cl, pH 7.4, 150 mM NaCl, 5 mM EDTA, 1% Triton X-100, 0.2% SDS, 10 mM NEM.

9. After the final wash, the residual buffer is removed with a gel-loading tip and 30 μl of 2× SDS–PAGE sample buffer are added.

10. The bound proteins are eluted by a 37 °C, 30 min incubation, and the supernatant is split after a brief centrifugation.

11. One-half of the supernatant (to detect ^{125}I-ubiquitin-modified protein) is analyzed using a 6 cm, 6% (denaturing) SDS–polyacrylamide gel. After fixation and drying, the gel is exposed to a phosphorimager plate.

12. The other half of the sample is analyzed first by SDS–PAGE but the gel is then blotted and used to detect the amount of unmodified Ste6p* with anti-HA antibody and the appropriate secondary antibody. We use the SuperSignal West Pico Chemiluminescent Substrate (Thermo Scientific) to visualize the signal.

3.2. Analysis of integral membrane protein retrotranslocation

The Cdc48p complex-dependent retrotranslocation of ubiquitinated Ste6p* can be followed by inserting a centrifugation step into the protocol described above. The involvement of the Cdc48p complex in this event was established through the use of cytosols prepared from a *cdc48-3* strain that had been shifted to 37 °C for 5 h and from an *npl4Δ* strain (Nakatsukasa *et al.*, 2008). Therefore, these materials serve as a negative control in the following experiments. A TAP-tagged version of Cdc48p was also shown to coprecipitate the retrotranslocated species (Nakatsukasa *et al.*, 2008), further implicating the Cdc48p complex in Ste6p* retrotranslocation.

1. Ubiquitination reactions are set up as described in Section 3.1, except that a final volume of 25 μl is achieved.

2. At the completion of the 1 h 23 °C incubation, the microsomes are pelleted in a refrigerated microcentrifuge at 18,000×g for 10 min.

3. The reaction tube is returned to ice and the supernatant (~20 μl), which contains the retrotranslocated, ubiquitinated Ste6p*, is removed with a gel-loading tip and placed in a new microcentrifuge tube on ice.

4. The pelleted microsomes are resuspended in 25 μl of ice-cold B88, and 20 μl of this suspension is placed in a new tube.

5. To analyze the amount of ubiquitinated Ste6p* in the supernatant (cytosol) and pellet (microsome) fractions, 80 μl of the SDS stop solution (see above) is added to the supernatant and resuspended microsomes, and the mixtures are briefly agitated on a Vortex mixer.

6. As above, the solution is incubated at 37 °C for 30 min, and then 400 μl of 50 mM Tris–Cl, pH 7.4, 150 mM NaCl, 5 mM EDTA, 2% Triton X-100 is added and placed on ice.

7. The Ste6p* in both fractions is immunoprecipitated with anti-HA antibody/Protein A–Sepharose and analyzed under denaturing conditions via SDS–PAGE.

REFERENCES

Ahner, A., et al. (2007). Small heat-shock proteins select DeltaF508-CFTR for endoplasmic reticulum-associated degradation. Mol. Biol. Cell **18**, 806–814.

Becker, J., et al. (1996). Functional interaction of cytosolic hsp70 and a DnaJ-related protein, Ydj1p, in protein translocation in vivo. Mol. Cell. Biol. **16**, 4378–4386.

Belden, W. J., and Barlowe, C. (2001). Role of Erv29p in collecting soluble secretory proteins into ER-derived transport vesicles. Science **294**, 1528–1531.

Brodsky, J. L., and Schekman, R. (1993). A Sec63p-BiP complex from yeast is required for protein translocation in a reconstituted proteoliposome. J. Cell Biol. **123**, 1355–1363.

Brodsky, J. L., et al. (1993). Reconstitution of protein translocation from solubilized yeast membranes reveals topologically distinct roles for BiP and cytosolic Hsc70. J. Cell Biol. **120**, 95–102.

Brodsky, J. L., et al. (1999). The requirement for molecular chaperones during endoplasmic reticulum-associated protein degradation demonstrates that protein export and import are mechanistically distinct. J. Biol Chem. **274**, 3453–3460.

Caplan, S., et al. (1991). Glycosylation and structure of the yeast MF alpha 1 alpha-factor precursor is important for efficient transport through the secretory pathway. J. Bacteriol. **173**, 627–635.

Chirico, W. J., et al. (1988). 70 K heat shock related proteins stimulate protein translocation into microsomes. Nature **332**, 805–810.

Cross, B. C., et al. (2009). Delivering proteins for export from the cytosol. Nat. Rev. Mol. Cell Biol. **10**, 255–264.

Deshaies, R. J., and Schekman, R. (1989). SEC62 encodes a putative membrane protein required for protein translocation into the yeast endoplasmic reticulum. J. Cell Biol. **109**, 2653–2664.

Deshaies, R. J., et al. (1988). A subfamily of stress proteins facilitates translocation of secretory and mitochondrial precursor polypeptides. Nature **332**, 800–805.

Gillece, P., et al. (1999). Export of a cysteine-free misfolded secretory protein from the endoplasmic reticulum for degradation requires interaction with protein disulfide isomerase. J. Cell Biol. **147**, 1443–1456.

Gnann, A., et al. (2004). Cystic fibrosis transmembrane conductance regulator degradation depends on the lectins Htm1p/EDEM and the Cdc48 protein complex in yeast. Mol. Biol. Cell **15**, 4125–4135.

Hampton, R. Y., et al. (1996). Role of 26S proteasome and HRD genes in the degradation of 3-hydroxy-3-methylglutaryl-CoA reductase, an integral endoplasmic reticulum membrane protein. Mol. Biol. Cell **7**, 2029–2044.

Hanna, J., and Finley, D. (2007). A proteasome for all occasions. FEBS Lett. **581**, 2854–2861.

Hansen, W., et al. (1986). In vitro protein translocation across the yeast endoplasmic reticulum: ATP-dependent posttranslational translocation of the prepro-alpha-factor. Cell **45**, 397–406.

Helmkamp, R. W., et al. (1960). High specific activity iodination of gamma-globulin with iodine-131 monochloride. *Cancer Res.* **20**, 1495–1500.

Hershko, A., and Heller, H. (1985). Occurrence of a polyubiquitin structure in ubiquitin–protein conjugates. *Biochem. Biophys. Res. Commun.* **128**, 1079–1086.

Hiller, M. M., et al. (1996). ER degradation of a misfolded luminal protein by the cytosolic ubiquitin–proteasome pathway. *Science* **273**, 1725–1728.

Hrizo, S. L., et al. (2007). The Hsp110 molecular chaperone stabilizes apolipoprotein B from endoplasmic reticulum-associated degradation (ERAD). *J. Biol. Chem.* **282**, 32665–32675.

Huyer, G., et al. (2004). Distinct machinery is required in *Saccharomyces cerevisiae* for the endoplasmic reticulum-associated degradation of a multispanning membrane protein and a soluble luminal protein. *J. Biol. Chem.* **279**, 38369–38378.

Jensen, T. J., et al. (1995). Multiple proteolytic systems, including the proteasome, contribute to CFTR processing. *Cell* **83**, 129–135.

Jentsch, S., and Rumpf, S. (2007). Cdc48 (p97): A "molecular gearbox" in the ubiquitin pathway? *Trends Biochem. Sci.* **32**, 6–11.

Johnston, J. A., et al. (1998). Aggresomes: A cellular response to misfolded proteins. *J. Cell Biol.* **143**, 1883–1898.

Jonikas, M. C., et al. (2009). Comprehensive characterization of genes required for protein folding in the endoplasmic reticulum. *Science* **323**, 1693–1697.

Julius, D., et al. (1984). Glycosylation and processing of prepro-alpha-factor through the yeast secretory pathway. *Cell* **36**, 309–318.

Kabani, M., et al. (2003). Dependence of endoplasmic reticulum-associated degradation on the peptide binding domain and concentration of BiP. *Mol. Biol. Cell* **14**, 3437–3448.

Kurjan, J., and Herskowitz, I. (1982). Structure of a yeast pheromone gene (MF alpha): A putative alpha-factor precursor contains four tandem copies of mature alpha-factor. *Cell* **30**, 933–943.

Latterich, M., et al. (1995). Membrane fusion and the cell cycle: Cdc48p participates in the fusion of ER membranes. *Cell* **82**, 885–893.

Lee, R. J., et al. (2004). Uncoupling retro-translocation and degradation in the ER-associated degradation of a soluble protein. *EMBO J.* **23**, 2206–2215.

Loayza, D., et al. (1998). Ste6p mutants defective in exit from the endoplasmic reticulum (ER) reveal aspects of an ER quality control pathway in *Saccharomyces cerevisiae*. *Mol. Biol. Cell* **9**, 2767–2784.

McCracken, A. A., and Brodsky, J. L. (1996). Assembly of ER-associated protein degradation in vitro: Dependence on cytosol, calnexin, and ATP. *J. Cell Biol.* **132**, 291–298.

McFarlane, A. S. (1958). Efficient trace-labelling of proteins with iodine. *Nature* **182**, 53.

Nakatsukasa, K., et al. (2008). Dissecting the ER-associated degradation of a misfolded polytopic membrane protein. *Cell* **132**, 101–112.

Nishikawa, S. I., et al. (2001). Molecular chaperones in the yeast endoplasmic reticulum maintain the solubility of proteins for retrotranslocation and degradation. *J. Cell Biol.* **153**, 1061–1070.

Otsu, M., et al. (1995). A possible role of ER-60 protease in the degradation of misfolded proteins in the endoplasmic reticulum. *J. Biol. Chem.* **270**, 14958–14961.

Pickart, C. M., and Cohen, R. E. (2004). Proteasomes and their kin: Proteases in the machine age. *Nat. Rev. Mol. Cell Biol.* **5**, 177–187.

Pilon, M., et al. (1997). Sec61p mediates export of a misfolded secretory protein from the endoplasmic reticulum to the cytosol for degradation. *EMBO J.* **16**, 4540–4548.

Rapoport, T. A., et al. (1999). Posttranslational protein translocation across the membrane of the endoplasmic reticulum. *Biol. Chem.* **380**, 1143–1150.

Rothblatt, J. A., and Meyer, D. I. (1986). Secretion in yeast: Reconstitution of the translocation and glycosylation of alpha-factor and invertase in a homologous cell-free system. *Cell* **44,** 619–628.

Schnell, D. J., and Hebert, D. N. (2003). Protein translocons: Multifunctional mediators of protein translocation across membranes. *Cell* **112,** 491–505.

Shen, S. H., *et al.* (1991). Primary sequence of the glucanase gene from Oerskovia xanthineolytica. Expression and purification of the enzyme from *Escherichia coli. J. Biol. Chem.* **266,** 1058–1063.

Sherman, M. Y., and Goldberg, A. L. (2001). Cellular defenses against unfolded proteins: A cell biologist thinks about neurodegenerative diseases. *Neuron* **29,** 15–32.

Sommer, T., and Jentsch, S. (1993). A protein translocation defect linked to ubiquitin conjugation at the endoplasmic reticulum. *Nature* **365,** 176–179.

Sorger, P. K., and Pelham, H. R. (1987). Purification and characterization of a heat-shock element binding protein from yeast. *EMBO J.* **6,** 3035–3041.

Su, K., *et al.* (1993). Pre-Golgi degradation of yeast prepro-alpha-factor expressed in a mammalian cell. Influence of cell type-specific oligosaccharide processing on intracellular fate. *J. Biol. Chem.* **268,** 14301–14309.

Travers, K. J., *et al.* (2000). Functional and genomic analyses reveal an essential coordination between the unfolded protein response and ER-associated degradation. *Cell* **101,** 249–258.

Vashist, S., and Ng, D. T. (2004). Misfolded proteins are sorted by a sequential checkpoint mechanism of ER quality control. *J. Cell Biol.* **165,** 41–52.

Vembar, S. S., and Brodsky, J. L. (2008). One step at a time: Endoplasmic reticulum-associated degradation. *Nat. Rev. Mol. Cell Biol.* **9,** 944–957.

Wahlman, J., *et al.* (2007). Real-time fluorescence detection of ERAD substrate retro-translocation in a mammalian in vitro system. *Cell* **129,** 943–955.

Ward, C. L., *et al.* (1995). Degradation of CFTR by the ubiquitin–proteasome pathway. *Cell* **83,** 121–127.

Waters, M. G., and Blobel, G. (1986). Secretory protein translocation in a yeast cell-free system can occur posttranslationally and requires ATP hydrolysis. *J. Cell Biol.* **102,** 1543–1550.

Werner, E. D., *et al.* (1996). Proteasome-dependent endoplasmic reticulum-associated protein degradation: An unconventional route to a familiar fate. *Proc. Natl. Acad. Sci. USA* **93,** 13797–13801.

Wiertz, E. J., *et al.* (1996). The human cytomegalovirus US11 gene product dislocates MHC class I heavy chains from the endoplasmic reticulum to the cytosol. *Cell* **84,** 769–779.

Wigley, W. C., *et al.* (1999). Dynamic association of proteasomal machinery with the centrosome. *J. Cell Biol.* **145,** 481–490.

Youker, R. T., *et al.* (2004). Distinct roles for the Hsp40 and Hsp90 molecular chaperones during cystic fibrosis transmembrane conductance regulator degradation in yeast. *Mol. Biol. Cell* **15,** 4787–4797.

Zhang, Y., *et al.* (2001). Hsp70 molecular chaperone facilitates endoplasmic reticulum-associated protein degradation of cystic fibrosis transmembrane conductance regulator in yeast. *Mol. Biol. Cell* **12,** 1303–1314.

Zhang, Y., *et al.* (2002). CFTR expression and ER-associated degradation in yeast. *Methods Mol. Med.* **70,** 257–265.

CHAPTER TWENTY-EIGHT

A Protein Transformation Protocol for Introducing Yeast Prion Particles into Yeast

Motomasa Tanaka*,†

Contents

Abstract

A range of methods for transforming organisms with nucleic acids has been established. However, techniques for introducing proteins, or particularly protein aggregates, into cells are less developed. Here, we introduce a highly efficient protocol for introducing protein aggregates such as prions into yeast. The protein transformation protocol allows one to infect yeast with amyloid fibers of recombinant fragments (Sup-NM) of Sup35p, the protein determinant of the yeast prion state [*PSI*⁺], or *in vivo* Sup35p prions. Infectivity is dependent on the concentration of Sup-NM fibers and approaches approximately 100% at high Sup-NM concentrations. We also describe a method to create distinct conformations of Sup-NM amyloids. Using the protein transformation protocol, infection of yeast with different Sup-NM amyloid conformations leads to distinct [*PSI*⁺] strains. This protein transformation procedure is readily adaptable to

* Tanaka Research Unit, RIKEN Brain Science Institute, Hirosawa, Wako, Saitama, Japan
† PRESTO, Japan Science and Technology Agency, Honcho, Kawaguchi, Saitama, Japan

Methods in Enzymology, Volume 470
ISSN 0076-6879, DOI: 10.1016/S0076-6879(10)70028-8

other prion proteins and makes it possible to bridge *in vitro* and *in vivo* studies and greatly helps to elucidate the principles of prion inheritance.

1. INTRODUCTION

Like mammalian prions, cytoplasmic genetic elements such as $[PSI^+]$ and $[URE3]$ are observed in yeast. Their responsible proteins, Sup35p and Ure2p, are called "yeast prions" (Wickner, 1994). Although mammalian and yeast prion proteins are unrelated in amino acid sequence, their prion states share common structural features such as β-sheet rich fibrillar aggregates, amyloids. The facile genetics of yeast together with the ability to create *de novo* infectious forms of proteins from pure material have greatly helped to elucidate the structural and mechanistic basis of prion inheritance as well as the role of cellular factors in facilitating and inhibiting prion replication. Here, we describe a procedure for generating distinct, self-propagating amyloid conformations of a prionogenic Sup35p fragment termed Sup-NM and a highly efficient protocol for transforming yeast cells with *in vitro* amyloid fibers or *in vivo* prion particles of Sup35p, an essential translation termination factor (Tanaka *et al.*, 2004). This protocol can be readily adapted to other proteins and as such represents a general tool for studying yeast prions (Alberti *et al.*, 2009; Brachmann *et al.*, 2005; Patel and Liebman, 2007; Patel *et al.*, 2009).

Sup35p forms self-propagating aggregates and leads to a nonsense suppression phenotype, resulting in a $[PSI^+]$ prion state (Chien *et al.*, 2004; Tessier and Lindquist, 2009; Tuite and Cox, 2003). In yeast containing a nonsense mutation in the *ade1* gene, $[PSI^+]$ colonies are white or pink and grow on media lacking adenine, while nonprion $[psi^-]$ colonies are red and require adenine (Chernoff *et al.*, 1995). $[PSI^+]$ propagation is mediated by an N-terminal Gln/Asn-rich sequence and, to a lesser extent, by a highly charged middle (M) domain (Bradley and Liebman, 2004; DePace *et al.*, 1998; Glover *et al.*, 1997; Liu *et al.*, 2002; Ter-Avanesyan *et al.*, 1994). Transient overexpression of the Sup-NM fragment (residues 1–254) leads to protein aggregation and *de novo* appearance of $[PSI^+]$. *In vitro*, Sup-NM is shown to be sufficient to form self-seeding amyloid fibers (Glover *et al.*, 1997; King *et al.*, 1997). Like mammalian prions and other yeast prions, $[PSI^+]$ exhibits a range of heritable phenotypic strain variants (Derkatch *et al.*, 1996). These strains differ in mitotic stability (Derkatch et al, 1996), dependence on the cellular chaperone machinery (Kushnirov *et al.*, 2000b), solubility and activity of Sup35 (Derkatch *et al.*, 1996; Kochneva-Pervukhova *et al.*, 2001; Uptain *et al.*, 2001; Zhou *et al.*, 1999). These differences in strain variants lead to distinct *ade1* color phenotype as well as efficiency and specificity of prion transmission.

To reconcile the existence of strains with the "protein-only" hypothesis of prion transmission, it has been proposed that a single protein can misfold into multiple distinct infectious forms, one for each different strain (Aguzzi and Haass, 2003; Collinge, 2001; Liebman, 2002; Prusiner *et al.*, 1998). Several studies have found correlations between strain phenotypes and conformations of prion particles (Bessen *et al.*, 1995; Chien and Weissman, 2001; Chien *et al.*, 2003; Peretz *et al.*, 2002; Telling *et al.*, 1996). Nonetheless, whether such differences cause or are simply a secondary manifestation of prion strains had remained unclear, largely due to the difficulty in creating an infectious protein from an *in vitro* source and introducing it into organisms (Aguzzi and Haass, 2003; Liebman, 2002).

An earlier study demonstrated that introduction of bacterially produced pure Sup-NM by liposome fusion induces the [*PSI*$^+$] state (Sparrer *et al.*, 2000). However, Sup-NM amyloid formation occurred after encapsulation within liposomes, making it difficult to control for or evaluate the conformation of the infectious form. Therefore, this liposome-based infection strategy was poorly suited to test the role of prion conformations in strain diversity. To overcome this limitation, we developed a novel transformation protocol by which *in vitro* preformed Sup-NM amyloid fibers or *in vivo* Sup35p prions are introduced into yeast spheroplasts with high efficiency. Here, we describe protocols to prepare different Sup-NM amyloid forms and *in vivo* prion particles as well as to efficiently introduce them into yeast.

2. PURIFICATION OF BACTERIALLY EXPRESSED SUP-NM

A plasmid of Sup-NM containing C-terminal polyhistidine tags under control of a T7 promoter (pAED4-Sup-NM) is transformed in *Escherichia coli* BL21 (DE3). The cells are grown at 37 °C until OD reaches 0.5 and the protein was expressed with 0.4 mM isopropyl-β-thiogalactoside for 4 h at 37 °C. Sup-NM was purified in a denatured condition by Ni agarose and cation-exchange columns, as follows (DePace *et al.*, 1998; Glover *et al.*, 1997).

2.1. Purification of Sup-NM

1. Add 25 ml buffer A (8 M urea, 25 mM Tris, 300 mM NaCl, pH 7.8) to bacterial cells harvested from 1 l LB media. Vortex vigorously to resuspend pellets. Sonicate the suspension (Sonic Dismembrator, Fischer Scientific, 40% intensity, tip diameter 3 mm) until the cells are fully disrupted (approximately 30 s).

2. Incubate the protein solution with gentle agitation for 1 h at room temperature. Spin at 15,000 rpm (Sorvall SS-34 rotor) for 30 min. Samples are prefiltered (Millex AP 20, Millipore), followed by filtration with 0.45 μm filters.

3. Pour the protein solution onto a Ni-NTA agarose column ($\sim$2 ml agarose gels per liter of cell culture) preequilibrated with buffer A. Wash the column with 4× column volumes of buffer B (8 M urea, 25 mM Tris, pH 7.8).

4. Elute bound proteins with buffer C (8 M urea, 25 mM Tris, pH 4.5). Collect 3–4 ml fractions and run them onto SDS–PAGE. Pool fractions with highest abundance of Sup-NM and store at −80 °C.

5. Load the partially purified Sup-NM from the Ni-NTA column onto a 6-ml Resource S column (GE) preequilibrated with buffer D (8 M urea, 50 mM MES, pH 6.0).

6. Elute Sup-NM with a 0–200-mM NaCl gradient in buffer D. Typically, we monitor absorbances at 229 and 280 nm, and collect 1 ml fractions.

7. Analyze purity of the fractions by SDS–PAGE. Pool fractions that contain >90% pure Sup-NM. Concentrate the fractions, exchange buffer D with 6 M guanidine hydrochloride (Gdn) containing 5 mM potassium phosphate (pH 7.4) and concentrate it again generally to more than 1 mM, using a VIVASPIN concentrator (Sartorius AG). Filter the concentrated protein with Microcon YM-100 (Millipore). Divide the filtered protein into $\sim$10 μl aliquots and store them at −80 °C until use.

3. Preparation of Different Conformations of *In Vitro* Sup-NM Amyloid

Sup-NM has the ability to misfold into multiple different amyloid conformations (DePace and Weissman, 2002). We found that the simplest method for controlling the conformation of Sup-NM amyloids is to alter the temperature at which the polymerization occurs. For preparation of 4 or 37 °C Sup-NM amyloid, guanidine hydrochloride-denatured Sup-NM was diluted by more than 200-fold into 5 mM potassium phosphate buffer (pH 7.4) containing 150 mM NaCl at 4 or 37 °C, respectively. It is important to use at least a 200-fold dilution of the Sup-NM stocks and preincubate the polymerization buffer at the proper temperatures to robustly obtain the 4 and 37 °C fibers. The Sup-NM solution (typically 2.5–5 μM) is immediately rotated at 8 rpm in an end-over-end manner overnight. Conformational differences in the two Sup-NM amyloid fibers were assessed by the following melting temperature analysis.

3.1. Melting temperature analysis of Sup-NM amyloid

1. Generate Sup-NM amyloid fibers at 4 or 37 °C. Add 6× SDS sample buffer (final 1.7% SDS) to the fiber solution and aliquot 20 μl into 500 μl PCR tubes.

2. Incubate the fiber solution at increasing temperatures from 25 to 95 °C in 10 °C intervals and 100 °C for 5 min in a PCR thermal cycler. Transfer each tube incubated at a specific temperature into a water bath ($\sim$10 °C) in order to cool the tubes quickly.

3. Run samples onto SDS–PAGE. Probe thermally solubilized Sup-NM monomer by western blotting with a polyclonal Sup-NM antibody (Santoso et al., 2000), followed by detection with chemiluminescence.

4. Quantitate band intensities by western blot with ImageJ (NIH) and fit them as a function of temperature with IgorPro (WaveMetrics Inc.), using the following equation: $y = A + B/(1 + 10^{(C - x)/D})$, where x, y, A, B, C, D indicate temperature, band intensity, band intensity at baseline, amplitude of band intensity, melting temperature (T_m), and width of the melting transition (W), respectively. 4 °C fibers showed $T_m = 56 \pm 2\,°C$ and $W = 27 \pm 2\,°C$, compared with values for 37 °C fibers of $T_m = 77 \pm 2\,°C$ and $W = 14 \pm 1\,°C$.

4. PREPARATION OF *IN VIVO* PRIONS FROM YEAST

For preparation of crude yeast extracts, yeast cells were spheroplasted with lyticase (see Section 6 for preparation of spheroplasts) or lysed with glass beads in the presence of a protease inhibitor cocktail (Roche), and sonicated on ice for 10 s (Sonic Dismembrator, Fischer Scientific, 20% intensity, tip diameter 3 mm) before use.

1. For preparation of partially purified prion particles, spheroplast yeast cells with lyticase ($\sim$250 μg for yeast cells cultured from 50 ml YPD (1% yeast extract, 2% bactopeptone, 2% dextrose)) in SCE-buffer (1 M sorbitol, 10 mM EDTA, 10 mM DTT, 100 mM sodium citrate, pH 5.8) containing protease inhibitors (protease inhibitor cocktail, Roche) (see the following paragraph for preparation of lyticase).

2. Lyse the spheroplasts by sequential addition of sodium deoxycholate to 0.5% (w/v) and Brij-58 to 0.5% (w/v) after 5 min (Uptain et al., 2001). After incubation on ice for 15 min, spin the lysate at 10,000×g for 5 min at 4 °C and ultracentrifuge the supernatant at 100,000×g for 30 min (TLA100.3 rotor, Beckman).

3. Resuspend the pellet with 1 M lithium acetate, incubate it on ice for 30 min with gentle agitation and spin again at 100,000×g for 30 min.

Resuspend the pellet with 5 mM potassium phosphate buffer including 150 mM NaCl and sonicate it on ice for 10 s (Sonic Dismembrator, Fischer Scientific). Determine concentration of total protein in the yeast extracts and partially purified prion particles by Bradford or BCA assay, using BSA as a standard.

5. Preparation of Lyticase

1. Transform $E.$ $coli.$ with pUV5-lyticase (Scott and Schekman, 1980), a bacterial plasmid that expresses periplasmically localized lyticase under control of $T7$ promoter into $E.$ $coli.$ BL21(DE3).
2. Inoculate a single colony into 100 ml of LB media including 100 μg/ml ampicillin and culture it at 37 °C until OD_{600} reaches $\sim$0.5. Transfer 20 ml of the culture media to 1 l of LB media and culture again at 37 °C.
3. Add isopropyl-β-thiogalactoside (final concentration of 0.5 mM) when OD_{600} reaches $\sim$0.5 to initiate protein expression. After 3 h, collect cells at 5000 rpm (Sorvall SLA-3000 rotor) for 20 min.
4. Resuspend cells in 20 ml of 25 mM Tris–HCl (pH 7.4), incubate at room temperature with gentle agitation for 30 min using a nutator and centrifuge at 7500 rpm (Sorvall SS-34 rotor) for 10 min.
5. Resuspend the pellet in 20 ml of 5 mM $MgCl_2$, incubate at 4 °C for 30 min and spin at 15,000 rpm for 30 min (Sorvall SS-34 rotor) to separate periplasmic components.
6. Dialyze the supernatant against 2 l of 50 mM sodium citrate buffer (pH 5.8) for 3 h twice, ultracentrifuge the solution at 100,000$\times g$ for 30 min (TLA100.3 rotor, Beckman) and store the supernatant at -80 °C. Determine the concentration of lyticase by the Bradford method.

6. Protein Transformation

Throughout, we used isogenic [psi^-] and [PSI^+] derivatives of 74D-694 [MATa, $his3$, $leu2$, $trp1$, $ura3$; suppressible marker $ade1$-14(UGA)] (Santoso et $al.$, 2000). Similar results have been obtained in a W303 background. In contrast to de $novo$ prion induction by Sup-NM overexpression, the infection efficiency of protein transformation does not depend on the [PIN] state of yeast (Tanaka et $al.$, 2004). The presence of a $URA3$ plasmid allows one to preselect for yeast that have successfully taken up material from the solution but is not absolutely required (Fig. 28.1A).

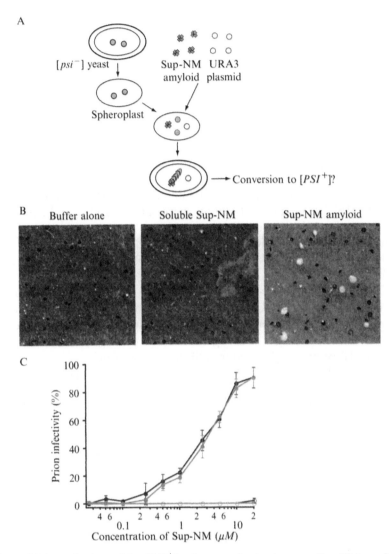

Figure 28.1 Induction of the [*PSI*⁺] prion state by *in vitro* pure Sup-NM amyloids. (A) Schematic of transformation procedure. The [*PSI*⁺] status is assessed by plating spheroplast mixture on SD-Ura containing trace amounts of adenine or on SD-Ura plates, followed by streaking transformants onto YPD plates. (B) Examples of yeasts transformed with the indicated materials on SD-Ura plates. Large and white colonies are yeasts that are converted to [*PSI*⁺] prion states. (C) Concentration-dependent and [*PIN*] state-independent infectivity by Sup-NM amyloid. The indicated concentration of fibrillar (filled circle) or soluble (open circle) Sup-NM was transformed into isogenic [*psi*⁻][*PIN*⁺] (black line) or [*psi*⁻][*pin*⁻] (grey line) strains. Throughout, values with error bars are expressed as mean ± S.D.

1. Grow yeast cells in 50 ml of YPD media to an OD_{600} of 0.5 and successively wash with 20 ml of sterile H_2O, 1 M sorbitol and SCE-buffer (1 M sorbitol, 10 mM EDTA, 10 mM DTT, 100 mM sodium citrate, pH 5.8).

2. Spheroplast cells with lyticase ($\sim$250 μg for yeast cells cultured from 50 ml YPD) in SCE-buffer at 30 °C for 30 min. Commercially available lyticase (Sigma, L-5263) is also used to prepare spheroplasts (King and Diaz-Avalos, 2004). Spheroplasts are then centrifuged (400$\times$g, 5 min) and successively wash with 20 ml of 1 M sorbitol and STC-buffer (1 M sorbitol, 10 mM $CaCl_2$, 10 mM Tris, pH 7.5).

3. Resuspend pelleted spheroplasts with 1 ml of STC-buffer and 100 μl of the spheroplast was mixed with sonicated Sup-NM amyloid fibers (2.5–10 μM) or in vivo prions (200–400 μg/ml), URA3 marked plasmid (pRS316) (20 μg/ml) and salmon sperm DNA (100 μg/ml). Incubate mixture for 30 min at room temperature.

4. Induce fusion by addition of 9 volumes of PEG-buffer (20% (w/v) PEG 8000, 10 mM $CaCl_2$, 10 mM Tris, pH 7.5) at room temperature for 30 min.

5. Collect the cells (400 $\times$ g, 5 min), resuspend with 150 μl of SOS-buffer (1 M sorbitol, 7 mM $CaCl_2$, 0.25% yeast extract, 0.5% bactopeptone), incubate at 30 °C for 30 min and plate on synthetic media lacking uracil overlaid with top agar (2.5% agar). Adenine (20 mg/l) is absent (procedure [1]) or present (procedure [2]) in the agar plate.

6. Incubate the plates at 30 °C for $\sim$10 days (procedure [1]) or 4–6 days (procedure [2]). For procedure [2], after the incubation, streak single colonies randomly chosen from the SD-URA plates onto modified YPD plates containing 1/4 of the standard amount of yeast extract (0.25%) to enhance color phenotypes of [PSI$^+$] and [psi$^-$] states. Include streaks of strong [PSI$^+$] and [psi$^-$] controls on each plate for comparison.

7. Determination of Prion Conversion Efficiency and Prion Strain Phenotypes

For procedure [1], [psi$^-$] yeast cells form small and intensely red colonies, whereas [PSI$^+$] yeast form large white colonies (Fig. 28.1B). Transformation of [psi$^-$] yeast with plasmid alone or with soluble Sup-NM does not lead to detectable formation of [PSI$^+$] colonies, whereas inclusion of preformed Sup-NM amyloid seeds leads to significant production of large, white Ade$^+$ colonies. The Ade$^+$ colonies exhibited the hallmarks of the [PSI$^+$] prion: the Ade$^+$ trait was inherited in a non–Mendelian manner, was readily cured by transient growth on medium containing 5 mM guanidine hydrochloride, and was associated with the formation of large Sup35p aggregates that are readily pelletable by high-speed centrifugation. We calculated a fraction of Ade$^+$ colonies from more than 200 total colonies on at least three different SD-URA plates containing trace amounts of adenine. Increasing the concentration of Sup-NM fibers resulted in a dose-dependent increase in Ade$^+$ convertants, with the fraction of Ade$^+$ colonies among the Ura$^+$ colonies approaching 100% at high Sup-NM concentrations (Fig. 28.1C). As expected for a prion, the efficiency of prion conversion was sensitive to proteinase but not nuclease treatment (Tanaka et al., 2004).

For procedure [2], the efficiency of conversion to prion state as well as the phenotypic strength of prion strains was examined by monitoring color with modified 1/4 YPD plates. Transformants remained [psi$^-$] (red) or converted to a weak [PSI$^+$] (pink) or strong [PSI$^+$] (white) state (Fig. 28.2A and B). After the YPD plates were incubated at 30 °C for a few days, each streak was classified into strong [PSI$^+$] (white), weak (pink) [PSI$^+$], or [psi$^-$] (red) strains. In quantification experiments, typically 56 colonies from at least three independent transformations are streaked on YPD plates. Ade$^+$ revertants, which are rare, were readily excluded from the statistics, as they showed brown color on 1/4 YPD plates.

Fibers formed at 4 °C had a high efficiency of infection as shown by the large majority of colonies with a strong (white) [PSI$^+$] strain phenotype. However, fibers formed at 37 °C had lower infectivity and produced almost exclusively weak (pink and/or sectored) [PSI$^+$] strains (Fig. 28.3A and B). The weak strains showed increased levels of soluble Sup35p and were more readily cured by Hsp104 overexpression (Tanaka et al., 2004). The strain phenotypes did not depend on the concentration of seed or the infection efficiency: 4 °C fibers yielded strong strains even when diluted 10-fold (infection rate $\sim$20%) and 37 °C fibers yielded weak strains even when the concentration was increased 10-fold (infection rate $\sim$80%) (Fig. 28.3C). Thus, the [PSI$^+$] strain is determined by conformation of infectious Sup-NM amyloid. These results establish that amyloid is an infectious form of prion protein and that conformational differences in amyloid determine prion strain variations.

In summary, we developed an efficient protein transformation protocol, which allows one to introduce in vitro amyloid or in vivo prions efficiently into yeast. Using the technique, we directly demonstrated that conformational differences of amyloid fibers constitute the physical foundation of the heritable differences in prion strains. Importantly, any protein and protein aggregate can be efficiently introduced into yeast by this procedure. Thus, this novel and versatile protein transformation technique will be a powerful tool in yeast biology.

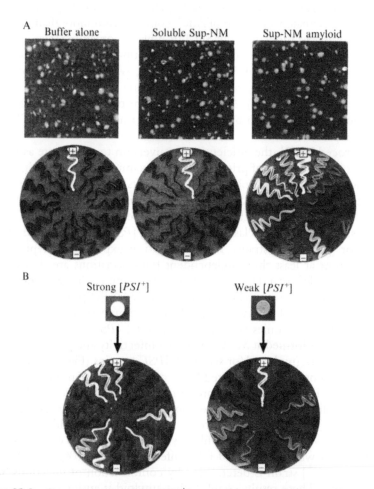

Figure 28.2 Generation of multiple [*PSI*⁺] strains by *in vitro* Sup-NM amyloid fibers and *in vivo* prions. (A) Infection of yeast without selection for the prion state. Following transformation with the indicated material, cells were recovered on SD-Ura (top) and randomly selected colonies were streaked on 1/4 YPD plates to identify [*PSI*⁺] convertants (bottom). *In vitro*-converted amyloid fibers induced a range of white to pink (grey color) [*PSI*⁺] strains. Throughout, the top (+) and bottom (−) streaks in a 1/4 YPD plate are strong [*PSI*⁺] and [*psi*⁻] controls, respectively. (B) Induction of prion state by partially purified prion particles derived from strong (white) or weak (grey) [*PSI*⁺] strains. Note that successful [*PSI*⁺] infectants (bottom) show the same strain phenotypes as the donor [*PSI*⁺] strain (top).

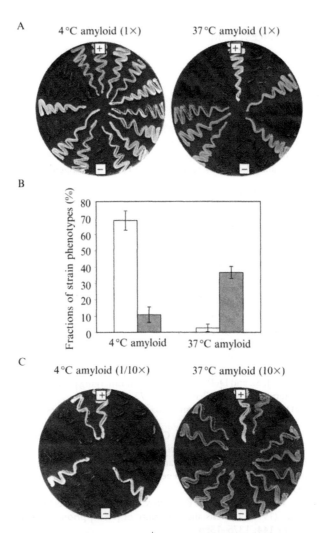

Figure 28.3 Generation of distinct [*PSI*⁺] strains by different conformations of *in vitro* Sup-NM amyloids. (A) Examples of [*PSI*⁺] strains resulting from infection with Sup-NM amyloid spontaneously formed at 4 or 37 °C. White and pink (grey color) and/or sectored colonies are strong and weak [*PSI*⁺] variants, respectively. (B) Quantification of frequency of [*PSI*⁺] strains induced by transformation with 4 or 37 °C Sup-NM amyloids. White and grey bars show fractions of strong and weak [*PSI*⁺] strains phenotypes. (C) Induction of [*PSI*⁺] states using 1/10th the amount of 4 °C Sup-NM amyloid or 10-fold the amount of 37 °C Sup-NM amyloid used in (A).

ACKNOWLEDGMENTS

I thank Jonathan Weissman and members of the Weissman laboratory (HHMI, UCSF) for their advice throughout the work. This study was partly supported by Uehara Memorial Foundation and JST PRESTO (M. T.).

REFERENCES

Aguzzi, A., and Haass, C. (2003). Games played by rogue proteins in prion disorders and Alzheimer's disease. *Science* **302,** 814–818.

Alberti, S., Halfmann, R., King, O., Kapila, A., and Lindquist, S. (2009). A systematic survey identifies prions and illuminates sequence features of prionogenic proteins. *Cell* **137,** 146–158.

Bessen, R. A., *et al.* (1995). Non-genetic propagation of strain-specific properties of scrapie prion protein. *Nature* **375,** 698–700.

Brachmann, A., Baxa, U., and Wickner, R. B. (2005). Prion generation in vitro: Amyloid of Ure2p is infectious. *EMBO J.* **24,** 3082–3092.

Bradley, M. E., and Liebman, S. W. (2004). The Sup35 domains required for maintenance of weak, strong or undifferentiated yeast [*PSI$^+$*] prions. *Mol. Microbiol.* **51,** 1649–1659.

Chernoff, Y. O., Lindquist, S. L., Ono, B., Inge-Vechtomov, S. G., and Liebman, S. W. (1995). Role of the chaperone protein Hsp104 in propagation of the yeast prion-like factor [*PSI$^+$*]. *Science* **268,** 880–884.

Chien, P., and Weissman, J. S. (2001). Conformational diversity in a yeast prion dictates its seeding specificity. *Nature* **410,** 223–227.

Chien, P., DePace, A. H., Collins, S., and Weissman, J. S. (2003). Generation of prion transmission barriers by mutational control of amyloid conformations. *Nature* **424,** 948–951.

Chien, P., Weissman, J. S., and DePace, A. H. (2004). Emerging principles of conformation-based prion inheritance. *Annu. Rev. Biochem.* **73,** 617–656.

Collinge, J. (2001). Prion diseases of humans and animals: Their causes and molecular basis. *Annu. Rev. Neurosci.* **24,** 519–550.

DePace, A. H., and Weissman, J. S. (2002). Origins and kinetic consequences of diversity in Sup35 yeast prion fibers. *Nat. Struct. Biol.* **9,** 389–396.

DePace, A. H., Santoso, A., Hillner, P., and Weissman, J. S. (1998). A critical role for amino-terminal glutamine/asparagine repeats in the formation and propagation of a yeast prion. *Cell* **93,** 1241–1252.

Derkatch, I. L., Chernoff, Y. O., Kushnirov, V. V., Inge-Vechtomov, S. G., and Liebman, S. W. (1996). Genesis and variability of [*PSI$^+$*] prion factors in *Saccharomyces cerevisiae*. *Genetics* **144,** 1375–1386.

Glover, J. R., *et al.* (1997). Self-seeded fibers formed by Sup35, the protein determinant of [*PSI$^+$*], a heritable prion-like factor of S. *cerevisiae*. *Cell* **89,** 811–819.

King, C. Y., and Diaz-Avalos, R. (2004). Protein-only transmission of three yeast prion strains. *Nature* **428,** 319–323.

King, C. Y., Tittmann, P., Gross, H., Gebert, R., Aebi, M., and Wuthrich, K. (1997). Prion-inducing domain 2-114 of yeast Sup35 protein transforms in vitro into amyloid-like filaments. *Proc. Natl. Acad. Sci. USA* **94,** 6618–6622.

Kochneva-Pervukhova, N. V., Chechenova, M. B., Valouev, I. A., Kushnirov, V. V., Smirnov, V. N., and Ter-Avanesyan, M. D. (2001). [*PSI$^+$*] prion generation in yeast: Characterization of the 'strain' difference. *Yeast* **18,** 489–497.

Kushnirov, V. V., Kryndushkin, D. S., Boguta, M., Smirnov, V. N., and Ter-Avanesyan, M. D. (2000). Chaperones that cure yeast artificial [PSI^+] and their prion-specific effects. *Curr. Biol.* **10**, 1443–1446.

Liebman, S. (2002). Progress toward an ultimate proof of the prion hypothesis. *Proc. Natl. Acad. Sci. USA* **99**, 9098–9100.

Liu, J. J., Sondheimer, N., and Lindquist, S. L. (2002). Changes in the middle region of Sup35 profoundly alter the nature of epigenetic inheritance for the yeast prion [PSI^+]. *Proc. Natl. Acad. Sci. USA* **99**, 16446–16453.

Patel, B. K., and Liebman, S. W. (2007). 'Prion-proof' for [PIN^+]: Infection with *in vitro*-made amyloid aggregates of Rnq1p-(132–405) induces [PIN^+]. *J. Mol. Biol.* **365**, 773–782.

Patel, B. K., Gavin-Smyth, J., and Liebman, S. W. (2009). The yeast global transcriptional co-repressor protein Cyc8 can propagate as a prion. *Nat. Cell Biol.* **11**, 344–349.

Peretz, D., *et al.* (2002). A change in the conformation of prions accompanies the emergence of a new prion strain. *Neuron* **34**, 921–932.

Prusiner, S. B., Scott, M. R., DeArmond, S. J., and Cohen, F. E. (1998). Prion protein biology. *Cell* **93**, 337–348.

Santoso, A., Chien, P., Osherovich, L. Z., and Weissman, J. S. (2000). Molecular basis of a yeast prion species barrier. *Cell* **100**, 277–288.

Scott, J. H., and Schekman, R. (1980). Lyticase: Endoglucanase and protease activities that act together in yeast cell lysis. *J. Bacteriol.* **142**, 414–423.

Sparrer, H. E., Santoso, A., Szoka, F. C. Jr., and Weissman, J. S. (2000). Evidence for the prion hypothesis: Induction of the yeast [PSI^+] factor by *in vitro*-converted sup35 protein. *Science* **289**, 595–599.

Tanaka, M., Chien, P., Naber, N., Cooke, R., and Weissman, J. S. (2004). Conformational variations in an infectious protein determine prion strain differences. *Nature* **428**, 323–328.

Telling, G. C., *et al.* (1996). Evidence for the conformation of the pathologic isoform of the prion protein enciphering and propagating prion diversity. *Science* **274**, 2079–2082.

Ter-Avanesyan, M. D., Dagkesamanskaya, A. R., Kushnirov, V. V., and Smirnov, V. N. (1994). The *SUP35* omnipotent suppressor gene is involved in the maintenance of the non-Mendelian determinant [PSI^+] in the yeast *Saccharomyces cerevisiae*. *Genetics* **137**, 671–676.

Tessier, P. M., and Lindquist, S. (2009). Unraveling infectious structures, strain variants and species barriers for the yeast prion [PSI^+]. *Nat. Struct. Mol. Biol.* **16**, 598–605.

Tuite, M. F., and Cox, B. S. (2003). Propagation of yeast prions. *Nat. Rev. Mol. Cell Biol.* **4**, 878–890.

Uptain, S. M., Sawicki, G. J., Caughey, B., and Lindquist, S. (2001). Strains of [PSI^+] are distinguished by their efficiencies of prion-mediated conformational conversion. *EMBO J.* **20**, 6236–6245.

Wickner, R. B. (1994). [URE3] as an altered URE2 protein: Evidence for a prion analog in *Saccharomyces cerevisiae*. *Science* **264**, 566–569.

Zhou, P., Derkatch, I. L., Uptain, S. M., Patino, M. M., Lindquist, S., and Liebman, S. W. (1999). The yeast non-Mendelian factor [ETA^+] is a variant of [PSI^+], a prion-like form of release factor eRF3. *EMBO J.* **18**, 1182–1191.

CHAPTER TWENTY-NINE

OVEREXPRESSION AND PURIFICATION OF INTEGRAL MEMBRANE PROTEINS IN YEAST

Franklin A. Hays,*,1 Zygy Roe-Zurz,† and Robert M. Stroud*,†,‡

Contents

Abstract

The budding yeast *Saccharomyces cerevisiae* is a viable system for the overexpression and functional analysis of eukaryotic integral membrane proteins (IMPs). In this chapter we describe a general protocol for the initial cloning, transformation, overexpression, and subsequent purification of a putative IMP and discuss critical optimization steps and approaches. Since expression and purification are often the two predominant hurdles one will face in studying this difficult class of biological macromolecules the intent is to outline the general workflow while providing insights based upon our collective experience. These insights should facilitate tailoring of the outlined protocol to individual IMPs and expression or purification routines.

* Department of Biochemistry and Biophysics, University of California at San Francisco, San Francisco, California, USA
† Membrane Protein Expression Center, University of California at San Francisco, San Francisco, California, USA
‡ Center for the Structure of Membrane Proteins, University of California at San Francisco, San Francisco, California, USA
1 Current addres: Department of Biochemistry and Molecular Biology, University of Oklahoma Health Sciences Center, Oklahoma City, Oklahoma, USA

Methods in Enzymology, Volume 470
ISSN 0076-6879, DOI: 10.1016/S0076-6879(10)70029-X

1. INTRODUCTION

Obtaining sufficient quantities of a purified integral membrane protein (IMP) for downstream experiments, such as structural or functional analysis, can be a daunting task. Common hurdles that one may encounter include obtaining sufficient IMP overexpression, extracting the IMP from cellular membranes with a detergent and purifying the IMP in functional form. Advances in addressing these bottlenecks should facilitate efforts by the broader scientific community in pursuing their own particular IMP of interest. One such advance is use of the budding yeast *Saccharomyces cerevisiae* to overexpress IMPs (Bill, 2001; Bonander *et al.*, 2005; Griffith *et al.*, 2003; Hays *et al.*, 2009; Li *et al.*, 2009; White *et al.*, 2007). When combined with a broad range of methods for *in vivo* functional characterization of IMPs in yeast, with its exhaustive genetic toolkit, one can appreciate the inherent power of using *S. cerevisiae* as an expression system. Thus, the objective of this chapter is to provide a general approach for overexpression of IMPs in the yeast *S. cerevisiae*. In addition, we will provide an introduction to purifying the IMP of interest following expression. To accomplish this, we will describe our approach to the task while highlighting critical steps within the protocol that may require heightened attention. It is important to note that overexpression and purification of functional IMPs is still a laborious endeavor fraught with problems. As with most difficult journeys, many small decisions often come together in dictating the outcome.

2. GENERAL CONSIDERATIONS

S. cerevisiae is an intensely studied eukaryotic organism. The approach we have taken with the current chapter is to outline the yeast expression protocol currently deployed within our research efforts. At almost every step throughout this chapter, an alternative method, vector, column, buffer, affinity tag, etc., could be employed with possibly better outcomes for the specific protein being studied. Our intent is to convey a generic strategy and, where possible, highlight alternatives that we feel the reader should be aware of. Since working with IMPs is often an endeavor replete with nuances and hurdles, it is our desire that this chapter will provide a foundation for those not familiar with membrane protein overexpression and purification.

This chapter is organized around the expression and purification of a putative integral membrane protein termed "POI" for "protein of interest." The intent is that a reader can substitute the membrane protein he/she is interested in for this target. As with most procedures, our approach is not the only viable strategy. It works very well in many cases though modifications can be customized to suit the system under study. We have made some key choices based upon our prior experience including: (1) yeast strain W303-Δpep4 (*leu2-3,112 trp1-1 can1-100 ura3-1 ade2-1 his3-11,15 Δpep4 MATα*) is used, (2) pRS423-GAL1-based inducible plasmid to drive expression, (3) a C-terminal [linker]-[3C-protease]-[10× His] tag fused to the expressed protein, and (4) solubilization in the detergent *n*-dodecyl-β-D-maltopyranoside (DDM). Each of these is a critical step and should be examined if the described procedure should fail. Finally, previously published protocols may also be of interest to the reader (Hays *et al.*, 2009; Li *et al.*, 2009; Newby *et al.*, 2009).

3. PROTOCOL—MOLECULAR BIOLOGY

Our experience is such that multiple expression plasmids, affinity tags, and fusion constructs should be tried when pursuing a specific protein. Vectors designed to better leverage *S. cerevisiae* as an expression system are often chimeric shuttle vectors with yeast and bacterial derived sequences. The yeast contribution to the vector sequence will determine the location of transformation: extrachromosomal ectopic expression or chromosomal integration in mitotically stable yeast strains (Boer *et al.*, 2007). Episomal expression requires that the cloned gene needs be free of introns for plasmid or genomic expression. If properly implemented, episomal overexpression in yeast can be rapidly deployed and often yields milligram quantities of IMPs (Li *et al.*, 2009; Mumberg *et al.*, 1995). This is accomplished through the autonomously replicating sequence from native yeast 2μ plasmid (Christianson *et al.*, 1992). Thus, for the current example, POI is cloned into a high-copy 2μ episomal expression vector containing a GAL1 promoter (Fig. 29.1). The GAL1 promoter is useful because it is tightly repressed in the presence of glucose and strongly induced by galactose allowing for stringent control of protein expression. Expression levels can be further manipulated by altering the copy number through the origin of replication and by swapping out the GAL1 promoter for constitutive (ADH1, TEF2) or other inducible promoters (e.g., MET25, PHO5) (Mumberg *et al.*, 1994). In our experience, constitutive promoters are not effective for IMP overexpression.

To facilitate the process of shuttling a gene between numerous expression vectors, and even expression systems, we use ligation independent cloning (LIC). We previously described in detail how LIC cloning is performed within our yeast system (Supplementary Information in Li *et al.*, 2009). Our experience has led us to prefer a C-terminal rhinovirus 3C cleavable poly-histidine tag as an initial choice when pursuing novel IMPs. Approximately 30% of IMPs contain an N-terminal signal peptide

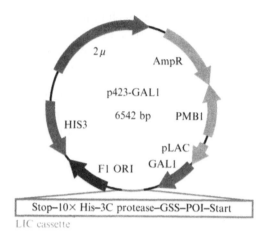

Figure 29.1   Schematic of the p423-GAL1 expression plasmid. Yeast 2μ-based expression plasmid containing a GSS-3C protease-$10\times$ His tag for the protein of interest using LIC. Plasmid contains yeast (shown in red), bacterial (shown in green), and phage (shown in blue) elements conducive to molecular cloning and transformation methodologies. (See Color Insert.)

involved in proper protein maturation and targeting to cellular membranes. Since N-terminal tags can interfere with this processing, the preference is to use C-terminal tags when available. In addition, C-terminal tags provide greater assurance that the protein being purified through initial steps is the full-length construct and free of truncation or degradation. If the POI does not contain a signal peptide, which is often difficult to ascertain for eukaryotic genes, then N-terminal tags provide greater flexibility in developing expression constructs. Whatever tag is chosen, care should be taken to ensure that it is either added to the design of synthetic primers during cloning or already present within the selected plasmid. Also, a critical step when including C-terminal tags is to ensure that the native stop codon is removed from the gene of interest. For the current discussion, we will clone POI into our p423-GAL1 expression plasmid containing the following design: Start-[POI]-[linker]-[3C site]-[$10\times$ His]-Stop. The choice of using a rhinovirus 3C protease for tag cleavage is described later.

A general protocol for cloning POI into this plasmid is as follows. Refer to Li *et al.* (2009) for a detailed protocol:

1. POI is PCR amplified with primers palindromic to our p423-GAL1 LIC vector.
2. Amplified POI and p423-GAL1 separately undergo T4-polymerase $3'-5'$ exonuclease digestion in the presence of dATP and dTTP, respectively.
3. The digested gene and plasmid are then combined at room temperature, annealed, and transformed into competent *Escherichia coli* cells.

4. Colony PCR is used to confirm POI insertion into p423-GAL1. Sequencing the plasmid with GAL1 and CYC1 primers validates POI identity.

4. Protocol—Cell Growth

The plasmid containing POI destined for transcription, translation, and proper membrane insertion must first be introduced into the yeast host through transformation. Although there are several methods to introduce genetic material into *S. cerevisiae*, including *Agarobacterium tumefaciens*-mediated transformation (Piers *et al.*, 1996), we use a lithium acetate transformation protocol with PEG 3350. The episomal vector p423-GAL1 contains the HIS3 gene needed by our strain, W303-Δpep4 (*leu2-3,112 trp1-1 can1-100 ura3-1 ade2-1 his3-11,15 Δpep4 MATα*), and must be cultured in synthetic complete media without histidine to maintain selection for the plasmid containing POI. Cultures are grown in 375 ml volumes containing SC-His with 2% glucose in 1L baffled flasks shaking at 220 rpm at 30 °C. Following a growth period of 24 h, the optical density at 600 nm ranges between 15 and 20 for most cultures with glucose concentration generally ≪0.1%. The culture is induced by adding 125 ml of 4× YPG (yeast extract, bactopeptone, and galactose) to each flask bringing the final volume to 500 ml. DMSO has previously been shown to improve the expression of certain IMPs and may be tried as a growth additive during induction (Andre *et al.*, 2006). Growths can easily transition from shaker flasks to the zymurgy route of large-scale fermentation as the choice of inducible promoter enables careful regulation and timing of expression. Cells are harvested after 16 h at 6000×g and resuspended in 30 ml of lysis buffer containing 50 mM Tris (pH 7.4, RT), 500 mM NaCl, and 20% glycerol (v/v) per half liter of growth. Ideally, we adjust the volume of growth culture to obtain a minimum of 2–3 mg purified protein per growth (200–300 μl at 10 mg/ml).

5. Protocol—Membrane Preparation and Solubilization

Once harvested, cells expressing POI are lysed mechanically using a microfluidizer or by bead beating in the presence of protease inhibitors. For bead beating, we use a 90-ml canister containing approximately 40 ml of resuspended cell pellet. Each canister is then filled to the top with 0.5 mm prechilled glass beads and lysed using four cycles of 1 min "on" and 1 min "off." We find that aggressive protease inhibition is not always needed and is target specific. Crude cell lysate is then centrifuged at $6000 \times g$ for 15 min. After this centrifugation, qualitative lysis efficiency can be determined by the debris pellet which will contain two layers: a bottom pink (strain dependent) layer of unlysed cells and top lighter layer of organelles and cellular debris from lysed cells. The ratio of the top lysed cells to the bottom unlysed cells is a qualitative indicator of lysis efficiency. We typically have >90% efficiency at this stage but >70% is considered acceptable. Collect cell lysate from the supernatant of the previous low-speed spin while being careful not to contaminate supernatant with cell debris, and spin the supernatant at $138,000 \times g$ (42,000 rpm using a Ti 45 rotor) for 2 h. Discard the supernatant from the high-speed spin. Occasionally, a loose upper layer is obtained following the high-speed spin that should be retained as it often contains a predominant portion of the expressed protein. Resuspend membranes in approximately 5 ml of membrane resuspension buffer (50 mM Tris (pH 7.4, RT), 200 mM NaCl, 10% (v/v) glycerol, and 2 mM fresh PMSF) per liter of culture growth with 10 μl HALT protease inhibitor cocktail (or your protease inhibitor cocktail of choice). Stir on ice for 30 min and flash freeze membranes in LN2 or use immediately.

We commonly use the following detergents for solubilizing membrane proteins leading to structural work: n-octyl-β-D-glucopyranoside (OG), n-nonyl-β-D-glucopyranoside (NG), n-decyl-β-D-maltopyranoside (DM), n-dodecyl-β-D-maltopyranoside (DDM), n-dodecyl-N,N-dimethylamine-N-oxide (LDAO), and n-dodecylphosphocholine (FC-12). Detergents are purchased in high-purity form (i.e., "ANAGRADE") from Anatrace. Numerous other detergents are possible depending on the individual experiment being performed. Once the POI-3C-10His has been expressed, it is important to access how well it can be extracted from the membrane with a detergent. This is generally accomplished through a broad screen of several detergents. The recommended concentrations, for detergents listed above, when solubilizing cellular membranes are 270 mM OG, 140 mM NG, 10 mM DM, 20 mM DDM, 200 mM LDAO, and 20 mM FC-12 ($10\times$ CMC for other detergents is a recommended starting point). A detailed protocol for performing this step is available in Box 1 of Newby *et al.* (2009). Generally, small aliquots of cellular membranes are mixed with an equal volume of buffer containing detergent at the above concentration and then stirred at 4 °C for 12–14 h. Unsolubilized cellular membranes will pellet at $200,000 \times g$, so the extent to which a given detergent is able to solubilize POI-3C-10His can be evaluated by the amount of protein left in the supernatant following a high-speed spin. When evaluating initial expression levels via western blots, one may observe several background bands specific to yeast that may be visible in an epitope dependent manner. For anti-His westerns IST2, a 946 amino acid polypeptide containing a stretch of seven histidines near the C-terminus, runs at around 100 kDa. When using anti-FLAG, an unidentified contaminant band often appears around 60 kDa. An HRP-conjugated Penta-His antibody (Invitrogen) works best for probing C-terminal poly-histidine tags in our experience. If available, functional assays to verify activity following detergent solubilization are highly informative.

6. PROTOCOL — PROTEIN PURIFICATION

Once the POI-3C-10His protein is extracted from cellular membranes in soluble form, it may be purified to obtain a sample that is Pure (free of other proteins and contaminants), Homogenous (single uniform population), Stable (typically over a week in concentrated form at 4 °C), and Free of protein-free detergent micelles (this combined state will be referred to as "PHSF"). To accomplish this, we employ a narrow range of techniques including immobilized metal affinity (IMAC), size-exclusion (SEC) and ion-exchange chromatography. These methods are synergistic, iterative, and employed to varying degrees depending on the target protein. For the current discussion, we will detail a standard approach of IMAC followed by cleavage of the expression tag, reverse-IMAC to remove uncleaved protein, and finally SEC to obtain the purified protein in diluted form. This sample will then be concentrated and analyzed prior to use. If this sample is intended for structure determination (i.e., crystallization) then special caution should be taken to avoid a significant concentration of protein-free detergent micelles (Newby et al., 2009).

Thus, we will continue with the theme of purifying our target protein, POI-3C-10His, which was solubilized in the previous section. Recommended detergent concentrations for SEC buffers are as follows: 40 mM OG, 12 mM NG, 4 mM DM, 1 mM DDM, 12 mM LDAO, and 4 mM FC-12 ($2\times$ CMC is a good starting point for most detergents). For the current example, we will use 1 mM DDM in all buffers (as determined in Section 5). The initial step to protein purification is a metal-affinity purification of the solubilized membranes; we generally use 125 μl of Ni-NTA agarose resin (Qiagen) per mg of expected protein yield. The selected IMAC resin should be prepared according to manufacturer's specifications and optimized as needed. The solubilized membranes should be incubated with IMAC resin at 4 °C with nutation for at least 1 h though generally not longer than 3 h. We have found that the degree of target protein binding to Ni-NTA resin does not increase substantially past 3 h though increased proteolysis and binding of contaminant proteins may occur. Following incubation, the Ni-NTA resin containing bound POI-3C-10×His protein should be transferred to a gravity flow column and washed with 20 column volumes of Buffer A (20 mM Tris (pH 7.4, RT), 200 mM NaCl, 10% (v/v) glycerol, 4 mM β-ME, 1 mM PMSF, and 1 mM DDM) containing 10 mM imidazole. If following the wash by absorbance, it is beneficial to wash until A_{280} nm returns to baseline. It is important at this point to obtain about 10 μl of initial flow-through for SDS–PAGE analysis. The above steps are repeated with Wash 2 (Buffer A with 25 mM imidazole) and Wash 3 (Buffer A with 40 mM imidazole) buffers. Finally, POI-3C-10×His is eluted from the column using the IMAC elution buffer (Buffer A with 300 mM imidazole). If possible, reduce the flow rate prior to elution to ensure the target protein elutes in a minimal volume. Be careful to observe the eluted sample for turbidity, especially over the ensuing several minutes as the protein may be unstable in the prescribed buffer and thus precipitate out of solution at this point. If precipitation occurs, one can make appropriate changes to the IMAC buffers (e.g., changing salt concentration or pH) to increase protein stability. It is also advisable to perform a buffer exchange immediately following elution into 20 mM HEPES (pH 7.4), 150 mM NaCl, 10% (v/v) glycerol, 4 mM β-ME, 1 mM PMSF, and 1 mM DDM (SEC buffer). This can be accomplished with a small desalting column such as the Econo-Pac 10 DG disposable chromatography column from Bio-Rad (cat. no. 732-2010). Following IMAC and buffer exchange, the POI-3C-His protein is ready for cleavage of the linker-3C-10×His expression tag.

There are a broad number of site-specific proteases for cleaving affinity tags, though care should be taken to ensure they are active in the prescribed detergent (Mohanty *et al.*, 2003). The human rhinovirus 3C protease and thrombin are both robust and efficient proteases that have worked very well for cleaving affinity tags attached to detergent solubilized membrane proteins. We have had great success with an MBP-3C fusion construct described previously (Alexandrov *et al.*, 2001). To cleave the POI-3C-His affinity tag, the protein should be incubated overnight at 4 °C with approximately a 1:5 ratio of protease to target protein in whatever volume of buffer is obtained in the desalting step above. Retain pre- and postcleavage 10 μl samples for SDS–PAGE gel analysis to evaluate cleavage. Following cleavage a reverse-IMAC purification (i.e., the flow-through is retained) is performed using metal-affinity resin to separate cleaved 3C-His tag and protease (which is also His tagged) from the target protein. This step entails a 1h incubation with IMAC resin, such as "Talon" metal-affinity resin, in batch at 4 °C. Following incubation, the flow-through should be retained—this contains the cleaved POI protein that will be purified in the next step. Elute resin bound protein from the column using the IMAC elution buffer and collect a 10-μl sample for analysis on a gel to ascertain if nonspecific binding of the target protein is occurring. Following completion of this step, the Ni-purified, 3C-cleaved POI protein is now ready for further purification.

Ion-exchange chromatography is a powerful technique that separates macromolecules based upon charge state at a given pH. Though not discussed within this chapter, we have often used this technique to purify difficult targets, concentrate protein, reduce protein-free detergent micelles, perform detergent exchanges, or obtain a pH stability profile. We generally use 1 or 5 ml disposable HiTrap sepharose Q or SP ion-exchange columns from GE Healthcare. Though not performed on every membrane protein, ion-exchange chromatography has proven to be a valuable technique and should be leveraged when needed.

The collective experience from the numerous IMP purifications that we have performed is that SEC is an essential step in the process of obtaining a PHSF sample. SEC allows one to rapidly evaluate the quality of the purified protein by analyzing the retention time, shape, and number of eluted peaks from the sample. Elution in the void volume of a properly sized SEC column (i.e., the void volume is significantly higher than the expected molecular weight of the target protein–micelle complex) is indicative of protein that is not stable under the prescribed solution conditions. Often this means a new solubilization buffer should be used with optimized parameters for detergent selection, pH, and salt concentration. If the target is present within the included volume then careful analysis of the peaks should be performed. Is the POI resident within a single, Gaussian shaped, peak or multiple peaks indicative of several oligomeric states? If the latter, it may shift to the void over time and, either way, is often indicative of stability issues within the specified buffer. Ideally, one will see a single well-defined peak within the included volume corresponding to (and verified by gels/blots) the POI. For a detailed discussion of membrane protein SEC characteristics refer to Figure 3 of Newby et al. (2009). Coupling fluorescence with SEC, termed fluorescence–detection SEC, is another approach that requires very small amounts of expressed protein and is therefore conducive for broad screens (Kawate and Gouaux, 2006). Troubleshooting is often required during the SEC purification step to ascertain the correct buffer conditions for stabilizing the protein in solution within a monodisperse peak. A standard approach to this process is varying pH (e.g., 5.5 in MES, 7.0 in HEPES, and 8.0 in Tris), salt concentration (e.g., 25 mM, 250 mM, and 500 mM NaCl), presence or absence of osmolytes (e.g., adding varying concentrations of glycerol or sucrose), and addition of putative or known ligands. It is important to note that when approaching this step one should be systematic and linear to clearly differentiate effects on protein stability and homogeneity.

In continuing with our example of expressing and purifying POI, we now have a Ni-purified and 3C-cleaved protein sample that has been purified away from cleaved affinity tag and protease. Next, we describe a general SEC purification step for this protein sample. There are a number of chromatography columns available and care should be taken to ensure that the column is appropriate for the desired task and will not interact with the detergent (e.g., TSK columns may interact with the detergent LDAO) or POI. We generally use a Superdex 200 10/300 GL column from GE

Healthcare (cat. no. 17-5175-01). This column has a separation molecular weight range of 10,000–600,000 that is ideally suited for most membrane proteins. The POI protein is now in the SEC buffer described above. It is important that the SEC column be equilibrated for a minimum of 3 h at 0.5 ml/min or overnight at 0.1 ml/min to ensure complete equilibration with the detergent (DDM in our example). Once equilibrated, the POI sample can be run in iterative rounds with a peak height of approximately one absorbance unit at 280 nm. The amount of loaded sample will vary depending on the presence of contaminating or oligomeric peaks. Generally, our approach is to use a chromatography station equipped with an auto-injector and fraction collector to enable automated runs, often overnight. Care should be taken that the column is not overloaded with sample, as this may mask secondary peaks and lead to incomplete purification. A common approach is to inject 0.5 ml of two OD A_{280}/ml sample per run, though the optimal injection amount is ultimately sample dependent.

Some general considerations regarding the purification step should be highlighted. In particular, when working with solubilized membrane proteins, the actual identity of a sample is a membrane protein with a detergent micelle surrounding it. This micelle will often contain endogenous lipids from the expression host. First, the protein–detergent–lipid complex (PDLC) will likely have a shorter retention time relative to a soluble protein of the same mass. Thus, it can be hard to ascertain the oligomeric state of a PDLC based upon SEC retention time alone. This holds true when comparing it to molecular weight standards because these standards are usually composed of small molecules and soluble proteins. In addition, IMPs tend to migrate slightly faster than expected on SDS–PAGE gels, giving the impression that your target is of a smaller mass than expected. Another common hurdle is that detergent micelles can occlude the protease recognition site when trying to remove an expression tag resulting in no, or attenuated, cleavage. Two common ways to avoid this potential problem are to add a short linker, often three additional amino acids, between the target protein and protease recognition site, or to move the expression tag to the other protein terminus. Moving the tag may lead to additional problems since approximately 30% of IMPs contain a signal peptide at the N-terminus. N-terminal tags can interfere with the processing of this signal peptide by the signal peptidase leading to retention of the membrane protein intracellularly and, as a result, decreased expression levels. Finally, when concentrating the purified protein, it should be remembered that the sample contains protein-free detergent micelles (since the detergent concentration is above the detergent CMC). Since these micelles can impact biophysical properties of the sample, such as crystallization, it is important that detailed notes be maintained regarding the concentration factor (i.e., starting volume relative to final volume postconcentration) for the sample. If possible, one should generally work to minimize the concentration factor and thereby minimize the protein-free detergent micelles.

7. PROTOCOL—PROTEIN CHARACTERIZATION

Separating the protein from detergent micelles: The lack of absorbance at 280 nm by detergent micelles means that to separate the protein containing micelles from those that do not requires other detection strategies. We have found that an in-line four-way detection scheme is useful in differentiating these species and separating them from each other. These detectors consist of UV absorbance and refractive index (RI) detectors for measuring concentration, a differential pressure or intrinsic viscosity detector that indicates properties of size and shape, and a right angle light scattering detector that indicates molecular mass. In concert, these allow one to (1) optimize detergent micelle concentration while maintaining PDLC homogeneity of concentrated protein and (2) measure the PDLC oligomeric state (mass), size (Rh), shape (IV), detergent: protein ratio, and rate of change of RI (dn/dc). For common detergents, we have measured size-exclusion retention volume (SERV), dn/dc, micelle molecular weight, and retention behavior on different molecular weight cut-off filters for empty micelles in various systems. These micelle parameters are dependent on buffer composition, column type, detergent concentration, and the presence of PDLCs. The goal is to minimize the detergent micelle concentration during purification and concentration. The micelle SERV relative to the PDLC SERV dictates whether the PDLC can be concentrated before SEC (as they often have different SERVs), and if SEC can be used to remove excess micelles after protein concentration. Detergent dn/dc is used to quantify excess [detergent micelle] after protein concentration, and the amount of detergent bound in the PDLC. To accurately measure the PDLC physical parameters, the PDLC peaks must be baseline-resolved, of adequate intensity, and Gaussian with no comigrating excess micelles or other buffer contaminants (i.e., single SEC peaks for all four detectors). A simpler approach is to include an in-line RI detector to measure solution viscosity. These detectors can be added to existing chromatography stations with minimal alterations and will identify SERV values for the specific solubilization detergent and buffer combination being used. Overall, characterization of the PDLC within the prescribed detergent using the above methods facilitates the development of a robust purification and protein concentration scheme conducive to downstream endeavors. One should view PDLC and empty detergent micelles as separate entities during purification and work to identify the latter early within purification to minimize as needed.

8. CONCLUSION

S. cerevisiae is a viable and powerful system for overexpressing IMPs as yeast is a genetically tractable and inexpensive expression system that can be easily manipulated experimentally and is conducive for high-, medium-, or low-throughput methodologies. Furthermore, being a eukaryotic organism it contains the necessary posttranslational modification and membrane targeting machinery to facilitate expression of many higher eukaryotic IMPs (Li *et al.*, 2009). The methods described in this chapter are focused on the overexpression and purification of a nominated IMP within yeast. Subsequent purification of these proteins can be accomplished if one takes appropriate caution and is aware of common hurdles. Whenever possible, functional assays should be incorporated into the purification protocol to ensure the POI being purified is in functional form. As the collective knowledge and experience in working with IMPs increases so have the rewards and novel biological insights. Indeed, the outlook is very positive (White, 2009). With so little known about the vast majority of IMPs, we are undoubtedly entering a period of dramatic growth in our understanding.

ACKNOWLEDGMENTS

We are grateful to Drs. Min Li, Franz Gruswitz, and John K. Lee for kind assistance in preparing the manuscript. This work was supported by the NIH Roadmap Center grant P50 GM073210 (R. M. S.), Specialized Center for the Protein Structure Initiative grant U54 GM074929 (R. M. S.), National Research Service Award F32 GM078754 (F. A. H.), and a Sandler Biomedical Research postdoctoral fellowship (F. A. H.).

REFERENCES

Alexandrov, A., Dutta, K., and Pascal, S. M. (2001). MBP fusion protein with a viral protease cleavage site: One-step cleavage/purification of insoluble proteins. *Biotechniques* **30,** 1194–1198.
Andre, N., Cherouati, N., Prual, C., Steffan, T., Zeder-Lutz, G., Magnin, T., Pattus, F., Michel, H., Wagner, R., and Reinhart, C. (2006). Enhancing functional production of G protein-coupled receptors in Pichia pastoris to levels required for structural studies via a single expression screen. *Protein Sci.* **15,** 1115–1126.
Bill, R. M. (2001). Yeast–A panacea for the structure-function analysis of membrane proteins? *Curr. Genet.* **40,** 157–171.
Boer, E., Steinborn, G., Kunze, G., and Gellissen, G. (2007). Yeast expression platforms. *Appl. Microbiol. Biotechnol.* **77,** 513–523.
Bonander, N., Hedfalk, K., Larsson, C., Mostad, P., Chang, C., Gustafsson, L., and Bill, R. M. (2005). Design of improved membrane protein production experiments: Quantitation of the host response. *Protein Sci.* **14,** 1729–1740.

Christianson, T. W., Sikorski, R. S., Dante, M., Shero, J. H., and Hieter, P. (1992). Multifunctional yeast high-copy-number shuttle vectors. *Gene* **110,** 119–122.

Griffith, D. A., Delipala, C., Leadsham, J., Jarvis, S. M., and Oesterhelt, D. (2003). A novel yeast expression system for the overproduction of quality-controlled membrane proteins. *FEBS Lett.* **553,** 45–50.

Hays, F. A., Roe-Zurz, Z., Li, M., Kelly, L., Gruswitz, F., Sali, A., and Stroud, R. M. (2009). Ratiocinative screen of eukaryotic integral membrane protein expression and solubilization for structure determination. *J. Struct. Funct. Genomics* **10,** 9–16.

Kawate, T., and Gouaux, E. (2006). Fluorescence-detection size-exclusion chromatography for precrystallization screening of integral membrane proteins. *Structure* **14,** 673–681.

Li, M., Hays, F. A., Roe-Zurz, Z., Vuong, L., Kelly, L., Ho, C. M., Robbins, R. M., Pieper, U., O'Connell, J. D. 3rd, Miercke, L. J., *et al.* (2009). Selecting optimum eukaryotic integral membrane proteins for structure determination by rapid expression and solubilization screening. *J. Mol. Biol.* **385,** 820–830.

Mohanty, A. K., Simmons, C. R., and Wiener, M. C. (2003). Inhibition of tobacco etch virus protease activity by detergents. *Protein Expr. Purif.* **27,** 109–114.

Mumberg, D., Muller, R., and Funk, M. (1994). Regulatable promoters of *Saccharomyces cerevisiae*: Comparison of transcriptional activity and their use for heterologous expression. *Nucleic Acids Res.* **22,** 5767–5768.

Mumberg, D., Muller, R., and Funk, M. (1995). Yeast vectors for the controlled expression of heterologous proteins in different genetic backgrounds. *Gene* **156,** 119–122.

Newby, Z. E., O'Connell, J. D. 3rd, Gruswitz, F., Hays, F. A., Harries, W. E., Harwood, I. M., Ho, J. D., Lee, J. K., Savage, D. F., Miercke, L. J., *et al.* (2009). A general protocol for the crystallization of membrane proteins for X-ray structural investigation. *Nat. Protoc.* **4,** 619–637.

Piers, K. L., Heath, J. D., Liang, X., Stephens, K. M., and Nester, E. W. (1996). *Agrobacterium tumefaciens*-mediated transformation of yeast. *Proc. Natl. Acad. Sci. USA* **93,** 1613–1618.

White, S. H. (2009). Biophysical dissection of membrane proteins. *Nature* **459,** 344–346.

White, M. A., Clark, K. M., Grayhack, E. J., and Dumont, M. E. (2007). Characteristics affecting expression and solubilization of yeast membrane proteins. *J. Mol. Biol.* **365,** 621–636.

CHAPTER THIRTY

BIOCHEMICAL, CELL BIOLOGICAL, AND GENETIC ASSAYS TO ANALYZE AMYLOID AND PRION AGGREGATION IN YEAST

Simon Alberti,* Randal Halfmann,*,† and Susan Lindquist*,†,‡

Contents

Abstract

Protein aggregates are associated with a variety of debilitating human diseases, but they can have functional roles as well. Both pathological and nonpathological protein aggregates display tremendous diversity, with substantial differences in aggregate size, morphology, and structure. Among the different aggregation types, amyloids are particularly remarkable, because of their high degree of order and their ability to form self-perpetuating conformational states. Amyloids form the structural basis for a group of proteins called prions, which have the ability to generate new phenotypes by a simple switch in protein conformation that does not involve changes in the sequence of the DNA. Although protein aggregates are notoriously difficult to study, recent technological developments and, in particular, the use of yeast prions as model systems, have been very instrumental in understanding fundamental aspects of aggregation. Here, we provide a range of biochemical, cell biological and yeast genetic methods that are currently used in our laboratory to study protein aggregation and the formation of amyloids and prions.

* Whitehead Institute for Biomedical Research, Cambridge, Massachusetts, USA
† Department of Biology, Massachusetts Institute of Technology, Cambridge, Massachusetts, USA
‡ Howard Hughes Medical Institute, Cambridge, Massachusetts, USA

Methods in Enzymology, Volume 470
ISSN 0076-6879, DOI: 10.1016/S0076-6879(10)70030-6

1. Introduction

More than 40 years ago an unusual nonsense suppressor phenotype was reported that was inherited in a non-Mendelian manner (Cox, 1965). This phenotype, [PSI+], was later found to be caused by a change in the conformation of the translation termination factor Sup35p (Patino et al., 1996; Paushkin et al., 1996). Shortly after the discovery of [PSI+], a different genetic element, [URE3], was isolated (Lacroute, 1971) and the causal agent was afterward identified as a conformationally altered form of the nitrogen catabolite repressor Ure2p (Wickner, 1994). Ure2p and Sup35p are the founding members of an intriguing class of yeast proteins that can act as protein-based epigenetic elements. Also known as prions, these proteins can interconvert between at least two structurally and functionally distinct states, at least one of which adopts a self-propagating aggregated state. A switch to the aggregated prion state generates new phenotypic traits, which increase the phenotypic heterogeneity of yeast populations (Alberti et al., 2009; Shorter and Lindquist, 2005).

Sup35p is a translation termination factor. When Sup35p switches into a prion state, a large fraction of the cellular Sup35p is sequestered into insoluble aggregates. The resulting reduction in translation termination activity causes an increase in ribosomal frame-shifting (Namy et al., 2008; Park et al., 2009) and stop codon read-through (Liebman and Sherman, 1979; Patino et al., 1996; Paushkin et al., 1996). The sequences downstream of stop codons are highly variable, and this, in turn, facilitates the sudden generation of new phenotypes by the uncovering of previously hidden genetic variation (Eaglestone et al., 1999; True and Lindquist, 2000; True et al., 2004). The other well-understood prion protein, Ure2p, regulates nitrogen catabolism through its interaction with the transcriptional activator Gln3p. Its prion state, [URE3], causes the constitutive activation of Gln3p and consequently gives cells the ability to utilize poor nitrogen sources in the presence of a rich nitrogen source (Aigle and Lacroute, 1975; Wickner, 1994).

The epigenetic properties of Sup35p and Ure2p reside in structurally independent prion-forming domains (PrDs) with a strong compositional bias for residues such as glutamine and asparagine (Edskes et al., 1999; Li and Lindquist, 2000; Santoso et al., 2000; Sondheimer and Lindquist, 2000). This observation stimulated the first sequence-based query for additional yeast prions (Sondheimer and Lindquist, 2000) that ultimately lead to the identification of an additional prion protein called Rnq1p. The prion form of Rnq1p, [RNQ+], was found to underlie the previously characterized non-Mendelian trait [PIN+], which facilitates the de novo appearance of [URE3] and [PSI+] (Derkatch et al., 2000, 2001). Rnq1p, however, not only induces benign conformational transitions of other prion proteins, it can also induce toxic conformational changes of glutamine-expansion (polyQ) proteins that cause Huntington's disease and several other late-onset neurodegenerative disorders. When polyQ fragments are expressed in yeast, they create a polyQ length-dependent toxicity that is accompanied by the formation of visible amyloid-containing aggregates in a [RNQ+]-dependent manner (Duennwald et al., 2006a; Krobitsch and Lindquist, 2000). These features have allowed yeast prions to be widely used as model systems for the study of protein aggregation in vivo. They were tremendously useful for unraveling several aspects of protein aggregation, including the formation of structural variants, known as prion strains, or the presence of transmission barriers between related prion proteins (Tessier and Lindquist, 2009).

Protein aggregation is a highly complex process that is determined by intrinsic factors, such as the sequence and structure of the aggregation-prone protein, as well as extrinsic factors, such as temperature, salt concentration, and chaperones (Chiti and Dobson, 2006; Rousseau et al., 2006). A particular type of aggregation underlies the self-perpetuating properties of prions. All biochemically well-characterized yeast prions adopt an amyloid conformation. Amyloid is a highly ordered fibrillar aggregate. The fibril core is a continuous sheet of β-strands that are arranged perpendicularly to the fibril axis. The exposed β-strands at the ends of the fibril allow amyloids to polymerize by the continuous incorporation of polypeptides of the same primary sequence. This extraordinary self-templating ability allows prions to generate and multiply a transmissible conformational state (Caughey et al., 2009; Ross et al., 2005; Shorter and Lindquist, 2005). Prions can spontaneously switch to this transmissible state from a default conformational state that is usually soluble.

In a recent systematic attempt to discover new prions in yeast we used the unusual amino acid biases of known PrDs to predict novel prionogenic proteins in the Saccharomyces cerevisiae proteome. We subjected 100 such prion candidates to a range of genetic, cell biological, and biochemical assays to analyze their prion- and amyloid-forming propensities, and determined that at least 24 yeast proteins contain a prion-forming domain (Alberti et al., 2009). Our findings indicate that prions play a much broader role in yeast biology and support previous assumptions that prions buffer yeast populations against environmental changes. The fact that prions are abundant in yeast suggests that prions also exist in other organisms. Moreover, many more examples of functional aggregation, which do not involve an epigenetic mechanism for the inheritance of new traits, are likely to be discovered. The methods and techniques described here have vastly increased our understanding of protein aggregation in yeast. They will allow us to identify additional aggregation-prone proteins in yeast and other organisms and will advance our understanding of the pathological and nonpathological functions of aggregates.

2. METHODS

2.1. Detecting protein aggregation in yeast cells

Protein aggregates are formed when large numbers of polypeptides cooper-
ate to form nonnative molecular assemblies. These structures are highly
diverse, with differences in the amount of β-sheet content, their overall
supramolecular organization and their ability to induce the coaggregation of
other proteins (Chiti and Dobson, 2006; Rousseau *et al.*, 2006). Notwith-
standing the multifactorial nature and complexity of protein aggregation,
protein aggregates can simplistically be classified as disordered (amorphous)
or ordered (amyloid-like). Amorphous aggregates are generally not well
characterized due to their tremendous structural plasticity. Recent studies,
however, indicate that the constituent proteins of some amorphous aggregates have a conformation that is similar to their native structure in solution
(Qin *et al.*, 2007; Vetri *et al.*, 2007). Ordered aggregates, on the other hand,
contain greater amounts of β-sheet content and form densely packed
amyloid fibers. Amyloid-like aggregates can be distinguished from disordered aggregates based on their resistance to physical and chemical perturbations that affect protein structure, such as increased temperature, ionic
detergents or chaotropes. In the following section, we provide a variety of
methods for the analysis of diverse protein aggregates in yeast cells.

2.1.1. Fluorescence microscopy and staining of amyloid-like aggregates

Aggregating proteins coalesce into microscopic assemblies that can be
visualized by fluorescence microscopy. This very convenient method for
following aggregation in cells has greatly expanded our understanding of
protein misfolding diseases and other nonpathological aggregation-based
phenomena such as prions (Garcia-Mata *et al.*, 1999; Johnston *et al.*, 1998;
Kaganovich *et al.*, 2008). Two different experimental approaches are available for the direct visualization of protein aggregates in cells: (1) The
aggregate-containing cells can be fixed and treated with an antibody specific
to the aggregation-prone protein, or (2) the aggregation-prone protein can
be expressed as a chimera with a fluorescent protein (FP). Either approach
has both advantages and disadvantages. Immunofluorescence microscopy
requires an extensive characterization of the antibody to determine its
specificity and to ensure that it is able to recognize the aggregated state of
the protein. The benefit, however, is that the aggregation-prone protein
can be expressed unaltered. In fact, tagging of an aggregation-prone protein
with an FP can severely interfere with its aggregation propensity by changing its overall solubility. The FP tag could also sterically interfere with the

formation of the amyloid structure. Moreover, some FPs are known to self-interact. Particularly problematic in this regard are older versions of DsRed, although we have observed that in rare cases GFP fusions can also cause spurious aggregation. Generally, GFP-driven aggregation does not react with thioflavin T (ThT) (discussed below) and forms a characteristic well-defined high molecular weight species when analyzed by semidenaturing detergent–agarose gel electrophoresis (SDD–AGE) (described in Section 2.1.3). Experiments with FP chimeras, therefore, require some caution in the interpretation of the results and we recommend performing additional assays to independently establish whether a protein is aggregation-prone or not.

Despite these drawbacks, using FP-chimeras to study protein aggregation has several advantages, particularly cost-effectiveness, ease of use, and rapid generation of results. Diverse yeast expression vectors are now available for the tagging of aggregation-prone proteins with FPs, most of which are suitable for determining the aggregation propensities of a protein (Alberti et al., 2007, 2009; Duennwald et al., 2006a; Krobitsch and Lindquist, 2000). The choice of the promoter and the copy number of the yeast plasmid are also important parameters that need to be considered carefully. Expression from high copy 2-μm plasmids is highly variable, thus allowing the sampling of a range of different protein concentrations. This property is desirable if the goal is to determine whether a protein is generally able to nucleate and enter an amyloid-like state in a cellular environment. Low-copy CEN-based plasmids and expression cassettes for integration into the genome, however, have more uniform expression levels, a property which can be useful if more consistent aggregation behavior is desired. We usually try to avoid constitutive promoters, as aggregation can be associated with toxicity and frequently triggers growth arrest or cell death. Inducible promoters like GAL1 are more suitable and transient expression for 6–24 h is usually sufficient to induce aggregation in a significant fraction of the cell population.

Aggregation-prone proteins form foci that can be visualized by fluorescence microscopy. Yeast cells expressing aggregation-prone proteins form two characteristic types of fluorescent foci (Fig. 30.1A): (1) ring-like structures that localize to the vacuole and/or just below the plasma membrane and (2) punctate structures that can be distributed all over the cytoplasm but preferentially reside close to the vacuole (Alberti et al., 2009; Derkatch et al., 2001; Ganusova et al., 2006; Taneja et al., 2007; Zhou et al., 2001). The fibrillar appearance of ring-like structures and their reactivity with amyloid-specific dyes suggests that they consist of laterally associated amyloid fibers. Punctate foci, on the other hand, do not always stain with amyloid-binding dyes, suggesting that these structures can also be of the nonamyloid or amorphous type (Douglas et al., 2008). The two types of aggregation can

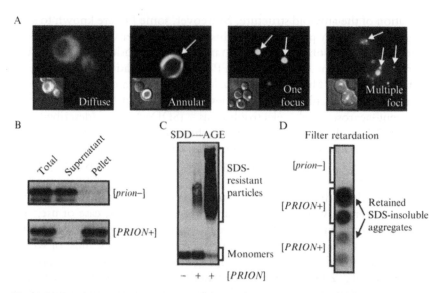

Figure 30.1 Diverse biochemical assays used to detect protein aggregation in yeast cell lysates. (A) Fluorescence microscopy was used to identify cellular aggregation of proteins that are expressed as fusions to GFP. GFP alone (left) is equally distributed throughout the cytosol and nucleus. Aggregation-prone proteins show annular or punctate fluorescent foci, resulting from the tight packing of amyloid fibrils in the cytosol. (B) Amyloid-containing fractions were isolated from yeast cell lysates by sedimentation in SDS-containing lysis buffer. The prion protein is detected in the total lysate, the SDS-soluble supernatant and the SDS-insoluble pellet fraction by immunoblotting with a specific antibody. (C) Lysates of yeast cells expressing a GFP-tagged prion protein were analyzed by SDD–AGE and Western blotting. The prion protein was detected by immunoblotting with an anti-GFP antibody. (D) [*prion*−] and [*PRION*+] cell lysates were subjected to a filter retardation assay. Aggregates retained on the membrane were detected by immunoblotting.

be distinguished based on their fluorescence intensity and their pattern of aggregation. Rings and large punctate foci with very bright fluorescence are indicative of highly ordered amyloid fibers, whereas multiple small puncta with low brightness usually do not react with amyloid-specific dyes and are therefore of the amorphous type.

To conclusively determine whether these foci result from amorphous or amyloid-like aggregation, we use a staining protocol with the amyloid-specific dye ThT. ThT has an emission spectrum that is red-shifted upon amyloid-binding, therefore, allowing colocalization with aggregation-prone proteins that are tagged with yellow FP. However, staining yeast cells with ThT can lead to high background levels and thus we recommend performing ThT costaining experiments only for proteins with relatively high expression levels. To grow the yeast cells for staining with ThT, a culture is inoculated in the appropriate selective medium for overnight growth, followed by dilution and regrowth until it reaches an OD_{600} of 0.25–1.0. Then 8 ml of the culture are transferred onto a 150-ml Nalgene bottle-top filter (45 mm diameter, 0.2 μM pores, SFCA membrane) and the solution is filtered by applying a vacuum. When the solution has passed through the filter, the vacuum is halted and 5 ml of freshly prepared fixing solution (50 mM H_2KPO_4, pH 6.5; 1 mM $MgCl_2$; 4% formaldehyde) is added to the cells on the filter. The cells are resuspended by swirling and the suspension is then transferred to a 15-ml tube. The cells are incubated at room temperature for 2 h and vortexed briefly every 30 min.

The fixed cells are collected by a brief centrifugation step (2 min at 2000 rcf) and the supernatant is removed carefully and completely. The cells are then resuspended in 5 ml buffer PM prepared freshly (0.1 M H_2KPO_4, pH 7.5; 1 mM $MgCl_2$) and collected again by centrifugation. After the supernatant is removed completely the cells are resuspended in buffer PMST (0.1 M H_2KPO_4, pH 7.5; 1 mM $MgCl_2$; 1 M sorbitol; 0.1% Tween 20; containing 1× EDTA protease inhibitor mix from Roche). The volume of the PMST buffer should be adjusted to generate a final cell density of 10 OD_{600}. One hundred microliters of the cell suspension is then transferred to a 0.5-ml Eppendorf tube. 0.6 μl of β-mercaptoethanol and 20 μl of 20,000 U/ml yeast lytic enzyme (ICN, or use zymolyase at 1 mg/ml) are added and the spheroplasted cells are incubated on a rotating wheel at room temperature for 15 min for spheroplasting. The spheroplasted cells are then resuspended gently in 100 μl PMST, collected by centrifugation and the resuspension step is repeated once. Subsequently, the cells are incubated in PBS (pH 7.4) containing 0.001% ThT for 20 min, washed three times with PBS and then used immediately for fluorescence microscopy.

The pattern of aggregation is protein-specific, dependent on the level and duration of expression and regulated by the physiological state of the cell. Many yeast prion proteins, for instance, proceed through a maturation pathway that includes an early stage with ring-like aggregation patterns and a later stage with punctate cytoplasmic distribution (Alberti et al., 2009). Other amyloidogenic proteins such as glutamine-expanded versions of huntingtin exon 1 almost exclusively form punctate foci, when expressed at comparable levels. Interestingly, toxic and nontoxic aggregates of glutamine-expanded huntingtin have distinct subcellular aggregation patterns. Toxic huntingtin forms multiple punctuate foci, whereas the nontoxic structural variant is present in a single cytosolic focus (Duennwald et al., 2006a,b).

Yeast cells have two different aggresome-like compartments to deal with aggregation-prone proteins such as huntingtin (Kaganovich et al., 2008).

These compartments, called JUNQ and IPOD, contain predominantly soluble or insoluble misfolded proteins, respectively. The JUNQ is believed to provide a subcellular location for the proteasome-dependent degradation of misfolded proteins, whereas the IPOD is enriched for chaperones like Hsp104 and seems to be a place for the sequestration of insoluble protein aggregates. Targeting to either of the compartments most likely influences the localization of aggregation-prone proteins, but the mechanisms of targeting and the factors involved in the maintenance of the compartments remain to be determined.

2.1.2. Sedimentation assay

In addition to their reactivity with amyloid-specific dyes like ThT, other criteria can be used experimentally to determine whether intracellular aggregates are amyloid-like, such as their unusual resistance to chemical solubilization. The detergent insolubility of amyloids has been used to isolate amyloid-containing fractions from yeast cell lysate by centrifugation (Bradley *et al.*, 2002; Sondheimer and Lindquist, 2000). In a typical amyloid sedimentation experiment 10 ml of yeast cells are grown to mid-logarithmic phase. The cells are collected by centrifugation and washed in water. The cell pellet is then resuspended in 300 μl of lysis buffer (50 mM Tris, pH 7.5, 150 mM NaCl, 2 mM EDTA, 5% glycerol). To inhibit proteolysis we supplement the lysis buffer with 1 mM phenylmethylsulphonyl fluoride (PMSF), 50 mM N-ethylmaleimide (NEM) and 1$\times$ EDTA-free protease inhibitor mix (Roche). The suspension is transferred to a 1.5-ml Eppendorf tube containing 300 μl of 0.5 mm glass beads. The cells are then lysed using a bead beater at 4 °C and are immediately placed on ice. Three hundred microliters of cold RIPA buffer (50 mM Tris, pH 7.0, 150 mM NaCl, 1% Triton X-100, 0.5% deoxycholate, 0.1% SDS) is added and the lysate is vortexed for 10 s. Subsequently, the crude lysate is centrifuged for 2 min at 800 rcf (4 °C) to pellet the cell debris. The sedimentation assay is performed by centrifuging 200 μl of the supernatant in a TLA 100-2 rotor for 30 min at 80,000 rpm and 4 °C using an Optima TL Beckman ultracentrifuge. Equal volumes of unfractionated (total) and supernatant samples are incubated in sample buffer containing 2% SDS and 2% β-mercaptoethanol at 95 °C for 5 min. The pellet is resuspended in 200 μl of a 1:1 mixture of lysis buffer and RIPA buffer containing protease inhibitors and boiled in sample buffer under the same conditions described above. The samples are then analyzed by SDS–PAGE and immunoblotting with an antibody specific to the aggregation-prone protein. If a putative prion is analyzed, it should predominantly be detectable in the supernatant of prion-free cells and in the pellet fraction of prion-containing cells (e.g., see Fig. 30.1B).

2.1.3. Semidenaturing detergent–agarose gel electrophoresis

The recent invention of SDD–AGE very conveniently allows the resolution of amyloid polymers based on size and insolubility in detergent (e.g., see Fig. 30.1C) (Bagriantsev et al., 2006). We adapted SDD–AGE for large-scale applications, allowing simultaneous detection of SDS-insoluble conformers of tagged proteins in a large number of samples (Halfmann and Lindquist, 2008). This advanced version of SDD–AGE enables one to perform high-throughput screens for novel prions and other amyloidogenic proteins.

As a first step, it is necessary to cast the detergent-containing agarose gel. Standard equipment for horizontal DNA electrophoresis can be used and the size of the gel casting tray and the comb should be adjusted according to the number and volume of the samples. We usually prepare a 1.5% agarose solution (medium gel-strength, low EEO) in 1× TAE. The agarose solution is heated in a microwave until the agarose is completely dissolved. Subsequently, SDS is added to 0.1% from a 10% stock. The agarose solution is then poured into the casting tray. After the gel has set, the comb is removed and the gel is placed into the gel tank. The gel is then completely submerged in 1× TAE containing 0.1% SDS.

The following lysis procedure is optimized for large numbers of small cultures processed in parallel, although it can be easily modified for individual cultures of larger volume. For high-throughput analysis of yeast lysates we use 2 ml cultures grown overnight with rapid agitation in 96-well blocks. The cells are harvested by centrifugation at 2000 rcf for 5 min and then resuspended in water. After an additional centrifugation step and removal of the supernatant, the cells are resuspended in 250 μl spheroplasting solution (1.2 M D-sorbitol, 0.5 mM MgCl$_2$, 20 mM Tris, pH 7.5, 50 mM β-mercaptoethanol, 0.5 mg/ml zymolyase, 100T) and incubated for 1 h at 30 °C. The spheroplasted cells are collected by centrifugation at 800 rcf for 5 min and the supernatant is removed completely. The pelleted spheroplasts are then resuspended in 60 μl lysis buffer (20 mM Tris, pH 7.5, 10 mM β-mercaptoethanol, 0.025 U/μl benzonase, 0.5% Triton X-100, 2× HALT protease inhibitor from Sigma-Aldrich). The 96-well block is covered with tape and vortexed at high speed for 1 min and then incubated for an additional 10 min at room temperature. The cellular debris is sedimented by centrifugation at 4000 rcf for 2 min and the supernatant is carefully transferred to a 96-well plate. As a next step, 4× sample buffer (2× TAE; 20% glycerol; 8% SDS; bromophenol blue to preference) is added to a final concentration of 1×, followed by brief vortexing to mix.

In SDS-containing buffers amyloid-like aggregates are stable at room temperature, but can be disrupted by boiling. Therefore, samples are incubated for an additional 10 min at room temperature, or, as a negative control, incubated at 95 °C. Most amyloids will be restored to monomers by the 95 °C treatment. The samples are then loaded onto the agarose gel. We usually also load one lane with prestained SDS–PAGE marker, enabling us to verify proper transfer and to estimate the size of unpolymerized SDS-soluble conformers. In addition, it is important to use protein aggregation standards. [psi−] and [PSI+] cell lysates or lysates of yeast cells overexpressing the huntingtin length variants Q25 (nonamyloid) and Q103 (amyloid) can be used for this purpose. The electrophoresis is performed at low voltage (≤ 3 V/cm gel length) until the dye front reaches ~ 1 cm from the end of the gel. It is important that the gel remains cool during the run since elevated temperatures can reduce the resolution.

For the blotting procedure, we prefer a simple capillary transfer using a dry stack of paper towels for absorption. One piece of nitrocellulose and eight pieces of GB002 blotting paper (or an equivalent substitute) are cut to the same dimensions as the gel. An additional piece of GB002, which serves as the wick, is cut to be about 20 cm wider than the gel. The nitrocellulose, wick, and four pieces of GB002 are immersed in $1\times$ TBS (0.1 M Tris–HCl, pH 7.5, 0.15 M sodium chloride). A stack of dry folded paper towels is assembled that is about 2 cm thick and the same length and width of the gel. On top of the stack of paper towels four pieces of dry GB002 are placed, then one piece of wet GB002, and finally the wet nitrocellulose. The gel in the casting tray is briefly rinsed in water to remove excess running buffer. It is then carefully moved from the tray onto the stack. We recommend adding extra buffer on the nitrocellulose to prevent bubbles from becoming trapped under the gel. The remaining three prewetted GB002 pieces are placed on top of the gel. To ensure thorough contact between all layers, a pipette should be rolled firmly across the top of the stack. The transfer stack is subsequently flanked with two elevated trays containing $1\times$ TBS and the prewetted wick is draped across the stack such that either end of the wick is submerged in $1\times$ TBS. Finally, the assembled transfer stack is covered with an additional plastic tray bearing extra weight (e.g., a small bottle of water) to ensure proper contact between all layers of the stack. The transfer should proceed for a minimum of 3 h, although we generally transfer over night. After the transfer the membrane can be processed by standard immunodetection procedures.

2.1.4. Filter retardation assay using yeast protein lysates

Another convenient method for analyzing aggregation is the size-dependent retention of aggregates on nonbinding membranes (Fig. 30.1D). This assay was initially developed to investigate amyloid formation of huntingtin in *in vitro* aggregation assays (Scherzinger et al., 1997), but it can also be used to detect aggregates in yeast cell lysates. Cells should be processed as for SDD–AGE (Section 2.1.3), except that the sample buffer is omitted. Instead, the lysates are treated with the desired detergent- or chaotrope-containing buffer. Generally, we use SDS at 0.1–2%. Samples are incubated at room temperature for 10 min, or, for a negative control, boiled in 2% SDS.

During this incubation period the vacuum manifold is prepared. First, a thin filter paper (GB002) that is soaked in water is placed on the manifold. Then, a cellulose acetate membrane (pore size 0.2 μm) is soaked in PBS containing 0.1% SDS and placed on top of the filter paper on the manifold. The manifold is closed and the samples are loaded into the wells of the manifold. The samples are filtered through the membrane by applying a vacuum and the membrane is washed five times with PBS containing 0.1% SDS. The cellulose acetate membrane can then be used for immunodetection with a protein-specific antibody. To demonstrate that equal amounts of protein were present in the samples, the same procedure can be repeated with a protein-binding nitrocellulose membrane. As with SDD–AGE, it is important for filter retardation experiments to include protein aggregation standards, such as [psi−] and [PSI+] cell lysates or lysates of yeast cells overexpressing the huntingtin length variants Q25 and Q103.

2.2. Assays for prion behavior

Prions are amyloids that are transmissible. Fragments of amyloid fibrils can be passed between cells or organisms, and the self-templating ability of amyloid results in the amplification of the structure, giving prions an infectious property. The prion properties of yeast prions reside in structurally independent PrDs. The PrDs of Sup35p and other prions are modular and can be fused to nonprion proteins, thereby creating new protein-based elements of inheritance (Li and Lindquist, 2000). We employ two assays which exploit this property of prions to experimentally determine whether a predicted PrD can confer a heritable switch in the function of a protein. The assays are based on the well-characterized prion phenotypes of the translation termination factor Sup35p and the nitrogen catabolite regulator Ure2p.

2.2.1. Sup35p-based prion assay

Sup35p consists of an N-terminal PrD (N), a highly charged middle domain (M) and a C-terminal domain (C), which functions in translation termination. Both the N and M domains are dispensable for the essential function of Sup35p in translation termination. The charged M domain serves to increase the solubility of the amyloid-forming N domain, thereby promoting the conformational bistability of the Sup35p protein. In the prion state, a large fraction of the cellular Sup35p is sequestered into insoluble aggregates, resulting in reduced translation termination activity and an increase in the read-through of stop codons. Premature stop codons in genes of the adenine synthesis pathway, which are present in lab strains such as 74D-694, provide a convenient way to monitor switching of Sup35p to the prion state

(Fig. 30.2A). In prion-free [*psi*−] cells, translation termination fidelity is high, leading to the production of truncated and nonfunctional Ade1p. As a consequence the [*psi*−] cells are unable to grow on adenine-free medium and accumulate a by-product of the adenine synthesis pathway that confers a red colony color when grown on other media such as YPD. Prion-containing [*PSI*+] cells, on the other hand, have a reduced translation termination activity that allows read-through of the *ade1* nonsense allele and the production of functional full-length Ade1p protein, resulting in growth on adenine-deficient medium and the expression of a white colony color.

The modular nature of prion domains enables the generation of chimeras between the C or MC domain of Sup35p and candidate PrDs that can then be tested for their ability to generate [*PSI*+]-like states (Fig. 30.2A). For several years plasmids have been available that allow the cloning of a candidate PrD N terminal to the C or MC domain of Sup35p (Osherovich and Weissman, 2001; Sondheimer and Lindquist, 2000). These plasmids can be integrated into the yeast genome to replace the endogenous *SUP35* gene. This procedure, however, is very laborious and has a low success rate, as the resulting strains frequently express the chimera at levels much lower than wild-type Sup35p. As a consequence, these strains aberrantly display a white colony color and show constitutive growth on medium lacking adenine, preventing their use in prion selection assays.

To overcome these difficulties, we recently developed a yeast strain (YRS100) in which a deletion of the chromosomal *SUP35* is covered by a Sup35p-expressing plasmid (Alberti *et al.*, 2009). When these cells are

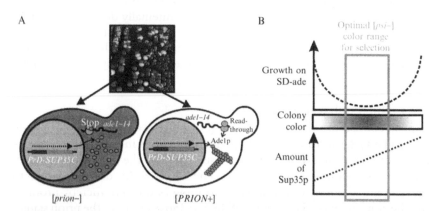

Figure 30.2 Using Sup35p to detect phenotype switching behavior. (A) Schematic representation of the Sup35p-based prion assay. The PrD of Sup35p is replaced with a candidate PrD and the resulting strains are tested for the presence of prion phenotypes, such as the switching between a red and a white colony color. (B) Relationship between the cellular concentration of Sup35p, the colony color in the [*psi*−] state and the ability to grow on synthetic medium lacking adenine. (See Color Insert.)

transformed with expression plasmids for *PrD-SUP35C* chimeras, a *URA3* marker on the covering *SUP35* plasmid allows it to be selected against in 5-FOA-containing medium (plasmid shuffle). The resulting strains contain PrD-Sup35C fusions as their only source of functional Sup35p. This plasmid-based expression system is more versatile than the previous versions, as it easily allows the use of different promoters and therefore a better control over the expression level of PrD-Sup35 proteins. We have generated vectors with four different promoters, which have the following relative strength of expression: *SUP35* < *TEF2* < *ADH1* < *GPD*. To minimize the time needed for strain generation, these vectors enable recombination-based cloning using the Gateway® system. In recent years, the Gateway® system has emerged as a very powerful cloning method that allows for the rapid *in vitro* recombination of candidate genes into diverse sets of expression vectors. We have generated hundreds of Gateway®-compatible yeast expression vectors, each with a different promoter, selectable marker or tag (Alberti *et al.*, 2007, 2009). This technological improvement now allows us to perform high-throughput testing of candidate PrD libraries for prion properties. All Gateway-compatible plasmids described here are available through the nonprofit plasmid repository Addgene (www.addgene.org).

The *SUP35C* and related fusion assays were instrumental in identifying new prions and prion candidates (Alberti *et al.*, 2009; Nemecek *et al.*, 2009; Osherovich and Weissman, 2001; Sondheimer and Lindquist, 2000). However, it can be difficult to work with PrD-Sup35 chimeras and it is therefore important to know about the shortcomings of this assay. Whether a Sup35p-based prion selection experiment will be feasible or not critically depends on the expression level of the PrD-Sup35C fusion protein and its functional activity in translation termination. Too low or too high levels of active Sup35p result in permanently elevated levels of stop codon read-through, with corresponding [*prion−*] strains that are able to grow on adenine-deficient medium and a colony color that is shifted to white (Fig. 30.2B). It is, therefore, important to find the appropriate window of expression to generate strains with adequate translation termination fidelity. In our lab, we obtained the best expression results with the *ADH1* promoter.

In general, a sufficient level of translation termination activity is indicated by a light to dark red colony color. Strains with pink or white colony colors usually have too high levels of translation termination activity and can thus not be used in prion selection assays that test for the ability to grow on adenine-deficient synthetic media (Fig. 30.2B). In rarer cases it is possible that a high level of read-through is caused by constitutive aggregation of the PrD and not by insufficient expression levels. To rule out that the PrD-Sup35C chimera is already present in an aggregated state, an SDD–AGE followed by immunoblotting with an anti-Sup35p antibody should be performed. It is also important to point out that the size of the PrD that is fused to Sup35p is a key factor that determines functionality of the chimera. Based on our experience, PrDs between 60 and 250 amino acids are well tolerated. PrDs above this threshold, however, tend to inhibit the translation termination activity of Sup35p. In some cases, it can be important to include a solubilizing domain such as the M domain between the PrD and the C domain, as the presence of M could slow down the aggregation kinetics, a property that is particularly desirable if a protein is very aggregation-prone.

To generate a PrD-Sup35C-expressing strain, we introduce the corresponding expression plasmid into the YRS100 strain and select the transformants on appropriate selective plates. The transformants are isolated, grown in liquid medium for a few hours and then plated on 5-FOA plates to counterselect the covering *SUP35* plasmid. Colonies growing on 5-FOA plates are streaked on YPglycerol to select for cells with functional mitochondria and eliminate petite mutants that change the colony color to white. Subsequently, the cells are transferred onto YPD plates to assess the colony color phenotype. In rare cases, strain isolates expressing the same construct can vary in colony color. We therefore recommend isolating a number of different colonies and using the isolates with the predominant colony colors for subsequent experiments.

At this stage, some strains might already show switching between a red and a white colony color on YPD plates, indicating that the PrD under investigation has prion properties. The strains can then be plated on SD medium lacking adenine to more thoroughly select for the prion state. However, switching rates of prions can be as low as 10^{-6}–10^{-7}. Therefore, in cases where spontaneous switching is not observed, conformational conversion to the prion state should be induced. Amyloid nucleates in a concentration-dependent manner. Thus, transient overexpression of the PrD can be used to increase the switching frequency of a prion. To do this, we usually introduce an additional plasmid for expression of a PrD-EYFP fusion under the control of a galactose-inducible promoter. Induction of the prion state is achieved by growing the resulting transformants with the *GAL1* plasmid in galactose-containing medium for 24 h. The cells are then plated on YPD and SD-ade plates at a density of 200 and 50,000 per plate, respectively. The same strains grown in raffinose serve as a control. A greater number of Ade$^+$ colonies under inducing conditions suggest that expression of cPrD-EYFP induced a prion switch. In these cases, the Ade$^+$ colonies should be streaked on YPD plates to determine if a colony color change from red to white or pink has occurred. Often several colony color variants can be observed in one particular *PrD-SUP35C* strain. This variation could be due to the presence of weak and strong prion variants, or "strains," that have been reported previously for other prions (Tessier and Lindquist, 2009). A single prion protein can generate multiple variants that differ in the strength of their prion phenotypes. The underlying basis for this phenomenon is the presence of initial structural differences in the amyloid-forming nucleus that are amplified and maintained through the faithful self-templating mechanism of amyloids.

After having isolated several putative prion strains that exhibit a change in colony color, it is important to determine whether these changes are based on a conformational conversion of the PrD-Sup35C protein. Known yeast prions critically depend on the chaperone disaggregase Hsp104p for propagation (Ross *et al.*, 2005; Shorter and Lindquist, 2005). Thus, deletion of the *HSP104* gene or repeated streaking of the putative prion strains on YPD plates containing 5 m*M* of the Hsp104p inhibitor guanidinium hydrochloride (GdnHCl) are convenient ways of testing whether a color change is due to a prion switch. However, as some prion variants can propagate in the absence of Hsp104p, we suggest testing those that are resistant to Hsp104p inactivation for the presence of aggregated PrD-Sup35C by SDD–AGE and immunoblotting with a Sup35p-specific antibody. We found that many strains are able to switch to a prion-like state that is not based on a conformational change in the PrD-Sup35 protein, but involve other genetic or epigenetic changes of unknown origin. To rule out these false positive candidates, it is very important to rigorously test putative prion strains for the presence of conformationally altered PrDs.

2.2.2. Ure2p-based prion assay

Ure2p is a 354-amino acid protein consisting of an N-terminal PrD and a globular C-terminal region. The C-terminal region of Ure2p shows structural similarity to glutathione transferases and is necessary and sufficient for its regulatory function. Ure2p regulates nitrogen catabolism through its interaction with the transcriptional activator Gln3p. [*URE3*], the prion state of Ure2p, results in the constitutive activation of Gln3p, and the prion-containing cells acquire the ability to utilize poor nitrogen sources in the presence of a rich nitrogen source. One of the genes activated by Gln3p is the *DAL5* gene. It codes for a permease that is able to transport ureidosuccinate (USA), an essential intermediate of uracil biosynthesis. This ability to take up USA has historically been used to monitor the presence of [*URE3*] (Wickner, 1994).

The N-terminal region of Ure2p is required for its prion properties *in vivo* and deletion of the N-terminal region has no detectable effect on the stability or folding of the C-terminal functional part of the protein. Analogous to the Sup35p assay described in the previous section, the Ure2p PrD can be replaced with a candidate PrD, and the resulting chimera can then be tested for its ability to create heritable phenotypes that mimic [*URE3*] (Nemecek *et al.*, 2009). Assaying [*URE3*] by selection on USA-containing plates has several disadvantages and for this reason we use a strain that contains an *ADE2* reporter gene that is placed under the control of the *DAL5* promoter (Brachmann *et al.*, 2006). In prion-free [*ure-o*] cells, the

ADE2 gene expression is repressed and as no functional Ade2p protein is produced, colonies are red and fail to grow on medium lacking adenine. Derepression of the *DAL5* promoter in [*URE3*] cells, however, results in a switch to a white colony color and the ability to grow on adenine-free medium (Fig. 30.3A).

We have also developed a yeast strain in which the chromosomal copy of the *URE2* gene was deleted in the BY334 background (Brachmann *et al.*, 2006). This strain was fully complemented by transformation of an expression plasmid for Ure2p. In order to use this strain for the detection of novel prions, we generated Gateway® vectors for the formation of chimeras

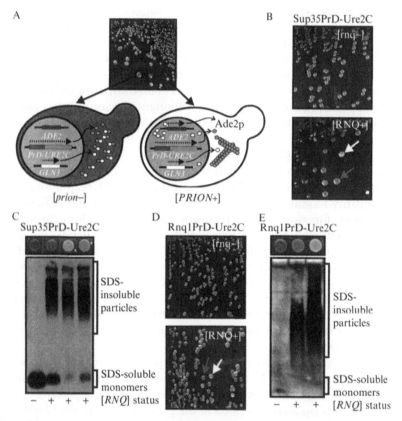

Figure 30.3 Using Ure2p to detect phenotype switching. (A) Schematic representation of a prion detection assay based on Ure2p. See text for further details. (B) A chimera between the PrD of Sup35p and the C-terminal region of Ure2p shows switching behavior in [*RNQ+*] cells. (C) SDD–AGE of different colony color isolates from a Sup35PrD-Ure2C strain. (D) A chimera between the Rnq1p PrD and Ure2C displays colony color switching in [*RNQ+*] cells. (E) SDD–AGE of different colony color isolates from the plates shown in (D).

between a candidate PrD and Ure2C (amino acids 66–354). The C terminus of Ure2p contains an HA tag that allows the detection of PrD-Ure2C chimeras with HA-specific antibodies. In addition, these vectors are available with three different promoters that can be used to express PrD-Ure2C chimeras at different levels (*TEF2* < *ADH1* < *GPD*). Using these vectors, we tested the PrDs of two well-characterized yeast prions, Sup35p and Rnq1p, for their ability to undergo prion switching when fused to the C-terminal domain of Ure2p. A chimera between the PrD domain of Sup35p and the C-terminal domain of Ure2p showed prion switching and formed weak and strong prion strains (Fig. 30.3B and C). A fusion between the PrD of Rnq1p and Ure2C also behaved as a prion, with at least two different color variants (Fig. 30.3D and E). Additionally, we found that even the full-length Rnq1p protein, when fused to Ure2C, resulted in a fully functional chimera that showed prion-dependent inactivation (data not shown). This finding suggests that the C domain of Ure2p is much more tolerant of larger PrDs than the C domain of Sup35p. We tested additional previously described PrDs for their ability to induce switching when fused to Ure2C and we found that many of these showed prion switching behavior (data not shown).

Despite its usefulness as a tool for the detection of prion switching behavior, there are disadvantages associated with the Ure2C-based selection system. The selection for the prion state of a PrD-Ure2C chimera on adenine-deficient media is usually very difficult due to high levels of background growth. Therefore, in those cases in which a prion state cannot be isolated by selection, we suggest identifying prion-containing strains based on colony color changes on media containing adenine. We noticed that the PrD-Ure2C fusions readily enter an aggregated state, which is probably due to the fact that the C domain of Ure2p does not have a solubilizing effect as strong as the C domain of Sup35p. In many cases, the high switching rates of PrD-Ure2C fusions allowed us to readily isolate several prion-containing strains from a single plate. Again, it is important to establish that the putative prion phenotypes are based on a conformationally altered state of the PrD-Ure2C chimera. The methods that are most convenient for testing are repeated streaking on plates containing GdnHCl or SDD–AGE followed by immunoblotting with an HA-specific antibody.

2.2.3. Prion selection assays

Although yeast prions described to date cause a loss-of-function when in the prion state, prion switches could also induce a gain-of-function phenotype, as has been described for the CPEB protein that is involved in long-term memory formation (Si *et al.*, 2003). To rigorously establish that a candidate protein operates as a prion in a physiologically relevant manner, robust prion selection assays have to be applied to isolate a prion-containing strain. The functional annotation of the yeast genome is a tremendously powerful resource for unraveling the biology of putative prions. A wealth of data from genome-wide deletion and overexpression screens as well as chemical and phenotypic profiling studies is now available to design functional prion assays (Cooper *et al.*, 2006; Hillenmeyer *et al.*, 2008; Sopko *et al.*, 2006; Zhu *et al.*, 2003).

The transcriptional activity of a putative prionogenic transcription factor, for example, can be monitored in a transcriptional reporter assay. We recently used such an approach to verify the prion properties of Mot3p (Alberti *et al.*, 2009). Mot3p is a globally acting transcription factor that modulates a variety of processes, including mating, carbon metabolism, and stress response. It tightly represses anaerobic genes, including *ANB1* and *DAN1*, during aerobic growth. To analyze Mot3p transcriptional activity, we created Mot3p-controlled auxotrophies by replacing the *ANB1* or *DAN1* ORFs with *URA3*. The resulting strains could not grow without supplemental uracil due to the Mot3p-mediated repression of *URA3* expression. However, *URA3* expression and uracil-free growth could be restored upon reduction of Mot3p activity by deletion of *MOT3* or by inactivation via prion formation.

To select for the Mot3p prion state, we transiently overexpressed the Mot3p PrD, via a galactose-inducible expression plasmid, and plated the cells onto media lacking uracil. We isolated Ura + strains whose phenotype persisted even after the inducing plasmid had been lost. These putative prion strains were then analyzed using a variety of prion tests, including curing by inactivation of Hsp104, testing for non–Mendelian inheritance in mating and meiotic segregation experiments and probing for an aggregated form of Mot3p in prion-containing cells. Candidate-tailored prion selection assays analogous to the one developed for Mot3p allow the study of prion in their natural context, thus providing valuable insights into the biological functions of prion conformational switches.

2.2.4. Methods to analyze prion inheritance

Yeast prions are protein-based epigenetic elements that are inherited in a non-Mendelian manner. The unusual genetic properties of yeast prions can be used to determine whether a phenotype is based on a prion (Ross *et al.*, 2005; Shorter and Lindquist, 2005). Diploid cells that result from a mating between [*prion*−] and [*PRION*+] cells are usually uniformly [*PRION*+]. The [*PRION*+] phenotype emerges from the self-perpetuating nature of prions: when a prion-containing cell fuses with a prion-free cell, the efficient prion replication mechanism rapidly consumes nonprion conformers until a new equilibrium is reached that is shifted in favor of the prion conformer. Diploid [*PRION*+] cells that undergo meiosis and sporulation normally generate [*PRION*+] progeny with a 4:0 inheritance pattern. However, stochastic deviations from the 4:0 pattern are possible, if the number of prion-replicating units or propagons is low, causing some progeny to receive no propagons and to correspondingly lose the prion state.

To establish whether a putative prion trait is inherited in a "protein only" cytoplasm-based manner through the cytoplasm, cytoduction can be performed. Cytoduction is an abortive mating in which the cytoplasms of a prion-free recipient cells and a prion-containing donor cell mix without fusion of the nuclei. Daughter cells with mixed cytoplasm but only one nucleus bud off from the zygote and can be selected (Conde and Fink, 1976). As a result, donor cells only contribute cytoplasm, whereas recipient cells contribute cytoplasm and nucleus to the progeny. The recipient cells we use are karyogamy-deficient and carry a mitochondrial petite mutation termed rho^0. As a consequence, recipient cells cannot fuse their nuclei with those of the donor cell and are unable to grow on a nonfermentable carbon source, such as glycerol, unless they receive wild-type mitochondria from the donor cytoplasm. Following cytoduction, haploid progeny are selected that retained the nuclear markers of the recipient strain but can also grow in medium containing glycerol. If the aggregated state of the candidate PrD was successfully transmitted through the cytoplasm, the selected cytoductants should display the prion phenotype under investigation. For a more detailed description of cytoduction we direct the reader to two recent articles on this topic (Liebman *et al.*, 2006; Wickner *et al.*, 2006).

2.2.5. Transformation of prion particles

The principal tenet of the prion hypothesis is that prions replicate in a protein-only manner, without the direction of an underlying nucleic acid template. The most rigorous proof for a prion, then, is to show that nucleic acid-free preparations of aggregated protein are "infectious" in and of themselves. That is, they have the capacity to convert cells to a stable prion state when they are introduced into those cells. Protein transformations have irrefutably established the protein-only nature of prion propagation for a number of yeast prions, and have also been used to show that the prion strain phenomenon results from conformational variations in the underlying amyloid structure.

Protein transformations are generally done by fusing prion particles with recipient cell spheroplasts using polyethylene-glycol (see Fig. 30.4). A selectable plasmid is typically cotransformed with the prion particles to allow for the determination of total transformation efficiency, as prion protein preparations have variable infectivities. The transformed spheroplasts are allowed to recover in agar and then analyzed for the prion phenotype. The putative prion particles to be used for protein transformations can be obtained either from [*PRION+*] cells or from recombinant protein that has been allowed to aggregate *in vitro*.

When using crude extracts as a source of prions, care must be taken to eliminate all viable nonlysed cells remaining in the extract (e.g., by

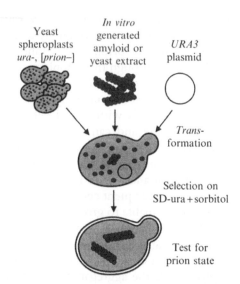

Figure 30.4 Protein transformations to examine transmissibility of protein aggregates (adapted from Tanaka and Weissman, 2006). *In vitro*-generated amyloid fibers, or alternatively, partially purified yeast extracts, can be used to transform cells to a stable prion state. The rigid cell wall of recipient cells is removed to generate competent spheroplasts. The spheroplasts are then incubated with a transformation mix containing prion particles and a selectable plasmid, followed by recovery of transformants on isotonic media that is selective for the plasmid. These transformants are then screened for the prion state using phenotypic or biochemical assays.

centrifugation or filtration), as they may otherwise appear as false positives. For this reason, we recommend performing control transformations without recipient cells to verify that there are no contaminating cells in the extract. Extracts from [PRION+] yeast can be generated either by spheroplasting (e.g., Section 2.1.3) or by glass bead lysis (Brachmann *et al.*, 2005; King *et al.*, 2006). Cleared lysates resulting from either of these procedures may be adequate for transformations without further manipulations in some cases. However, there are also a number of techniques that can be used to enrich prion particles relative to other cellular components and thereby improve transformation efficiencies. We direct the reader to the corresponding references concerning these procedures, which include: partial purification of aggregated protein by sedimentation (Tanaka and Weissman, 2006), sedimentation followed by affinity purification of the prion protein (King *et al.*, 2006), and amplification of prion particles in cell extracts by seeding the conversion of exogenously added recombinant prion protein (Brachmann *et al.*, 2005; King *et al.*, 2006).

The most rigorous proof that a protein is a prion is to transform cells to the [PRION+] state using solely recombinant protein from a heterologous host that has been converted to the putative prion form *in vitro*. This procedure avoids potential confounding factors that are present in yeast extracts, and also allows one to easily generate highly concentrated infectious preparations without the need for labor-intensive enrichment of prion particles from [*PRION+*] yeast cells.

We routinely purify yeast PrDs from *Escherichia coli* and convert them to amyloid fibers in a near-physiological buffer (Alberti *et al.*, 2009). To form infectious amyloids *in vitro*, denatured proteins are diluted from a GdnHCl stock to a final concentration of 10 μM in 1 ml assembly buffer (5 mM K$_2$HPO$_4$, pH 6.6, 150 mM NaCl, 5 mM EDTA, 2 mM TCEP) and rotated end-over-end for at least 24 h at room temperature. The formation of amyloid is most easily monitored by ThT fluorescence (450 nm excitation, 482 nm emission), added at 20-fold molar excess, to aliquots taken from the assembly reaction. Following amyloid conversion, the reaction is centrifuged at maximum speed (20,000 rcf) in a table top centrifuge for 30 min at room temperature, and the pellet of aggregated protein resuspended in 200 μl PBS. The protein is then sonicated with a tip sonicator at the lowest setting for 10 s. Sonication shears amyloid fibers into smaller pieces, thereby greatly enhancing their infectivity.

Proper negative controls are essential for interpreting protein transformations. Prions arise spontaneously at a low frequency and this frequency increases after cells are exposed to stress (Tyedmers *et al.*, 2008). Consequently, the efficiency of transformation to [*PRION+*] must be normalized against mock transformations, such as freshly diluted (soluble) prion protein, amyloid fibers of other prions, or [*prion−*] cell extract.

Recipient cells are prepared for protein transformations by a gentle enzymatic removal of the cell wall (spheroplasting), using a protocol adapted from Tanaka and Weissman (2006). Many prions have an increased rate of appearance in yeast cells harboring the [*PIN+*] prion. For this reason, we recommend using [*pin−*] yeast to reduce background from the spontaneous appearance of the prion state of interest. Yeast are grown to an OD of 0.5 in YPD, harvested by centrifugation, and washed twice in sterile distilled water. The cells are then washed with 1 ml SCE (1 M sorbitol, 10 mM EDTA, 10 mM DTT (added just before use), 100 mM sodium citrate, pH 5.8) and then resuspended in 1 ml SCE. Sixty microliters lyticase solution (4.2 mg/ml lyticase (Sigma); 50 mM sodium citrate, pH 5.8) is added and the cells are incubated for 20–30 min at 30 °C while shaking at 300 rpm. It is very important that this step not be allowed to proceed for too long, or cells will lose viability. We recommend standardizing this step using identical aliquots of lyticase solution prepared from a single lot, which are stored at −80 °C. The progress of spheroplasting can be monitored during this incubation by placing 2 μl of cells in 20 μl of 1% SDS and observing them under a microscope. Spheroplasts with SDS should lyse and be invisible or appear as ghost cells.

The spheroplasts are harvested at 500 rcf for 3 min at room temperature, followed by washing twice with 1 ml of STC buffer (1 M sorbitol, 10 mM CaCl$_2$, 10 mM Tris–HCl, pH 7.5). Finally, the spheroplasts are resuspended in 0.5 ml of STC buffer. Spheroplasts are sensitive to shear forces and consequently must be handled gently during all manipulations. To resuspend spheroplasts, we use a 1-ml plastic pipette tip that has ~1 cm of the tip removed.

We add 100 μl of spheroplasts to 4 μl of 10 mg/ml salmon sperm DNA, 25 μl of 0.1 mg/ml selectable plasmid (e.g., pRS316 for a URA3-marked plasmid) and 33 μl of the protein solution to be transformed. The final protein concentration of amyloid fibers should be ~10 μM, or if using yeast extract, 200–400 μg/ml total protein. The samples are tapped gently to mix and incubated for 30 min at room temperature. Next, proteins are fused to spheroplasts by adding 1.35 ml PEG-buffer (20% (w/v) PEG 8000, 10 mM CaCl$_2$, 10 mM Tris–HCl, pH 7.5) and incubating for 30 min at room temperature. Note that the optimal concentration and molecular weight of PEG used in this step may vary depending on the transformed protein (Patel and Liebman, 2007). Spheroplasts are collected at 500 rcf for 3 min at room temperature, and resuspended in 0.5 ml of SOS buffer (1 M sorbitol, 7 mM CaCl$_2$, 0.25% yeast extract, 0.5% bactopeptone), followed by incubation for 1 h at 30 °C with 300 rpm shaking. Meanwhile, 8 ml aliquots of spheroplast recovery media are prepared in 15 ml tubes and maintained in a 48 °C water bath. The spheroplast recovery media needs to be selective for the plasmid (e.g., SD-ura) and for the prion state if desired (see below), and is supplemented with 1 M sorbitol and 2.5% agar. Each transformation reaction is diluted into one aliquot of media, mixed by gentle inversion, and overlayed immediately onto the appropriate selective plates that have been prewarmed to 37 °C.

Plates are incubated at 30 °C under high humidity until colonies develop (up to 1 week). Colonies can then be picked out of the agar and scored for [PRION+] phenotypes. If transformation efficiencies are low, it is especially important that putative [PRION+] transformants are verified by secondary assays like SDD–AGE, to distinguish them from genetic revertants or other background colonies.

We and others (Brachmann et al., 2005) have found that selecting directly for the [PRION+] state of some prions (e.g., [MOT3+] and [URE3]) during spheroplast recovery increases conversion to the [PRION+] state relative to delaying selection until after the cells have recovered. Newly induced prion states are often initially unstable and seem to be lost at a high frequency under nonselective conditions. Consequently, applying an immediate mild selective pressure during the spheroplast recovery step can improve the apparent transformation efficiency by preventing many prion-containing spheroplasts from losing the prion state during colony formation in the sorbitol-containing media. However, stringent selective conditions can also inhibit spheroplast recovery resulting in drastically reduced transformation efficiencies. For instance, [*URE3*] spheroplasts recover at a low frequency when plated directly to USA containing media (Brachmann *et al.*, 2005), and we have observed that [*PSI+*] spheroplasts generally recover poorly in adenine–deficient media. For each prion and selection scheme, there is likely to be an optimum window of selection stringency that maximizes the number of [*PRION+*] transformants recovered.

 ## 3. CONCLUDING REMARKS

Aggregation has been suggested to be a generic property of proteins (Chiti and Dobson, 2006), but most proteins aggregate only under conditions that fall outside of the normal physiological range. Studies of proteins that aggregate under nonphysiological experimental conditions have provided important insights into general aspects of protein aggregation. Yet proteins that aggregate under physiological conditions are much more interesting from a biological point of view. Recent studies show that misfolding and aggregation propensities are likely to be a dominant force in the evolution of protein sequences (Chen and Dokholyan, 2008; Drummond and Wilke, 2008). This hypothesis is further underscored by the presence of complex quality control mechanisms that govern the abundance and structure of protein aggregates. Studies that identify large numbers of aggregation–prone proteins under physiological conditions will be necessary to understand how aggregation propensities shape the sequence and structure of proteins and the composition of proteomes. These studies will also allow us to generate comprehensive inventories of proteins that are capable of forming functional or pathological aggregates. Such inventories will be an important asset, as their analysis will facilitate the identification of sequence determinants that drive aggregation behavior. A growing toolbox of scalable and adaptable protein aggregation assays now enables rapid identification and characterization of the repertoire of aggregation–prone proteins in yeast. Moreover, they place yeast at the vanguard of new technological developments that have a tremendous impact on our understanding of fundamental aspects of biology.

REFERENCES

Aigle, M., and Lacroute, F. (1975). Genetical aspects of [URE3], a non–mitochondrial, cytoplasmically inherited mutation in yeast. *Mol. Gen. Genet.* **136**(4), 327–335.
Alberti, S., Gitler, A. D., and Lindquist, S. (2007). A suite of Gateway cloning vectors for high–throughput genetic analysis in *Saccharomyces cerevisiae*. *Yeast* **24**(10), 913–919.

Alberti, S., Halfmann, R., King, O., Kapila, A., and Lindquist, S. (2009). A systematic survey identifies prions and illuminates sequence features of prionogenic proteins. *Cell* **137**(1), 146–158.

Bagriantsev, S. N., Kushnirov, V. V., and Liebman, S. W. (2006). Analysis of amyloid aggregates using agarose gel electrophoresis. *Methods Enzymol.* **412**, 33–48.

Brachmann, A., Baxa, U., and Wickner, R. B. (2005). Prion generation in vitro: Amyloid of Ure2p is infectious. *EMBO J.* **24**(17), 3082–3092.

Brachmann, A., Toombs, J. A., and Ross, E. D. (2006). Reporter assay systems for [URE3] detection and analysis. *Methods* **39**(1), 35–42.

Bradley, M. E., Edskes, H. K., Hong, J. Y., Wickner, R. B., and Liebman, S. W. (2002). Interactions among prions and prion "strains" in yeast. *Proc. Natl Acad. Sci. USA* **99** (Suppl. 4), 16392–16399.

Caughey, B., Baron, G. S., Chesebro, B., and Jeffrey, M. (2009). Getting a grip on prions: Oligomers, amyloids, and pathological membrane interactions. *Annu. Rev. Biochem.* **78**, 177–204.

Chen, Y., and Dokholyan, N. V. (2008). Natural selection against protein aggregation on self-interacting and essential proteins in yeast, fly, and worm. *Mol. Biol. Evol.* **25**(8), 1530–1533.

Chiti, F., and Dobson, C. M. (2006). Protein misfolding, functional amyloid, and human disease. *Annu. Rev. Biochem.* **75**, 333–366.

Conde, J., and Fink, G. R. (1976). A mutant of *Saccharomyces cerevisiae* defective for nuclear fusion. *Proc. Natl. Acad. Sci. USA* **73**(10), 3651–3655.

Cooper, A. A., Gitler, A. D., Cashikar, A., Haynes, C. M., Hill, K. J., Bhullar, B., Liu, K., Xu, K., Strathearn, K. E., Liu, F., Cao, S., Caldwell, K. A., *et al.* (2006). Alpha-synuclein blocks ER-Golgi traffic and Rab1 rescues neuron loss in Parkinson's models. *Science* **313**(5785), 324–328.

Cox, B. S. (1965). PSI, a cytoplasmic suppressor of the super-suppressor in yeast. *Heredity* **121**(20), 505–521.

Derkatch, I. L., Bradley, M. E., Masse, S. V., Zadorsky, S. P., Polozkov, G. V., Inge-Vechtomov, S. G., and Liebman, S. W. (2000). Dependence and independence of [PSI(+)] and [PIN(+)]: A two-prion system in yeast? *EMBO J.* **19**(9), 1942–1952.

Derkatch, I. L., Bradley, M. E., Hong, J. Y., and Liebman, S. W. (2001). Prions affect the appearance of other prions: The story of [PIN(+)]. *Cell* **106**(2), 171–182.

Douglas, P. M., Treusch, S., Ren, H. Y., Halfmann, R., Duennwald, M. L., Lindquist, S., and Cyr, D. M. (2008). Chaperone-dependent amyloid assembly protects cells from prion toxicity. *Proc. Natl. Acad. Sci. USA* **105**(20), 7206–7211.

Drummond, D. A., and Wilke, C. O. (2008). Mistranslation-induced protein misfolding as a dominant constraint on coding-sequence evolution. *Cell* **134**(2), 341–352.

Duennwald, M. L., Jagadish, S., Giorgini, F., Muchowski, P. J., and Lindquist, S. (2006a). A network of protein interactions determines polyglutamine toxicity. *Proc. Natl. Acad. Sci. USA* **103**(29), 11051–11056.

Duennwald, M. L., Jagadish, S., Muchowski, P. J., and Lindquist, S. (2006b). Flanking sequences profoundly alter polyglutamine toxicity in yeast. *Proc. Natl. Acad. Sci. USA* **103**(29), 11045–11050.

Eaglestone, S. S., Cox, B. S., and Tuite, M. F. (1999). Translation termination efficiency can be regulated in *Saccharomyces cerevisiae* by environmental stress through a prion-mediated mechanism. *EMBO J.* **18**(7), 1974–1981.

Edskes, H. K., Gray, V. T., and Wickner, R. B. (1999). The [URE3] prion is an aggregated form of Ure2p that can be cured by overexpression of Ure2p fragments. *Proc. Natl. Acad. Sci. USA* **96**(4), 1498–1503.

Ganusova, E. E., Ozolins, L. N., Bhagat, S., Newnam, G. P., Wegrzyn, R. D., Sherman, M. Y., and Chernoff, Y. O. (2006). Modulation of prion formation, aggregation, and toxicity by the actin cytoskeleton in yeast. *Mol. Cell. Biol.* **26**(2), 617–629.

Garcia-Mata, R., Bebok, Z., Sorscher, E. J., and Sztul, E. S. (1999). Characterization and dynamics of aggresome formation by a cytosolic GFP-chimera. *J. Cell Biol.* **146**(6), 1239–1254.

Halfmann, R., and Lindquist, S. (2008). Screening for amyloid aggregation by semi-denaturing detergent-agarose gel electrophoresis. *J. Vis. Exp.* (17), DOI: 10.3791/838.

Hillenmeyer, M. E., Fung, E., Wildenhain, J., Pierce, S. E., Hoon, S., Lee, W., Proctor, M., St Onge, R. P., Tyers, M., Koller, D., Altman, R. B., Davis, R. W., *et al.* (2008). The chemical genomic portrait of yeast: Uncovering a phenotype for all genes. *Science* **320** (5874), 362–365.

Johnston, J. A., Ward, C. L., and Kopito, R. R. (1998). Aggresomes: A cellular response to misfolded proteins. *J. Cell Biol.* **143**(7), 1883–1898.

Kaganovich, D., Kopito, R., and Frydman, J. (2008). Misfolded proteins partition between two distinct quality control compartments. *Nature* **454**(7208), 1088–1095.

King, C. Y., Wang, H. L., and Chang, H. Y. (2006). Transformation of yeast by infectious prion particles. *Methods* **39**(1), 68–71.

Krobitsch, S., and Lindquist, S. (2000). Aggregation of huntingtin in yeast varies with the length of the polyglutamine expansion and the expression of chaperone proteins. *Proc. Natl. Acad. Sci. USA* **97**(4), 1589–1594.

Lacroute, F. (1971). Non-Mendelian mutation allowing ureidosuccinic acid uptake in yeast. *J. Bacteriol.* **106**(2), 519–522.

Li, L., and Lindquist, S. (2000). Creating a protein-based element of inheritance. *Science* **287** (5453), 661–664.

Liebman, S. W., and Sherman, F. (1979). Extrachromosomal psi + determinant suppresses nonsense mutations in yeast. *J. Bacteriol.* **139**(3), 1068–1071.

Liebman, S. W., Bagriantsev, S. N., and Derkatch, I. L. (2006). Biochemical and genetic methods for characterization of [PIN+] prions in yeast. *Methods* **39**(1), 23–34.

Namy, O., Galopier, A., Martini, C., Matsufuji, S., Fabret, C., and Rousset, J. P. (2008). Epigenetic control of polyamines by the prion [PSI(+)]. *Nat. Cell Biol.* **10**(9), 1069–1075.

Nemecek, J., Nakayashiki, T., and Wickner, R. B. (2009). A prion of yeast metacaspase homolog (Mca1p) detected by a genetic screen. *Proc. Natl. Acad. Sci. USA* **106**(6), 1892–1896.

Osherovich, L. Z., and Weissman, J. S. (2001). Multiple Gln/Asn-rich prion domains confer susceptibility to induction of the yeast [PSI(+)] prion. *Cell* **106**(2), 183–194.

Park, H. J., Park, S. J., Oh, D. B., Lee, S., and Kim, Y. G. (2009). Increased-1 ribosomal frameshifting efficiency by yeast prion-like phenotype [PSI+]. *FEBS Lett.* **583**(4), 665–669.

Patel, B. K., and Liebman, S. W. (2007). "Prion-proof" for [PIN+]: Infection with in vitro-made amyloid aggregates of Rnq1p-(132–405) induces [PIN+]. *J. Mol. Biol.* **365**(3), 773–782.

Patino, M. M., Liu, J. J., Glover, J. R., and Lindquist, S. (1996). Support for the prion hypothesis for inheritance of a phenotypic trait in yeast. *Science* **273**(5275), 622–626.

Paushkin, S. V., Kushnirov, V. V., Smirnov, V. N., and Ter-Avanesyan, M. D. (1996). Propagation of the yeast prion-like [psi+] determinant is mediated by oligomerization of the SUP35-encoded polypeptide chain release factor. *EMBO J.* **15**(12), 3127–3134.

Qin, Z., Hu, D., Zhu, M., and Fink, A. L. (2007). Structural characterization of the partially folded intermediates of an immunoglobulin light chain leading to amyloid fibrillation and amorphous aggregation. *Biochemistry* **46**(11), 3521–3531.

Ross, E. D., Minton, A., and Wickner, R. B. (2005). Prion domains: Sequences, structures and interactions. *Nat. Cell Biol.* **7**(11), 1039–1044.

Rousseau, F., Schymkowitz, J., and Serrano, L. (2006). Protein aggregation and amyloidosis: Confusion of the kinds? *Curr. Opin. Struct. Biol.* **16**(1), 118–126.

Santoso, A., Chien, P., Osherovich, L. Z., and Weissman, J. S. (2000). Molecular basis of a yeast prion species barrier. *Cell* **100**(2), 277–288.

Scherzinger, E., Lurz, R., Turmaine, M., Mangiarini, L., Hollenbach, B., Hasenbank, R., Bates, G. P., Davies, S. W., Lehrach, H., and Wanker, E. E. (1997). Huntingtin-encoded polyglutamine expansions form amyloid-like protein aggregates in vitro and in vivo. *Cell* **90**(3), 549–558.

Shorter, J., and Lindquist, S. (2005). Prions as adaptive conduits of memory and inheritance. *Nat. Rev. Genet.* **6**(6), 435–450.

Si, K., Lindquist, S., and Kandel, E. R. (2003). A neuronal isoform of the aplysia CPEB has prion-like properties. *Cell* **115**(7), 879–891.

Sondheimer, N., and Lindquist, S. (2000). Rnq1: An epigenetic modifier of protein function in yeast. *Mol. Cell.* **5**(1), 163–172.

Sopko, R., Huang, D., Preston, N., Chua, G., Papp, B., Kafadar, K., Snyder, M., Oliver, S. G., Cyert, M., Hughes, T. R., Boone, C., and Andrews, B. (2006). Mapping pathways and phenotypes by systematic gene overexpression. *Mol. Cell.* **21**(3), 319–330.

Tanaka, M., and Weissman, J. S. (2006). An efficient protein transformation protocol for introducing prions into yeast. *Methods Enzymol.* **412**, 185–200.

Taneja, V., Maddelein, M. L., Talarek, N., Saupe, S. J., and Liebman, S. W. (2007). A non-Q/N-rich prion domain of a foreign prion, [Het-s], can propagate as a prion in yeast. *Mol. Cell.* **27**(1), 67–77.

Tessier, P. M., and Lindquist, S. (2009). Unraveling infectious structures, strain variants and species barriers for the yeast prion [PSI+]. *Nat. Struct. Mol. Biol.* **16**(6), 598–605.

True, H. L., and Lindquist, S. L. (2000). A yeast prion provides a mechanism for genetic variation and phenotypic diversity. *Nature* **407**(6803), 477–483.

True, H. L., Berlin, I., and Lindquist, S. L. (2004). Epigenetic regulation of translation reveals hidden genetic variation to produce complex traits. *Nature* **431**(7005), 184–187.

Tyedmers, J., Madariaga, M. L., and Lindquist, S. (2008). Prion switching in response to environmental stress. *PLoS Biol.* **6**(11), e294.

Vetri, V., Canale, C., Relini, A., Librizzi, F., Militello, V., Gliozzi, A., and Leone, M. (2007). Amyloid fibrils formation and amorphous aggregation in concanavalin A. *Biophys. Chem.* **125**(1), 184–190.

Wickner, R. B. (1994). [URE3] as an altered URE2 protein: Evidence for a prion analog in *Saccharomyces cerevisiae. Science* **264**(5158), 566–569.

Wickner, R. B., Edskes, H. K., and Shewmaker, F. (2006). How to find a prion: [URE3], [PSI+] and [beta]. *Methods* **39**(1), 3–8.

Zhou, P., Derkatch, I. L., and Liebman, S. W. (2001). The relationship between visible intracellular aggregates that appear after overexpression of Sup35 and the yeast prion-like elements [PSI(+)] and [PIN(+)]. *Mol. Microbiol.* **39**(1), 37–46.

Zhu, H., Bilgin, M., and Snyder, M. (2003). Proteomics. *Annu. Rev. Biochem.* **72**, 783–812.

OTHER FUNGI

CHAPTER THIRTY-ONE

GENETICS AND MOLECULAR BIOLOGY IN *CANDIDA ALBICANS*

Aaron D. Hernday,* Suzanne M. Noble,* Quinn M. Mitrovich,[†] *and* Alexander D. Johnson*,[‡]

Contents

* Department of Microbiology and Immunology, University of California at San Francisco, San Francisco, California, USA
[†] Department of Biochemistry and Biophysics, University of California at San Francisco, San Francisco, California, USA
[‡] Department of Biochemistry and Biophysics, University of California at San Francisco, San Francisco, California, USA

Methods in Enzymology, Volume 470   © 2010 Elsevier Inc.
ISSN 0076-6879, DOI: 10.1016/S0076-6879(10)70031-8   All rights reserved.

Abstract

Candida albicans is an opportunistic fungal pathogen of humans. Although a normal part of our gastrointestinal flora, *C. albicans* has the ability to colonize nearly every human tissue and organ, causing serious, invasive infections. In this chapter we describe current methodologies used in molecular genetic studies of this organism. These techniques include rapid sequential gene disruption, DNA transformation, RNA isolation, epitope tagging, and chromatin immunoprecipitation. The ease of these techniques, combined with the high-quality *C. albicans* genome sequences now available, have greatly facilitated research into this important pathogen.

Candida albicans is a normal resident of the human gastrointestinal tract; it is also the most common fungal pathogen of humans, causing both mucosal and systemic infections, particularly in immune compromised patients. *C. albicans* and *Saccharomyces cerevisiae* last shared a common ancestor more than 900 million years ago; in terms of conserved coding sequences, the two species are approximately as divergent as fish and humans. Although *C. albicans* and *S. cerevisiae* share certain core features, they also exhibit many significant differences. This is not surprising as *C. albicans* has the ability to survive in nearly every niche of a mammalian host, a property not shared by *S. cerevisiae*. Research into *C. albicans* is important in its own right, particularly with regards to its ability to cause disease in humans; in addition, comparison with *S. cerevisiae* can reveal important insights into evolutionary processes.

Many of the methodologies developed for use in *S. cerevisiae* have been adapted for *C. albicans*, and we describe some of the most common. Although alternative procedures are described in the literature, we have found those described below to be the most convenient. Because the *C. albicans* parasexual cycle is cumbersome to use in the laboratory, genetics in this organism has been based almost entirely on directed mutations. Because the organism is diploid, creating a deletion mutant requires two rounds of gene disruption. We describe a rapid method for creating sequential disruptions, one which can be scaled up to create large collections of *C. albicans* deletion mutants. We also describe a series of additional techniques including DNA transformation, mRNA isolation, epitope tagging, and chromatin immunoprecipitation (ChIP). The ease of these techniques, combined with the high-quality *C. albicans* genome sequences now available, has greatly increased the quality and pace of research into this important pathogen.

1. HOMOZYGOUS GENE DISRUPTION IN *C. ALBICANS*

Creating gene knockout mutants in *C. albicans* typically involves two rounds of transformation (to disrupt both alleles of a given gene) with a linear fragment of DNA bearing a selectable marker as well as sequences identical (or nearly identical) to those sequences flanking the target gene

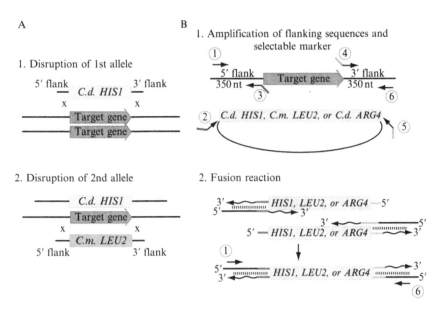

Figure 31.1 (A) Homozygous gene disruption by two rounds of transformation and homologous recombination. (B) Fusion PCR method.

(Fig. 31.1A). Approximately 60 nucleotides of flanking sequence on each side of the selectable marker approaches the minimum necessary for successful targeting, and the efficiency appears to improve with increasing lengths. The disruption cassette can be created by PCR or by traditional cloning, and the available selectable markers include multiple auxotrophic markers (such as *HIS1*, *LEU2*, *ARG4*, and *URA3*) and an antibiotic resistance gene (*SAT1*); note that the *URA3* marker should be used with care, because important *C. albicans* phenotypes such as morphogenesis and virulence are strongly dependent on the levels of *URA3* expression. Transformants are selected on appropriate media and then screened for integration of the disruption cassette at the correct genomic locus. Following disruption of the second allele, verification that the target ORF is truly deleted (achieved by PCR or Southern blotting) is crucial, as extra copies of chromosomes can arise during the tranformation procedures.

Following is a streamlined protocol based on fusion PCR that results in a high efficiency of gene disruption (Noble and Johnson, 2005; Fig. 31.1B). Maximal efficiency is achieved by the use of auxotrophic markers from non-*albicans Candida* species or bacterial antibiotic resistance genes as markers, these strategies reduce integration events at "off-target" locations in the *C. albicans* genome. The first round of PCR consists of three reactions: two to amplify DNA upstream and downstream of the target gene, and a third to amplify the selectable marker. Importantly, certain primers (indicated in Fig. 31.1 and described in detail in the protocol) contain complementary tails. In the second round of PCR, the three products of the first round of PCR serve as an aggregate template, resulting in a single product. Note that specific reagents and kits are recommended, but alternates can be substituted. We have found that Ex Taq (Takara) and Klentaq LA (DNA Polymerase Technology) yield better results than other commercial enzymes for fusion PCR.

1.1. Homozygous gene disruption by fusion PCR

The following auxotrophic markers are available as cloned genes from non-*albicans Candida* species (Noble and Johnson 2005):

C. dubliniensis HIS1 = pSN52
C. maltosa LEU2 = pSN40
C. dubliniensis ARG4 = pSN69

1. Design the following PCR primers:

Gene disruption primers (Fig. 31.1B):

1—primer to gene of interest (beginning of 5′ flank top, ~350 bp upstream of ORF)
2*—CCGCTGCTAGGCGCGCCGTG-selectable marker (5′ marker top)
3—CACGGCGCGCCTAGCAGCGG-gene of interest (end of 5′ flank bottom)
4—GTCAGCGGCCGCATCCCTGC-gene of interest (beginning of 3′flank top)
5*—GCAGGGATGCGGCCGCTGAC-selectable marker (3′ marker bottom)
6—primer to gene of interest (end of 3′ flank bottom, ~350 bp downstream of ORF)

* If using auxotrophic markers in the pSN series (Noble and Johnson), sequences of primer 2 and primer 5 are:

Primer 2	ccgctgctaggcgcgccgtgACCAGTGTGATGGATATCTGC
Primer 5	gcagggatgcggccgctgacAGCTCGGATCCACTAGTAACG

Knockout verification primers (Fig. 31.2):

Target Check Left—just upstream of primer 1
Target Check Right—just downstream of primer 6
ORF Left—internal to the deleted ORF
ORF Right—internal to the deleted ORF
Marker Check Left*—toward end of selectable marker that is near primer 2
Marker Check Right*—toward end of selectable maker that is near primer 5

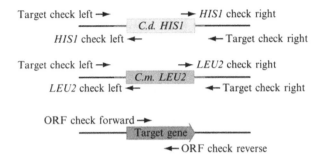

Figure 31.2 Primers for verification PCR.

★If using auxotrophic markers in the pSN series, one can use the following Marker Check primers:

C. dubliniensis HIS1 Check Left	ATTAGATACGTTGGTGGTTC
C. dubliniensis HIS1 Check Right	AACACAACTGCACAATCTGG
C. maltosa LEU2 Check Left	AGAATTCCCAACTTTGTCTG
C. maltosa LEU2 Check Right	AAACTTTGAACCCGGCTGCG
C. dubliniensis Check Left	TTCAACCTTTCAAACGATGC
C. dubliniensis Check Right	TCGATACATTTGCGGTACAG

2. Set up reactions for *PCR Round I* on ice, and run the PCR:

   Reaction 1 = primers 1 + 3 using genomic DNA as template
   Reaction 2 = primers 4 + 6 using genomic DNA as template
   Reaction 3 = primers 2 + 5 using auxotrophic marker as template,
      for example, pSN52 = *C. dubliniensis* HIS1

   PCR reaction (50 μl):

   5 μl 10× Ex Taq buffer
   4 μl 2.5 m*M* dNTPs
   36 μl H$_2$O
   <1 μl GENOMIC DNA (reactions 1 and 2) or <1 μl plasmid DNA
      (reaction 3)
   2 μl 1st primer (5 μ*M*)
   2 μl 2nd primer (5 μ*M*)
   0.5 μl Ex Taq polymerase

   PCR conditions:

   94 °C 5 min
   35 cycles 94 °C 30 s, 45 °C 45 s, 72 °C 1 min (flanks) or 4 min (marker)
   72 °C 10 min
   10 °C forever

3. Run out 5 μl PCR products on a 1% agarose gel to confirm successful synthesis.

4. Optional—Gel purify the product of Reaction 3 (marker fragment).
Run the PCR reaction on a 1% agarose gel and cut out the correct
sized band under long wave ultraviolet light. Recover DNA with a
Qiagen QIAquick gel extraction kit (eluting in 50 μl H_2O or Buffer
EB). By isolating the correct PCR product and eliminating contaminants, this step increases the efficiency of the fusion reaction for certain
targets and allows for stable storage of the marker fragment (at −20 °C).
5. Optional—Purify the products of Reactions 1 and 2 with a Qiagen
QIAquick PCR purification kit. Use as directed and elute in 50 μl H_2O
or Buffer EB.
6. Set up reaction for *PCR Round II* on ice, and run PCR.

PCR reaction (100 μl):

10 μl 10× Ex Taq buffer
8 μl 2.5 mM dNTPs
0.75 μl Ex Taq polymerase
67 μl H_2O
1.5 μl reaction 1 PCR product (5′ flank)
1.5 μl reaction 2 PCR product (3′ flank)
2 μl reaction 3 PCR product (Marker)
4 μl primer 1 (5 μM)
4 μl primer 6 (5 μM)

Fusion PCR conditions:

94 °C 5 min (Hot start: i.e., wait until the PCR block heats to ∼80 °C
    before introducing PCR reactions)
35 cycles 94 °C 30 s, 50 °C 45 s, 72 °C 4.5 min
72 °C 10 min
10 °C forever

7. Run out 5 μl PCR product on a 1% agarose gel to confirm success of
fusion reaction.
    The fusion product should be ∼3–4 kb, depending on the auxotrophic marker chosen and the length of flanking sequences. There is
typically a mix of full sized product, with a variable amount of a
minority shorter product.
    Note: If the fusion reaction is unsuccessful, a variation is to include
Betaine in the reaction; that is, add 20 μl of 5 M betaine (final 1.3 M)
and just 47 μl H_2O in the fusion reaction mix. If betaine is used,
decrease the PCR denaturation temperature to 92.5 °C.
8. Optional—Purify the fusion PCR product with a Qiagen QIAquick
PCR purification kit.
    Use kit as directed and elute in 30 μl of H_2O or Buffer EB.

9. Transform 10 μl of disruption fragment into fresh competent SN152 (or any strain that is auxotrophic for the selectable marker). Plate cells on the appropriate dropout plate, for example, -His. Incubate plates at 30 °C for 2 days, or until individual colonies are visible.

   Note: If selecting for Nourseothricin resistance, allow cells to recover by growing in YPD liquid media without selection for at least 5 hours at 30 °C prior to plating on selective media.

10. Patch ~10 transformants onto fresh medium, and perform colony PCR to verify the correct 5' and 3' junctions of the disrupted allele (Fig. 31.2). In separate PCR reactions, use the primer pairs Target Left + Marker Left and Target Right + Marker Right. Correct integrants should have PCR products of the expected size (~0.5 kb) with each primer set.

11. Pick at least two confirmed heterozygous knockout candidates and streak for single colonies on fresh medium.

    Note: Because unlinked mutations can be acquired during strain construction and because two rounds of transformation are required to create a homozygous deletion, it is wise to obtain at least 2 independent isolates of any *C. albicans* knockout.

12. Repeat the fusion PCR step, using the same flank products (PCR reactions 1 and 2) but a different selectable marker (e.g., pSN40 = *C. maltosa* LEU2).

13. Transform 10 μl of the new disruption fragment into two independent heterozygous knockout strains and plate on doubly selective medium (e.g., -His, -Leu).

14. Patch ~10 transformants of each strain onto fresh medium, and perform colony PCR as above to confirm the appropriate 5' and 3' junctions of the second disrupted allele.

15. For candidates with expected disruptions of both target alleles, perform a final PCR verification that there are no remaining copies of the target ORF. This step is necessary because aneuploidies or translocations commonly result in an extra copy of the target gene. Remember to test as a positive control a strain that retains a copy of the target gene (e.g., wild type or the heterozygous knockout). One should see a PCR product of the expected size in the positive control and no PCR product in the desired homozygous deletion strain.

2. *C. ALBICANS* DNA TRANSFORMATION

The following is a basic protocol for DNA transformation with *C. albicans*. Because stable extrachromosomal plasmids have yet to be fully developed for use in *C. albicans*, this protocol is typically used for transformation and stable integration of linear DNA fragments into the *C. albicans* genome. For efficient homologous recombination to occur, a minimum of 60 bp of sequence identical (or nearly identical) to the genomic target locus is required on either end of the DNA fragment that is to be transformed.

1. Inoculate a 5-ml culture in YEPD and grow overnight at 30 °C.
2. Dilute 300 μl of the overnight culture into 10 ml of fresh YEPD and grow at 30 °C for 4–6 h (or until OD is around 0.5–1.0).
3. Centrifuge for 2 min at $\sim 1000 \times g$ and discard supernatant.
4. Resuspend in 900 μl LiOAc/TE and transfer to a microcentrifuge tube.
5. Pellet for 1 min at $\sim 1000 \times g$ and discard supernatant.
6. Wash two more times with 900 μl LiOAc/TE then resuspend in ~ 400 μl final volume with LiOAc/TE.
7. In a separate microfuge tube mix (in order) the following:
   a. 10 μl 10 mg/ml denatured Herring Sperm (or Salmon Sperm) DNA
      i. Prepare by boiling ~ 2 min then snap cooling in ice water
   b. ≥ 1 μg of DNA fragment to be transformed (or ~ 20–50 μl of PCR product)
      ii. Highest transformation efficiencies are achieved if the DNA is NOT purified following enzymatic reactions (i.e., PCR products or plasmid digests)
   c. 200 μl washed cells in LiOAc/TE
   d. 1 ml PEG mix
8. Incubate overnight at room temperature.
9. Heat shock 1 h at 42 °C (or 44 °C for 15 min).
10. Pellet for 1 min at $\sim 1000 \times g$ and discard supernatant.
11. Wash one time with 1 ml sterile water.
12. Resuspend in ~ 150 μl final volume with sterile water.
    a. For selection of Nourseothricin resistance, transfer cells to 5 ml YEPD and recover by incubation at 30 °C for at least 5 h prior to plating on selective media.
13. Plate on selective media and incubate at 30 °C for 2–3 days.

2.1. Transformation buffers

LiOAc mix:

10 ml 1 M LiOAc
200 μl 0.5 M EDTA
1 ml 1 M Tris–HCl, pH 7.5
H_2O to 100 ml
Filter sterilize

PEG mix:

80 ml 50% PEG-3350
10 ml 1 M LiOAc

200 μl 0.5 M EDTA
1 ml 1 M Tris–HCl, pH 7.5
H_2O to 100 ml
Filter sterilize

3. *C. ALBICANS* TOTAL RNA PURIFICATION

Purifying total cellular RNA from liquid cultures of *C. albicans* is comparable in most regards to purifications from *S cerevisiae*. As with *S. cerevisiae*, lysing the *C. albicans* cell wall requires a more vigorous procedure than does lysis of animal cells. The procedure outlined below includes organic extractions in Phase Lock tubes (Eppendorf) for removal of proteins and other cellular material. Because *C. albicans* cellular debris tends to disrupt the Phase Lock gel matrix, the first organic extraction is performed in conventional rather than Phase Lock tubes; if desired, subsequent extraction steps may also be performed in conventional tubes. Purified RNA is suitable for most applications, including microarray analysis, quantitative RT-PCR and Northern hybridization.

1. Grow a 10-ml liquid culture of *C. albicans* cells to an appropriate concentration (e.g., OD_{600} of 1.0–1.5). For other volumes and cell densities, all steps may be scaled proportionally.
2. Collect cells by centrifugation (2000×g, 5 min, 4 °C) in a 15-ml polypropylene conical tube. Remove as much liquid as possible, and freeze by immersing tube in liquid nitrogen. Store frozen cell pellet at − 80 °C.
3. Transfer frozen tube containing cell pellet to ice, working quickly to avoid thawing prior to the addition of phenol.
4. To frozen pellet, first add 2 ml phenol, then 2 ml extraction buffer (50 mM sodium acetate [from pH 5.3 stock], 10 mM EDTA and 1% SDS). The use of acidic (pH ∼4.5) rather than neutral phenol will reduce, but not eliminate, DNA contamination. *While working with phenol and chloroform, use appropriate protective equipment (goggles, gloves, lab coat, fume hood) and dispose of hazardous waste appropriately.*
5. Ensure that each tube is well-capped, then mix by vortexing. Transfer tube to 65 °C water bath. Incubate for 10 min, removing to vortex vigorously every minute or so.
6. Transfer tube to ice for 5 min. Keep samples on ice for all subsequent steps, except where noted.
7. Add 2 ml chloroform to tube, cap securely, then mix well by vortexing.
8. Spin tube in tabletop centrifuge (2000×g, 5 min, 4 °C) to separate phases, along with an empty 15 ml Heavy Phase Lock Gel tube (Eppendorf) for use in the next step.

9. Carefully remove aqueous (top) phase by pipetting, avoiding as much material at the interface as possible. Transfer to the prespun Phase Lock tube along with 2 ml phenol:chloroform. (Use an equal volume mixture of phenol and chloroform; this can be either neutral or acidic, with or without isoamyl alcohol.) Cap tube and shake by hand, but do not vortex, as this may disrupt the gel matrix.

10. Spin tube in tabletop centrifuge ($1500 \times g$, 5 min, 4 °C). Organic phase should partition below the gel matrix.

11. Add 2 ml chloroform, then shake and spin as before. (This step removes residual phenol from the sample.)

12. From this point on, ensure that all reagents and containers are free of RNases. Pour aqueous phase into a fresh conical tube. Add 200 μl 3 M sodium acetate (pH 5.3) and 2 ml isopropyl alcohol. Shake or vortex briefly, then incubate at room temperature for 10 min.

13. Pellet RNA in tabletop centrifuge at maximum speed (20 min, 4 °C). A substantial white pellet of RNA should be visible. Pour off supernatant and let drain briefly with tube inverted on a Kimwipe.

14. Use 800 μl 70% ethanol to transfer pellet by pipet to a 1.5-ml microfuge tube, breaking up pellet if necessary. Spin in 4 °C microcentrifuge at maximum speed for 5 min.

15. Carefully remove as much liquid as possible from pellet with pipet tip, then air dry for a few minutes. Do not allow RNA pellet to become too dry, as this will make resuspension difficult.

16. Resuspend RNA in ~200 μl RNase-free water. Pipetting up and down will help to resuspend the RNA; if necessary, the tube can also be incubated at 50 °C for 10 min.

17. To determine the concentration of RNA in solution, measure its absorbance in an ultraviolet spectrophotometer. If using a NanoDrop (Thermo Scientific), 2 μl of solution can be measured directly. Otherwise, dilute 1:100 to measure. The concentration of the measured solution (in ng/μl) is given by the absorbance at 260 nm multiplied by 40. Expected yield is roughly 400 μg.

18. RNA can be stored at −80 °C, then thawed slowly on ice for use. For downstream applications that may be compromised by contaminating DNA (such as quantitative RT-PCR), RNA should first be treated with an RNase-free DNase (e.g., RQ1 DNase from Promega).

4. C-TERMINAL EPITOPE TAGGING IN *C. ALBICANS*

This protocol relies on homologous recombination to integrate the coding sequence for a C-terminal epitope tag in place of the stop codon for any gene at its endogenous locus. The pADH34 vector contains the coding sequence for a 13× myc repeat, while pADH52 encodes a 6-His/FLAG tandem affinity purification (TAP) tag. As both of these constructs use the same linker sequence, either tag can be amplified with a single pair of PCR primers. Briefly, long oligonucleotides (typically 90–120 bp total) are used to amplify a 4.8-kb DNA fragment which, when integrated into the genome, will replace the stop codon of the target gene with the epitope tag coding sequence, followed by the SAT1/flipper cassette. Upon confirmation of integration at the desired locus in nourseothricin resistant colonies, the SAT1/flipper cassette is excised, leaving only the epitope tag coding sequence and a minimally disruptive FLP recombinase target sequence behind. The SAT1/flipper cassette and marker excision procedure was developed by Reuss *et al.* (2004).

4.1. Primer design

Synthesize a "forward knock-in primer" encompassing the sense strand sequence of the target gene up to, but not including, the stop codon. In place of the stop codon, add the sequence "CGGATCCCCGGGT-TAATTAACGG" to the 3' end of the forward knock-in primer. To generate the "reverse knock-in primer," take the reverse complement of the sequence immediately downstream of the stop codon and add the sequence "GGCGGCCGCTCTAGAACTAGTGGATC" to the 3' end.

4.2. PCR conditions

Perform 30–35 cycles of amplification with pADH34 or pADH52 and the knock-in primers using Ex Taq (Takara) or a similar increased fidelity/high-activity thermostable polymerase. Use a three-step program for the first five cycles, with annealing at 58 °C and 5 min extensions at 72 °C. For the remaining cycles perform a 2-step program, eliminating the annealing step. To minimize the chances of acquiring PCR generated mutations in the knock-in cassette, perform three independent PCR reactions for each target gene and pool the reactions following amplification.

4.3. Transformation

Directly transform ~20–50 µl of the knock-in cassette PCR product (without purification) into the target strain using standard *C. albicans* transformation methods as described above. Following transformation, but prior to selection, wash the cells twice with YEPD then split to two independent 5 ml cultures (to insure isolation of independent clones) and recover for at least 5 h at 30 °C. Pellet and plate the entire culture onto YEPD + 400 µg/ml nourseothricin. *Note*: addition of adenine and/or uridine to the growth medium, even with prototrophic strains, can increase the efficiency of several steps in this protocol.

4.4. Integration confirmation

Screen nourseothricin resistant colonies by colony PCR with the following primers:

Upstream flank check: Use a primer that hybridizes ∼500 bp upstream of the target gene stop codon (extending toward the stop codon) and AHO300. (CCGTTAATTAACCCGGGGATC). AHO300 anneals to the linker sequence, which is common to both pADH34 and pADH52, and extends into the tagged ORF.
Downstream flank check: Use a primer that hybridizes ∼500 bp downstream of the target gene stop codon (extending toward the stop codon) and AHO301 (GGAACTTCAGATCCACTAGTTCTAGAGC), which anneals to both pADH34 and pADH52.

4.5. SAT1 marker excision

To induce excision of the SAT1/flipper cassette, culture nourseothricin resistant strains in YEP-maltose (2%) for at least 5 h (or overnight) at 30 °C and plate ∼100 cells/plate on YEPD +25 μg/ml nourseothricin. (Note that some mutant strains are hypersensitive to nourseothricin, and lower concentrations (≤5 μg/ml) may be necessary.) Following 1–2 days of growth at 30 °C, small, medium, and large colonies should be observed. Patch small and medium sized colonies on to YEPD +400 μg/ml nourseothricin and onto YEPD without selection to screen for nourseothricin sensitive colonies. To confirm excision of the SAT1/FLP cassette, perform colony PCR with either AHO302 (TCACTAGTGAATTCGCGCTCGAG, for myc tagging with pADH34) or AHO405 (TAAATAATGAATTCGCGCTCGAG, for TAP tagging with pADH52) and the downstream flank check primer described above.

4.6. Tag sequence confirmation

To confirm that the target ORF and the epitope tag are free of mutations, perform colony PCR using a high fidelity polymerase and the following primers: AHO283 (GGCGGCCGCTCTAGAACTAGTGGATC, common to both pADH34 and pADH52) and the upstream flank check primer designed above. AHO283 anneals to the 3′ end of the residual SAT1/FLP cassette sequence (including the FRT) and extends toward the tagged gene. Following colony PCR amplification from the tagged strain, purify the PCR product and sequence with AHO283 as the sequencing primer.

Note: The pADH34 myc tagging construct inserts 629 bp in place of the original stop codon, while the pADH52 TAP tagging construct inserts 152 bp. To determine the expected size of the PCR product for sequencing, add this number to the distance between the upstream flank check primer and the original stop codon of the target gene.

4.7. Schematic of the 13× myc tagging procedure

Note: The following figures outline the myc tagging protocol, which uses pADH34. Refer to the text above for a description of the minor variations in this process that are specific to TAP tagging with pADH52.

A Schematic diagram of pADH34, including the regions to which the forward and reverse knock-in PCR primers hybridize

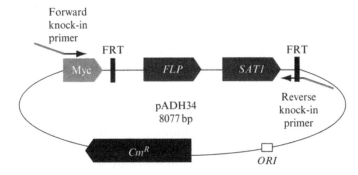

B Schematic of the integrated knock-in cassette and locations of flank check primers

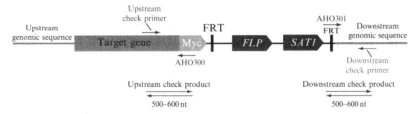

C Schematic of the tagged gene following excision of the SAT1/FLP cassette

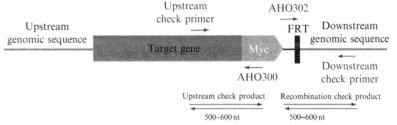

5. *C. ALBICANS* CHROMATIN IMMUNOPRECIPITATION

Chromatin immunoprecipitation (ChIP) procedures with *C. albicans* are comparable overall to those used with *S. cerevisiae* and mammalian cells, and the following protocol is based on standard ChIP methods (For example, see Lee *et al.*, 2006). We have found, however, that the methods used for cell lysis and DNA shearing are critical for performing high-resolution genome-wide ChIP (ChIP-chip) experiments with *C. albicans*. The following protocol has been used successfully, with reproducible results, to perform high-resolution ChIP-chip experiments with planktonic cultures of *C. albicans*, *Kluyveromyces lactis*, and *S. cerevisiae* (Tuch *et al.*, 2008). This protocol has also been used, with increased lysis times, to perform ChIP with *C. albicans* biofilms (Nobile *et al.*, 2009) and with both yeast and mycelial forms of *Histoplasma capsulatum* (Nguyen and Sil 2008; Webster and Sil, 2008). We also describe a rapid method for amplification of ChIP DNA samples and hybridization to high-density oligonucleotide tiling arrays.

In the previous section, we described a method for C-terminal epitope tagging in *C. albicans* that can be used to rapidly tag genes of interest for ChIP. Although the use of affinity-purified polyclonal antibodies raised against a unique peptide within a protein of interest is arguably a less disruptive method of immunoprecipitation, there are several drawbacks to such an approach. These "peptide antibodies" are costly, take time to produce, and often require extensive optimization for ChIP experiments. To control for cross-reactivity, which is often a problem with peptide antibodies, a viable gene deletion strain is required as a negative control, making ChIP results with essential genes much more difficult to validate. Lastly, at least two different peptides from each protein of interest should be used to raise antibodies, as it is not unusual to have one or both sets of antibodies fail completely in ChIP experiments. C-terminal epitope tagging and immunoprecipitation with commercially available, high-specificity monoclonal antibodies offers a rapid, economical, and effective method to circumvent many of these problems.

5.1. Chromatin immunoprecipitation protocol

Step 1: Culture growth and cross-linking

1. Grow 200–400 ml of cells to OD = 0.4

    Note: ∼200 ml of OD = 0.4 is sufficient for one batch of lysate, which is sufficient material for as many as 10 individual ChIPs.

2. Add fresh formaldehyde (if previously opened, use within 1 month) to final concentration of 1% (stock is 37%) and cross-link 15 min at room temperature with occasional mixing.
3. Quench cross-linking by adding 2.5 M glycine (make fresh) to a final concentration of 125 mM and incubate for 5 min at RT.
4. Collect cells by centrifugation for 10 min at 1000×g in a fixed-angle centrifuge rotor.
5. Decant and resuspend pellets in 10 ml ice-cold TBS and transfer to 15 ml Falcon tubes, pellet, decant, and repeat wash 1 more time, then resuspend pellet in 2 ml ice-cold TBS. Split cell suspension to two 2 ml Sarstaedt tubes, pellet, decant, and freeze pellets in liquid nitrogen. Store at − 80 °C or proceed to step 2 (skip freezing).

Step 2: Cell lysis and immunoprecipitation

1. Thaw cell pellets on ice, weigh the pellet (tare scale w/empty tube) and resuspend pellets in 700 μl ice-cold lysis buffer with protease inhibitors (Add protease inhibitors immediately prior to use at the following final concentrations: 1 mM PMSF, 1 mM benzamadine, 1 μg/ml each leupeptin, pepstatin, and bestatin; alternatively, Roche complete protease inhibitor cocktail (EDTA-free, catalog #11836170001) can be used; mix 1 Roche tablet with 10 ml lysis buffer).
2. Transfer cell suspension to a fresh 1.75-ml microfuge tube filled to the 500 μl mark with 0.5 mm glass beads.
3. Place in an Eppendorf mixer (part #5432), clamped vortex genie, or equivalent for ∼2 h at 4 °C.
   Note: Mixing times may vary, depending on cell type and growth conditions. For example, this technique has been used successfully with biofilms by extending the mixing time to >4 h on a vortex mixer.
4. Check cell lysis under the microscope and if >90% of cells are lysed, proceed to step 5.
   Note: Cells should appear as a mixture of "ghosts" and fragmented cell debris by phase contrast microscopy.
5. *Recover the lysate*: Invert the tubes containing lysate/beads and wipe with 70% ethanol. Allow to dry then pierce the bottom of the tube with a 26-gauge needle. Open the tube and place it (right side up) into a 5-ml falcon tube and pierce the falcon tube (above the level of the bottom of the microfuge tube) with an 18-guage needle attached to a vacuum line. The lysate should flow through to the bottom of the falcon tube (alternate: recover by centrifugation into a larger tube). Recover the lysate and transfer 300 ml to each of two fresh 1.75 ml microfuge tubes (for Bioruptor shearing) or transfer entire lysate to one fresh 1.75 ml tube (for microtip sonication).

6. Shear chromatin by sonication in a Diagenode Bioruptor™ (15 min, high setting, 30 s on, 1 min off) or with a microtip sonicator (5×20 s at level 2, 100% duty cycle, with 1 min on ice between each pulse).

   Note: Shearing with a Bioruptor yields smaller fragment sizes, tighter shear distribution, and greater consistency than the tip sonication method and is highly recommended for ChIP-chip applications.

7. Pellet cell debris for 5 min at 14,000 rpm at 4 °C and transfer the supernatant (lysate) to a fresh tube.

8. Remove 50 μl of the lysate and transfer to 200 μl TE/1% SDS. This is the "input DNA" sample which can be stored at −20 °C until the end of step 3 when it is processed along with the immunoprecipitated DNA.

9. Aliquot and dilute sheared lysate according to the number of IPs to be perfomed. For each IP, use 50–500 μl of crude lysate in 500 μl (final volume) lysis buffer (with fresh protease inhibitors). The relative amounts of lysate in each IP can be equalized between strains or samples by normalizing against the mass of each cell pellet.

10. Add antibody (typically 5 μg of affinity-purified polyclonal antibody or 2 μg of monoclonal anti-myc antibody) and incubate overnight at 4 °C on a nutator.

11. The next day, add 50 μl of a 50% slurry of protein-A or protein-G Sepharose beads (washed two times with TBS and three times with lysis buffer) and incubate at least 2 h at 4 °C on a nutator.

Step 3: Recovery of immunoprecipitated DNA

1. Wash beads as follows:
   Pellet 1 min at 1000×g and draw off the supernatant with an 18-gauge needle on a vacuum line. Wash with the buffers indicated below for 5 min each with mixing on a nutator:
   2× with 1 ml lysis buffer
   2× with 1 ml lysis buffer w/500 mM final NaCl
   2× with 1 ml Wash buffer
   1× with 1 ml TE

   Note: Although Wash buffer temperatures, incubation temperatures, and incubation times can all be optimized for each antibody, we have found that ice-cold buffers and 5 min incubations at room temperature work best for most antibodies.

2. After the last wash, draw off TE and add 110 μl elution buffer, vortex, and incubate 10 min at 65 °C, mixing every 2 min.

3. Pellet 30 s at 14,000 rpm at room temperature and remove 100 μl to a fresh tube.

4. Add 150 μl TE + 0.65% SDS and vortex vigorously. Pellet and remove 150 μl and pool with previous eluate (250 μl final).

5. Incubate IP and "input DNA" samples (from step 2) for ∼16 h at 65 °C.

Step 4: Cross-link reversal and DNA cleanup

1. Add 250 μl of proteinase K mix (for each sample: 238 μl TE, 2 μl 5 mg/ml glycogen, 10 μl 10 mg/ml proteinase K) and incubate 2 h at 37 °C.

   Note: Make a fresh proteinase K solution each time from lyophilized powder.

2. Add 55 μl 4 *M* LiCl and 500 μl phenol:chloroform:isoamyl alcohol (25:24:1), pH 8.0. Vortex briefly, then spin 1 min at $> 10,000 \times g$ and remove 500 μl of the aqueous layer to a fresh tube.

   Note: AMRESCO Biotechnology Grade phenol:chloroform:isoamyl alcohol (code 0883-100 ml) has provided reliable performance in this protocol.

3. Add 1 ml ice-cold 100% ethanol and incubate at -20 °C overnight or at least 1 h at -80 °C.

4. Centrifuge 30 min at $> 10,000 \times g$ at 4 °C and decant carefully with a 1-ml pipette.

5. Wash pellet 1× with 70% EtOH, spin 5–10 min, decant, spin briefly, and remove residual EtOH.

6. Air dry the pellets and resuspend. Use 25 μl TE for IP samples and 100 μl TE + 100 μg/ml RNaseA for input DNA samples.

7. Incubate input DNA/RNaseA solution for 1 h at 37 °C, then store at -20 °C.

   Note A: An optional DNA cleanup step could be performed on the input DNA following this step (i.e., a commercial DNA cleanup kit), however, this adds an additional variable (relative to the IP'd DNA) and may actually contribute to "spiky" data in ChIP-chip experiments. It is probably safest to leave the RNaseA in the input DNA sample and avoid any cleanup steps prior to amplification.

   Note B: Although chromatin shearing with the Bioruptor is highly reproducible (assuming cell lysis is $>90\%$), it is advisable to monitor sheer distribution of the input DNA sample prior to proceeding with subsequent analysis of ChIP samples. Test the sheer distribution by running ~ 200–500 ng of purified input DNA (purify an aliquot with a DNA purification mini column) on a 2% agarose gel at ~ 5 V/cm. The average sheer size from the Bioruptor is typically ~ 200 bp, with most fragments distributed between 100 and 400 bp.

5.2. Chromatin immunoprecipitation buffers

Be sure to use autoclaved ddH$_2$O and baked glassware when making buffers to avoid DNA contamination. This caution is especially important for the final Wash buffers and post-elution steps.

TBS: 20 m*M* Tris/HCl (pH 7.5), 150 m*M* NaCl

Lysis buffer: 50 mM HEPES/KOH (pH 7.5), 140 mM NaCl, 1 mM EDTA, 1% Triton X-100, 0.1% Na-deoxycholate

Lysis buffer w/500 mM NaCl: same as above, increase total NaCl concentration to 500 mM

Wash buffer: 10 mM Tris/HCl (pH 8.0), 250 mM LiCl, 0.5% NP-40, 0.5% Na-deoxycholate, 1 mM EDTA

Elution buffer: 50 mM Tris/HCl (pH 8.0), 10 mM EDTA, 1% SDS

TE/0.67% SDS: 10 mM Tris/HCl (pH 8.0), 1 mM EDTA, 0.67% SDS

TE/1% SDS: 10 mM Tris/HCl (pH 8.0), 1 mM EDTA, 1% SDS

4 M LiCl

2.5 M glycine (fresh)

10 mg/ml proteinase K in TE (fresh)

5 mg/ml glycogen (in TE)

5.3. Strand displacement amplification of ChIP samples

The following protocol uses high concentration exo$^-$ Klenow (New England Biolabs #M0212M) and random DNA nonamers to perform strand displacement amplification of the input and IP DNA samples from ChIP experiments. Prior to amplification, input and IP DNA concentrations are normalized by dilution of the input DNA for each corresponding IP based on the qPCR values for a nonenriched locus, such as the ADE2 ORF (primers AHO294: GTTGTCAGATCATTAGAAGGGGAAG and AHO295: AAGTATCTGGGATCCTGGCA). Input and IP samples are amplified separately, in parallel, and should yield similar amounts of product following each round of amplification. Typically, three rounds of amplification are required prior to dye coupling and hybridization of the ChIP samples. If the IP DNA concentration is sufficient, Round B amplification can be omitted. Since this is a nonspecific amplification, all DNA will be amplified by this approach; all amplification steps should be performed with clean gloves, filter tips, autoclaved ddH$_2$O and dedicated reagents which are free of any contaminating DNA.

- *Round A (primary amplification):*
  1. Mix:
     - 12 μl of IP sample or diluted input (diluted in TE)
       Note: equalize the input and IP samples based on qPCR values for a nonenriched locus.
     - 12 μl H$_2$O
     - 20 μl 2.5× SDA buffer
  2. Incubate 95 °C, 5 min then immediately transfer to an ice water bath for 5 min.
  3. Add 5 μl dNTP mix (1.25 mM each nucleotide).
  4. Add 1 μl 50 U/μl exo$^-$ Klenow (NEB).

5. Incubate 2 h at 37 °C with heated lid in a thermal cycler.

   Note: If needed, let the reactions sit up to ~2 h at 10 °C following amplification or add 5 μl 0.5 M EDTA and store at −20 °C.

6. Purify product with Zymo[25] columns (Zymo Research):

   Add at least 3 volumes of binding buffer, bind, wash one time with 200 μl binding buffer, two times with 200 μl Wash buffer, spin 1 min at 10,000×g to dry, and elute with 30 μl H_2O into a fresh tube.

7. Check 1.5 μl on a NanoDrop spectrophotometer (Thermo Scientific). If ≥400 ng total, skip to Round C. Otherwise continue with Round B.

- *Round B (secondary amplification)*:
  1. Mix:
     - 24 μl Round A DNA
     - 20 μl 2.5× SDA buffer
  2. Repeat steps 2–7 of Round A, but elute with 50 μl H_2O.
- *Round C (aminoallyl-dUTP incorporation and final amplification)*:

Preferred approach:

Perform 100 μl reactions with 1–2 μg total Round B DNA for each sample. Yields only ~2.5- to 3-fold amplification, but dye coupling is still very efficient.

1. Mix:
   - 1–2 μg of Round B DNA + H_2O to 48 μl total
   - 40 μl 2.5× SDA
2. Incubate 95 °C, 5 min then immediately transfer to an ice water bath for 5 min.
3. Add 10 μl 1.25 mM aminoallyl-dNTP mix (1:10 dilution of stock solution).
4. Add 2 μl 50 U/μl exo⁻ Klenow.
5. Incubate 2 h at 37 °C with heated lid in a thermal cycler

   Note: If needed, let the reactions sit up to ~2 h at 10 °C following incubation or add 5 μl 0.5 M EDTA and store at −20 °C.

6. Purify Round C product with Zymo[25] columns: Add at least 3 volumes of binding buffer, bind, wash one time with 200 μl binding buffer, two times with 200 μl Wash buffer, spin 1 min at 10,000×g to dry, and elute with 50 μl H_2O into a fresh tube.

7. Check 1.5 μl on NanoDrop; the yield should be ~5 μg of total DNA per reaction.

Alternate Round C approach:

If Round B yields less than 1 μg total DNA, set up 2× 100 μl Round C reactions for each sample, using 200–400 ng of Round B DNA per tube. Perform amplification and cleanup as described for the preferred approach, but pool the two independent reactions prior to the Zymo[25] column purification.

5.4. Strand displacement amplification solutions

2.5× SDA mix: (best if made fresh, but can be kept at −20 °C for up to 1 month)
- 125 mM Tris–HCl, pH 7.0
- 12.5 mM MgCl$_2$
- 25 mM BME
- 750 μg/ml random DNA nonamers (dN9)

10× aminoallyl-dNTP stock solution:
- 12.5 mM dATP
- 12.5 mM dCTP
- 12.5 mM dGTP
- 5 mM dTTP
- 7.5 mM aa-dUTP

5.5. Dye coupling

1. Speed-vac amplified input and IP reactions from Round C to ≤9 μl each, or until dry.
2. Resuspend or QS to 9 μl with H$_2$O and add 1 μl of fresh 1 M Na bicarbonate, pH 9.0.
   Note: Prepare Na bicarbonate on the day of labeling and carefully pH using a pH meter.
3. Immediately add 1.25 μl Cy3 (input sample) or Cy5 (IP sample)
   Note: We use Amersham monoreactive dye packs (Cat. #PA23001 and PA25001). Each tube contains sufficient die for eight labeling reactions. Resuspend the dye in 10 μl DMSO and use 1.25 μl of dye per labeling reaction. If fewer than eight reactions will be performed, either decrease the volume of DMSO to use the entire tube or aliquot and speed-vac the unused dye. Store any unused dye under desiccation at 4 °C, protected from light.
4. Incubate labeling reactions for 1 h at room temperature in darkness.
5. Purify dye-coupled DNA with Zymo[25] columns (Zymo Research):
   Add 800 μl of Zymo DNA binding buffer to each sample and load onto a Zymo[25] column. Wash one time with 200 μl DNA binding buffer, two times with 200 μl Wash solution, spin 1 min at 10,000×g to dry, then elute with 50 μl H$_2$O. Check 1.5 μl on a NanoDrop spectrophotometer using the "microarray" setting to quantitate the total yield and dye-coupling efficiency (Typically > 20 picomoles of dye per microgram of DNA).
6. Proceed to array hybridization, following the array manufacturer's guidelines. Equalize the input and IP samples to 5 μg each for a 1×244 K format Agilent microarray.

Note: We have found Agilent custom oligonucleotide arrays, hybridization buffers, and Wash buffers to consistently yield high-quality data. The following hybridization protocol was adapted from the Agilent oligo aCGH/ChIP-on-Chip hybridization kit.

5.6. ChIP-chip hybridization protocol (adapted from the Agilent oligo aCGH/chip-on-chip hybridization kit)

This protocol is for competitive hybridization of amplified, dye-coupled ChIP and input DNA using Agilent 1×244 K format oligonucleotide tiling array. While a newer version of this protocol can be found on the Agilent website, we include this protocol and notes for convenience. Please follow manufacturer's guidelines for other array formats.

1. Mix 5 μg each (input and IP) sample and bring volume to 150 μl with H_2O.
   Note: Less DNA can be used; as little as 1 μg each of input and IP samples have been successfully hybridized and scanned with no significant decrease in data quality.
2. Add 50 μl of 1 mg/ml Human Cot-1 DNA (Invitrogen).
3. Add 50 μl of $10 \times$ Agilent blocking agent.
4. Add 250 μl of Agilent hybridization buffer.
5. Mix thoroughly then quick-spin to collect.
6. Incubate 3 min at 95 °C then transfer immediately to 37 °C for 30 min.
7. Spin 1 min at full speed in microcentrifuge then carefully remove 490 μl, load onto gasket slide, cover with array slide, and assemble hybridization chamber.
8. Hybridize for $\sim$40 h at 65 °C in an Agilent microarray hybridization oven with the rotation speed set at "20."
9. Disassemble the array and wash using Agilent Wash buffers.
   a. Incubate array 5 min with mixing in Agilent oligo aCGH/ChIP-on-Chip Wash Buffer 1 at 25 °C.
   b. Incubate 5 min with mixing in Agilent oligo aCGH/ChIP-on-Chip Wash Buffer 2 at 32 °C.
   c. Incubate 1 min with mixing in acetonitrile at 25 °C.
   d. Incubate 30 s with agitation in Agilent drying and stabilization solution.
      Note: To ensure even drying, very slowly remove the slide holder such that $\sim$10 s elapse prior to complete removal from the solution.
      Note: For disassembly, hold the microarray/gasket slide "sandwich" submerged in Wash buffer 1 while gently gripping sides of the microarray slide. Carefully pry the gasket slide off of the array by inserting the tip of a plastic forceps between the outer edge of the two slides and lightly twisting the forceps. The gasket slide will fall away, while the array should remain in your hands. Be sure to avoid any contact with the printed array surface.

REFERENCES

Lee, T. I., Johnstone, S. E., and Young, R. A. (2006). Chromatin immunoprecipitation and microarray-based analysis of protein location. *Nat. Protoc.* **1,** 729–748.

Nguyen, V. Q., and Sil, A. (2008). Temperature-induced switch to the pathogenic yeast form of *Histoplasma capsulatum* requires Ryp1, a conserved transcriptional regulator. *Proc. Natl. Acad. Sci. USA* **105,** 4880–4885.

Nobile, C. J., Nett, J. E., Hernday, A. D., Homann, O. R., Deneault, J., Nantel, A., Andes, D. R., Johnson, A. D., and Mitchell, A. P. (2009). Biofilm matrix regulation by *Candida albicans* Zap1. *PLoS Biol.* (in press).

Noble, S. M., and Johnson, A. D. (2005). Strains and strategies for large-scale gene deletion studies of the diploid human fungal pathogen *Candida albicans*. *Eukaryot. Cell* **4,** 298–309.

Reuss, O., Vik, A., Kolter, R., and Morschhauser, J. (2004). The SAT1 flipper, an optimized tool for gene disruption in *Candida albicans*. *Gene* **341,** 119–127.

Tuch, B. B., Galgoczy, D. J., Hernday, A. D., Li, H., and Johnson, A. D. (2008). The evolution of combinatorial gene regulation in fungi. *PLoS Biol.* **6,** e38.

Webster, R. H., and Sil, A. (2008). Conserved factors Ryp2 and Ryp3 control cell morphology and infectious spore formation in the fungal pathogen *Histoplasma capsulatum*. *Proc. Natl. Acad. Sci. USA* **105,** 14573–14578.

CHAPTER THIRTY-TWO

MOLECULAR GENETICS OF SCHIZOSACCHAROMYCES POMBE

Sarah A. Sabatinos *and* Susan L. Forsburg

Contents

Department of Molecular and Computational Biology, University of Southern California, Los Angeles, California, USA

Methods in Enzymology, Volume 470
ISSN 0076-6879, DOI: 10.1016/S0076-6879(10)70032-X

Abstract

In this chapter we present basic protocols for the use of *Schizosaccharomyces pombe*, commonly known as fission yeast, in molecular biology and genetics research. Fission yeast is an increasingly popular model organism for the study of biological pathways because of its genetic tractability and as a model for metazoan biology. It provides an alternative and complimentary approach to *Saccharomyces cerevisiae* for addressing questions of cell biology, physiology, genetics, and genomics/proteomics. We include details and considerations for growing fission yeast, information on crosses and genetics, gene targeting and transformation, cell synchrony and analysis, and molecular biology protocols.

1. INTRODUCTION

Schizosaccharomyces pombe, or fission yeast, is an archaeascomycete that is evolutionarily remote from budding yeast (approximately 10^9 years separation) (Heckman *et al.*, 2001). While the genomes are similar in size (13.8 Mb for *S. pombe*, 12.1 Mb for *Saccharomyces cerevisiae*), they are neither related nor syntenic. Fission yeast is much less likely than budding yeast to have duplicated genes (Hughes and Friedman, 2003), although it is more likely to have introns (Wood *et al.*, 2002). The fission yeast genome is divided between just three chromosomes, with large and complex replication origins and centromeres that are models for metazoan structures (Forsburg, 1999). Additionally, it shares some genes with humans that are missing from budding yeast (Aravind *et al.*, 2000), which makes it a complementary experimental system to budding yeast.

S. pombe offers the traditional yeast strengths of excellent genetics and an easily manipulated genome. The tools for fission yeast are conceptually similar to those for budding yeast, although their biological details are specific to the *S. pombe* system. The fission yeast community is smaller than the budding yeast community, and has traditional strengths in cell growth and division, differentiation, DNA replication and repair, and chromosome dynamics. More recently, investigators have used *S. pombe* to study a variety of other cell biology problems including signal transduction, RNA splicing, and cell morphology. An international fission yeast conference is held every 2–3 years, alternating between Europe, Japan, and North America. There is also a list-serv ("pombelist") and a number of Web-based resources (Wood and Bahler, 2002) (refer to Appendix).

In this chapter we provide essential methods for *S. pombe* growth and manipulation suitable for a beginning lab. Other excellent reviews on *S. pombe* biology and methods may be found in Egel (2007), Forsburg (2003a), and Forsburg and Rhind (2006).

2. Biology, Growth, and Maintenance of Fission Yeast

Fission yeasts are rod-shaped cells that divide by medial fission. Wild-type cells are typically 8–14 μm in length and 4 μm wide, and grow primarily in length, not in width. Diploids are proportionately larger in all dimensions. The cell cycle is divided into distinct G1 (10%), S (10%), G2 (70%), and M (10%) phases. In exponentially growing wild-type cells, the nuclear division cycle is staggered relative to cell division and newly divided nuclei enter the next cell cycle and undergo G1 and S phase prior to cytokinesis. Thus, upon completion of cell division, the newborn cell is already in G2 phase. For this reason, a single cell particle almost always has a 2C DNA content, either because of the extended G2 phase, or the binucleate G1 or S phase cells. Because of the strict control of cell size, overall length is an effective metric for the position in the cell cycle.

The essentials of *S. pombe* culture are identical to those used for *S. cerevisiae*; they differ in growth rate and media selection. Cultures are usually initiated on rich medium and replica plated to test markers and ploidy. Media formulations are detailed in Tables 32.1 and 32.2. Yeast-extract medium, YE or YES (+supplements), is rich but poorly defined, and is used for initial growth and/or recovery. Synthetic and well-defined media such as Edinburgh minimal media 2 (EMM2) or its derivative, pombe minimal glutamate (PMG) is used in experiments requiring stable physiological conditions or maintenance of auxotrophic markers. Appropriate supplements are added as needed, and single marker dropouts used to determine auxotrophy. Cells will grow on *S. cerevisiae* media, but *S. pombe* media is optimized for this species, and is recommended. Mating media are described in Table 32.2.

2.1. Other media supplements—Phloxin B

Phloxin B (PB; Magdala Red, Sigma P4030) is a vital stain used in agar plates to identify compromised cells. Healthy cells are efficient at eliminating PB, making pale pink colonies. Sick cells retain more PB, leading to a dark pink or red color. This is useful for rapid screening of replica plates, for example, screening for drug sensitivity, isolating temperature sensitive mutations, or identifying auxotrophs on medium lacking supplements. Additionally, diploid cells are also less efficient at removing PB and thus diploid colonies are darker pink than haploid. Assessment is based upon visual inspection of colonies, as the color differences are not apparent in single cells. It is useful to have a known, healthy haploid streaked alongside test strains for comparison. PB is mildly toxic and is only used for short-term

Table 32.1 Growth media for fission yeast, *Schizosaccharomyces pombe*

Name	Use	Recipe	Concentration	Notes
Rich media: Poorly defined because of variations in yeast extract and preparation, this media supports vigorous growth with added glucose. YE is a naturally low-adenine medium. Additional supplements are added to maximize growth rate				
Yeast extract (YE)	Vegetative growth, thiamine repression[a]	5 g/l yeast extract 30 g/l glucose	0.5% (w/v) 3.0% (w/v)	• Inhibits conjugation and sporulation
Yeast extract + supplements (YES)[b,c]	Vegetative growth; maintaining diploids	YE base plus: 225 mg/l each of adenine, L-histidine, L-leucine, uracil, and L-lysine		• Medium of choice for most nonphysiological growth; inhibits conjugation and sporulation
YES + Phloxin B	• Screening diploids • Temperature or drug sensitivity detection	YES plus: 5 mg/l phloxin B[d] (Sigma, P4030)		• Make 2000× PB stock as 10 g/l in water and filter sterilize. Store at room temperature in the dark • Store plates in dark, room temperature, up to 1 month. Discard upon color loss or browning
Synthetic media: All these media are derived from the buffered defined media called EMM. Variations in carbon source, nitrogen source, and additional supplements may be required for different uses. EMM-glutamate, called PMG, is used as a standard medium in some labs				
Edinburgh minimal medium 2 (EMM)[b]	Vegetative growth	3 g/l potassium hydrogen pthallate 2.2 g/l Na₂HPO₄ 5 g/l NH₄Cl 20 g/l glucose 20 ml/l salts (50× stock) 1 ml/l vitamins (1000× stock) 0.1 ml/l minerals (10,000×)	14.7 mM 15.5 mM 93.5 mM 111 mM	• Defined growth medium • Add supplements (ade, arg, his, leu, lys, ura) from stock solutions, described below

50× salt stock		52.2 g/l MgCl₂·6H₂O	0.26 M	• Made in water and autoclaved. Store stock at 4 °C indefinitely

Name	Description	Composition	Concentration	Notes
50× salt stock		52.2 g/l MgCl$_2$·6H$_2$O	0.26 M	• Made in water and autoclaved. Store stock at 4 °C indefinitely
		0.735 g/l CaCl$_2$·2H$_2$O	4.99 mM	
		50 g/l KCl	0.67 M	
		2 g/l Na$_2$SO$_4$	14.1 mM	
1000× vitamin stock		1 g/l pantothenic acid	4.2 mM	• Made in water and autoclaved. Store stock at 4 °C indefinitely
		10 g/l nicotinic acid	81.2 mM	
		10 g/l inositol	55.5 mM	
		10 mg/l biotin	40.8 μM	
10,000× mineral stock		5 g/l boric acid	80.9 mM	• Made in water and autoclaved. Store stock at 4 °C indefinitely
		4 g/l MnSO$_4$	23.7 mM	
		4 g/l ZnSO$_4$·7H$_2$O	13.9 mM	
		2 g/l FeCl$_2$·6H$_2$O	7.40 mM	
		1 g/l KI	6.02 mM	
		0.4 g/l molybdic acid	2.47 mM	
		0.4 g/l CuSO$_4$·5H$_2$O	1.60 mM	
		10 g/l citric acid	47.6 mM	
Supplements,[e] stock solutions (EMM + supplements)	For auxotrophic markers	7.5 g/l adenine, histidine, leucine, lysine, arginine 3.75 g/l uracil	50–225 mg/l	• Made in water and autoclaved. Add to EMM or PMG as required[f] • Heat uracil gently to solubilize
Minimal low adenine[g]	Screening *ade6⁻* alleles	EMM + required supplements + 7.5 mg/l adenine		• Adenine concentrations from 7.5 to 30 mg/l may be used
Minimal – N	G1 arrest at 25 °C	As for EMM, but omit NH$_4$Cl		• Starvation medium used for G1 arrest
Minimal – N – G	G2 arrest at 25 °C; storing spores	As for EMM, but omit NH$_4$Cl and glucose		• Starvation medium, used to arrest diploids in G2

(continued)

Table 32.1 (continued)

Name	Use	Recipe	Concentration	Notes
Minimal + thiamine[d]	nmt1 promoter repression[a]	As for EMM, add: 5 µg/ml thiamine before use	15 µM thiamine	• 2000× thiamine stock solution is 10 mg/ml in water. Filter sterilize and store in the dark, room temperature • Thiamine can be titrated from 0.01 to 20 µM to regulate expression (Javerzat et al., 1996)
Pombe minimal glutamate (PMG)[h]	Vegetative growth, and screening heterochromatic markers	As for EMM, but omit NH$_4$Cl and add: 3.75 g/l L-glutamic acid, monosodium salt (Sigma G 5889)	22.1 mM	• More even growth between Ura+/− strains • Compatible with G418 selection[b] • Add supplements as appropriate[g]

Media are prepared in bulk batches, and autoclaved at 121 °C, 30.5 psi for 20 min to avoid overcaramelizing sugars. Some fission yeast media are commercially available, and require only mixing powder and water.

Solid media are made by adding 2% Difco Bacto Agar.

[a] When transforming a construct using the nmt1 promoter, we recommend inclusion of thiamine in growth and maintenance steps to prevent toxicity from gene overexpression. This has an added benefit of promoting growth.

[b] Drug plates are typically made with YES to eliminate limiting nutrient effects. When made in EMM, the drug concentration is different and often higher than the effective dose on YES. The drug G418 is most often used in YES, and never in EMM. However, G418 resistance can be studied selectively using PMG.

[c] Rich medium (YE/YES) contains thiamine, as does SD medium from budding yeast, and cannot be used for high-level protein expression in S. pombe under the nmt1 promoter.

[d] Phloxin B and thiamine are always added to cool (60 °C) molten agar.

[e] Supplements are typically added once agar has cooled to 60 °C. Dropout media, in which all possible supplements are added except one, is not widely used in fission yeast, simply because of habit.

[f] Anecdotal evidence suggests that lower supplement concentrations are required in PMG medium, typically 75 mg/l. This is likely due to more efficient uptake of supplements in the glutamate medium than in EMM.

[g] Low adenine medium allows the development of red color in Ade− strains, and can be used to help discriminate ade6 alleles visually. PMG may require a lower amount of adenine for color discrimination (<7.5 mg/l).

[h] PMG may cause unwanted sporulation in some diploid strains. YES is the first choice for growing diploid cultures, followed by EMM for physiological experiments. PMG is preferred when screening heterochromatic silencing markers or G418 selection. Some labs use PMG exclusively.

Table 32.2 Media for conjugation and sporulation

Name	Recipe	Concentration	Notes
Malt extract (ME)	30 g/l bacto–malt extract 225 mg/l each: adenine, histidine, leucine, and uracil Adjust to pH 5.5 with NaOH	3% (w/v)	• Complex nitrogen-limiting medium; not well defined • Contains thiamine • Some batch-to-batch variation
Sporulation agar with supplements (SPAS)	10 g/l glucose 1 g/l KH_2PO_4 1 ml/l $1000\times$ vitamin stock 45 mg/l each: adenine, histidine, leucine, uracil, and lysine hydrochloride	1% (w/v) 7.3 mM 1/5 normal concentrations	• Stringent synthetic mating medium • Benefit: consistency batch-to-batch • Drawback: may produce inefficient mating in sickly strains

Solid media is made by adding 2% agar. Autoclave media for 15–20 min, 121 °C, 30 psi.

testing and not long-term storage. Cells on PB plates will die rapidly if refrigerated, and colonies should be used or transferred to YES agar within a few days of plating on YES + PB. PB plates are light sensitive and should be stored in the dark.

2.2. Other media supplements—Adenine and low-Ade media

The $ade6^+$ gene, the ortholog of $ADE2$ in budding yeast, provides a colorimetric system for plasmid and chromosome maintenance. On conditions of low adenine (Table 32.1), cells induce the adenine biosynthetic pathway and $ade6$ mutants accumulate a red/pink intermediate. Growth on full levels of adenine represses the pathway and colonies remain white.

2.3. Storage of fission yeast

Frozen glycerol stocks are made by mixing 1 volume of a late-logarithmic YES culture with 1 volume of sterile 50% glycerol (v/v in YES or water) in a cryovial, and freezing at $-80\,°C$. Once frozen, strains can be maintained at $-80\,°C$ indefinitely, although there is some loss of viability over long periods (10 years or more).

2.4. Growth in liquid media

A wild-type culture in liquid YES at the optimal temperature of $32\,°C$ has a doubling time of 2–3 h (3–4 h at 25 °C); in EMM/PMG this is slightly longer, usually by 0.5–1 h at $32\,°C$ (+1 h or more at 25 °C). A starter culture is made in a small volume of liquid YES and grown overnight at an appropriate temperature. The starter culture is diluted into selective medium to obtain a physiologically reproducible population for study, using Eq. (32.1):

$$\text{vol}_{SC} = \frac{\text{vol}_C \times OD_{des}}{OD_{SC} \times 2^{(n-1)}} \quad \text{and} \quad n = \frac{\text{time}}{t_D} \tag{32.1}$$

where vol_{SC} is the volume of starter used to inoculate, vol_C is the volume of the overnight culture, OD_{des} is the desired OD_{600} of the culture at a given time, OD_{SC} is the measured OD_{600} for the starter culture at the time of inoculation, and t_D is the doubling time for a given strain. If the starter culture is late-log or stationary $2^{(n-1)}$ is used, but if the culture is in mid-exponential growth this may be changed to 2^n as the first-generation "lag time" will not be as pronounced.

Growth rates can be measured by charting OD_{600} over time in an exponentially growing culture. Growth curves are particularly important for conditional mutants such as temperature sensitive strains, in which the cells are shifted to the restrictive temperature to determine how quickly they stop growing. For most temperature sensitive mutants, temperature arrest will occur within 1–2 generations (4–6 h). However OD_{600} measures cell mass, which may not be the same as cell number in cell cycle mutants that continue growth without dividing. For these phenotypes, accurate cell numbers are determined using a hemocytomer or a Coulter Counter.

3. GENETICS AND PHYSIOLOGY

The *S. pombe* life cycle is predominantly haploid (Fig. 32.1), and in contrast to budding yeast, fission yeast cells only reluctantly make diploids when starved for nitrogen. In normal conditions these diploids are transient, and the zygotes immediately proceed through meiosis and sporulation. Diploids can be recovered in the laboratory by selection for complementing markers and maintained through vegetative growth, but they are unstable and prone to sporulate (see Section 3.6). Azygotic asci produced from

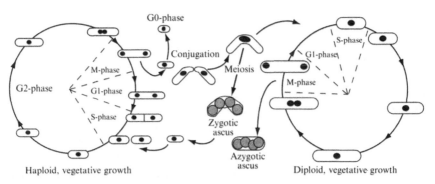

Figure 32.1 Life cycle of *Schizosaccharomyces pombe*. The fission yeast life cycle is predominantly haploid, and divided into distinct phases: G1 (10%), S (10%), G2 (70%), and M (10%). The nuclear division cycle is offset relative to cell division such that newly divided nuclei undergo G1 and S-phase prior to cytokinesis. Consequently, septation index is a metric for the proportion of S-phase cells in a population. Cell length is regulated, and increases throughout G2, and is an indicator of cell cycle position. *S. pombe* cells can be induced to enter a quiescent (G0) state by nitrogen starvation. Starvation also induces expression of mating-type factors that permit conjugation in the presence of a partner of the opposite mating type. Meiosis is quickly initiated in the transient diploid, producing zigzag- or banana-shaped zygotic asci. However, diploid cells may be trapped through complementing markers and induced to enter a diploid vegetative cycle by maintenance on rich medium and at higher temperatures. Upon starvation, diploid cells will quickly complete meiosis and sporulate as linear azygotic asci.

diploids are small and linear. Zygotic asci from coupled mating–sporulation are banana or zigzag-shaped.

Fission yeast are induced to mate by nitrogen starvation, which synchronizes cells in G1 and induces expression of P (Plus, h^+) or M (Minus, h^-) information from the *mat1* locus (Beach *et al.*, 1982; Willer *et al.*, 1995). Homothallic wild-type strains (h^{90}) are capable of switching mating type between h^+ and h^- every second generation with information from silent loci mating loci. The mechanism of switching is conceptually similar, but molecularly distinct from that in budding yeast (Klar, 1992). Approximately half of the cells in an h^{90} population will be h^+ and half will be h^- at a given time. For convenience, common lab strains are generally mating-type stable, or heterothallic, and require a partner of the opposite mating type for conjugation. While most h^- lab strains have lost P-information, h^+ lab strains retain both M- and P-information in a rearranged configuration and are capable of switching at a low frequency in the population from h^+ to h^{90} ($< 10^{-3}$ per generation) (Klar, 1992; Klar *et al.*, 1991).

These differences in mating and sporulation make fission yeast genetics different in practice than budding yeast. Because mating and meiosis are coupled, generally it is not necessary to isolate diploids first; simply cross two strains and let them proceed to sporulation. However, this also means that complementation tests are less convenient than linkage analysis. Some effort is required to maintain diploids, which are always primed to sporulate. Additionally, haploid-specific genes are not shut off in diploids, meaning that diploids express both mating types and can mate with each other to form tetraploids at low but discernable frequencies.

Mating and sporulation are also intrinsically temperature sensitive. Efficient mating requires that cells arrest in the G1 phase of the cell cycle under nitrogen starvation; however, efficiency of G1 arrest declines with increasing temperatures with a higher fraction of cells arresting in G2. Additionally, sporulation is inhibited by high temperature, so maintaining a diploid at temperatures > 30 °C helps reduce unwanted sporulation. Thus, mating and sporulation should optimally be performed at 25 °C and no higher than 29 °C.

3.1. Performing genetic crosses

1. Take a single colony of one strain and smear into a small patch onto an ME- or SPAS-agar plate. With a fresh sterile toothpick, take a single colony of the second strain of opposite mating type, and smear alongside the first.
2. Place 10 μl of sterile water onto the yeast and mix with a fresh sterile toothpick. Allow the patch to dry briefly before inverting and incubating at 25 °C for 2–3 days.

3. Confirm mating microscopically; strains that have mated successfully will produce banana-shaped or zigzag zygotes and asci, and will stain positively with iodine (Section 3.2).

3.2. Testing mating type and sporulation by iodine staining

Successful conjugation and sporulation may be assessed by exposing a mating mixture on plates to iodine vapor, which stains the starch in the ascus walls but not the vegetative cells. However, iodine kills both vegetative cells and spores, so it must be performed on a duplicate plate, and any further analysis performed on the strains from the untreated plate.

1. Patch strains to be tested onto YES in a line. Include h^+ and h^- controls. On a separate YES plate, patch a long line each of h^+ and h^- control (Fig. 32.2). Grow plates for 2–3 days.

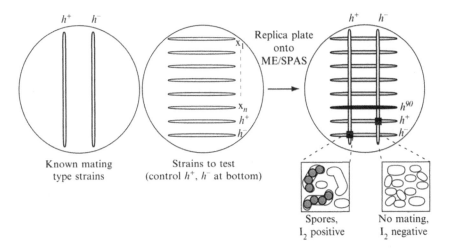

Figure 32.2 Mating-type testing by replica plating. To screen large numbers of unknown strains, a plate of known mating-type strains (one h^+, one h^-) is made and grown on YES (A). At the same time, a plate with unknowns plus a control strain of each mating type is made with lines of culture across the plate, and grown on YES or suitable medium (B). Once the plates have grown, they are replica plated onto ME or SPAS agar, and incubated for 2–3 days at 25 °C. The lines from each plate are perpendicular to each other to create intersecting areas with both known mating-type streaks (C). When exposed to iodine vapor, intersections with sporulated yeast will turn black (shown for the control strains); in the case of a h^{90} strain, the parent will also mate and stain black (line labeled h^{90}). Since mating may be poor or incomplete, patches should be examined microscopically for the presence of zygotes and asci (left box). Unmated cells will be round and starved (right). If replica plated onto *two* ME plates, one plate is used to iodine stain cells (causing death), while the second is used to recover spores.

2. Replica plate the control h^+ and h^- plate onto a ME-agar plate. Change filter. Replica plate the strains for testing onto the same ME-agar plate perpendicularly to the control lines (Fig. 32.2). Incubate plates at 25 °C for 2–3 days.
3. Invert the agar plate over a dish of iodine crystals (Sigma I3380) for 5–10 min. Sporulated patches will stain black, fading to purple with time. Un-sporulated yeast will be orange/cream colored. Strains with switching or mating defects may show a gray, or speckled phenotype. The iodine crystals can be "capped" with an empty plate, and stored in a fume hood for repeated use, adding fresh crystals as needed. Iodine is toxic and volatile, and should only be used in a fume hood.

3.3. Random spore analysis

Random spore analysis (RSA) is the simplest method to analyze a large number of sporulation products. Asci from the cross are incubated in a dilute solution of glusulase (snail gut enzyme) to digest the ascus wall and liberate spores, while killing any vegetative cells that did not mate. This is highly efficient in fission yeast and makes RSA a practical option for many strain constructions, reducing the need for more time-consuming tetrad analysis. Its success depends on the ability to unambiguously identify double mutants, and on screening sufficient numbers for robust statistics to determine appropriate segregation ratios. When in doubt, pull tetrads.

1. Prepare 1 ml of 0.5% glusulase solution (PerkinElmer #NEE154001) in water for each cross. With a sterile toothpick, scoop up some of the mating patch from the mating plate, and resuspend in the enzyme solution. Vortex well and incubate overnight at 25 °C.
2. Count spore concentration in each cross using a hemocytometer. Spores are small, round, and highly refractive, while vegetative cells will have a typical *S. pombe* rod shape and appear darker or fuzzy from the glusulase treatment. Plate no more than 500 spores onto a YES plate. Remaining spores can be washed in water or EMM − N − G for up to a month (survival is better in EMM − N − G).

Grow spores at 25–36 °C for 4–6 days, then replica plate to test markers and determine candidates for analysis.

3.4. Tetrad dissection

Tetrad analysis is performed similarly to *S. cerevisiae* with only a few *S. pombe*-specific considerations. Because the *S. pombe* ascus breaks down very easily, no pretreatment with glusulase is required; however, tetrads must be dissected within a day or two of mating. Cells from a mating mixture are spread onto a YES plate and inspected. Ripe asci with refractive spores are identified and moved to a unique position on the plate using a micromanipulator. Plates are incubated until asci degrade, releasing the spores. This may be carried out at room temperature overnight, or as high as 36 °C (depending on the strain and time requirements) to accelerate ascus breakdown. Upon inspection, the four spores will have popped out of the ascus and can be easily dissected into appropriate positions on the plate. Following colony formation, the tetrad plate may be replica plated to score segregation of different markers.

3.5. Bulk spore germination

For lethal disruptions that have no conditional alleles, a convenient way to assess phenotype is by a mass spore germination experiment where spores from the disruption diploid are inoculated in liquid culture under conditions where only the disrupted spores with the selectable marker can germinate. Their phenotype can then be monitored in a timecourse using morphological analysis, flow cytometry, protein characterization, or other methods.

1. Inoculate a sporulation-competent diploid colony in 10 ml YES and grow to OD_{600} of 0.8–0.9, shaking at 32 °C (25 °C if temperature sensitive).
2. Inoculate 100 ml of liquid ME with 10 ml of the culture from step 1, and shake at 25 °C for 3 days. Check microscopically for sporulation.
3. Harvest cells in a sterile tube. Centrifuge $500 \times g$, 5 min and decant. Resuspend the cells in 10 ml of 2% glusulase (v/v in water), and shake at 25 °C, 150 rpm overnight.
4. Check microscopically that asci and vegetative cells have been digested. Harvest cells in a sterile tube, and centrifuge as above. Wash pelleted cells five times with 15 ml of EMM − N − G, and resuspend in 5 ml EMM − N − G.
5. Prepare 40 ml of sterile 25% glycerol (v/v in water) in 50 ml conical tubes. Overlay the spores. Centrifuge 10 min, $2000 \times g$, room temperature and then carefully remove supernatant. (This step removes cell debris from digestion.)
6. Wash pellet three times with 15 ml EMM − N − G. Resuspend pellet in 10 ml EMM − N − G, and examine. Repeat step 5 to further clean spores if necessary. Store at 4 °C.
7. Plate a small amount onto YES to test markers. Inoculate spores into minimal selective media and sample hourly time points for germination and DNA replication. At 32 °C, S phase occurs between approximately 6 and 8 h.

3.6. Isolating diploids

Some situations require isolating and propagating vegetative diploids. For example, outcrossing a homothallic h^{90} strain (which would otherwise mate with itself), performing synchronous meiosis experiments, or any other experiment requiring the presence of homologous chromosomes depends upon diploid isolation. Diploids are isolated by selecting for complementing markers, and are maintained during growth on YES. It is important to remember that diploids are de-repressed for sporulation, forming azygotic asci upon starvation that may occur even within a single colony. The banana-shaped zygotic asci from mating cells may be distinguished morphologically from the linear azygotic asci produced by sporulating diploids. Diploids maintained for a prolonged time are prone to accumulate mutations that block meiosis and sporulation. For this reason, diploids are generally made freshly when needed and not stored for extended periods.

The intragenic complementing markers *ade6-M210* and *ade6-M216* are an ideal tool for diploid selection. These mutations confer an Ade^+ phenotype if present in the same (diploid) cell, while individually in haploids they are Ade^-. Using these as the complementing markers of choice avoids potential complications from recombinant haploid offspring during strain isolation. If these are not available, any pair of complementing markers may be used, but care must be taken to ensure that the selected strain is a diploid and not a recombinant haploid offspring.

Importantly, because diploids are de-repressed for sexual differentiation, any conditions that lead to nitrogen limitation can lead to unwanted sporulation. For example, colonies on minimal media have likely sporulated in the middle. Thus final screening of isolated diploids on YES medium is essential. High temperatures (>30 °C) also suppress the sporulation response of diploids.

1. Mate strains with complementing markers on ME/SPAS agar (as in Section 3.1).
2. Remove a patch of the mating mixture and streak on selective media at intervals following mating (typically 16 and 26 h to start). Incubate at 32 °C for 4–5 days (unless temperature sensitive) until small colonies form. These colonies will have sporulated in the middle, so actual diploid cells must be further purified.
3. Streak several colonies onto YES + PB, including a haploid control for comparison. Diploid cells are longer and wider than haploids, and stain dark pink on PB. Identify appropriate colonies and streak on YES. They can be stored for no longer than 3–4 weeks at 4 °C. It is recommended to make diploids fresh when needed if possible. Prolonged storage even at -80 °C leads to loss of sporulation capacity. Before use, verify sporulation ability by streaking to single colonies on YES and replica plating to ME for iodine screening. Diploids will sporulate fully within

12 h of plating on ME/SPAS media. Ploidy may be confirmed by sporulation competence or by FACS.

4. On rare occasions, a vegetative haploid will skip mitosis and form a stable homozygous diploid, a process called endo-reduplication. These strains may be identified by phloxin B staining, examination of cell size, and flow cytometry to confirm DNA content. Such diploids are unable to sporulate as they are mating-type homozygous and will behave like a haploid. However, expression of the missing mating-type information from an ectopic plasmid can induce sporulation, allowing some recovery of haploid offspring (Willer et al., 1995). Haploidization can also be induced by treating with drugs that promote chromosome loss such as mFPA (Kohli et al., 1977); because aneuploidy is not tolerated in fission yeast, loss of a single chromosome in a diploid is accompanied by rapid loss of the remaining chromosomes to reduce the complement to haploid.

3.7. Cell synchrony in fission yeast cultures

In an asynchronous, exponentially growing wild-type fission yeast culture, approximately 70% of cells will be in G2 phase at any given time with the remaining 30% equally divided between G1, S, and M phases. Cytokinesis occurs after the peak of S phase. The historically favored method to isolate a synchronous population is elutriation, which requires an elutriation rotor and specially modified centrifuge (e.g., Beckman J26XPI, product #393134) capable of differential sedimentation and regulating media flow during separation and collection (Walker, 1999). Fortunately, fission yeast cultures can also be synchronized by growth arrest, cell cycle mutations, or size selection, which do not require specialized equipment. Using media lacking nitrogen or glucose, or adding hydroxyurea, takes advantage of fission yeast metabolism to synchronize populations in G1, G2, or S phases, respectively. Cell cycle mutant alleles use temperature shift and are tight arrests. Finally, an alternative size-based separation method using lactose gradient can separate a G2 population without special equipment. To confirm synchrony and monitor progression, samples can be monitored in real time for septation index, or fixed in ethanol for later analysis by morphology or flow cytometry. Septation index is obtained by counting the number of cells with a septum that have not yet invaginated. Septa may be viewed under phase microscopy or by staining with calcofluor (see Section 5.4). Since septation occurs concurrently with the next S phase, septation index is a measure of the next cell cycle, as well as an indicator of degree of synchrony, ideally 60–80% in the population for an effective block and release experiment.

3.7.1. S-phase arrest by hydroxyurea block and release

Hydroxyurea (HU) depletes ribonucleotide reductase, causing nucleotide depletion and S-phase stalling. HU treatment has the potential to disturb cells with checkpoint or replication recovery defects, or alter S-phase events. We find that filtration is the most efficient and most gentle way to wash cells, but low-speed centrifugation may also be used to remove HU ($500 \times g$, 5 min).

1. Grow a culture in minimal medium to mid-log phase (OD_{600} 0.3–0.6, approximately 1–2×10^7 cells/ml).
2. Add HU to 12 mM, and shake for 4 h at 25–32 °C. HU (FW $= 76.05$ g/mol) stock solution is made up freshly at 1 M in water and filter sterilized.

Note: 10–15 mM HU will arrest cells; 20–25 mM HU will cause an irreversible arrest and significant toxicity to wild-type cells. HU will also cause cell elongation over prolonged incubation.

3. Set up a vacuum filtration unit, with a fresh, sterile glass fiber filter (Whatman #18220555). Turn on vacuum line, and pour culture into filter bell. Once the liquid has passed through leaving cells on the filter paper, wash twice with 0.5 culture volume of fresh EMM.
4. Remove filter, and scrape cells into 1 volume of fresh EMM + supplements. Keep as sterile as possible. This will yield approximately 50% synchrony for the next S phase.

3.7.2. Lactose gradient centrifugation

Centrifuging cultures through a lactose gradient separates the cells based on volume/size. This allows enrichment for the smallest fission yeast cells, which are "newborn" G2 cells immediately after cytokinesis, and avoids cell cycle disruption (Carr *et al.*, 1995). Because it relies on cell size, only strains with uniform morphology can be synchronized in this way; strains with irregular shapes or elongation phenotypes are not good candidates for this method.

1. Grow a 100-ml culture in minimal media to mid-log phase, OD_{600} of 0.5–1.0. Harvest 100 ODs of cells and resuspend in 3 ml media.
2. Prepare 45 ml of 10–40% lactose gradient (in media) using a gradient maker.

Alternatively, 45 ml of 20% lactose can be frozen at -80 °C for 4 h, and then thawed without disturbing for 3 h at 30 °C, spontaneously generating a 10–30% gradient. The 40% solution must be warmed to solubilize the lactose, and both 40% and 10% solutions are made in YES or EMM, autoclaved, and stored at room temperature.

3. Layer the cells on top of the lactose gradient.
4. Centrifuge at 500×g, 5 min in a swinging bucket rotor. Remove 2 ml fractions from the top of the gradient, and examine for uniformly small cells (usually in fractions 1–5).
5. Pool fractions and centrifuge 500×g, 5 min to pellet cells. Wash cells once with fresh medium and then resuspend in 0.2 volumes of medium.

This method isolates newly divided G2 cells, representing about 5–10% of the total culture amount (fewer than 1% should be septated). If desired, a second (10 ml) gradient may be used.

3.7.3. Block and release using temperature sensitive alleles

The key to temperature shift is rapid increase and decrease of temperature. Standard blocks for *S. pombe* are *cdc10* (G1/START), *cdc25* (G2 immediately before mitosis), and *nuc2* (M), by temperature shift to 36 °C for 4 h, followed by rapid temperature shift in an ice–water bath to 25 °C for release. Incubation at higher temperatures, or for longer than 4 h, will cause cell death. The *cdc10-V50* allele causes cells to accumulate in G1, and enter into S phase ~30 min after release, completing replication approximately 90 min post release as assessed by flow cytometry. The *cdc25-22* allele is tightly temperature sensitive and delivers a G2-arrested population that releases quickly and synchronously (~90%) into the next S phase. Cells grow very long during the *cdc25-22* arrest as cell volume increases in G2 and flow cytometry for DNA content using whole cells may be variable due to the variability in cell size and shape. Therefore, septation index is critical to assess block and release kinetics in a *cdc25-22* block and release.

1. Grow cultures in EMM with supplements to early/mid-log phase (OD_{600} of 0.3–0.5, approximately 1×10^7 cells/ml). Sample the asynchronous culture for comparison.
2. Temperature shift the cultures to 36 °C in a prewarmed water bath and shake at 225 rpm, 36 °C for 4 h. At the end of the incubation, sample the synchronized culture ($t = 0$).
3. Shift back to 25 °C by swirling flasks in ice/water, for approximately 5 min. A thermometer cleaned with ethanol is useful to monitor the culture temperature to 25 °C.
4. Put flasks into a prewarmed 25 °C water shaker (225 rpm) and assess synchrony by sampling for septation index/flow every 20 min.

3.7.4. G1/G0 arrest of fission yeast by nitrogen starvation

Nitrogen starvation at low temperature is used to enrich a population for G1 cells, particularly for release into synchronous meiosis using the *pat1-114* allele. However, cells require approximately 2 h of lag time to reenter the cell cycle. This reentry is not uniform, so cells shows only modest synchrony compared to other methods.

1. Grow a culture in EMM to OD 0.8–1.0 (approximately 2–3×10^7 cells/ml). Using a flask $5\times$ larger than the culture volume will enhance aeration and starvation at all points of the experiment.
2. Harvest cells in a sterile conical tube by centrifuging $500 \times g$, 5 min, at room temperature. Decant supernatant and vortex pellet.
3. Wash cells twice in 0.5 culture volume EMM − N medium, vortexing well.
4. Resuspend cells in 1 volume EMM − N. If cells are Ade auxotrophes, add adenine stock solution to 7.5 mg/l (1/1000 dilution). Other supplements are not required.
5. Shake cells in a 25 °C air shaker for 16–18 h. Past 20 h is not recommended, as cells will arrest in G0 and not release. Temperature above 25 °C will lead to G2, rather than G1 arrest.
6. To release cells, add 1 culture volume of EMM(+N) and supplements at 150 mg/l. The final concentration of NH_4Cl in solution will be 2.5 g/l, half of the usual amount in EMM. Adding an equal volume of EMM (+N + supplements) avoids the medium shock that could occur if a large volume of NH_4Cl solution were added.

Harvest and fix cells in 70% ethanol for flow and morphology (Section 5.1), including an asynchronous culture sample (prepared before starving cells), a G1 starved sample ($t = 0$), and every 15–30 min after release.

Alternates: While temperatures > 25 °C will lead to a predominantly G2 cell population, this is not the recommended method for achieving uniform G2 arrest. The tightest G2 synchrony is achieved using the *cdc25-22* temperature sensitive allele (for basic protocol, refer to Section 3.7.3). As an alternative, useful for cells with two temperature sensitive alleles or diploid cultures (which are sensitive to nitrogen depletion), we recommend a variation on the nitrogen starvation protocol above. Removing both glucose and nitrogen (EMM − N − G) enriches a population for G2 cells, and is particularly useful to arrest diploid cells for release into meiosis in the absence of nitrogen.

4. MOLECULAR ANALYSIS

4.1. Fission yeast plasmids

Fission yeast plasmids were originally derived from budding yeast vectors, but these are maintained poorly in *S. pombe* due to inefficient replication, high frequency of rearrangement, and poor complementation of *S. pombe* auxotrophs by most *S. cerevisiae* markers. Generally, modern fission yeast episomes contain a *S. pombe* replication origin (*ars*) and an *S. pombe*-specific nutritional or drug-resistance marker. The exception is the *S. cerevisiae LEU2* marker, which efficiently complements *S. pombe leu1* mutations, and is still common. Drug selection markers including hygromycin, G418, nourseothricin, and phleomycin are available under *S. pombe*-specific promoters (Sato *et al.*, 2005).

Plasmid copy number varies from a few to 20 (Patterson *et al.*, 1995). There are no *S. pombe* single-copy or centromere vectors, because the fission yeast centromere is too large to encompass on a simple plasmid. Without a centromere, lower copy number plasmids may be lost frequently, and in some cases 10–20% of a population of cells even under positive selection may lack the plasmid of interest. Additionally, plasmids are unstable through meiosis and are recovered typically in only 10–20% of spores. One solution to the stability issue is to target integration into the genome at a known sequence such as *leu1* (Keeney and Boeke, 1994).

Some plasmids may also carry counter-selectable markers including *ura4+* (which can be selected against with 5-FOA, and selected for by minus-uracil media), *can1+* (selected against with the toxic arginine analogue canavanine; Fantes and Creanor, 1984), or *adh-thymidine kinase* (which renders cells sensitive to the thymidine analogue FudR (Kiely *et al.*, 2000)). Of course, any cloned sequence associated with a growth defect is functionally a counter-selectable marker. The *sup3-5* marker, a tRNA that suppresses the nonsense allele *ade6-704*, is an example of this that has been exploited to monitor successful integration. There is a modest toxicity associated with this marker on high-copy plasmids due to inappropriate insertion of the wrong tRNA. This leads to a subpopulation of cells that lose the plasmid, and are pink on low-adenine media due to the *ade6* deficiency. Conversely, if integrated into the genome, the *sup3-5* marker is stable, there is no Ade⁻ subpopulation, and the colonies are white on low-adenine media (Grallert *et al.*, 1993).

Expression of a cloned sequence from a plasmid relies on a cloned promoter. The two most common constitutive promoters are *adh1* (constitutive high expression; Broker and Bauml, 1989) and CaMV (constitutive low expression, engineered to be repressed with tetracycline; Gmunder and Kohli, 1989). Unfortunately, inducible expression options in *S. pombe* are more limited than in budding yeast. The most popular inducible promoter is the *nmt1* promoter (*no message in thiamine*), which is repressed in the presence of thiamine and is one of the strongest inductions in fission yeast (Maundrell, 1993). The *nmt1* promoter is repressed by the addition of thiamine stock solution (2000×, 10 mg/ml) to 15 μM or 5 μg/ml just before use. Rich medium (YE/YES) already contains thiamine, as does SD medium for budding yeast, and therefore cannot be used for high-level protein expression in *S. pombe* under the *nmt1* promoter. Thus, synthetic minimal medium (EMM or PMG) is required, in which thiamine can be added as necessary. When transforming a construct using the *nmt1* promoter, we recommend including thiamine in growth and maintenance steps to prevent unwanted toxicity from gene overexpression. This has an added benefit of promoting growth, because thiamine is modestly growth limiting and its presence accelerates growth rate.

To activate *nmt1*-driven expression, liquid cultures with thiamine are washed to remove excess thiamine, by filtration or centrifugation ($500 \times g$, 5 min to pellet cells). Cells are washed twice in medium without thiamine, then resuspended in an equal volume of medium and grown at appropriate temperature for a minimum of 16 h to completely activate the *nmt1* promoter. The long induction reflects the considerable pools of intracellular thiamine that must be depleted before the promoter becomes active. However, *nmt1* still has low expression in repressive conditions (+thiamine); therefore, only unstable proteins may be reliably "switched off" by adding thiamine. Some fine tuning of this system has occurred by isolating mutations that attenuate both induced and repressed levels making medium- and low-level derivatives (*nmt** (Tommasino and Maundrell, 1991) and *nmt*** (Basi *et al.*, 1993; Forsburg, 1993)), or titrating thiamine concentration to generate intermediate levels of repression (Javerzat *et al.*, 1996; Tommasino and Maundrell, 1991). There are numerous versions of *nmt1* containing plasmids with different epitope tags and promoter derivatives (Craven *et al.*, 1998; Forsburg and Sherman, 1997; Tasto *et al.*, 2001).

Additional regulated promoters have been characterized including *fbp1* (glucose; Hoffman and Winston, 1989), *inv1* (invertase; Iacovoni *et al.*, 1999), *urg1* (uracil; Watt *et al.*, 2008), and *ctr4* (copper; Bellemare *et al.*, 2001). Although these have not become as popular as the *nmt1* promoter, which is well characterized and widely employed, some of these promoters have the advantage of a faster induction time. However, the constructs available using these alternative promoters are not as diverse as those available for *nmt1*, and the kinetics of transcript induction and repression from these promoters must be determined by the user for individual proteins.

4.2. Integrations in fission yeast

The *S. pombe* genome is easily modified by integration via homologous recombination, at which it is very proficient, although it is also capable of performing nonhomologous recombination. Ectopic integration into an auxotrophic locus may be accomplished with any plasmid that contains no *ars* and has sufficient sequence to target the event. For example, a plasmid containing a construct of interest may be integrated into *leu1-32* by linearizing the integrating plasmid in the cloned *leu1*[+] gene to target the insertion (Keeney and Boeke, 1994).

Integrations for precise gene disruptions or replacements (knock-outs and knock-ins) require additional planning. Maximum integration efficiency with minimal background occurs using homologous sequences >300 base pairs (bp) (Krawchuk and Wahls, 1999). Short tracts of homology <100 bp may be used (Bahler et al., 1998) but can generate a high background of nonhomologous events compared to S. cerevisiae. All homologous integration success is dependent upon the specific gene and chromatin context. Additionally, fragments or plasmids without replication origins can sometimes be maintained as unstable concatamers. Thus, appropriate screening for locus-specific integration must be carried out, including verification of stable integration (2:2 segregation through meiosis). This is easily determined using random spore analysis.

4.3. Isolation and analysis of novel mutations

Genome mutagenesis can be performed with nitrosoguanidine or ethyl methane sulfonate (EMS), although recently nonchemical methods have been considered, for example, ultraviolet (UV) radiation or insertional mutagenesis (Chua et al., 2000; Wang et al., 2004). The general methods of chemical or radiation-induced mutagenesis are the same for budding and fission yeasts. This requires determination of dose and lethality, screening or selection, and subsequent genetic analysis including backcrossing, dominance and recessiveness testing, complementation, and linkage analysis. For detailed considerations to setting up a genetic screen, several excellent references are Barbour and Xiao (2006), Barbour et al. (2006), Boone et al. (2007), and Forsburg (2001).

If mutagenizing a plasmid for point mutant analysis of a specific gene, the cDNA is first put into an expression plasmid with a marker that will complement the strain in question. Mutagenesis can be performed in vitro using hydroxylamine (Liang and Forsburg, 2001; Rose and Fink, 1987), via mutagenic PCR, or using mutagenic bacteria during amplification (Rasila et al., 2009). The mutagenized plasmid is transformed into cells with a mutant copy of the allele (deletion or temperature sensitive mutant) and screened for rescue under conditions that would kill cells. Optimizing the screen ahead of time is critical to success, and Phloxin B staining helps distinguish colonies that are attenuated for growth under appropriate conditions.

4.4. Transformation of DNA into fission yeast

4.4.1. Electroporation of fission yeast

We find electroporation is the fastest and most efficient method for DNA transformation into fission yeast. Cells are made electrocompetent by washing in cold sorbitol, although sorbitol is removed following electroporation as it retards growth during selection.

1. Grow 25–50 ml of fission yeast cells in minimal medium to mid-log phase (OD_{600} 0.4–1.0), approximately $1–2.5 \times 10^7$ cells/ml.
2. Harvest cells in a sterile tube and centrifuge at $500 \times g$, 5 min, 4 °C. Keep cells and solutions on ice in subsequent steps.
3. Wash cells once in 0.5 volume of chilled, sterile water. Centrifuge as above.
4. Wash cells once in 0.25 volume of chilled, sterile 1 M sorbitol, and centrifuge as above. (*Note*: 15 min incubation of cells in 1 M sorbitol + 25 mM DTT may increase electrocompetence; Suga and Hatakeyama, 2001.)
5. Resuspend cells at $1–5 \times 10^9$ cells/ml in chilled, 1 M sorbitol. Aliquot 40 μl to prechilled microfuge tubes and add 100 ng DNA. Mix gently and incubate 5 min on ice.
6. Transfer mixture to prechilled 0.5 cm-gap electroporation cuvettes. Electroporate cells according to manufacturer. For a BioRad instrument: 1.5 kV, 200 Ω, 25 μF.
7. Immediately add 0.9 ml of cold, 1 M sorbitol, and remove from cuvette into a microfuge tube on ice. Keep tubes on ice while all electroporations are performed.
8. Centrifuge tubes ($3000 \times g$, 5 min, 4 °C) and remove sorbitol. Wash cells once with 1 ml of chilled sterile water. Plate cells as soon as possible onto minimal selective medium. Transformants appear in 3–5 days at 32 °C, 5–7 days at 25 °C.

4.4.2. Rapid lithium acetate transformation

This method is quick, independent of special equipment, and uses small culture volumes, resulting in transformation efficiency of approximately 10^3 transformants per microgram of DNA (Kanter-Smoler *et al.*, 1994).

1. Use a fresh colony from a plate and resuspend in 150 μl of 0.1 M lithium acetate (LiOAc), pH 4.9. Incubate 1 h at room temperature.
2. Add 1 μg DNA and mix (less DNA may be used, but 1 μg is recommended to start due to the lower transformation efficiency).
3. Add 350 μl of 50% polyethylene glycol 8000 (PEG8000). Mix, and incubate hour at room temperature.
4. Centrifuge cells $3000 \times g$, 5 min, at room temperature to pellet. Wash cells once with 1 ml of sterile water, then resuspend in a small volume of sterile water to plate onto selective medium.

A more stringent version of this protocol produces higher transformation efficiency, and is better when transforming PCR products for homologous integration (Bahler *et al.*, 1998; Forsburg, 2003b).

4.5. Harvesting DNA, RNA, and protein from fission yeast

The methods of isolating nucleic acids and protein are similar to that in *S. cerevisiae*, although the cell wall of *S. pombe* can be difficult to crack. The lysis step is critical to success, and can be accomplished with enzymes or mechanically. In large-scale preparations, fission yeast may be frozen in liquid nitrogen and ground (e.g., Retsch RM100 Mortar Grinder; www.retsch.com), broken (e.g., Retsch Ball Mill, RMM301) or microfluidized (e.g., Microfluidics M110-S; www.microfluidicscorp.com); these are more efficient but more expensive.

4.5.1. DNA preparation from fission yeast

This small-scale preparation should deliver approximately 1 μg of DNA per OD harvested. Scale up as required. This protocol can also be used to isolate plasmids from yeast for transformation into *E. coli* (Hoffman and Winston, 1987).

1. Grow a 5–10 ml culture to OD_{600} of 0.8 or higher. Harvest, and wash once with 1× PBS or water, transferring cells to a screw-cap tube. Pellets may be stored at $-80\ ^\circ$C.
2. Add 250 μl of DNA lysis buffer (see Table 32.3), 250 μl of acid-washed glass beads and 250 μl of phenol:chloroform:isoamyl alcohol (PCI; 25:24:1). Vortex for 1 min, rest 1 min, and repeat.
3. Centrifuge for 5 min at $>13,000\times g$ and then move the aqueous top phase to a fresh microfuge tube, avoiding the interface. Estimate the lysate volume.
4. Add 250 μl TE (total ~ 450 μl), and 1 volume of PCI. Vortex well and centrifuge 5 min. Recover aqueous layer to a new tube. Repeat PCI extraction if necessary (e.g., cloudy interface).
5. Add 1 ml 100% ethanol, vortex and chill for 15 min on ice or at $-20\ ^\circ$C. Centrifuge at top speed for 5 min.
6. Wash pellet once with 1 ml 70% ethanol (v/v in water). Air dry pellet briefly.
7. Resuspend pellet in 100 μl TE + 20 mg/ml RNaseA (boiled to inactivate DNases), and store at 4 or $-20\ ^\circ$C.

4.5.2. Colony PCR

This method uses zymolyase enzyme to digest the cell walls, and then spheroplasts are directly used in PCR reactions with gene-specific primers. The number of cycles may be increased to detect product from this method.

1. Aliquot 10 μl of zymolyase solution (Table 32.3) to fresh microfuge or PCR tubes.

Table 32.3 Buffers for biochemistry

	Composition	Notes
DNA lysis buffer	10 mM Tris, pH 8.0 100 mM NaCl 1 mM EDTA, pH 8.0 1% SDS	• For isolation of fission yeast genomic DNA or episomal plasmid isolation • Variation: addition of 1% Triton X-100 to buffer
Zymolyase solution (colony PCR)	2.5 mg/ml zymolyase 20T[a] 1.2 M sorbitol 0.1 M sodium phosphate, pH 7.4	• Aliquots may be stored at −20 °C for >6 months, but should not be refrozen
RNA lysis buffer	50 mM sodium acetate, pH 4.3 10 mM EDTA 1% SDS	• Due to 1% SDS, this is essentially RNase-free and does not need to be treated with DEPC
Soluble protein lysis buffer 1	50 mM Tris, pH 7.5 150 mM NaCl 5 mM EDTA 10% glycerol 1 mM phenylmethylsulfonyl-fluoride (PMSF)	• Add PMSF fresh before use • May add additional protease or phosphatase inhibitors if desired
"STOP" buffer	0.2% sodium azide 150 mM NaCl 10 mM EDTA, pH 8.0 Use ice-cold	• May also be made as a 10× stock • Store at 4 °C for less than 2 months
Laemmli buffer part A (1× buffer)	2.5% SDS 60 mM Tris base (un-pH'd) 6.25 mM EDTA, pH 8.0 Make up in water, store at room temperature	• Acetone-washed pelleted protein is solubilized by boiling in this • Add fresh protease inhibitors (Sigma P8125) • Add phosphatase inhibitors if desired
Laemmli buffer, part B (5× buffer)	50% glycerol 100 mM Tris base 0.1–0.05% (w/v) Bromophenol blue 400 mM dithiothretol 10% 2-mercaptoethanol (v/v) optional	• 1 ml aliquots made as needed; can be stored at −20 °C for several months • Add fresh 1 M DTT to 50 mM in samples if desired

[a] Zymolyase 20T from Seikagaku Biobusiness product #120491.

2. Use a clean plastic pipette tip to pick up a single yeast colony. Avoid toothpicks, which may interfere with the PCR reaction. Touch colony onto a numbered patch of a plate (appropriate medium), to keep as a master copy.
3. Transfer the majority of the colony to the Zymolyase solution (step 1), and rinse the yeast from the tip into the solution.
4. Incubate 10 min at 37 °C. Use 2 μl of spheroplasts per 50 μl PCR reaction.

4.5.3. RNA preparation from fission yeast

This small-scale preparation should yield approximately 2 μg of RNA per OD harvested. Scale up as required. Alternatively, reagents such as TRIzol (Invitrogen) may be used to purify RNA. After phenol/chloroform steps, take care that solutions, tubes, and tips are all RNase free.

1. Grow 20–50 ml culture to OD_{600} of 0.5–1.0. Harvest approximately 20 ODs of cells.
2. Resuspend pellet in 250 μl of RNA lysis buffer (see Table 32.3), and transfer to a screw-cap tube.
3. Add 250 μl of acid-washed glass beads (0.5 mm) and 250 μl of acidic phenol (pH 4.7). Vortex for 1 min, rest 1 min, and then vortex 1 min.
4. Incubate 10 min at 65 °C. Centrifuge tubes for 5 min at $> 13,000 \times g$. Move aqueous top phase to a fresh microfuge tube, avoiding the interface.
5. Add 250 μl TE or RNA lysis buffer, and estimate volume of lysate ($\sim$450 μl). Repeat organic extraction with 1 volume acidic phenol, taking aqueous phase.
6. Add 1 volume of phenol/chloroform and vortex. Centrifuge and move the aqueous layer to a fresh tube.
7. Add 1 ml 100% ethanol, vortex and chill for 15 min on ice or at -20 °C. Centrifuge at top speed for 10 min, 4 °C.
8. Wash pellet once with 1 ml 70% ethanol (v/v in water). Air dry pellet briefly.
9. Resuspend pellet in appropriate buffer (i.e., RNase-free water). Store samples at -80 °C and do not refreeze. For long-term storage, keep in 70% ethanol at -80 °C.

4.5.4. Soluble protein extract

This protocol is a starting point for known soluble proteins, which is suitable for direct Western blots or immunoprecipitation. A whole-extract preparation using TCA follows (Section 4.5.5). Treating cells with a "STOP buffer" containing sodium azide will prevent proteolysis or other changes during preparative steps, and "stopped" cell pellets in screw-cap lysate tubes may be frozen at $-80\,°C$ until all samples are ready to prepare together (at step 3). Solutions and samples should be kept at $4\,°C$ as much as possible during lysate preparation.

1. Grow yeast to mid-log phase and harvest 5–20 ODs in a conical tube. Add $10\times$ STOP buffer to $1\times$, or 0.1 volume of 2% sodium azide solution, and incubate culture on ice for 5 min. Centrifuge 5 min, $500\times g$, $4\,°C$, and decant supernatant.
2. Wash cells once with $1\times$ PBS. Centrifuge as above. Transfer pellet to a screw-cap lysate tube in a small volume of PBS and centrifuge to remove PBS ($3000\times g$, 5 min, $4\,°C$).
3. Resuspend pellet in 200 μl lysate buffer. Add approximately 500 μl acid-washed glass beads (0.5 mm) to the level of the buffer meniscus.
4. Vortex for 5 min at $4\,°C$ at top power. Alternately, use a bead beater, or disruptor (e.g., Fastprep); typically 4–6 pulses, 5.5 m/s, 45 s each, resting for 2 min on ice in between. Check microscopically for cell breakage, and repeat disruption if required.
5. Puncture the top and bottom of the screw-cap tube with a red-hot needle and place in a collection tube. Centrifuge tubes to collect lysate, 10 s at top speed in a microfuge or $500\times g$ in a swinging bucket rotor. Collect lysate into a fresh tube.
6. Spin lysate for 5 min, $>13,000\times g$, $4\,°C$, then remove cleared lysate to a fresh tube.
7. Quantitate protein concentration and store remaining lysate at $-80\,°C$.

4.5.5. Whole cell protein extract with trichloroacetic acid (TCA)

TCA protein extraction and precipitation is an alternative to soluble lysates, and is used to assess insoluble proteins, such as those that remain chromatin bound. Consequently, TCA lysis better represents the *total* protein pool within cells but is not suitable for subsequent immunoprecipitation.

1. Prepare cells as in the preceding protocol to step 2. Add 400 μl of 20% TCA (v/v in water) to pellet and approximately 500 μl of acid-washed glass beads. Disrupt as in step 4 above, and then puncture tube top/ bottom with a needle and place in a collection tube.
2. Centrifuge tubes to collect lysate, 10 s at top speed in a microfuge or $500\times g$ in a swinging bucket rotor.
3. Add 800 μl of 5% TCA (v/v in water) to wash beads. Repeat step 2.
4. Harvest all TCA lysate (now 10% TCA, v/v). Centrifuge $>13,000\times g$, 5 min, room temperature, to pellet protein. Remove supernatant.
5. Add 1 ml of acetone and vortex well, and then rock 5 min at room temperature. Centrifuge $>13,000\times g$ in a microfuge, 3 min, room temperature, and remove acetone wash. Repeat twice, for a total of three acetone washes.

6. Air-dry protein in a fume hood for 10–30 min until completely dry.
7. Add 200 µl of Laemmli buffer A (see Table 32.3) with fresh protease inhibitors. Boil 3–5 min in a heat block at 100 °C. TIP-tubes will pop open, close caps with holders.
8. Centrifugate >13,000×g, 3 min then remove lysate to a fresh tube. Repeat if lysate is cloudy.
9. Quantitate protein by BCA assay and store lysates at −80 °C. For SDS–PAGE, make up equal protein amounts in Laemmli buffer A, adding 5× Laemmli buffer B (Table 32.3) to 1×. Heat samples at 95 °C for 5 min and then centrifuge for 5 min >13,000×g before loading for electrophoresis.

5. CELL BIOLOGY

5.1. Preparation and analysis of cell populations

In this section, we describe methods to fix and prepare cells for DNA quantitation by flow cytometry, nucleus and septum staining, and basic immunofluorescence.

5.1.1. Ethanol fixation of cells for flow cytometry (FACS) and septation index

Fission yeast cells fixed in 70% ethanol may be stored at 4 °C for >1 year. The simplest and quickest method to fix cells, best for a timecourse with closely spaced collection points, is to prepare 1 ml of 100% ethanol into labeled microfuge tubes and chill to 4 °C ahead of time. Then, 420 µl of culture is added directly to the tube and vortexed, bringing ethanol to ~70%. For larger culture volumes:

1. Remove appropriate amount of culture. Centrifuge at 500×g, 5 min. Decant.
2. Vortex the pellet well to resuspend cells.
3. Add an equal volume of cold 70% ethanol (v/v in water, prepared ahead and stored at −20 °C) very slowly and vortexing continually, then store samples at 4 °C, minimally 15 min.

5.2. Flow cytometry (whole cells)

This is the most straightforward method to assess DNA content in fission yeast samples. However, the cell cycle of S. pombe must be considered when interpreting the data: cells do not separate until after S phase, meaning that an asynchronous haploid population displays a predominantly 2C DNA content, and S phase is commonly detected as a shift from 2C to 4C. The key to data interpretation is including control samples that are prepared with other samples at the same time, including asynchronous (G2 = 2C), nitrogen starved (G1 = 1C), and diploid (2C/4C) controls. Samples may be prepared directly in flow cytometer tubes, or in microfuge tubes and transferred to flow tubes later. For additional protocols and considerations, refer to Sabatinos and Forsburg (2009).

1. Remove 300 μl of cells fixed in 70% ethanol, and pellet. Wash cells twice with 1–3 ml of 50 mM sodium citrate (made in water from a 0.5-M sodium citrate autoclaved stock solution), vortexing well with each wash.
2. Resuspend cells in 0.5 ml of 50 mM sodium citrate + 0.1 mg/ml RNase A (prepared as a 10 mg/ml stock and boiled to kill DNases). Vortex. Incubate for 2 h at 37 °C.
3. Add 0.5 ml of 50 mM sodium citrate + 1 μM Sytox Green. Vortex and incubate for 15 min or longer at 4 °C or on ice. Samples may be sealed and stored at 4 °C for several months, protected from light, and rerun at a later date.
4. Just before running, sonicate samples with a microtip sonicator for 5 s on low to medium power. Samples are acquired using FL1 for green fluorescence. On a Becton Dickinson FACScan flow cytometer with 488 nm excitation, guideline settings are FSC voltage E00, gain 1.36, linear mode; SSC voltage 300, gain 2.47, linear mode; FL1 (Sytox fluorescence) voltage 455, gain 1.0, logarithmic mode; FL1-Area gain 3.64 gain, linear mode; FL2-Wide gain 3.60 gain, linear mode.

5.3. Flow cytometry (nuclear "ghosts")

Due to the replication and division cycle of fission yeast, the cellular volume, shape, and mitochondrial DNA content may affect DNA profiles, making it difficult to distinguish a single cell particle with a 2C nucleus from a single cell particle with two 1C nuclei (Sazer and Sherwood, 1990). By digesting fixed cells with enzymes, the cell wall and membrane are removed, leaving just the nuclei (Carlson et al., 1997). Appropriate controls are the same as for whole cell FACS, but must be prepared as nuclei alongside samples.

1. Place 1×10^7 ethanol fixed cells in a microfuge tube (typically 1 ml of fixed culture) and centrifuge samples at 3000×g, 5 min to pellet cells.
2. Wash cells once with 1 ml of 0.6 M KCl. Vortex, and centrifuge.
3. Resuspend cells in 1 ml of 0.6 M KCl + 0.5 mg/ml zymolyase 20 T + 1.0 mg/ml lysing enzymes. Vortex, and incubate for 30 min at 37 °C.

4. Centrifuge cells as above, and resuspend pellet in 1 ml of 0.6 M KCl + 0.1% Triton X-100. Rotate samples for 5 min at room temperature.
5. Centrifuge cells, and wash once in 1 ml of 1× PBS.
6. Resuspend washed cells in 1 ml PBS + 0.1 mg/ml RNaseA. Vortex and incubate at 37 °C for 2 h to overnight.
7. Sonicate cells to release nuclei without fragmenting them. This will require optimization for each sonicator; we use 50% amplitude for 5 s with a digital Branson 5 mm microtip sonicator. Nuclei may be stored at 4 °C for several months.
8. Stain by mixing 100 μl of nuclei with 400 μl of PBS + 1 μM Sytox Green. Vortex and incubate 15–30 min in the dark, then run samples. Guideline settings are similar to those for whole cells (Section 5.2) with the following exceptions to account for the smaller size of nuclei: FSC voltage E00, gain 5.74, linear mode; SSC voltage 304, gain 6.55, linear mode.

5.4. DNA and septum staining in fixed cells

Septation index may be monitored in real time, or using fixed cells and calcofluor staining. This is a good metric for cell synchrony as well as an index of S phase in a population, since S phase corresponds with the peak of cytokinesis in exponentially growing cells. Staining fixed cells is useful for cell morphology and is essential to monitor mitotic index, chromosome segregation, and meiotic progression using DAPI staining.

1. Resuspend ethanol fixed cells, and remove 100 μl to a fresh microfuge tube. Wash cells with 1 ml water to rehydrate, then resuspend in 50–100 μl water.
2. Pipette 5 μl of cells onto a poly-lysine coated or charged slide. Heat fix cells on a slide warmer set at low for a few minutes until barely dry around the edges, or let air-dry. Do not over dry!
3. Add 5 μl of mounting solution (50% glycerol in water (v/v) with 1 mg/ml p-phenylenediamine (PPD), 1 μg/ml DAPI, 0.2 mg/ml calcofluor; will keep a few days, but best prepared fresh). Cover with a slip, seal with VALAP or nail polish, and visualize using a microscope with UV excitation source.
4. Alternately, resuspend rehydrated cells (from step 1) in 100 μl of water with 1 μg/ml DAPI and 0.2 mg/ml calcofluor. Mix for 5 min at room temperature in the dark, then wash cells three times with 1 ml water. Resuspend cells in 50–100 μl water and mount as in step 2. Mount solution is 50% glycerol in water (v/v) with 1 mg/ml PPD as antifade.

Calcofluor staining can be difficult to optimize, and is often very bright. If this is the case, let the sample photobleach briefly and reexpose to photograph nuclei and septa. Calcofluor is prepared in 50 mM sodium citrate, 100 mM sodium phosphate, pH 6.0, for long-term storage at $-20\,°C$ (centrifuge aliquots to remove particulates before use). VALAP is a (1:1:1, w/w/w) mixture of petroleum jelly, lanolin, and paraffin that are combined in a beaker, and melted on low heat to form a suspension used to seal slides.

5.5. Fission yeast whole-cell immunofluorescence

Protocols for *S. pombe* immunofluorescence are similar to other cells, although the fission yeast cell wall is somewhat resistant to digestion. The critical parameter is getting samples into fixative quickly, and consistency of fixation between experiments. The choice of fixative depends on the purpose and the epitope stability. For additional considerations on microscopy and immunofluorescence using fission yeast, refer to Green *et al.* (2009).

1. Harvest cells, and centrifuge 500×g, 5 min to pellet cells. Decant. Vortex cell pellet and add an equal volume of fixative solution (refer to Table 32.4). Incubation time and temperature is fixative-dependent. Centrifuge cells, and wash pellet twice in 1× PBS. Store cells up to 6 months at 4 °C in 1× PBS. Ethanol/methanol cells may be stored in fixative at 4 °C, and washed well before step 2.
2. Wash cells once with 1 ml 1× PBS, and then 1 ml 1× PEM (100 mM PIPES, pH 6.9, 1 mM EGTA, 1 mM MgSO$_4$; may be made as a 10× stock, filter sterilized and diluted to 1× as required).
3. Resuspend cells in 1 ml of PEMS (100 mM PIPES, 1 mM EGTA, 1 mM MgSO$_4$, 1.2 M sorbitol. pH to 6.5–6.9; filter sterilize and store at room temperature) with digesting enzymes 0.2 mg/ml lysing enzymes (Sigma #L1412) and 0.5 mg/ml zymolyase 20 T (Seikagaku Biobusiness #120491). Incubate for 10–30 min at room temperature. Check digestion microscopically (cells will lose refractive halo), and by adding a drop of 1% SDS to the slide (to induce lysis). Complete digestion is not required, merely enough to allow antibodies to enter/exit.
4. Wash cells three times in 1 ml PEMS.
5. Block cells in 1 ml PEMBAL (100 mM PIPES, 1 mM EGTA, 1 mM MgSO$_4$, 3% BSA, 0.1% NaN$_3$, 100 mM lysine hydrochloride) >30 min at room temperature on a rocking platform.

   Note: blocking agent, BSA may be changed to fetal calf serum, or a combination of BSA and FCS. NaN$_3$ inhibits fungal growth and permits extended incubations at room temperature, but is optional if solutions are made fresh and used quickly, and if overnight incubations are performed at 4 °C. Lysine is optional, and reduces background staining by blocking negative charges, particularly in the nucleus. The

Table 32.4 Fixatives commonly used in fission yeast immunofluorescence

Fixative[a]	Composition	Conditions[b]	Recommended uses
70% ethanol	70% ethanol (v/v) in water	Store at −20 °C. Add while vortexing cells to reduce clumping[c]	Fast, efficient fixative for DNA quantitation (e.g., nuclei with DAPI or flow cytometry). May destroy fluorescent protein fluorescence (see methanol) Good fixative for tubulin immunofluorescence
Methanol	100% methanol (v/v)	Store at −20 °C, and add to harvested cells while vortexting. Fix for 15 min or longer and store at 4 °C[c]	Preservative for cytoplasmic microtubule and actin architecture. May destroy GFP fluorescence; use anti-GFP antibody. Long-term incubation may destroy nuclear architecture
Methanol/ formaldehyde[d]	3.7% formaldehyde (v/v) 10% methanol (v/v) 0.1 M potassium phosphate, pH 6.5	Use 1 culture volume of fixative, rocking for 15–30 min. Store fixed cells in PBS at 4 °C long-term[c] Store leftover fixative at room temperature, in the dark	Good starting fixative for immunofluorescence experiments. May destroy GFP fluorescence (use anti-GFP antibody) Good for tubulin staining and nuclear morphology
Methanol/acetic acid (MAA)	25% methanol (v/v) 75% glacial acetic acid (v/v)	Resuspend harvested cells in fixative and incubate for 15 min. Replacing fixative with fresh MAA after 5 min may enhance fixation[c]	Reported to be a good chromosome structure fixative, but less successful for cytoplasmic staining.

(continued)

Table 32.4 (continued)

Fixative[a]	Composition	Conditions[b]	Recommended uses
Paraformaldehyde[c] (4%PFA/PBS)	4% paraformaldehyde (w/v) made up in water or 1× PBS[f,g] Best if made and used fresh, although unused aliquots may be stored at −20 °C wrapped in foil[h]	Fix for 15 min to start (up to 30 min), rocking at room temperature Remove PFA/PBS, wash once with 1 volume of PBS, then incubate in 0.2% Triton X-100 (v/v in water) to permeabilize cells, 15 min, room temperature	General fixative 4%PFA/PBS may fix enough of GFP structure while retaining GFP fluorescence for flow cytometry/immunofluoresence. This requires titration of PFA concentration, from 0.1% to 4%

[a] General word of caution: fixatives are dangerous, use carefully and dispose of waste as dictated by regional guidelines. Methanol, PFA, and formaldehyde are all toxic. PFA and formaldehyde are potential carcinogens.

[b] The length of exposure of the fixative to the cells may cause epitope destruction. Fixation conditions must be optimized for each experiment/epitope and recorded for reproducibility between experiments.

[c] Permeabilization is generally not required after alcohol fixation.

[d] For 500 ml of methanol/formaldehyde buffer mix: 16.9 ml 1 M K_2HPO_4, 33.1 ml KH_2PO_4, 50 ml 100% methanol, 50 ml 37% formaldehyde and bring to 500 ml with water.

[e] Quenching aldehydes is generally not required for fixatives other than glutaraldehyde.

[f] 4% PFA solution is made by mixing 2 g of PFA powder in 45 ml water + 5 ml of 10× PBS stock. Add 10 μl of 1 N NaOH, and stir in a double boiler (60–80 °C) to bring the powder completely into solution. Do not overheat! Once in solution, take pH of solution—optimal range is 7.2–8.0. PFA is toxic and caustic—perform all steps in a fume hood and dispose of waste appropriately.

[g] PFA may be made in 1× PBS, as well as 1× PEM buffer, or HEPES.

[h] PFA solution stored at −20 °C may precipitate upon thawing. If this is the case, heat to 60 °C in a water bath or heat block, and vortex precipitate into solution. Do not refreeze aliquots.

choice of blocking reagent may be changed depending on the protein under study (suggested alternates: 10% FCS in 1× PBS, or 5% FCS in PEMBAL).

6. Split samples into tubes for primary antibody incubation, and centrifuge to pellet cells. Prepare primary antibody solution in blocking buffer, 200–500 μl for each sample if kept in microfuge tube. Incubate for 2 h to overnight on a rotator.

7. Wash cells 3× 1 ml PEMBAL, rotating for 10 min at room temperature with each wash.

8. Add secondary antibody in PEMBAL (200–500 μl volume per tube), and incubate 1–2 h in the dark at room temperature. We typically use a dilution of 1:250–1:500 for secondary antibodies under these incubation conditions, but the optimal concentration should be determined by the user and checked for background and nonspecific signal using appropriate controls. We do not recommend overnight incubations in secondary antibody.

9. Wash 2× 1 ml PEMBAL, incubating 10 min on a rotator in the dark with each wash. Wash once in 1 ml PEMBAL with 1 μg/ml DAPI, freshly prepared. Rotate 10 min in the dark then harvest cells.

10. Resuspend the cells in 1 ml PEMBAL to briefly wash then centrifuge to pellet cells. Add a small volume of sterile water, and spot 5–10 μl onto a poly-L-lysine treated coverslip. Heat fix briefly, until liquid starts to evaporate, and wick away excess with a tissue. Mount the coverslip on a slide with mount (50% glycerol, 1 mg/ml PPD). Seal slides with nail varnish or VALAP. Store slides at − 20 to 4 °C, protected from light.

6. CONCLUSION

S. pombe methods use microbiology techniques common to S. cerevisiae, with unique S. pombe modifications. This chapter, along with a friendly community and numerous online resources, makes it easy for novices to begin analysis of this system and "go fission."

APPENDIX: SCHIZOSACCHAROMYCES POMBE ONLINE RESOURCES

General fission yeast resources include PombeNet (www.pombe.net) and the Sanger Centre web site for genome data. There are now many strain resources for complementing and studying fission yeast phenotypes. The FYSSION collection, housed at the University of Sussex, UK, curates libraries of temperature sensitive mutants and nonessential deletion mutants (Armstrong *et al.*, 2007). This also includes the Bioneer fission yeast deletion library, which is available commercially. Another useful resource is the GFP-tagged ORFeome; all *S. pombe* ORFs tagged with GFP, to track protein localization in live cells response to conditions available from the RIKEN centre (Matsuyama *et al.*, 2006). Resources for analysis such as ortholog mapping (YOGI; Wood and Bahler, 2002) and interactions (e.g., BioGRID; Breitkreutz *et al.*, 2008) are increasingly powerful, and many are housed or linked to the Sanger Genome Center fission yeast resources.

Name	URL
General information, links, resources	
PombeNet: Protocols, resources, people, general information (Forsburg Lab)	http://www.pombe.net/
Sanger Centre	http://www.sanger.ac.uk/Projects/ S_pombe/links.shtml
NIH *S. pombe* page	http://www.nih.gov/science/models/ Schizosaccharomyces/index.html
Genome information	
GeneDB (Genome browser and database)	http://genedb.org/genedb/pombe/
Epigenome home page (Grewal lab)	http://pombe.nci.nih.gov/genome/
Sequences of related species Fungal genomes at the Broad Institute	http://broadinstitute.org/science/projects/ fungal-genome-initiative/
Expression data	
Bähler lab data, UCL (UK)	http://www.bahlerlab.info/
Gene Expression Viewer	http://www.bahlerlab.info/cgi-bin-SPGE/ geexview
Transcriptome Viewer	http://www.bahlerlab.info/Transcriptome Viewer/
Software links (disruption, tagging)	http://www.bahlerlab.info/resources
Strains and plasmids	
ATCC	http://www.atcc.org/ Schizosaccharomycespombe/tabid/680/ Default.aspx
Bioneer (commercial gene deletions)	http://www.bioneer.com

National Collection of Yeast Cultures (Norwich, UK)	http://www.ncyc.co.uk/
RIKEN, GFP-tagged ORFeome	http://yeast.lab.nig.ac.jp/nig/index_en.html
Yeast Resource Centre, Japan	http://yeast.lab.nig.ac.jp/nig/english/index_en.html
Email list (list-serv)	
Pombelist (community list-serv)	http://lists.sanger.ac.uk/mailman/listinfo/pombelist

REFERENCES

Aravind, L., Watanabe, H., Lipman, D. J., and Koonin, E. V. (2000). Lineage-specific loss and divergence of functionally linked genes in eukaryotes. *Proc. Natl. Acad. Sci. USA* **97,** 11319–11324.

Armstrong, J., Bone, N., Dodgson, J., and Beck, T. (2007). The role and aims of the FYSSION project. *Brief. Funct. Genomic. Proteomic.* **6,** 3–7.

Bahler, J., Wu, J. Q., Longtine, M. S., Shah, N. G., McKenzie, A. 3rd, Steever, A. B., Wach, A., Philippsen, P., and Pringle, J. R. (1998). Heterologous modules for efficient and versatile PCR-based gene targeting in *Schizosaccharomyces pombe*. *Yeast* **14,** 943–951.

Barbour, L., and Xiao, W. (2006). Synthetic lethal screen. *Methods Mol. Biol.* **313,** 161–169.

Barbour, L., Hanna, M., and Xiao, W. (2006). Mutagenesis. *Methods Mol. Biol.* **313,** 121–127.

Basi, G., Schmid, E., and Maundrell, K. (1993). TATA box mutations in the *Schizosaccharomyces pombe* nmt1 promoter affect transcription efficiency but not the transcription start point or thiamine repressibility. *Gene* **123,** 131–136.

Beach, D., Nurse, P., and Egel, R. (1982). Molecular rearrangement of mating-type genes in fission yeast. *Nature* **296,** 682–683.

Bellemare, D. R., Sanschagrin, M., Beaudoin, J., and Labbe, S. (2001). A novel copper-regulated promoter system for expression of heterologous proteins in *Schizosaccharomyces pombe*. *Gene* **273,** 191–198.

Boone, C., Bussey, H., and Andrews, B. J. (2007). Exploring genetic interactions and networks with yeast. *Nat. Rev. Genet.* **8,** 437–449.

Breitkreutz, B. J., Stark, C., Reguly, T., Boucher, L., Breitkreutz, A., Livstone, M., Oughtred, R., Lackner, D. H., Bahler, J., Wood, V., Dolinski, K., and Tyers, M. (2008). The BioGRID interaction database: 2008 update. *Nucleic Acids Res.* **36,** D637–D640.

Broker, M., and Bauml, O. (1989). New expression vectors for the fission yeast *Schizosaccharomyces pombe*. *FEBS Lett.* **248,** 105–110.

Carlson, C. R., Grallert, B., Bernander, R., Stokke, T., and Boye, E. (1997). Measurement of nuclear DNA content in fission yeast by flow cytometry. *Yeast* **13,** 1329–1335.

Carr, A. M., Moudjou, M., Bentley, N. J., and Hagan, I. M. (1995). The chk1 pathway is required to prevent mitosis following cell-cycle arrest at 'start'. *Curr. Biol.* **5,** 1179–1190.

Chua, G., Taricani, L., Stangle, W., and Young, P. G. (2000). Insertional mutagenesis based on illegitimate recombination in *Schizosaccharomyces pombe*. *Nucleic Acids Res.* **28,** E53.

Craven, R. A., Griffiths, D. J., Sheldrick, K. S., Randall, R. E., Hagan, I. M., and Carr, A. M. (1998). Vectors for the expression of tagged proteins in *Schizosaccharomyces pombe*. *Gene* **221,** 59–68.

Egel, R. (2007). The Molecular Biology of *Schizosaccharomyces pombe*: Genetics, Genomics and Beyond, p. 450. Springer, Heidelberg.

Fantes, P. A., and Creanor, J. (1984). Canavanine resistance and the mechanism of arginine uptake in the fission yeast *Schizosaccharomyces pombe*. *J. Gen. Microbiol.* **130,** 3265–3273.

Forsburg, S. L. (1993). Comparison of *Schizosaccharomyces pombe* expression systems. *Nucleic Acids Res.* **21,** 2955–2956.

Forsburg, S. L. (1999). The best yeast? *Trends Genet.* **15,** 340–344.

Forsburg, S. L. (2001). The art and design of genetic screens: Yeast. *Nat. Rev. Genet.* **2,** 659–668.

Forsburg, S. L. (2003a). Growth and manipulation of *S. pombe*. *Curr. Protoc. Mol. Biol.* Chapter 13, Unit 13.16.

Forsburg, S. L. (2003b). Introduction of DNA into *S. pombe* cells. *Curr. Protoc. Mol. Biol.* Chapter 13, Unit 13.17.

Forsburg, S. L., and Rhind, N. (2006). Basic methods for fission yeast. *Yeast* **23,** 173–183.

Forsburg, S. L., and Sherman, D. A. (1997). General purpose tagging vectors for fission yeast. *Gene* **191,** 191–195.

Gmunder, H., and Kohli, J. (1989). Cauliflower mosaic virus promoters direct efficient expression of a bacterial G418 resistance gene in *Schizosaccharomyces pombe*. *Mol. Gen. Genet.* **220,** 95–101.

Grallert, B., Nurse, P., and Patterson, T. E. (1993). A study of integrative transformation in *Schizosaccharomyces pombe*. *Mol. Gen. Genet.* **238,** 26–32.

Green, M. D., Sabatinos, S. A., and Forsburg, S. L. (2009). Microscopy techniques to examine DNA replication in fission yeast. *In* "DNA Replication. Methods and Protocols," (S. Venegrova and J. Z. Dalgaard, eds.), Vol. 521, pp. 463–482. Humana Press.

Heckman, D. S., Geiser, D. M., Eidell, B. R., Stauffer, R. L., Kardos, N. L., and Hedges, S. B. (2001). Molecular evidence for the early colonization of land by fungi and plants. *Science* **293,** 1129–1133.

Hoffman, C. S., and Winston, F. (1987). A ten-minute DNA preparation from yeast efficiently releases autonomous plasmids for transformation of *Escherichia coli*. *Gene* **57,** 267–272.

Hoffman, C. S., and Winston, F. (1989). A transcriptionally regulated expression vector for the fission yeast *Schizosaccharomyces pombe*. *Gene* **84,** 473–479.

Hughes, A. L., and Friedman, R. (2003). Parallel evolution by gene duplication in the genomes of two unicellular fungi. *Genome Res.* **13,** 1259–1264.

Iacovoni, J. S., Russell, P., and Gaits, F. (1999). A new inducible protein expression system in fission yeast based on the glucose-repressed inv1 promoter. *Gene* **232,** 53–58.

Javerzat, J. P., Cranston, G., and Allshire, R. C. (1996). Fission yeast genes which disrupt mitotic chromosome segregation when overexpressed. *Nucleic Acids Res.* **24,** 4676–4683.

Kanter-Smoler, G., Dahlkvist, A., and Sunnerhagen, P. (1994). Improved method for rapid transformation of intact *Schizosaccharomyces pombe* cells. *Biotechniques* **16,** 798–800.

Keeney, J. B., and Boeke, J. D. (1994). Efficient targeted integration at leu1–32 and ura4–294 in *Schizosaccharomyces pombe*. *Genetics* **136,** 849–856.

Kiely, J., Haase, S. B., Russell, P., and Leatherwood, J. (2000). Functions of fission yeast orp2 in DNA replication and checkpoint control. *Genetics* **154,** 599–607.

Klar, A. J. (1992). Developmental choices in mating-type interconversion in fission yeast. *Trends Genet.* **8,** 208–213.

Klar, A. J., Bonaduce, M. J., and Cafferkey, R. (1991). The mechanism of fission yeast mating type interconversion: Seal/replicate/cleave model of replication across the double-stranded break site at mat1. *Genetics* **127,** 489–496.

Kohli, J., Hottinger, H., Munz, P., Strauss, A., and Thuriaux, P. (1977). Genetic mapping in *Schizosaccharomyces pombe* by mitotic and meiotic analysis and induced haploidization. *Genetics* **87,** 471–489.

Krawchuk, M. D., and Wahls, W. P. (1999). High-efficiency gene targeting in *Schizosaccharomyces pombe* using a modular, PCR-based approach with long tracts of flanking homology. *Yeast* **15,** 1419–1427.

Liang, D. T., and Forsburg, S. L. (2001). Characterization of *Schizosaccharomyces pombe* mcm7(+) and cdc23(+) (MCM10) and interactions with replication checkpoints. *Genetics* **159,** 471–486.

Matsuyama, A., Arai, R., Yashiroda, Y., Shirai, A., Kamata, A., Sekido, S., Kobayashi, Y., Hashimoto, A., Hamamoto, M., Hiraoka, Y., Horinouchi, S., and Yoshida, M. (2006). ORFeome cloning and global analysis of protein localization in the fission yeast *Schizosaccharomyces pombe*. *Nat. Biotechnol.* **24,** 841–847.

Maundrell, K. (1993). Thiamine-repressible expression vectors pREP and pRIP for fission yeast. *Gene* **123,** 127–130.

Patterson, T. E., Stark, G. R., and Sazer, S. (1995). A strategy for quickly identifying all unique two-hybrid or library plasmids within a pool of yeast transformants. *Nucleic Acids Res.* **23,** 4222–4223.

Rasila, T. S., Pajunen, M. I., and Savilahti, H. (2009). Critical evaluation of random mutagenesis by error-prone polymerase chain reaction protocols, Escherichia coli mutator strain, and hydroxylamine treatment. *Anal. Biochem.* **388,** 71–80.

Rose, M. D., and Fink, G. R. (1987). KAR1, a gene required for function of both intranuclear and extranuclear microtubules in yeast. *Cell* **48,** 1047–1060.

Sabatinos, S. A., and Forsburg, S. L. (2009). Measuring DNA Content by Flow Cytometry in Fission Yeast. *In* "DNA Replication. Methods and Protocols," (S. Venegrova and J. Z. Dalgaard, eds.), Vol. 521, pp. 449–462. Humana Press.

Sato, M., Dhut, S., and Toda, T. (2005). New drug-resistant cassettes for gene disruption and epitope tagging in *Schizosaccharomyces pombe*. *Yeast* **22,** 583–591.

Sazer, S., and Sherwood, S. W. (1990). Mitochondrial growth and DNA synthesis occur in the absence of nuclear DNA replication in fission yeast. *J. Cell Sci.* **97**(Pt 3), 509–516.

Suga, M., and Hatakeyama, T. (2001). High efficiency transformation of *Schizosaccharomyces pombe* pretreated with thiol compounds by electroporation. *Yeast* **18,** 1015–1021.

Tasto, J. J., Carnahan, R. H., McDonald, W. H., and Gould, K. L. (2001). Vectors and gene targeting modules for tandem affinity purification in *Schizosaccharomyces pombe*. *Yeast* **18,** 657–662.

Tommasino, M., and Maundrell, K. (1991). Uptake of thiamine by *Schizosaccharomyces pombe* and its effect as a transcriptional regulator of thiamine-sensitive genes. *Curr. Genet.* **20,** 63–66.

Walker, G. M. (1999). Synchronization of yeast cell populations. *Methods Cell Sci.* **21,** 87–93.

Wang, L., Kao, R., Ivey, F. D., and Hoffman, C. S. (2004). Strategies for gene disruptions and plasmid constructions in fission yeast. *Methods* **33,** 199–205.

Watt, S., Mata, J., Lopez-Maury, L., Marguerat, S., Burns, G., and Bahler, J. (2008). urg1: A uracil-regulatable promoter system for fission yeast with short induction and repression times. *PLoS ONE* **3,** e1428.

Willer, M., Hoffmann, L., Styrkarsdottir, U., Egel, R., Davey, J., and Nielsen, O. (1995). Two-step activation of meiosis by the mat1 locus in *Schizosaccharomyces pombe*. *Mol. Cell. Biol.* **15,** 4964–4970.

Wood, V., and Bahler, J. (2002). Website review: How to get the best from fission yeast genome data. *Comp. Funct. Genomics* **3,** 282–288.

Wood, V., Gwilliam, R., Rajandream, M. A., Lyne, M., Lyne, R., Stewart, A., Sgouros, J., Peat, N., Hayles, J., Baker, S., Basham, D., Bowman, S., *et al.* (2002). The genome sequence of *Schizosaccharomyces pombe*. *Nature* **415,** 871–880.

CHAPTER THIRTY-THREE

APPLYING GENETICS AND MOLECULAR BIOLOGY TO THE STUDY OF THE HUMAN PATHOGEN *CRYPTOCOCCUS NEOFORMANS*

Cheryl D. Chun *and* Hiten D. Madhani

Contents

Abstract

The basidiomycete yeast *Crytococcus neoformans* is a prominent human pathogen. It primarily infects immunocompromised individuals producing a meningoencephalitis that is lethal if untreated. Recent advances in its genetics and molecular biology have made it a model system for understanding both the Basidiomycota phylum and mechanisms of fungal pathogenesis. The relative ease of experimental manipulation coupled with the development of murine models for human disease allow for powerful studies in the mechanisms of

University of California, San Francisco, California, USA

Methods in Enzymology, Volume 470
ISSN 0076-6879, DOI: 10.1016/S0076-6879(10)70033-1

virulence and host responses. This chapter introduces the organism and its life cycle and then provides detailed step-by-step protocols for culture, manipulation of the genome, analysis of nucleic acids and proteins, and assessment of virulence and expression of virulence factors.

1. INTRODUCTION

Although members of the Ascomycota phylum, particularly *Sacchromyces cerevisiae*, are the most studied fungi, there are 80,000 known species of the Fungi kingdom. There is a great deal of diversity in the kingdom, ranging from small harmless unicellular yeast such as *S. cerevisiae* to the great plant pathogen *Armillaria ostoyae*, one of the largest organisms in the world. This latter species is a member of the Basidiomycota phylum, a phylum less well understood than Ascomycota.

While no basidiomycete species has been studied in as much detail as *S. cerevisiae*, it is a fascinating and diverse group of organisms. Basidiomycetes produce many interesting secondary metabolites used in medicine, industry, and research. Members of the phylum account for about 10% (40 species) of known human fungal pathogens (Morrow and Fraser, 2009). With the onset of the AIDS epidemic, one basidiomycete in particular, *Cryptococcus neoformans*, has risen from a little-known pathogen to one of the top fungal killers of immunocompromised patients.

C. neoformans is primarily found as a haploid yeast, and is widely present in the environment worldwide, including in avian excreta, soil, and tree bark. Studies have shown that humans come into frequent contact with *C. neoformans*: individuals with no history of cryptococcosis possess antibodies against the yeast (Chen *et al.*, 1999), and most children appear to have been exposed by the age of five (Goldman *et al.*, 2001). This suggests that the majority of individuals encounter *C. neoformans* in the environment, most likely through inhalation into the lungs. Immunocompetent individuals are usually able to control and contain the infection, often leading to an asymptomatic latent state of infection. If the patient's immune system becomes compromised at a later date, the latent infection can reactivate. In the case of the immunocompromised individual, pulmonary infection can lead to pneumonia followed by dissemination via the bloodstream to other organs. *C. neoformans* is one of only a few fungal species known to cross the blood–brain barrier and infect the brain (Kim, 2006), leading to meningitis that is fatal if left untreated. When the AIDS epidemic began in the 1980s, there was a concomitant surge in cryptococcosis cases worldwide. In recent years, the increased usage of antiretroviral therapy and antifungals has reduced the overall incidences of fatal cryptococcal meningitis. Yet in areas where access to treatment is limited, *C. neoformans* remains an important concern in the care of the immunocompromised, including AIDS, cancer, and organ transplant patients. In addition, recent outbreaks of cryptococcosis in immunocompetent individuals in the Pacific Northwest raise concerns about the risk of cryptococcal infection even in otherwise healthy individuals (Bartlett *et al.*, 2008; Hoang *et al.*, 2004).

As a haploid yeast cell, *C. neoformans* is amenable to many of the extensive protocols that have been developed for *S. cerevisiae*, requiring in most cases only a few adjustments. However, having diverged from the ascomycete lineage some 400 million years ago (mya) (Taylor and Berbee, 2006), there are significant differences in its cellular machinery and life cycle (see below). Comparative genomics promises to yield rich information about the evolution of shared and diverged genes, proteins, and pathways, as well as offering insight into the differences between species that allow one yeast to exist as a benign saprophyte and another to cause lethal infection in a mammalian host.

2. SEROTYPES, STRAINS, AND SEQUENCES

C. neoformans is classified into four different serotypes based on its reactivity with monoclonal antibodies to surface capsular polysaccharide (Kabasawa *et al.*, 1991). These serotypes have historically been further classified into three different varieties: var. *neoformans* (serotype D), var. *grubii* (serotype A), and var. *gattii* (serotypes B and C). However, in recent years, var. *gattii* has been proposed to comprise its own species as *Cryptococcus gattii*, based on morphological and biochemical evidence (Kwon-Chung and Varma, 2006). *C. neoformans* var. *neoformans* and var. *grubii* primarily infect immunocompromised individuals, with var. *grubii* causing ∼99% of cryptococcal infections in HIV-infected patients (Mitchell and Perfect, 1995). *C. gattii* has the ability to infect immunocompetent individuals, as evidenced by an emergent outbreak in the Pacific Northwest that has resulted in hundreds of human and veterinary infections. Based on analysis of mutation frequency in conserved genes, it is thought that *C. neoformans* and *C. gattii* diverged about 37 mya, while *C. neoformans* var. *neoformans* and var. *grubii* split 18.5 mya, and within *C. gattii*, serotypes B and C diverged 9.6 mya (Xu *et al.*, 2000). To date, the genomes of five strains of *C. neoformans* and *C. gattii* have been sequenced to at least 6× coverage: JEC21 (serotype D), B-3501 (serotype D), H99 (serotype A), WM276 (serotype B), and R265 (serotype B). H99 and R265 are clinical isolates, while WM276 was isolated from the environment. JEC21 and B-3501 are laboratory-derived strains, where JEC21 was derived from B-3501 through a series of crosses and backcrosses (Heitman *et al.*, 1999), and their genomes are 99.5% identical (Loftus *et al.*, 2005).

Online resources for *C. neoformans* genome sequences

Strain	URL
JEC21	http://www.tigr.org/tdb/e2k1/cna1/
B-3501	http://www-sequence.stanford.edu/group/C.neoformans/
H99	http://www.broad.mit.edu/annotation/fungi/ cryptococcus_neoformans
WM276	http://www.bcgsc.ca/gc/cryptococcus/
R265	http://www.broad.mit.edu/annotation/fungi/ cryptococcus_neoformans_b/

The genomes of *C. neoformans* and *C. gattii* contain about 19 Mb of DNA spread over 14 chromosomes with about 7000 predicted protein-coding genes. The genomic sequence is relatively GC-rich (48% GC content) when compared to the genome of *S. cerevisiae* (38% GC content). Nucleic acid enzymatic protocols from *S. cerevisiae* laboratories that have been adapted for use with *C. neoformans* take this into account with the addition of DMSO (5% final concentration) or betaine (1.3 M final concentration) to resolve secondary structures resulting from the higher GC content.

Unless otherwise noted, the use of "*C. neoformans*" in the text of this chapter refers to *C. neoformans* var. *neoformans* and var. *grubii*. Although many of the same techniques are applicable to *C. gattii*, their usage is less well documented and may require additional adaptations.

 ## 3. LIFE CYCLE

C. neoformans and *C. gattii* primarily exist as haploid yeast cells that reproduce asexually through budding. They also possess a bipolar mating system, with mating types **a** and α. The mating (*MAT*) locus regulates the sexual cycle and encodes for more than 20 genes, including genes for cell type identity and the production and sensing of pheromone. Similar to *S. cerevisiae*, *MAT***a** cells produce MF**a** pheromone that is sensed by *MAT*α cells. In response to pheromone, *MAT*α cells produce a conjugation tube (Fig. 33.1A). Likewise, *MAT***a** cells respond to the MFα pheromone produced by *MAT*α cells, although the response of *MAT***a** cells is to form large swollen cells that can then fuse to the conjugation tubes of the *MAT*α cells. The *MAT***a** and *MAT*α nuclei divide, and the *MAT*α nuclei travel through the conjugation tube into the *MAT***a** cell. *MAT***a** and *MAT*α nuclei move into the hypha formed by the *MAT***a** cell, and a septum forms between the hypha and the *MAT***a** cell. The hypha may then elongate through cell

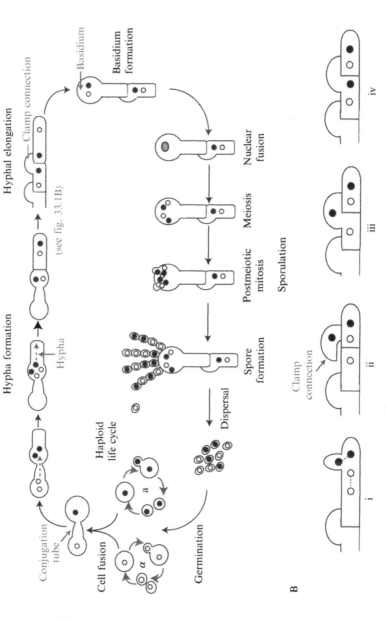

Figure 33.1 Life cycle of *C. neoformans*.

growth and division. During hyphal elongation, the nuclei divide mitoti-cally (Fig. 33.1B). One nucleus divides in such an orientation as to enter into a bulge in the cell wall that will later form a clamp connection (i). Septa form between the posterior cell wall, the tip of the hypha, and the clamp connection, leaving one nucleus in the posterior cell, two nuclei in the tip of the hypha, and one nucleus in the clamp (ii). The clamp fuses back to merge with the posterior cell (iii), allowing the nucleus present in the clamp to join the nucleus in the posterior cell (iv). During sporulation (Fig. 33.1A), a basidium forms at the tip of the hypha. In the basidium, the *MATa* and *MATα* nuclei fuse and undergo meiosis. The new *MATa* and *MATα* nuclei then undergo repetitive rounds of mitosis, eventually forming four chains of spores that emerge from the basidium. The spores are then dispersed, and germinate into haploid yeast cells (Bovers *et al.*, 2008; McClelland *et al.*, 2004).

In the laboratory, mating is achieved through nitrogen starvation on V8 medium, consisting of 5% (v/v) V8 juice, 3 mM KH$_2$PO$_4$, 4% (w/v) agar, pH 5.0. For technical details, several excellent studies have been performed examining mating conditions; we refer you to these (Escandon *et al.*, 2007; Nielsen *et al.*, 2003; Xue *et al.*, 2007).

Haploid fruiting has also been observed, where cells of one mating type become diploid and form hyphae. These monokaryotic hyphae are char-acterized by unfused clamp connections. Similar to mating, monokaryotic hyphae also form basidia, undergo meiosis, and sporulate (Lin *et al.*, 2005; Tscharke *et al.*, 2003; Wickes and Edman, 1995).

4. TECHNIQUES FOR BASIC CULTURE

C. neoformans is classified as a Biosafety Level 2 (BSL-2) organism, and as such does not require elaborate biohazard safety facilities. Current pre-caution recommendations include the use of a Class I or Class II biological safety cabinet for manipulation of environmental samples or spore forms. Incidences of infection in laboratory personnel are rare, limited in the literature to skin puncture accidents with needles heavily contaminated with *C. neoformans* (Casadevall *et al.*, 1994).

C. neoformans may be cultured using similar medium to that used in *S. cerevisiae* cultivation. Common media used include YPAD and YNB. For some assays (e.g., see Sections 6.1.6 and 6.3.2), Sabouraud dextrose medium is used for culturing for historic reasons and its promotion of yeast growth over bacterial growth.

	Composition
YPAD (yeast peptone adenine dextrose)	
1% bacto yeast extract (Becton Dickinson, Cat. No. 212720)	10 g
2% bacto peptone (Beckton Dickinson, Cat. No. 211820)	20 g
2% glucose	20 g
0.73 mM L-tryptophan (Sigma, Cat. No. T8941)	0.15 g
0.27 mM adenine (Sigma, Cat. No. A2786)	0.037 g
Water	to 1 l
YNB (yeast nitrogen base)	
0.15% YNB w/o amino acids, w/o dextrose, w/o ammonium sulfate (BIO 101, Cat. No. 4027-032)	3 g
75 mM ammonium sulfate	10 g
2% glucose	20 g
Water	to 1 l
Sabouraud dextrose	
3% Sabouraud dextrose broth (Becton Dickinson, Cat. No. 238210)	30 g
Water	to 1 l

Standard growth is performed in YPAD medium, typically at 30 °C, with the alternative use of the defined medium YNB. During logarithmic growth in YPAD medium at 30 °C, the doubling time of wild-type *C. neoformans* is approximately 110 min. Consistent with its role as a human pathogen, *C. neoformans* also grows robustly at 37 °C. Unlike *S. cerevisiae*, *C. neoformans* does not perform fermentation, and therefore requires a minimal amount of oxygen for growth. *C. neoformans* is sensitive to alkaline pH, growing poorly at pH 9. However, it is insensitive to acidic pH, exhibiting normal doubling times in conditions as low as pH 3.

Frozen stocks of *C. neoformans* can be maintained in 15% glycerol solution at −80 °C. These stocks may be revived following transfer by sterile applicator stick to a YPAD plate.

4.1. Dominant drug selection markers

Our laboratory and others have used resistance to nourseothricin (NAT), G418, and hygromycin for selection in *C. neoformans*.

In the plasmids pHL001-STM-# and pJAF1, the genes encoding for proteins conferring resistance to NAT and G418 have been inserted in between the promoter element of *C. neoformans ACT1* and the terminator element of *C. neoformans TRP1* (both of these sequences were derived from the H99 strain) (Table 33.1). In the plasmid pHYG7-KB1, the gene

Table 33.1 Dominant drug selection markers

Drug selection	Plasmid name	Structure	Reference
Nourseothricin	pHL001-STM-#	$CnACT1$ promoter-NAT^R -$CnTRP1$ term	Gerik *et al.* (2005)
G418	pJAF1	$CnACT1$ promoter-NEO^R -$CnTRP1$ term	Fraser *et al.* (2003)
Hygromycin	pHYG7-KB1	$CnACT1$ promoter-HYG^R -$CnGAL7$ UTR	Hua *et al.* (2000)

encoding for resistance to hygromycin was inserted between the promoter element of *C. neoformans ACT1* and the untranslated region (UTR) of *C. neoformans GAL7* (where the *ACT1* sequence was derived from H99 and the *GAL7* sequence was derived from JEC21).

For selection of yeast containing the appropriate drug resistance cassette, we use YPAD agar plates made with 0.1 mg/ml nouseothricin (clonNAT, Werner BioAgents), 0.2 mg/ml G418 (VWR, Cat. No. 45000-626), and/ or 0.3 mg/ml hygromycin (Sigma, Cat. No. H7772).

5. BASIC MOLECULAR BIOLOGY TECHNIQUES

5.1. Fusion polymerase chain reaction

Manipulation of the genomic sequence of a species is a powerful tool for analyzing the importance of specific genes in the function of the organism. Homologous recombination, or the integration of an exogenous DNA construct into the genome, is a crucial step in the site-directed mutagenesis of a target gene. *C. neoformans* performs homologous recombination at relatively low frequencies when compared with other fungi (1–4% as compared to nearly 100% in *S. cerevisiae*) but transformation with linear constructs flanked by a significant amount (0.3–1 kb) of sequence homologous to the genome creates stable integrants reproducibly (Davidson *et al.*, 2000; Nelson *et al.*, 2003). For example, a linear construct to target a gene for deletion might contain an antibiotic resistance cassette flanked on the 5'- and 3'-ends with 1 kb sequences homologous to the 5'- and 3'-ends of the targeted gene.

Construction of linear constructs for homologous recombination uses a procedure known as fusion PCR or PCR overlap (Davidson *et al.*, 2002).

In this process, two or more DNA fragments are joined together during the polymerase chain reaction (PCR) by virtue of a shared region of homology. This region of homology is engineered during previous PCR steps using primers containing linker sequences that are then shared between the two fragments to be fused together (Fig. 33.2).

5.1.1. Fusion PCR for targeted gene deletion

As mentioned previously, a linear construct for targeted gene deletion is designed to contain an antibiotic resistance cassette flanked by 1 kb sequences homologous to the targeted sequence. To create this construct, an antibiotic resistance cassette, such as resistance to NAT, is first amplified with primers containing 22 bp of homology to the 5′ and 3′ ends of the cassette, and 21 bp of linker sequence which is different for the 5′ and 3′ primers (these primers are designated primers 3 and 4 in Table 33.2 and Fig. 33.2A). Then, from genomic DNA we amplify 1 kb of sequence upstream and downstream of the ORF targeted for deletion. We term these sequences the 5′ and 3′ flanks for the targeted gene deletion construct. For the 5′ flank, the forward primer (1) is 22 bp of exact homology to the genomic sequence. The reverse primer (2) is 21 bp of linker sequence that is antiparallel to the linker sequence in the forward primer (3) for amplifying the antibiotic resistance cassette, followed by 22 bp of homology to the genomic sequence. For the 3′ flank, the forward primer (5) is 21 bp of linker sequence that is antiparallel to the linker sequence in the reverse primer (6) for amplifying the antibiotic resistance cassette, followed by 22 bp of homology to the genomic sequence. The reverse primer is 22 bp of exact homology to the genomic sequence. Table 33.2 contains the linker sequences we use to design these primers.

Primers 1–6 are used first for amplification of the 5′- and 3′-flanks and the antibiotic resistance cassette. Then primers 1 and 6 are used in the fusion PCR to amplify the full-length linear construct.

5.1.1.1. Conditions for amplification of 5′ and 3′ flanks and antibiotic resistance cassette *(50 μl final volume)*: 400 nM each primer, 0.25 mM dNTPs, 20 mM Tris–HCl (pH 8.8), 2 mM MgSO$_4$, 10 mM KCl, 10 mM (NH$_4$)$_2$SO$_4$, 0.1% (v/v) Triton X-100, 0.01% (w/v) BSA (98% electrophoresis grade, Sigma, Cat. No. A7906), 5% (v/v) DMSO, 2.5 U Pfu polymerase, 0.5 μl of template DNA (genomic DNA at 1 μg/μl or plasmid bearing antibiotic resistance cassette at 30 ng/μl).

We maintain at 4 °C a 10× stock of PCR buffer that contains 200 mM Tris–HCl (pH 8.8), 20 mM MgSO$_4$, 100 mM KCl, 100 mM (NH$_4$)$_2$SO$_4$, 1% (v/v) Triton X-100, 0.1% (w/v) BSA. We then add the primers, DMSO, dNTPs, Pfu, and template DNA separately.

PCR conditions, performed on a PTC-200 Peltier Thermal Cycler (MJ Research): 93 °C for 3 min, followed by 35 cycles of (93 °C for 30 s, 45 °C

A   Targeted gene deletion

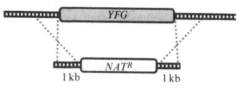

Targeted gene deletion construct generation

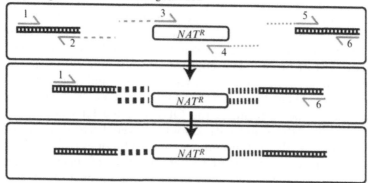

B   Epitope-tagging

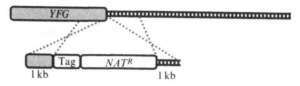

C   Promoter replacement

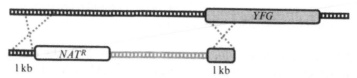

Figure 33.2   (A) *Targeted gene deletion*: The construct contains an antibiotic resistance cassette flanked on the 5′- and 3′-ends with 1 kb regions of homology upstream and downstream of the targeted gene. (B) *Epitope-tagging*: The construct contains the epitope tag and an antibiotic resistance cassette flanked on the 5′-end with a 1-kb region of homology to the 3′-end of the targeted gene, and on the 3′-end with a 1-kb region of homology to the region immediately downstream of the targeted gene. (C) *Promoter replacement*: The construct contains an antibiotic resistance cassette and the desired promoter region flanked on the 5′-end with a 1-kb region homologous to the sequence upstream of the promoter to be replaced, and on the 3′-end with a 1-kb region homologous to the 5′-end of the targeted gene. NAT^R, nourseothricin resistance cassette; YFG, targeted gene.

Table 33.2 Primers for construction of targeted gene deletion construct by fusion PCR

Primer	Sequence[a,b]
1	Forward primer to 5′ flank: 22 bp of sequence 1 kb upstream of ORF
2	Reverse primer to 5′ flank: CACGGCGCGCCTAGCAGCGGA-22 bp of sequence immediately upstream of ORF
3	Forward primer to antibiotic resistance cassette: CCGCTGCTAGGCGCGCCGTGA-22 bp of sequence at 5′ end of antibiotic resistance cassette
4	Reverse primer to antibiotic resistance cassette: GCAGGGATGCGGCCGCTGACA-22 bp of sequence at 3′ end of antibiotic resistance cassette
5	Forward primer to 3′ flank: GTCAGCGGCCGCATCCCTGCA-22 bp of sequence immediately downstream of ORF
6	Reverse primer to 3′ flank: 22 bp of sequence 1 kb downstream of ORF

[a] The linker sequences are not exactly antiparallel with each other; you will note that all linker sequences in primers 2′–5′ end in an adenine prior to the 22 bp of homologous sequence. Taq polymerase exhibits terminal transferase activity, which adds an additional adenosine onto the 3′ ends of PCR products. Therefore, the extra adenine in the primers allows for perfect homology between the linker sequences during the actual fusion PCR that fuses the three fragments together into the targeted gene deletion construct. The sequences for these primers are listed 5′–3′.

[b] The linker sequences were adapted from previous work Reid *et al.* (2002).

for 30 s, 72 °C for 3.5 min or appropriate amount of time for the length of your antibiotic resistance cassette), followed by 72 °C for 5 min.

Purify the PCR products by running the PCR out on a 0.8% agarose gel. Cut out the appropriate size band in the gel, and purify using a QIAquick Gel Extraction kit (Qiagen, Cat. No. 28704), following the manufacturer's instructions. In the final step, elute from the column with 30 μl of elution buffer (EB), letting the column stand for 1 min then centrifuging for 1 min at 13,000 rpm.

5.1.1.2. Conditions for fusion PCR *(50 μl final volume)*: 400 nM each primer (primers 1 and 6), 0.25 mM dNTPs, 10 mM Tris–HCl (pH 8.3), 50 mM KCl, 2 mM MgCl$_2$, 1.3 M betaine, 1 U Taq polymerase, 0.25 U Pfu polymerase, 50 nmol each of 5′ flank, 3′ flank, and antibiotic resistance cassette (roughly equal to 2 μl of the eluted volume from the QIAquick Gel Extraction).

We maintain at 4 °C a 10× stock of PCR buffer that contains 100 mM Tris–HCl (pH 8.3), 500 mM KCl, 20 mM MgCl$_2$. We then add the betaine, dNTPs, Pfu, and Taq polymerases separately. Betaine is maintained as a 5 M stock at 4 °C.

PCR conditions, performed on a PTC-200 Peltier Thermal Cycler (MJ Research): 72 °C for 10 min, 92.5 °C for 3.5 min, followed by 35 cycles of (92.5 °C for 12 s, 52 °C for 12 s, 72 °C for 7 min or appropriate amount of time for the length of the full targeted gene deletion construct), followed by 72 °C for 5 min.

Purify the PCR product by running the PCR out on a 0.8% agarose gel. Cut out the appropriate size band in the gel, and purify the DNA in the gel slice using a QIAquick Gel Extraction kit, following the manufacturer's instructions. In the final step, elute from the column with 50 μl of EB, letting the column stand for 1 min, then centrifuging the column for 1 min at 13,000 rpm. Add another 30 μl of EB to the column, let stand for 1 min, then centrifuge for 1 min at 13,000 rpm.

With modifications, fusion PCR may be utilized for a variety of genetic manipulations. By selecting different sequences for amplification, we have successfully used fusion PCR to introduce epitope tags into the 5'- and 3'-ends of genes (Fig. 33.2B), and to replace the promoters of genes (e.g., for overexpression of genes or placing genes under the control of inducible promoters) (Fig. 33.2C).

5.2. Transformation

The preferred method for transformation into *C. neoformans* is biolistic delivery. While studies have shown that *C. neoformans* is transformable by electroporation, these transformations have been low efficiency, resulting in some stable ectopic transformants but also many unstable transformants harboring extrachromosomal DNA material (Edman, 1992; Edman and Kwon-Chung, 1990). In addition, electroporation of different *C. neoformans* strains has varying degrees of success; strain H99 (serotype A) is much less tractable to transformation in this way than strain B-3501 (serotype D). In contrast, both serotypes A and D are readily transformed with relatively high efficiency by biolistic delivery (Davidson *et al.*, 2000; Toffaletti *et al.*, 1993).

In biolistic delivery, DNA is introduced into the yeast cell using a biolistic particle delivery system (PDS-1000/He, Bio-Rad) hooked up to a vacuum pump (Maxima C Plus M6C, Fisher Scientific) and compressed helium tank (Fig. 33.3). The targeted gene deletion constructs generated by fusion PCR (see above) are deposited onto gold bead microcarriers that are then positioned on a macrocarrier disk in the main chamber of the biolistic PDS. The air in the biolistic PDS is removed by a vacuum pump, and helium is pumped into a small chamber (termed the gas acceleration tube) positioned above the macrocarrier disk, separated from the main chamber

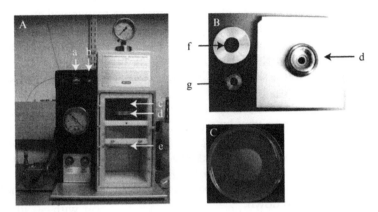

Figure 33.3 (A) Example PDS-1000/He biolistic particle delivery system (Bio-Rad): (a) "VENT/HOLD/VAC" toggle switch, (b) "FIRE" toggle switch, (c) gas acceleration tube/retaining cap, (d) microcarrier launch assembly, and (e) plate holder. (B) Close-up of microcarrier launch assembly for biolistic particle delivery system: (d) microcarrier launch assembly, (f) Top to microcarrier launch assembly, (g) Macrocarrier holder. (C) Example of a YPAD plate immediately following biolistic transformation. Note the scattering of microcarriers in the center of the patch of *C. neoformans* cells.

by a pressure-calibrated rupture disk. At a high enough pressure, the rupture disk breaks and the helium blasts into the main chamber of the biolistic PDS, propelling the macrocarrier disk downward against a metal stopping screen. The force of the impact against the stopping screen propels the DNA-coated microcarriers off the macrocarrier disk at high velocity and into *C. neoformans* cells that have been plated onto an agar plate and positioned below.

5.2.1. Protocol for biolistic transformation
Preparation of constructs (may be done anytime prior to the day of transformation)

1. Transfer 30 μl of the purified construct from fusion PCR (see above) into a microcentrifuge tube or into one well of a 96-well skirted PCR plate (Fisher, Cat. No. 055068).
2. Concentrate the DNA to the bottom of the tube by removing all moisture by SpeedVac.
3. Add 2.5 μl water, pipetting up and down and around the walls of the tube/well multiple times to resuspend the DNA.
4. Add 12.5 μl of microcarriers (0.6 μm gold beads, Bio-Rad, Cat. No. 1652262, resuspended in water to 60 mg/ml and maintained at 4 °C).
5. Add 12.5 μl of 2.5 M CaCl$_2$ (maintained at 4 °C). Pipette up and down to mix.

6. Add 5 μl of 0.1 M spermidine (1 M stocks are maintained at $-80\,°C$ and diluted to 0.1 M in water prior to use).
7. Mix on a vortexer at low speed for 4 min. We use a Vorex Genie 2 (Fisher, Cat. No.12-812) with a platform attachment, set to Vortex level 1. If using a skirted PCR plate, cover with a plastic plate seal (Qiagen, Cat. No. 1018104).
8. Collect the microcarriers by centrifugation at 500 rpm for 10 s.
9. Remove the supernatant by pipette.
10. Wash microcarriers by adding 50 μl 70% ethanol and immediately removing by pipette, being careful to not disturb the microcarrier pellet.
11. Add 50 μl 100% ethanol and immediately remove by pipette.
12. Resuspend the DNA-coated microcarriers in 12.5 μl 100% ethanol by pipetting up and down.
13. Transfer all 12.5 μl of microcarriers onto the center of a macrocarrier disk (Bio-Rad, Cat. No. 1652335) deposited in a 6-well culture dish (Falcon, Cat. No. 35-3224).
14. Dry the disk until all ethanol has evaporated, leaving a dark gold residue on the surface of the macrocarrier. We dry the disk by placing the 6-well culture dish in a desiccator hooked up to the house vacuum.

Day one

1. Inoculate *C. neoformans* from plate stock into 50 ml liquid YPAD medium in a 250 ml flask. Grow with aeration at 30 °C for 2–3 days.

Day three

1. Collect *C. neoformans* culture by centrifugation at 3000 rpm for 10 min.
2. Resuspend cell pellet in 5 ml regeneration medium (see recipe below).
3. Pipette 140 μl of resuspended cells onto the center of a YPAD plate, one plate for each transformation to be performed. Use a spreader device (Marsh Brand, Cat. No. KG-5P) to spread the cells into a circular patch, 4–5 cm in diameter.
4. Let the plates dry with lids ajar at 30 °C for 20–30 min.
5. Perform transformation with biolistic PDS.
   a. Open valve of compressed helium tank.
   b. Turn on vacuum pump and biolistic delivery system.
   c. Unscrew retaining cap at the end of the gas acceleration tube.
   d. Dip rupture disk briefly (2–3 s) in 70% isopropanol, to sterilize the disk and aid in its retention in the retaining cap following rupture.
   e. Place the rupture disk in the retaining cap of the biolistic delivery system, screw retaining cap back into place.
   f. Press microcarrier-coated macrocarrier (microcarrier-side up) in the macrocarrier holder.
   g. Place stopping screen at bottom of the microcarrier launch assembly of the biolistic delivery system.

   h. Place the macrocarrier holder into the microcarrier launch assembly, above the stopping screen (microcarrier-side down). Screw top on microcarrier launch assembly.

   i. Load microcarrier launch assembly in the first slot from the top in the main chamber of the biolistic delivery system.

   j. Load plate holder in the third slot from the top in the main chamber of the biolistic delivery system.

   k. Place a YPAD plate with patch of *C. neoformans* cells from step 3 on the plate holder. Remove its lid.

   l. Close the door of the biolistic delivery system, flipping the switch of the biolistic delivery system to "VAC" to start drawing air out of the chamber.

   m. When the pressure gauge reads more than 27 in. Hg vacuum, flip the switch of the biolistic delivery system to "HOLD."

   n. Hold down the "FIRE" button until you hear the rupture disk break—it will sound like a loud pop—and the helium pressure drops down to zero.

   o. Release the "FIRE" button and flip the switch of the biolistic delivery system to "VENT" to release the vacuum from the chamber.

6. Incubate transformed plates at 30 °C for 4 h to allow for recovery.
7. Resuspend cells in 800 μl phosphate-buffered saline (PBS) using a spreader device, then transfer by pipette to a plate containing selective medium. Spread the cells over the surface of the plate using a spreader device.
8. Dry the plates at 30 °C with the lids ajar for 30 min or until dry.
9. Cover plates and incubate 2–3 days. Colonies should be visible by the end of the next day and of a pickable size ($\sim$0.5 mm diameter) by the second day. These colonies should be picked and patched out onto selective medium plates for confirmation of their genotype by PCR. We typically patch out the colonies for a verification of the 5′-junction. The colonies that show successful integration of the construct at the 5′-junction are streaked out to single colonies on selective medium, and new colonies are patched out for verification of the 3′-junction.

Notes

1. For concentrating the transformation construct, if using microcentrifuge tubes we run them for 30–45 min in a Savant SC100 SpeedVac on high drying rate. If using a skirted PCR plate, we use a Savant AES2010 SpeedVac outfitted with plate holders, set to run for 6 h with 45 min of radiant cover heating on high drying rate.
2. Sterilize the macrocarriers and stopping screens prior to use by washing in 70% ethanol and drying in a sterile environment. We find a 15-cm Petri dish to work well for this purpose.

3. We use rupture disks rated between 1100 and 1350 psi (Bio-Rad, Cat. Nos. 1652329 and 1652330). Both have given good transformation results.

4. Following successful transformation, it is usually possible to see a spattering of gold beads embedded into the YPAD plate (Fig. 33.3C). If this is not visible, it is likely that not enough gold beads were used in the macrocarrier setup.

5. We find it best to pick the colonies by the end of the second day, because one obtains a higher rate at that time of successful transformants that test positive for the integration of the drug selection cassette at the targeted locus. Waiting until the third day or later allows for false positive colonies to catch up in size with true positives. In general, we find it best to pick the largest colonies on the plate, although for disruptions in genes that positively regulate cell growth, these knockouts can be slower growing than some false positives. We typically pick and patch out 6–8 colonies per transformation, although more may be picked for transformations with a lower success rate.

6. We have observed varying transformation success rates among strains that are theoretically genetically identical (i.e., H99 strains from different laboratory sources). We hypothesize that in the process of passaging these strains, mutations have been acquired that affect homologous recombination efficiency.

Regeneration medium	Composition
0.9% YNB w/o ammonium sulfate w/o dextrose w/o ammonium sulfate (BIO 101, Cat. No. 4027-032)	9 g
1 M sorbitol	182 g
1 M mannitol	182 g
2.6% glucose	26 g
0.267% bacto yeast extract (Becton Dickinson, Cat. No. 212720)	0.27 g
0.054% bacto peptone (Beckton Dickinson, Cat. No. 211820)	0.54 g
0.133% Gelatin (Sigma, Cat. No. G-8150)	1.33 g
Water	to 1 l

5.3. Colony PCR

The genotype of the transformed strain is verified through PCR-based detection of the expected 5'- and 3'-junctions of the resistance marker with the genomic DNA. While verifying the genotypes of many transformations, it is easiest and fastest to perform colony PCR.

1. Patch out colonies from the transformation into 48- or 96-well grid format on plates containing selective medium.
2. Grow at 30 °C for 2 days.
3. Using a 48- or 96-well pin replicator, transfer a generous amount of cells into 7 μl of water in each well of a PCR plate.
4. Seal the PCR plate with PCR plate thermal adhesive sealing film.
5. Flash freeze the PCR plate in liquid nitrogen, then immediately transfer to a PCR block set at 100 °C. Incubate for 2–5 min.
6. Perform PCR as below.

   (50 μl final volume): 400 nM each primer, 0.25 mM dNTPs, 20 mM Tris–HCl (pH 8.8), 2 mM MgSO$_4$, 10 mM KCl, 10 mM (NH$_4$)$_2$SO$_4$, 0.1% (v/v) Triton X-100, 0.01% (w/v) BSA (98% electrophoresis grade, Sigma, Cat. No. A7906), 5% (v/v) DMSO, 0.15 U Pfu polymerase, 0.5 U Taq polymerase.

   We maintain at 4 °C a 10× stock of PCR buffer that contains 200 mM Tris–HCl (pH 8.8), 20 mM MgSO$_4$, 100 mM KCl, 100 mM (NH$_4$)$_2$SO$_4$, 1% (v/v) Triton X-100, 0.1% (w/v) BSA. We then add the primers, DMSO, dNTPs, Pfu, and Taq polymerases separately.

   PCR conditions, performed on a PTC-200 Peltier Thermal Cycler (MJ Research): 92.5 °C for 3 min, followed by 35 cycles of 92.5 °C for 15 s, 45 °C for 15 s, 72 °C for 1 min 45 s or appropriate amount of time for the length of targeted amplicon), followed by 72 °C for 5 min.

   Notes

1. For adequate DNA recovery, there should be a visible amount of cells in the 7 μl of water in the PCR plate prior to flash freezing.
2. We use a fixed solid pin replicator (V&P Scientific, Cat. No. VP 408H) both to mark the selection medium plates on which colonies are patched and for the transfer of cells into a PCR plate. We use thin well PCR plates from RPI (Research Products International, Cat. No. 141314) and TempPlate Sealing Film (USA Scientific, Cat. No. 2921-000) for the PCR.
3. Colony PCR may also be performed in single tube reactions, using a sterile toothpick in this case to transfer an appropriate number of cells into 7 μl water in a PCR tube.
4. To verify successful gene deletion, we use primers designed to amplify DNA sequences of approximately 1 kb. The verification primers target the sequences outside of the region amplified as 5'- and 3'-flanks for the targeted gene deletion construct, and are paired with common primers internal to the gene encoding for the drug resistance.
5. As an additional test for successful gene replacement, it is often useful to perform a PCR to the ORF of the targeted gene to confirm its absence in the transformed strain.

5.4. Genomic DNA extraction

1. Inoculate 50 ml of YPAD with a *C. neoformans*. Culture at 30 °C until saturation (1–2 days).
2. Harvest cells by centrifugation in a 50-ml Falcon tube at 3000 rpm for 10 min.
3. Remove supernatant, add 30 ml water.
4. Vortex to mix, then harvest cells by centrifugation at 3000 rpm for 10 min.
5. Remove supernatant and flash freeze cell pellet in liquid nitrogen.
6. Transfer the conical tube containing the cell pellet into a lyophilizer vessel.
7. Attach to a lyophilizer (FreeZone 4.5 Liter Benchtop Freeze Dry System, Labconco) connected to a vacuum pump (Maxima C Plus M6C, Fisher Scientific).
8. Lyophilize cell pellet overnight, or until all the liquid has sublimated and a dry powdery pellet is left.
9. Add 3–5 ml of 3 mm glass beads and vortex vigorously until a fine powder is created.
10. Add 10 ml CTAB extraction buffer (100 mM Tris (pH 7.5), 0.7 M NaCl, 10 mM EDTA, 1% (w/v) CTAB (hexadecyltrimethylammonium bromide, Sigma, Cat. No. H6269), 1% (v/v) beta-mercaptoethanol) and mix.
11. Incubate at least 30 min at 65 °C.
12. Add an equal volume of chloroform and mix gently.
13. Pellet cell debris to the interphase by centrifugation at 3000 rpm for 10 min.
14. Transfer aqueous phase to a fresh tube.
15. Add an equal volume of isopropanol and mix gently.
16. Pellet DNA by centrifugation at 3000 rpm for 10 min.
17. Wash DNA pellet with 70% ethanol.
18. Aspirate out supernatant. Invert tube and allow pellet to dry overnight.
19. Resuspend DNA in 500 μl TE (100 mM Tris (pH 7.5), 1 mM EDTA) and transfer to a 1.5-ml microcentrifuge tube.
20. Add 1 μl RNase (1 mg/ml stock solution), and incubate at least 30 min at 37 °C.
21. Add 5 μl proteinase K (20 mg/ml stock solution), and incubate 2 h at 55 °C.
22. Add 500 μl equilibrated phenol (Sigma, Cat. No. P4557). Separate phases by centrifugation at 14,000 rpm for 10 min.
23. Transfer aqueous phase to a new 1.5-ml microcentrifuge tube. Add 500 μl chloroform. Vortex briefly to mix. Separate phases by centrifugation at 14,000 rpm for 10 min.

24. Transfer aqueous phase to a new 1.5-ml microcentrifuge tube. Add 1/10 volume 3 M NaOAc and 2–3 volumes 100% EtOH. Briefly vortex or flick to mix. Incubate at $-20\,°C$ for 2 h or $-80\,°C$ for 0.5–1 h.
25. Centrifuge at 14,000 rpm for 10 min. Aspirate out the supernatant.
26. Dry the pellet in a SpeedVac concentrator for 2 min.
27. Resuspend DNA in 500 μl TE.

Note

Lyophilization greatly enhances recovery of nucleic acid from *C. neoformans*. Dessication may weaken the structure of the polysaccharide capsule and cell wall, allowing greater disruption in later steps.

5.5. RNA extraction

1. Culture *C. neoformans* cells in the conditions desired for harvesting RNA.
2. Harvest cultures by centrifugation.
3. Remove medium and flash freeze cell pellet in liquid nitrogen.
4. Lyophilize cell pellet until dry.
5. Resuspend cell pellet in 1 ml TRIzol Reagent (Invitrogen, Cat. No. 15596018) and transfer to a 2-ml screw-cap microcentrifuge tube (Sarstedt, Cat. No. 72.693.005) containing ~200 μl volume of 0.5 mm zirconia/silica beads (Bio-Spec Products, Cat. No. 11079105z).
6. Bead-beat at least twice for 2.5-min intervals in a Mini-BeadBeater-8 (BioSpec Products).
7. Centrifuge samples at 12,000×g for 10 min at 4 °C.
8. Transfer cleared lysate to a 1.5-ml microcentrifuge tube and add 200 μl chloroform.
9. Vortex for 15 s to mix.
10. Centrifuge samples at 12,000×g for 10 min at 4 °C.
11. Transfer the aqueous phase to a new 1.5-ml microcentrifuge tube and add 500 μl isopropanol.
12. Briefly vortex and allow to sit at 4 °C for at least 15 min.
13. Centrifuge samples at 12,000×g for 10 min at 4 °C.
14. Remove supernatant and wash pellets with 1 ml 75% ethanol (prepared with RNase-free water).
15. Vortex to mix and centrifuge sample 10,000×g for 5 min at 4 °C.
16. Remove supernatant and dry pellet by spinning in a SpeedVac concentrator (Savant, Model SC100) for 2 min.
17. Resuspend pellet in 100 μl RNase-free water if performing DNase treatment, or 500 μl if not DNase-treating the sample.

18. *Optional*: You may DNase-treat the RNA at this step to remove contaminating DNA.
   a. Add 10 μl of 10× DNase buffer (0.1 M Tris (pH 7.5), 25 mM MgCl$_2$, 5 mM CaCl$_2$, made with RNase-free water and stored at $-20\,°C$), and 5 μl (50 U) of DNase I (Roche, Cat. No. 047 716 728 001).
   b. Incubate at 37 °C for 30 min.
   c. Incubate at 75 °C for 5 min to heat-inactivate the DNase I.
   d. Add 400 μl RNase-free water.
   e. Add 500 μl acid-equilibrated phenol:chloroform (Sigma, Cat. No. P1944). Vortex 15 s to mix.
   f. Let stand at room temperature until phases have separated ($\sim$10 min), then spin at 14,000 rpm for 10 min at 4 °C.
   g. Transfer aqueous phase to a new microcentrifuge tube.
   h. Add 500 μl chloroform. Vortex 1 min to mix.
   i. Spin at 14,000 rpm for 10 min at 4 °C.
   j. Transfer aqueous phase to a new tube.
19. Add 500 μl chloroform to the samples.
20. Vortex for 1 min.
21. Centrifuge samples at 12,000×g for 10 min at 4 °C.
22. Transfer the aqueous phase to a new microcentrifuge tube and add 15 μl 3 M NaOAc and 900 μl isopropanol. Briefly vortex and allow to sit at -20 °C for at least 15 min.
23. Centrifuge samples at 12,000×g for 10 min at 4 °C.
24. Remove supernatant and wash pellets with 1 ml 75% ethanol (prepared with RNase-free water).
25. Vortex to mix and centrifuge sample 10,000×g for 5 min at 4 °C.
26. Remove supernatant and dry pellet in a SpeedVac concentrator for 2 min.
27. Resuspend pellet in 100 μl RNase-free water.

Notes

1. Depending on the growth conditions being assayed, it may be necessary to increase the number and length of intervals in the Mini-BeadBeater-8. For example, conditions of increased capsule synthesis require upward of five 10-min intervals. Experimentation may be required in order to determine optimal durations for your growth conditions.
2. DNase treatment is optional but highly recommended, especially if the RNA will later be reverse-transcribed for use in quantitative PCR (qPCR). This step appears to be less critical for microarray analysis of transcript level, but is nonetheless recommended.
3. The second round of chloroform extractions (step 19 onward) has in our hands led to cleaner RNA extractions that offer greater yields of cDNA following reverse transcription.

5.6. Protein extraction for SDS–PAGE

1. Harvest cells in mid-logarithmic growth phase corresponding to $OD_{600} = 2$ (e.g., if cells are at $OD = 0.5$, harvest 4 ml) by centrifugation.
2. Remove medium and resuspend cells in 500 μl ice-cold H_2O.
3. Transfer to 2 ml screw-cap microcentrifuge tube (Sarstedt, Cat. No. 72.693.005).
4. Centrifuge samples at 12,000×g for 5–10 min.
5. Remove supernatant and flash freeze cell pellet in liquid nitrogen.
6. Lyophilize until pellet is dry.
7. Resuspend pellet in 1 ml ice-cold water.
8. Add 150 μl NaOH/beta-mercaptoethanol mixture (1.85 N NaOH, 7.5% (v/v) beta-mercaptoethanol) to each sample.
9. Incubate on ice with occasional vortexing for 30 min.
10. Add 150 μl trichloracetic acid (TCA, 55% (w/v) in water kept at 4 °C in a foil-wrapped bottle) to each sample.
11. Incubate on ice with occasional vortexing for 30 min.
12. Centrifuge samples at 12,000×g for 10–20 min at 4 °C.
13. Remove most of the supernatant.
14. Optional (when harvesting >1 OD of cells): Add 100 μl ice-cold acetone to optimize removal of residual TCA.
15. Centrifuge samples at 12,000×g for 1 min at 4 °C.
16. Remove the remaining supernatant.
17. Resuspend pellet in 50 μl HU buffer (200 mM sodium phosphate buffer (pH 6.8), 8 M urea, 5% (w/v) SDS, 1 mM EDTA, bromophenol blue. Store at − 20 °C and add 100 mM DTT immediately before use).

Notes

1. To load samples in HU buffer, care should be taken not to boil them. Instead, the samples should be heat-denatured at 65–70 °C for 10–15 min or at 37 °C for 30 min.
2. If HU buffer in the resuspended protein pellet turns yellow due to residual TCA, add 10–20 μl of 1 M Tris (pH 6.8).

6. METHODS FOR ASSAYING PATHOGENESIS

6.1. Murine model of infection

Mice are relatively susceptible to *C. neoformans* infection, when compared with other mammalian hosts such as rats and rabbits. Immunocompetent murine strains will succumb to pulmonary infection, and will experience dissemination to other organs including the brain, similar to human crypto-coccosis. A murine model of cryptococcal infection offers several advantages over other species, including the consistency of susceptibility within a given strain, the availability of genetically modified strains (useful for examining host factors that may be involved in infection), as well as their small size and low cost. Our laboratory utilizes two routes of infection for introducing *C. neoformans* into a murine model of infection: intranasal and intravenous.

6.1.1. Considerations of murine strain and age

Inbred mouse strains may vary in their susceptibility to *C. neoformans* infection. For example, in some studies, BALB/c mice have been demonstrated to be more resistant to *C. neoformans* infection than C57BL/6 mice, as evidenced by both fungal load in the lungs following infection (Chen et al., 2008; Huffnagle et al., 1998) and degree of dissemination to other organs (Chen et al., 2008), although in some survival curve analyses, C57BL/6 mice survive slightly longer than BALB/c mice following infection with *C. neoformans* (Nielsen et al., 2005). There are varying theories as to the source of the differences in susceptibility in mouse strains to *C. neoformans*: studies have linked relative resistance of a mouse strain to the production of Th1-type cytokines, where their production is associated with pulmonary clearance (Huffnagle et al., 1998), or the presence of complement protein C5 (Rhodes et al., 1980). Our laboratory performs murine infections with 5- to 6-week-old A/J mice, which are C5-deficient and therefore slightly more susceptible to *C. neoformans* infection than C5-sufficient mouse strains (e.g., BALB/c and C57BL/6 mice) (Nielsen et al., 2005; Wormley et al., 2007). It bears noting that the inocula listed below have been determined by our laboratory for use with our strain of H99 *C. neoformans* in A/J mice. Use of other mouse strains may necessitate adjustment of the inocula to a higher or lower dosage of *C. neoformans* cells. Additionally, derivatives of the H99 strain (i.e., H99 stocks maintained by different laboratories) appear to have varying levels of virulence.

Studies have also shown that the age of the mice may affect their relative susceptibility to infection. Older C57BL/6 mice (e.g., 17-week-old) are better able to clear an intratracheal infection from their lungs, brains, and spleens than younger (e.g., 5-week-old) mice (Blackstock and Murphy, 2004). We have found it best to infect mice of a consistent age to reduce variability in our data.

6.1.2. Intranasal infection

An intranasal infection is thought to more closely mimic the natural course of infection, beginning with the inhalation of *C. neoformans* cells leading to pulmonary disease followed by dissemination to other organs. The mice are first anesthetized with a mixture of ketamine hydrochloride (Orion Pharma

Animal Health) and medetomidine hydrochloride (Domitor®, Orion Pharma Animal Health) via intraperitoneal injection. For all murine injections, we use ½ cc insulin syringes with 28G½ needles (Becton Dickinson, 329461). The anesthetic is mixed to contain 18.75 mg/ml ketamine hydrochloride and 0.625 mg/ml medetomidine hydrochloride. We administer 30–50 µl of this formulation to each mouse (10–15 g), leading to doses of 50–60 mg/kg ketamine hydrochloride and 1.5–2.0 mg/kg medetomidine hydrochloride. The mice usually succumb to the anesthesia after 5–10 min, at which point they are weighed, their ears notched for later identification, and ointment (Artifical Tears, Webster Veterinary, Cat. No. 07-841-4071) applied to their eyes to prevent them from drying out. A silk thread (50 Denier Weight, obtainable from a sewing supply store) is strung between two supports—we use ring stands for this purpose. The mice are then suspended by their incisors upon this silk thread (Fig. 33.4A). The inoculum

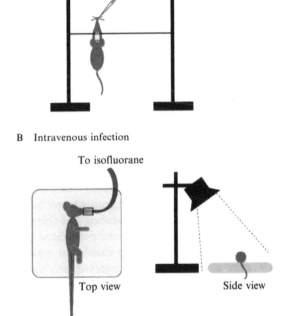

Figure 33.4 (A) Intranasal infection. A silk thread is tied across two supports (such as ring stands). The anesthetized mouse is suspended from its front incisors on the thread. The inoculum of yeast cells is pipetted down one nare. (B) Intravenous infection. The mouse is anesthetized with isofluorane administered by face mask (top view), while the tail vein is dilated through a combination of a sodium acetate heating pad from below and a heating lamp from above (side view). When the mouse is laid on its side, the lateral tail vein of the mouse will be at the top of the tail (top view).

of *C. neoformans* cells (5×10^5 cells/mouse in 50 μl) is slowly pipetted directly into one nare using a pipette fitted with a filter tip. Take care to allow for complete dispension of the inoculum into the nare; if signs of struggling are seen in the mouse, pipetting should be suspended until the mouse no longer shows signs of discomfort. We typically anesthetize and inoculate batches of five mice at a time. Following completion of inoculation, the mice remain suspended for 10 min, to allow for complete aspiration into the lungs, before being lowered and the anesthesia reversed via intraperitoneal injection of atipamezole hydrochloride (Antisedan®, Orion Pharma Animal Health). We administer 40–50 μl of 1 mg/ml atipamezole hyrochloride per mouse, leading to a dose of 2.5–3.5 mg/kg. It typically takes 10–15 min following injection with atipamezole hydrochloride to see signs of stirring in the mice, and another 15–20 min before the mice begin to walk around again.

6.1.3. Inocculum preparation

Inocula are prepared by growing *C. neoformans* in liquid YPAD overnight at 30 °C. Cells are counted by hemocytometer and, for an intranasal infection, 1×10^7 cells are washed twice with PBS and resuspended in 1 ml of PBS. Fifty microliters of this inoculum are used per mouse (5×10^5 cells). For an intravenous infection, 2×10^7 cells are washed in PBS and resuspended in 1 ml of PBS. One hundred microliters of this inoculum is used per mouse (2×10^6 cells). Inocula concentrations are confirmed by plating appropriate dilutions onto YPAD plates and counting the colony forming units (CFU) after 2 days growth at 30 °C.

6.1.4. Intravenous infection

An intravenous infection, via the lateral tail vein, leads to more uniform dissemination to the organs. The mice are weighed prior to infection and marked by ear notching for later identification. They are anesthetized via inhalation of 3% isofluorane in oxygen, administered by face mask, then remain on a sodium acetate rechargeable heating pad (Heat Solution, Prism Enterprises) beneath a heating lamp during the procedure (see Fig. 33.4B) in order to dilate the vein so that it is more visible for easier injection. The inoculum (2×10^6 cells in 100 μl PBS) is injected into the lateral tail vein. Following successful inoculation, the mice are immediately removed to their cage where they will rapidly recover from the anesthesia.

6.1.5. Monitoring disease progression

Mice are weighed prior to infection, and then monitored every 2–3 days postinfection. Signs of disease progression include hunched posture, abnormal gait, weight loss, and decreased grooming as indicated by ruffled fur. Our laboratory uses two endpoints for assessing time of survival: the point at which the mouse has lost 15% of its initial weight, or 25% of its peak weight.

We find the latter to be more consistent when the mice were infected at a younger age (e.g., close to 4 weeks in age) and are hence smaller at the initial time point.

6.1.6. Murine infection evaluations

"Time-to-endpoint" survival curve analysis monitors the infection of 8–10 mice with a single strain of *C. neoformans*, until their endpoints (as defined above). In this manner, mice infected with less virulent strains of *C. neoformans* survive longer than mice infected with more virulent strains (Fig. 33.5A).

This analysis gives a gross determination of the virulence of a single strain on the entire host system. A more specific analysis might address questions such as the initial rate of colonization to a specific organ, the rate of proliferation and/or rate of killing by the host immune cells, or the rate of dissemination to other organs. This additional analysis may be performed by assessing fungal load in the organs at various time points following infection. We typically examine fungal loads in the lungs, brain and spleen, although we have also examined the liver and kidneys. To measure organ loads after the animal is euthanized, the selected organs are removed by dissection, and placed on ice in 17×100 mm polypropylene sterile tubes (Evergreen Scientific, Cat. No. 222-2393-080), one tube per organ per mouse. Take care to wash the dissecting tools in water and ethanol between organs to eliminate carryover of yeast from organ to organ. Each organ is homogenized in 5 ml sterile PBS (we use a PRO200 tissue homogenizer, PRO Scientific, Oxford, CT), then serial dilutions in PBS are plated on Sabouraud dextrose agar plates (made with Sabouraud dextrose agar, Becton Dickinson, Cat. No. 211661) containing 40 μg/ml gentamycin and 50 μg/ml carbenicillin to discourage bacterial growth. CFU are assessed, and comparisons can be made for a single strain in different organs, rate of growth in different organs over time, or between multiple strains for relative fitness.

6.1.7. Signature-tagged mutagenesis screening

Evaluation of infectivity and virulence for a large number of *C. neoformans* strains through single-strain infections as described above can quickly add up in terms of both time and cost, as many mice must be used for each strain. Pooling mutant strains into a single infection allows rapid assessment of multiple strains in a single mouse. This can be easily and effectively performed using a technique known as signature-tagged mutagenesis (STM) screening. Each mutant contains a signature tag, or a unique sequence similar to a barcode, in its DNA. When pooled together in a group, individual strains can still be identified through qPCR of pooled genomic DNA using signature-tag-specific primers. By identifying relative representation in the pool of genomic DNA before and after infection, relative rates of infectivity can be assessed rapidly and reproducibly for multiple mutants in a single infection. Using 48 unique signature tag sequences, this technique has been

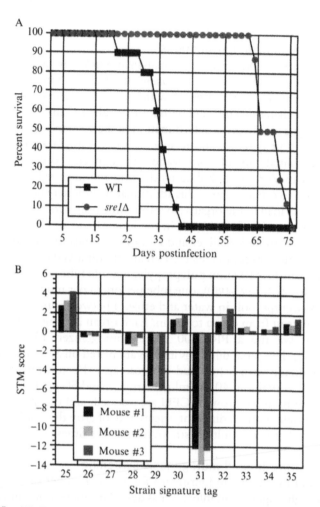

Figure 33.5 (A) Example of a survival curve. Mice were inoculated via tail-vein injection with 2×10^5 cells/mouse of either WT (H99) or *sre1Δ* strains of *C. neoformans*. On average, mice infected with *sre1Δ* survived 30 days longer, indicating *sre1Δ* that is a hypovirulent strain. (B) Example of STM score data. Forty-eight signature-tagged strains were grown individually in liquid YPAD medium in a 96-well deep pocket plate, then pooled together to generate the inoculum. Three mice were inoculated with 5×10^5 cells/mouse via intranasal infection. The mice were monitored to the disease endpoint, at which point they were sacrificed. Shown are a subset of the data from the lungs, following qPCR and calculation of the STM score for each signature tag in each mouse.

employed by our laboratory for the production and quantitative analysis of a library containing ~1200 targeted gene deletion strains (available without restriction from the Fungal Genetic Stock Center or the American Type Culture Collection (ATCC)) (Liu *et al.*, 2008).

In detail, to analyze a group of 48 signature-tagged strains, the group is first grown up in liquid YPAD in 96-well deep-pocket plates (Grenier Bio-One, Cat. No 780270), one strain per well, at 30 °C without shaking for 3 days. Two hundred microliters of each culture is pooled together, and the number of cells assessed by hemocytometer. 2×10^7 cells (for tail vein injection) or 1×10^7 cells (for intravenous infection) are washed twice in sterile PBS and resuspended in 1 ml sterile PBS. This pool is used as the inoculum to infect three mice, either by intranasal infection (5×10^5 cells/mouse) or tail vein injection (2×10^6 cells/mouse). Fifty microliters (5×10^5 cells) of this pool is also plated in triplicate on Sabouraud dextrose agar plates containing 40 μg/ml gentamycin and 50 μg/ml carbenicillin, which are then incubated at 30 °C for 2 days. The resulting colonies are scraped off each plate, resuspended in water, flash frozen in liquid nitrogen and then lyophilized. Genomic DNA is prepared from these samples as described above. This DNA constitutes the "input DNA" for later analysis.

After monitoring and sacrifice of the animals, the organs of interest are removed and homogenized in 5 ml sterile PBS. Serial dilutions in triplicate are made in sterile PBS and plated on Sabouraud dextrose agar plates containing 40 μg/ml gentamycin and 50 μg/ml carbenicillin. These plates are incubated at 30 °C for 2 days. The resulting colonies are scraped off each plate, resuspended in water, flash frozen in liquid nitrogen, and then lyophilized. The genomic DNA that is prepared from these samples constitutes the "output DNA" of the experiment.

The input and output DNA are analyzed using qPCR using a common primer targeted to the drug resistance marker that has replaced the targeted gene, coupled with signature tag-specific primers.

6.1.7.1. STM qPCR conditions

(50 μl final volume): 400 nM each primer, 0.25 m*M* dNTPs, 10 m*M* Tris–HCl (pH 8.3), 50 m*M* KCl, 2 m*M* MgCl$_2$, 1.3 *M* betaine, 1 U Taq polymerase, 0.25 U Pfu polymerase, 1–4 μg genomic DNA, 2 μl 2× Sybr Green I (Molecular Probes, Cat. No. S-7563).

We maintain at 4 °C a 10× stock of PCR buffer that contains 100 m*M* Tris–HCl (pH 8.3), 500 m*M* KCl, 20 m*M* MgCl$_2$. We then add the betaine, dNTPs, Sybr Green I, Pfu, and Taq polymerases separately. Betaine is maintained as a 5 *M* stock at 4 °C. Sybr Green I is kept at −20 °C as a 100× stock in DMSO, and diluted 1:50 in TE buffer to 2× stock immediately prior to addition to the PCR mix.

PCR conditions, performed on a DNA Engine Opticon (MJ Research): 93 °C for 4 min, followed by 40 cycles of (93 °C for 45 s, 52 °C for 25 s, 72 °C for 1 min, then a plate read by the machine), followed by 72 °C for 5 min.

6.1.7.2. Calculating STM score

The threshold cycle (C_T), or cycle number where the amplified target reaches a fixed threshold, of each primer pair is used to calculate an STM score, using a variation of the $2^{-\Delta\Delta C_T}$ method for quantitation analysis (Livak and Schmittgen, 2001). For each signature tag, a ΔC_T is calculated by subtracting the C_T of the specific primer pair from the median C_T for all 48 pooled strains to ($\Delta C_T = C_{T\text{-median}} - C_{T\text{-tag}}$). The ΔC_T values for each of the three independent input DNA samples are averaged to calculate the $\Delta C_{T\text{-input}}$ value. The ΔC_T values for each of the three independent output DNA samples is similarly calculated by subtracting the median C_T for all 48 pooled strains to the C_T of the specific primer pair. However, ΔC_T values ($\Delta C_{T\text{-output}}$) for the three independent output DNA samples are not averaged. The value $\Delta\Delta C_T$ is then calculated, where ($\Delta\Delta C_T = \Delta C_{T\text{-output}} - \Delta C_{T\text{-input}}$). The STM score is then equal to $\Delta\Delta C_T$. The STM scores from the three mice (i.e., each of the three output DNA samples) are then averaged to determine a final STM score for each mutant. Strains with reduced levels of persistence in the organ have STM scores less than 0, while strains with increased levels of persistence have STM scores greater than 0 (Fig. 33.5B). The STM score correlates with the relative fold change in persistence of a strain with respect to wild type.

This method of analysis makes the basic assumption in its normalization of the data that most of the signature-tagged strains in the pooled infection will have phenotypes similar to wild type. If you desire to assay a significant number of strains that you believe to have different survival rates than wild-type *C. neoformans* in the mouse, you may need to also seed the inoculum pool with signature-tagged strains that are known to have a wild-type phenotype to prevent skewing during the normalization process. We frequently use knockouts in the gene *SXI1* (CNAG_06814 in the H99 sequence database of the Broad Institute) in this manner, as *SXI1* is required for mating but dispensable for virulence (Hull *et al.*, 2004).

It is important to note that this screen assays for relative persistence of a strain within a particular organ. It is not a true test of virulence *per se*, as it is conceivable that a strain may persist in large numbers in a tissue but fail to cause disease in the host. However, in our experience (Liu *et al.*, 2008), the STM screen, when used to assay persistence of mutant strains in the lungs of 5-week-old A/J mice following intranasal infection, has resulted in STM scores that are both reproducible from mouse-to-mouse and pool-to-pool, but are also to a certain degree quantitative, by which we mean that the relative value of the STM score correlates with relative hypo- and hypervirulent phenotypes of the strains when assayed by survival curve analysis.

STM screens also cannot avoid *in trans* effects from mixing of strains; theoretically a wild-type strain may complement the phenotype of a mutant strain, allowing for a false negative result. Single-strain infections bypass this limitation of the STM screen approach.

6.2. Tissue culture

Although analysis of virulence in the host organismal level offers obvious correlations between a particular genotype and its efficacy at disease development, it is often problematic to determine the specific host–pathogen interactions responsible for a certain virulence phenotype. It is therefore useful to examine in closer detail the interaction of *C. neoformans* with a particular host tissue or cell type.

Many studies of the virulence of *C. neoformans* have focused on its interactions with the immune system, and, in particular, its interactions with macrophages. Alveolar macrophages are thought to be the first line of defense against pulmonary cryptococcal infection. Macrophages and macrophage-derived cells have been observed in the periphery of cryptococcal-containing granuloma formations in the lungs during latent infection of immunocompetent hosts. Additionally, depletion of macrophages from the murine host through the administration of silica has proven to be detrimental to fungal clearance (Monga, 1981). *C. neoformans* mutants that are more susceptible to killing by macrophages are hypovirulent in "time-to-endpoint" survival curves (e.g., *FHB1* which encodes for flavohemoglobin; de Jesus-Berrios *et al.*, 2003). For these and other reasons, it is of interest to examine the interaction of macrophages and macrophage-like cells with *C. neoformans* yeast.

Unopsonized *C. neoformans* cells are rarely taken up by macrophages in the absence of activation by cytokines such as IFN-γ or potent antigens such as lipopolysaccharide (LPS). Therefore, phagocytosis assays and assessments of killing by macrophages are commonly done in the presence of both opsonins (such as anti-*C. neoformans* antibodies or murine or human sera) and activating agents. We most commonly use the murine macrophage-like cell line RAW264.7 (American Type Culture Collection, No. TIB-71), and have had better success using anti-*C. neoformans* antibody than sera as an opsonizing agent.

6.2.1. Assay for killing of *C. neoformans* by macrophages

1. Seed RAW264.7 macrophages overnight into 96-well tissue culture plates (Corning, Cat. No. 3598) in 200 μl RAW cell medium (high-glucose DMEM (UCSF Cell Culture Facility, Cat. No. CCFAA005), 20 mM HEPES/NaOH buffer (pH 7.4) (UCSF Cell Culture Facility, Cat. No. CCFGL001), 20 mM glutamine (UCSF Cell Culture Facility, Cat. No. CCFGB002)) with IFN-γ (100 U/ml, Millipore, Cat. No. 005) at a density of 5×10^5 cells/well.
2. Culture the strain(s) of *C. neoformans* in 5 ml YPAD medium overnight at 30 °C.
3. The following day, wash an aliquot of the overnight *C. neoformans* culture 3× in sterile PBS.

4. Resuspend the *C. neoformans* cells to a concentration of 10^7 cells/ml.
5. Remove the RAW medium from the macrophages and replace with 200 μl of fresh RAW medium with 30 ng/ml LPS (Sigma, Cat. No. L4391), 100 U/ml IFN-γ, and anti-*C. neoformans* antibody.
6. Add 10 μl of the *C. neoformans* cells (10^5 cells) to the macrophages, and 10 μl to a well containing only 200 μl of RAW medium. Incubate for 24 h.
7. Remove supernatant from the wells, and retain for plating.
8. Lyse the macrophages by adding 0.01% SDS to each well. Wait 15 min, then remove and add to the supernatant previously removed. Repeat at least three times.
9. Check for complete lysis of the macrophages via a microscope.
10. Dilute the collected supernatants and plate for CFU. Determine the rate of killing by the macrophages by comparing the CFU from the wells with macrophages to the CFU from the wells without macrophages.

6.3. Assays for characterized virulence factors

C. neoformans has a number of characteristics previously shown to be involved in its virulence. These include (1) ability to grow at 37 °C (2) melanization, thought to aid in resistance to host killing (Nosanchuk and Casadevall, 2003), and (3) polysaccharide capsule formation, thought to be involved in host immune system evasion (Del Poeta, 2004), (Monari *et al.*, 2006). Below are methods to test the relative efficiency of a strain for production of the virulence factors melanin and capsule.

6.3.1. Melanization

Melanization, or the ability for the yeast to form dark pigment compounds from catecholamine substances such as L-DOPA (3,4-dihydroxy-L-phenyl-alanine) by the enzyme laccase (Lac1), has long been associated with *C. neoformans* virulence. It has been hypothesized that melanin protects the yeast from oxidative or nitrosative damage originating from the host cells. To test strains for melanization, we utilize plates containing 100 ng/ml L-DOPA.

L-DOPA plates	Composition
2% Difco Bacto Agar (Becton-Dickinson, Cat. No. 214030)	20 g
7.6 mM L-asparagine monohydrate	1 g
5.6 mM glucose	1 g
22 mM KH$_2$PO$_4$	3 g
1 mM MgSO$_4$·7H$_2$O	250 mg
0.5 mM L-DOPA (Sigma, Cat. No. D9628)	100 mg

0.3 mM thiamine–HCl	1 mg
20 nM biotin	5 μg
Water	to 1 l

To make 1 l of L-DOPA plate medium, autoclave 20 g of Difco Bacto Agar in 900 ml water so that it dissolves. In 100 ml water, add L-asparagine, glucose, KH_2PO_4, $MgSO_4 \cdot 7H_2O$, and L-DOPA in the amounts indicated in the above recipe. Add phosphoric acid to the medium to pH 5.6, then add thiamine–HCl and biotin. Mix with the dissolved agar, and pour into plates.

6.3.1.1. Melanization test protocol

1. Inoculate cultures into YPAD from colonies on a plate for growth overnight.
2. Measure the optical density (OD) by spectrophotometer for each culture to be tested.
3. Dilute the cultures to the equivalent of $OD_{600} = 0.6$ with PBS and array in a 96-well assay plate.
4. Spot 4–6 μl of each diluted strain onto an L-DOPA plate.
5. Incubate for 2–5 days at 30 or 37 °C, under observation (Fig. 33.6A).

Note

1. We have observed that the kinetics of melanization differ between growth at 30 and 37 °C, with a greater range of phenotypes visible at 30 °C. If screening a large numbers of strains, growth at 37 °C may be useful to highlight the mutants with more extreme defects in melanization. In addition, it may be useful to monitor the degree of melanization

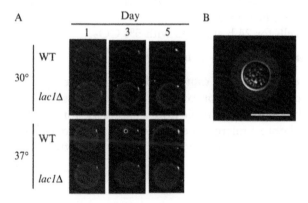

Figure 33.6 (A) *Melanin assay*: The kinetics of melanization varies depending on the incubation temperature. *lac1*Δ is deficient in the primary laccase enzyme responsible for melanization in *C. neoformans*. (B) *Capsule formation assay*: WT (H99) cell grown under capsule inducting conditions (DMEM, 37 °C, 5% CO_2) and visualized with India ink. Bar denotes 10 μm.

at early time points, as we have observed some strains that begin to melanize later than wild type, but reach a similar final level of melanization after 3 days.

6.3.2. Capsule formation

Secretion of a polysaccharide capsule is one of the major virulence factors of *C. neoformans*. When *C. neoformans* is mixed with India ink, the particles of the ink are excluded by a network of capsule fibers, thereby producing a characteristic halo around the yeast cell (Fig. 33.6B). Capsule is produced at low levels in typical YPAD culture—for best visualization, capsule production must be induced. Capsule can be induced through two different methods, described as follows.

Capsule induction via low nutrient conditions:

1. Inoculate *C. neoformans* from a colony on a plate into liquid Sabouraud dextrose medium.
2. Grow overnight at 30 °C.
3. Dilute the culture 1/100 in 10% Sabouraud dextrose medium buffered to pH 7.3 with 50 m*M* MOPS.
4. Grow cultures at 30 °C for 2 days in a rotating drum.

Capsule induction via carbon dioxide exposure:

1. Inoculate *C. neoformans* from a colony on a plate in liquid YNB medium.
2. Grow overnight at 30 °C.
3. Count cells on hemocytometer.
4. Wash 2×10^7 cells three times with PBS.
5. Resuspend the cells in 2.5 ml DMEM in a 6-well tissue culture dish (Falcon, Cat. No. 35-3224).
6. Culture cells for 24 h at 37 °C with 5% CO_2.

Visualization of capsule by India ink staining:

1. Collect cells grown in capsule-inducing conditions. Concentrate the cells by centrifugation for ease of viewing if necessary.
2. Add 4 μl India ink (obtainable from a stationery store) to 20 μl of culture.
3. Drop 2 μl onto a microscope slide, mount with coverslip glass.
4. Visualize on a microscope with 60×–100× objective.

7. Concluding Remarks

While the methods described here are by no means all inclusive for what can be accomplished with *C. neoformans*, we hope they provide a guide for working with this basidiomycete, as well as a starting point for adapting other techniques.

ACKNOWLEDGMENTS

We thank Joseph Heitman, Jennifer Lodge, Gary Cox, Tamara Doering, John Perfect, Peter Williamson, Christina Hull, Andrew Alspaugh, James Kronstad, Thomas Kozel, Arturo Casadevall, June Kwon-Chung, and Suzanne Noble for sharing strains, protocols, reagents, and general expertise. We thank Jessica Brown and Oliver Liu for critical reading of this manuscript. The work was supported by an Opportunity Grant from the Herb and Marion Sandler Foundation and a grant from the NIAID (R01AI065519). C. D. C. was supported by a National Science Foundation Predoctoral Fellowship.

REFERENCES

Bartlett, K. H., Kidd, S. E., and Kronstad, J. W. (2008). The emergence of *Cryptococcus gattii* in British Columbia and the Pacific Northwest. *Curr. Infect. Dis. Rep.* **10,** 58–65.

Blackstock, R., and Murphy, J. W. (2004). Age-related resistance of C57BL/6 mice to *Cryptococcus neoformans* is dependent on maturation of NKT cells. *Infect. Immun.* **72,** 5175–5180.

Bovers, M., Hagen, F., and Boekhout, T. (2008). Diversity of the *Cryptococcus neoformans–Cryptococcus gattii* species complex. *Rev. Iberoam. Micol.* **25,** S4–S12.

Casadevall, A., Mukherjee, J., Yuan, R., and Perfect, J. (1994). Management of injuries caused by *Cryptococcus neoformans*—Contaminated needles. *Clin. Infect. Dis.* **19,** 951–953.

Chen, L. C., Goldman, D. L., Doering, T. L., Pirofski, L., and Casadevall, A. (1999). Antibody response to *Cryptococcus neoformans* proteins in rodents and humans. *Infect. Immun.* **67,** 2218–2224.

Chen, G. H., McNamara, D. A., Hernandez, Y., Huffnagle, G. B., Toews, G. B., and Olszewski, M. A. (2008). Inheritance of immune polarization patterns is linked to resistance versus susceptibility to *Cryptococcus neoformans* in a mouse model. *Infect. Immun.* **76,** 2379–2391.

Davidson, R. C., Cruz, M. C., Sia, R. A., Allen, B., Alspaugh, J. A., and Heitman, J. (2000). Gene disruption by biolistic transformation in serotype D strains of *Cryptococcus neoformans*. *Fungal Genet. Biol.* **29,** 38–48.

Davidson, R. C., Blankenship, J. R., Kraus, P. R., de Jesus Berrios, M., Hull, C. M., D'Souza, C., Wang, P., and Heitman, J. (2002). A PCR-based strategy to generate integrative targeting alleles with large regions of homology. *Microbiology* **148,** 2607–2615.

de Jesus-Berrios, M., Liu, L., Nussbaum, J. C., Cox, G. M., Stamler, J. S., and Heitman, J. (2003). Enzymes that counteract nitrosative stress promote fungal virulence. *Curr. Biol.* **13,** 1963–1968.

Del Poeta, M. (2004). Role of phagocytosis in the virulence of *Cryptococcus neoformans*. *Eukaryot. Cell* **3,** 1067–1075.

Edman, J. C. (1992). Isolation of telomerelike sequences from *Cryptococcus neoformans* and their use in high-efficiency transformation. *Mol. Cell. Biol.* **12,** 2777–2783.

Edman, J. C., and Kwon-Chung, K. J. (1990). Isolation of the URA5 gene from *Cryptococcus neoformans* var. *neoformans* and its use as a selective marker for transformation. *Mol. Cell. Biol.* **10,** 4538–4544.

Escandon, P., Ngamskulrungroj, P., Meyer, W., and Castaneda, E. (2007). In vitro mating of Colombian isolates of the *Cryptococcus neoformans* species complex. *Biomedica* **27,** 308–314.

Fraser, J. A., Subaran, R. L., Nichols, C. B., and Heitman, J. (2003). Recapitulation of the sexual cycle of the primary fungal pathogen *Cryptococcus neoformans* var. *gattii*: Implications for an outbreak on Vancouver Island, Canada. *Eukaryot. Cell* **2,** 1036–1045.

Gerik, K. J., Donlin, M. J., Soto, C. E., Banks, A. M., Banks, I. R., Maligie, M. A., Selitrennikoff, C. P., and Lodge, J. K. (2005). Cell wall integrity is dependent on the PKC1 signal transduction pathway in *Cryptococcus neoformans*. *Mol. Microbiol.* **58**, 393–408.

Goldman, D. L., Khine, H., Abadi, J., Lindenberg, D. J., Pirofski, L., Niang, R., and Casadevall, A. (2001). Serologic evidence for *Cryptococcus neoformans* infection in early childhood. *Pediatrics* **107**, E66.

Heitman, J., Allen, B., Alspaugh, J. A., and Kwon-Chung, K. J. (1999). On the origins of congenic MATalpha and MATa strains of the pathogenic yeast *Cryptococcus neoformans*. *Fungal Genet. Biol.* **28**, 1–5.

Hoang, L. M., Maguire, J. A., Doyle, P., Fyfe, M., and Roscoe, D. L. (2004). *Cryptococcus neoformans* infections at Vancouver Hospital and Health Sciences Centre (1997–2002): Epidemiology, microbiology and histopathology. *J. Med. Microbiol.* **53**, 935–940.

Hua, J., Meyer, J. D., and Lodge, J. K. (2000). Development of positive selectable markers for the fungal pathogen *Cryptococcus neoformans*. *Clin. Diagn. Lab. Immunol.* **7**, 125–128.

Huffnagle, G. B., Boyd, M. B., Street, N. E., and Lipscomb, M. F. (1998). IL-5 is required for eosinophil recruitment, crystal deposition, and mononuclear cell recruitment during a pulmonary *Cryptococcus neoformans* infection in genetically susceptible mice (C57BL/6). *J. Immunol.* **160**, 2393–2400.

Hull, C. M., Cox, G. M., and Heitman, J. (2004). The alpha-specific cell identity factor Sxi1alpha is not required for virulence of *Cryptococcus neoformans*. *Infect. Immun.* **72**, 3643–3645.

Kabasawa, K., Itagaki, H., Ikeda, R., Shinoda, T., Kagaya, K., and Fukazawa, Y. (1991). Evaluation of a new method for identification of *Cryptococcus neoformans* which uses serologic tests aided by selected biological tests. *J. Clin. Microbiol.* **29**, 2873–2876.

Kim, K. S. (2006). Microbial translocation of the blood–brain barrier. *Int. J. Parasitol.* **36**, 607–614.

Kwon-Chung, K. J., and Varma, A. (2006). Do major species concepts support one, two or more species within *Cryptococcus neoformans*? *FEMS Yeast Res.* **6**, 574–587.

Lin, X., Hull, C. M., and Heitman, J. (2005). Sexual reproduction between partners of the same mating type in *Cryptococcus neoformans*. *Nature* **434**, 1017–1021.

Liu, O. W., Chun, C. D., Chow, E. D., Chen, C., Madhani, H. D., and Noble, S. M. (2008). Systematic genetic analysis of virulence in the human fungal pathogen *Cryptococcus neoformans*. *Cell* **135**, 174–188.

Livak, K. J., and Schmittgen, T. D. (2001). Analysis of relative gene expression data using real-time quantitative PCR and the 2(-Delta Delta C(T)) method. *Methods* **25**, 402–408.

Loftus, B. J., Fung, E., Roncaglia, P., Rowley, D., Amedeo, P., Bruno, D., Vamathevan, J., Miranda, M., Anderson, I. J., Fraser, J. A., Allen, J. E., Bosdet, I. E., *et al.* (2005). The genome of the basidiomycetous yeast and human pathogen *Cryptococcus neoformans*. *Science* **307**, 1321–1324.

McClelland, C. M., Chang, Y. C., Varma, A., and Kwon-Chung, K. J. (2004). Uniqueness of the mating system in *Cryptococcus neoformans*. *Trends Microbiol.* **12**, 208–212.

Mitchell, T. G., and Perfect, J. R. (1995). Cryptococcosis in the era of AIDS–100 years after the discovery of *Cryptococcus neoformans*. *Clin. Microbiol. Rev.* **8**, 515–548.

Monari, C., Kozel, T. R., Paganelli, F., Pericolini, E., Perito, S., Bistoni, F., Casadevall, A., and Vecchiarelli, A. (2006). Microbial immune suppression mediated by direct engagement of inhibitory Fc receptor. *J. Immunol.* **177**, 6842–6851.

Monga, D. P. (1981). Role of macrophages in resistance of mice to experimental cryptococcosis. *Infect. Immun.* **32**, 975–978.

Morrow, C. A., and Fraser, J. A. (2009). Sexual reproduction and dimorphism in the pathogenic basidiomycetes. *FEMS Yeast Res.* **9**, 161–177.

Nelson, R. T., Pryor, B. A., and Lodge, J. K. (2003). Sequence length required for homologous recombination in *Cryptococcus neoformans*. *Fungal Genet. Biol.* **38**, 1–9.

Nielsen, K., Cox, G. M., Wang, P., Toffaletti, D. L., Perfect, J. R., and Heitman, J. (2003). Sexual cycle of *Cryptococcus neoformans* var. *grubii* and virulence of congenic a and alpha isolates. *Infect. Immun.* **71**, 4831–4841.

Nielsen, K., Cox, G. M., Litvintseva, A. P., Mylonakis, E., Malliaris, S. D., Benjamin, D. K. Jr., Giles, S. S., Mitchell, T. G., Casadevall, A., Perfect, J. R., and Heitman, J. (2005). *Cryptococcus neoformans* alpha strains preferentially disseminate to the central nervous system during coinfection. *Infect. Immun.* **73**, 4922–4933.

Nosanchuk, J. D., and Casadevall, A. (2003). The contribution of melanin to microbial pathogenesis. *Cell. Microbiol.* **5**, 203–223.

Reid, R. J., Lisby, M., and Rothstein, R. (2002). Cloning-free genome alterations in *Saccharomyces cerevisiae* using adaptamer-mediated PCR. *Methods Enzymol.* **350**, 258–277.

Rhodes, J. C., Wicker, L. S., and Urba, W. J. (1980). Genetic control of susceptibility to *Cryptococcus neoformans* in mice. *Infect. Immun.* **29**, 494–499.

Taylor, J. W., and Berbee, M. L. (2006). Dating divergences in the Fungal Tree of Life: Review and new analyses. *Mycologia* **98**, 838–849.

Toffaletti, D. L., Rude, T. H., Johnston, S. A., Durack, D. T., and Perfect, J. R. (1993). Gene transfer in *Cryptococcus neoformans* by use of biolistic delivery of DNA. *J. Bacteriol.* **175**, 1405–1411.

Tscharke, R. L., Lazera, M., Chang, Y. C., Wickes, B. L., and Kwon-Chung, K. J. (2003). Haploid fruiting in *Cryptococcus neoformans* is not mating type alpha-specific. *Fungal Genet. Biol.* **39**, 230–237.

Wickes, B. L., and Edman, J. C. (1995). The *Cryptococcus neoformans* GAL7 gene and its use as an inducible promoter. *Mol. Microbiol.* **16**, 1099–1109.

Wormley, F. L. Jr., Perfect, J. R., Steele, C., and Cox, G. M. (2007). Protection against cryptococcosis by using a murine gamma interferon-producing *Cryptococcus neoformans* strain. *Infect. Immun.* **75**, 1453–1462.

Xu, J., Vilgalys, R., and Mitchell, T. G. (2000). Multiple gene genealogies reveal recent dispersion and hybridization in the human pathogenic fungus *Cryptococcus neoformans*. *Mol. Ecol.* **9**, 1471–1481.

Xue, C., Tada, Y., Dong, X., and Heitman, J. (2007). The human fungal pathogen Cryptococcus can complete its sexual cycle during a pathogenic association with plants. *Cell Host Microbe* **1**, 263–273.

CHAPTER THIRTY-FOUR

THE FUNGAL GENOME INITIATIVE AND LESSONS LEARNED FROM GENOME SEQUENCING

Christina A. Cuomo *and* Bruce W. Birren

Contents

Abstract

The sequence of *Saccharomyces cerevisiae* enabled systematic genome-wide experimental approaches, demonstrating the power of having the complete genome of an organism. The rapid impact of these methods on research in yeast mobilized an effort to expand genomic resources for other fungi. The "fungal genome initiative" represents an organized genome sequencing effort to promote comparative and evolutionary studies across the fungal kingdom. Through such an approach, scientists can not only better understand specific organisms but also illuminate the shared and unique aspects of fungal biology that underlie the importance of fungi in biomedical research, health, food production, and industry. To date, assembled genomes for over 100 fungi are available in public databases, and many more sequencing projects are underway. Here, we discuss both examples of findings from comparative analysis of fungal sequences, with a specific emphasis on yeast genomes, and on the analytical approaches taken to mine fungal genomes. New sequencing methods are accelerating comparative studies of fungi by reducing the cost and difficulty of sequencing. This has driven more common use of sequencing applications, such as to study genome-wide variation

Broad Institute of MIT and Harvard, Cambridge, Massachusetts, USA

Methods in Enzymology, Volume 470
ISSN 0076-6879, DOI: 10.1016/S0076-6879(10)70034-3

in populations or to deeply profile RNA transcripts. These and further technological innovations will continue to be piloted in yeasts and other fungi, and will
expand the applications of sequencing to study fungal biology.

1. INTRODUCTION: YEAST GENOMES AND BEYOND

Saccharomyces cerevisiae was the first eukaryote to be sequenced (Goffeau
et al., 1996). The ensuing genome-wide functional studies (Aparicio *et al.*,
2004; Boone *et al.*, 2007; DeRisi *et al.*, 1997; Washburn *et al.*, 2001; Zhu
et al., 2001) using this sequence established the power of genomic
approaches in providing a comprehensive understanding of higher organisms.
Subsequent sequencing of a number of close relatives of *S. cerevisiae* demonstrated the additional value of comparative analysis in identifying genes, gene
families, regulatory elements, signatures of selection, and the molecular
mechanisms that underlie genome evolution. When advances in "whole
genome shotgun" sequencing methods reduced the cost and organizational
barriers that had precluded sequencing the larger genomes of filamentous
fungi, comparative studies of the fungal kingdom began.

The fungal genome initiative represents a concerted effort by fungal
researchers and sequencing groups around the world to use genome
sequencing and comparative analysis to understand individual fungi that
are important in research, health, and industry and the biological innovations
that support fungal lifestyles. The fungal kingdom has now been extensively
sampled by sequencing with ~110 genomes assemblies representing 80
species submitted to NCBI to date (Table 34.1) and another ~60 species
targeted for sequencing (http://www.genomesonline.org).

Comparative genome analysis has provided insights into the genetic
differences between groups of fungi and the genomic changes that shaped
their evolution. Comparing genome sequences between fungi over a wide
range of phylogenetic distances permits identification of ancient and very
recent molecular innovations, along with core conserved gene sets among
all fungi or specific subsets of related fungi. In this chapter, we review what
has been learned from fungal genome sequencing, with a focus on comparisons that have informed our understanding of *Saccharomyces* and the general
evolutionary principles that have helped define the fungal kingdom. With
the advent of next-generation sequencing methods and the resulting acceleration in genome sequencing, we expect comparative genomics will
become an increasingly common tool in yeast genetic analysis.

Within the fungal kingdom, the Saccharomycotina are the most highly
sequenced subphylum (Table 34.1). The focus on sequencing this group of
organisms reflects the small size of these genomes relative to other fungi and
other model systems, the carefully described phylogeny of closely related

Table 34.1 Fungal genome assemblies in GenBank and EMBL-Bank (Data updated from www.broadinstitute.org, www.jcvi.org, www.genolevures.org, www.genome.jgi-psf.org, www.genome.wustl.edu, www.ebi.ac.uk, www.yeastgenome.org, www.candidagenome.org, agd.vital-it.ch, and www.aspergillusgenome.org.)

Genome	Size (Mb)	Genes	Strains	Phylum	Subphylum
Ashbya gossypii	9.6	4726	1	Ascomycota	Saccharomycotina
Candida albicans	14.3–14.4	6094–6160	2	Ascomycota	Saccharomycotina
Candida glabrata	12.3	5202	1	Ascomycota	Saccharomycotina
Candida guilliermondii	10.6	5920	1	Ascomycota	Saccharomycotina
Candida lusitaniae	12.1	5941	1	Ascomycota	Saccharomycotina
Candida parapsilosis	13.1	5733	1	Ascomycota	Saccharomycotina
Candida tropicalis	14.6	6258	1	Ascomycota	Saccharomycotina
Debaryomyces hansenii	12.2	6272	1	Ascomycota	Saccharomycotina
Kluyveromyces lactis	10.6	5076	1	Ascomycota	Saccharomycotina
Kluyveromyces waltii	10.9	10,721	1	Ascomycota	Saccharomycotina
Lodderomyces elongisporus	15.5	5802	1	Ascomycota	Saccharomycotina
Pichia stipitis	15.4	5841	1	Ascomycota	Saccharomycotina
Pichia pastoris	9.4	5313–5450	2	Ascomycota	Saccharomycotina
Saccharomyces bayanus	10.2–11.9	11,992	2	Ascomycota	Saccharomycotina
Saccharomyces castellii	11.4	4700	1	Ascomycota	Saccharomycotina
Saccharomyces cerevisiae	10.7–12.3	5196–5904	7	Ascomycota	Saccharomycotina
Saccharomyces kluyveri	11.3	5321	1	Ascomycota	Saccharomycotina
Saccharomyces kudriazevii	11.4	nd	1	Ascomycota	Saccharomycotina
Saccharomyces mikatae	12.6	10,311	1	Ascomycota	Saccharomycotina
Saccharomyces paradoxus	12.2	10,554	1	Ascomycota	Saccharomycotina
Saccharomyces pastorianus	22.4	14,152	1	Ascomycota	Saccharomycotina
Vanderwaltozyma polyspora	14.7	5652	1	Ascomycota	Saccharomycotina

(*continued*)

Table 34.1 (continued)

Genome	Size (Mb)	Genes	Strains	Phylum	Subphylum
Yarrowia lipolytica	20.5	6448	1	Ascomycota	Saccharomycotina
Alternaria brassicicola	30.3	10,688	1	Ascomycota	Pezizomycotina
Ascosphaera apis	21.5	nd	1	Ascomycota	Pezizomycotina
Aspergillus clavatus	27.9	9125	1	Ascomycota	Pezizomycotina
Aspergillus flavus	37.1	12,074	1	Ascomycota	Pezizomycotina
Aspergillus fumigatus	28.8–29.2	9631–9906	2	Ascomycota	Pezizomycotina
Aspergillus nidulans	30.1	10,560	1	Ascomycota	Pezizomycotina
Aspergillus niger	33.9	14,165	1	Ascomycota	Pezizomycotina
Aspergillus oryzae	37.1	12,079	1	Ascomycota	Pezizomycotina
Aspergillus terreus	29.3	10,406	2	Ascomycota	Pezizomycotina
Blastomyces dermatitidis	66.6–75.4	9522–9555	2	Ascomycota	Pezizomycotina
Blumeria graminis	161.4	nd	1	Ascomycota	Pezizomycotina
Botrytis cinerea	42.7	16,448	1	Ascomycota	Pezizomycotina
Chaetomium globosum	34.9	11,124	1	Ascomycota	Pezizomycotina
Coccidioides immitis	27.7–29.0	10,355–10,608	4	Ascomycota	Pezizomycotina
Coccidioides posadasii	25.5–28.6	9897–10,060	11	Ascomycota	Pezizomycotina
Colletotrichum graminicola	51.6	nd	1	Ascomycota	Pezizomycotina
Fusarium graminearum	36.5	13,332	1	Ascomycota	Pezizomycotina
Fusarium oxysporum	61.4	17,735	1	Ascomycota	Pezizomycotina
Fusarium verticillioides	41.8	14,179	1	Ascomycota	Pezizomycotina
Histoplasma capsulatum	30.4–38.9	9233–9532	5	Ascomycota	Pezizomycotina
Magnaporthe grisea	41.7	11,074	1	Ascomycota	Pezizomycotina
Microsporum canis	23.2	8765	1	Ascomycota	Pezizomycotina
Microsporum gypseum	23.2	8876	1	Ascomycota	Pezizomycotina
Nectria haematococca	54.4	15,707	1	Ascomycota	Pezizomycotina

Neosartorya fischeri	32.6	10,407	1	Ascomycota	Pezizomycotina
Neurospora crassa	39.2	9826	1	Ascomycota	Pezizomycotina
Paracoccidioides brasiliensis	29.1–32.9	7876–9136	3	Ascomycota	Pezizomycotina
Penicillium marneffei	28.5	nd	1	Ascomycota	Pezizomycotina
Pyrenophora tritici-repentis	37.8	12,171	1	Ascomycota	Pezizomycotina
Sclerotinia sclerotiorum	38.3	14,522	1	Ascomycota	Pezizomycotina
Stagonospora nodorum	37.2	16,597	1	Ascomycota	Pezizomycotina
Talaromyces stipitatus	35.7	12,449	1	Ascomycota	Pezizomycotina
Trichoderma atroviride	36.1	11,100	1	Ascomycota	Pezizomycotina
Trichoderma reesei	34.1	9129	1	Ascomycota	Pezizomycotina
Trichoderma virens	38.8	11,643	1	Ascomycota	Pezizomycotina
Trichophyton equinum	24.1	8560	1	Ascomycota	Pezizomycotina
Trichophyton tonsurans	23.0	nd	1	Ascomycota	Pezizomycotina
Uncinocarpus reesii	22.3	7798	1	Ascomycota	Pezizomycotina
Verticillium dahliae	33.8	10,535	1	Ascomycota	Pezizomycotina
Verticillium albo-altrum	32.8	10,221	1	Ascomycota	Pezizomycotina
Schizosaccharomyces japonicus	11.3	4814	1	Ascomycota	Taphrinomycotina
Schizosaccharomyces octosporus	11.2	4925	1	Ascomycota	Taphrinomycotina
Schizosaccharomyces pombe	12.6	4962	1	Ascomycota	Taphrinomycotina
Coprinus cinereus	36.3	13,392	1	Basidiomycota	Agaricomycotina
Cryptococcus neoformans	17.5–19.1	6210–6967	4	Basidiomycota	Agaricomycotina
Laccaria bicolor	64.9	20,614	1	Basidiomycota	Agaricomycotina
Moniliophthora perniciosa	26.7	16,329	1	Basidiomycota	Agaricomycotina
Phanerochaete chrysosporium	35.1	10,048	1	Basidiomycota	Agaricomycotina
Postia placenta	90.9	17,173	1	Basidiomycota	Agaricomycotina
Puccinia graminis f. sp. *tritici*	88.6	20,567	1	Basidiomycota	Pucciniomycotina
Malassezia globbosa	8.9	4285	1	Basidiomycota	Ustilaginomycotina

(*continued*)

Table 34.1 *(continued)*

Genome	Size (Mb)	Genes	Strains	Phylum	Subphylum
Ustilago maydis	19.7	6522	1	Basidiomycota	Ustilaginomycotina
Batrachochytrium dendrobatidis	23.7	8794	1	Chytridiomycota	Chytridiomycota
Spizellomyces punctatus	24.1	8804	1	Chytridiomycota	Chytridiomycota
Rhizopus oryzae	46.1	17,467	1	Zygomycota	Mucormycotina
Encephalitozoon cuniculi	2.9	1997	1	Microsporidia	
Enterocytozoon bieneusi	3.9	3632	1	Microsporidia	
Nosema ceranae	7.9	2614	1	Microsporidia	

Notes: Genomes listed are taken from GenBank list of submitted assemblies and EMBL–Bank, as of 09-09-09. nd, no data; the Zygomycota phyla was not included in the recent "AFTOL classification" of fungi (Hibbett *et al.*, 2007).

species, the wealth of genetic resources for *S. cerevisiae*, and the importance of *Candida albicans* as a pathogen. This group comprises two predominant subclades: the *Saccharomyces* group and the *Candida* group (Fig. 34.1), which diverged ~200–400 mya (Hedges *et al.*, 2006; Taylor and Berbee, 2006). These groups include species specialized to live on different carbon sources

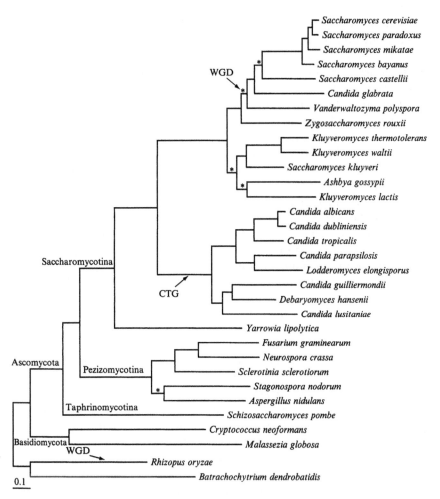

Figure 34.1 Phylogeny of sequenced yeasts and other fungi. Species phylogeny relationship of the sequenced yeasts in Saccharomycotina and selected other fungi based on 20,000 randomly sampled sites from 294 orthologous protein alignments. Tree topology and branch lengths were inferred with RAxML (Stamatakis, 2006); branches are well supported (>80% of bootstrap replicates) except as indicated with an asterisk. The location of whole genome duplication (WGD) events and change in CTG decoding are shown.

as well as to live as human commensals and pathogens. Draft genome assemblies have been released for 23 Saccharomycotina species (Table 34.1). Several were specifically targeted to trace the evolutionary history of a whole genome duplication (WGD) event in this group. Also recent work described several genomes related to *C. albicans* which share the translation of CTG codons as Serine instead of Leucine (Butler *et al.*, 2009), and representatives of other major *Saccharomyces* groups which did not undergo a WGD (Souciet *et al.*, 2009).

Genome sequencing has also targeted additional species within the Ascomycota phylum, of which the Saccharomycotina are one of three subphyla (Fig. 34.1). Ascomycota is the largest phylum in the fungal kingdom, and these species share a common morphology of forming asci around meiotic spores (James *et al.*, 2006). Many species from the Pezizomycotina subphylum of filamentous fungi have also been sequenced, including many animal pathogens (including dimorphic fungi, dermatophytes, *Aspergillus fumigatus*), plant pathogens (*Fusarium graminearum*, *Magnaporthe grisea*, *Stagonospora nodorum*), as well as model systems (*Aspergillus nidulans* and *Neurospora crassa*). Sequenced genomes from the basally branching Taphrinomycotina subphylum include the *Schizosaccharomyces* (*S. pombe*, *S. octosporus*, and *S. japonicus*) and *Pneumocystis*. Despite morphological similarities, DNA sequence analysis shows that the *Schizosaccharomyces* are more diverged in evolutionary distance from *S. cerevisiae* than the filamentous Ascomycetes (Fig. 34.1).

Recent classification of fungi suggests that there are seven phyla in the fungal kingdom, including Ascomycota, Basidiomycota, Glomeromycota, Microsporidia, and three phyla of Chytrids, as well as four subphyla of the Zygomycota group of fungi that are not placed in any phylum (Hibbett *et al.*, 2007). Genome sequencing outside of Ascomycota has been more limited, but has sampled the other major groups of the fungal kingdom (Table 34.1, Fig. 34.1). Among these groups, the Basidiomycetes are most closely related to Ascomycetes and include fungi that span a wide range of life cycles. Sequenced Basidiomycetes include long-studied human pathogens, such as the yeasts *Cryptococcus neoformans* and *Malassezia globosa*, the plant pathogens *Ustilago maydis* and *Puccinia graminis*, and important models such as the biodegrading fungus *Phanerochaete chyrsosprorium* and the mushroom *Coprinus cinereus*. Sequencing of *Glomus intraradices*, a representative of the Glomeromycetes, which form symbiotic interactions with plants, has been undertaken though the data have proven very difficult to assemble (Martin *et al.*, 2008). Sequencing of the basal Zygomycota and Chytrid polyphyletic groups of fungi have been sparse to date. Sequencing of Zygomycotas has been limited to just two genomes from the Mucorales subphylum; these include *Rhizopus oryzae*, the major cause of mucormycosis, and *Phycomyces blakesleeanus* (http://genome.jgi-psf.org/Phybl1/Phybl1.home.html). Chytrid fungi, which are characterized by having flagella, are subdivided into three phyla; two species from the

Chytridiomycota phylum have been sequenced; the fungal pathogen of amphibians *Batrachochytrium dendrobatidis* and the saprobe *Spizellomyces puncatus*. Genome sequencing for microsporidia, a group diverging from within basal fungi (James *et al.*, 2006), has sampled major pathogens including the human pathogen *Encephalitozoon cuniculi* and the bee pathogen *Nosema ceranae*. Given the tremendous diversity of the Basidiomycetes, Zygomycotas and Chytrids, many more sequenced representatives will be required to describe and understand the origin or lineage-specific evolution of these groups.

2. COMPUTATIONAL PREDICTION OF GENES AND NONCODING ELEMENTS

A primary goal of genome sequencing is a complete gene list for an organism. New tools for genome annotation make use of a variety of supporting data to improve prediction not only of protein coding genes but also RNA-encoding genes and other functional elements. As the genes for the Saccharomycotina primarily consist of a single exon, early gene prediction methods for sequenced genomes emphasized identification of ORFs above a certain size limit. Introns were incorporated into these gene predictions based on the presence of conserved splice signals. Although effective in identifying the majority of genes, this approach is confounded by small genes, which can be impossible to distinguish from randomly occurring small ORFs. Simple ORF identification also suffers from the inability to correctly distinguish between alternative closely spaced start sites. To identify small genes with greater specificity, as well as to confirm the more complex gene structures in filamentous fungi where the majority of genes are spliced, gene prediction has benefited from evidence of transcription by EST sequences and hybridization to microarrays. Transcriptional data are also crucial in revealing alternative splicing in fungal multiexon genes. Analysis of transcriptional data has shown that retained introns are the primary mechanism of generating alternate transcripts (McGuire *et al.*, 2008). Application of new sequencing technologies that allow the direct sequencing of RNA (RNA-seq) is likely to supplant other methods for identifying transcripts and thereby greatly increase the depth and resolution of transcript mapping (Nagalakshmi *et al.*, 2008; Wilhelm *et al.*, 2008) and gene prediction.

New gene prediction methods are also being developed to exploit comparative genomic data. Because genes and other functional elements show different patterns of sequence conservation than nonfunctional sequences, comparison of closely related species can help pinpoint conserved sequences without prior knowledge of their function, including both genes and regulatory elements (Hardison, 2003). Comparative annotation requires additional sequences from several species that are each closely enough related to the genome being annotated to contain the same genes and regulatory mechanisms and allow unambiguous alignment of the sequences, while having diverged sufficiently to have accumulated base changes in nonselected regions. A fine example of this approach was used in the analysis of the *S. cerevisiae* genome. To identify species at an appropriate evolutionary distance for comparative annotation of *S. cerevisiae*, light sequencing was used to evaluate a number of candidate species and an informative set was selected for subsequent deeper sequencing (Cliften *et al.*, 2001; Souciet *et al.*, 2000). Species from the *senso stricto* group, which can form viable diploids with *S. cerevisiae*, were sequenced to deeper draft coverage, including *S. paradoxus*, *S. mikatae*, *S. bayanus*, and *S. kudriavzevii* (Cliften *et al.*, 2003; Kellis *et al.*, 2003). Additional more distant species important for these comparisons were also sequenced, including *S. castellii* and *S. kluyveri* (Cliften *et al.*, 2003).

Through this comparative approach, ~40 new *S. cerevisiae* genes were predicted based on their conservation among the sequenced species. All of the newly predicted genes were < 100 amino acids in length, highlighting the value of identifying patterns of conservation as an indication of which of the many small ORFs are functional and under selective pressure. Further, each of these comparative analyses demonstrated that ~500 predicted genes in *S. cerevisiae* are likely dubious protein coding genes because they are not well conserved among closely related species, and in addition not supported by experimental data available at the time.

Comparative annotation methods continue to undergo improvement. For example, methods developed for *Drosophila* genomes (Lin *et al.*, 2007) were recently used to reannotate the diploid *C. albicans* (Butler *et al.*, 2009). For whole genome alignments across the *Candida* clade, measuring codon substitution frequencies in addition to reading frame conservation in genic regions provided evidence for 91 new or updated genes, and revealed 222 dubious genes. Dubious genes were manually curated to determine if any experimental evidence suggested a functional role; 80% have no current evidence. Comparative analysis also identified 226 well-conserved regions in *C. albicans* where an ORF appeared interrupted by a nonsense mutation or frameshift. Creating a consensus assembly from sequence of a diploid organism that is heterozygous at many loci, can lead to incorrect merging of the sequences of alternative alleles. One consequence of these regions of mixed origin can be problems with gene structures such as nonsense mutations or frameshifts. Additional analysis of sequencing reads for this strain of *C. albicans* found 80% of these predicted errors to be fixable using existing data, which led to revision of the consensus sequence and a more complete gene set.

Increasingly, methods have sought to identify the entire transcriptome of the yeast genome by direct sequencing or hybridization to tiling micorarrays. As high transcript coverage using RNA-seq is now achievable at low cost, these data can be used to annotate gene structures more completely, including upstream and downstream untranslated regions. *De novo* gene predictions from RNA sequencing data can achieve high coverage of the annotated genes in *S. cerevisiae*, although overlapping genes and small ORFs still pose challenges to these methods. Gene predictions based on transcriptional data suggest that some dubious ORFs are transcribed (Yassour *et al.*, 2009), although they are not broadly conserved between species. Recent analysis of large-scale transcript and proteomic data has also found that many nonconserved ORFs are transcribed (Li *et al.*, 2008), though many may be noncoding RNAs.

Population level variation within a species can also provide evidence of conservation needed to support gene models. For example, using SNPs within a species as a measure of divergence, genes are more highly conserved than intergenic regions, although they are also less well conserved between strains than species conserved genes (Li *et al.*, 2008). Such experiments suggest that multiple sources of high-throughput data may be required to identify all the ORFs in a genome.

In addition to refining gene annotation, comparative analysis can identify conserved nongenic sequence, including regulatory elements. By examining whole genome alignments between orthologous regions, "phylogenetic footprints" highlight functionally constrained regions. These elements are far more difficult to identify than genes as they are short, typically only 6–11 bases in length, contain degenerate bases, and can vary in distance from the genes they regulate. For example, computational searches in *S. cerevisiae* identified a small number of conserved motifs in intergenic regions, and found them enriched upstream of genes (Cliften *et al.*, 2003; Kellis *et al.*, 2003); subsequent analysis connected predicted motifs to binding site data for individual transcription factors (Harbison *et al.*, 2004). As variation in the sequence of such motifs between species may contribute to differences in expression and phenotype (Borneman *et al.*, 2007), detection of binding site occupancy in different species by chromatin immunoprecipitation followed by either hybridization to tiling microarrays (ChIP-chip) or direct sequencing (ChIP-seq) may uncover recent differences in regulatory element usage.

3. MECHANISMS OF GENOME EVOLUTION

Dramatic differences in genome size and structure within the fungal kingdom reflect the large span of time that led to the emergence of different species and as well as a variety of genetic mechanisms. Saccharomycotina genomes are small compared to other fungi, but display surprisingly large variations in size between related species. The sequenced *Saccharomyces* have a median genome size of 12 Mb, and a median of ~5900 protein coding genes. By contrast, the related Pezizomycotina have a median genome size of 36 Mb and contain a median of ~10,600 protein coding genes. The large differences in genome size, such as the 50% variation between *Candida guilliermondii* and *Lodderomyces elongisporus*, are due variation in intergenic size rather than the number of protein coding genes (Butler *et al.*, 2009). Even within species, genome size can vary widely; the two sequenced strains of *Blastomyces dermatitidis* differ by 9 Mb (Table 34.1). Among fungi, the proportion of the genome comprising repetitive sequences can also vary substantially; as little as 0.1% of the genome is repetitive in *F. graminearum* whereas about 20% of *R. orzyae* consists of transposable elements. Among strains of *S. cerevisiae*, repetitive elements can differ in number by 10-fold, and some strains with higher repeat content appear prone to genome instability (Scheifele *et al.*, 2009). Compared to the yeasts which have few DNA transposons (Dujon, 2006), the genomes of filamentous fungi contain a larger variety of different transposable element types.

Genome evolution in the *Saccaromyces* clade has been shaped by an ancient WGD event. Initial analysis of the *S. cerevisiae* genome revealed numerous instances in which the same pairs of genes were found next to each other at different locations across the genome. These recurring pairs were suggested to have arisen at the same time from a WGD event (Wolfe and Shields, 1997). Although WGD was suggested to have doubled the number of chromosomes from 8 to 16, most duplicated genes were subsequently lost, with only 12% retained in duplicate in *S. cerevisiae*. The sequence from additional yeast genomes which did not undergo a WGD, *Kluyveromyces waltii* and *Ashbya gossypii*, allowed comparison of each WGD paralogous gene pair to the ancestral organization of orthologs and provided evidence to unequivocally validate this hypothesis (Brachat *et al.*, 2003; Kellis *et al.*, 2004). Alignments of *S. cerevisiae* to either *A. gossypii* or *K. waltii* revealed a "two-to-one" mapping pattern in which two distinct regions from *S. cerevisiae* map to a single region in each of these other genomes (Dietrich *et al.*, 2004; Kellis *et al.*, 2003). In all, 90% of the *S. cerevisiae* genome falls into recognizable blocks that map in this manner. Within these duplicated blocks in *S. cerevisiae*, genes were conserved in the same order and orientation between the corresponding region in *A. gossypii* or *K. waltii*. When the presence of genes in either of the duplicated blocks is compared to the ancestral version represented by *A. gossypii* or *K. waltii*, it is clear that differential gene loss has occurred between the blocks in *S. cerevisiae*. Recognizing such a pattern of interleaved gene loss, as well as a clear 2:1 pairing of centromeres between *S. cerevisiae* and the preduplication species, strongly supports the origin of paired regions by a WGD.

Evolutionary pressures may act differently on each member within the pairs of genes retained in duplicate after a WGA. In Ohno's model of evolution after duplication, one gene maintains the ancestral role whereas the other duplicate is free to diverge in sequence and function (Ohno, 1970). In *S. cerevisiae*, nearly all cases of accelerated evolution involve only one of the two paralogs of retained duplicate pairs (Kellis *et al.*, 2004). Such computational analyses support a model of neo-functionalization, where one of the paralogs retains the ancestral function and the other evolves a new function, over a model of subfunctionalization, in which each new copy assumes part of the ancestral function (Lynch and Force, 2000).

Examining additional postduplication species has shown that the same genes are often maintained as WGD duplicates, but the pattern of gene loss differs. Analysis of additional loss in *C. glabrata* and *S. castelli* compared to *S. cerevisiae* revealed that while one copy of most WGD duplicates was resolved prior to speciation, reciprocal and independent loss events were also observed. Analysis of *Kluyveromyces polysporus*, a WGD yeast most diverged from *S. cerevisiae*, identified a similar fraction of genes retained as WGD pairs. Roughly half the gene pairs are found in both species, and they are enriched for protein kinases, cell wall organization, and carbohydrate metabolism suggesting that subsequent to WGD there was a general advantage to retaining these genes. For the many genes retained in single copy the retained paralog differs, suggesting that the retention of different copies of duplicated genes may have played a role in speciation (Scannell *et al.*, 2006).

WGD events have been described in other eukaryotes and are likely to have occurred numerous times in the long history of fungal evolution. Recently WGD has been documented within the *Mucorales* group of basal fungi (Ma *et al.*, 2009). *R. oryzae*, the primary cause of mucormycosis, contains 9% of genes in duplicated pairs distributed across the genome. Comparison to a related *Mucorales* species, *P. blakesleeanus*, established a 2:1 relationship for 78% of the *R. oryzae* gene pairs. While a similar number of genes were retained in duplicate after the WGD events in the *Saccharomyces* and the *Mucorales*, the genes retained as duplicates comprise quite different functional groups. In *R. oryzae*, the duplicates include nearly all the individual subunits of the proteosome and mitochondrial ATPase. Finer dating and analysis of the WGD in *R. oryzae* will require identification and sequencing of a more closely related preduplication genome.

Duplication of individual genes is also a primary mechanism by which new functions evolve. Unlike genes originating with a WGD event, which are all created at the same time, individual gene duplications can arise over time leading to gene families in which the genes display a wide range of sequence similarity. As mentioned above, following gene duplication, alternative fates including neo-functionalization or subfunctionalization can create new gene functions. Individual gene duplications played a major role in enlarging and diversifying gene families important for pathogenesis in *Candida* species (Butler *et al.*, 2009). Variation in gene copy number is found more widely in genes involved in transport, the cell wall, and stress response across the Ascomycetes (Wapinski *et al.*, 2007), suggesting greater variation in these processes. Strikingly, some genomes are limited in the preservation of gene duplicates by a genome defense process called repeat induced point mutation (RIP) in which duplicated genes are highly mutated during meiosis. As RIP renders duplications nonfunctional, there is a strong selection against duplication in this organism. The *N. crassa* genome has been profoundly shaped by RIP, with a dearth of both gene duplications and transposons (Galagan *et al.*, 2003). A similar process appears to constrain gene duplication and transposons in *F. graminearum* (Cuomo *et al.*, 2007). This lack of opportunity for gene innovation via duplication suggests that these species may be more constrained in adapting to changing environments.

Related species of *Saccharomyces* have largely collinear genomes in which gene order has been highly conserved in large blocks, termed syntenic regions. Within the *senso stricto* group, only a few inversion and translocation events differentiate the sequenced species (Kellis *et al.*, 2003). The breaks in synteny are often accompanied by the presence of repetitive sequences, suggesting rearrangements between similar repeats underlie these few large-scale structural changes. However, in some species particular genomic regions, such as subtelomeres, often display a faster evolutionary rate than the rest of the genome. In *Saccharomyces*, differences between the *senso stricto* species were highest at the subtelomeres (Kellis *et al.*, 2003). Among the subtelomeric gene families undergoing rapid change, in both copy number and sequence, are FLO genes and other surface markers that govern how cells are recognized by each other and the environment (Reynolds and Fink, 2001; Verstrepen *et al.*, 2005).

In other species, higher rates of evolution are not confined to the subtelomeres. In *F. graminearum*, a pathogen of wheat, genome-wide diversity between two strains was high at the subtelomeres, but also at discrete internal regions (Cuomo *et al.*, 2007). Chromosome number in *F. graminearum* is reduced compared to related species, and whole genome alignment with *F. verticillioides* supports a hypothesis that high diversity has been maintained at sites of chromosome fusion. *Aspergillus* sp. also contain genomic regions of lineage-specific sequence, which include many subtelomeres (Fedorova *et al.*, 2008; Galagan *et al.*, 2005). As in *S. cerevisiae*, particular gene families are found in the faster evolving regions of the filamentous fungal genomes. These include secreted and other proteins implicated in plant interactions in *F. graminearum* (Cuomo *et al.*, 2007), and secondary metabolite gene clusters, transporters, and proteins involved in metabolism and detoxification in *A. fumigatus* (Fedorova *et al.*, 2008). As a practical matter, any genomic region with a large concentration of a given gene family or families is likely to be poorly represented in a typical draft genome assembly. The software used to assemble shotgun sequence data is confounded by repeated sequences. Thus, the regions and gene families that are of greatest interest in defining recent evolutionary events, or population-based differences, are often underrepresented or entirely absent from draft assemblies. The fidelity of subtelomeric regions in draft assemblies is likely to further degrade as genome sequencing increasingly relies on new technologies that produce shorter read lengths from shorter DNA fragments. For some genomes, targeted sequencing was undertaken to fill in genes missing from the subtelomeric regions in draft assemblies (Wu *et al.*, 2009).

The availability of whole genome sequences allows for a comprehensive search and phylogenetic analysis of potential horizontal gene transfer events. Analysis of the fungal genomes sequenced to date indicates that while horizontal transfer within fungi is not as frequent as in bacteria, such events can be very important for altering phenotype. One such case was seen in the analysis of the genome of the wheat pathogen *S. nodorum*. An ortholog of the ToxA toxin gene from *Pyrenophora tritici-repentis* was identified in the *S. nodorum* genome; the ToxA orthologs share 99.7% similarity (Friesen *et al.*, 2006) whereas best bidirectional hit orthologs show a median of 71% similarity. The presence of an adjacent hAT transposon suggested that transposons may contribute to inter species transfer. Horizontal transfer has also been detected between fungi and other eukaryotes, including transfer from filamentous fungi to oomycetes (Richards *et al.*, 2006), between plants and fungi (Richards *et al.*, 2009), and recently between viruses and fungi (Frank and Wolfe, 2009). Clusters of linked genes may also be transferred as a group, such as those which produce secondary metabolites (Khaldi *et al.*, 2008). An extreme case of horizontal transfer may involve conditionally dispensable chromosomes in *Fusaria* spp., which are enriched for genes that do not follow the species phylogeny, display atypical GC content and codon usage, and contain a high percentage of transposable elements (Coleman *et al.*, 2009; Ma *et al.*, 2010).

 ## 4. GENOMIC POTENTIAL FOR SEX

Completed genome sequences provide the catalog needed to investigate the presence of entire pathways or complex biological processes. The conservation of genes required for mating and meiosis is sought as evidence for whether sexual cycles can occur, and how similar they may be in different fungi. While mating and meiosis has been well described in *Saccharomyces*, sexual cycles in other fungi can be less standard, often lacking some features common to most eukaryotes. *C. albicans* can mate through a parasexual cycle which is not thought to involve meiosis, and some components of meiosis were found to be missing from the genome sequence

(Tzung *et al.*, 2001). Further analysis of these genes in related *Candida* clade genomes demonstrated that they are missing in all *Candida*. Strikingly, the sexual *Candida* species are missing additional meiotic and mating genes, including proteins involved in synaptonemal complex formation and recombination (Butler *et al.*, 2009). Some of these proteins are also missing in other eukaryotes, such as *Drosophila* and *Caenorhabditis elegans*, suggesting considerable plasticity of the genes involved in meiosis. Innovation of meiotic genes within the *Saccharomyces* was supported by finding many meiosis and sporulation genes unique to the *Saccharomyces* clade among Ascomycetes, including the meiotic regulator IME1 (Wapinski *et al.*, 2007). For other species analysis of the genome sequence has provided the first evidence of a sexual cycle. In Aspergilli, mating was inferred to occur in *A. fumigatus* from the genome sequence (Galagan *et al.*, 2005), and subsequently shown experimentally (O'Gorman *et al.*, 2009).

5. Gene Family Conservation and Evolution

Examining the conservation of genes between species can help predict the potential functional capacity of an organism. Such comparative genomic analysis permits tracing the history of gene gain and loss events, which suggest changes in functional capacity along specific lineages. Protein similarity is the basis for clustering genes into families, which can then be refined based on synteny or phylogeny to establish orthology, which sequence divergence can obscure over time.

Comparative analysis of *Saccharomyces* genomes suggests that most genes are part of a common core set. *A. gossypii* has a small genome and gene content compared to other *Saccharomyces*, and nearly all (95%) of genes from *A. gossypii* have homologs in *S. cerevisiae*. This conservation suggests that the core genome is ~4500 genes, of which most are present in syntenic regions. The noncore gene set for each species includes examples of specialization, for example, the loss of the galactose utilization pathway in *S. kudriavzevii* (Hittinger *et al.*, 2004).

The divergence between the *Candida* and *Saccharomyces* clades is apparent in the considerable variation seen in gene families. In a total of 6209 gene families examined for seven *Candida* clade species and nine *Saccharomyces* clade species only about half (3923) were conserved between the two groups (Butler *et al.*, 2009; http://www.broadinstitute.org/annotation/genome/candida_group/MultiHome.html). By contrast 3765 gene families are specific to the *Saccharomyces* clade and 1521 are specific to the *Candida* clade. Several families important for pathogenesis were identified as expanded or unique to the *Candida* genomes compared to *Saccharomyces*. These include gene families that encode cell surface proteins, including the Als adhesins and the Iff/Hyr GPI-linked proteins; other cell wall families such as the FLO genes are highly specific to *Saccharomyces*. The Als and Flo families differ in sequence, but have a very similar structure. Each contains intragenic tandem repeats, which can vary in copy number and thereby contribute to phenotypic differences (Verstrepen *et al.*, 2005). In addition to variation at the cell surface, differences in nutrient acquisition are found with some proteins involved in transport overrepresented in the *Candida*.

While *S. cerevisiae* is separated from the other well-studied yeast *S. pombe* by a long evolutionary distance spanning the filamentous Ascomycetes, these species share some similar genomic features. Some genes appear to have duplicated in parallel in these species, allowing for adaptation to similar environments (Hughes and Friedman, 2003). Gene families specific to the yeasts *S. cerevisiae* and *S. pombe* included four nuclear pore proteins, which were presumably lost from the ancestor of the filamentous fungi (Cornell *et al.*, 2007).

By examining gene families built across the Ascomycetes, innovations within specific subgroups can be identified. A large fraction (84%) of *S. cerevisiae* genes are conserved in orthology groups which also are conserved in filamentous fungi (Wapinski *et al.*, 2007). Although many essential yeast proteins are in this conserved set, a significant fraction of spindle pole body components are specific to yeast (Cornell *et al.*, 2007; Wapinski *et al.*, 2007). The larger genomes of filamentous fungi contain expansions of specific gene families involved in transport of molecules in and out of cells and DNA binding proteins. The expansion of DNA binding proteins may suggest more complex regulation of transcription required to control a larger number of genes in the filamentous fungi. Within the filamentous plant pathogens, there is particular innovation of genes involved in secondary metabolite production, including cytochrome p450 and polyketide synthases (Soanes *et al.*, 2008). Such analysis in other more basal fungal phyla will be increasingly possible as the number of sequenced genomes increases.

6. IMPACT OF NEXT-GENERATION SEQUENCING

The advent of new sequencing technologies that produce dramatically less expensive data at vastly higher throughput is elevating the use of sequencing as a general purpose tool for studying genomes and biology. Several different sequencing technologies such as Illumina, 454, and ABI SOLiD are now in routine use, with more expected soon. Each uses amplification of DNA fragments and does not rely on propagation of bacterial clone libraries; sequencing chemistries differ, as do other properties such as read length and quality profiles which rapid development continues to improve. Each of these methods has distinct advantages, preferred applications, and cost structures, though the ongoing rapid changes in these platforms means that the relative differences are likely to shift.

New laboratory methods and assembly algorithms are still being optimized to address the inherent difficulty in assembling genomes from the shorter sequence reads produced by these new methods. However, early results indicate that high-quality genome assemblies can be produced by these new technologies. Resequencing a previously finished genome provides the most rigorous validation of any new sequencing strategy. We have resequenced *N. crassa* using several new technologies, along with genomes that have draft sequences but have higher amounts of repetitive DNA. After testing a variety of coverage depths, sequencing fragment and read length, and technology type, assemblies can be produced that rival or greatly surpass the quality of those generated by traditional Sanger sequencing, according to measures such as genome coverage, contig length, and consensus base quality (C. Nusbaum, personal communication). Further, new genome sequencing methods can capture genomic sequences that are typically hard to obtain by older methods that relied on cloning genomic fragments in bacteria. For example, sequencing *N. crassa* genome with 454 captured over 1 Mb of additional genome sequence that was missing from the original assembly based on Sanger-sequencing. The sequences obtained only by 454 were found to be much higher in AT content than average and were absent from the bacterial clone libraries. Finally, it is important to note that the overall quality of the consensus sequence found in draft assemblies produced from either 454 or Illumina is extremely high. In fact, the consensus quality within contigs of these *draft* assemblies matches that of *finished* sequence from prior sequencing data. Assembling short read sequencing data from fungal genomes generally yields assemblies with short contig lengths. Providing additional linking information, for example, by including Sanger paired-end sequence from 40 kb Fosmids, or end sequences from jumping fragments, will improve the continuity of the assembled sequence. Thus, *N. crassa* assemblies generated from a hybrid of 454 and Sanger sequence from Fosmid ends are comparable to the previous Sanger-only in terms of genome coverage and contig size. For a more repetitive genome, such as *P. brasiliensis*, contigs are about fourfold smaller on average than an assembly based on Sanger sequence alone (though again, the hybrid assembly combining both data types covered more of the genome). More recent work demonstrates that draft sequences can be assembled for 40 Mb fungal genomes from Illumina data, which points to even further reductions in cost, if solutions to the assembly of highly repetitive or polymorphic genomes are found.

In contrast to the large number of draft genome assemblies that have been produced, only a few have been fully finished, that is, have been completely sequenced from telomere to telomere. The paucity of finished sequence reflects the much greater cost and effort required compared to generating draft sequences. Although certain key fungi will warrant the expense of fully finished sequence, large-scale efforts are likely to continue to focus on draft genomes, especially as the cost differential between draft and finished sequence is likely to grow with the adoption shotgun sequencing methods that use new technologies. Hence, it is important to consider potential consequences of this distinction. Systematic losses of specific classes of sequence from draft assemblies routinely impact genome analysis. Repetitive sequences are frequently missing from draft genome assemblies or are greatly underrepresented. For example, repeat elements at centromeres and telomeres, or local gene clusters with highly repeated domains are rarely found intact in genome assemblies, and the highly repeated rDNA genes are underrepresented in draft assemblies. Repeated gene families that lie in subtelomeric regions are also often incorrectly assembled or missing from draft assemblies. Because these subtelomeric regions often harbor rapidly evolving genes that hold clues to an organism's most recent history and ecological specialization, draft assemblies are a poor substrate for comparisons of very recently diverged relatives, strains, or isolates. As a rule, multiple lines of evidence, including experimental data, are needed to confirm genes suspected to be missing from a genome based on draft assemblies.

The amount and location of sequencing missing from a draft assembly can be estimated by comparison to an independent genome map, such as a physical, genetic or optical map, or to a reference gene set, such as from ESTs. Optical maps, which are based on automated restriction fragment measurements (Samad et al., 1995), have proven practical and highly useful for fungal genomes; for the highly repetitive genome of R. oryzae, an optical map was instrumental in evaluating and improving the large-scale accuracy of the assembly (Ma et al., 2009). Alignment of genome sequence to the map not only allows sequence contigs to be correctly anchored and ordered along the chromosomes but also provides a measure of the size of sequence gaps. In addition to estimates of missing sequence based on optical maps or known genome sizes, the extent of missing genes can be determined by aligning a reference gene set or a set of ESTs to the assembly, and assessing the ones which do not align for genic potential. Researches with interests in specific regions of the genome that are challenging for whole genome sequencing strategies, such as repetitive regions and segmental duplications, need to consider targeted sequencing efforts to better deal with the challenges for representing these regions.

7. FUTURE DIRECTIONS

The growing ease and lower costs of genome sequencing will continue to profoundly alter fungal research. In the past, genome sequencing was the domain of large specialized centers, but in the future individual investigators will be encouraged to sequence new fungi as part of their research agenda. This change is likely to produce sequence from a far greater range of organisms and these resources will further enable research on a range of fungi.

Thus, while genome sequencing has previously emphasized a single Phylum, the Ascomycota, recent efforts seek to increase coverage of the basally branching groups of fungi. Survey sequencing for use in building multigene species phylogenies is becoming more cost-effective. Greater sequence coverage of basal fungi and their closest relatives is essential to define the commonalities of all fungi. Additionally, sequencing of additional early diverging eukaryotes will allow more detailed comparison of genomic changes that differentiate fungi from their nearest eukaryotic relatives. Further sequencing of microsporidia, a group placed near or within the fungal kingdom, should help shed light on the evolution of this group and clarify its phylogenetic relationship to other fungi.

As sequencing becomes cheaper, the ability to sequence multiple strains from a population, or even populations of fungi, becomes feasible. While initial sequencing efforts targeted single representatives of a species, the new methodologies will permit a more comprehensive understand of the genome structure of an organism, by understanding diversity within the population and the population's response to selective pressures. Future studies will be able to sequence hundreds of strains or isolates for what a single genome cost just a few years ago. Two recent studies examined polymorphism in collections of ∼70 *S. cerevisiae* strains each, from diverse geographic and source of origin (Liti *et al.*, 2009; Schacherer *et al.*, 2009). Sequencing multiple strains allow maps of the diversity within a species, which in turn can be used to identify genes and regions under heightened selective pressures. For example, one could identify rapidly evolving genes that may be important in the recent evolution of emerging pathogens, or highly constrained sequences that may be important targets for diagnostics, vaccines, or therapeutic measures. Finally, fungal researchers are likely to benefit from the power of new sequencing technologies to quantify expression. The combination of high dynamic range, base pair resolution, and rapidly falling costs make this avenue highly attractive. Already direct RNA sequencing using new technologies is adding much needed precision to the task of genome annotation. A single lane of Illumina sequence represents a comparable cost to a microarray experiment, and barcoding of samples allows greater throughput. Sequencing by these new approaches also has the advantage of simultaneously detecting variation in transcripts, which could be used to assay mutations between samples or differences in expression of alleles.

ACKNOWLEDGMENTS

We thank the National Human Genome Research Institute for initial support of the fungal genome initiative (FGI) through a series of white papers and the leadership of the FGI steering committee for guidance in selecting initial genome sequencing targets.

REFERENCES

Aparicio, O., Geisberg, J. V., and Struhl, K. (2004). Chromatin immunoprecipitation for determining the association of proteins with specific genomic sequences *in vivo*. *Curr. Protoc. Cell Biol.* Chapter 17, Unit 17.7.

Boone, C., Bussey, H., and Andrews, B. J. (2007). Exploring genetic interactions and networks with yeast. *Nat. Rev. Genet.* **8**(6), 437–449.

Borneman, A. R., *et al.* (2007). Divergence of transcription factor binding sites across related yeast species. *Science* **317**(5839), 815–819.

Brachat, S., *et al.* (2003). Reinvestigation of the *Saccharomyces cerevisiae* genome annotation by comparison to the genome of a related fungus: *Ashbya gossypii. Genome Biol.* **4**(7), R45.

Butler, G., *et al.* (2009). Evolution of pathogenicity and sexual reproduction in eight *Candida* genomes. *Nature* **459**(7247), 657–662.

Cliften, P. F., *et al.* (2001). Surveying Saccharomyces genomes to identify functional elements by comparative DNA sequence analysis. *Genome Res.* **11**(7), 1175–1186.

Cliften, P., *et al.* (2003). Finding functional features in Saccharomyces genomes by phylogenetic footprinting. *Science* **301**(5629), 71–76.

Coleman, J. J., *et al.* (2009). The genome of *Nectria haematococca*: Contribution of supernumerary chromosomes to gene expansion. *PLoS Genet.* **5**(8), e1000618.

Cornell, M. J., *et al.* (2007). Comparative genome analysis across a kingdom of eukaryotic organisms: Specialization and diversification in the fungi. *Genome Res.* **17**(12), 1809–1822.

Cuomo, C. A., *et al.* (2007). The *Fusarium graminearum* genome reveals a link between localized polymorphism and pathogen specialization. *Science* **317**(5843), 1400–1402.

DeRisi, J. L., Iyer, V. R., and Brown, P. O. (1997). Exploring the metabolic and genetic control of gene expression on a genomic scale. *Science* **278**(5338), 680–686.

Dietrich, F. S., *et al.* (2004). The *Ashbya gossypii* genome as a tool for mapping the ancient *Saccharomyces cerevisiae* genome. *Science* **304**(5668), 304–307.

Dujon, B. (2006). Yeasts illustrate the molecular mechanisms of eukaryotic genome evolution. *Trends Genet.* **22**(7), 375–387.

Fedorova, N. D., *et al.* (2008). Genomic islands in the pathogenic filamentous fungus *Aspergillus fumigatus. PLoS Genet.* **4**(4), e1000046.

Frank, A. C., and Wolfe, K. H. (2009). Evolutionary capture of viral and plasmid DNA by yeast nuclear chromosomes. *Eukaryot. Cell* **8**(10), 1521–1531.

Friesen, T. L., *et al.* (2006). Emergence of a new disease as a result of interspecific virulence gene transfer. *Nat. Genet.* **38**(8), 953–956.

Galagan, J. E., *et al.* (2003). The genome sequence of the filamentous fungus *Neurospora crassa. Nature* **422**(6934), 859–868.

Galagan, J. E., *et al.* (2005). Sequencing of *Aspergillus nidulans* and comparative analysis with *A. fumigatus* and *A. oryzae. Nature* **438**(7071), 1105–1115.

Goffeau, A., *et al.* (1996). Life with 6000 genes. *Science* **274**(5287), 546563–567.

Harbison, C. T., *et al.* (2004). Transcriptional regulatory code of a eukaryotic genome. *Nature* **431**(7004), 99–104.

Hardison, R. C. (2003). Comparative genomics. *PLoS Biol.* **1**(2), E58.

Hedges, S. B., Dudley, J., and Kumar, S. (2006). TimeTree: A public knowledge-base of divergence times among organisms. *Bioinformatics* **22**(23), 2971–2972.

Hibbett, D. S., *et al.* (2007). A higher-level phylogenetic classification of the Fungi. *Mycol. Res.* **111**, 509–547.

Hittinger, C. T., Rokas, A., and Carroll, S. B. (2004). Parallel inactivation of multiple GAL pathway genes and ecological diversification in yeasts. *Proc. Natl. Acad. Sci. USA* **101**(39), 14144–14149.

Hughes, A. L., and Friedman, R. (2003). Parallel evolution by gene duplication in the genomes of two unicellular fungi. *Genome Res.* **13**(5), 794–799.

James, T. Y., *et al.* (2006). Reconstructing the early evolution of Fungi using a six-gene phylogeny. *Nature* **443**(7113), 818–822.

Kellis, M., *et al.* (2003). Sequencing and comparison of yeast species to identify genes and regulatory elements. *Nature* **423**(6937), 241–254.

Kellis, M., Birren, B. W., and Lander, E. S. (2004). Proof and evolutionary analysis of ancient genome duplication in the yeast *Saccharomyces cerevisiae*. *Nature* **428**(6983), 617–624.

Khaldi, N., *et al.* (2008). Evidence for horizontal transfer of a secondary metabolite gene cluster between fungi. *Genome Biol.* **9**(1), R18.

Li, Q. R., *et al.* (2008). Revisiting the *Saccharomyces cerevisiae* predicted ORFeome. *Genome Res.* **18**(8), 1294–1303.

Lin, M. F., *et al.* (2007). Revisiting the protein-coding gene catalog of *Drosophila melanogaster* using 12 fly genomes. *Genome Res.* **17**(12), 1823–1836.

Liti, G., *et al.* (2009). Population genomics of domestic and wild yeasts. *Nature* **458**(7236), 337–341.

Lynch, M., and Force, A. (2000). The probability of duplicate gene preservation by subfunctionalization. *Genetics* **154**(1), 459–473.

Ma, L. J., *et al.* (2009). Genomic analysis of the basal lineage fungus *Rhizopus oryzae* reveals a whole-genome duplication. *PLoS Genet.* **5**(7), e1000549.

Ma, L.-J., *et al.* (2010). Comparative genomics reveals mobile pathogenicity chromosomes in *Fusarium oxysporum*. *Nature* (submitted).

Martin, F., *et al.* (2008). The long hard road to a completed *Glomus intraradices* genome. *New Phytol.* **180**(4), 747–750.

McGuire, A. M., *et al.* (2008). Cross-kingdom patterns of alternative splicing and splice recognition. *Genome Biol.* **9**(3), R50.

Nagalakshmi, U., *et al.* (2008). The transcriptional landscape of the yeast genome defined by RNA sequencing. *Science* **320**(5881), 1344–1349.

O'Gorman, C. M., Fuller, H. T., and Dyer, P. S. (2009). Discovery of a sexual cycle in the opportunistic fungal pathogen *Aspergillus fumigatus*. *Nature* **457**(7228), 471–474.

Ohno, S. (1970). Evolution by Gene and Genome Duplication. Springer, New York.

Reynolds, T. B., and Fink, G. R. (2001). Bakers' yeast, a model for fungal biofilm formation. *Science* **291**(5505), 878–881.

Richards, T. A., *et al.* (2006). Evolution of filamentous plant pathogens: Gene exchange across eukaryotic kingdoms. *Curr. Biol.* **16**(18), 1857–1864.

Richards, T. A., *et al.* (2009). Phylogenomic analysis demonstrates a pattern of rare and ancient horizontal gene transfer between plants and fungi. *Plant Cell* **21**(7), 1897–1911.

Samad, A., *et al.* (1995). Optical mapping: A novel, single-molecule approach to genomic analysis. *Genome Res.* **5**(1), 1–4.

Scannell, D. R., *et al.* (2006). Multiple rounds of speciation associated with reciprocal gene loss in polyploid yeasts. *Nature* **440**(7082), 341–345.

Schacherer, J., *et al.* (2009). Comprehensive polymorphism survey elucidates population structure of *Saccharomyces cerevisiae*. *Nature* **458**(7236), 342–345.

Scheifele, L. Z., *et al.* (2009). Retrotransposon overdose and genome integrity. *Proc. Natl. Acad. Sci. USA* **106**(33), 13927–13932.

Soanes, D. M., *et al.* (2008). Comparative genome analysis of filamentous fungi reveals gene family expansions associated with fungal pathogenesis. *PLoS ONE* **3**(6), e2300.

Souciet, J., *et al.* (2000). Genomic exploration of the hemiascomycetous yeasts: 1. A set of yeast species for molecular evolution studies. *FEBS Lett.* **487**(1), 3–12.

Souciet, J. L., *et al.* (2009). Comparative genomics of protoploid Saccharomycetaceae. *Genome Res.* **19**(10), 1696–1709.

Stamatakis, A. (2006). RAxML-VI-HPC: Maximum likelihood-based phylogenetic analyses with thousands of taxa and mixed models. *Bioinformatics* **22**(21), 2688–2690.

Taylor, J. W., and Berbee, M. L. (2006). Dating divergences in the Fungal Tree of Life: Review and new analyses. *Mycologia* **98**(6), 838–849.

Tzung, K. W., *et al.* (2001). Genomic evidence for a complete sexual cycle in *Candida albicans*. *Proc. Natl. Acad. Sci. USA* **98**(6), 3249–3253.

Verstrepen, K. J., *et al.* (2005). Intragenic tandem repeats generate functional variability. *Nat. Genet.* **37**(9), 986–990.

Wapinski, I., *et al.* (2007). Natural history and evolutionary principles of gene duplication in fungi. *Nature* **449**(7158), 54–61.

Washburn, M. P., Wolters, D., and Yates, J. R. 3rd (2001). Large-scale analysis of the yeast proteome by multidimensional protein identification technology. *Nat. Biotechnol.* **19**(3), 242–247.

Wilhelm, B. T., *et al.* (2008). Dynamic repertoire of a eukaryotic transcriptome surveyed at single-nucleotide resolution. *Nature* **453**(7199), 1239–1243.

Wolfe, K. H., and Shields, D. C. (1997). Molecular evidence for an ancient duplication of the entire yeast genome. *Nature* **387**(6634), 708–713.

Wu, C., *et al.* (2009). Characterization of chromosome ends in the filamentous fungus *Neurospora crassa*. *Genetics* **181**(3), 1129–1145.

Yassour, M., *et al.* (2009). Ab initio construction of a eukaryotic transcriptome by massively parallel mRNA sequencing. *Proc. Natl. Acad. Sci. USA* **106**(9), 3264–3269.

Zhu, H., *et al.* (2001). Global analysis of protein activities using proteome chips. *Science* **293**(5537), 2101–2105.

CHAPTER THIRTY-FIVE

ULTRADIAN METABOLIC CYCLES IN YEAST

Benjamin P. Tu

Contents

Abstract

Budding yeast are capable of displaying various modes of oscillatory behavior. Such cycles can occur with a period ranging from 1 min up to many hours, depending on the growth and culturing conditions used to observe them. This chapter discusses the robust oscillations in oxygen consumption exhibited by high-density yeast cell populations during continuous, glucose-limited growth in a chemostat. These ultradian metabolic cycles offer a view of the life of yeast cells under a challenging, nutrient-poor growth environment and might represent useful systems to interrogate a variety of fundamental metabolic and regulatory processes.

1. INTRODUCTION

Many cyclic and oscillatory phenomena can be observed in nature. Perhaps the most recognized of such periodic phenomena are the circadian rhythms present in virtually all kingdoms of life. Although the budding yeast

Department of Biochemistry, University of Texas Southwestern Medical Center, Dallas, Texas, USA

Methods in Enzymology, Volume 470   © 2010 Elsevier Inc.
ISSN 0076-6879, DOI: 10.1016/S0076-6879(10)70035-5

Saccharomyces cerevisiae does not seem to display any *bona fide* circadian behavior, yeast cells have long been known to be capable of exhibiting other modes of oscillatory behavior, typically with a period much shorter than 24 h (Richard, 2003).

Among the first observed of such oscillations were reduced pyridine nucleotide oscillations that occurred with a period of ~ 1 min both in cell-free extracts and intact cells (Chance *et al.*, 1964a,b; Ghosh and Chance, 1964; Hommes, 1964). Subsequently, oscillations in oxygen consumption and other physical parameters were observed during continuous culture of yeast cells using a chemostat (Parulekar *et al.*, 1986; Porro *et al.*, 1988; Satroutdinov *et al.*, 1992; von Meyenburg, 1969). These sustained oscillations displayed periods ranging anywhere from ~ 40 min to over 10 h, depending on the strain as well as the cultivation protocol.

Long-period oscillations of oxygen consumption on the order of several hours were first reported by several groups during continuous, glucose-limited growth (Kuenzi and Fiechter, 1969; Parulekar *et al.*, 1986; Porro *et al.*, 1988; von Meyenburg, 1969). Later on, short-period ~ 40-min oscillations during continuous culture growth were observed under higher glucose concentrations (Satroutdinov *et al.*, 1992). Collectively, these studies demonstrated that under steady-state, nutrient-poor growth conditions, synchronous changes in a variety of metabolic parameters could be observed in yeast cells that repeat over time (Kuenzi and Fiechter, 1969; Porro *et al.*, 1988; Satroutdinov *et al.*, 1992; von Meyenburg, 1969). Consequently, the use of chemostats to create controlled growth environments has revealed interesting aspects of the behavior of yeast cells that would otherwise be difficult to observe using traditional batch culturing methods. In essence, these ultradian cycles illustrate how yeast cells might cope with a demanding growth environment that is not replete with glucose and other nutrients.

2. INDUCTION OF ULTRADIAN CYCLES OF OXYGEN CONSUMPTION USING A CHEMOSTAT

Several prototrophic strains have been observed to exhibit ultradian cycles of oxygen consumption during continuous growth in a chemostat (Parulekar *et al.*, 1986; Porro *et al.*, 1988; Satroutdinov *et al.*, 1992; Tu *et al.*, 2005; von Meyenburg, 1969). Our preferred strain to study such cycles is the CEN.PK prototroph (van Dijken *et al.*, 2000). The CEN.PK strain grows more rapidly than the common S288C laboratory strain and is genetically tractable (Fig. 35.1). The use of a chemostat to achieve steady-state growth conditions is critical for the observation of the robust oscillations in oxygen consumption that are a hallmark of these cycles. The growth media and general procedure to induce long-period (~ 4–5 h) cycles

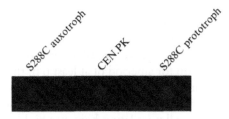

Figure 35.1 Comparison of colony size between CEN.PK and S288C. Note that the prototrophic CEN.PK cycling strain displays more robust growth than the common laboratory strain S288C. Cells were grown on YEPD plates at 30 °C for 48 h.

Table 35.1 Chemostat media recipe

$(NH_4)_2SO_4$	5 g/l
KH_2PO_4	2 g/l
$MgSO_4$	0.5 g/l
$CaCl_2$	0.1 g/l
$FeSO_4$	0.02 g/l
$ZnSO_4$	0.01 g/l
$CuSO_4$	0.005 g/l
$MnCl_2$	0.001 g/l
Yeast extract	1 g/l
Glucose	10 g/l
70% H_2SO_4	0.5 ml/l
Antifoam	0.5 ml/l

using a chemostat is detailed below (Table 35.1). In order to observe short-period (~40 min) oscillations as described by Kuriyama, Klevecz, Murray, and colleagues, the same general procedure is used except the glucose concentration is 20 g/l (2%) and use of the polyploid strain IFO0233 is recommended (Klevecz *et al.*, 2004; Satroutdinov *et al.*, 1992).

2.1. Chemostat setup

Following sterilization, the chemostat is set to maintain a constant temperature of 30 °C and a pH of 3.4. The chemostat vessel is aerated with house air at ~1 l/min and agitated at ~450 rpm to maintain sufficient aeration for the cell population. The exact agitation and aeration settings will be dependent on the chemostat setup; the key is to provide sufficient aeration so that molecular oxygen never becomes limiting in the growth environment, which can be an issue at high cell densities. The pH is kept constant at 3.4 by regulated dosing of 1 *M* sodium hydroxide. Cycles can be observed at a range of pH values (3–5.5); the acidic conditions help minimize contamination of the growth media.

Once the pH and temperature have stabilized, the chemostat vessel is seeded with an overnight starter culture and the cells are allowed to grow up to density in batch mode ($OD_{600} \sim 10\text{--}12$). A return to $\sim 100\%$ dO_2 levels signifies that the cells have exhausted available carbon sources in the media (Fig. 35.2). After starvation for a period of at least 6 h, the cell population is continuously fed the same media at a dilution rate of ~ 0.09 h^{-1}, which leads to the production of long-period, $\sim 4\text{--}5$ h cycles. Shortly after the start of continuous mode, the cell population becomes highly synchronized and exhibits robust oscillations in oxygen consumption (Fig. 35.2). The cycles will persist as long as media is continuously supplied to the cells. We have observed up to ~ 100 consecutive metabolic cycles over the course of 3 weeks without significant loss of synchrony.

2.2. Comments

The growth media contains nitrogen, phosphate and sulfur sources, essential metals (Ca^{2+}, Mg^{2+}, Fe^{2+}, Zn^{2+}, Cu^{2+}, Mn^{2+}) and glucose as the carbon source, and limiting nutrient. A little bit of yeast extract (0.1%) supplies remaining trace vitamins and cofactors. Yeast nitrogen base without amino acids and ammonium sulfate can supplement for the yeast extract. The sulfuric acid acidifies the media which minimizes contamination.

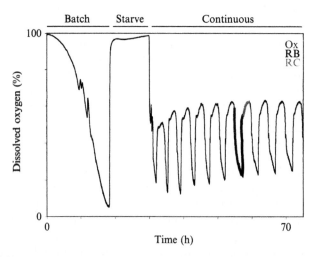

Figure 35.2 Long-period metabolic cycles during continuous, glucose-limited growth. During batch mode, the cells are grown to a high density and then starved for a short period. During continuous mode, media containing glucose is introduced to the chemostat culture at a constant dilution rate (~ 0.09 h^{-1}). These metabolic cycles are comprised three phases: Ox, RB, and RC.

Long-period cycles have also been observed to occur at 25 and 37 °C, the period length is approximately the same at different temperatures; however, the shape of the oscillations in oxygen consumption is different (Tu, unpublished data). In this respect, the cycles exhibit temperature compensation. Higher dilution rates lead to shorter cycles and lower dilution rates lead to longer cycles. Both haploid and diploid strains can cycle, although under the same culturing conditions diploid cycles are ~25% longer (Tu, unpublished data). Thus far, we have not been able to observe robust, long-period cycles using prototrophic derivatives of the common laboratory strains S288C and W303.

3. LONG-PERIOD CYCLES

Long-period cycles have a period ranging anywhere from 2 h to upward of 10 h depending on the chemostat dilution rate. These cycles are characterized by phases of rapid oxygen consumption that alternate with phases of slower oxygen consumption. Interestingly, a variety of additional parameters such as budding index, storage carbohydrate content, ethanol levels, and carbon dioxide production have been observed to oscillate as a function of these cycles, although not necessarily in phase with the dissolved oxygen oscillation (Kuenzi and Fiechter, 1969; Porro et al., 1988; Tu et al., 2005, 2007; Wang et al., 2000; Xu and Tsurugi, 2006).

Microarray analysis of gene expression during ~5 h long-period cycles of a CEN.PK diploid has revealed the nature of the cellular events that occur as a function of such cycles (Tu et al., 2005). Over half of yeast genes were found to exhibit robust cyclic expression. Gene products with functions associated with energy and metabolism and those localized to the mitochondria tended to be expressed periodically. For these reasons, these long-period cycles of oxygen consumption were given the name "yeast metabolic cycle" or "YMC" (Tu et al., 2005).

Analysis of the long-period expression dataset revealed three superclusters of gene expression, which were used to define three major phases: Ox (oxidative, respiratory), RB (reductive, building), and RC (reductive, charging) (Tu et al., 2005). Different categories of genes peak during each phase, and cells traverse each of these three phases in every metabolic cycle. The Ox phase represents the peak of respiration and is associated with a transient induction of ribosomal genes and many genes involved in growth. Cell division and the upregulation of genes that encode mitochondrial proteins occur during the RB phase, when the rate of oxygen consumption begins to decrease. Then in the RC phase, many genes associated with starvation and stress-associated responses (e.g., ubiquitin-proteasome, vacuole, autophagy, heat shock proteins) are activated prior to the next Ox phase (Tu *et al.*, 2005; Brauer *et al.*, 2008; Fig. 35.2).

The extensive orchestration of gene expression around bursts of respiration indicates that many essential cellular and metabolic processes such as cell division, mitochondria biogenesis, ribosome biogenesis, fatty acid oxidation, and autophagy tend to occur during specified temporal windows of these cycles (Tu *et al.*, 2005). Temporal compartmentalization of such processes might enable cells to execute a variety of processes in a more coordinated and efficient fashion and help minimize futile reactions.

The many oscillating gene expression patterns that occur as a function of these cycles predict oscillatory changes in the metabolic state of yeast cells. Both liquid chromatography–tandem mass spectrometry (LC–MS/MS) and 2-D gas chromatography/time-of-flight mass spectrometry (GCxGC–TOFMS) metabolite profiling methods were used to monitor the intracellular concentrations of over ~150 common metabolites at different time intervals of the metabolic cycles (Tu *et al.*, 2007). The results of these surveys show that many metabolites including amino acids, nucleotides, and carbohydrates, oscillate in abundance with a periodicity precisely matching that of the cycles. Consequently, many fundamental biological processes can be predicted to be intimately coupled to these cyclic changes in cellular metabolic state.

4. SHORT-PERIOD CYCLES

The majority of studies on short-period (~40-min) ultradian cycles have been carried out with the polyploid IFO0233 strain. These cycles are also defined by robust oscillations in oxygen consumption. However, these oscillations occur on a more rapid timescale, perhaps due to higher cell densities that result from higher glucose concentrations in the media. Low-amplitude oscillations of transcription as well as metabolite levels were reported to occur during these 40-min cycles (Klevecz *et al.*, 2004; Murray *et al.*, 2007). Furthermore, gating of DNA replication was observed in these short-period cycles, despite the cell division time exceeding the period length of these cycles (Klevecz *et al.*, 2004). Surprisingly, there appears to be minimal phase correlation between the set of short-period cyclic transcripts and the set of long-period cyclic transcripts (Tu and McKnight, 2006). How cycles of transcription, translation, mRNA degradation, and protein turnover can occur on such a short timescale remains a fascinating but outstanding question. More information on these short-period oscillations can be found in the following reviews on the topic (Klevecz *et al.*, 2004; Lloyd and Murray, 2005; Murray *et al.*, 2001).

5. Significance of Ultradian Cycles

The ability to undergo robust metabolic cycles during continuous growth appears to be a feature of prototrophic yeast strains that do not require supplementation of amino acids or nucleobases for growth. Moreover, there are several significant reasons to favor the use of prototrophic strains as opposed to more convenient auxotrophic strains for studies of this sort. Mutations or deletions in metabolic genes present in auxotrophic strains might compromise the output of numerous cellular and metabolic pathways and elicit compensatory responses that are not typical of wild strains of yeast. For example, an auxotrophic strain that cannot synthesize adenine, uracil, or methionine will be absolutely dependent on supplementation of these metabolites for growth, which will undoubtedly alter flux through key metabolic pathways and hence the regulation of particular cellular processes (Boer *et al.*, 2008). Moreover, under nutrient-poor conditions, these supplemented amino acids and nucleobases might be interconverted to other metabolites and become limiting for growth once again. Therefore, it is not surprising that auxotrophic strains might not exhibit robust cycles under conditions that necessitate glucose being the sole limiting nutrient.

How do the growth conditions in the chemostat that lead to cycles of oxygen consumption compare to more commonly used exponential phase growth conditions? Microarray analysis of gene expression during long-period cycles has shown that many genes that encode proteins with starvation and stress-associated functions that are normally not expressed during log phase are actually induced during particular temporal windows (Tu *et al.*, 2005). Moreover, periodic bursts of mitochondrial respiration and peroxisomal β-oxidation are a hallmark of such cycles, whereas in exponential phase fermentation is the predominant mode of energy production. Indeed, many mitochondrial and peroxisomal proteins are not present in high quantities during log phase growth (Ghaemmaghami *et al.*, 2003). Therefore, it is clear that during cycling yeast cells utilize a more extensive assortment of metabolic and regulatory strategies than compared to log phase (Tu and McKnight, 2006; Tu *et al.*, 2005). Thus, the chemostat conditions that bring about such cycles can be likened to a glucose-limited and nutrient-poor growth environment.

While the use of nutrient-rich, log phase growth conditions is convenient and often informative, they might not be typical of conditions in the wild, which are likely to be nutrient-poor. Moreover, yeast typically grow in colonies, where the cell density is very high and the distance between cell neighbors is small. Thus, the dense population of cells in the chemostat that undergo ultradian metabolic cycles to a first approximation might resemble a colony exposed to a nutrient-poor environment.

The high degree of synchrony exhibited by yeast cell populations under-going ultradian cycles is self-evident. Following starvation, cells usually rapidly self-synchronize upon initiation of continuous mode. The precise mechanisms that lead to the establishment and maintenance of synchrony are not entirely clear; however, some secreted metabolites have been implicated in the process (Murray et al., 2003; Porro et al., 1988). Regardless, this metabolically achieved synchrony is remarkably stable, having lasted up to 100 cycles ($\sim$20 days) (Tu, unpublished data). Since cells undergoing long-period metabolic cycles are also highly synchronized with respect to the cell cycle, the microarray expression data have enabled the high resolution timing of cell-cycle regulated gene expression to a previously unachievable resolution of $\sim$2–3 min (Rowicka et al., 2007). Thus, these cycling systems might facilitate the study of almost any temporally regulated process or pathway.

Perhaps the most fascinating aspect of these ultradian metabolic cycles is the manner by which so many metabolic outputs are precisely orchestrated about bursts of mitochondrial respiration. One particularly striking example is the observation that DNA replication and cell division are precisely gated to temporal windows when oxygen consumption decreases (Chen et al., 2007; Klevecz et al., 2004; Kuenzi and Fiechter, 1969; Porro et al., 1988; Tu et al., 2005; von Meyenburg, 1969). The gating of cell division is reminiscent of the circadian gating of cell division previously observed in cyanobacteria, mouse liver, and cultured fibroblasts (Matsuo et al., 2003; Mori et al., 1996; Nagoshi et al., 2004). Moreover, yeast cells secrete ethanol, a product of glycolytic metabolism and become much less dependent on mitochondrial respiration as they enter the cell cycle, which is reminiscent of cancer cell division and the Warburg effect (Tu et al., 2005, 2007; Warburg, 1956). How cell division as well as other fundamental cellular processes are intimately coordinated with the metabolic state of a cell remain important open questions.

In conclusion, a chemostat enables the maintenance of constant pH, temperature, aeration, and nutrient levels thereby creating steady-state growth conditions that are not easily achievable in batch cultures. In turn, it becomes possible to observe the behavior of yeast cell populations with minimal interference from external variables. Based on the precise, coordinated expression of many groups of genes in a manner that makes biological sense, these ultradian metabolic cycles offer a view of the diverse metabolic and regulatory strategies undertaken by a yeast cell under glucose-limited, nutrient-poor conditions. It is hopeful that the study of these prototrophic yeast strains undergoing synchronized metabolic oscillations will contribute toward our understanding of biological cycles as well as complex metabolic diseases such as cancer and aging. Leo Szilard, one of the inventors of the chemostat, fittingly predicted almost 60 years ago: "A study of this slow-growth phase by means of the chemostat promises to yield information on some value on metabolism, regulatory processes, adaptations, and mutations of microorganisms" (Novick and Szilard, 1950).

ACKNOWLEDGMENTS

The author thanks Peter Kotter, Robert Klevecz, and Andrew Murray for sharing yeast strains. B. P. T. is a W. W. Caruth, Jr. Scholar in Biomedical Research at UT Southwestern Medical Center. This work was also supported by a Burroughs Wellcome Fund Career Award in Biomedical Sciences and a Welch Foundation Research Grant.

REFERENCES

Boer, V. M., Amini, S., and Botstein, D. (2008). Influence of genotype and nutrition on survival and metabolism of starving yeast. *Proc. Natl. Acad. Sci. USA* **105,** 6930–6935.

Brauer, M. J., Huttenhower, C., Airoldi, E. M., Rosenstein, R., Matese, J. C., Gresham, D., Boer, V. M., Troyanskaya, O. G., and Botstein, D. (2008). *Mol. Biol. Cell* **19**(1), 352–367.

Chance, B., Estabrook, R. W., and Ghosh, A. (1964a). Damped sinusoidal oscillations of cytoplasmic reduced pyridine nucleotide in yeast cells. *Proc. Natl. Acad. Sci. USA* **51,** 1244–1251.

Chance, B., Schoener, B., and Elsaesser, S. (1964b). Control of the waveform of oscillations of the reduced pyridine nucleotide level in a cell-free extract. *Proc. Natl. Acad. Sci. USA* **52,** 337–341.

Chen, Z., Odstrcil, E. A., Tu, B. P., and McKnight, S. L. (2007). *Science* **316**(5833), 1916–1919.

Ghaemmaghami, S., Huh, W. K., Bower, K., Howson, R. W., Belle, A., Dephoure, N., O'Shea, E. K., and Weissman, J. S. (2003). Global analysis of protein expression in yeast. *Nature* **425,** 737–741.

Ghosh, A., and Chance, B. (1964). Oscillations of glycolytic intermediates in yeast cells. *Biochem. Biophys. Res. Commun.* **16,** 174–181.

Hommes, F. A. (1964). Oscillatory Reductions of Pyridine Nucleotides during Anaerobic Glycolysis in Brewers' Yeast. *Arch. Biochem. Biophys.* **108,** 36–46.

Klevecz, R. R., Bolen, J., Forrest, G., and Murray, D. B. (2004). A genomewide oscillation in transcription gates DNA replication and cell cycle. *Proc. Natl. Acad. Sci. USA* **101,** 1200–1205.

Kuenzi, M. T., and Fiechter, A. (1969). Changes in carbohydrate composition and trehalase-activity during the budding cycle of *Saccharomyces cerevisiae. Arch Mikrobiol.* **64,** 396–407.

Lloyd, D., and Murray, D. B. (2005). Ultradian metronome: Timekeeper for orchestration of cellular coherence. *Trends Biochem. Sci.* **30,** 373–377.

Matsuo, T., Yamaguchi, S., Mitsui, S., Emi, A., Shimoda, F., and Okamura, H. (2003). Control mechanism of the circadian clock for timing of cell division in vivo. *Science* **302,** 255–259.

Mori, T., Binder, B., and Johnson, C. H. (1996). Circadian gating of cell division in cyanobacteria growing with average doubling times of less than 24 hours. *Proc. Natl. Acad. Sci. USA* **93,** 10183–10188.

Murray, D. B., Roller, S., Kuriyama, H., and Lloyd, D. (2001). Clock control of ultradian respiratory oscillation found during yeast continuous culture. *J. Bacteriol.* **183,** 7253–7259.

Murray, D. B., Klevecz, R. R., and Lloyd, D. (2003). Generation and maintenance of synchrony in *Saccharomyces cerevisiae* continuous culture. *Exp. Cell Res.* **287**, 10–15.

Murray, D. B., Beckmann, M., and Kitano, H. (2007). Regulation of yeast oscillatory dynamics. *Proc. Natl. Acad. Sci. USA* **104**, 2241–2246.

Nagoshi, E., Saini, C., Bauer, C., Laroche, T., Naef, F., and Schibler, U. (2004). Circadian gene expression in individual fibroblasts: Cell-autonomous and self-sustained oscillators pass time to daughter cells. *Cell* **119**, 693–705.

Novick, A., and Szilard, L. (1950). Description of the chemostat. *Science* **112**, 715–716.

Parulekar, S. J., Semones, G. B., Rolf, M. J., Lievense, J. C., and Lim, H. C. (1986). Induction and elimination of oscillations in continuous cultures of *Saccharomyces cerevisiae*. *Biotechnol. Bioeng.* **28**, 700–710.

Porro, D., Martegani, E., Ranzi, B. M., and Alberghina, L. (1988). Oscillations in continuous cultures of budding yeast: A segregated parameter analysis. *Biotechnol. Bioeng.* **32**, 411–417.

Richard, P. (2003). The rhythm of yeast. *FEMS Microbiol. Rev.* **27**, 547–557.

Rowicka, M., Kudlicki, A., Tu, B. P., and Otwinowski, Z. (2007). High-resolution timing of cell cycle-regulated gene expression. *Proc. Natl. Acad. Sci. USA* **104**, 16892–16897.

Satroutdinov, A. D., Kuriyama, H., and Kobayashi, H. (1992). Oscillatory metabolism of *Saccharomyces cerevisiae* in continuous culture. *FEMS Microbiol. Lett.* **77**, 261–267.

Tu, B. P., and McKnight, S. L. (2006). Metabolic cycles as an underlying basis of biological oscillations. *Nat. Rev. Mol. Cell Biol.* **7**, 696–701.

Tu, B. P., Kudlicki, A., Rowicka, M., and McKnight, S. L. (2005). Logic of the yeast metabolic cycle: Temporal compartmentalization of cellular processes. *Science* **310**, 1152–1158.

Tu, B. P., Mohler, R. E., Liu, J. C., Dombek, K. M., Young, E. T., Synovec, R. E., and McKnight, S. L. (2007). Cyclic changes in metabolic state during the life of a yeast cell. *Proc. Natl. Acad. Sci. USA* **104**, 16886–16891.

van Dijken, J. P., Bauer, J., Brambilla, L., Duboc, P., Francois, J. M., Gancedo, C., Giuseppin, M. L., Heijnen, J. J., Hoare, M., Lange, H. C., et al. (2000). An interlaboratory comparison of physiological and genetic properties of four *Saccharomyces cerevisiae* strains. *Enzyme Microb. Technol.* **26**, 706–714.

von Meyenburg, H. K. (1969). Energetics of the budding cycle of *Saccharomyces cerevisiae* during glucose limited aerobic growth. *Arch. Microbiol.* **66**, 289–303.

Wang, J., Liu, W., Uno, T., Tonozuka, H., Mitsui, K., and Tsurugi, K. (2000). Cellular stress responses oscillate in synchronization with the ultradian oscillation of energy metabolism in the yeast *Saccharomyces cerevisiae*. *FEMS Microbiol. Lett.* **189**, 9–13.

Warburg, O. (1956). On respiratory impairment in cancer cells. *Science* **124**, 269–270.

Xu, Z., and Tsurugi, K. (2006). A potential mechanism of energy-metabolism oscillation in an aerobic chemostat culture of the yeast *Saccharomyces cerevisiae*. *FEBS J.* **273**, 1696–1709.

Author Index

Subject Index

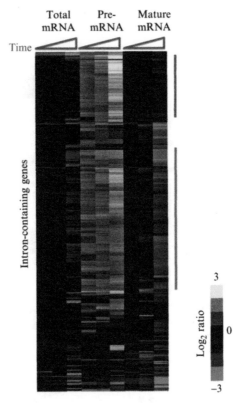

Maki Inada and Jeffrey A. Pleiss, Figure 3.4 Genome-wide changes in pre-mRNA splicing. Results are presented from an experiment comparing a strain containing a cold-sensitive *prp16-302* mutation with a matched wild-type strain as both were shifted to the nonpermissive temperature. Data are shown from unshifted samples (grown at 30 °C), as well as after 10 and 60 min of incubation at 16 °C. Each horizontal line represents the behavior of a single intron-containing gene during this time course. Notice that some genes (indicated with a red bar) show a dramatic increase in pre-mRNA level with very little change in mature mRNA level, whereas other genes (indicated with a green bar) show a strong increase in pre-mRNA level concomitant with a strong decrease in mature mRNA level.

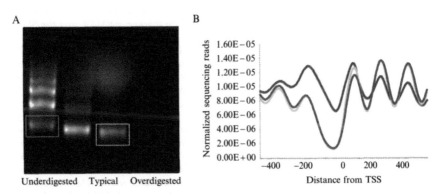

Oliver J. Rando, Figure 5.2 Effect of digestion level on nucleosome maps. A. Gel, as in Figure 5.1, showing an MNase titration from which 3 mononucleosome bands have been excised (indicated with boxes), corresponding to under-, well, and over-digested chromatin, from left to right. B. Chromatin maps differ depending on digestion level. Deep sequeucning data for the three nucleosome preps in A were normalized, and data for all genes aligned by transcription start site (TSS) are averaged for each dataset.

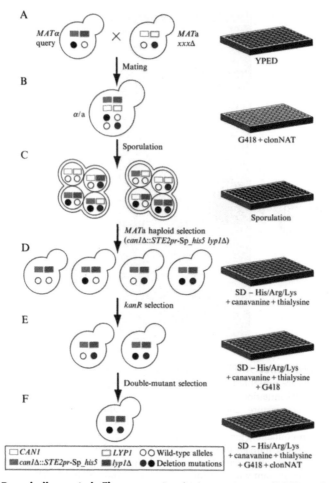

Anastasia Baryshnikova et al., Figure 7.1 Synthetic genetic array (SGA) methodology. (A) A *MATα* strain carries a query mutation linked to a dominant selectable marker (filled black circle), such as the nourseothricin-resistance marker, *natMX4*, and the SGA reporter, *can1Δ::STE2pr-Sp_his5* (in which *STE2pr-Sp_his5* is integrated into the genome such that it deletes the open reading frame (ORF) of the *CAN1* gene, which normally confers sensitivity to canavanine). The query strain also lacks the *LYP1* gene. Deletion of *LYP1* confers resistance to thialysine. This query strain is crossed to an ordered array of *MATa* deletion mutants (*xxxΔ*). In each of these deletion strains, a single gene is disrupted by the insertion of a dominant selectable marker, such as the kanamycin-resistance (*kanMX4*) module (the disrupted gene is represented as a filled red circle). (B) The resulting heterozygous diploids are transferred to a medium with reduced carbon and nitrogen to induce sporulation and form haploid meiotic spore progeny. (C) Spores are transferred to a synthetic medium that lacks histidine, which allows selective germination of *MATa* meiotic progeny owing to the expression of the SGA reporter, *can1Δ::STE2pr-Sp_his5*. To improve this selection, canavanine and thialysine, which select *can1Δ* and *lyp1Δ* while killing *CAN1* and *LYP1* cells, respectively, are included in the selection medium. (D) The *MATa* meiotic progeny are transferred to a medium that contains kanamycin which selects single mutants equivalent to the original array mutants and double mutants. (E, F) An array of double mutants is selected on a medium that contains both nourseothricin and kanamycin.

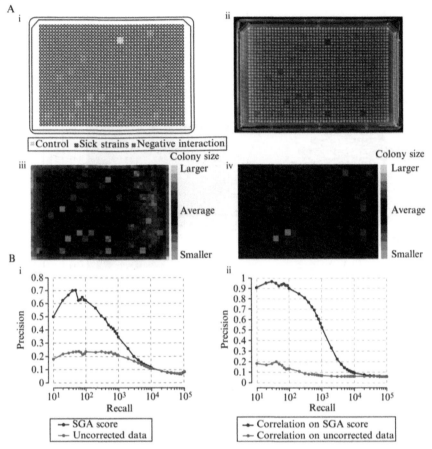

Anastasia Baryshnikova *et al.*, Figure 7.3 Systematic effect correction. (A) A schematic of a typical double mutant array plate shows the systematic biases affecting colony sizes. (i, ii) A typical double mutant array plate contains control spots (gray circles), strains with low fitness (blue circles), negative interactions (red circles), and positive interactions (not shown). On visual inspection, all three cases appear as small colonies or empty spots. (iii) Quantification of colony areas shows a distinctive spatial pattern affecting opposite sides of the plate (bigger colonies on the right, smaller colonies on the left) that was not obvious on visual inspection. Failure to correct for this spatial pattern will result in false-positive interactions. (iv) Corrects spatial patterns, eliminates false positives, and highlights true genetic interactions. (B) Precision–recall curves on genetic interaction scores (i) and genetic profile similarity (ii) show the increased functional prediction capacity of genetic data after correcting for systematic biases. A set of 1712 genome-wide SGA screens (Costanzo *et al.*, in press) were processed using the SGA score (Baryshnikova *et al.*, manuscript in preparation) and a version of the SGA score without systematic effect correction. Both direct genetic interactions and genetic profile similarities, as measured by Pearson correlations, were assessed for function by calculating precision and recall of functionally related gene pairs as described in the study of Myers *et al.* (2006). As a measure of functional relatedness, we used coannotation to the same Gene Ontology term.

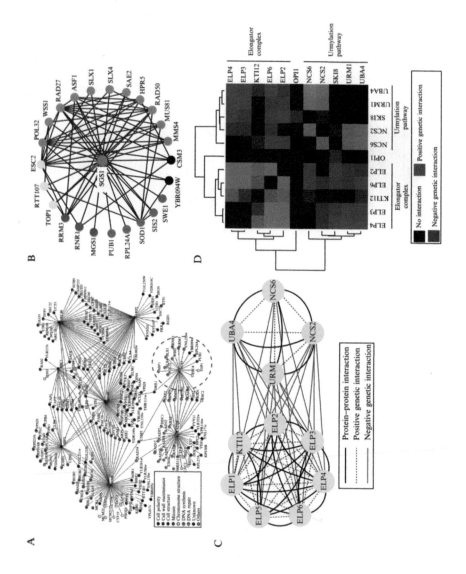

Anastasia Baryshnikova _et al._, Figure 7.4 Properties of genetic interactions. (A) Example of a yeast synthetic lethal network. The synthetic lethal network is a sparse network, indicating that genetic interactions are rare. The frequency of true synthetic lethal interactions (blue lines) is less than 1%. A detailed description of how this initial network was generated can be found elsewhere (Tong _et al._, 2001). (B) Functional neighborhood corresponding to indicated region (dashed gray circle) in (A). Despite being rare, synthetic lethal interactions (blue lines) occur frequently among genes that are functionally related, such as those involved in DNA replication and repair shown here. The frequency of synthetic lethal interaction between functionally related genes ranges from 18% to 25%. (C) Orthogonal relationships. Negative interactions tend to occur between nonessential complexes and pathways. Positive interactions overlap significantly with physical interactions and tend to connect members of the same pathway or complex. (D) Grouping genes according to patterns of genetic interactions revealed a functional relationship between the elongator complex and the urmylation pathway, which act in concert to modify specific tRNAs.

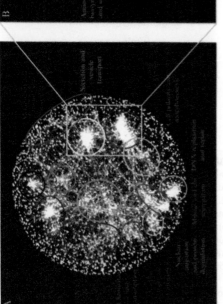

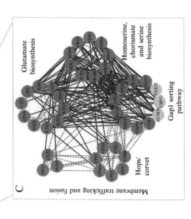

Anastasia Baryshnikova _et al._, Figure 7.5 The genetic landscape of the cell. (A) A correlation-based network connecting genes with similar interaction profiles (Costanzo _et al._, in press). Genetic profile similarities were measured for all gene pairs by computing Pearson correlation coefficients (PCC) from the complete genetic interaction matrix. Gene pairs whose profile similarity exceeded a PCC > 0.2 threshold were connected in the network. An edge-weighted, spring-embedded network layout, implemented in Cytoscape (Shannon _et al._, 2003), was applied to determine node position based on genetic profile similarity. This resulted in the unbiased assembly of a network whereby genes sharing similar patterns of genetic interactions are proximal to each other in two-dimensional space, while less-similar genes are positioned further apart. Circled regions correspond to gene clusters enriched for the indicated biological processes. (B) Magnification of the functional map resolves cellular processes with increased specificity and enables precise functional predictions. A subnetwork corresponding to the indicated region of the global map is shown. Node color corresponds to a specific biological process; amino acid biosynthesis and uptake (dark green); Signaling (light green); ER/Golgi (light purple); endosome and vacuole sorting (dark purple); ER-dependent protein degradation (yellow); protein glycosylation, cell wall biosynthesis and integrity (red); tRNA modification (fuchsia); cell polarity and morphogenesis (pink); autophagy (orange); uncharacterized (black). (C) Individual genetic interactions contributing to genetic profiles revealed by (B). A subset of genes belonging to the amino acid biosynthesis and uptake region of the network in (B). Nodes are grouped according to profile similarity and edges represent negative (red) and positive (green) genetic interactions. Nonessential (circles) and essential (diamonds) genes are colored according to the biological process indicated in (B) and uncharacterized genes are depicted in yellow.

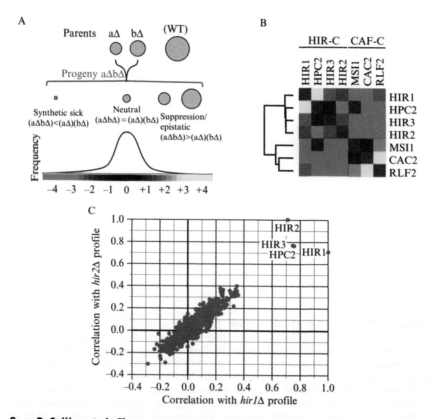

Sean R. Collins et al., Figure 9.1 Epistatic interactions within and between chromatin assembly complexes. (A) The entire spectrum of genetic interactions. Quantitative genetic analysis can identify negative ((aΔbΔ) < (aΔ)(bΔ)), positive ((aΔbΔ) > (aΔ) (bΔ)), and neutral ((aΔbΔ) = (aΔ)(bΔ)) genetic interactions. (B) Genetic interactions between and within the HIR and CAF chromatin assembly complexes. Using the E-MAP approach (Collins et al., 2007a), strong negative interactions were detected between components of the HIR-C (*HIR1*, *HPC2*, *HIR3*, and *HIR2*) and the CAF-C (*MSI1*, *CAC2*, and *RLF2*), which are known to function in parallel pathways to ensure efficient chromatin assembly. Conversely, positive genetic interactions were observed between components within each complex. Blue and yellow interactions correspond to negative and positive genetic interactions, respectively. (C) Plot of correlation coefficients generated from comparison of the genetic profiles from *hir1Δ* and *hir2Δ* to all other ~750 profiles from the chromosome biology E-MAP (Collins et al., 2007a). Note the high pairwise correlations with *HIR1*, *HIR2*, *HIR3*, and *HPC2*.

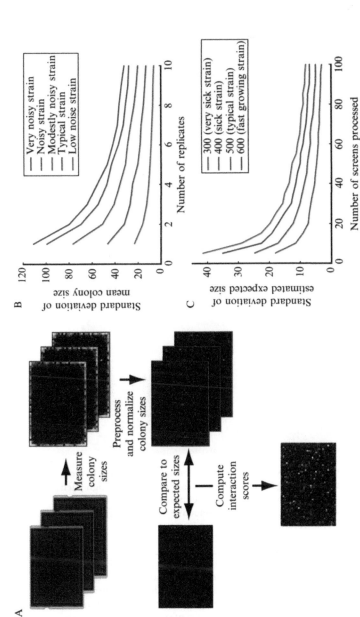

Sean R. Collins et al., Figure 9.4 Overview of the data processing procedure. (A) A flowchart describing the data processing procedure for a single screen is shown. The first images are digital photographs of arrays of yeast colonies. In the following images heatmaps of either measured colony sizes or genetic interaction scores are shown. In colony size heatmaps, blue represents small colonies, black represents average-sized colonies, and yellow represents large colonies. In the genetic interaction heatmap, blue represents negative interactions, black represents neutral interactions, yellow represents positive interactions, and gray represents missing data (or data filtered out during quality control). (B) The variability in measured growth phenotype (mean colony size over the replicate measurements) is shown as a function of number of experimental replicates. The curves shown were generated using data from 36 replicates of a control screen run while generating the early secretory pathway E-MAP (Schuldiner et al., 2005). On a given curve, the point corresponding to N replicates was generated by randomly drawing N of the 36 replicates for a particular strain and computing the mean normalized colony size. This process was repeated

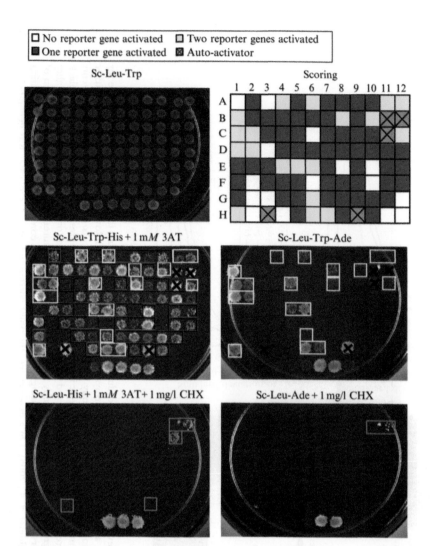

Matija Dreze et al., Figure 12.5 Phenotyping plates and scoring. First, autoactivators are identified and crossed out. The stringency of autoactivator detection is high such that even slight growth on the CYH control plates leads to elimination of the respective candidate. Subsequently, growth is evaluated on the selective –His and –Ade plates using the six controls (Fig. 12.1) as reference.

1000 times, and the standard deviation over these 1000 repeats was plotted. Each curve represents data for a different strain. Five different representative strains with different levels of measurement variability were chosen for analysis. (C) The variability of the empirically determined expected double–mutant phenotype was estimated as a function of the number of screens analyzed in parallel. In this case, sets of screens of the indicated size were drawn at random from a set of 329 screens completed at approximately the same time from the early secretory pathway E-MAP (Schuldiner et al., 2005). For each point on each curve, 1000 random draws were completed, and each time expected colony size values were computed. The standard deviation of the expected colony sizes (over the 1000 random draws) was plotted for four representative strains with different single mutant phenotypes.

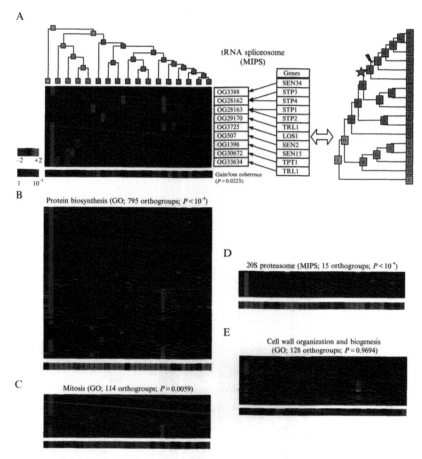

Ilan Wapinski and Aviv Regev, Figure 18.11 Coherent evolution of functionally related proteins. (A) Phylogenetic coherence of the tRNA spliceosome orthogroup class. The set of *S. cerevisiae* tRNA spliceosome genes (MIPS) was mapped to the set of orthogroups that contain these genes (black arrows, middle panel). Some orthogroups (e.g., #28162) contain multiple paralogs from the gene set. The ECVPs of all the orthogroups in the set are compiled into a matrix (left panel). Each row denotes one profile, and each column the copy number changes in one species (red, increase; blue, decrease; black, no change). The species (extant and ancestral) are ordered according to the order of the nodes in the species tree (top). The bottom row shows the coherence score for each column (purple, coherent), as evaluated by comparing the number of events to the distribution of events in the specific node within a random set of orthogroups of the same size. The overall significance of the coherence is reported in a *P*-value. (B–E) Phylogenetic coherence of the protein biosynthesis, Mitosis, 20S proteasome, and cell wall organization and biogenesis orthogroup sets. Copy-number variation coherence is presented as in (A). The copy-number changes observed in the protein biosynthesis (B), Mitosis (C), and 20S proteasome (D) orthogroup classes are coherent. Those in the cell wall organization and biogenesis orthogroup set are not coherent (E). Orthogroup classes are projected from the GO gene classes.

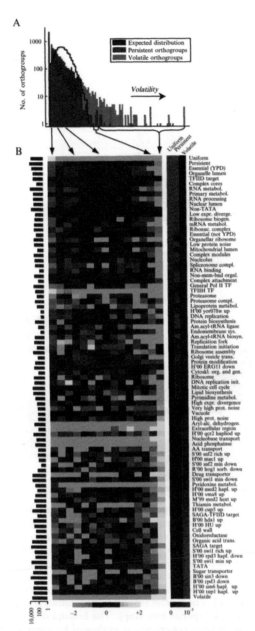

Ilan Wapinski and Aviv Regev, Figure 18.13 A functional dichotomy of uniform, persistent, and volatile orthogroups. (A) Orthogroup volatility. The histogram shows the number of orthogroups in each bin of volatility scores. The expected distribution when sampling random orthogroups from the evolutionary model is shown as a black line. Uniform orthogroups (leftmost blue column) are the lowest scoring orthogroups. Persistent orthogroups usually receive low scores as well (blue columns). All orthogroups with a score above three standard deviations from the expected mean are volatile (red columns). (B) Gene class annotations enriched in uniform, persistent

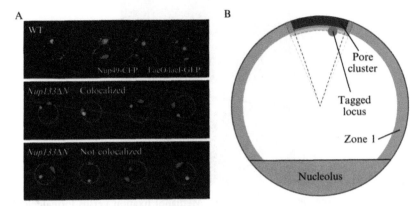

Peter Meister, *et al.*, Figure 21.5 Determining the significance of colocalization. (A) Nuclear pores tagged with Nup49-GFP (red) and a LacI-tagged locus (green). The two left images in the upper panel are not deconvolved, all other images are. In the *nup133ΔN* mutant, the nuclear pores form a cluster (Schober *et al.*, 2009). (B) The expected colocalization for a randomly positioned spot and the pore cluster can be calculated as a ratio of volumes (see text). The figure shows a cut through the nucleus. The pore cluster is modeled as a conical layer shown in red. The spot is considered as colocalizing if it at least touches the pore cluster, which results in the colocalization zone (green).

and volatile orthogroups. Annotations that were significant (at most $P < 10^{-5}$ after FDR correction, purple significance) among either uniform, persistent or volatile orthogroups are shown (the size of each annotation is shown on the left). The functional and mechanistic dichotomy between volatile and nonvolatile orthogroups largely reproduces along the full range of volatility scores (columns—bins of orthogroups with similar volatility scores; Rows—significant annotations. Yellow/blue higher/lower relative enrichment compared to the expected enrichment in the class).

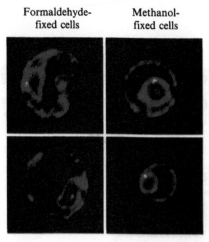

Donna Garvey Brickner et al., Figure 22.2   Methanol fixation versus formaldehyde fixation. Shown are representative examples of cells fixed using either methanol (as described in the protocol) or 4% formaldehyde (twice for 30 min). The ER/nuclear envelope was stained with Sec63-myc (red) and the chromosomal locus (in this case, the *INO1* gene) was stained with the GFP-Lac repressor (green). Note the nonspherical shape of the nucleus in the formaldehyde-fixed cells.

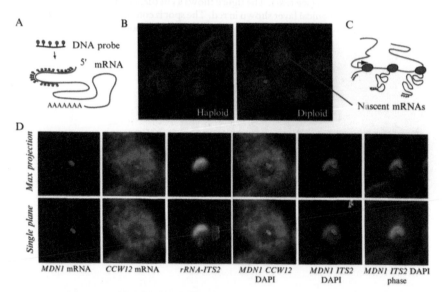

Daniel Zenklusen and Robert H. Singer, Figure 26.1   Single mRNA sensitivity fluorescent *in situ* hybridization (FISH). (A) Schematic diagram of the FISH protocol. A mix of four 50 nt DNA oligonucleotides, each labeled with five fluorescent dyes, is hybridized to paraformaldehyde fixed yeast cells to obtain single transcript resolution. (B) Single mRNA FISH for *MDN1* mRNA. Single mRNAs are detected in the cytoplasm, higher intensity spot in the nucleus. Haploid and diploid yeast cells are shown. Probes hybridize to the 5' of the mRNA. *MDN1* mRNA (red), DAPI (blue), and DIC. (C) Cartoon illustrating that the number of nascent mRNAs at the site of transcription is used to determine the polymerase loading on a gene using probes to the 5' end of *MDN1*. (D) Nascent transcripts of neighboring genes colocalize at the site of transcription. Diploid cells are hybridized with probes against *MDN1* (red) and *CCW12* (green). Nucleolus is stained with probes against the ITS2 spacer of the rRNA precursor (yellow). Maximum projection of 3D-dataset and single plane containing the transcription sites are shown (Zenklusen *et al.*, 2008).

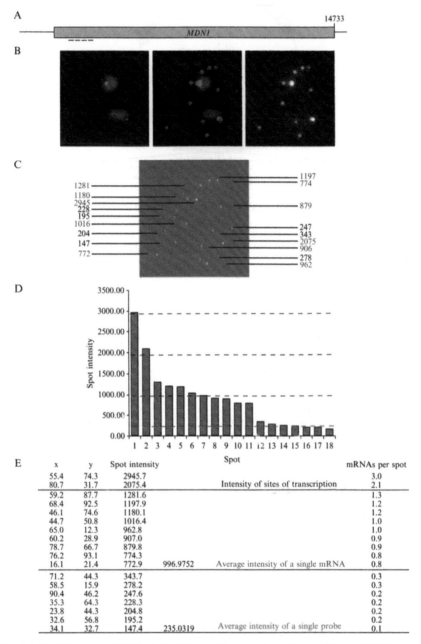

Daniel Zenklusen and Robert H. Singer, Figure 26.3 Quantifying mRNA signal intensities. Intensity of a single mRNA can be calculated by determining the fluorescence intensity emitted from a single probe. (A) FISH probes hybridizing to the 5′ end of *MDN1* mRNA were used for hybridization shown in (B). A small number of single probes tend to hybridize unspecifically to the cells and can be visualized by changing the contrast levels (B, compare left and middle panel). The intensity is determined using a

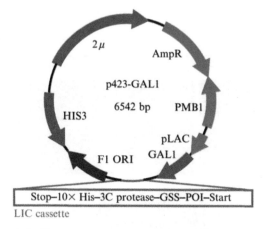

Stop–10× His–3C protease–GSS–POI–Start

LIC cassette

Franklin A. Hays et al., Figure 29.1 Schematic of the p423-GAL1 expression plasmid. Yeast 2μ-based expression plasmid containing a GSS-3C protease-10× His tag for the protein of interest using LIC. Plasmid contains yeast (shown in red), bacterial (shown in green), and phage (shown in blue) elements conducive to molecular cloning and transformation methodologies.

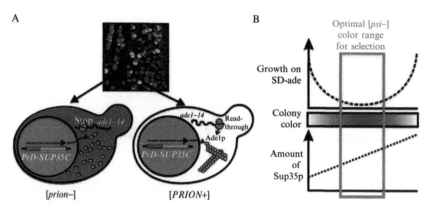

Simon Alberti et al., Figure 30.2 Using Sup35p to detect phenotype switching behavior. (A) Schematic representation of the Sup35p-based prion assay. The PrD of Sup35p is replaced with a candidate PrD and the resulting strains are tested for the presence of prion phenotypes, such as the switching between a red and a white colony color. (B) Relationship between the cellular concentration of Sup35p, the colony color in the [psi–] state and the ability to grow on synthetic medium lacking adenine.

spot detection porgram. (C) Signal intensity of each spot corresponding to a single DNA probe is shown in black, signal intensities of single mRNA and sites of transcription are shown in red. Consistent with the four probes used in the hybridization (A), intensity of single mRNA signals in the cytoplasm is four times the intensity of a single probe (D, E). Nascent mRNAs at the site of transcription are two and three times the intensity of a single mRNA in the cytoplasm (Zenklusen et al., 2008).

Printed and bound by CPI Group (UK) Ltd, Croydon, CR0 4YY

14/10/2024

01774195-0001